Lecture Notes in Physics

Springer
Berlin
Heidelberg
New York
Barcelona
Hong Kong
London
Milan
Paris
Singapore
Tokyo

The Editorial Policy for Monographs

The series Lecture Notes in Physics reports new developments in physical research and teaching - quickly, informally, and at a high level. The type of material considered for publication in the monograph Series includes monographs presenting original research or new angles in a classical field. The timeliness of a manuscript is more important than its form, which may be preliminary or tentative. Manuscripts should be reasonably self-contained. They will often present not only results of the author(s) but also related work by other people and will provide sufficient motivation, examples, and applications.
The manuscripts or a detailed description thereof should be submitted either to one of the series editors or to the managing editor. The proposal is then carefully refereed. A final decision concerning publication can often only be made on the basis of the complete manuscript, but otherwise the editors will try to make a preliminary decision as definite as they can on the basis of the available information.
Manuscripts should be no less than 100 and preferably no more than 400 pages in length. Final manuscripts should preferably be in English, or possibly in French or German. They should include a table of contents and an informative introduction accessible also to readers not particularly familiar with the topic treated. Authors are free to use the material in other publications. However, if extensive use is made elsewhere, the publisher should be informed. Authors receive jointly 50 complimentary copies of their book. They are entitled to purchase further copies of their book at a reduced rate. As a rule no reprints of individual contributions can be supplied. No royalty is paid on Lecture Notes in Physics volumes. Commitment to publish is made by letter of interest rather than by signing a formal contract. Springer-Verlag secures the copyright for each volume.

The Production Process

The books are hardbound, and quality paper appropriate to the needs of the author(s) is used. Publication time is about ten weeks. More than twenty years of experience guarantee authors the best possible service. To reach the goal of rapid publication at a low price the technique of photographic reproduction from a camera-ready manuscript was chosen. This process shifts the main responsibility for the technical quality considerably from the publisher to the author. We therefore urge all authors to observe very carefully our guidelines for the preparation of camera-ready manuscripts, which we will supply on request. This applies especially to the quality of figures and halftones submitted for publication. Figures should be submitted as originals or glossy prints, as very often Xerox copies are not suitable for reproduction. For the same reason, any writing within figures should not be smaller than 2.5 mm. It might be useful to look at some of the volumes already published or, especially if some atypical text is planned, to write to the Physics Editorial Department of Springer-Verlag direct. This avoids mistakes and time-consuming correspondence during the production period.
As a special service, we offer free of charge LaTeX and TeX macro packages to format the text according to Springer-Verlag's quality requirements. We strongly recommend authors to make use of this offer, as the result will be a book of considerably improved technical quality.
Manuscripts not meeting the technical standard of the series will have to be returned for improvement.
For further information please contact Springer-Verlag, Physics Editorial Department II, Tiergartenstrasse 17, D-69121 Heidelberg, Germany.

Friedrich H. Busse   Stefan C. Müller   (Eds.)

Evolution of Spontaneous Structures in Dissipative Continuous Systems

Springer

Editors

Friedrich H. Busse
Physikalisches Institut
Universität Bayreuth
D-95440 Bayreuth, Germany

Stefan C. Müller
Fakultät für Naturwissenschaften
Institut für Experimentelle Physik
Otto-von-Guericke-Universität Magdeburg
Universitätsplatz 2
D-39106 Magdeburg, Germany

Library of Congress Cataloging-in-Publication Data.

Die Deutsche Bibliothek - CIP-Einheitsaufnahme

Evolution of spontaneous structures in dissipative continuous systems / Friedrich Busse ; Stefan C. Müller (ed.). - Berlin ; Heidelberg ; New York ; Barcelona ; Hong Kong ; London ; Milan ; Paris ; Singapore ; Tokyo : Springer, 1998
(Lecture notes in physics : N.s. M, Monographs ; 55)
ISBN 3-540-65154-3

ISSN 0940-7677 (Lecture Notes in Physics. Monographs)
ISBN 3-540-65154-3 Springer-Verlag Berlin Heidelberg New York

Typesetting: Camera-ready by the authors/editors
Cover design: *design & production*, Heidelberg

SPIN: 10552627 55/3144 - 5 4 3 2 1 0 – Printed on acid-free paper

Preface

In the past decades the subject of the spontaneous formation of structures in systems far from equilibrium has grown into a major branch of physics research with strong ties to neighboring disciplines. It has become evident that a diverse range of phenomena can be understood within a common mathematical framework which has been called nonlinear dynamics of continuous systems. This name emphasizes the close relationship to the field of nonlinear dynamics of systems with few degrees of freedom which has evolved into a mathematically mature subject in the recent past. Many dynamical features of continuous systems can actually be described through a reduction to a few degrees of freedom and properties of the latter type of systems continue to inspire the study of continuous systems.

The goal of this book is to demonstrate through numerous examples the opportunities that exist for the study of nonlinear phenomena through the use of common tools of mathematical analyses and dynamical interpretations. Instead of providing a comprehensive overview of the rapidly evolving field, the contributors to this book are trying to communicate to a wide scientific audience the essence of what they have learnt about the formation of spontaneous structures in dissipative continuous systems and about the competition between order and chaos that characterizes these systems. It is hoped that the book will be helpful even to those scientists whose disciplines are not represented by the contributing authors.

The first chapter introduces the mathematical foundations of the subject and some of the mathematical methods used in its study. This chapter clearly plays a central role even though it is not a prerequisite for understanding the following chapters. All chapters of the book are self-contained. But the first one challenges most strongly the mathematical abilities of the reader.

Fluid dynamics traditionally has played a leading role in the development of the nonlinear dynamics of continuous systems. The basic equations of motion are well known in this case and the mathematical analysis has proceeded further into regimes of spatio-temporal chaos than has been possible in other continuous systems. Even the mathematical analysis of the dynamics of complex fluids such as liquid crystals has advanced sufficiently such that quantitative comparisons with observations can be achieved. The seventh and eighth chapters focus on recent progress in this area.

The examples of structure formation in simple Newtonian fluids are far too numerous to be surveyed in the present volume. Several books have been devoted to the classical cases of spontaneous formation of patterns such as Rayleigh–Bénard convection in a fluid layer heated from below and the Taylor–Couette system. Recent progress in the understanding of unusual dynamical features of the latter system is described in the second chapter. Convection driven by centrifugal buoyancy and surface tension induced instabilities are other topics of this chapter. In the third chapter a modified Rayleigh–Bénard problem is discussed where the pure fluid is replaced by a mixture of two fluids. Also addressed in this chapter is the influence of a mean shear flow on problems of pattern formation. Examples of purely hydrodynamic instabilities in shear layers and the subsequent transition to turbulence are considered in the fourth chapter. In comparison with the closed systems treated in most of the other chapters, the formation of patterns cannot be exhibited as easily in open systems and sophisticated methods of data analysis are often required to identify "coherent structures".

In contrast to the usually assumed time independence of the external conditions, the Faraday experiment depends on time periodic variations of the effective gravity acting on a fluid layer with a free surface. A fascinating variety of standing surface waves is observed in this case. Recent theoretical developments together with experimental observations are reported in the fifth chapter. Periodic conditions can also be imposed in the spatial domain. The resulting dynamics gains complexity through the interaction of modes with forced and spontaneous structures. These and other heterogeneity effects are discussed in the sixth chapter. But examples can also be found in other chapters as, for example, in the eighth chapter where the influence of noise on the onset of electrohydrodynamic instability in liquid crystals has been analyzed.

After hydrodynamical systems, chemical oscillations are the next best understood pattern forming systems. Although the details of the chemical reactions are often too complex to be modelled accurately, quantitative comparisons with measurements are still possible on the basis of models with reaction–diffusion equations for just the major chemical species. Particularly interesting examples of pattern dynamics have been found in recent years in chemical reactions on catalytic surfaces, a topic which is presented in the ninth chapter. Because of buoyancy driven flows caused by concentration gradients, chemical reactions in the fluid phase are intimately connected with convection phenomena as is discussed in the tenth chapter. The wide area of pattern formation in flames could be added here if sufficient space were available.

It is remarkable to what extent the reaction–diffusion type equations can be applied for the description of structure forming processes in semiconductors. Especially useful are coupled activator and inhibitor equations. The activator part corresponds to ionisation or to electron–hole pair creation, while the buildup of space charges acts as an inhibitor. In the eleventh chapter a wide variety of experimentally observed phenomena are described within this framework. The basic coherent structure is the current filament. It appears as a single entity or in regular patterns and its dynamics can be influenced by magnetic fields or with electron or laser beams. The numerous electronic techniques available will undoubtedly ensure increasing applications of the formations of structures in semiconductors.

The last three chapters are devoted to subjects that are further removed from the central theme of the book. Granular media such as ordinary sand exhibit a variety of different dynamical patterns when shaken or rotated. But a quantitative continuum description is not yet possible in spite of the similarity between sandripples and wave phenomena. The theoretical interpretations in the chapter on granular materials thus rely primarily on computational simulations of the molecular dynamics type which have become a highly successful tool in studying the dynamics of collections of little solid spheres. The dynamo problem outlined in the next to last chapter, on the other hand, represents a typical bifurcation problem in the dynamics of electrically conducting fluids. There exists little opportunity for pattern formation, however, since the structure of the generated magnetic field will be governed by the outer boundary conditions. This property is caused by the huge magnetic diffusivity of the order of $1 m^2/s$ of typical liquid metals. The formation of spontaneous magnetic structures thus becomes a problem of cosmic dimensions and only with special efforts it may be realized in the laboratory.

Finally, the last chapter presents an example of the formation of spatio-temporal structures in biology and demonstrates the extent to which the development of multicellular organisms can be understood on the basis of reaction–diffusion type equations together with equations of motion. It is fascinating to see how the morphogenetic mechanisms in the evolution of a slime mould can be explained on the basis of rather simple physical and chemical principles.

Acknowledgement: The authors of the chapters of this volume are grateful to the Deutsche Forschungsgemeinschaft for its support of the research focus on Evolution of Spontaneous Structures in Dissipative Continuous Systems. They also wish to thank the evaluation committee and especially its chairman, Prof. S. Großmann, for the guidance it has provided throughout the past seven years.

Bayreuth and Magdeburg, October 1998

F.H. Busse
S.C. Müller

Contents

Mathematical Tools for Pattern Formation

Gerhard Dangelmayr[1,2] and Lorenz Kramer[3]

[1] Universität Tübingen, Institut für Theoretische Physik, Morgenstelle, D-72076 Tübingen, FRG
[2] Permanent Adress: Department of Mathematics, Colorado State University, Fort Collins, Co 80523, USA
[3] Physikalisches Institut, Universität Bayreuth, D-95440 Bayreuth, FRG

1 Introduction

1.1 Overview

We give an introduction and overview of some basic mathematical tools for the description and analysis of pattern formation in spatio-temporally continuous systems. Since we are dealing with macro- or mesoscopic phenomena a deterministic description on a hydrodynamic level is usually adequate. Thus the starting point is a set of nonlinear partial differential equations (PDEs) that represents the evolution of the relevant variables of the physical system, together with boundary conditions. Typical examples can be found in various chapters of this book. These systems are always dissipative, although the conservative limit is sometimes useful, in particular for wave phenomena.

For physical reasons the basic equations should guarantee that the variables always remain bounded, although in some instances it may also prove useful to study limiting cases that produce singularities. Only situations are considered where the external constraints or "controls" are time independent, or in some cases time periodic. For an example where the constraint is stochastic in time ("external noise") see [1]. Thus our systems display a continuous, or in some cases discrete, time translation symmetry. One is mostly interested in solutions that describe the long-time behavior. For finite dimensional systems governed by ordinary differential equations these are categorized in stationary, periodic and quasiperiodic solutions, and chaotic solutions showing complicated persistent (recurrent) dynamics. In contrast, for the PDEs considered here this categorization is not sufficient. The present state of research is in fact far from a satisfactory classification of the variety of possible complex spatio-temporal behaviours observed in simulations and experiments. Nonetheless we make the hypothesis that solutions describing the long-time behaviour of the PDEs exist and refer to these in a heuristic manner as "states". One calls a state an attractor if it belongs to some region (or basin) of attraction, i.e. a set of initial conditions. Although the general initial-value problem is never solvable in the cases of interest here, one can study systematically the evolution of small excursions away from such a state by analyzing the linear perturbational equations. The condition for a nonchaotic state to be an attractor is the condition of asymptotic stability, i.e. small excursions die out in the future time evolution of the system. This is guaranteed if all stability exponents of the linearized operator are negative.

Usually one first tries to identify a simple stationary attractor that serves as a "basic state" for the search for further attractors. In many cases of interest the basic state will loose stability as a control parameter Λ crosses some critical value Λ_c. Thus, at Λ_c the perturbational equations allow a nontrivial solution (critical mode) that does not decay and signifies a bifurcation of one (or more) long-time solutions from the basic state ("primary bifurcation"). It is important to appreciate the intimate connection between instabilities and bifurcations.

Beyond Λ_c the system has to settle down in a new attractor, i.e. there is a transition to a new state that typically exhibits qualitative new behavior. There are now various possibilities and it is useful to start a classification that relates to those symmetries of the basic state that are lifted, or "spontaneously broken", in the bifurcation. If no symmetry breaking occurs the bifurcating solution is also stationary and will simply take over the role of the previous basic state (transcritical bifurcation, see Sec. 3.1). Since the other cases involve the breaking of a discrete or continuous symmetry in the system, more than one symmetry-degenerate solution bifurcates. In particular one has the breaking of time-translation invariance which represents the Hopf (or Andronov-Hopf) bifurcation, see Sec. 3.2. The other symmetries we are concerned with here relate in particular to spatial degrees of freedom. Thus, typically, the basic state is translationally invariant and the bifurcation breaks this invariance. For a study of pattern formation in an inhomogeneous environment see [2]. For our systems, with periodic boundary conditions (or an infinite system where boundedness is demanded) one has spatially periodic bifurcating solutions. There is also the possibility that the bifurcating solution is unstable and hence no stable solution emerges above Λ_c from the basic state (subcritical bifurcation). Then the system has to undergo a discontinuous jump to another attractor, which may have arizen via a so called saddle-node bifurcation. This is the third kind of simple, generic bifurcation that can occur without additional symmetry breaking, see Sec. 3.1. At a saddle-node bifurcation stable and unstable solutions are generated. In a rather typical scenario the unstable branches connect with the unstable solutions of the previous subcritical bifurcation. Thus one can (in principle) pursue a systematic search for nonchaotic attractors by following solution branches and crossing bifurcations.

The interesting feature about bifurcations is that the long-time behavior in their vicinity, i.e. Λ near Λ_c and the solutions near the corresponding attractors, can be decribed by simplified or reduced equations for amplitudes or "order parameters", possibly coupled to other equations describing neutral or "Goldstone" modes if the pre-bifurcation state is continuously degenerate as is often the case in secondary (or higher) bifurcations. Moreover, these equations have the same form for different systems if the symmetries involved are the same. This expresses the universality found in these systems. The main aim of this article is to derive and discuss such "normal-form" equations.

The case where the attractors are not only time independent, but thermodynamic equilibrium states (possibly metastable if fluctuations are included) is not excluded. Examples are buckling instabilities in elastic media and vortex states in type II superconductors. Then, if the system remains during transitions between such states near thermal equilibrium, the dynamics can be decribed as the gradient flow of an appropriate minimizing thermodynamic potential, which gives an evolution criterion. Interestingly, even when the final states are stationary non-equilibrium states one often has a variational dynamics in a limited parameter range near the primary bifurcation. However, in general, one has nonvariational effects, and the search for useful evolution criteria has been unsuccessful.

We will not discuss the influence of (thermal) fluctuations which may become important in the vicinity of bifuractions and which can be incorporated in the basic equations in the form of stochastic Langevin forces. Some discussion on this topic can be found in [1, 3]. In general the state selected by the system will depend on random fluctuations as well as symmetry breaking imperfections.

Phase transitions also describe abrupt qualitative changes of physical systems and it is useful to set them in relation to bifurcations as considered here. Phase transitions pertain to molecular rearrangements, so fluctuations are usually important. They become sharp only in the thermodynamic limit (above the critical dimension). In experimental systems exhibiting bifurcations one is, in that sense, not in the thermodynamic limit, so the inclusion of fluctuations leads to rounding of the transitions. Some additional remarks on this topic can be found in Subsecs 6.3 and 8.1 (see also [5]).

The organization of this article reflects two distinct approaches towards the analysis of instabilities. We exemplify the techniques in terms of a simple system of reaction diffusion equations which will be introduced below. The first approach may be called the "finite dimensional approach" and leads to a description of instabilities in terms of ordinary differential equations (ODEs). It is the subject of Sections 2 - 5 for which the first author (G.D.) is responsible. The second approach may be called the "modulation approach" and leads to normal form equations in the form of PDEs. This approach is the subject of Sections 6 - 8 for which the second author (L.K.) is responsible. The two parts can in fact be read largely independent.

In the first approach the spatial domain is taken as bounded or periodic boundary conditions are imposed. Then any instability of a stationary state can be described by a finite-dimensional system of ODEs which may be derived from the basic PDEs by means of a technique called "center manifold reduction". The form of a reduced ODE-system is mainly determined by the type of instability and the symmetries. Therefore, a short introduction into the relevant symmetries and symmetry groups is given in Section 2. In Subsections 3.1 - 3.2 we introduce, discuss and interpret the reduced ODE's for "generic" instabilities for some simple cases, without explicitly relating them to underlying partial differential equations. This discussion is straightforward; advanced group theoretical terminology is avoided as much as possible. In Subsection 3.3 we describe generalizations and in particular introduce and exemplify some useful, more advanced group theoretical concepts which will be used in later Sections. In Subsection 3.4 the gap between the reaction diffusion system and the ODE-systems is closed by relating the latter to the former by formal computations. In Section 4 we review classification results for two-dimensional spatial and wave patterns and in Section 5 an overview of further bifurcation scenarios involving more complicated spatio-temporal dynamics is presented.

Then, in Section 6 the approach where an infinite spatial domain is taken as the starting point is presented. The reduction then leads to PDEs of Ginzburg-Landau type. In systems that are extended in only one dimension (1D) the procedure is rather straightforward and for primary bifurcations there is in fact a large degree of universality, see Subsections 6.2 - 6.4. In 2D various complications may arise and in fact a reduction to one or two PDEs is strictly possible only in the presence of anisotropy, see Subsection 6.5. In Section 7 results on some important classes of attractors of several of the equations are briefly presented with some emphasis on the complex Ginzburg-Landau equation. In Section 8 we discuss complications and generalizations (Subsection 8.1), phase equations valid also away from threshold (Subsection 8.2), and finally remark very briefly on secondary instabilities.

1.2 Example: A Reaction Diffusion Equation

A prototype example for a spatio-temporal evolution system is provided by the following reaction diffusion equation,

$$\frac{\partial u}{\partial t} = [D(\Lambda)\nabla^2 + K(\Lambda)]u + F(u, \Lambda) \tag{1}$$

for fields $u(t, x) = (u_1(t, x), u_2(t, x))^T$ [1] describing deviations of chemical concentrations $c = (c_1, c_2)^T$ from homogenous equlibrium values $c^*(\Lambda)$. Here, t is time, x denotes the space variables which will be specified below, $\Lambda \in \mathbf{R}^s$ is a set of parameters and the diffusion is described by a positive definite 2×2–diffusion matrix $D(\Lambda)$ and the Laplacean ∇^2. Assuming that the chemical kinetics is determined by a function $f(c, \Lambda)$ such that $f(c^*(\Lambda), \Lambda) = 0$, the reaction term is split into $f(c^* + u, \Lambda) = K(\Lambda)u + F(u, \Lambda)$ with its linear part governed by the 2×2–matrix $K(\Lambda)$ and a nonlinear part F whose Taylor expansion starts at second order. A typical form for $f = (f_1, f_2)^T$ is that of the

[1] A superscript T indicates that this quantity is actually a column vector.

trimolecular reaction of Lefever and Prigogine, known as the "Brusselator" [6]. Here, in suitable dimensionless units, $f_1 = A - c_1(1+B) + c_1^2 c_2$ and $f_2 = Bc_1 - c_1^2 c_2$, leading to $c^* = (A, B/A)^T$ and

$$K = \begin{bmatrix} B-1 & A^2 \\ -B & -A^2 \end{bmatrix}, \quad F = (\frac{B}{A}u_1^2 + 2Au_1u_2 + u_1^2 u_2)\begin{pmatrix} 1 \\ -1 \end{pmatrix}.$$

More realistic chemical or biochemical models such as that of the Belusov Zhabotinsky reaction (see Five, [6]) may involve more concentrations. Interesting effects arise from the coupling to hydrodynamic motion. We mention that reaction-diffusion models are also used to describe transport in semiconductors, see [4].

1.3 Linear Stability Analysis

Clearly, $u = 0$ is a solution of (1) for all values of the parameters Λ. We refer to this solution as the basic state and investigate its transition to instability (this analysis will be generalized in Section 6.4). In general, the stability of a stationary state is determined by the linearized system, which in our case is simply obtained by omitting the nonlinear terms, i.e.,

$$\dot{u} = \mathcal{L}(\Lambda)u\,, \tag{2}$$

where

$$\mathcal{L}(\Lambda) = D(\Lambda)\nabla^2 + K(\Lambda) \tag{3}$$

is the linear operator. When the special form of $\mathcal{L}$ as a matrix differential operator is emphasized we make this explicit by writing $\mathcal{L}(-\nabla^2, \Lambda)$. The basic state is stable if all allowed solutions of (2) decay exponentially. Since $\mathcal{L}(\Lambda)$ does not depend on t, any solution of (2) is a superposition of solutions of the form $u(t,x) = e^{\sigma t}v(x)$ where, by substitution in (2), v and σ satisfy the eigenvalue problem $\mathcal{L}(\Lambda)v = \sigma v$. Thus the basic state is stable if the linearized operator has only eigenvalues with negative real parts. Since $\mathcal{L}$ varies with Λ so do the eigenvalues, hence a transition to instability occurs at critical parameter values Λ_c for which some of the eigenvalues are on the imaginary axis while the remaining ones still have negative real parts. We refer to such a point as a primary instability or bifurcation point. To characterize instabilities further we need to specialize the spatial domain in more detail because this is crucial for the class of allowed solutions of (1) and (2) and hence for the above eigenvalue problem.

Bounded Domains. Assume first that the reaction diffusion process takes place in a one-, two- or three-dimensional bounded domain Ω ($x \in \Omega$). In this case we need to supplement (1) by appropriate boundary conditions in order to have a well defined evolution equation. The standard choices are homogenous boundary conditions of Dirichlet ($u = 0$), Neumann (normal derivative $\partial u/\partial n = 0$) or Robin ($Au + B\partial u/\partial n = 0$) type, where A and B are here assumed to be positive constants (this restriction will be relaxed in Section 3.2). For all of these choices u_1 and u_2 satisfy the same equation at the boundary, thus we can consider the scalar eigenvalue problem $-\nabla^2\phi = \mu\phi$ subject to the given boundary conditions. Without specializing these and the domain further we know [7] that $-\nabla^2$ has a discrete, unbounded set of real, nonnegative eigenvalues and each eigenvalue μ has finite multiplicity, that is, there exists only a finite number of linearly independent eigenfunctions. Since for arbitrary $p \in \mathbf{R}^2$ we can write

$$\mathcal{L}(-\nabla^2, \Lambda)[p\phi] = [\mathcal{L}(\mu, \Lambda)p]\phi \tag{4}$$

whenever $(\nabla^2 + \mu)\phi(x) = 0$, the eigenvalues of $\mathcal{L}(\Lambda)$ are determined by those of the discrete set of 2×2-matrices $\mathcal{L}(\mu, \Lambda)$ associated with the eigenvalues of $-\nabla^2$. Recall that for any 2×2-matrix the eigenvalues are $\mathrm{Tr}/2 \pm [(\mathrm{Tr}/2)^2 - \det]^{1/2}$ where Tr and det are its trace and determinant, hence both real parts are negative if $\mathrm{Tr} < 0$ and $\det > 0$.

A primary bifurcation occurs when some eigenvalues of $\mathcal{L}(\Lambda)$ have zero real parts and all other eigenvalues have negative real parts. This imposes conditions on the matrices $\mathcal{L}(\mu, \Lambda)$. The generic situation is that one of these, associated with a fixed, "critical" eigenvalue μ_c, has either a simple zero eigenvalue or a pair of imaginary eigenvalues and that all other eigenvalues μ of $\mathcal{L}$ have negative real parts. The first case leads to the conditions

$$\det\mathcal{L}(\mu_c, \Lambda) = 0 \tag{5}$$

$$\text{with } \mathrm{Tr}\mathcal{L}(\mu, \Lambda) < 0 \text{ for all } \mu \text{ and } \det\mathcal{L}(\mu, \Lambda) > 0 \text{ if } \mu \neq \mu_c\,. \tag{6}$$

The second case yields

$$\mathrm{Tr}\mathcal{L}(\mu_c, \Lambda) = 0 \tag{7}$$

$$\text{with } \det\mathcal{L}(\mu, \Lambda) > 0 \text{ for all } \mu \text{ and } \mathrm{Tr}\mathcal{L}(\mu, \Lambda) < 0 \text{ if } \mu \neq \mu_c\,. \tag{8}$$

A point Λ_c in parameter space for which (5) and (6) hold is referred to as a generic stationary instability or bifurcation point. Locally, for Λ near Λ_c, the parameter space is divided into two regions distinguished by $\det\mathcal{L}(\mu_c, \Lambda) > 0$ and $\det\mathcal{L}(\mu_c, \Lambda) < 0$. In the region with positive determinant the basic state is linearly unstable against stationary perturbations proportional to $p\phi(x)$ where $\mathcal{L}(\mu_c, \Lambda_c)p = 0$. Analogously, a point Λ_c for which (7) and (8) hold is referred to as a generic oscillatory instability (or Hopf bifurcation) point. As in the stationary case, these points provide a local division of the parameter space such that in the region where $\sigma_r = \mathrm{Tr}\mathcal{L}(\mu_c, \Lambda) > 0$ the linearized system shows growing oscillations determined by $\sigma = \sigma_r \pm \imath\omega$ with $\omega^2 = \det\mathcal{L}(\mu_c, \Lambda)$. Both types of primary instabilities are called bifurcations of codimension one because the parameters have to satisfy the single equation (5) or (7).

As an example consider the Brusselator with the diffusion matrix $D = \mathrm{diag}(1, A^2/d^2)$ and Neumann boundary conditions at the ends of an interval

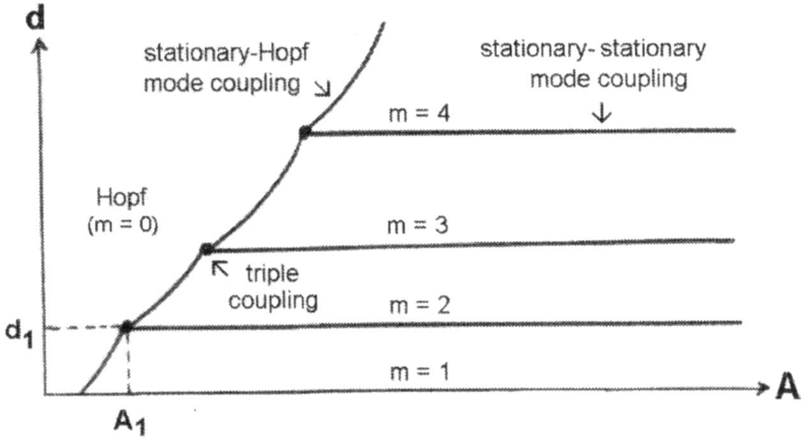

Fig. 1. Division of (A, d)-plane for the Brusselator. (schematically). In the regions labeled $m = 1$, $m = 2$ etc. the m-th mode encounters a stationary primary instability when B is increased through $B_m(d)$. In the region labeled "Hopf" one finds a Hopf bifurcation involving the constant mode when B increases through $B = A^2+1$. The boundaries between the regions are the straight lines $A > A_m \equiv m^2 + m + 1$, $d = d_m \equiv m(m+1)$ and the curve segments $A^2 + 1 = B_m(d)$ $(d_{m-1} < d < d_m)$ that connect the triple couplings.

normalized to $0 \le x \le \pi$. The eigenvalues of $-d^2/dx^2$ are $\mu = n^2$ with eigenfunctions $\phi_n = \cos nx$ where $n = 0, 1, 2, \ldots$. There are three parameters $\Lambda = (B, A, d)$ and the matrices $\mathcal{L}_n \equiv \mathcal{L}(n^2, \Lambda)$ become

$$\mathcal{L}_n = \begin{bmatrix} B - 1 - n^2 & A^2 \\ -B & -A^2(1 + n^2/d^2) \end{bmatrix} .$$

The trace of $\mathcal{L}_0$ vanishes if $B = 1 + A^2$. Moreover, if $\mathrm{Tr}\mathcal{L}_m = 0$ for $m > 0$ we have $\mathrm{Tr}\mathcal{L}_0 > 0$, hence only oscillatory instabilities involving the mode $m = 0$ are possible leading to spatially homogenous and temporally oscillating solutions (bulk oscillations). For $m > 0$ the determinant of $\mathcal{L}_m$ vanishes if $B = B_m(d) \equiv (1/m^2)(m^2 + 1)(m^2 + d^2)$. Some inspection of the inequality constraints (6), (8) shows that the primary bifurcations depend on (A, d) as shown in Figure 1.

Multiple Bifurcations. If two different equations defining primary bifurcations are satisfied simultaneously we find a so called multiple bifurcation of codimension two. When these equations involve different eigenvalues $\mu_{c1} \neq \mu_{c2}$, the multiple bifurcation point Λ_c is called a mode coupling. One distinguishes stationary-stationary, stationary-Hopf and Hopf-Hopf mode couplings, this notation refers to the two types of primary bifurcations that occur at the multiple bifurcation point. If the pair of equations involves the same eigenvalue, $\det\mathcal{L} = \mathrm{Tr}\mathcal{L} = 0$, $\mathcal{L}(\mu_c, \Lambda_c)$ has a zero eigenvalue with algebraic multiplicity two. A bifurcation at such a point is commonly referred to as a Takens-Bogdanov bifurcation because it was considered first by these authors [8]. It can be viewed as the limit of a Hopf bifurcation when the frequency goes to zero. If more than two concentrations are present in (1), there is also the possibility of codimension two bifurcation points defined by the simultaneous appearance of a zero and an imaginary and two imaginary eigenvalues for the same eigenvalue μ_c, respectively, see Section 6.4. In the Brusselator only stationary-stationary mode couplings involving neighboring eigenvalues, $\mu_{c1} = m^2$, $\mu_{c2} = (m+1)^2$, and stationary-Hopf mode couplings involving $m > 0$ and $m = 0$ occur as is apparent in Figure 1.

One can continue this way and define multiple bifurcations of codimension 3,4 etc. until the dimension of the parameter space is reached. For example, in the Brusselator stationary-stationary-Hopf triple couplings occur. In general, the dynamical behaviour observed in the vicinity of a bifurcation point shows increasing complexity when the codimension increases. Isolated bifurcation points in multi-parameter systems, where the codimension coincides with the dimension of the parameter space, are sometimes called "organizing centers".

Neutral Stability Curves. When the size of Ω increases, the distance between neighboring eigenvalues of $-\nabla^2$ becomes smaller and smaller. In the case of a stationary instability, for example, there are thus many other modes besides the critical ones governed by μ_c that become unstable for Λ slightly above Λ_c. To include also these modes in the bifurcation analysis, one usually idealizes Ω by an infinitely extended domain. In this limit the distance between neighboring eigenvalues goes to zero and we encounter a continuous spectrum. We consider specifically $\Omega = \mathbf{R}$ or $\Omega = \mathbf{R}^2$. The eigenvalues μ of $-\nabla^2$ may then be identified with $\mu = |k|^2 \geq 0$, corresponding to Fourier modes $\exp(\imath k \cdot x)$ with continuous wave numbers $k \in \mathbf{R}$ or $\mathbf{R}^2$. Thus the discrete family of 2×2-matrices becomes now a continuous family $\mathcal{L}(|k|^2, \Lambda)$. As before, neutral linear stability occurs when

$$\det\mathcal{L}(|k|^2, \Lambda) = 0, \quad \mathrm{Tr}\mathcal{L}(|k|^2, \Lambda) \leq 0 \tag{9}$$

$$\text{or} \quad \mathrm{Tr}\mathcal{L}(|k|^2, \Lambda) = 0, \quad \det\mathcal{L}(|k|^2, \Lambda) \geq 0\,. \tag{10}$$

For simplicity let us assume that $\Lambda \in \mathbf{R}$. The equations (9) and (10) define curves in the $(|k|, \Lambda)$-plane on which the determinant and the trace change signs. The curve (9) is called a stationary neutral stability curve, referred to as C_s, and (10) is called an oscillatory neutral stability curve, referred to as C_o. The typical situation is that for $(|k|, \Lambda)$ below these curves the determinant is positive and the trace is negative, respectively, and both C_s and C_o possess a minimum with respect to Λ. Moreover, C_o terminates on C_s at a point where the trace and the determinant vanish simultaneously. Such a point is also often referred to as a Takens-Bogdanov point. The lower

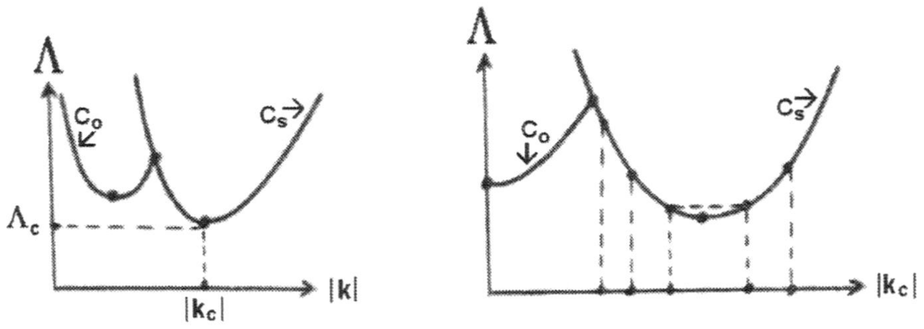

Fig. 2. Typical forms of neutral stationary (C_s) and oscillatory (C_o) stability curves. The second diagram also illustrates the appearance of a stationary mode coupling when the spectrum is discrete.

of the two minima of C_s and C_o marks the transition to instability. When further parameters are varied the curves are deformed and one may achieve that either of these minima is the lower one.

Typical forms for C_s, C_o are sketched in Figure 2. In both diagrams of that figure the transition to instability occurs at the minimum $(|k_c|, \Lambda_c)$ of C_s. The difference between these diagrams concerns the location of the C_o-minimum. It is in the first diagram at some $|k| \neq 0$ (nonuniform modes) and in the second diagram at $|k| = 0$ (constant mode). In the Brusselator the second type of diagram is encountered: replacing n^2 by k^2 and identifying Λ with B we find $C_s:\ B = k^{-2}(k^2+1)(k^2+d^2)$ with minimum at $k = d$ and $C_o:\ B = 1 + A^2 + k^2(1+d^2)$ with minimum at $k = 0$.

The neutral stability curves are also useful for investigating instabilities in bounded domains as illustrated in the second diagram of Figure 2. Here the points marked on the $|k|$-axis indicate the discrete eigenvalues in the case of a bounded domain. A stationary mode coupling occurs when two neighboring eigenvalues are at the same height on C_s as shown in the figure. Obviously, stationary triple couplings cannot occur when C_s has precisely one minimum.

1.4 Nonlinear Approaches

Depending on the spatial domain, there are two distinct approaches towards a (weakly) nonlinear analysis of (1) near a primary instability point Λ_c. In case of a bounded domain or with periodic boundary conditions imposed there is always, for any type of instability, a finite number of "critical" (or "marginally stable") modes corresponding to the eigenvalues of $\mathcal{L}(\Lambda_c)$ on the imaginary axis. All other modes have eigenvalues in the left half plane at finite distance from the imaginary axis. When Λ varies about Λ_c some eigenvalues may cross the imaginary axis and develop positive real parts, but the number of modes that become unstable (exponentially increasing solutions of (2)) cannot exceed the number of critical modes for $|\Lambda - \Lambda_c|$ sufficiently small. The basic technique is then to reduce the full PDE to a system of ODE's for the amplitudes of the critical modes alone, whereby the (infinitely many) stable modes, corresponding to eigenvalues of $\mathcal{L}(\Lambda_c)$ with negative real parts, are "slaved by" (become functions of) the critical modes. This procedure is called a center manifold reduction and its technical aspects, including specific computations, will be discussed in Section 3.4. The main objective of Sections 3 and 4 is to describe the structure of the reduced systems and their solutions. The boundedness of the domain is crucial here because it guarantees that $-\nabla^2$ and hence $\mathcal{L}$ have only isolated eigenvalues.

When the domain is unbounded we find beyond a primary instability a continuous family of unstable modes because $-\nabla^2$ has a continuous spectrum (arbitrary wave vectors $\boldsymbol{k}$). The appropriate approach to analyzing these instabilities is in terms of modulation equations of the Ginzburg-Landau type, canonical systems of PDE's for slowly varying envelopes. This approach and extensions of it will be discussed in Sections 6 – 8. A simpler approach, where one restricts the analysis to spatially periodic solutions, consists in artificially imposing periodic boundary conditions with basic wave numbers given by $|\boldsymbol{k}_c|$, see Section 4 for a treatment in $2D$.

2 Symmetries and Symmetry Groups

We summarize first basic symmetry concepts that will be used in subsequent sections. In mathematical terms symmetries are governed by group theory. Abstractly, a group Γ consists of a set of elements $\gamma \in \Gamma$ with an associative composition law $(\gamma_1, \gamma_2) \to \gamma_1\gamma_2 \in \Gamma$, an identity element and a unique inverse γ^{-1}.The group is called Abelian or commutative if $\gamma_1\gamma_2 = \gamma_2\gamma_1$ holds for all $\gamma_1, \gamma_2 \in \Gamma$.

The groups we consider are derived from transformations γ acting on x as $x \to \gamma x \equiv \gamma_o x + x_0$ where γ_o is an orthogonal matrix (rotation or reflection) and x_0 is a translation vector, hence $\gamma^{-1}x = \gamma_o^{-1}(x - x_0)$. Transformations of that kind "induce" operations $u \to \gamma u$ on (scalar or vector) functions $u(x)$, defined by $(\gamma u)(x) \equiv u(\gamma^{-1}x)$; the use of γ^{-1} is a convention. This class of transformations is the largest class that commutes with the Laplacean, that is, $\nabla^2(\gamma u) = \gamma(\nabla^2 u)$ where it is understood that in the first expression x is first replaced by $\gamma^{-1}x$ and then ∇^2 is applied whereas in the second expression one computes first $\nabla^2 u(x)$ and then replaces x by $\gamma^{-1}x$.

Concerning the reaction diffusion system (1), let us denote the full nonlinear operator on the right hand side in a short hand notation by $\mathcal{F}(u)$. Since the Laplacean is the only differential operator contained in $\mathcal{F}$ we find $\mathcal{F}(\gamma u) = \gamma\mathcal{F}(u)$ which has to be understood in the same manner as before. Hence $\mathcal{F}$ also commutes with all transformations γ defined above which together constitute the Euclidean group $E(n)$, $n = 1, 2$ or 3. However, not all of these transformations are permitted if (1) is posed in a bounded domain or subject to periodic boundary conditions. The set of transformations admitted in these cases forms a group Γ (a subgroup of $E(n)$) and the above commutation property of the nonlinear operator $\mathcal{F}$, restricted to $\gamma \in \Gamma$, is referred to as (Γ-) equivariance, another term is covariance.

An immediate consequence of equivariance is that $\gamma u(t,x)$ is a solution of (1) whenever $u(t,x)$ is a solution. If both are the same, i.e., u is left invariant, $u = \gamma u$, one calls the transformation γ a symmetry of u. Invariances provide an important scheme to characterize spatio-temporal patterns. Moreover, symmetries in general dictate to a large extent the form of a reduced system that describes an instability as well as its solutions. In the following we introduce the basic symmetry groups that will be exploited further in Sections 3 and 4.

When (1) is posed in a bounded domain Ω, we can admit only the group of transformations which leave Ω invariant, i.e., $\gamma x \in \Omega$ whenever $x \in \Omega$. The simplest case occurs when Ω is an interval, $0 \le x \le l$. The basic symmetry transformation here is reflection about the midpoint, denoted as $\kappa : x \to l - x$. Reflecting twice brings us back, i.e., $\kappa^2 \equiv \kappa\kappa = id$, the identity, hence κ and id together form the simplest possible group consisting only of two elements, which is called the cyclic group C_2 of order 2.

For annuli or circular discs the allowed symmetry operations are rotations and reflections. In terms of polar coordinates (r, φ), a (clockwise) rotation of x by an angle ψ simply corresponds to a translation of the angular coordinate, $\psi : \varphi \to \varphi - \psi$, where φ and $\varphi + 2k\pi$ as well as ψ and $\psi + 2k\pi$ are identified when k is an integer (translation along a circle). All translations together define the rotation group $SO(2)$. The reflection is again denoted by $\kappa : \varphi \to -\varphi$. It corresponds in Cartesian coordinates to $(x_1, x_2) \to (x_1, -x_2)$ and generates a reflection group, denoted by C_2 as before. The combined operation $\psi\kappa : \varphi \to -\varphi - \psi$ corresponds to a reflection of (x_1, x_2) about a line through the origin which has an angle $-\psi/2$ with the x_1-axis. All translations ψ and reflections $\psi\kappa$ together constitute $O(2)$, the group of orthogonal linear transformations in the plane which is also referred to as the "symmetry group of a circle". Note, however, that $\kappa\psi : \varphi \to -\varphi + \psi$ reflects about the line with angle $\psi/2$, hence $\kappa\psi = (-\psi)\kappa$.

In general, one can combine two groups Γ_1, Γ_2 to the so called product group if the set of all combinations $\gamma_1\gamma_2$ with $\gamma_1 \in \Gamma_1$, $\gamma_2 \in \Gamma_2$ is closed under composition. The product group is called direct, written as $\Gamma_1 \times \Gamma_2$, if always $\gamma_1\gamma_2 = \gamma_2\gamma_1$ and semidirect if these operations do not commute in general in which case we write $\Gamma_1 \times_s \Gamma_2$. In that sense, $O(2)$ is the semidirect product $C_2 \times_s SO(2)$. A simple example for a direct product is the dihedral group $D_2 = C_2 \times C_2$, the symmetry group of a rectangle. Here, the two C_2-factors are generated by two independent reflections, e.g. $\kappa_1 : (x_1, x_2) \to (-x_1, x_2)$ and

$\kappa_2 : (x_1, x_2) \to (x_1, -x_2)$. The combined reflection $\kappa_1\kappa_2 = \kappa_2\kappa_1$ is inversion about the origin which is the same as rotation by $\pi \in SO(2)$.

All symmetry groups of two-dimensional bounded domains are subgroups of $O(2)$. Besides C_2 (one reflection axis) and D_2 (two reflections axes) there is the cyclic subgroup C_m of $SO(2)$ of order m, defined by restricting the allowed translations ψ to angles that are multiples of $2\pi/m$. If supplemented by the reflection κ, this defines $D_m = C_2 \times_s C_m$, the symmetry group of a "regular m-gon" (equilateral triangle for $m = 3$, square for $m = 4$ etc.).

Analogously to the two-dimensional case, a symmetry group of a three-dimensional bounded domain is necessarily a subgroup of $O(3)$, the group of orthogonal linear transformations in $\mathbf{R}^3$. We do not discuss problems that involve this more complicated symmetry group here, a short summary of the basic results obtained so far is given in Section 5. A simple extension of the two-dimensional case is provided by domains which are invariant under reflections about a mid plane. If no further true space symmetry is present, the basic symmetry group has the form $\Gamma \times C_2$, where Γ is the symmetry group of the two-dimensional cross section of Ω in the reflection plane, for example $O(2)$ if Ω is a cylinder or an ellipsoid of revolution.

If $x \in \mathbf{R}$ and periodic boundary conditions are imposed, $u(t, x + 2\pi) = u(t, x)$ (choosing an appropriate length scale for x), the relevant symmetry group is again $O(2)$ acting as above on $\varphi \equiv x$ (the real line is "compactified" to a circle). In the case of periodic boundary conditions in a plane we have to choose a lattice. As in crystallography, a planar lattice is specified by two linearly independent generating vectors e_1, e_2 and consists of all points of the form $n_1x_1 + n_2x_2$ with integers n_1, n_2. Periodicity with respect to the lattice means $u(t, x + x_l) = u(t, x)$ whenever x_l is a lattice point. There are now two independent translation invariances $x \to x - (\psi_i/2\pi)e_i$ $(i = 1, 2)$, each of them defining a $SO(2)$-translation group because of the periodicity. Since they commute, their product is direct and is called the "torus group" $T^2 = SO(2) \times SO(2)$. In addition to these translation symmetries there is a discrete subgroup D_m of $O(2)$ (called the point group) which leaves the lattice as a whole invariant. Depending on the angle between and relative sizes of e_1, e_2, there are three possibilities for m, namely D_2 (oblique, rhombic or rectangular lattice), D_4 (square lattice) and D_6 (hexagonal lattice). The full symmetry group under which (1) is equivariant is $D_m \times_s T^2$ which is a semidirect product because these operations do not commute generally. We specialize the generating vectors and the corresponding symmetry operations in more detail in Section 4. For further details on group and group representation theory we refer to [9, 10].

Besides spatial symmetries, a PDE may possess further symmetries involving the dependent variables. The simplest case is equivariance under a reflection operation, for example $\kappa_0 : u \to -u$ if F is odd in u. For more general PDE's, in particular in fluid mechanics, transformations of the dependent variables may have to be combined with space transformations.

3 Generic Bifurcations with Discrete and Circular Symmetries

3.1 Stationary Bifurcations

Assume that, for $\Lambda = \Lambda_c$, (1) encounters a generic primary, stationary instability involving the eigenvalue μ_c of $-\nabla^2$, i.e., $\det\mathcal{L}(\mu_c, \Lambda_c) = 0$ and $\mathrm{Tr}\mathcal{L}(\mu_c, \Lambda_c) < 0$. The critical modes are then of the form $p\phi(x)$ where $p \in \mathbf{R}^2$ satisfies $\mathcal{L}(\mu_c, \Lambda_c)p = 0$ and $\nabla^2\phi + \mu_c\phi = 0$. Due to our assumptions on the spatial domain there are finitely many, say n, linearly independent eigenfunctions $\phi_j(x)$ $(1 \leq j \leq n)$ of $-\nabla^2$ corresponding to μ_c. In the cases considered below $n = 1$ or $n = 2$, in Section 4 $n \leq 12$ and in case of $O(3)$-symmetric domains n can be any odd number, cf. Section 5. Introducing (real) amplitudes X_j, an arbitrary critical mode can be written as a superposition $\sum_{j=1}^{n} X_j \Phi_j(x)$ of the "basic critical modes" $\Phi_j \equiv p\phi_j$. Hence the critical modes form an n-dimensional space which is called the center eigenspace, $\mathcal{E}_c$, of $\mathcal{L}(\Lambda_c)$ and the X_j play the role of coordinates in this space. Setting $X = (X_1, \ldots, X_n)$ we can identify $\mathcal{E}_c$ with the Euclidean space $\mathbf{R}^n$.

Let now Γ be the symmetry group of the spatial domain. A general result of group representation theory is that the eigenspaces of the Laplacean are invariant under Γ, i.e., $\gamma\phi_j = \sum_{i=1}^{n} A_{ij}(\gamma)\phi_i$ with certain orthogonal $n \times n$-matrices $A = \{A_{ij}\}$ which satisfy $A(\gamma_1\gamma_2) = A(\gamma_1)A(\gamma_2)$. This induces a group operation on the amplitude vectors X, defined by $\gamma(\sum_j^n X_j\Phi_j) = \sum_i^n (\gamma X)_i \Phi_i$, where $(\gamma X)_i = \sum_j^n A_{ij}(\gamma)X_j$ is the i-th coordinate of the transformed amplitude vector γX. In general, a collection of orthogonal matrices $A(\gamma)$ as above is called a representation (or action) of Γ in $\mathbf{R}^n$. Since in our case it was derived by restricting transformations $u \to \gamma u$ to critical modes, we refer to it as the "induced representation in the center eigenspace".

Consider first the simple case where Ω is the interval $0 \leq x \leq \pi$ and Dirichlet boundary conditions are imposed. The eigenvalues and eigenfunctions of $-d^2/dx^2$ are $\mu = k^2$ and $\phi_k(x) = \sin kx$ with k a positive integer. Assume $\mu_c = m^2$, thus there is only one critical mode $p \sin mx$ with amplitude $X \in \mathbf{R}$. The reflection $\kappa : x \to \pi - x$ applied to $Xp\sin mx$ yields $(-1)^{m+1}Xp\sin mx$, inducing the representation $X \to -X$ if m is even and $X \to X$ (trivial representation) if m is odd. These two representations exhaust also in general all possibilities if there is only one critical eigenfunction $\phi(x)$ associated with μ_c. In particular, if Ω admits one or several reflection invariances and $\phi(x)$ is odd under at least one reflection κ, $\kappa\phi = -\phi$, we find $X \to -X$ as resulting symmetry operation in $\mathcal{E}_c$ whereas no nontrivial symmetry operation remains if ϕ is even under all reflections.

A slightly more complicated situation occurs if (1) is posed on the real line with 2π-periodic boundary conditions. In this case the eigenvalues and eigenfunctions of $-d^2/dx^2$ are, in complex representation, given by $\mu = k^2$ and $e^{\pm ikx}$ with nonnegative integers k. Assuming again that $\mu_c = m^2 > 0$, there are now two linearly independent eigenfunctions, thus two real amplitudes are needed. These are most conveniently represented by a single complex amplitude Z, yielding $(Ze^{imx} + \bar{Z}e^{-imx})p$ as parametrization of the center eigenspace. A translation $x \to x - \psi$ then corresponds to a phase shift $Z \to e^{im\psi} Z$ and the reflection $x \to -x$ leads to $Z \to \bar{Z}$ as induced symmetry operations on Z where the overbar denotes complex conjugation. The same induced symmetries occur in case of two-dimensional circular domains such as annuli or discs. Here the eigenfunctions of the Laplacean are of the form $\phi_k(r)e^{\pm ik\varphi}$ and translations and reflections of φ have the same effect on complex amplitudes as before. If Ω is a cylinder with circular base it is left invariant also by reflection about the mid plane which may induce $Z \to -Z$ as additional symmetry operation.

As mentioned earlier, the center manifold reduction yields a closed ODE-system for the amplitudes of the critical modes or, in other words, the dynamics of (1) is governed by a vector field in the center eigenspace. We write this ODE-system as $\dot{X} = g(X, \lambda)$ where $\lambda \equiv \Lambda - \Lambda_c$ describes deviations of Λ from criticality. Since X and λ are small one can work with a truncated Taylor series which is referred to as a "normal form" and contains certain "relevant terms". These are basically determined by the Γ-equivariance of g, that is, $g(\gamma X, \lambda) = \gamma g(X, \lambda)$ for all γ in the domain's symmetry group Γ, where $X \to \gamma X$ is the induced representation in $\mathcal{E}_c$. Moreover, $g(0, \lambda) = 0$ because $X = 0$ is always a solution that corresponds to the basic state. In addition, the linearized matrix $d_X g(0, \lambda)$ represents the critical eigenvalue in $\mathcal{E}_c$ and hence vanishes for $\lambda = 0$.

(A) Elementary Bifurcations. The simplest case occurs if only one eigenfunction $\phi(x)$ is present so that $X \in \mathbf{R}$. As discussed before, there are then only two possible symmetry operations, namely the trivial one, $X \to X$, and the reflection $X \to -X$. In the first case there are no symmetry restrictions whereas in the second case g is odd, $g(-X, \lambda) = -g(X, \lambda)$, thus the lowest order expansions read, respectively,

$$\dot{X} = \alpha X + aX^2 , \tag{11}$$

$$\dot{X} = \alpha X + aX^3 . \tag{12}$$

Here, $\alpha \equiv \alpha(\lambda)$ is the linear, λ-dependent coefficient which represents the linearized part of (1) in $\mathcal{E}_c$, hence $\alpha(0) = 0$. Usually, α is computed only at linear order in λ, an explicit formula will be given in Subsection 3.4. We consider α as an independent, essential small parameter in which the parameter dependence of the original system is encapsulated. The quadratic or cubic coefficient a may also vary with λ, however, since there are no further restrictions we will generically find $a(0) \neq 0$. Thus in practice a is computed at criticality, i.e., for $\lambda = 0$. Its non-vanishing provides also a justification for the omission of higer order terms.

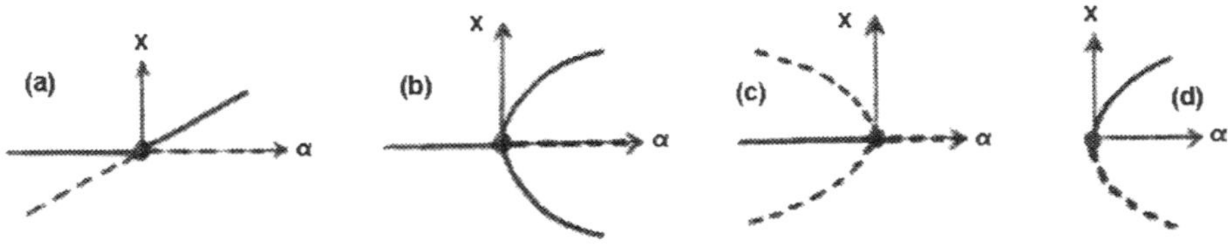

Fig. 3. Generic stationary bifurcation diagrams. (a): transcritical bifurcation, eq. (11), with $a < 0$. (b) and (c): super- ($a < 0$) and subcritical ($a > 0$) pitchfork bifurcations, eq. (12). (d): saddle node bifurcation, eq. (13), with $a < 0$. Solid/dashed branches are stable/unstable.

The equations (11) and (12) are called the normal forms for a transcritical and a pitchfork bifurcation, respectively. Being one-dimensional, their behaviour is simple, the analysis involves merely an investigation of the fixed points, i.e., solutions X of the equation $\dot{X} = g = 0$. Their dependence on α is usually shown graphically in the form of a bifurcation diagram X versus α which includes also the information about the stability (sign of $d_X g$ along a solution of $g = 0$). These diagrams are sketched in Figures 3(a)-(c). As is apparent from Figure 3(a), (11) describes a stability exchange between the basic state $X = 0$ and a nonlinear branch, $\alpha + aX = 0$, with linear dependence on α (at leading order). In case of (12) we find besides $X = 0$ a pair of nontrivial, symmetry-related ($X \to -X$) solutions, $\alpha + aX^2 = 0$, with square root dependence on α. They are stable ($d_X g < 0$) in the supercritical case $a < 0$ where they are created after the basic state has become unstable, and unstable ($d_X g > 0$) in the subcritical case $a > 0$ where they coexist with the stable basic state. For both normal forms a nontrivial steady state X corresponds to a spatial solution of (1) which is approximately given by $Xp\phi(x)$, where $p\phi$ is the critical mode. We note that the *full* solution of (1) involves additional terms coming from the stable modes, however, these depend at least quadratically on X and can, therefore, be neglected in the small amplitude limit. The pitchfork is the simplest and most basic bifurcation describing "spontaneous symmetry breaking" which here refers to the fact that $Xp\phi(x)$ is not left invariant under a reflection κ, e.g. $x \to \pi - x$, in contrast to the basic state from which it bifurcates.

Besides (11) and (12) there is a further basic normal form describing bifurcations with a one-dimensional center eigenspace, namely

$$\dot{X} = \alpha + aX^2 , \tag{13}$$

which follows from (11) when the linear term is replaced by a X-independent term. Equation (13) is called the normal form for a saddle node bifurcation and describes the coalescence and disappearance of a stable and an unstable solution, as shown in Figure 3(d). The main difference to (11) is that it does not admit $X = 0$ as basic solution for $\alpha \neq 0$ and hence does not describe bifurcation from a basic state. Instead, a bifurcation governed by this normal form typically takes place on a nontrivial branch of stationary solutions. A typical scenario is that the unstable branch created in a subcritical pitchfork "bends back" in a saddle node bifurcation. This scenario is described by a "degenerate" bifurcation, with normal form $\dot{X} = \alpha X + \beta X^3 + bX^5$. Here, α, β are two small parameters and the nonlinear coefficient b is negative.

(B) Circular Symmetry. If instead of sines the harmonic modes $e^{\pm \imath m x}$ (periodic boundary boundary conditions on a line) or two-dimensional modes with angular parts $e^{\pm \imath m \varphi}$ (circular disc or annulus) are critical, this instability is described by a complex amplitude Z. As pointed out before, the induced symmetry operations are $Z \to e^{\imath m \psi} Z$ and $Z \to \bar{Z}$. The reduced ODE $\dot{Z} = g(Z, \bar{Z}, \lambda)$ has to be equivariant under these operations, that is, $g(e^{\imath m \psi} Z, e^{-\imath m \psi} \bar{Z}, \lambda) = e^{\imath m \psi} g(Z, \bar{Z}, \lambda)$ and $g(\bar{Z}, Z, \lambda) = \bar{g}(Z, \bar{Z}, \lambda)$. It is easy to see that a general function g with these properties has the form $g = Zh(|Z|^2, \lambda)$ with h real valued, hence up to cubic order the resulting normal form reads

$$\dot{Z} = \alpha Z + aZ^2 \bar{Z}, \tag{14}$$

where α and a are real and have the same meanings as before. In terms of polar coordinates, $Z = Re^{\imath \Psi}$, (14) becomes $\dot{R} = \alpha R + aR^3$, $\dot{\Psi} = 0$, thus the equation for R reduces to the pitchfork normal form (12). Generically we find, therefore, again the sub- and supercritical bifurcation diagrams of Figure 3(b),(c), with $X = R$ and restricted to $R \geq 0$. A nontrivial steady state corresponds, however, to a full circle of solutions of (14) because the phase Ψ is arbitrary. The normal form (14) may, therefore, be also called a "pitchfork of revolution".

A point (R, Ψ) on the steady state circle corresponds to a spatial, nonhomogeneous stationary solution of (1), given approximately by $2Rp\cos(mx + \Psi)$ and $2Rp\phi_m(r)\cos(m\varphi + \Psi)$ in the one- and two-dimensional cases. Rotations along the circle correspond to translations (or rotations) of the pattern. The new state shows now only an m-fold translation symmetry, i.e., the continuous translation symmetry of the basic state is "spontaneously broken". Besides these discrete translations, the new state is also left invariant under the reflection $x \to -x + 2\Psi/m$. Taken together, the m-fold translations and the reflection generate D_m as basic symmetry group under which the new pattern stays invariant.

(C) $\mathbf{D_m}$-Symmetric Secondary Bifurcation. Assume now that a primary, stationary instability leads to the appearance of a stable D_m-symmetric solution pattern $u(x)$ or $u(r, \varphi)$ as described before with $m \geq 3$. This solution is called a "primary solution" (or "primary branch") because it is generated from the basic state. If one follows such a pattern into global parameter regimes (usually numerically, the above representation is valid only near the primary instability), this pattern may itself become unstable in a so called secondary instability or bifurcation. To determine such an instability one has to expand (1) about u and inspect the corresponding linearized operator. This expansion shows no longer equivariance under the full, original symmetry $O(2)$ but only under the discrete symmetry D_m of the pattern. Accordingly, a generic instability is characterized by critical modes which form a representation space for D_m. In the case of a stationary instability, this space is typically spanned by a single complex mode $\Phi(x)$ ($x \in \mathbf{R}$, 2π-periodic boundary conditions) or $\Phi(r, \varphi)$ (annulus or disc). We focus on the one-dimensional case, the two-dimensional case is basically the same.

Invariance under D_m means $\Phi(x + 2\pi/m) = e^{2\pi i l/m}\Phi(x)$ and $\Phi(-x) = \bar{\Phi}(x)$ for some integer l, for simplicity we assume $l = 1$. Note that this implies $\phi(x + \pi) = -\phi(x)$ if m is even. As before we introduce a complex amplitude Z and write $Z\Phi + \bar{Z}\bar{\Phi}$ to parametrize a general critical mode in the center eigenspace. Also here it is possible to derive a reduced ODE-system $\dot{Z} = g(Z, \bar{Z}, \lambda)$ for Z which is equivariant under $Z \longrightarrow e^{i\psi}Z$ and $Z \longrightarrow \bar{Z}$, however, ψ is now restricted to multiples of $2\pi/m$. The corresponding normal form is obtained by adding to (14) the lowest order term that breaks the continuous to an m-fold rotation symmetry. This leads to

$$\dot{Z} = \alpha Z + aZ^2\bar{Z} + b\bar{Z}^{m-1}\,, \tag{15}$$

with a further (real) nonlinear coefficient b. Equation (15) plays the role of the normal form for a generic, D_m-equivariant stationary bifurcation [10]. In polar coordinates (15) becomes

$$\dot{R} = \alpha R + aR^3 + bR^{m-1}\cos m\Psi\,, \quad \dot{\Psi} = -bR^{m-2}\sin m\Psi\,,$$

thus the steady states admit no longer an arbitrary phase. There are two sets of stationary solutions with phases $\Psi_+ = 2\pi k/m$ and $\Psi_- = (2k+1)\pi/m$ where k is an arbitrary integer. Solutions distinguished by different values of k are symmetry related (the group theoretical jargon is "conjugate"), hence we may set $\Psi_+ = 0$ and $\Psi_- = \pi$ (π/m) for m odd (even). Substituting into $\dot{R} = 0$ yields $\alpha + aR^2 \pm bR^{m-2} = 0$, giving a transcritical solution branch for $m = 3$, distinct pitchforks for $m = 4$ and pitchforks with same leading terms which are split apart at higher order for $m \geq 5$. Inspection of the eigenvalues of the Jacobean evaluated along these solutions shows that for $m = 3$ both are unstable and for $m \geq 4$ at most one can be stable. The patterns $u_\pm(x)$ corresponding to $\Psi_\pm$ have the form $u_+(x) = 2R\Re[\Phi(x)]$ and $u_-(x) = -u_+(x)$ for m odd whereas for m even $u_-(x) = 2R\Re[e^{i\pi/m}\Phi(x)]$, with $\Re$ refering to the real parts. These patterns have to be superimposed on the underlying D_m-equivariant solution $u(x)$. Both of them show reflection invariances, $u_\pm(-x) = u_\pm(x)$ for m odd and $u_+(-x) = u_+(x)$, $u_-(-x - 2\pi/m) = u_-(x)$ as well as $u_\pm(x + \pi) = -u_\pm(x)$ for m even. In any case, we encounter a further spontaneous symmetry breaking because the discrete translation invariance of $u(x)$ is broken, leaving only the above reflection invariances as remaining symmetries of the patterns.

3.2 Hopf Bifurcation

The system (1) encounters a primary Hopf bifurcation when $\mathcal{L}(\mu_c, \Lambda_c)$ has imaginary eigenvalues $\pm \imath\omega_c$ and $\mathrm{Tr}\mathcal{L}(\mu, \Lambda_c) < 0$ for all noncritical eigenvalues μ. Let $p \in \mathbf{C}^2$ be the corresponding eigenvector, $\mathcal{L}(\mu_c, \Lambda_c) = \imath\omega_c p$. The eigenfunctions of $\mathcal{L}(\Lambda_c)$ corresponding to $\imath\omega_c$ are complex modes of the form $\Phi(x) = p\phi(x)$ where as in the preceeding subsection $\phi(x)$ is a (real) eigenfunction of the Laplacean, $(\nabla^2 + \mu_c)\phi = 0$. When n linearly independent eigenfunctions $\Phi_j = p\phi_j$ of $\mathcal{L}(\Lambda_c)$ corresponding to $\imath\omega_c$ are present, one associates to each of them a complex amplitude Z_j and represents an (arbitrary real) critical mode as $\sum_j(Z_j\Phi_j + \bar{Z}_j\bar{\Phi}_j)$. These critical modes form a $2n$-dimensional space which is again called the center eigenspace, $\mathcal{E}_c$, and the Z_j play the role of complex coordinates in that space. Setting $Z = (Z_1, \ldots, Z_n)$ we may identify $\mathcal{E}_c$ with $\mathbf{C}^n$ or, if real coordinates $Z_j = X_j + \imath Y_j$ are used, with $\mathbf{R}^{2n}$. The symmetry group Γ comes into play in the same way as in the preceeding subsection: transformations $\gamma\phi_j$ lead to orthogonal matrices $A(\gamma)$ which induce a representation $Z \rightarrow \gamma Z$ on the amplitude vector. Being real, these operations act in the same way on the real and imaginary parts of Z. In a group theoretical jargon this fact is referred to by saying "$\mathcal{E}_c$ is the direct sum of two copies of the same representation space".

The center manifold reduction leads here to a reduced, Γ-equivariant ODE-system $\dot{Z} = g(Z, \bar{Z}, \lambda)$ where again $g(0, 0, \lambda) = 0$ and the linearized part at criticality ($\lambda = \Lambda - \Lambda_c = 0$) is $\dot{Z} = \imath\omega_c Z$. This linear part admits periodic solutions of the form $Z(t) = e^{\imath\omega_c t}Z_0$, hence a time translation $t \rightarrow t+t_0$ can be absorbed in the initial condition by transforming $Z_0 \rightarrow e^{\imath\theta}Z_0$ where $\theta = \omega_c t_0$. As shown in many texts on dynamical systems, e.g. [11], this fact can be used to obtain a significant simplification of the full nonlinear ODE-system. Namely, one can perform a so called near identity transformation on $(Z, \bar{Z})$ such that the transformed vector field $\dot{Z} = g_N(Z, \bar{Z}, \lambda)$ attains the additional equivariance $g_N(e^{\imath\theta}Z, e^{-\imath\theta}\bar{Z}, \lambda) = e^{\imath\theta}g_N(Z, \bar{Z}, \lambda)$ for arbitrary phases θ. The underlying symmetry operations $Z \rightarrow e^{\imath\theta}Z$ constitute a $SO(2)$-translation group which is commonly referred to as S^1 (identifying rotations with points on a circle) to distinguish it from spatial rotation groups. The transformation that brings g to the form g_N is of the form $Z \rightarrow Z + \text{h.o.t.}$ where h.o.t. represents terms of higher order, including λ-dependent linear terms which vanish for $\lambda = 0$. These and the resulting terms of g_N can be computed successively, order by order. Since the leading term of the transformation is Z (the transformation is "near the identity"), we still can describe solutions of (1) approximately by solutions $Z(t)$ of the transformed ODE in terms of the representation of $\mathcal{E}_c$ which typically leads to wave patterns.

As in the preceeding subsection we will set up and discuss suitably truncated series expansions of g_N and refer to these as "normal forms". The relevant terms are inferred from the equivariance of g_N under the extended symmetry group $\Gamma \times S^1$. This product is direct because "temporal" phase shifts θ and "spatial" transformations γ commute.

Concerning parameter variations, we note that in all cases considered here and in the next section the center manifold reduction yields a linear part of the form $\dot{Z} = (\imath\omega_c + \alpha_1(\lambda))Z + \alpha_2(\lambda)\bar{Z}$ where $\alpha_1(0) = \alpha_2(0) = 0$. Since the second term is incompatible with the normal form symmetry, one effect of the near identity transformation is to remove it leaving $\dot{Z} = (\imath\omega_c + \alpha(\lambda))Z$, with $\alpha_1 = \alpha$ at leading order. Obviously $\imath\omega_c + \alpha(\lambda)$ is the complex eigenvalue of $\mathcal{L}(\Lambda)$ into which $\imath\omega_c$ evolves when Λ is varied about Λ_c. Its imaginary part, $\alpha_i(\lambda)$, describes small frequency variations about ω_c whereas the real part, $\alpha_r(\lambda)$, is the basic critical parameter which marks the transition to instability. Analogously as in the case of stationary bifurcations we consider α_r as essential, independent parameter in terms of which bifurcating solutions are analyzed whereas α_i plays the role of another small parameter that determines frequencies. If only one parameter Λ is varied, α_i is to be considered as function of α_r.

We note that the normal form for a Hopf bifurcation is a special case of a so called Poincare Birkhoff normal form. The basic idea behind these normal forms is to exploit the linear part of a vector field to remove certain nonlinear Taylor terms. A systematic approach to general normal form theory is developed in [12].

(A) Elementary Hopf Bifurcation. We begin again with the simplest case where only one critical eigenfunction $\phi(x)$ with corresponding critical mode $\Phi = p\phi$ is present. Accordingly there is a single complex amplitude Z and the normal form is equivariant under $Z \to e^{\imath\theta}Z$. Since for $\theta = \pi$ this yields $Z \to -Z$, spatial reflection symmetries need not be considered.

Any equivariant g has the form $g(Z, \bar{Z}, \lambda) = Zh(|Z|^2, \lambda)$ where, in contrast to the $O(2)$-symmetric case, h is complex valued. Thus the lowest order terms of the normal form are

$$\dot{Z} = (\imath\omega_c + \alpha + a|Z|^2)Z \,. \tag{16}$$

In polar coordinates, $Z = Re^{\imath\Psi}$, (16) becomes

$$\dot{R} = \alpha_r R + a_r R^3 \,, \quad \dot{\Psi} = \omega_c + \alpha_i + a_i R^2 \,, \tag{17}$$

where the subscripts r and i refer to real and imaginary parts. The R-equation reduces to a pitchfork bifurcation and governs the essential behaviour. Its stationary points R correspond to bifurcating periodic orbits $Z = Re^{\imath\omega t}$ with $\omega = \dot{\Psi}$ close to ω_c. Thus Figures 3(b) or (c) (with $X = R$, $\alpha = \alpha_r$, $a = a_r$) describe the local bifurcation behaviour when $a_r \neq 0$. We find a unique stable and unstable periodic orbit if $a_r < 0$ (supercritical Hopf bifurcation) and $a_r > 0$ (subcritical Hopf bifurcation), respectively. This periodic orbit induces a standing wave solution of (1) which is approximately given by $2R\Re[pe^{\imath\omega t}]\phi(x)$.

(B) Circular Symmetry. We consider first the one-dimensional case in which the two linearly independent complex modes $pe^{\pm imx}$ are critical. Accordingly two amplitudes (Z_1, Z_2) are needed, hence the center eigenspace is represented as

$$(Z_1 e^{imx} + Z_2 e^{-imx})p + cc, \tag{18}$$

where cc refers to the complex conjugate expression. The transformations induced by translations $\psi : x \to x - \psi$ and the reflection $\kappa : x \to -x$ are here

$$\psi : (Z_1, Z_2) \to (e^{im\psi} Z_1, e^{-im\psi} Z_2), \quad \kappa : (Z_1, Z_2) \to (Z_2, Z_1), \tag{19}$$

and in addition we have the normal form symmetry

$$\theta : (Z_1, Z_2) \to (e^{i\theta} Z_1, e^{i\theta} Z_2). \tag{20}$$

It can be shown [13] that a general vector field that is equivariant under the operations (19),(20) has the form $\dot{Z}_1 = g(|Z_1|^2, |Z_2|^2, \lambda) Z_1$, with a complex valued function g and the $\dot{Z}_2$-equation following by interchanging Z_1 and Z_2. Up to cubic order one obtains, therefore,

$$\dot{Z}_1 = (i\omega_c + \alpha + a|Z_1|^2 + b|Z_2|^2) Z_1, \tag{21}$$

with two nonlinear complex coefficients a, b. In polar coordinates, $Z_j = R_j e^{i\Psi_j}$ $(j = 1, 2)$, (21) becomes

$$\dot{R}_1 = (\alpha_r + a_r R_1^2 + b_r R_2^2) R_1, \quad \dot{\Psi}_1 = \omega_c + \alpha_i + a_i R_1^2 + b_i R_2^2, \tag{22}$$

and the equations for (R_2, Ψ_2) follow by interchanging the indices 1 and 2. Since the phases decouple, one can look for stationary solutions of the (R_1, R_2)-system. There are two basic types of solutions, denoted by TW: $(R_1, R_2) = (R, 0)$ or $(0, R)$ which are related to each other by the reflection, and SW: $(R_1, R_2) = (R, R)$. Both are determined by pitchforks, $\dot{R} = (\alpha_r + AR^2)R = 0$ where $A = a_r$ for TW and $A = a_r + b_r$ for SW. Some inspection shows that these are in fact the only solutions of the coupled equations $\dot{R}_1 = \dot{R}_2 = 0$ if $a_r \neq 0$ and $a_r^2 - b_r^2 \neq 0$. To determine their stabilities one has to analyze the eigenvalues of the Jacobean of $(\dot{R}_1, \dot{R}_2)$. By checking trace and determinant it is easy to see that TW is stable if $b_r < a_r < 0$ and SW is stable if $a_r < 0$ and $a_r^2 > b_r^2$, otherwise both are unstable. Thus there is one stable solution if and only if $a_r < 0$ and $a_r + b_r < 0$ which means that both pitchforks are supercritical. Moreover, the solution which is stable in this case is the one with the larger total amplitude $R \equiv (R_1^2 + R_2^2)^{1/2}$ as indicated in the bifurcation diagram of Figure 4.

The two types of stationary solutions of the (R_1, R_2)-system correspond to periodic solutions of (21) of the form TW: $(Z_1, Z_2) = (Re^{i\omega t}, 0)$ and SW: $(Z_1, Z_2) = Re^{i\omega t}(e^{i\psi}, e^{-i\psi})$ with an arbitrary phase ψ. The letter ω refers here to a generic frequency close to ω_c. The precise form can be determined from the $\dot{\Psi}_j$-equations and differs usually along the two solution types. When substituted into the representation of $\mathcal{E}_c$ we find the patterns TW:

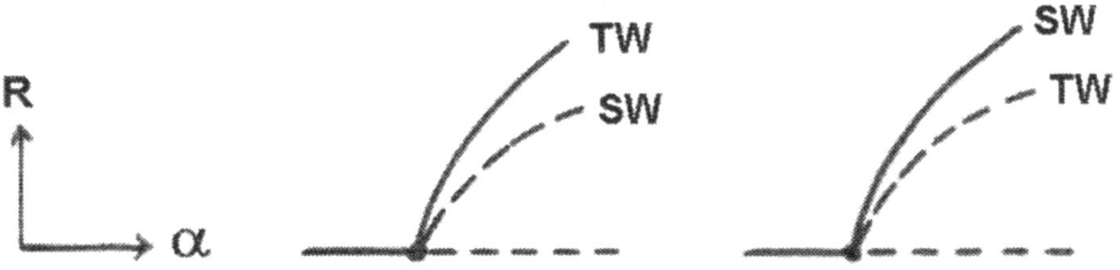

Fig. 4. Bifurcation diagrams for (21) when (a) $b_r < a_r < 0$ (stable SW) and (b) $a_r < 0$ and $a_r^2 - b_r^2 > 0$ (stable TW).

$u(t,x) = 2R\Re[p\exp(\imath\omega t + mx)]$ and SW: $u(t,x) = 4R\Re(pe^{\imath\omega t})\cos(mx+\psi)$. Obviously TW corresponds to a left-travelling wave (if $R_1 = 0, R_2 = R$ it is a right-travelling wave) and SW corresponds to a (reflection invariant) standing wave. The appearance of an arbitrary phase ψ is a consequence of the continuous $SO(2)$-subgroup of $O(2)$ and corresponds to translations of the pattern. In a dynamical systems jargon one says that "the SW-solutions form a 2-torus which is foliated by periodic orbits". The term 2-torus refers to the presence of two phases, a temporal phase ωt and a parameter phase ψ.

Rotating versus Spiral Waves. As in the stationary case the normal form (21) also comes into play if (1) is posed in two-dimensional annuli or discs. The critical modes are here of the form $v_c(r)e^{\pm\imath m\varphi}$ where $v_c(r) = p\phi_m(r)$ with $\phi_m(r)$ being the radial part of the critical eigenfunction $\phi_m(r)e^{\pm\imath m\varphi}$ of the Laplacean. The resulting wave pattern for TW can be written as

$$u(t,r,\varphi) = 2R\Re[v_c(r)\exp\{\imath(\omega t + m\varphi)\}]\,. \tag{23}$$

Analogously one can write down an expression for the SW-pattern, but we are here mainly interested in the form in which the TW-pattern will be observed. Recall first that in Section 1.3 we made the restriction that the components u_1, u_2 of u satisfy the same boundary conditions. It is this restriction which made it possible to decouple the linearized problem into 2×2-matrices governed by the eigenvalues and eigenfunctions of the Laplacean. As a consequence the phases of both components of the radial part $v_c(r) = p\phi_m(r)$ of the critical modes are determined by p and are, therefore, constant. Thus (23) will represent a rotating wave, analogous to the one-dimensional case, because the harmonic motion involves only φ. Under more general conditions the decomposition of the linearized problem in 2×2-problems is, however, no longer possible. This happens in particular if u_1, u_2 satisfy different boundary conditions, for example $u_j = -A_j\partial u_j/\partial n$ with $A_1 \neq A_2$. Nevertheless, owing to the $O(2)$-symmetry the angular part still can be split off because $\mathcal{L}[v(r)e^{\pm\imath n\varphi}] = [\mathcal{L}_n v(r)]e^{\pm\imath n\varphi}$ where here $\mathcal{L}_n$ is the radial differential operator

$$\mathcal{L}_n(\Lambda) = (\frac{1}{r}\frac{d}{dr}r\frac{d}{dr} - \frac{n^2}{r^2})D(\Lambda) + K(\Lambda)\,, \tag{24}$$

acting on $v(r) = (v_1(r), v_2(r))^T$. In this setting, a generic oscillatory instability occurs if $\mathcal{L}(\Lambda_c)$ has a simple imaginary eigenvalue $\imath\omega_c$ with eigenfunction $v_c(r)$ and all other eigenvalues of $\mathcal{L}$ have negative real parts as usual. Now $v_c(r) = (v_{c1}(r), v_{c2}(r))^T$ is a complex eigenfunction and we may split off the phases by setting $v_{cj}(r) = V_j(r)e^{\imath\theta_j(r)}$. The components of (23) can then be written as

$$u_j(t, r, \varphi) = 2RV_j(r)\cos[\omega t + m\varphi + \theta_j(r)]\,. \tag{25}$$

The wave form as which (25) is observed depends crucially on $\theta_j(r)$. When $\theta_j(r)$ is monotonically decreasing or increasing (25) defines a spiral wave and when $\theta_j = const$ (as before) it defines a rotating wave. The fact that a simple Hopf bifurcation can lead to spiral waves in bounded systems has been discovered just recently [14], although these patterns are the subject of extensive research since several years. In [14] a representation of the critical eigenfunctions in terms of Bessel functions of complex arguments is presented.

(C) $\mathbf{D_m}$-Symmetric Secondary Hopf Bifurcation. As in the preceeding subsection we assume now that a (primary) D_m-invariant stationary solution of an $O(2)$-symmetric system becomes unstable in a secondary instability, but now with imaginary eigenvalues $\pm\imath\omega_c$ of the linearized operator. We confine the discussion again to the one-dimensional case. For $m \geq 3$ a generic possibility for this symmetry is that a complex critical eigenmode $\Phi(x)$ corresponding to $\imath\omega_c$ occurs such that $\Phi(-x)$ and $\Phi(x)$ are linearly independent and $\Phi(x + 2\pi/m) = e^{2\pi\imath/m}\Phi(x)$ as in the stationary case. Thus again two complex amplitudes (Z_1, Z_2) are needed to represent an arbitrary critical mode in the center eigenspace as

$$Z_1\Phi(x) + Z_2\Phi(-x) + cc\,.$$

This representation is analogous to (18), hence we find again the symmetry operations (19), (20) but with the spatial translations ψ in (18) restricted to multiples of $2\pi/m$. The ODE-system for (Z_1, Z_2) is, therefore, obtained by adding the lowest terms to (21) which break the continuous translation symmetry to a discrete one, yielding [15]

$$\dot{Z}_1 = (\imath\omega_c + \alpha + a|Z_1|^2 + b|Z_2|^2)Z_1 + c\bar{Z}_1^{m'-1}Z_2^{m'}\,, \tag{26}$$

where $m' = m$ $(m/2)$ if m is odd (even) and c is another complex coefficient. In polar coordinates (26) becomes

$$\dot{R}_1 = (\alpha_r + a_r R_1^2 + b_r R_2^2)R_1 + R_1^{m'-1}R_2^{m'}(c_r\cos m'\Psi - c_i\sin m'\Psi) \tag{27}$$

$$\begin{aligned}\dot{\Psi} = &(R_2^2 - R_1^2)[a_i - b_i - c_i(R_1R_2)^{m'-2}\cos m'\Psi] \\ &-c_r(R_1R_2)^{m'-2}(R_1^2 + R_2^2)\sin m'\psi\,,\end{aligned} \tag{28}$$

where $\Psi = \Psi_2 - \Psi_1$. The equation for R_2 follows by interchanging R_1 and R_2 and replacing Ψ by $-\Psi$. The fourth equation is an ODE for the total phase $\Psi_1 + \Psi_2$ which basically determines frequencies along periodic solutions, i.e., $\Psi_{1,2} = \omega t \mp \Psi/2$ with ω close to ω_c. In contrast to the $O(2)$-symmetric case the phase difference Ψ is now coupled to $\dot{R}_1, \dot{R}_2$. The case $m = 4$ ($m' = 2$) is special because then the last term in (26) or (27) is also cubic, thus we confine the discussion to $m \neq 4$ where this term is of higher order than the cubic $O(2)$-symmetric terms. As in the $O(2)$-case we find here again a periodic solution TW determined by $R_1 = R$, $R_2 = 0$. When substituted in the $\mathcal{E}_c$-representation we obtain the pattern $u(t,x) = 2R\Re[e^{\imath\omega t}\Phi(x)]$. This pattern has the invariance property $u(t,x) = u(t - T/m, x + 2\pi/m)$, where $T = 2\pi/\omega$ is the period, and hence can be interpreted as a "discrete version" of a travelling wave. Its stability properties are determined by the Jacobean of the (R_1, R_2)-system (Ψ is not defined) and are the same as in the $O(2)$-case: TW is stable if $b_r < a_r < 0$.

Other periodic solutions follow from the steady states of the (R_1, R_2, Ψ)-system. Again one finds generically (if $a_r \neq 0$, $a_r^2 \neq b_r^2$, $c_r \neq 0$) only solutions with $R_1 = R_2 = R$, but this time the phase Ψ is no longer arbitrary because $\sin m'\Psi = 0$. As in the stationary case we have two types of solutions, $\Psi_+ = 2\pi k/m'$ and $\Psi_- = (2k+1)\pi/m'$ with solutions distinguished by different integers k being equivalent (symmetry-related), hence we set again $\Psi_+ = 0$ and $\Psi_- = \pi$ $(-\pi/m')$ if m' is odd (even). The amplitudes R along these solutions satisfy $\alpha_r + (a_r + b_r)R^2 \pm c_r R^{2m'-2} = 0$. Thus at leading order both branches are identical pitchforks which are split apart at higher order. As in the $O(2)$-case this leads to a stable solution if $a_r < 0$ and $b_r^2 < a_r^2$, but only one of the two $\Psi_\pm$-solutions can be stable, which one depends on the Ψ-eigenvalue whose sign is determined by $-c_r \cos m\Psi = \mp c_r$. Transforming back to $\Psi_{1,2}$ we find that the periodic solution following from Ψ_+ is given by $(Z_1, Z_2) = Re^{\imath\omega t}(1,1)$ and for Ψ_- we obtain $Re^{\imath\omega t}(-\imath, \imath)$ if m' is odd and $Re^{\imath\omega t}(e^{\imath\pi/m}, e^{-\imath\pi/m})$ if m' is even.

To represent the corresponding wave patterns we set $2\Phi_+(x) = \Phi(x) + \Phi(-x)$ for any m and $\Phi_-(x) = \Phi(x) - \Phi(-x)$ if m' is odd while for m' even $2\Phi_-(x) = \rho\Phi(x) + \bar{\rho}\Phi(-x)$ with $\rho = e^{\imath\pi/m}$. The wave patterns corresponding to $\Psi_\pm$ can then be compactly written as $u_+(t,x) = 4R\Re[e^{\imath\omega t}\Phi_+(x)]$ and $u_-(t,x) = 4R\Im[e^{\imath\omega t}\Phi_-(x)]$ if m' is odd whereas for m' even the imaginary part has to be replaced by the real part. All of these patterns are standing waves showing the following invariance properties: $u_+(t,-x) = u_+(t,x)$ for all m, $u_-(t+T/2,-x) = u_-(t,x)$ if m' is odd (m odd or $m-2$ a multiple of 4) and $u_-(t,-x-2\pi/m) = u_-(t,x)$ if m' is even (m a multiple of 4). In addition, if m is even both patterns show the invariance $u_\pm(t+T/2, x+\pi) = u_\pm(t,x)$. For all of these cases the D_m-symmetry of the underlying spatial pattern is broken down to a spatial or spatio-temporal reflection symmetry.

Concerning $m = 4$ the basic periodic solutions have the same invariances as described above for m' even, however, the branching behaviour and the stabilities are different. Moreover, in [16] it is shown that in certain ranges of the coefficients the normal form also admits quasiperiodic solutions which can be stable.

3.3 Generalizations

Spontaneous Symmetry Breaking. In the preceeding subsections we have set up normal forms for stationary and Hopf bifurcations and determined the basic solutions as well as the corresponding spatial and wave patterns. In addition we characterized the patterns in terms of their invariance properties. The common feature was that the new patterns showed always less symmetry than the underlying state from which they bifurcated. This is commonly referred to as "spontaneous symmetry breaking". Typically, when a parameter is varied, the basic state becomes unstable in a primary bifurcation giving rise to one (or several) so called "primary solution branches" in which some of the invariances of the basic state are broken, for example, a reflection invariance in case of a pitchfork bifurcation. A primary branch may itself become unstable in a secondary bifurcation leading to new solutions which may be called secondary branches and are characterized by further broken invariances and so forth.

Amplitude-Phase Coupling. If a nontrivial stationary solution becomes unstable one can always utilize a weakly nonlinear analysis and investigate the bifurcating patterns in terms of ODE's for critical mode amplitudes (provided the spatial domain is bounded). The resulting symmetry equivariance is that of the underlying pattern that becomes unstable. As an example we discussed bifurcation from a D_m-symmetric state. In this analysis we have, however, cheated a bit because a D_m-symmetric solution in a circular symmetric system is not isolated but is embedded in a continuous family obtained by spatial translations, i.e., phase shifts $\Phi(x + \psi)$. This fact has not been accounted for in our discussion. To include it one has to supplement the ODE-system by a further equation for the phase which is, however, decoupled from the amplitude equations (note that ψ is *not* a phase of an amplitude). The formal procedure behind that approach is the so called "center bundle reduction" [17]. In the case of a D_m-symmetric stationary bifurcation the lowest order terms through which Z is coupled to the phase are of the form $\dot{\psi} = \imath(Z^m - \bar{Z}^m)$ which vanish along the stationary solutions, hence there is no effect. In contrast, if the underlying pattern shows only a reflection symmetry we find generically a real critical mode and the lowest order term that couples X to ψ is of the form $\dot{\psi} = X$ which induces a slow drift, i.e., the symmetry broken states are revealed as travelling waves with small speed. This instability is commonly referred to as "parity breaking instability", referring to the broken reflection invariance of the bifurcating states. In [18] the effect of small symmetry breaking imperfections is studied on the basis of an extension of the center bundle reduction. In [19] the center bundle reduction is utilized to describe instabilities of spiral waves. We note that in infinitely extended systems also phase instabilities may occur. These are described by partial differential equations for ψ which are commonly referred to as "phase equations", see Section 8.

Isotropy Subgroups and Fixed Point Subspaces. The characterization of spatial or wave patterns through invariances usually proceeds in terms of subgroups of the full symmetry group of the system. In group representation theory these are called isotropy subgroups [10]. Recall that $\Sigma \subset \Gamma$ is a subgroup of Γ if it is closed under composition. Let $X \to \gamma X$ be a representation of Γ in $\mathbf{R}^n$ such as our induced representation in $\mathcal{E}_c$. The fixed point subspace, $Fix(\Sigma)$, of Σ is the space of all vectors X which are fixed by all transformations $\sigma \in \Sigma$, i.e., $\sigma X = X$. Σ is an isotropy subgroup if it is the largest subgroup which has $Fix(\Sigma)$ as fixed point subspace. For example, in the representation $Z \to e^{i\psi} Z$, $Z \to \bar{Z}$ of $O(2)$ the fixed point subspace of the reflection κ is the real axis, $Z = X \in \mathbf{R}$. In general, one calls two subgroups Σ, Σ' conjugate if $\Sigma' = \gamma \Sigma \gamma^{-1}$ for some $\gamma \in \Gamma$. The fixed point subspaces of conjugate subgroups are isomomorphic because $Fix(\Sigma)$ is mapped under γ to $Fix(\Sigma')$. In the $O(2)$-example the nontrivial isotropy subgroups are all conjugate reflection groups and their fixed point subspaces are straight lines through the origin. A general result of group representation theory is that the isotropy subgroups can be classified (up to conjugacy) and ordered by inclusion. In turn, the corresponding fixed point subspaces are nested/isomorphic subspaces of $\mathbf{R}^n$. Loosely: the larger Σ the smaller is $Fix(\Sigma)$.

For Γ-equivariant vector fields $\dot{X} = g(X)$ the importance of fixed point subspaces relies on the fact that these are flow-invariant which follows simply from $g(X) = g(\sigma X) = \sigma g(X)$. The term flow-invariant means that a solution $X(t)$ remains in such a space whenever the initial condition $X(0)$ is in that space, for example, $\dot{Z} = Zh(|Z|^2)$ leads to $\dot{X} = Xh(X^2)$ along the straight line $Z = e^{i\psi} X$. The presence of nested flow-invariant subspaces has strong implications on the possible dynamical behaviour. In particular the possibility of persistent heteroclinic cycles is a consequence of this structure. Roughly speaking, a heteroclinc cycle is a collection of fixed points or periodic orbits located in (usually low-dimensional) fixed point subspaces which have connecting orbits in higher-dimensional fixed point subspaces such that a cycle is formed [20]. In non-symmetric systems such cycles would be destroyed by allmost every perturbation whereas in a symmetric system they "survive" small perturbations that respect the symmetry.

Isotropy subgroups are extremely useful for bifurcation analyses. Intuitively one expects solutions with high symmetry to be more likely created in an instability than solutions with low symmetry. This fact is accounted for in equivariant branching lemmas. In case of a generic stationary bifurcation it tells [21] that one always finds solutions living in one-dimensional fixed point subspaces. Their isotropy subgroups are necessarily maximal, i.e., there is no larger isotropy subgroup containing them (except the full group Γ). In complicated systems one can, therefore, focus on the analysis of these maximally symmetric solutions which leads to a significant simplification because they are determined by a single bifurcation equation for a real variable. Typically one finds here transcritical or pitchfork bifurcations, for example $\dot{X} = \alpha X + aX^3$ if one sets $Z = X$ in the $O(2)$-equivariant stationary bifurcation.

In case of a Hopf bifurcation we encountered the extended symmetry group $\Gamma \times S^1$ with its phase shift operations $Z \to e^{i\theta} Z$. One can show here [13] that any isotropy subgroup Σ of $\Gamma \times S^1$ consists of a "spatial" part $\Sigma' \subset \Gamma$ such that to each spatial symmetry operation $\sigma \in \Sigma'$ there belongs a phase shift $\theta(\sigma)$. If $\theta(\sigma) = 0$ for all σ the isotropy subgroup is called spatial, if there are nontrivial phase shifts one calls it "twisted" or spatio-temporal. One can further show that the dimension of $Fix(\Sigma)$ is always even, hence this space can also be parametrized by complex variables (amplitudes). Moreover, if two-dimensional, the dynamics in $Fix(\Sigma)$ is at leading order governed by the simple Hopf normal form (16) and the "equivariant Hopf branching lemma" states that there is always a periodic solution in such a space. For example, the isotropy subgroup of the TW-solution of the $O(2)$-equivariant Hopf bifurcation consists of combined spatial and temporal translations $m\psi = -\theta$ (a time translation is compensated by a spatial translation) which constitutes an $SO(2)$-subgroup of $O(2) \times S^1$, denoted as $\widetilde{SO(2)}$. In contrast, the isotropy subgroup of the SW-solution is just the spatial reflection (ignoring phase shifts by $2\pi/m$ which leave any Z invariant). The fixed point subspaces of these symmetries are $(Z_1, Z_2) = (Z, 0)$ for $\widetilde{SO(2)}$ and $(Z_1, Z_2) = (Z, Z)$ for the reflection. Substituting this into (21) we find $\dot{Z} = \alpha Z + AZ^2\bar{Z}$ where $A = a$ in the first case and $A = a + b$ in the second, i.e., in both subspaces we arrive at the elementary Hopf normal form (16). Similarly one can determine and interpret the solutions of the D_m-equivariant Hopf bifurcation. In general, the approach to the analysis of bifurcating solutions via restriction to fixed point subspaces simplifies computations considerably, but requires knowlege of the isotropy subgroups and fixed point subspaces.

Irreducible Representations. Let us first recall that a representation of a group Γ in $\mathbf{R}^n$ (or more generally in a finite dimensional vector space) is irreducible if no proper invariant subspace U such that $\gamma X \in U$ whenever $X \in U$ exists. An irreducible representation is called absolutely irreducible if the only linear operators (matrices) that commute with all group transformations are multiples of the identity. Concerning bifurcations, the important point is that in case of a stationary bifurcation the center eigenspace generically forms an absolutely irreducible representation space. In case of a Hopf bifurcation it is generically the direct sum of two copies of the same absolutely irreducible representation space which is revealed in the complex amplitudes one normally uses. All the representations we encountered in Subsections 3.1

and 3.2 had these properties. Mode couplings as described in Section 1.3 may be interpreted as couplings of different irreducible representation spaces.

3.4 Normal Form Computations

We describe here some details of the computation of the coefficients up to cubic order that appeared in the normal forms for the primary bifurcations discussed in Subsections 3.1 and 3.2. Let p be the (real or complex) critical eigenvector of $\mathcal{L}(\mu_c, \Lambda_c)$, $\mathcal{L}(\mu_c, \Lambda_c)p = 0$ or $\mathcal{L}(\mu_c, \Lambda_c)p = \imath\omega_c p$, and let p^* be the corresponding adjoint eigenvector, a row vector such that $p^*\mathcal{L}(\mu_c, \Lambda_c) = 0$ or $p^*\mathcal{L}(\mu_c, \Lambda_c) = \imath\omega_c p^*$. As normalization we choose $p^* \cdot p = 1/C$ with $C = \int_\Omega |\phi|^2 dx$ where $\phi(x)$ is the critical eigenfunction of ∇^2. In the periodic case Ω is the interval $0 \le x \le 2\pi$ and $\phi = e^{\imath m x}$, thus $C = 1/2\pi$.

In all cases considered the linear coefficient α describes the change of the critical eigenvalue 0 or $\imath\omega_c$ when Λ varies about Λ_c. Thus at linear order in λ, $\alpha(\lambda)$ is given by

$$\alpha(\lambda) = C \sum_{j=1}^{s} (p^* \cdot \frac{\partial \mathcal{L}(\mu_c, \Lambda_c)}{\partial \Lambda_j} p)\, \lambda_j \,.$$

Concerning the nonlinear coefficients, these are all computed at criticality, hence we can set $\lambda = 0$. Consider first the case of a zero eigenvalue involving only one critical eigenfunction $\phi(x)$ of the Laplacean. In the nonlinear regime the critical mode has to be supplemented by a contribution from the stable modes which we write as $u(X;x) = Xp\phi(x) + X^2 w_2(x) + X^3 w_3(x) + O(X^4)$. Here, w_2, w_3 are "coefficient functions" which have to be determined from the condition that, to the order considered, $u(X(t);x)$ should present a solution of (1) when $X(t)$ is a solution of the reduced ODE for X. Moreover, w_2 is a superposition of the "stable modes", the eigenfunctions of $\mathcal{L}(\Lambda_c)$ corresponding to eigenvalues with negative real parts. This means that $w_2(x)$ is orthogonal to the adjoint critical mode, $\int_\Omega \phi p^* \cdot w_2 dx = 0$ and analogously w_3 etc. Since we are at criticality we can write $\dot{X} = a_2 X^2 + a_3 X^3 + O(X^4)$. Substituting $u(X;x)$ as above and $\dot{X}$ into (1) and comparing powers in X yields at second order $-\mathcal{L}_c w_2 = F_{2c}[p,p]\phi^2 - a_2 p\phi$. Here, $\mathcal{L}$ is the linear operator (3) and the nonlinear part is expanded as $F(u) = F_2[u,u] + F_3[u,u,u] + \ldots$ with symmetric, bilinear and cubic mappings F_2 and F_3 in $\mathbf{R}^2$. A subscript "c" denotes here and in the following evaluation at criticality, $\Lambda = \Lambda_c$. The above equation for w_2, subject to the orthogonality constraint, has a unique solution if and only if the right hand side is orthogonal to $p^*\phi$. This (Fredholm) condition yields $a_2 = p^* \cdot F_{2c}[p,p] \int_\Omega \phi^3 dx$ and determines the coefficient $a \equiv a_2$ in (11). At cubic order in X, (1) yields

$$-\mathcal{L}_c w_3 = -2a_2 w_2 + 2F_{2c}[p, w_2]\phi + F_{3c}[p,p,p]\phi^3 - a_3 p\phi \,.$$

The solvability condition for w_3 then leads to

$$a_3 = 2p^* \cdot \int_\Omega F_{2c}[w_2, p]\phi^2 dx + p^* \cdot F_{3c}[p,p,p] \int_\Omega \phi^4 dx \,.$$

We note that when ϕ is odd under a reflection ($a_2 = 0$), w_2 is even but usually nonzero. Thus the computation of the cubic coefficient $a \equiv a_3$ in (12) requires to solve the nonhomogenous, linear boundary value problem for w_2. In the special case when Ω is the interval $0 \le x \le \pi$ and Dirichlet boundary conditions are imposed, we have $\phi(x) = \sin mx$ for some even integer m. The coefficient function $w_2(x)$ can be computed straightforward in this case whereas in two (and higher) dimensions one usually has to resort to numerical computations.

The complex coefficient a in (16) involves a normal form transformation. Writing the reduced system (before the normal form transformation) as $\dot{Z} = \imath\omega_c Z + \sum_{ij} g_{ij} Z^i \bar{Z}^j$ with $g_{ij} \in \mathbf{C}$, explicit normal form transformation yields (see [22])

$$a = g_{21} + \frac{\imath}{\omega_c}(g_{20}g_{11} - |g_{11}|^2 - \frac{2}{3}|g_{02}|^2) \,,$$

hence all quadratic and one of the cubic terms of $\dot{Z}$ are needed. The computation of these coefficients proceeds in the same manner as before. The critical modes are supplemented by quadratic terms, giving the representation $u = (Zp\phi + Z^2 w_{22} + cc) + Z\bar{Z} w_{20} + \ldots$, with complex and real coefficient functions $w_{22}(x)$ and $w_{20}(x)$. Expansion of (1) then shows that these functions satisfy the following boundary value problems,

$$\begin{aligned}(2\imath\omega_c - \mathcal{L}_c)w_{22} &= F_{2c}[p,p]\phi^2 - (g_{20}p + \bar{g}_{02}\bar{p})\phi \\ -\mathcal{L}_c w_{20} &= 2F_{2c}[p,\bar{p}]\phi^2 - (g_{11}p + \bar{g}_{11}\bar{p})\phi \,,\end{aligned}$$

subject to two orthogonality constraints. These constraints can be satisfied only if

$$(g_{20}, g_{11}, g_{02}) = (p^* \cdot F_{2c}[p,p], 2p^* \cdot F_{2c}[p,\bar{p}], p^* \cdot F_{2c}[\bar{p},\bar{p}]) \int_\Omega \phi^3 dx \,.$$

A further solvability condition for a cubic coefficient function finally leads to the following expression for g_{21} that involves the solutions of the above boundary value problems,

$$g_{21} = \int_\Omega 2p^* \cdot \{F_{2c}[w_{22}(x), \bar{p}] + F_{2c}[w_{20}(x), p]\}\phi^2 dx + 3p^* \cdot F_{3c}[p,p,\bar{p}] \int_\Omega \phi^4 dx \,.$$

In a similar way one can determine the coefficients of the $O(2)$-equivariant normal forms in the case of 2π-periodic boundary conditions. We use the notation W_2 to denote the second order contribution of the representation of u. The individual terms of W_2 are here Fourier modes. For example, the coefficient function of Z^2 involves $(e^{\imath mx}Z)^2$ and hence is given by $e^{2\imath mx}$ times a vector to be determined. In case of a stationary bifurcation we find $W_2 =$

$(Z^2 q_2 e^{2\imath mx} + cc) + Z\bar{Z}q_0$, with real vectors q_2, q_0 defined by $\mathcal{L}(4m^2, \Lambda_c)q_2 = K_c q_0/2 = -F_{2c}[p,p]$, yielding $a = 2\pi p^* \cdot \{2F_{2c}[q_0 + q_2, p] + 3F_{3c}[p,p,p]\}$.

In case of a Hopf bifurcation we obtain

$$W_2 = (Z_1\bar{Z}_1 + Z_2\bar{Z}_2)q_{00} + V_2 + \bar{V}_2 \quad \text{with}$$

$$V_2 = (Z_1^2 e^{2\imath mx} + Z_2^2 e^{-2\imath mx})q_{22} + Z_1\bar{Z}_2 e^{2\imath mx} q_{02} + Z_1 Z_2 q_{20}\,.$$

Here the vectors q_{kj} satisfy $[\imath k\omega_c - \mathcal{L}(j^2m^2, \Lambda_c)]q_{kj} = p_{kj}$, where $2p_{22} = p_{20} = 2F_{2c}[p,p]$ and $p_{02} = p_{00} = 2F_{2c}[p,\bar{p}]$. The coefficients a, b occuring in (21) can then be computed as

$$a = 2\pi p^* \cdot \{3F_{3c}[p,p,\bar{p}] + 2F_{2c}[q_{22},\bar{p}] + 2F_{2c}[q_{00},p]\}\,,$$
$$b = 4\pi p^* \cdot \{3F_{3c}[p,p,\bar{p}] + F_{2c}[q_{02} + q_{00}, p] + F_{2c}[q_{20},\bar{p}]\}\,.$$

These formulae hold also for the generalized problem posed in circular discs or annuli if scalar multiplication of a vector q by p^* is replaced by the radial inner product $\int_{r_0}^{r_1} v_c^*(r) \cdot q(r) r\, dr$ and the linear equations for vectors q are reinterpreted as radial boundary value problems defined in terms of (24).

We conclude this discussion by noting that in geometrical terms the above nonlinear "amplitude-representations", which are generally of the form $u = \sum_j X_j \Phi_j + W(X, \lambda)$ with W orthogonal to to the adjoint critical modes, form a parametrization of a flow-invariant manifold in the infinite dimensional phase space in which (1) is posed. This manifold is called the "center manifold" and the ODE-system $\dot{X} = g(X, \lambda)$ for the critical mode amplitudes can be considered as the "projection" of the vector field that determines the flow in the center manifold onto the center eigenspace, hence the notion "center manifold reduction". Under some technical assumptions, which assure also uniqueness of the Taylor series, the existence of a center manifold is guaranteed [23].

4 Generic Bifurcations for Lattice Symmetries

Here we discuss generic bifurcations when (1) is posed in $2D$ subject to periodic boundary conditions with respect to a lattice. Let the lattice be generated by two basis vectors e_1, e_2. The dual lattice in Fourier space is generated by the dual basis $k_1, k_2 \in \mathbf{R}^2$, defined by $k_i \cdot e_j = 2\pi\delta_{ij}$. To any dual lattice vector $k = n_1 k_1 + n_2 k_2$, labelled by an integer pair (n_1, n_2), there corresponds an eigenfunction $\phi_{n_1 n_2} = e^{\imath k \cdot x}$ for $-\nabla^2$ with eigenvalue $\mu_{n_1 n_2} = |k|^2$. The requirement that the lattice's symmetry commutes with the Laplacean restricts us to rectangular, square and hexagonal lattices. In the first two cases, with suitable normalizations of x_1, x_2, we may choose $e_1 = (2\pi, 0)$, $e_2 = (0, 2\pi/s)$, where $s \neq 1$ for the rectangular and $s = 1$ for the square lattices. Periodicity with respect to these lattices then corresponds to imposing 2π- and $2\pi/s$-periodicity along the x_1- and x_2-axes, thus the eigenfunctions and eigenvalues of $-\nabla^2$ are

$$\phi_{n_1n_2}(x_1, x_2) = \exp\{\imath(n_1x_1 + sn_2x_2)\}\,, \quad \mu_{n_1n_2} = n_1^2 + s^2n_2^2\,, \tag{29}$$

In the case of the hexagonal lattice we choose as generating vectors $e_1 = 2\pi(1/\sqrt{3}, 1)$, $e_2 = 4\pi(1/\sqrt{3}, 0)$. The dual vectors are $k_1 = (0, 1)$, $k_2 = (\sqrt{3}/2, -1/2)$, hence the eigenvalues are $\mu_{n_1n_2} = n_1^2 + n_2^2 - n_1n_2$, with eigenfunctions

$$\phi_{n_1n_2}(x_1, x_2) = \exp\{\tfrac{\imath}{2}[(\sqrt{3}n_2x_1 + (2n_1 - n_2))x_2]\}\,. \tag{30}$$

When μ_c is a critical eigenvalue, its multiplicity depends on the number of integer solutions of the equation $\mu_c = m_1^2 + s^2m_2^2$ or $\mu_c = m_1^2 + m_2^2 - m_1m_2$. Note that number theoretical degeneracies may occur leading to reducible eigenspaces. In the discussion below we exclude such degeneracies and confine ourselfes to situations where the eigenspace of μ_c is irreducible under the PDE's symmetry group, which is the semidirect product $D_m \times_s T^2$ with $m = 2, 4$ and 6 for the rectangular, square and hexagonal cases, respectively. In terms of angles $0 \le \psi_1, \psi_2 < 2\pi$, the action of $(\psi_1, \psi_2) \in T^2$ is induced by $x \to x + (\psi_1e_1 + \psi_2e_2)/2\pi$. The action of D_m is induced by its standard action on $x \in \mathbf{R}^2$ and will be described below.

In the following Subsections we describe the generic bifurcations for these symmetries individually. We confine the discussion to maximally symmetric solutions whose existence is guaranteed by the equivariant branching lemmas, referring to the literature for further details. In most cases the various solution branches are of pitchfork type, thus bifurcation diagrams are omitted.

A variety of distinct spatial and spatiotemporal (wave) patterns have been classified through maximal isotropy subgroups. For reasons of space we do not include pictures of these patterns. Instead, analytical expressions in terms of trigonometric functions are presented. These reveal to a large extent the basic features of the patterns and, moreover, can be used by the reader to generate his/her own pictures, e.g. via MAPLE.

4.1 Rectangular Lattice

One motivation for posing (1) on a rectangular lattice is that this provides the simplest way to describe anisotropy effects in a large spatial domain. The relevant symmetry group $D_2 \times_s T^2$ coincides with $O(2) \times O(2)$, where the two $O(2)$-components simply account for reflections and translations in the two x-directions. Letting $\mu_c = m_1^2 + s^2m_2^2$, we assume that $m_1 \neq 0$ and $m_2 \neq 0$ and, for simplicity, that m_1, m_2 are relatively prime which means that there is no smaller lattice supporting the critical modes. This leads generically to a four dimensional irreducible eigenspace of μ_c with basis functions $\phi_{\pm m_1, \pm m_2}$. Accordingly, the center eigenspace in case of a stationary bifurcation at Λ_c can be represented in terms of complex coordinates (Z_1, Z_2) as $\mathcal{E}_c:\ (Z_1\phi_{m_1m_2} + Z_2\phi_{-m_1m_2} + \mathrm{cc})p$, and the actions of Γ are

$$(\psi_1,\psi_2):(Z_1,Z_2)\rightarrow(e^{i(m_1\psi_1+m_2\psi_2)}Z_1,e^{i(m_2\psi_2-m_1\psi_1)}Z_2)\,, \quad (31)$$
$$\kappa_1:(Z_1,Z_2)\rightarrow(Z_2,Z_1)\,,\quad \kappa_2:(Z_1,Z_2)\rightarrow(\bar{Z}_2,\bar{Z}_1)\,. \quad (32)$$

Here, $(\psi_1,\psi_2)\in T^2$ and (31) describes the effect of translations $(x_1,x_2)\rightarrow(x_1+\psi_1,x_2+\psi_2/s)$ on the mode amplitudes (Z_1,Z_2). Similarly, the actions (32) follow from the reflections $\kappa_j : x_j \rightarrow -x_j$ $(j=1,2)$. Observe that the (ψ_1,ψ_2)- and κ_1-actions have the same effect (independent phase shifts and permutation) like the group operations that were encountered in the Hopf bifurcation with $O(2)$-symmetry. Thus to cubic order the generic equivariant vector field reads

$$\dot{Z}_1=(\alpha+a|Z_1|^2+b|Z_2|^2)Z_1\,, \quad (33)$$

and the $\dot{Z}_2$-equation follows by applying κ_1, however, equivariance under the second reflection κ_2 enforces α, a, b to be real. The maximal isotropy subgroups are $\widetilde{O(2)}$ and D_2 with one-dimensional fixed point subspaces $(X,0)$ and (X,X), respectively. The "mixed" subgroup $\widetilde{O(2)}$ of Γ is generated by the combined translation and reflection $(\psi_1,\psi_2)=(m_1\psi,-m_2\psi)$ and $\kappa_1\kappa_2$. A solution in $Fix(\widetilde{O(2)})$ corresponds to a translation and reflection invariant ("stripe"- or "roll"-) pattern of the form $2Xp\cos(m_1x_1+m_2sx_2)$, whereas a solution in $Fix(D_2)$ corresponds to $4Xp\,\cos(m_1x_1)\,\cos(m_2sx_2)$ which is a rectangular pattern. Both $\widetilde{O(2)}$ and D_2 solutions are governed by (rotated) pitchforks and their amplitudes satisfy the same equations as those of the TW- and SW-solutions of the $O(2)$-equivariant Hopf bifurcation if a_r and b_r are replaced by a and b. For example, the two diagrams of Figure 4 (a) and (b) with the labels TW and SW replaced by $\widetilde{O(2)}$ and D_2 occur when $b<a<0$ and $b<0$, $a^2-b^2>0$, respectively.

The Hopf bifurcation with $O(2)\times O(2)$ has been introduced and discussed independently in [24] and [25]. The center eigenspace is here eight-dimensional and may be represented in the form

$$\mathcal{E}_c:\ (Z_1\phi_{m_1m_2}+Z_2\phi_{-m_1m_2}+Z_3\phi_{-m_1-m_2}+Z_4\phi_{m_1-m_2})p+cc\,, \quad (34)$$

with complex variables $\mathbf{Z}\equiv(Z_1,Z_2,Z_3,Z_4)$. The induced actions of $(\psi_1,\psi_2)\in T^2$ and κ_1, κ_2 are as before inferred from this representation: a translation $(\psi_1,\psi_2)\in T^2$ induces the phase factor $Z_1\rightarrow\phi_{m_1m_2}(\psi_1,\psi_2)Z_1$ and analogously for Z_2,Z_3,Z_4. The reflections act as

$$\kappa_1:\ \mathbf{Z}\rightarrow(Z_2,Z_1,Z_4,Z_3)\,,\quad \kappa_2:\ \mathbf{Z}\rightarrow(Z_4,Z_3,Z_2,Z_1)\,. \quad (35)$$

In addition to these symmetries, the normal form shows equivariance under the S^1-action $\mathbf{Z}\rightarrow e^{i\theta}\mathbf{Z}$ corresponding to time translations $t\rightarrow t+\theta/\omega$ of a periodic solution. Up to cubic order a vector field that is equivariant under this $O(2)\times O(2)\times S^1$-action can be written as

names	symmetries	$Fix(\Sigma)$	wave pattern
TTW (*TR*)	$\tilde{\psi}_1$, $\tilde{\psi}_2$	$(Z,0,0,0)$	$2\cos(m_1x_1 + sm_2x_2 + \omega t)$
STW (*TRe*1)	κ_1, $\tilde{\psi}_2$	$(Z,Z,0,0)$	$4\cos(m_1x_1)\cos(sm_2x_2 + \omega t)$
TSW (*TRe*2)	$\tilde{\psi}_1$, κ_2	$(Z,0,0,Z)$	$\cos(sm_2x_2)\cos(m_1x_1 + \omega t)$
SSW (*SRe*)	κ_1, κ_2	(Z,Z,Z,Z)	$8\cos(m_1x_1)\cos(sm_2x_2)\cos(\omega t)$
SW (*SR*)	ψ_{12}, $\kappa_1\kappa_2$	$(Z,0,Z,0)$	$4\cos(m_1x_1 + sm_2x_2)\cos(\omega t)$
AW (*AR*)	$\tilde{\kappa}_1, \tilde{\kappa}_2$	$(Z,\imath Z,Z,\imath Z)$	$4\cos(m_1x_1 + sm_2x_2)\cos(\omega t)$ $-4\cos(m_1x_1 - sm_2x_2)\sin(\omega t)$

Table 1. Maximally symmetric periodic solutions of (36) ($Z = Re^{\imath\omega t}$) and their wave patterns. The second column summarizes the symmetry generators of the isotropy subgroups under which these solutions are invariant. Here, $\tilde{\psi}_j$ ($j = 1, 2$) and ψ_{12} are defined by $\theta = -m_j\psi_j$ and $m_2\psi_1 + m_1\psi_2 = 0$ and $\tilde{\kappa}_1$, $\tilde{\kappa}_2$ denote the spatial reflections κ_1, κ_2 combined with a temporal phase shift by $\theta = \pi$. When multiplied by $R|p_1|$, the expressions in the last column represent the leading terms of the wave patterns $u_1(t, x)$, for suitable choice of the phases.

$$\dot{Z}_1 = (\imath\omega_c + \alpha + \sum_{j=1}^{4} a_j|Z_j|^2)Z_1 + a_5Z_2\bar{Z}_3Z_4\,, \tag{36}$$

with complex coefficients a_j ($j = 1..5$). The equations for $\dot{Z}_2$, $\dot{Z}_3$ and $\dot{Z}_4$ can be obtained by applying the reflections κ_1, $\kappa_1\kappa_2$ and κ_2, respectively, to (36).

As explained in Subsection 3.3, the maximally symmetric periodic solutions of a Γ-equivariant Hopf normal form are classified in terms of the isotropy subgroups of the $\Gamma \times S^1$-action which have two-dimensional fixed point subspaces. Such a subspace is parametrized by a single complex variable Z and its branching behaviour is governed by the standard Hopf normal form (21). In the case of (36) with $\Gamma = O(2) \times O(2)$, the group theoretical classification leads to six distinct isotropy subgroups which are summarized in Table 1. In this table, the second column contains the group operations which fix the solution. The third column shows the fixed point subspaces; the periodic solutions in these spaces are all of the form $Z = Re^{\imath\omega t}$. The last column summarizes the leading terms of the wave patterns associated with these solutions. Observe that the first four isotropy subgroups are of the form $\Sigma_1 \times \Sigma_2$ where both Σ_1 and Σ_2 are one of the maximal isotropy subgroups $\widetilde{SO}(2)$ or C_2 encountered in the Hopf bifurcation with $O(2)$-symmetry. Accordingly, the solutions are denoted in [25] *TTW*, *STW*, *TSW* and *SSW* where "*T*" refers to an $\widetilde{SO}(2)$-component (travelling wave) and "*S*" refers to a C_2-component (standing wave). For example, the isotropy subgroup of *TTW* is $\widetilde{SO}(2) \times \widetilde{SO}(2)$ which induces a pure travelling wave. Similarly the *STW*-solution has isotropy subgroup $C_2 \times \widetilde{SO}(2)$ and is observed as a travelling wave in the x_2-direction and as a standing wave in the x_1-direction. The solutions SSW and SW induce both standing waves, SSW with a rectangular shape and SW with a spatial translation invariance. The (purely spatial) subgroup for SW is $\widetilde{O(2)}$ which was already defined in the case of a stationary bifurcation. The names in parantheses in Table 1 are taken from [24]. Here, T and S refer again to travelling and standing waves. The letter R stands for rolls (or stripes), indicating a spatial translation invariance, whereas Re refers to a rectangular shape. There are two different types of travelling rectangles that have to be distinguished. The most interesting pattern follows from the AW/R-solution (alternating wave/rolls). Its temporal behaviour is characterized by a "switching" between differently oriented (perpendicular for $m_1 = m_2 = 1$) spatially translation invariant patterns.

The stabilities and branching directions of these solutions depend on certain combinations of the real parts of the five nonlinear coefficients. We do not go into details here (see [24, 25]) but mention that each of them can be stable and that there are eight combinations of two different solutions that can be simultanously stable. Among these are, for example, (SSW, AW) and (TTW, AW).

Besides the above maximal isotropy subgroups there exist eight distinct submaximal isotropy subgroups with four-dimensional fixed point subspaces. The full dynamics of (36) is not completely understood, but some partial results have been obtained so far. For example it has been shown [24, 25], using the results of [16], that stable quasiperiodic solutions exist in one of the four-dimensional fixed point subspaces. Moreover, the existence of heteroclinic cycles connecting three and four maximally symmetric solutions has been verified in [24] and [25], respectively. One of the possible 3-cycles is $STW \to SSW \to AW$ and an example for a 4-cycle is $TTW \to STW \to SSW \to SW$. Both cycles can be attracting for appropriate values of the parameters.

4.2 Square Lattice

The D_4-part of the relevant symmetry group $D_4 \times_s T^2$ is most conveniently generated by the two noncommuting reflections $\kappa_1 : x_1 \to -x_1$ and $\kappa_d : (x_1, x_2) \to (x_2, x_1)$ in terms of which the rotation by $\pi/2$ is represented by $\rho_{\pi/2} = \kappa_1 \kappa_d$. The torus group T^2 acts as in Subsection 4.1 via $(\psi_1, \psi_2) : (x_1, x_2) \to (x_1 + \psi_1, x_2 + \psi_2)$. When $m_1 = m_2$, the critical eigenfunctions are $\exp[\pm\imath(m_1(x_1 + x_2)]$, thus κ_d acts trivial and we are basically left with an $O(2)$-action. Excluding this case, we have to distinguish the cases $m_1 > 0 = m_2$ and $m_1 > m_2 > 0$. For simplicity we will confine ourselfes to actions that are translation free [26], that is, there are no smaller lattices supporting the critical modes. If $m_1 > 0 = m_2$ this restricts us to $m_1 = 1$ and if $m_1 > m_2 > 0$ we have to assume that m_1, m_2 are relatively prime.

In the first case, a simple stationary bifurcation lives in a four-dimensional center eigenspace, parametrized by $(Z_1, Z_2) \equiv \mathbf{Z} \in \mathbf{C}^2$ as $\mathcal{E}_c : \ (Z_1 e^{\imath x_1} + Z_2 e^{\imath x_2} + cc)p$, giving rise to the actions

$$(\psi_1,\psi_2) : \mathbf{Z} \to (e^{\imath\psi_1}Z_1, e^{\imath\psi_2}Z_2), \quad \kappa_1 : \mathbf{Z} \to (\bar{Z}_1, Z_2), \quad \kappa_d : \mathbf{Z} \to (Z_2, Z_1)\,.$$

To cubic order the resulting normal form has again the form (33) as in the $O(2) \times O(2)$-case discussed before. Thus we find the same solutions and stabilities with, however, different group theoretical meanings. The former $\widetilde{O(2)}$-symmetric solution with fixed point subspace $(X, 0)$ corresponds now to the $O(2) \times C_2$-invariant stationary pattern $2Xp\cos x_1$, which is translation and reflection invariant in the x_2-direction (the $O(2)$-factor) and in addition reflection invariant in the x_1-direction (the C_2-factor). The former D_2-symmetric solution with fixed point subspace (X, X) leads here to the pattern $4xp\cos x_1 \cos x_2$ showing a D_4-invariance.

The Hopf bifurcation for the case $(m_1, m_2) = (1, 0)$ leads to the eight-dimensional center eigenspace

$$\mathcal{E}_c : \ (Z_1 e^{\imath x_1} + Z_2 e^{\imath x_2} + +Z_3 e^{-\imath x_1} + Z_4 e^{-\imath x_2})p + cc\,, \tag{37}$$

with the following symmetry actions on $\mathbf{Z} = (Z_1, Z_2, Z_3, Z_4)$,

$$(\psi_1,\psi_2) : \mathbf{Z} \to (e^{\imath\psi_1}Z_1, e^{\imath\psi_2}Z_2, e^{-\imath\psi_1}Z_3, e^{-\imath\psi_2}Z_4) \tag{38}$$

$$\kappa_1 : \mathbf{Z} \to (Z_3, Z_2, Z_1, Z_4)\,, \quad \kappa_2 : \mathbf{Z} \to (Z_2, Z_1, Z_4, Z_3)\,, \tag{39}$$

and the usual normal form equivariance $\mathbf{Z} \to e^{\imath\theta}\mathbf{Z}$.

A comprehensive study of this case is presented in [27]. The normal form up to cubic order is again given by (36), with the restriction $a_3 = a_4$ by virtue of the higher symmetry. We summarize in Table 2 the information about the maximally symmetric periodic solutions, their symmetries and their wave patterns. The names stand here for travelling rolls (TR), travelling squares (TS), standing rolls (SR), standing squares (SS) and alternating waves (AW). The meanings of these patterns are similar as in Subsection 4.1 (see Table 1), but have to be understood in the context of the square symmetry instead of a rectangular symmetry. We note that, as in the $O(2) \times O(2)$-case, $(Z, Z, 0, 0)$ and $(Z, 0, 0, Z)$ are fixed point subspaces, however, the underlying isotropy subgroups are conjugate to each other and both resulting patterns are travelling squares. Moreover, the symmetry group for the alternating wave is $\tilde{D}_4 \times C_2^\pi$, where $\tilde{D}_4$ is the spatio-temporal symmetry generated by $\tilde{\kappa}_d$ and κ_1 and C_2^π is generated by spatial rotation by π.

In [27] a detailed stability analysis of the solutions of Table 2 is carried out. This analysis shows that each of these solutions can occur stably for appropriate choices of the nonlinear coefficients. There are 13 different cases which lead to at least one stable solution. Moreover, there is the possibility that one or even two subcritical branches coexist with a stable branch. For example, if SS is subcritical, SW, AW or TW can be stable and, conversely, if TW and/or SW is subcritical SS can be stable. Finally, SS can be simultaneously stable with one solution among AW, TW or SW.

name	symmetries	$Fix(\Sigma)$	wave pattern
TR	$\psi_2,\ \kappa_2,\ \theta=-\psi_1$	$(Z,0,0,0)$	$2\cos(x_1+\omega t)$
TS	$\kappa_d,\ \theta=-\psi_1=-\psi_2$	$(Z,Z,0,0)$	$4\cos[(x_1-x_2)/2]\cos[(x_1+x_2)/2+\omega t)]$
SR	$\kappa_\pi,\ \kappa_1,\ \kappa_2,\ \psi_2$	$(Z,0,Z,0)$	$4\cos(x_1)\cos(\omega t)$
SS	$\kappa_\pi,\ \kappa_1,\ \kappa_d$	(Z,Z,Z,Z)	$8\cos[(x_1-x_2)/2]\cos[(x_1+x_2)/2]\cos(\omega t)$
AW	$\kappa_\pi,\ \tilde{\kappa}_d,\ \kappa_1$	$(Z,-\imath Z,Z,-\imath Z)$	$4\cos(x_1)\cos(\omega t)+4\cos(x_2)\sin(\omega t)$

Table 2. Same as Table 1 for the $D_4 \times_s T^2$-equivariant Hopf normal form, eq. (36) with $a_3 = a_4$. κ_π is defined by the translation $(\pi,\pi) \in T^2$ combined with spatial rotation by π. $\tilde{\kappa}_d$ is defined by κ_d combined with a temporal phase shift by π and the translation $\psi_2 = \pi$.

Further analysis carried out in [27] is devoted to the dynamics in four-dimensional fixed point subspaces, in particular with respect to the existence of heteroclinic cycles. In certain ranges of the nonlinear coefficients cycles between TS, SS, AW and between TS, SS, SR, TR could be shown to occur.

In the second case, $m_1 > m_2 > 0$ with m_1, m_2 relatively prime, a generic stationary bifurcation is posed in an eight-dimensional center eigenspace. A group theoretical classification of the stationary solutions as well as a stability analysis of the generic branches is given in [28]. One has to distinguish here between $m_1 + m_2$ even or odd; only the latter case is considered in [28]. We outline these results using a slightly different notation. The center eigenspace is parametrized as

$$\mathcal{E}_c :\ (Z_1\phi_{m_1m_2} + Z_2\phi_{m_1,-m_2} + Z_3\phi_{m_2m_1} + Z_4\phi_{m_2,-m_1} + cc)p\,, \tag{40}$$

where the $\phi_{n_1n_2}$ are defined in (29) with $s = 1$. (The representation used in [28] follows from (40) when (Z_2, Z_4) is replaced by $(\bar{Z}_4, \bar{Z}_2)$.) The induced actions of $D_4 \times_s T^2$ on $\mathbf{Z} \equiv (Z_1, Z_2, Z_3, Z_4)$ are as usual inferred from (40): the translations (ψ_1, ψ_2) act as phase factors $Z_1 \to \phi_{m_1m_2}(\psi_1, \psi_2)Z_1$ etc. and the D_4-generators act as

$$\kappa_1\ :\mathbf{Z} \to (\bar{Z}_2, \bar{Z}_1, \bar{Z}_4, \bar{Z}_3)\,,\quad \kappa_d\ :\mathbf{Z} \to (Z_3, \bar{Z}_4, Z_1, \bar{Z}_2)\,. \tag{41}$$

The "relevant" terms of a vector field that is equivariant under this action is given by

$$\dot{Z}_1 = (\alpha + \sum_{j=1}^{4} a_j|Z_j|^2)Z_1 + b_1\bar{Z}_1^{m_2-1}\bar{Z}_2^{m_2}(Z_3Z_4)^{m_1} + b_2\bar{Z}_1^{m_1-1}Z_2^{m_1}(Z_3\bar{Z}_4)^{m_2}\,, \tag{42}$$

with real coefficients α, a_j, b_j. The equations for $\dot{Z}_2, \dot{Z}_3, \dot{Z}_4$ are obtained by applying $\kappa_1, \kappa_d, \kappa_1\kappa_d$ to (41). Observe that the cubic truncation of (41) shows

name	symmetries	$Fix(\Sigma)$	spatial pattern
R	$\rho_\pi,\ \tilde{\psi}$	$(X,0,0,0)$	$2\cos(m_1x_1+m_2x_2)$
SuS	$\kappa_1,\ \kappa_d$	(X,X,X,X)	$4\cos(m_1x_1)\cos(m_2x_2)+4\cos(m_1x_2)\cos(m_2x_1)$
AnS	$\tilde{\kappa}_1,\ \tilde{\kappa}_d$	$(X,-X,-X,X)$	$4\sin(m_2x_1)\sin(m_1x_2)-4\sin(m_1x_1)\sin(m_2x_2)$
SiS	$\rho_{\pi/2},\ S_s$	$(X,0,0,X)$	$2\cos(m_1x_1+m_2x_2)+2\cos(m_2x_1-m_1x_2)$
Rha	$\rho_\pi,\ \kappa_1,\ S_a$	$(X,X,0,0)$	$2\cos(m_1x_1+m_2x_2)+2\cos(m_1x_1-m_2x_2)$
Rhb	$\rho_\pi,\ \kappa_d,\ S_b$	$(X,0,X,0)$	$2\cos(m_1x_1+m_2x_2)+2\cos(m_2x_1+m_1x_2)$

Table 3. Maximally symmetric solution types ($X \in \mathbf{R}$) and their spatial patterns for (42). $\tilde{\kappa}_1, \tilde{\kappa}_d$ are defined by κ_1, κ_d combined with $Z \to -Z$. $\tilde{\psi}$ denotes translations in T^2 with $\psi_1/m_2 = -\psi_2/m_1$. The symmetries S_s, S_a, S_b (called $S_{1,2}, S_{1,3}, S_{1,4}$ in [28]) are generated by two discrete translations $\psi^{(j)} = (\psi_1^{(j)}, \psi_2^{(j)}) \in T^2$, $j = 1, 2$, which are summarized in Table 4 When multiplied by Xp, the expressions in the last column represent the leading terms of the solution patterns $u(t, x)$.

an $O(2)^4$-symmetry. The terms described by b_1, b_2 are the lowest order terms that break this high symmetry down to $D_4 \times_s T^2$.

In Table 3 we summarize the maximally symmetric solutions ($X = X(\alpha) \in \mathbf{R}$), their fixing symmetries and the resulting spatial patterns. The symmetries S_s, S_a, S_b for the last three patterns and their actions on $\mathbf{Z}$ are summarized in Table 4. The letters in the first column stand for rolls, super squares, anti squares, simple squares, type-a-rhombs and type-b-rhombs. These names reveal already the basic forms of these patterns, details about their fixing symmetries are summarized in the second column. For example, R stands for a translation invariant pattern, SuS has a full D_4-symmetry, AnS has a D_4-symmetry whose generating reflections are combined with a phase shift (π, π) and SiS shows a cyclic C_4-symmetry together with a discrete translation invariance generated by S_s. Of special interest are the two rhombic patterns Rha, Rhb. The wave numbers of these patterns are $k_1 = (m_1, m_2)$, $k_2 = (m_1, -m_2)$ and $k_1 = (m_1, m_2)$, $k_2 = (m_2, m_1)$ with angles $\cos\gamma = (m_1^2 - m_2^2)/(m_1^2 + m_2^2)$ and $\cos\gamma = 2m_1m_2/(m_1^2+m_2^2)$ between k_1, k_2 for Rha and Rhb, respectively. Thus the rhombic patterns are supported by the finer lattice generated by $e_1 = \psi^{(1)}$, $e_2 = \psi^{(2)}$. Their basic rectangles possess aspect ratios m_2/m_1 and $(m_1 - m_2)/(m_1 + m_2)$, respectively, and both have basic wavelength $l_c = (m_1^2 + m_2^2)^{-1/2}$. In [28] it is argued that the normal form analysis of (42) can also be interpreted in the following way: one may assume that an instability occurs with a critical wavenumber $k_c = 2\pi/l_c$ and consider the two rhombic patterns as basic solutions characteristic of that wavelength. Then the other solutions of Table 3 have periodicity $l >> l_c$, thus the normal form (42) provides a stability analysis of the rhombic patterns versus solutions with much larger speriod than that predicted by k_c.

	$\psi^{(1)}$	$\psi^{(2)}$	α	β	$\psi^{(1)}$ – action	$\psi^{(2)}$ – action
S_s	$\frac{2\pi(m_1,m_2)}{m_1^2+m_2^2}$	$\frac{2\pi(-m_2,m_1)}{m_1^2+m_2^2}$	$\frac{2\pi(m_1^2-m_2^2)}{m_1^2+m_2^2}$	$\frac{2\pi m_1 m_2}{m_1^2+m_2^2}$	$e^{\imath\alpha}Z_2,\ e^{\imath\beta}Z_3$	$e^{\imath\beta}Z_2,\ e^{\imath\alpha}Z_3$
S_a	$(\frac{\pi}{m_1}, \frac{\pi}{m_2})$	$(-\frac{\pi}{m_1}, \frac{\pi}{m_2})$	$\frac{\pi(m_1^2+m_2^2)}{m_1 m_2}$	$\frac{\pi(m_1^2-m_2^2)}{m_1 m_2}$	$e^{\imath\alpha}Z_3,\ e^{-\imath\beta}Z_4$	$e^{\imath\beta}Z_3,\ e^{-\imath\alpha}Z_4$
S_b	$\frac{2\pi(m_1,-m_2)}{m_1^2-m_2^2}$	$\frac{2\pi(-m_2,m_1)}{m_1^2-m_2^2}$	$\frac{2\pi(m_1^2+m_2^2)}{m_1^2-m_2^2}$	$\frac{4\pi m_1 m_2}{m_1^2-m_2^2}$	$e^{\imath\alpha}Z_2,\ e^{\imath\beta}Z_4$	$e^{-\imath\beta}Z_2,\ e^{-\imath\alpha}Z_4$

Table 4. Generators $\psi^{(1)}, \psi^{(2)} \in T^2$ of the discrete translation groups S_s, S_a, S_b occuring in Table 3. The induced actions are rotations of two components of $\mathbf{Z}$ by angles α, β.

The stabilities of the basic solutions depend as usual on the nonlinear coefficients appearing in (42). We refer to [28] for details and mention only some basic facts. As in previous cases, any of the maximally symmetric solutions can bifurcate stably and, if so, it must be supercritical. Moreover, at least one solution is stable if all branches are supercritical. Conversely, if any solution is subcritical, R, SuS and AnS are unstable and if all of these three branches are subcritical, then all maximally symmetric solutions are unstable. Concerning multiple stable solutions, there can be at most two solutions that are simultaneously stable, namely any combination of two states of SiS, Rha, Rhb.

The Hopf bifurcation corresponding to an eight-dimensional irreducible representation involves a sixteen-dimensional center eigenspace. A purely group theoretical analysis of this more complicated case is carried out in [29] and provides a variety of new types of waves.

4.3 Hexagonal Lattice

Here the translations (ψ_1, ψ_2) act as $x \to x + ((\psi_1 + 2\psi_2)/\sqrt{3}, \psi_1)$. A convenient choice for the generators of D_6 is provided by the reflection $\kappa_1 : x \to (-x_1, x_2)$, the inversion $\rho_\pi : x \to -x$ and the $2\pi/3$-rotation $\rho_{2\pi/3} : x \to (-\frac{1}{2}(x_1 + \sqrt{3}x_2), \frac{1}{2}(\sqrt{3}x_1 - x_2))$. We also will need the reflection $\kappa_2 : x \to (x_1, -x_2)$ in terms of which the cyclic generator $\rho_{\pi/3}$ can be expressed as $\rho_{\pi/3} = \kappa_1 \rho_{2\pi/3} \kappa_2$.

Given an integer pair (wave vector) (m_1, m_2), one can generate an irreducible representation space for $D_6 \times_s T^2$ by applying all group elements of D_6 to an eigenfunction $\phi_{m_1 m_2}(x_1, x_2)$ as defined in (30). When m_1, m_2 are both nonzero the resulting space is twelve-dimensional. This case has been first examined in [30] and later more systematically in [26, 28]. We confine ourselfes here to the six-dimensional case where $m_2 = 0$ and $m_1 = 1$ for simplicity.

As mentioned in Section 2, the governing PDE may possess an additional reflection symmetry κ_0 involving the dependent variables, for example $\kappa_0 : u \to -u$ if (1) is odd in u. For the rectangular and square lattices the presence of an additional reflection symmetry does, however, not lead to further restrictions on the normal forms near a generic bifurcation because these are always odd in $\mathbf{Z}$. In contrast, in case of a stationary bifurcation on a hexagonal lattice generic quadratic terms are present which are forced to vanish in the presence of an additional reflection symmetry. Since quadratic terms usually induce transcritical bifurcations whereas odd normal forms lead to pitchforks, one finds different branching behaviors and stabilities. It is, therefore, of much interest to study bifurcations on a hexagonal lattice also in the presence of an additional reflection symmetry.

Generic stationary bifurcations for the six-dimensional irreducible representation of $D_6 \times T^2$ have been analyzed in [31] without and in [32] with an additional reflection symmetry. The Hopf bifurcation for this representation was first analyzed in [33]; an additional C_2-symmetry does not have an effect here. In the following we give an overview of these results.

In case of a stationary bifurcation (with $\mu_c = m_1^2 = 1$) the center eigenspace is parametrized as $\mathcal{E}_c$: $(Z_1\phi_{10} + Z_2\phi_{-1,-1} + Z_3\phi_{01} + cc)p$, and the symmetry generators act on $\mathbf{Z} \equiv (Z_1, Z_2, Z_3)$ in the form

$$(\psi_1, \psi_2) : \mathbf{Z} \to (e^{i\psi_1} Z_1, e^{-i(\psi_1+\psi_2)} Z_2, e^{i\psi_2} Z_3) \tag{43}$$

$$\rho_{2\pi/3} : \mathbf{Z} \to (Z_3, Z_1, Z_2), \quad \kappa_1 : \mathbf{Z} \to (Z_1, Z_3, Z_2), \quad \rho_\pi : \mathbf{Z} \to \bar{\mathbf{Z}}. \tag{44}$$

Note that $(\pi, 0) \in T^2$ operates as $\mathbf{Z} \to (-Z_1, -Z_2, Z_3)$. Similarly, $\psi_2 = \pi$ induces a sign change of Z_2, Z_3 and both together a sign change of Z_1, Z_3, however, a sign change of a single component with the remaining two fixed is not contained in the action of $D_6 \times_s T^2$. In contrast, if the additional C_2-symmetry $\kappa_0 : \mathbf{Z} \to -\mathbf{Z}$ is present, the enlarged symmetry group $D_6 \times_s T^2 \times C_2$ contains the sign change $\hat{\kappa}_3 : \mathbf{Z} \to (Z_1, Z_2, -Z_3)$ derived from simultaneous operation by $\psi_1 = \pi$ and κ_0. Similarly one can generate the analogously defined reflections $\hat{\kappa}_1$ and $\hat{\kappa}_2$.

The relevant terms of the normal form without a κ_0-reflection are given by (see [31])

$$\dot{Z}_1 = (\alpha + a|Z_1|^2)Z_1 + c\bar{Z}_2\bar{Z}_3\,, \tag{45}$$

with real coefficients α, a, c. The equations for $\dot{Z}_2, \dot{Z}_3$ follow from cyclic permutations due to the $\rho_{2\pi/3}$-action. The group theoretical classification of the isotropy subgroups of the above action shows that there are two maximal isotropy subgroups with the one-dimensional fixed point subspaces $(X, 0, 0)$ and (X, X, X). The first subspace is left invariant by $(0, \psi_2) \in T^2$ and the D_2-group generated by ρ_π, κ_1 and hence corresponds to a roll (stripe) pattern, given by $2Xp\cos x_2$. Points in the subspace (X, X, X) are left invariant by all D_6-operations and so induce a hexagonal pattern, which can be written as

$$u(x_1, x_2) = 8Xp\cos(x_2/2)\cos[(\sqrt{3}x_1 + x_2)/4]\cos[(\sqrt{3}x_1 - x_2)/4] - 2Xp\,.$$

name	symmetries	$Fix(\Sigma)$	spatial pattern
R	ψ_2, κ_1, ρ_π, $\hat{\kappa}_3$	$(X,0,0)$	$2\cos(x_2)$
H	κ_1, $\rho_{2\pi/3}$, ρ_π	(X,X,X)	$8\Pi_{j=1}^3 \cos(k_j \cdot x/2) - 2$
PQ	κ_1, ρ_π, $\hat{\kappa}_1$	$(0,X,X)$	$4\cos(\sqrt{3}x_1/2)\cos(x_2/2)$
Tr	κ_1, $\rho_{2\pi/3}$, $\kappa_0\rho_\pi$	$(\imath X, \imath X, \imath X)$	$8\Pi_{j=1}^3 \sin(k_j \cdot x/2)$

Table 5. Same as Table 3 for (46). The wave numbers $k_1 = (0,1)$, $k_2 = (\sqrt{3}/2, -1/2)$ are the dual basis vectors of e_1, e_2, and $k_3 = -k_1 - k_2$.

Observe that (45) restricted to $(X,0,0)$ has the form of a pitchfork bifurcation whereas restricted to (X,X,X) it becomes a transcritical normal form. The stability analysis shows that both solutions are always unstable (similarly as for the stationary bifurcation with D_3-symmetry). In [31], therefore, also various normal forms of codimension two, with a or c "small" and higher order terms included, have been analyzed. In these extended analyses one of the unstably generated primary branches is stabilized in a secondary bifurcation, e.g., the transcritical hexagonal branch can gain stability in a saddle node bifurcation.

With additional reflection symmetry the relevant terms of the normal form are

$$\dot{Z}_1 = [\alpha + a|Z_1|^2 + b(|Z_2|^2 + |Z_3|^2)]Z_1 + c\bar{Z}_1 \bar{Z}_2^2 \bar{Z}_3^2 \,, \tag{46}$$

with again real coefficients. The basic solutions and their fixing symmetries are summarized in Table 5. The names in the first column of this table stand for rolls (stripes), hexagons, patchwork quilt and (regular) triangles. The meanings of the first two names are obvious from their symmetries. The PQ-solution corresponds to a rectangular pattern showing besides the reflection invariances of a rectangle an additional reflection invariance under translation of x_1 by $2\pi/\sqrt{3}$ combined with the inversion $u \longrightarrow -u$. Analogously the Tr-solution corresponds to a triangular pattern that exhibits the additional reflection invariance $u(x) = -u(-x)$. Concerning stabilities, the analysis of [32] shows that generically all branches are pitchforks and that the bifurcation diagram possesses at most one stable branch which can be any of the four solutions of Table 5. Necessary and sufficient conditions for a stable branch to occur is that all branches are supercritical. The fifth order term in (46) needs to be taken into account in order that both amplitudes and stabilities of the H versus Tr solutions can be distinguished.

In the case of a Hopf bifurcation involving the six-dimensional irreducible representation of $D_6 \times_s T^2$, the center eigenspace is parametrized as

$$\mathcal{E}_c : \ (Z_1\phi_{10} + Z_2\phi_{-1,-1} + Z_3\phi_{01} + Z_4\bar{\phi}_{10} + Z_5\bar{\phi}_{-1,-1} + Z_6\bar{\phi}_{01})p + cc \,. \tag{47}$$

name	symmetries	$Fix(\Sigma)$	wave pattern
SR	$\kappa_1,\ \rho_\pi,\ \psi_2,\ \tilde{\kappa}$	$(Z,0,0,Z,0,0)$	$4\cos(\omega t)\cos(x_2)$
TR	$\kappa_1,\ \psi_2,\ \tilde{\psi}_1$	$(Z,0,0,0,0,0)$	$2\cos(\omega t + x_2)$
$WR1$	$\rho_\pi,\ \tilde{\rho}_{\pi/2}$	$(0,Z,\imath Z,0,Z,\imath Z)$	$4\sum_{j=2}^{3}\sin(\omega t + j\pi/2)\cos(k_j\cdot x)$
$WR2$	$\tilde{\rho}^{(1)}_{\pi/3}$	$(Z,\rho Z,\rho^2 Z,$ $-Z,-\rho Z,-\rho^2 Z)$	$4\sum_{j=0}^{2}\sin(2\pi j/3 - \omega t)\sin(k_{j+1}\cdot x)$
$TRe1$	$\kappa_1,\ \tilde{\psi}_{12}$	$(0,Z,Z,0,0,0)$	$4\cos(\omega t - x_2/2)\cos(\sqrt{3}x_1/2)$
$TRe2$	$\rho_\pi,\ \tilde{\psi}_2$	$(0,Z,0,0,0,Z)$	$4\cos(\omega t - \sqrt{3}x_1/2)\cos(x_2/2)$
$TwPQ$	$\kappa_2,\ \tilde{\rho}^{(2)}_{\pi/3}$	$(Z,\rho^2 Z,\rho Z,$ $Z,\rho^2 Z,\rho Z)$	$4\sum_{j=1}^{3}\cos(\omega t + 2\pi j/3)\cos(k_{j+1}\cdot x)$
SPQ	$\rho_\pi,\ \kappa_2,\ \tilde{\pi}$	$(0,Z,Z,0,Z,Z)$	$8\cos(\omega t)\cos(\sqrt{3}x_1/2)\cos(x_2/2)$
STr	$\kappa_2,\ \tilde{\rho}^{(3)}_{\pi/3}$	$(Z,Z,Z,$ $-Z,-Z,-Z)$	$16\sin(\omega t)\Pi_{j=1}^{3}\sin(k_j\cdot x/2)$
OTr	$\rho_{2\pi/3},\ \kappa_1,\ \tilde{\rho}_{2\pi/3}$	$(Z,Z,Z,0,0,0)$	$2\sum_{j=1}^{3}\cos(\omega t + k_j\cdot x)$
SH	$\rho_\pi,\ \rho_{2\pi/3},\ \kappa_1$	(Z,Z,Z,Z,Z,Z)	$16\cos(\omega t)\Pi_{j=1}^{3}\cos(k_j\cdot x/2) - 4$

Table 6. Same as Table 1 for the $D_6 \times_s T^2$-equivariant Hopf normal form (48). Wavenumbers k_j are as in Table 5 and $\rho = e^{2\pi\imath/3}$.

The translations (ψ_1, ψ_2) act on the first three components (Z_1, Z_2, Z_3) of $\mathbf{Z}$ as in (43) and on (Z_4, Z_5, Z_6) by the complex conjugate phase factor. The generators $\rho_{2\pi/3}$, κ_1 and ρ_π permute the indices $(1,2,3,4,5,6)$ according to $(3,1,2,6,4,5)$, $(1,3,2,4,6,5)$ and $(4,5,6,1,2,3)$, respectively, whereas κ_2 and $\rho_{\pi/3}$ permute as $(4,6,5,1,3,2)$ and $(5,6,4,2,3,1)$. In addition there is the usual Hopf normal form symmetry $\mathbf{Z} \to e^{\imath\theta}\mathbf{Z}$. Up to "relevant" order, the normal form for this action is given by

$$\dot{Z}_1 = (\imath\omega + \alpha + \sum_{j=1}^{6} a_j |Z_j|^2) Z_1 + b Z_4(\bar{Z}_3^2 Z_5^2 + \bar{Z}_2^2 Z_6^2) + c\bar{Z}_2\bar{Z}_3 Z_4 Z_5 Z_6\,, \quad (48)$$

with $a_2 = a_3$ and $a_5 = a_6$ by virtue of the κ_1-equivariance. In [33] eleven distinct maximally symmetric solutions have been classified which are summarized in Table 6. In this table additional group generators marked by a tilde correspond to twisted symmetry operations, that is, spatial symmetries combined with a temporal phase shift θ.

The operations $\tilde{\kappa}$, $\tilde{\rho}_{\pi/2}$, $\tilde{\rho}_{2\pi/3}$ and $\tilde{\rho}^{(j)}_{\pi/3}$ $(j = 1,2,3)$ generate cyclic twisted subgroups of orders $2, 4, 3$ and 6, respectively. Specifically, $\tilde{\kappa}$ is a combined spatial and temporal phase shift by π, $\tilde{\rho}_{\pi/2}$ is a temporal phase shift by $\pi/2$ combined with a reflection about the x_1-axis and a spatial phase shift by π, $\tilde{\rho}_{2\pi/3}$ corresponds to simultaneous spatial and temporal translations in both directions by $2\pi/3$ and in $\tilde{\rho}^{(j)}_{\pi/3}$ the rotation by $\pi/3$ is combined with the temporal phase shifts $\pi/3$, $2\pi/3$, π for $j = 1, 2, 3$, respectively. The operations $\tilde{\psi}_1$, $\tilde{\psi}_2$ and $\tilde{\psi}_{12}$ form continuous symmetries of travelling waves, defined by $(\psi_1, \psi_2) = (-\theta, 0)$, $(0, \theta)$ and $(2\theta, -\theta)$, respectively. Finally, $\tilde{\pi}$ is a standard symmetry of a standing wave, defined by $\psi_1 = 0$, $\psi_2 = \theta = \pi$.

Concerning the individual patterns in Table 6, a first letter S or T stands for a standing or travelling wave. The second letters R, PQ, Tr and H refer to the spatial form of a roll (stripe), patchwork quilt, (regular) triangle and hexagon as classified in Table 5. In addition, Re refers to a rectangular shape. As usual, temporally periodic roll patterns are characterized by a full spatial translation invariance (we have chosen ψ_2-invariance in the table), additional discrete invariances and a full spatiotemporal translation invariance if it is a travelling roll. The two types of "travelling rectangles" $TRe1$and $TRe2$ show a spatial reflection invariance and again a full spatiotemporal translation invariance. The meanings of the names standing triangle (STR) and standing hexagon (SH) is obvious from the symmetries and the wave forms. Note that STr has besides the D_3-invariance an additional reflection symmetry, similarly as in the stationary case. In [33] it is, therefore, more accurately denoted as standing regular tiangle. For an application of these results to magnetoconvection see [34].

The spatial shape of the patchwork quilt PQ (see Table 5) occurs in two different wave patterns distinguished by SPQ and $TwPQ$. The SPQ is a standing wave, whereas $TwPQ$ ("twisted PQ") refers to the $\tilde{\rho}^{(2)}_{\pi/3}$-twist symmetry shown by this pattern. Finally, OTr refers to a rotating (oscillating) pattern of triangular shape. Its spatiotemporal symmetry $\tilde{\rho}_{2\pi/3}$ is also characteristic for the discrete travelling wave occuring in the Hopf bifurcation with D_3-symmetry.

The stabilities of the patterns of Table 6 depend as usual on the nonlinear coefficients a_j, b_j. It turns out here that the travelling rectangle $TRe1$ is always unstable while all other patterns can occur stably for appropriate values of the coefficients. It should be noted that the cubic truncation of the normal form (48) shows an additional continuous symmetry, namely equivariance against $(Z_2, Z_5) \rightarrow (e^{i\psi_3} Z_2, e^{-i\psi_3} Z_5)$ (with Z_j for $j \neq 2, 5$ fixed). As a consequence, at cubic order the STr and SH as well as the $WR2$ and $TwPQ$ solutions show the same stabilities at cubic order. The two quintic terms are needed to distinguish the stability properties of the solutions in each of these two pairs. Their effect is to change a zero eigenvalue of the cubic truncation to a nonzero value.

Both the stationary and the Hopf normal form involving a twelve-dimensional irreducible representation of $D_6 \times_s T^2$ (or $D_6 \times_s T^2 \times C_2$) are rather complicated and we refer to [28, 29] for details. In the stationary case a similar, though richer solution structure is found as in the case of the square lattice involving the eight-dimensional irreducible representation (Subsection 4.2).

In particular, also here generic rhombic patterns can occur that are supported by finer lattices.

5 Further Developments

In Sections 3 and 4 we restricted attention to (i) generic stationary and Hopf bifurcations and (ii) symmetries which are characteristic of PDE's posed in one- and two-dimensional domains. A large variety of further results has been obtained so far concerning other symmetries, degenerate and multiple bifurcations as well as the effect of symmetry breaking imperfections. In this section we present a summary of further work related to the analysis of instablities in an appropriate symmetry context without, however, claiming to be complete regarding the large number of papers published in that area.

Bifurcations with Spherical Symmetry. The interst in PDE's posed in spheres or spherical shells arises from various kinds of convection problems occuring in the astrophysical and geophysical sciences [35] such as convection in the earth mantle which is believed to be the driving force for plate tectonics [36]. The "simplest" model equations describing such problems are the Boussinesq equations for a self-gravitating spherical fluid layer heated at its inner boundary. When the gravitational field and the heat source distribution are isotropic and the Coriolis force can be neglected, these equations stay invariant under arbitrary orthogonal transformations in $\mathbf{R}^3$ which constitute the spherical symmetry group $O(3)$. Since the basic (heat conduction) state depends only on $r = |x|$, the convective instability is governed by a primary bifurcation in an $O(3)$-equivariant system.

The study of steady state bifurcations with $O(3)$-symmetry started in the seventies, both mathematically [37] and in the context of specific applications including besides convection [38] also the buckling of spherical shells [39]. The irreducible representation spaces are the $2l + 1$-dimensional spaces of spherical harmonics $Y_{lm}(\vartheta, \varphi)$ of order l. The lattices of isotropy subgroups for these representations have been classified in [40] and show increasing complexity when l increases. The symmetries that a primary stationary branch can possess depend on l and include the space symmetries of the Platonic bodies (octahedral, isocahedral and tetrahedral symmetries), an $O(2)$-symmetry (axisymmetric patterns) and all dihedral symmetries. When l is odd, stable supercritical branches can appear, for example patterns with six-fold rotation symmetry or with octahedral symmetry when $l = 3$. For l even the primary branches are typically transcritical and unstable because here a single quadratic term exists in the reduced equations (which is not present for l odd). Stabilized primary as well as secondary and tertiary branches for l even can be determined to some extent near a "codimension two bifurcation" for which the quadratic term is small and higher order terms are included. This kind of analysis has been first pursued in [41] for $l = 2$, however, it took 16 years until computer algebra and sufficient computational power permitted the normal form computations for the Boussinesq equations [42]. The persistent stable branches are here all axisymmetric. In contrast, for $l = 4$ the persistent stable branches show either a four-fold rotation symmetry or an octahedral symmetry and these cases occur in the Boussinesq equations for small and large values of the Prandtl number, respectively [43]. Numerical simulations indicate that these patterns persist to ranges of the Rayleigh number far beyond the convective instability. Further results concerning generic and degenerate stationary bifurcations with $O(3)$-symmetry can be found in [44].

Hopf bifurcations with that symmetry have not been studied yet in so much detail. A partial classification of isotropy subgroups is given in [13] and a special case is analyzed in [45].

Mode Couplings. A variety of different mode couplings have been analyzed so far, concerning on the one hand their types and on the other hand different symmetries as well as different representations for a given symmetry. A basic example is provided by stationary-stationary mode couplings involving two single modes which are odd under different reflections. Up to cubic order such an instability is described by the following system of ODE's for two real amplitudes (X_1, X_2),

$$\dot{X}_1 = \alpha X_1 + aX_1^3 + bX_1X_2^2\,, \quad \dot{X}_2 = \beta X_2 + cX_1^2X_2 + dX_2^3\,, \tag{49}$$

where a, b, c, d are nonlinear coefficients and (α, β) are small parameters which are varied about $(0,0)$. The basic features of (49) are "pure mode solutions" $(X_1, 0)$, $(0, X_2)$, created in distinct primary bifurcations, and "mixed mode solutions" with X_1, X_2 both nonzero, created in secondary bifurcations from the pure modes. In certain ranges of a, b, c, d also "tertiary" Hopf bifurcations from secondary solutions may occur. Their analysis requires to take fifth order terms into account. In the context of two-dimensional confined convection systems, (49) describes the situation where the number of convection rolls generated at the convective instability increases by one when the aspect ratio passes through certain critical values [46].

The appearance of the above types of solutions is a basic characteristic feature of all mode couplings. In case of "simple" Hopf-stationary and Hopf-Hopf mode couplings described by amplitudes (Z, X) (assuming that X belongs to an odd mode) and (Z_1, Z_2), the leading terms of the corresponding normal forms written in polar coordinates are also governed by (49), with (X_1, X_2) replaced by (R, X) and (R_1, R_2), respectively, because the phases decouple. Here pure mode solutions with a radial variable R or R_j nonzero correspond to periodic orbits in the (Z, X)- or (Z_1, Z_2)-systems. If R_1, R_2 are both nonzero (secondary solution) we find a 2-torus (two-frequency motion) in the (Z_1, Z_2)-system. Moreover, a periodic orbit of (49), created in a tertiary bifurcation, induces a 2-torus in the (Z, X)-system and a 3-torus in the (Z_1, Z_2)-system. A carefull analysis [47], in which the normal form symmetry is regarded as an approximate rather than an exact geometrical symmetry, shows that the Hopf-stationary and Hopf-Hopf mode couplings lead to chaotic dynamics if (49) admits tertiary Hopf bifurcations. In particular the 3-torus breaks up into 2-tori and chaotic attractors reflecting the general result that three-frequency motions are structurally unstable [48].

When more complicated symmetries are present the dimension of the phase space (center eigenspace) becomes larger. For example, a mode coupling involving two stationary modes e^{imx} and e^{inx}, referred to as "spatial (m, n)-resonance", is governed by an ODE-system for two complex amplitudes (Z_1, Z_2) of the form [49] (see, also, [50])

$$\begin{aligned}\dot{Z}_1 &= (\alpha + a|Z_1|^2 + b|Z_2|^2)Z_1 + c\bar{Z}_1^{n-1}Z_2^m \\ \dot{Z}_2 &= (\beta + d|Z_1|^2 + e|Z_2|^2)Z_2 + fZ_1^n\bar{Z}_2^{m-1}\,.\end{aligned}$$

The branching behaviour of this system depends on the nonlinear coefficients as well as on the mode numbers m, n and admits tertiary bifurcations to both travelling and standing waves. The case of the $(1, 2)$-resonance is special, here also heteroclinic cycles may appear [51] in terms of which some of the complex dynamical behaviour observed in simulations of the Kuramoto Sivashinsky equation could be explained [52]. The $(1, 2)$-resonance plays also a role for interacting convection rolls [53] as well as for wave-vector restrictions of axisymmetric Taylor vortex flows [54] and MHD-flows [55].

Further research on stationary-stationary mode couplings has been pursued for the case of an $O(3)$-symmetry. The $(l = 1, l = 2)$-coupling was first analyzed, mainly numerically, in [56] and in [57] the existence of heteroclinic cycles was shown. In recent work [58] it was proved that for any spherical $(l, l + 1)$-mode coupling a heteroclinic cycle can exist for generic values of the nonlinear coefficients and, moreover, does appear if these coefficients are derived from the Boussinesq equations.

Hopf-stationary mode couplings in the presence of $O(2)$-symmetry are analyzed in [59]. One of the characteristic features of this interaction is the possibility of invariant 3-tori which are identified with various types of wavy vortex flow patterns. Normal form computations for this type of mode coupling in case of the Taylor-Couette problem are performed in [60]. The stability diagram of the parameter space of the normal form shows here a remarkable similarity with experimentally determined stability boundaries [61]. An extension to triple couplings, where two Hopf modes interact with a stationary mode, is studied in [62]. Although the phase space for this problem is ten-dimensional, the use of group theoretical methods led to considerable insight in the dynamics and made it possible to classify a variety of new types of wavy vortex patterns.

Concerning Hopf-Hopf mode couplings with higher symmetries, also here the research focused mainly on $O(2)$-equivariant systems. In [63] basic normal forms were set up and analyzed using group theoretical methods. The main result is that 3-tori as well as 4-tori can appear in the normal form dynamics, although 4-tori are not structurally stable in $O(2)$-symmetric systems.

Takens-Bogdanov Bifurcation. One way to view this kind of multiple bifurcation is as a Hopf bifurcation in the limit of zero frequency or infinite period. Since a branch of periodic orbits whose period goes to infinity typically terminates in a homoclinic or heteroclinic loop, one can expect such loops to occur generically near a Takens-Bogdanov (TB) bifurcation. The lowest possible dimension for a TB bifurcation is two and the corresponding normal form can be written as a second order ODE for a single variable X as $\ddot{X} + \dot{X}g(X) + f(X) = 0$, where $f(X)$ and $g(X)$ describe conservative and friction terms. The forms of f and g depend on the symmetries. For example, in the presence of a reflection symmetry $X \rightarrow -X$ we have $f = \alpha X + aX^3$ and $g = \beta + bX^2$, where (α, β) and a, b are the usual small parameters and nonlinear coefficients. The reflection symmetric version has been set up and analyzed in [64] in the context of double diffusion convection. The various solutions which can be found here are nontrivial steady states and periodic orbits with large periods, the latter corresponding to convection patterns in the form of standing waves (periodic reversal of the circulation direction in the convection cells). In the case of an $O(2)$-symmetry the real variable X has to be replaced by a complex variable Z. The corresponding normal form was analyzed in [65] and applied in [66] to magnetoconvection and laser systems. Here also travelling and modulated waves (revealed as tori) occur.

Hidden Symmetries. To illustrate the basic ideas consider the reaction diffusion system (1) on the interval $0 \leq x \leq \pi$ with Neumann boundary conditions. The concept of a hidden symmetry comes into play if we consider the same equation with 2π-periodic boundary conditions. The basic point is that any solution of the Neumann problem can be extended via even reflection and periodic extension to a reflection invariant solution of the periodic problem. Conversely, any (smooth) reflection invariant (under $x \rightarrow -x$) solution of the periodic problem satisfies Neumann boundary conditions at $x = 0$ and $x = \pi$. Consequently, the Neumann problem is a subproblem of the periodic problem, defined by restricting the latter to the subspace of reflection invariant functions. In particular this implies that a bifurcation analysis for the Neumann problem may be pursued by restricting a bifurcation analysis of the periodic problem to solutions that are reflection invariant. Now a generic stationary bifurcation for the periodic problem involving a mode e^{imx} is governed by the normal form $\dot{Z} = \alpha Z + aZ^2\bar{Z}$ with $Z \in \mathbf{C}$ (cf. Section 3.1(B)). Restricting this to reflection invariant solutions simply means $Z = \bar{Z} = X$ and so leads to $\dot{X} = \alpha X + aX^3$ as generic equation for a stationary bifurcation involving a $\cos mx$-mode. The important point here is that we always find a pitchfork (symmetry $X \rightarrow -X$), irregardless of the mode number m. While for m odd this is what one expects because $\cos mx$ is odd under $x \rightarrow \pi - x$, one would expect for m even generically a transcritical bifurcation because then $\cos mx$ is an even mode. The fact that also for m even a pitchfork rather than a transcritical bifurcation occurs is a consequence of the $O(2)$-symmetry of the enlarged (periodic) problem. This $O(2)$-symmetry is interpreted as a "hidden symmetry" for the Neumann problem. In [67] (see, also, [50]) further implications of this hidden symmetry, in particular for stationary-stationary mode couplings are discussed. The result is that the normal forms have a special structure that depends on the mode numbers and leads to unexpected bifurcation sequences.

For Dirichlet boundary conditions similar arguments apply, however, an additional reflection symmetry is needed in order that hidden symmetries have an effect [68]. Further research on hidden symmetries has been carried out in [69] for general n-dimensional rectangles, in [70] for hemispheres (the relevant hidden symmetry group is here $O(3)$) and in [71] for the case of a square which is embedded in a square lattice. In the latter case the hidden symmetry group is $D_4 \times_s T^2$ and its presence requires the domain to have the shape of a perfect square. The effect of this hidden symmetry could be observed experimentally in the Faraday problem [72]: perturbations in the shape of the container that preserved the square symmetry but destroyed the perfect square shape altered the bifurcation scenarios significantly.

Forced Symmetry Breaking. Perfect symmetries are rare in macroscopic systems, thus any symmetry based approach to study pattern formation should be considered as an idealization. This observation has stimulated the investigation of the effects of small symmetry breaking imperfection terms on equivariant normal forms. Such effects are most severe when a continuous symmetry is broken. Consider for example (1) with periodic boundary conditions on a line. In a primary stationary bifurcation described by a single, complex mode amplitude Z one finds a family of nontrivial stationary solutions $u(x+\Psi)$ because the phase Ψ of Z is arbitrary. The main effect when the continuous translation symmetry is broken, for example due to inhomogeneities or distant side walls, is that the phase is no longer arbitrary. Instead, only a discrete set of phases is permitted, which ones depending on the remaining symmetries of the imperfections. This selection of discrete phases from a continuous family is referred to as "phase pinning".

More severe phenomena can appear in time dependent problems. In [73] the effect of breaking the translation symmetry on the $O(2)$-equivariant Hopf normal form has been studied and the resulting perturbed normal form was related to coupled complex Ginzburg Landau equations (cf. Section 6) describing dissipative waves in confined systems. The main results are that (i) travelling waves are absorbed at the boundary and (ii) a kind of modulated wave appears that describes a "switching" between left and right travelling waves. Such waves have been observed as convection states in binary fluids [74], where they were called "blinking states". Further numerical analysis of the dynamics of these waves has been carried out in [75].

Bursts and Intermittency. Recently started research in equivariant dynamics is directed towards a geometrical/dynamical description of bursts and intermittency, thereby extending previous work on heteroclinic cycles. Symmetries are crucial here because they typically lead to nested invariant subspaces. The basic idea is that regular or chaotic attractors reside inside such subspaces, however, there are positive Liapunov exponents in directions transverse to them. If global reinjection mechanisms are present, one may find excursions between different subspaces which should be observed as bursts. Some basic scenarios have been described and related to complex spatio-temporal dynamics observed in experiments and simulations of dynamos, lasers and magnetoconvection [76]. It is very likely that these scenarios can be found in a variety of equivariant normal forms, in particular in those describing mode couplings.

6 Constructing Ginzburg-Landau Equations

6.1 Beyond Finite-Dimensional Bifurcation Analysis

As demonstrated in Subsec. 1.3 the multiplicity of solutions generated in a pattern-forming instability develops into a quasi-continuum when the size of the system becomes large. In fact, in the simplest case of a quasi one-dimensional (1D) system of length L, i.e. a system that is extended and homogeneous in one direction and short in the other (transverse) directions, one typically has solutions with wavenumber $k_m \sim 2\pi m/L$. [2] As the system becomes longer one also expects the effect of distant sidewalls to becomes weak in many situations. The previous approach that focuses on the bifurcations of solutions that satisfy specific boundary conditions (modes), may then become inconvenient.

Let us approximate the neutral stability curve introduced in Subsec. 1.3 in the vicinity of its minimum (the threshold of instability) by a parabola

$$\Lambda_0 \approx \Lambda_c(1 + \xi^2(k - k_c)^2), \tag{50}$$

which defines the coherence length ξ. The Ginzburg-Landau approach to be discussed in the following is useful when one has many unstable modes on the marginal branch (50), while all other modes remain well damped. The first condition holds for $\epsilon >> \epsilon_{min} := (2\pi\xi/L)^2$, where we have introduced the reduced control parameter

$$\epsilon = \Lambda/\Lambda_c - 1. \tag{51}$$

Then bulk effects should be separable from surface effects. For lower values of ϵ the modal approach is sufficient, although the GL theory remains applicable as long as $L >> \xi$ holds. The other condition restricts ϵ from above. In

[2] From now on we will adopt a physical length scale.

fact the approach is restricted to the "weakly nonlinear regime" where the amplitude(s) of the pattern is/are small and (50) is a good approximation to the neutral curve. Further above threshold a complementary approach making use of phase equations still applies, see Subsec. 8.2. For other reviews on physically oriented derivations of Ginzburg-Landau equations in nonequilibrium systems, see e.g. [77, 78, 5], and references therein. A more mathematically oriented treatment is found in [79].

6.2 Construction of the GL Equations in 1D

As mentioned before one usually starts from a linear stability analysis of the basic unstructured state taken to be homogeneous in x. We here assume that the basic state is not degenerate. This excludes e.g. situations with Galilean invariance as would arise in convection with free-slip boundary conditions and many secondary instabilities, see Subsec. 8.1. One is led to modal solutions of the form

$$\mathbf{u} = e^{\sigma t} e^{ikx} \mathbf{p}(k). \tag{52}$$

The eigenfuction $\mathbf{p}(\mathbf{k})$ usually includes variations in the transverse directions. We assume that transverse dimensions are short in the sense that "higher" transverse eigenmodes play no role (condition of quasi-onedimensionality). This may exclude e.g. wave phenomena over deep water. Some experimentally important systems are driven periodically in time, see [1]. Then $\mathbf{p}$ is time periodic and the (complex) growth rate

$$\sigma = \sigma_r + i\omega \tag{53}$$

with σ_r and ω real, is actually a Floquet exponent. Solving the linear problem leads to a dispersion relation in the form of an implicit function $f(\sigma, k; \Lambda) = 0$ (often this relation is accessible only numerically). Here Λ stands for the "main" control parameter that carries the system across the instability upon increase. Secondary control parameters will be indicated where necessary. In principle one can solve this relation for the growth rate $\sigma(k; \Lambda)$. If there are several branches, the one(s) with the largest σ_r is/are to be chosen (in view of the transverse dimensions there usually exist in fact infinitely many branches). Setting $\sigma_r = 0$ determines the neutral curve $\Lambda_0(k)$ and the frequency $\omega_0(k)$ at the neutral curve.

The spatial symmetries of the system typically manifest themselves in the dispersion relation, and therefore in the neutral curve. Translation invariance has already been assumed and the only remaining question in 1D is that of reflection symmetry. In most cases one considers a reflection symmetric situation, and then the dispersion relation depends only on k^2. We will adopt this condition here and make some remarks on the non-reflection symmetric case later on. From the previous discussion it is clear that one then has four generic types of bifurcations at $\Lambda = \Lambda_c$ where the growth rate reaches zero for the first time (we denote the Hopf frequency at the minimum k_c of the neutral curve by ω_c):

iB) Homogeneous and stationary instability ($\omega_c = 0$, $k_c = 0$, see Subsec. 3.1 (A)).
iiB) Stationary finite-k instability ($\omega_c = 0$, $k_c \neq 0$, see Subsec. 3.1 (B)).
iiiB) Homogeneous Hopf bifurcation ($\omega_c \neq 0$, $k_c = 0$, see Subsec. 3.2 (A)).
ivB) Finite-k Hopf bifurcation ($\omega_c \neq 0$, $k_c \neq 0$, see Subsec. 3.2 (B)).

The lowest-order description in terms of GL equations can be obtained by starting from the respective truncated normal forms. In case iB) one has a real center-space variable $X(t)$ and the normal form is given either by (11) or (12). Some remarks on this case will be made below. In the other cases one has one or two complex center-space variables $Z_i(t)$ and their dynamical equations are of the form

$$\begin{aligned} \dot{Z}_1 &= (\sigma_r(k,\epsilon) + \imath\,\omega(k,\epsilon))Z_1 + (\tilde{g}|Z_1|^2 + \tilde{h}|Z_2|^2)Z_1 \\ \dot{Z}_2 &= (\sigma_r(-k,\epsilon) + \imath\omega(-k,\epsilon))Z_2 + (\tilde{g}|Z_2|^2 + \tilde{h}|Z_1|^2)Z_2. \end{aligned} \tag{54}$$

(Note the new notation for the nonlinear coefficients.) The second equation arises only in the case ivB), see (21) (otherwise $Z_2 = 0$). We have included explicitly the fact that Z_1 and Z_2 correspond to opposite wavenumbers. One has $\sigma_r(\pm k_c, 0) = 0$ and $\omega(\pm k_c, 0) = \omega_c$. For stationary bifurcations ω is zero and the nonlinear coefficient is real.

The wavenumber k can usually not be prescribed externally. When the system is carried across the instability one may expect a state with $k = k_c$ (or at least with k near k_c) to develop. We now wish to allow for general functions $\mathbf{u}(t,x)$, with the restriction that the spatial variations represent slow modulations of the "critical" periodic solution with $k = k_c$. In Fourier space this means that we allow for functions that are represented by narrow wavepackets centered around k_c. (Actually, for $k_c \neq 0$, the real fields $\mathbf{u}$ will contain two peaks centered around k_c and $-k_c$, but the complex center-space representation picks out one of them, see Subsecs. 3.1 (B) and 3.2 (B) where m is replaced by k). Thus the dynamics in this generalized center space is now described by many (in principle infinitely many) variables (indexed by k), each of which satisfies (54). The function $\mathbf{u}$ is then given by

$$\mathbf{u}(t,x) = \int (Z_1(t,k)e^{\imath kx} + Z_2(t,k)e^{-\imath kx})\mathbf{p}dk/2\pi + cc + \text{hot}. \tag{55}$$

where we have introduced hot for higher-order terms.

Here one may argue that the nonlinear terms in eq.(54) should be replaced by a convolution integral allowing for interaction between all wavenumbers. This is indeed a delicate point that has to be refined in higher dimensions. The problem is with contributions from $k \sim 0$ (for $k_c \neq 0$) which are excited at quadratic order and contribute at cubic order to wavenumbers near k_c. It appears that with the assumptions made these contributions can be eliminated analytically. Then it is consistent to keep the approximation (54) with $\tilde{g}$ and $\tilde{h}$ evaluated at $\epsilon = 0$ and $k = k_c$. Similarly, one may neglect the k dependence of the eigenfunction $\mathbf{p}$ and take it at $k = k_c$.

We may now expand $\sigma_r(k,\epsilon)$ and $\omega(k,\epsilon)$ around $\epsilon = 0$ and $k^2 = k_c^2$, which leads to expressions of the form

$$\sigma_r(\pm k,\epsilon) \approx \tau^{-1}(\epsilon - \xi^2(k-k_c)^2), \tag{56}$$
$$\omega(\pm k,\epsilon) \approx \omega_c + \tau^{-1}(\epsilon a \pm \tau v_g(k-k_c) - \xi^2 b(k-k_c)^2). \tag{57}$$

The first relation defines the relaxation time τ and involves the coherence length ξ introduced in (50). They depend on the specific choice of the control parameter, but the diffusivity ξ^2/τ is unique. Since the expansion is about the minima of the neutral curve, which correspond to maxima of σ_r, no terms linear in $k - k_c$ arise. The second relation defines the linear frequency shift $\tau^{-1}\epsilon a$ and the dispersion coefficient b. The group velocity term can only arise in case (ivB) because ω must have an extremum at $k = 0$ (under the assumption of reflection symmetry). Then the two components have opposite sign.

At this point it is convenient to switch from Fourier to real space. It is also convenient to separate out the frequency at band center $\omega_\epsilon := \omega_c + \tau^{-1}\epsilon a$. Thus we define the (generalized) center-manifold variables in real space by

$$\tilde{A}_{1,2}(t,x) := e^{-i\omega_\epsilon t}\int Z_{1,2}(t,k)e^{\mp ikx}dk/2\pi. \tag{58}$$

Then (55) goes over into

$$\mathbf{u}(t,x) = \left(\tilde{A}_1(t,x)e^{ik_cx} + \tilde{A}_2(t,x)e^{-ik_cx}\right)e^{i\omega_\epsilon t}\mathbf{p}(k_c) + c.c. + \text{hot} \tag{59}$$

and (54) transforms into the following equation for the "amplitudes" $\tilde{A}_j$

$$\tau(\partial_t - v_{gr}\partial_x)\tilde{A}_1 = (\epsilon + \tilde{D}_C)\tilde{A}_1 - g\left((1+ic)|\tilde{A}_1|^2 + h(1+id)|\tilde{A}_2|^2\right)\tilde{A}_1,$$
$$\tau(\partial_t + v_{gr}\partial_x)\tilde{A}_2 = (\epsilon + \tilde{D}_C)\tilde{A}_2 - g\left((1+ic)|\tilde{A}_2|^2 + h(1+id)|\tilde{A}_1|^2\right)\tilde{A}_2 \tag{60}$$

where $\tilde{D}_C = \xi^2(1+ib)\partial_x^2$ and $g(1+ic) = \tau\tilde{g}$, $gh(1+id) = \tau\tilde{h}$. The real and imaginary parts of the coefficients are exhibited explicitly, so the new coefficients are all real. We will call these equations (or their rescaled versions, see below) the (1D) Coupled Complex GL equations (coupled CGLE). The restriction to one amplitude A is called the Complex GL equation (CGLE). Even in case (ivB) the restriction is relevant when $h > 0$, since then the counterpropagating waves tend to suppress each other. By contrast, when $0 > h > -1$ the amplitudes superpose to give standing waves.

6.3 Remarks

Here are a number of remarks on eqs.(60):

1. *Interpretation.* The complex amplitude (or "envelope", "modulation", "order parameter") function describes a slow modulation of a periodic signal.

In the terminology of signal processing it combines amplitude with phase (or "frequency") modulation. The fact that the two aspects appear here on the same level is a consequence of being near the bifurcation, where the amplitude is a "soft" degree of freedom. Further away from threshold only the phase remains soft. Then slow modulations can still be described by phase equations, see Subsec. 8.2. The different terms allow simple physical interpretations: the first term on the rhs describes growth/decay of the critical mode and the second term diffusion and dispersion. The nonlinear terms incorporate saturation and nonlinear frequency shift. We remind that a linear frequency shift has been absorbed in ω_ϵ.

2. *Rescaling.* Scaling time by τ and length by ξ removes these quantities from eqs. (60). One can go further by introducing the rescaling

$$t = (\tau/|\epsilon|)T, \; x = (\xi/\sqrt{|\epsilon|})X, \; \tilde{A}_j = \sqrt{|\epsilon|/g|}A_j, \; u = \tau v_{gr}/(\xi\sqrt{|\epsilon|}), \tag{61}$$

leading to

$$(\partial_T - u\partial_X)A_1 = (\pm 1 + D_C)A_1 - \left((1+ic)|A_1|^2 + h(1+id)|A_2|^2\right)A_1,$$
$$(\partial_T + u\partial_X)A_2 = (\pm 1 + D_C)A_2 - \left((1+ic)|A_2|^2 + h(1+id)|A_1|^2\right)A_2 \tag{62}$$

where $D_C = (1+ib)\partial_X^2$ (for simplicity we here assumed $g > 0$). The $\pm$ sign pertains to positive and negative ϵ, respectively. One sees that in the absense of the group velocity term there is uniform scaling of the coupled CGLE in ϵ. Otherwise u diverges for $\epsilon \longrightarrow 0$. In fact, in the limit $\epsilon \longrightarrow 0$, keeping the system length large but finite, a further reduction of the equations is possible. This point will be discussed in Subsec. 7.2.

3. *Derivation by multiple scale method.* The slow time and length scales introduced above indicate that near threshold the relevant dynamics and spatial modulations become slow and thereby separate from the fast scales of the critical mode. This shows that an alternative perturbational approach employing multiple time and space scales in real space is possible. This was the traditional method to derive modulational equations, as first done in the context of Rayleigh-Bénard convection by Newell, Whitehead and Segel [80].

4. *Derivation by renormalisation group.* An alternate perturbative derivation proposed some years ago was called the renormalization group approach [81]. To our knowledge the method has not produced new results, but it supposedly requires less assumptions and makes connection with other asymptotic methods for global analysis. The method has also been proposed for the derivation of phase equations [82].

5. *Real GL equation.* For a stationary bifurcation all imaginary coefficients vanish in (60),(62), i.e. $b = c = d = 0$, and there is just one amplitude. The resulting equation is usually called the real GL equation (RGLE) although A, of course, remains complex (we are here not considering case iB). In dimensionless units the RGLE is free of parameters demonstrating universality for all stationary, supercritical, finite-k bifurcations in 1D. The RGLE will be discussed in Subsec.7.1. We have mentioned already that the RGLE describes

a purely relaxational gradient dynamics governed by a Lyapunov potential. It shows many similarities with equilibrium systems.

6. *Nonlinear Schroedinger Equation.* Another limiting case, which is opposite to the above one, is obtained when the real coefficients vanish (or become vanishingly small). In the case of a single amplitude one is left with conservative equations describing the modulation of weakly nonlinear waves. It is convenient to rescale differently (and in fact not uniquely) to obtain the celebrated nonlinear Schroedinger equation (NLS)

$$\dot{A}(t'', x'') = \imath(\partial_{x'}^2 \pm |A|^2)A. \qquad (63)$$

The NLS is completely integrable and has been studied extensively, see e.g. [85]. It represents the universal description of (weakly nonlinear) waves in 1D with "strong" dispersion where nonlinearity is balanced by dispersion, like e.g. in gravity driven surface waves in deep water. The NLS is particularly important in nonlinear optics where it describes soliton propagation ("envelope solitons") in optical fibres, see e.g. [84]. From that application the names "focussing case" (+ sign in (63)) exhibiting "bright solitons" with pulse-like envelopes and defocussing case (− sign in (63)) exhibiting "dark" (and "gray") solitons with hole-like envelopes derive.

7. *Bifurcations with $k_c = 0$, $\omega_c = 0$ (case iB).* For stationary and homogeneous instabilities the amplitude/order parameter is *real* and quadratic terms occur generically, unless excluded by some additional symmetry, see Subsec. 3.1 (A). Then

$$\tau \partial_t \tilde{A} = (\epsilon + \tilde{D}_R)\tilde{A} + g_2 \tilde{A}^2 - g\tilde{A}^3 \qquad (64)$$

($\tilde{D}_R = \xi^2 \partial_x^2$) is the minimal equation that ensures with $g > 0$ boundedness of solutions. It is sometimes called the Nonlinear Diffusion equation. If, by an additional symmetry, the quadratic term vanishes, the equation scales with ϵ, like the CGLE. If the quadratic term cannot be considered as a small perturbation, one has to be aware of other higher-order terms, e.g. proportional to $(\partial_x \tilde{A})^2$.

8. *Long-wave instabilities* There are other classes of instabilities, where in principle $k_c = 0$, but the homogeneous mode is excluded by a conservation law, like interfacial instabilities with conserved volume. For an overview, see e.g. [86]. They also arise in the context of phase equations and some of them will be briefly considered in Subsec. 8.2.

9. *Validity.* The question of the validity of the coupled CGLE is in particular the question wether initial conditions chosen from the properly generalized center space remain there for all times. All experience indicates that this is indeed the case in 1D as long as $g \geq 0$, $1 + h \geq 0$ (and $\xi^2 > 0$). For the RGLE this is easy to show. Studies on the validity may be found in [87].

10. *Relation to phase transitions.* Local bifurcations in spatially extended systems exhibit similarities with continuous *phase transitions*. In fact, phase transitions in the so-called mean-field description are described by such bifurcations. The concept has become popular with the success of the (static)

phenomenological GL theory of superconductivity [83]. This theory has a particular resemblance with the modulational theories of pattern-forming systems because the order parameter here is also complex, although for a very different reason: superconductivity being a macroscopic quantum phenomenon requires an order parameter that has the symmetries of a wave function. One has a number of analogies like that of a stable family of current-carrying states limited in thin filaments by an intrinsic critical current. The critical current corresponds to the Eckhaus instability limiting the stable band of periodic solutions. In both systems one has phase slip processes and topological defects (vortices), although the systems differ in detail since currents in extended superconductors couple to a magnetic field (this leads to the mixed state in Type-II materials) and the dynamics in superconductors is simple only in dirty materials (short mean-free path of electrons) and in the vicinity of the critical temperature. In superconductors the time dependence is rendered non-variational through the coupling to the electric field, a mechanism that presumably has no analog in pattern-forming systems. Phase transitions can be captured beyond the mean-field approximation by including additive $\delta-$correlated noise terms.

11. *Absence of reflection symmetry.* Without reflection symmetry finite-k instabilities are expected to always exhibit a nonzero frequency, i.e. in general the patterns drift. In that situation the separation between stationary and Hopf bifurcations (with finite k) disappears. All finite-k instabilities are described by a single CGLE *with* a group velocity term, which influences the stability properties in finite systems. In fact, below the threshold for "absolute stability" $\epsilon_{abs} = \frac{\tau^2 v_{gr}^2}{4\xi^2(1+b^2)}$, where small, localized fluctuations are convected away faster than they grow, the basic state may persist. The bifurcation point then corresponds to a so-called convective instability. In open flow systems one always faces this situation and here the CGLE has been applied as well [88]. In the convectively unstable regime one may have noise-induced structures. which have been studied fairly intensively, see [89].

6.4 Example

As an example consider the two-component reaction-diffusion system (1) whose linear stability was discussed in Subsec. 1.3. We denote the eigenvalues σ of $-\nabla^2 = -\partial_x^2$ by k^2, so that the matrix governing the linear evolutions is $\mathcal{L}(k, \Lambda) = K(\Lambda) - D(\Lambda)k^2$, where K is the reaction and D the diffusion matrix. The relevant linear growth rate is

$$\sigma = \sigma_r + \imath\omega = \frac{1}{2}\mathrm{Tr}\mathcal{L} + \sqrt{\frac{1}{4}(\mathrm{Tr}\mathcal{L})^2 - \det\mathcal{L}}. \tag{65}$$

where one should choose the sign of the square root with positive real part. Stability requires

$$\det\mathcal{L} = \det K - Sk^2 + (\det D)k^4 > 0 \tag{66}$$
$$and \quad \mathrm{Tr}\mathcal{L} = \mathrm{Tr}K - (\mathrm{Tr}D)k^2 < 0 \tag{67}$$

where $S := D_{11}K_{22} + D_{22}K_{11} - (D_{12}K_{21} + D_{21}K_{12})$. The model is only reasonable if it is stable for large wavenumbers. Thus we always assume $\det D > 0$ and $\mathrm{Tr}D > 0$.

Breakdown of (66) (while $\mathrm{Tr}K < 0$) leads to a stationary bifurcation ($\omega_c = 0$) where one of the two real eigenvalues passes through zero. If $S < 0$ then $\det\mathcal{L}$ passes through zero at $k_c = 0$. If $S > 0$ then $\det\mathcal{L}$ first passes through zero at $k_c^2 = \frac{S}{2\mathrm{Tr}D}$ (> 0) when $S^2 = 4(\det K)(\det D)$ holds, which defines the critical point $\Lambda = \Lambda_c$. Let $L_c' := -d(\det\mathcal{L})/d\Lambda$ at the critical point. Then

$$\sigma_r = \frac{L_c'\Lambda_c}{-\mathrm{Tr}\mathcal{L}_c}(\epsilon - \frac{2S_c}{L_c'\Lambda_c}(k - k_c)^2). \tag{68}$$

Comparing this expression with (56) one reads off τ and ξ. This is a manifestation of the celebrated Turing (or diffusive) instability. Disregarding cross diffusion (which is usually a good approximation) one sees from the conditions $S > 0$ and $\mathrm{Tr}K < 0$ that one needs $K_{11} > 0$ and $K_{22} < -K_{11} < 0$ as well as $D_{11} > D_{22}$ (or vice versa). Moreover one has from the threshold condition $K_{12}K_{21} < 0$. Thus, the necessary ingredients for a Turing bifurcation are an autocatalytically growing and rapidly diffusing "activator" (substance 1) that stimulates growth of the slowly diffusing "inhibitor", which in turn suppresses the growth of the activator. The growth rate of the activator remains below the damping rate of the inhibitor.

Breakdown of (67) leads to a Hopf bifurcation where the real part of two complex eigenvalues passes through zero. This occurs when $\mathrm{Tr}K$ becomes positive at $k_c = 0$ and $\omega_c^2 = \det K$. Let $K_c' := d(\mathrm{Tr}K)/d\Lambda$ at the critical point. Then

$$\sigma_r = \frac{1}{2}K_c'\Lambda_c(\epsilon - \frac{\mathrm{Tr}D}{K_c'\Lambda_c}k^2), \tag{69}$$

$$\omega = \omega_c + \epsilon\frac{\Lambda_c}{2\omega_c}\frac{d(\det K)}{d\Lambda} - \frac{S_c}{2\omega_c}k^2. \tag{70}$$

Comparing this with (56,57) one reads off τ, ξ, a and b. Here one also needs an activator and an inhibitor. The growth rate of the activator must overcome the damping rate of the inhibitor and the product of the cross reaction terms must overcome the product of the direct terms.

Thus the two-component reaction-diffusion model exhibits all the generic types of bifurcations except the case $k_c \neq 0$, $\omega_c \neq 0$. In particular the crossover from the Turing to the Hopf bifurcation can be studied, see e.g. [90], as well as the processes $k_c \to 0$ in a stationary bifurcation and $\omega_c \to 0$ in a spatially homogeneous bifurcation. These can be described by generalizations of the GL equations presented.

Let us briefly demonstrate that with three components also the fourth type of instability is encountered. The dispersion relation is now a cubic polynomial in the growth rate σ

$$\sigma^3 - (\mathrm{Tr}K)\sigma^2 + B\sigma - \det\mathcal{L} = 0, \tag{71}$$

$$B = (K_{22} - D_2k^2)(K_{33} - D_3k^2) - K_{23}K_{32} + cycl., \tag{72}$$

(we disregard cross diffusion). A Hopf bifurcation occurs when $r := B\ \mathrm{Tr}K - \det\mathcal{L}$ changes from negative to positive while $B > 0$ and $\mathrm{Tr}K < 0$, and then $\omega_c^2 = B$. The special choice $K_{11} = -K_{22} = -K_{33} > 0$ and $K_{23} = K_{32} = 0$ simplifies the algebra. Then one has $r = [(D_{11} + D_{22})K_{12}K_{21} + (D_{11} + D_{33})K_{13}K_{31}]k^2 + O(k^4)$ and $B = -(K_{11}^2 + K_{12}K_{21} + K_{13}K_{31}) + 2D_{11}K_{11}k^2 + O(k^4)$. Clearly one can have $B(k = 0) > 0$ and still choose the term in square brackets to have either sign. Thus a crossover from $k_c = 0$ to $k_c \neq 0$ while $\omega_c \neq 0$ can be produced.

Whereas in reaction-diffusion systems one needs three components to obtain a primary instability with $k_c \neq 0$, $\omega_c \neq 0$, two are sufficient in hydrodynamical sytems (e.g. thermal convection in binary fluids). This is presumably related to the fact that in hydrodynamic systems the finite-k condition is usually enforced by the geometry together with mass conservation, whereas in reaction-diffusion systems the "diffusive-instability" condition has to be met intrinsically.

Nevertheless, reaction-diffusion systems offer a simpler anlysis of the instabilities since one can stay within 1D, whereas in hydrodynamic systems on has to include at least one transverse dimension. With realistic (rigid) boundary conditions analytic progress if often possible (or efficient) only by introducing suitable "test functions" for the variation along the transverse direction, amounting to a lowest-order Galerkin expansion, see e.g. [1].

The nonlinear terms are calculated as in Sec. 3.4. We conclude that the additional effort to obtain the GL equations after the ODE normal form has been calculated is actually quite small.

6.5 Construction of GL Equations in 2D

The system is assumed to be homogeneous in the x-y plane and one considers the linear growth rate $\sigma(\mathbf{k}; \Lambda) = \sigma_r + \imath\omega$ in Fourier space. The condition $\sigma_r = 0$ now gives a *neutral surface* $\Lambda_0(\mathbf{k})$. The symmetries with respect to rotation and reflection (the point groups) of the basic state, which we again assume to be nondegenerate, come into play. It is straightforward to generalize the treatment to two dimensions, as long as the conditions detailed in Subsections 6.1 and 6.2 are satisfied (quasi-onedimensionality is to be replaced by quasi-twodimensionality). Since isotropic systems undergoing a finite-k instability exhibit a continuous degeneracy of the critical mode they are actually more complicated to treat than anisotropic ones. Also, one maybe has to make a stronger proviso with regard to the problem of the analytical elimination of the contributions from $k \sim 0$ arizing at quadratic order (singular mean flow), although apparently in the presence of reflection symmetry of the basic state they do not appear at lowest order, see Subsection 8.1. Otherwise each critical mode gives rise to an amplitude $\tilde{A}_l(x, y, t)$ and the function $\mathbf{u}$ is obtained from

$$\mathbf{u}(t, \mathbf{x}) = \tilde{A}_1(t, \mathbf{x}) e^{\imath \mathbf{k}_c \mathbf{x} + \omega_0 t} \mathbf{p}_1(\mathbf{k}_c) + c.c. + \tilde{A}_2(t, \mathbf{x}) e^{\cdots} + \ldots + \text{hot} \tag{73}$$

We consider the most important cases.

Isotropic Systems Then on has the symmetry $(x, y) \to (x \cos\phi + y \sin\phi, -x \sin\phi + y \cos\phi)$. This includes reflection symmetry $(x, y) \to -(x, y)$. We also assume reflection symmetry with respect to an axis $(x, y) \to (-x, y)$ (chiral symmetry). Then σ only depends on $|\mathbf{k}|^2$. When $k_c = 0$ the neutral surface is a parabola of revolution and one simply has to replace in eqs.(60), (62) or (64) the derivatives ∂_x^2 by the (two-dimensional) Laplacian Δ ($v_{gr} = u = 0$). If $k_c \neq 0$ the situation is more complicated because the system becomes critical on a whole circle in k space (continuous degeneracy of the critical mode!). Then no full substitute of the GL scheme appears to exist because the generalized center space includes wavenumbers of all directions. On the linear level the growth rate would be approximated by $\sigma_r(k, \epsilon) \approx \tau^{-1}(\epsilon - \xi^2(k^2 - k_c^2)^2/(4k_c^2))$ (compare (56)). The problem appears on the nonlinear level, where one is left with a convolution integral that cannot in general be transformed into an expression that is local in real space. Locality requires additional approximations, leading in particular (after rescaling) to the often used Swift-Hohenberg equation, see [5]

$$\tau \partial_t U = \epsilon U - (\xi/(2k_c))^2 (\Delta + k_c^2)^2 U - U^3 \tag{74}$$

with real U, which reproduces the growth rate. Note that U still contains the fast variations. Equation (74) is not general. In particular it leads to rolls at threshold, which is not the only possible planform (there can also be squares and hexagons). To accomodate this one can admit other nonlinear terms that involve derivatives and include quadratic nonlinearities and which can be derived from the original non-local cubic term.

The GL scheme works for solution classes that can be represented by slow modulations of a periodic solution with a wavevector of *fixed* direction (here chosen in the x direction). One may then expand $(k^2 - k_c^2)^2 = [2k_c(k_x - k_c) + (k_x - k_c)^2 + k_y^2]^2 \approx 4k_c^2(k_x - k_c + \frac{k_y^2}{2k_c})^2$. Inserting this into the growth rate and transforming back to real space as done before, one arrives for stationary bifurcations at the celebrated Newell-Whitehead-Segel equation (NWSE) [80]

$$\tau \partial_t \tilde{A} = \left(\epsilon + \xi^2 (\partial_x - \frac{\imath}{2k_c} \partial_y^2)^2 \right) \tilde{A} - g |\tilde{A}|^2 \tilde{A}. \tag{75}$$

By rescaling as in (61) and $y = (\xi/(2k_c))^{1/2} |\epsilon|^{-1/4} Y$ one sees that the NWSE scales uniformly in ϵ and is free of parameters. For generalization to Hopf bifurcations, see [91, 78].

Chiral symmetry can be broken in particular by rotating a fluid layer around the z axis. Systems with broken chiral symmetry exhibit many interesting properties, but there is no room to discuss them here, see e.g. [92].

Anisotropic Systems The most important case is that of *axial anisotropy* where one has reflection symmetry with respect to an axis taken to be the x axis: $(x, y) \rightarrow (x, -y)$, together with reflection symmetry $(x, y) \rightarrow -(x, y)$. When $k_c = 0$ the neutral surface is a paraboloid, at lowest order in ϵ. For $\omega_c \neq 0$ one recovers eqs. (60) with $v_{gr} = 0$ and $\tilde{D}_C = \xi_x^2(1 + \imath b_x)\partial_x^2 + \xi_y^2(1 + \imath b_y)\partial_y^2$. Scaling x and y as in (61), but each with its own coherence length one arrives at (62) with $D_C = (1 + \imath b_x)\partial_X^2 + (1 + \imath b_y)\partial_Y^2$ and $U = 0$. The equation remaining when only one amplitude survives is encountered frequently (see below) and may be called the anisotropic CGLE (ACGLE). For $\omega_c = 0$ one obtains the RGLE with ∂_x^2 replaced by Δ, which is again free of parameters. For $\mathbf{k}_c \neq 0$ there are two generic situations:

Normal rolls. The minimum of the neutral curve (maximum of σ_r) occurs at a symmetry point $k_{xc} \neq 0$, $k_{yc} = 0$ (or vice versa). Then one obtains a straightforward generalization of (60) with (again) $\tilde{D}_C = \xi_x^2(1 + \imath b_x)\partial_x^2 + \xi_y^2(1 + \imath b_y)\partial_y^2$. The group velocity (for $\omega_c \neq 0$) is oriented along x. If only one mode survives one is led in the moving frame to the ACGLE (for $\omega_c = 0$ the RGLE).

Oblique rolls. The minimum occurs at $k_{xc} \neq 0$ and $k_{yc} \neq 0$. Then one has two degenerate critical wavevectors $k_{xc}, \pm k_{yc}$ corresponding to two systems of oblique rolls ("zig" and "zag"). For a stationary bifurcation one has two coupled real GL equations obtained from (60) with $v_{gr} = c = b = 0$ and $\tilde{D}_C = \xi_x^2\partial_x^2 \pm 2a\xi_x\xi_y\partial_x\partial_y + \xi_y^2\partial_y^2$, $|a| < 1$. If $0 > h > -1$ the rolls superpose to give rectangles. If only one mode persists one recovers the RGLE after a linear transformation of the spatial coordinates. Clearly for the case of a Hopf bifurcation one has four degenerate modes. We will not write down the equations, see e.g. [93]. If only one mode persists one is led in the frame moving with the group velocity, and after a linear transformation, with the ACGLE.

Crossover. As an example for a codimension-2 point related to the modulational parts in the equation we consider the crossover between normal and oblique rolls. To tune this situation a second control parameter w is needed. The crossover point is characterized by the fact that, coming from the normal-roll side one of the coherence lengths (say ξ_y) passes through zero. It is called a *Lifshitz Point* in analogy to similar points in the theory of phase transitions. One then has to expand the growth rate further, leading for a stationary bifurcation to $\tilde{D}_C = \xi_x^2\partial_x^2 + w\xi_2^2\partial_y^2 - \imath\bar{Z}\xi_x\xi_2^2\partial_x\partial_y^2 - \xi_2^4\partial_y^4$ with $|\bar{Z}| \leq 2$ and $\xi_2^4 > 0$ (if these relations are not satisfied, the crossover would be of a different type). Clearly, for $w = 0$ and $\bar{Z} = 2$ the NWSE (75) is recovered with $1/(2k_c)$ replaced by ξ_2^2/ξ_x. For $w > 0$ the threshold is at $\epsilon = 0$, $\mathbf{q} = (q_{c0}, 0)$, whereas for $w < 0$ one has $\epsilon_c = -|w|r$, $\mathbf{q} = (q_{c0} - \bar{Z}r/\xi_x, \pm(2r)^{1/2}/\xi_2)$ with

$r := |w|/(4 - \bar{Z}^2)$. Writing $w = \epsilon^{1/2}W$ with $W = O(1)$ and adopting otherwise a rescaling analog to the NWSE one can see that the resulting equation scales uniformly in ϵ. It is a simple matter to write down the corresponding equations for a Hopf bifurcation.

If there is *only reflection symmetry* $(x, y) \to -(x, y)$ the system is anisotropic but no axis is singled out. Then σ_r has maxima (Λ minima) either at $k_c = 0$ or at some some $\pm\mathbf{k}_c \neq 0$. Examples are systems that are influenced by effects that single out two axes at an oblique angle, like e.g. planarly aligned liquid crystal layers (this singles out a director axis) driven electrically or thermally into convection in the presence of an in-plane magnetic field oriented obliquely to the director axis, see [1]. Application of a linear shear flow along an oblique axis would lead to the same situation.

7 Solutions of the GL Equations

In this section we will deal with the respective equations in the scaled versions such that only the relevant parameters occur (primes etc. are omitted). We consider the regime above threshold and the nonlinearities are chosen to be saturating (unless stated otherwise). The systems are assumed to be extended and usually we do not consider the influence of distant boundaries. The differential operators Δ and ∇ are to be interpreted in the respective dimensions.

7.1 Stationary Bifurcations, $\omega_c = 0$

Homogeneous Case, $k_c = 0$. Here we comment very briefly on the equation

$$\partial_t A = (1 + \Delta)A + \alpha A^2 - A^3 \tag{76}$$

with real A ($\alpha = g_2/\sqrt{g\epsilon}$, compare (64)). It is not surprising that the only stable stationary bulk solutions are spatially constant (in arbitrary dimension). To see this it is useful to introduce the potential (Lyapunov functional)

$$\mathcal{V} = \int U d^D r, \qquad U = [-\frac{1}{2}A^2 - \frac{1}{3}\alpha A^3 + \frac{1}{2}A^4 + \nabla A^2], \tag{77}$$

so that $\partial_t A = -\delta\mathcal{V}/\delta A$, where the last term denotes the variational (Frechet) derivative. Since $\mathcal{V}$ is bounded from below the attractors of eq. (76) correspond to its (local) minima. The only minima are the homogeneous states A_1 (< 0) and A_2 (> 0) corresponding to the minima of U.

An ordering according to relative stability can be established: A_2 is more stable than A_1 if $U(A_2) < U(A_1)$, i.e. $\alpha > 0$. This can be seen by considering the evolution of *finite*, localized fluctuations around A_i. If the fluctuation around A_1 (the "higher" state) leads in the direction of A_2, i.e. if it is positive, and large enough, it will initialise a transition of the whole system to A_2. In fact, in 1D the fluctuation will evolve into two domain walls (fronts) moving apart with constant velocity c. The front is a solution of the equation

$$A'' - cA' = \partial_A U(A), \tag{78}$$

where now A depends only on $\zeta := x - ct$ and the prime denotes derivatives with respect to ζ. This equation allows an interpretation in terms of the mechanical analog of a damped particle (coordinate A, time ζ) in the potential $-U$. Thus, for a particular, positive damping c_s there exists a heteroclinic orbit solution connecting the two maxima of the mechanical potential $-U$ such that the particle moves from the higher to the lower maximum. This means that the front moves such that the system evolves from the higher minimum of U to the lower one. Here $c_s = \alpha/\sqrt{2}$.

Clearly one also has front solutions that connect a minimum of U with the maximum. In the mechanical analog they correspond to a particle released at a maximum of $-U$ and ending in the minimum, as is the case for $|c| > |c_s|$. Thus one has a *continuous family* of fronts describing the transition from an unstable to a stable state. One is confronted with a genuine selection problem that has captured the interest of mathematicians and physicists. For some recent work on this subject, see e.g [94] (also [5]).

The localized fluctuation necessary to trigger the transition from A_1 to A_2 can be characterized quantitatively in terms of a "critical fluctuation", which is the homoclinic orbit solution of (78) with $c = 0$ localized around A_1. It indeed exists if $U(A_2) < U(A_1)$ and corresponds to the undamped particle starting at A_1, falling down and climbing up towards A_2 until the motion reverts. It is an unstable solution with one unstable direction (saddle point). For $\alpha \rightarrow 0$ the critical fluctuation becomes infinitely large and degenerates into the stationary fronts expected to exist then.

One can generalize these concepts to higher dimensions (the critical fluctuation becomes the critical droplet known from the theory of first-order phase transitions). Then fronts can become curved and one can connect to the concept of surface tension. One can also generalize to more complicated models (e.g. RD models), in particular those that are not derivable from a potential. Then one can, in principle, use the concepts of critical fluctuations and front motion to *define* relative stability, but one has to be aware of situations with multiple critical fluctuations and fronts mediating transitions in opposing directions [95]. One may conclude that in models that only involve local interactions (no global coupling terms) there is at most one homogeneous state, which is stable under general localized, nonlinear fluctuations (globally stable state).

The Real GL Equation (RGLE)

$$\partial_t A = (1 + \Delta)A - |A|^2 A \tag{79}$$

describes a relaxational dynamics governed by the (real) potential (Lyapunov functional)

$$\mathcal{V} = \int U d^D r, \qquad U = \frac{1}{2}[-|A|^2 + \frac{1}{2}|A|^4 + |\nabla A|^2]. \tag{80}$$

One has $\partial_t A = -2\delta\mathcal{V}/\delta\bar{A}$ and one easily sees that $\mathcal{V}$ decreases under the dynamics. Since $\mathcal{V}$ is bounded from below the attractors of the RGLE correspond to its (local) minima. The only candidates for such attractors (apart from defect solutions, which will be considered below) are the periodic solutions

$$A = Fe^{\imath \mathbf{Q}\cdot\mathbf{x}+\phi_0}, \; F = \sqrt{1-Q^2}, \; Q^2 \leq 1. \tag{81}$$

When inserted into (73) one sees that they correspond to periodic solutions of the physical system with a displacement from band center $\mathbf{k}_c$ proportional to $\mathbf{Q}$ ($Q^2 = 1$ gives the neutral surface in scaled units).

The stability of these solutions can be studied directly by a linear stability analysis which will be discussed in the context of the CGLE. An alternate method is given in the next subsection. The critical modes correspond to long-wave modulations with wavenumbers parallel to $\mathbf{Q}$ (Eckhaus instability). Thus a 1D analysis is sufficient. The stable wavenumber band is given by $|Q| < Q_E := 1/\sqrt{3}$ Actually all stationary 1D solutions can be obtained by writing A in terms of amplitude and phase $A = Re^{\imath\theta}$ which leads to (we write $\partial_t X = X_{,t}$ etc.)

$$R_{,t} = R_{,xx} + (1 - \theta_{,x}^2 - R^2)R, \qquad R^2\theta_{,t} = \partial_x(R^2\theta_{,x}). \tag{82}$$

In a static situation this reduces to $R^2\theta_{,x} = h$ with an integration constant h and

$$R_{,xx} = \partial_R U \text{ where } U = \frac{1}{2}[-R^2 + \frac{1}{2}R^4 - h^2/R^2]. \tag{83}$$

This allows to invoke the mechanical analog of a point particle (position R, time x) moving in the potential $-U(R)$. On easily sees that for $|h| < h_E = \sqrt{4/27} \approx 0.385$ the potential $-U$ has extrema at $R_{1,2}^2\sqrt{1-R_{1,2}^2} = |h|$. Clearly they correspond to the periodic solutions with $F_{1,2} = R_{1,2}$ and $|Q_{1,2}| = \sqrt{1-R_{1,2}^2}$ $(= |\theta_{,x}|)$. The maximum corresponds to $|Q_1| < 1/\sqrt{3}$. It is stable since it correspond to a minimum of $\mathcal{V}$. The solution $|Q_2| > 1/\sqrt{3}$ is unstable. In this way the Eckhaus instability is recovered. From the mechanical analog one infers that the unstable solutions are embedded in (unstable) periodically modulated solutions corresponding to the particle oscillating in the potential well. They can be expressed in terms of elliptic functions and correspond to quasiperiodic solutions of the form

$$A = e^{\imath Qx+\phi_0}B(Kx) \tag{84}$$

with $2\pi-$periodic B. They bifurcate subcritically from the unstable periodic solutions at $K^2 = 2(3Q^2 - 1)$. To understand their significance it is useful to note that the space of quasiperiodic functions of the form (84) represents an invariant manifold of the RGLE. The quasiperiodic solutions separate the domains of attraction of the periodic solutions with wavenumbers Q and $Q \pm K$ on this manifold. For $K \to 0$ the solutions approach homoclinic orbits starting and ending at the maximum of $-U$. The homoclinic orbits are written down easily [96]. They separate periodic solutions differing by one "periodic unit on a length $L \to \infty$" and thus represent the saddle points that mediate wavelength-changing processes. They bifurcate from the periodic solutions at Q_E. Actually, for $Q \to \pm Q_E$, i.e. $h \to h_E$, all modulated (and constant) solutions collapse to $R = \sqrt{2/3}$, $Q = \pm 1/\sqrt{3}$. The scenario can be described efficiently with phase equations, see Subsec. 8.2.

In Fig. 5 we demonstrate a simulation of Eq. (79) of a wavelength changing process in 2D occuring when the system is initialised in an unstable periodic state with small stochastic perturbations. At early times one sees quasi-1D modulations with the characteristic fastest-growing wavenumber $K_m = \sqrt{(3Q^2 - 1)(Q^2 + 1)/4Q^2}$. Then defect pairs (zeros of A, see below) are created by pinching of rolls. The defects move and annihilate in such a way that the final wavenumer lies in the stable range (see [97]). The corresponding 1D process has also been studied [96].

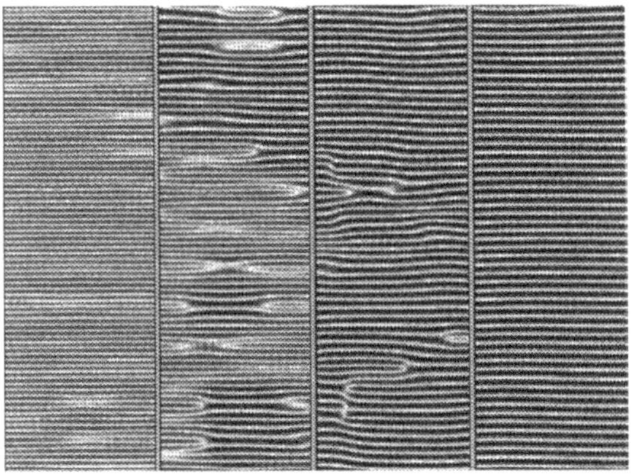

Fig. 5. Time series of the evolution of an Eckhaus unstable state into a stable state in 2D. Times from left to right $t = 120$, 140, 200, and 700. Shown is $\Re(A(x,y)e^{k_c x} + cc)$ in grey scale.
Initial wavenumber $Q = 0.7$; final wavenumber $Q \approx 0.14$; $L_x = 2L_y = 128$, periodic boundary conditions.

An important new feature in 2D is the existence of topological point defects corresponding to the zeros of A. In polar coordinates r, θ the stationary isolated defect (vortex) is of the form $A = R(r)e^{\pm i\theta}$, i.e. for $r \to \infty$ it connects to the band center solution with "background wavevector" $\mathbf{Q} = 0$. The $\pm$ sign signifies the two topological charges ± 1. The ODE for R, to be solved with boundary conditions $R \to 0$ for $r \to 0$ and $R \to 1$ for $r \to \infty$, are written down easily. It is interesting to consider the corresponding solution for a nonzero background wavevector. It can be shown [98, 99] that these solutions are of the form $A(\mathbf{x} - \mathbf{V}t)$ with the velocity $\mathbf{V}$ perpendicular to $\mathbf{Q}$ and $V \ln(V_0/V) = 2Q(1 - 0.35Q^2)$ with $V_0 = 3.29$. The term in the brackets on the rhs has been fitted to numerical results. The formula can be used throughout the stable range and is consistent with experiments in nematics [98]. The logarithmic singularity at $Q = 0$ is connected with the fact that the far-field of the moving defect cannot be treated adiabatically. The motion of a defect across the system carries away one periodic unit (roll pair) and changes the wavevector by $\pm(2\pi/L)\hat{z} \times \mathbf{V}$. Thus the wavevector change is parallel to $\mathbf{Q}$ and in fact reduces its magnitude. Therefore, a system that is offered defects will eventually end at band center, which corresponds to the absolute minimum of the potential $\mathcal{V}$. Finite-ϵ corrections often bring in nonvariational effects which change this simple behavior.

The interaction beween defects can be studied along similar lines and was compared with experiments on electroconvection in nematics [98, 99]. It is repulsive/attractive for equally/oppositely-charged pairs and decays monotonically with distance. Here again the far-field cannot be treated adiabatically [99]. For an oppositely-charged defect pair in an environment with not too large background wavevector $\mathbf{Q}$ there exists an equilibrium separation $\mathbf{R}$ such that the interaction is compensated by the "background force". Clearly $\mathbf{R} \perp \mathbf{Q}$. Moreover $R \to \infty$ for $Q \to 0$ (in fact, then $R \sim 1/|Q|$). For increasing $|Q|$ the equilibrium separation decreases, the zeros of A annihilate, and the solution goes over into the saddle-point solution for the homogeneous nucleation of defect pairs and thereby mediates the wavevector-changing processes in 2D. At Q_E the solution bifucates subcritically from the periodic solution and for $|Q| \to Q_E$ one has simple scaling behavior that can be derived from the appropriate 2D phase equation [98]. In real systems defects (or defect pairs) may form at weak spots or boundaries, so scanning out the ideal stable wavevector band is difficult.

The Newell-Whitehead-Segel Equation (NWSE)

$$\partial_t A = [1 + (\partial_x - \imath\partial_y^2)^2]A - |A|^2 A. \tag{85}$$

allows solutions of the form (81) with $F = \sqrt{1 - (Q_x + Q_y^2)}$. It suffices to investigate the stabiliy of the solutions with Q_y near 0. We will study their stability by a method, which is generally applicable to potential cases, by inspecting the "energy" $\mathcal{V}(Q_x, Q_y)$ per unit length. Adapting (80) one finds $U(Q_x, Q_y) = -\frac{1}{2}[1 - (Q_x + Q_y^2)^2]^2$. Consider slow modulations, which can be expressed as slow variations of Q_x and Q_y under the condition that their spatial average remains unchanged. In fact, a change of the local wavevector in one region has to be compensated by an opposite change somewhere else, since otherwise periodic units would have to be destroyed or created. Therefore in the expansion of $\mathcal{V}$ the terms linear in the variations δQ_x and δQ_y vanish and for stability one has to demand positivity of

$$\partial^2 U/\partial Q_x^2 = 2(1 - 3Q_x^2), \qquad \partial^2 U/\partial Q_y^2 = 4Q_x(1 - Q_x^2). \tag{86}$$

(Here we have set $Q_y = 0$). This recovers the (longitudinal) Eckhaus instability and moreover shows that the solution is stable with respect to transverse modulations only if $Q_x > 0$, which exhibits the universal zigzag instability against undulatory (or wavy) modulations.

Defects in the NWSE are complicated to treat, because there is no (rigorous) way to separate variables and thereby reduce the problem to ODEs. We will not deal with this subject here, see e.g. [78].

The Lifshitz Point Equation. As discussed in Subsec. 6.5 the equation valid near the (simplest type of) crossover between stationary normal and oblique rolls is in the scaling appropriate for the immediate vicinity of the Lifshitz point

$$\partial_T A = [\pm 1 + \partial_X^2 + W\partial_Y^2 - i\bar{Z}\partial_X\partial_Y^2 - \partial_Y^4]A - |A|^2 A. \tag{87}$$

For $W < 0$ this equation describes a neutral surface that has minima at $Q = ZW/(4 - Z^2)$, $P = \pm\sqrt{2|W|/(4 - Z^2)}$ corresponding to oblique rolls arising at slightly negative ϵ (therefore we have also allowed for the $-$sign in eq. (87)). The bands of stationary, periodic solutions and their stability have been discussed [100]. One also has linearly stable undulated rolls (normal and oblique) that develop out of the zigzag instability. The equation includes as limiting cases: for large and positive W the RGLE, for large and negative W the coupled GL equations for oblique rolls (near the Lifshitz point they do not superpose to give rectangles), and for $W = 0$, $Z = 2$ the NWSE. The undulated roll solutions also exist in the NWSE [101].

7.2 The Complex GL Equation (CGLE)

General considerations. Here we deal with the equation

$$\partial_t A = [1 + (1 + ib)\Delta - (1 + ic)|A|^2]A \tag{88}$$

which has been studied intensely during the last years. Sometimes it is useful to consider the following transformation

$$A(\mathbf{x}, t) = e^{i(\mathbf{Q}\cdot\mathbf{x} - \Omega t)} B(\mathbf{x} - \mathbf{V}t, t), \qquad \text{leading to} \tag{89}$$

$$\partial_t B = [a + \mathbf{v}_z \cdot \nabla + (1 + ib)\Delta - (1 + ic)|B|^2]B \tag{90}$$

with $a = 1 + i\Omega - (1 + ib)Q^2$, $\mathbf{v}_z = \mathbf{V} - i(1 + ib)\mathbf{Q}$. Separating out an "outer wavevector" $\mathbf{Q}$ is useful if the gradient of the phase of B, integrated over the system, is zero (or at least small). Most known solutions are either of the "coherent-structure" type, where B depends only on its first argument (i.e. with properly chosen Ω it is time independent in a moving frame), or appear disordered in the sense of spatio-temporal chaos. Coherent structures can be localized or extended.

With a little bit of algebra one can discover a useful *similarity transformation* that connects coherent structures along the lines

$$(b-c)/(1+bc) = const. \tag{91}$$

in parameter space. In this transformion time, length and the amplitude are rescaled. Solutions with $\mathbf{V} = 0$ remain stationary (for $b < \infty$) [102], otherwise the velocity is transformed as well [103]. The similarity line $b = c$ (vanishing group velocity, see below) includes the real case and is particularly simple. Note that the stability limits of coherent states are expected to conform with the similarity transform only in special cases. In the limit $b \to \infty$ the coefficient c tends to a constant and the transformation turns into a Galilei transformation, i.e. in this limit coherent structures appear as continuous families moving at arbitrary velocity. Note that by also letting $c \to \infty$ one gains in addition the scale invariance well-known from the NLS equation.

The simplest coherent structures are the plane-wave solutions

$$A = \sqrt{1-Q^2}e^{i(\mathbf{Q}\mathbf{x}-\Omega(Q)t)}, \qquad F^2 = 1-Q^2, \qquad \Omega(Q) = c(1-Q^2)+bQ^2. \tag{92}$$

which exist for $Q^2 < 1$. To test their stability one may determine the dispersion relation $\lambda(\mathbf{k})$ of the linear perturbations, which have the form of side band modulations. Destabilisation occurs for $\mathbf{k} \parallel \mathbf{Q}$ and one finds

$$\begin{aligned} &\lambda^2 + 2(F^2 + 2iQbk + k^2)\lambda \\ &+ik\{(1+b^2)(4Q^2-k^2)ik - 2F^2[(1+bc)ik - 2(b-c)Q]\} = 0. \end{aligned} \tag{93}$$

By expanding this equation for small k one finds

$$\lambda = \pm iV_g k - D_{||}k^2 + 0(k^3),\ V_g = 2(b-c)Q,\ D_{||} = 1+bc-2(1+c^2)Q^2/(1-Q^2) \tag{94}$$

Clearly the solutions (92) are stable as long as the phase diffusion constant $D_{||}$ is positive, which gives the generalized Eckhaus criterion. Thus one has a stable range of wave numbers embedding the homogeneous ($Q = 0$) state as long as the "Newell criterion" $1 + bc > 0$ holds, which signalizes the "Benjamin-Feir (BF) instability". This criterion is in line with the similarity transform (91). From (93,94) one sees that for $b - c \neq 0$ and $Q \neq 0$ the destabilizing modes have a group velocity $V_g = \partial_Q \Omega = 2(b-c)Q$, so the Eckhaus instability now is of convective nature and does not necessarily lead to destabilization of the pattern. The absolute stability threshold can be calculated from Eq.(93) [104]. In Fig. 6 the BF limit as well as the limit up to which absolutely stable plane waves exist (AI), are plotted (thick curves).

The (convective) Eckhaus instability signalizes bifurcations to quasiperiodic solutions (including the solitary limit), which are of the form (89) with periodic function $B(\mathbf{x}-\mathbf{V}t)$. Well away from the BF instability (on the stable side) the bifurcation is subcritical (as for the RGLE). However, it becomes supercritical before the BF line is reached [105] and remains so in the unstable range. The bifurcation is captured most easily in the long-wave limit by

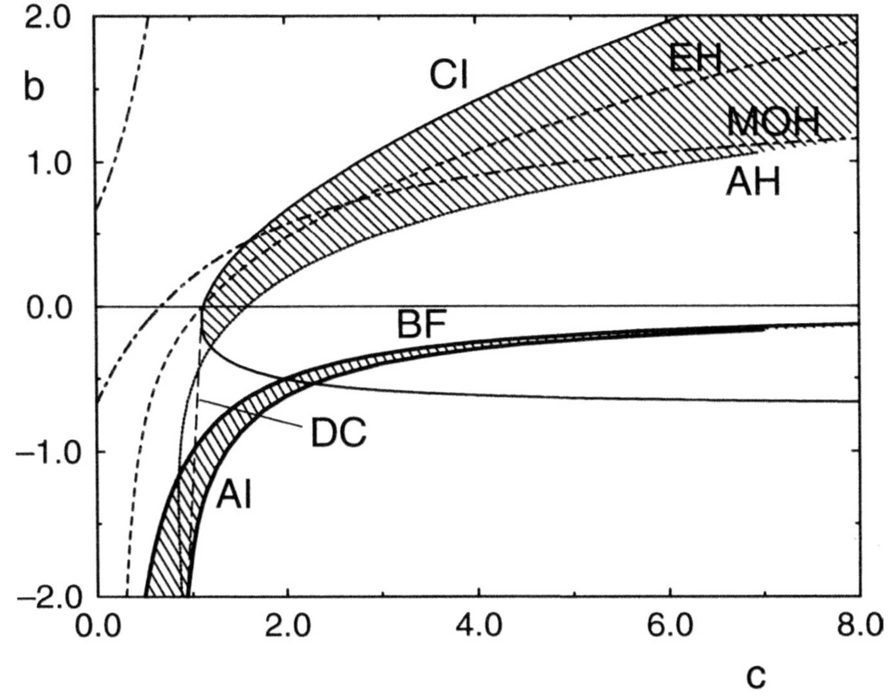

Fig. 6. Phase diagram of the 1D CGLE. BF = BF instability line up to which one has stable plane-wave solutions; AI= absolute instability line up to which one has convectively unstable plane-wave solutions; DC = boundary of existence of DC towards small $|c|$. The other lines pertain to Nozaki-Bekki hole solutions: CI=core instability line, EH = convective (Eckhaus) instability of the emitted plane waves, AH = absolute instability of the emitted plane waves, MOH = boundary between monotonic and osciallatory interaction.

phase equations, see Subsec. 8.2. The supercritical nature of the bifurcation allows to understand the existence of "phase chaos" (PC) that is found when the BF line is slightly crossed (in 1D approximately in the lower dashed region between curves BF and AI in Fig. 6). In that state $|A|$ remains saturated (typically above about 0.7) so there is long-range phase coherence and one has a well-defined background wavevector.

1D CGLE. Localized coherent structures connect asymptotically to plane-wave states or to the trivial solution. In 1D they satisfy three coupled ODEs and can be classified as *pulses*, *fronts* and *domain boundaries* [107, 5]. We will not discuss pulses (see e.g. [108]) and fronts (see [107, 5]). Domain boundaries are often *sources* (also called *holes*) or *sinks* (also called *shocks*) depending on wether in the co-moving frame the group velocity is directed outward or inward, respectively. They represent heteroclinic solutions connecting plane-wave states. Thus counting arguments are useful to determine their generic multiplicity [107, 5].

Whereas sinks indeed conform with these arguments (there is a continuous family), the sources pose a paradox. According to counting arguments there should be only a discrete number, including in particular the symmetric standing hole solution that has a zero at its center and emits plane waves of a definite wave number (for given b, c). However, the standing hole is embedded in a continuous family of *analytic* moving sources, the Nozaki and Bekki hole solutions. They are characterized by a localized dip in $|A|$ that moves with constant velocity v and emits plane waves with wave numbers $Q_1 \neq Q_2$ for nonzero velocity. This puzzle was (partially) resolved by the discovery that the moving holes do not survive a (general) perturbation of the CGLE, e.g. a small quintic term, so they are not structurally stable [110]. The holes then either accelerate or decelerate, depending on the sign of the perturbation. Thus only one sign leads to stable standing holes, whereas in the opposite case the standing holes accelerate spontaneously and disappear. Clearly the 1D cubic CGLE possesses a "hidden symmetry" and has retained some remnant of integrability from the NLS. Apparently this non-genericity has no consequences for other solutions and in higher dimensions.

The interest in the hole solutions comes in particular from the fact that they are (dynamically) stable in some range of parameters. The stability boundary for the standing hole can be calculated essentially analytically [110]. Roughly speaking the core region of the solution is stabilized by large $|c|$ and destabilized by increasing $|b|$, see curve CI (continuous) in Fig. 6. Destabilization occurs via a stationary, supercritical bifurcation [109] that transforms into a Hopf bifurcation when the equation is perturbed in the stabilizing direction. Then one has oscillating holes in a narrow parameter range. The scenario is described by a simple normal form. For holes to be stable also the emitted waves have to be in the stable or convectively unstable range. In Fig. 6 the convective (EH, dashed) and the absolute (AH, dotted) stability limits are included. Standing holes can be maintained stably in the upper dashed region.

Since the plane-wave states on the two sides of a hole are different, an array of holes has shocks in between. Above the curve MOH (dash-dotted) these shocks have the tendency to stay away from the holes, whereas below the curve the shocks may be associated with the holes. In fact, the curve MOH separates the region where the asymptotic interaction between localizes structures decays in space monotonically (above) and oscillatorily (below). This interaction will be discussed below in the 2D case. Thus in the oscillatory range one may have bound hole-shock pairs that travel at a fixed velocity. Their behavior can be understood largely analytically [110]. These objects are neither sources nor sinks. The plane-wave states on the two sides are the same and therefore they represent homoclinic connections. Possibly they join smoothly with the unstable "homoclinic holes" studied recently in a region below the curve AH [111] (see below).

Beyond the BF instability one has first PC (for background wavenumber $Q = 0$ approximately in the lower dashed region) and then defect chaos (DC) (or amplitude chaos) where phase slips (space-time defects of A) occur at some (mean) rate. For $|b|$ larger than about 1.8 (near the lower edge of Fig. 6) the transition between PC and DC is continuous (then the phase slip rate goes to zero at the crossover) and hysteretic otherwise. Thus there is a region where PC and DC coexist ("bichaos", lower dashed region to the right of line DC), and in fact DC even persists into the BF stable range (the limit towards small c is approximated by the dashed line DC in Fig. 6) [112, 113]. There the DC takes on the form of spatio-temporal intermittency [113]. From above the chaos terminates either where holes become stable (when there is stabilizing perturbation of the CGLE), i.e curve AH, or else it persists somewhat further. In fact, for small destabilizing perturbations the Eckhaus curve EH appears to give a good approximation of this upper bound.

The homoclinic holes seem to play an important role in the intermittency region of DC [111] (except very near to its boundary of existence, where the heteroclinic Nozaki-Bekki holes become dominant). They also seem to characterize the localized objects seen in PC, in particular with background wavenumber $Q \neq 0$ in Eq. (89) ("wound-up" PC) that has been studied recently in detail [115, 114]. We mention that one has this kind of PC in a band of wavenumbers Q that shrinks to zero at the boundary of existence of PC. Thus, as in the Eckhaus instability for plane-wave solutions, the state with $Q = 0$ is the last one to loose stability. Actually one expects wound up PC to persist also in the BF stable range. The homoclinic holes can also be looked upon as the localized limit of the quasiperiodic solutions studied first in [105]. Interestingly, there also exist stable, stationary, nonperiodic arrangements of homoclinic holes [114]. This can presumably be understood in terms of an oscillatory interaction. PC, quasiperiodic, and spatially disordered solutions can coexist.

In summary we may say that, although many aspects of the behavior of the cubic CGLE in 1D have been clarified, there remain important open questions. One example is a better understanding of the limit of existence of PC.

2D CGLE. The topological defects (zeros of A) that corrrespond to vortices in the real case (and more generally in the case $b = c$) now become rotating spirals [116, 102, 117]. The isolated spiral is a localized coherent structure of the form

$$A(r, \theta) = F(r)exp[i(-\Omega t \pm \theta + \psi(r))] \tag{95}$$

where (r, θ) are polar coordinates. The functions F and ψ have the following asymptotic behavior $F(r) \sim r$, $\psi'(r) \sim r$ for $r \to 0$ and $F(r) \to \sqrt(1 - Q^2)$, $\psi'(r) \to Q$ for $r \to \infty$ and Ω as in Eq.(92). Q is the wavenumber of the

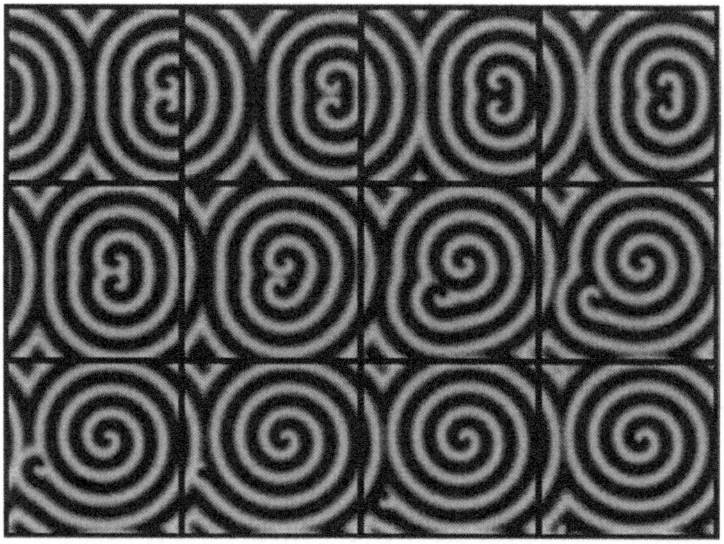

Fig. 7. Snapshots of the evolution of an oppositely charged spiral pair into a stable symmetry broken state [120]. $b = 0$, $c = 1.0$, $L = 100$, periodic boundary conditions. Time runs from top left to bottom right; $\Delta t = 500$. $\Re A(x, y)$ is shown in gray scale.

(locally) plane waves emitted asymptotically by the spiral. It is determined uniquely for given b, c, but in general has to be determined numerically [102]. Here the similarity transform (91) is very useful. In Fig. 7 a well-developed spiral can be seen in the lower row. In the upper row the evolution of a spiral pair is exhibited.

Spirals can, in a sense, be looked upon as the 2D analogs of the 1D standing hole solutions. In this interpretation the second dimension acts like a (massive) stabilizing perturbation (in the sense introduced in the previous subsection). There is the analog of the core instability, but in 2D large values of $|b|$ are needed ($|b| > 6$) [118]. At the critical value of b one has a Hopf bifurcation leading to spontaneous acceleration of the core along a spiraling trajectory. The instability is on the linear level analogous to a well-known instability which leads to meandering motion of the spiral core in excitable media [119], but in the CGLE the instability is not saturated by the nonlinearities. Instead, it leads to spatio-temporally chaotic states (see below).

Of particular importance is the interaction between spirals, i.e. the motion of one spiral that results from the presence of another spiral. The interaction is of short range, in contrast to the real case (and $b = c$). The motion can be decomposed into a radial (along the line connecting the spiral cores) and a tangential component (perpendicular to that line). From simple symmetry arguments follows that for an oppositely-charged spiral pair the tangential velocities of the two spirals are equal, so there is a common drift (see Fig.

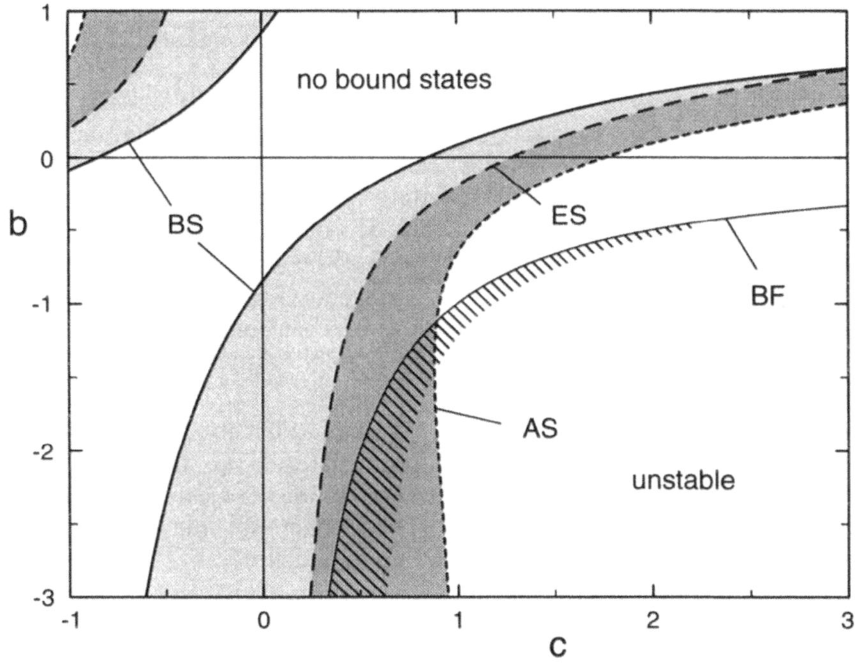

Fig. 8. Phase diagram for the 2D CGLE. PC exists approximately in the dashed region to the right of the BF instability line. The other lines pertain to the plane waves emitted by spirals: ES = convective (Eckhaus) instability, AS = absolute instability, BS = boundary between monotonic and oscillatory interaction.

7), whereas for equally-charged spirals the tangential velocities are equal but opposite, so there is a rotation around the "center of mass". A bound state is a (stable) configuration with zero radial velocity, so only the tangential motion is left.

One has two parameter regions with qualitative different behavior of the interaction, namely oscillatory and monotonic [120]. This can be seen already from the asymptotic behavior for $r \longrightarrow \infty$ of (stationary) perturbations of a localized structure, which can be obtained from the dispersion relation (93) of plane waves with $\lambda = 0$ and Q the asymptotic wavenumber. It is then convenient to set $ik = p$ so that $\delta A \sim exp(pr)$. The root $p = 0$ of the resulting real polynomial corresponds to the translation mode. There is always one real negative root p_3. The other two roots $p_{1,2}$ are either complex-conjugated with positive real part or real and positive, depending on $|c-b|/(1+bc)$ being larger or smaller than $c_{cr} \approx 0.845$, respectively. The limiting curve is shown as BS in Fig. 8. Clearly $p_{1,2}$ are asymptotically leading and one therefore expects oscillatory behavior when $p_{1,2}$ are complex and monotonic behavior otherwise. A matched asymptotic expansion leads to the result $Im[p_1 X_l + \mu \ln X_l] = -\phi + \pi l$ for the half-distance X_l between the spirals. Here $l = 1, 2, 3...$, $\phi = -arg[(\delta C_y p_1^{1/2})]$, μ and δ are functions of Q and p, and C_y is a constant determined from the asymptotics of the perturbation [120]. The first (and only relevant) bound state corresponds to $l = 2$, as is verified by numerical simulations ($l = 1$ corresponds to an unstable equilibrium position).

For an equally-charged spiral pair the radial velocity behaves similar. The bound state can then be interpreted as a two-armed (or doubly charged) spiral. In numerical simulations also multiple-bound states corresponding to multi-armed spirals were observed [120]. In the monotonic range one appears to have always asymptotic repulsion and no bound state.

The above results relate to symmetric/antisymmetric states, where the emitted waves collide on the symmetry line between the spirals. Here a "shock structure" is formed. The region on either side of the shock is organized by the respective spiral. It turns out that in the oscillatory range, i.e. the range where one has the bound states, such bound pairs are actually unstable against spontaneous breaking of the symmetry [120]. One spiral increases its range of influence by pushing away the shock which moves from its symmetric position towards the core of the other spiral. In Fig. 7 this process is shown in a simulation. An analytic description of this spontaneous symmetry breaking can again be done by a matching procedure. The fact that symmetry breaking goes hand in hand with oscillatory interaction between spirals should be interpreted as expressing the fact that one has the same qualitative behavior of the interaction between shocks and spirals as between spirals. For oppositely-charged spiral pairs the final state in not too large square regions with periodic boundary conditions corresponds to a stationary lattice of spirals where only one population of spirals (e.g. the positively charged ones) emit waves. The others are reduced to their cores trapped inside the shocks bounding the unit cell of the lattice (in particular in the corners). The asymmetric spiral lattice can be numerically stable in regions containing many elementary cells. In the monotonic range symmetric spiral lattices were found to be stable [120].

The trapped defects found in symmetry-broken lattices represent a typical defect structure where the group velocity, instead of pointing outward as in spirals, points inward, so they are sinks instead of sources. They are passive objects reacting completely to the incoming waves. There exists a continuous family of the ideal, rotationally symmetric solution, corresponding essentially

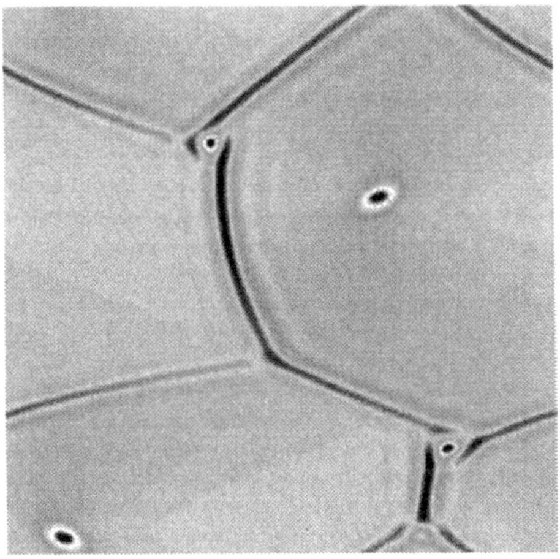

Fig. 9. Greyscale plot of $|A(x, y)|$ for a vortex glass for the anisotropic CGLE [106]. $b_1 = -2.0$, $c = -b_2 = 0.7$, $L = 130$, periodic boundary conditions. The structure drifts very slowly, which is not the case for the isotropic CGLE.

to wavenumbers with group velocity larger than that of the spiral (for smaller group velocities the defect will become active and evolve into spiral). They play an important role in the disordered, frozen states to be discussed below.

Slightly beyond the BF instability there is a range of PC, similar to, but narrower than in 1D [121]. DC [122] exists in the b,c range where the wavenumber emitted by the spirals is absolutely unstable, i.e. to the right of the curve AS in Fig. 8. In DC the distribution of the number n of defects present in the system is described by $p(n) \sim exp[-(n - \bar{n})^2/\bar{n}]$ where $\bar{n}$ is the average number of defects. The formula follows in the limit $n \to \infty$ from treating defect *pairs* as statistically independent entities [123]. In the region between the Eckhaus instability (curve ES) and AS one has "transient DC" [104, 124, 125, 126]. Actually in large systems the transient times may become prohibitively large close to the absolute stability boundary. Eventually, in large systems an extended, spatially disordered, coherent state with symmetry-broken spirals, surrounded by shocks is found which may be called "vortex glass" or "frozen state". The effect can presumably be understood in terms of the non-monotonic interaction (together with the symmetry breaking). In Fig. 9 an example is shown. Actually this is from a simulation of the anisotropic CGLE to be introduced below, but the pattern is very similar. The shape of the shock lines making up the cellular structure can be understood by simple geometric arguments [127]. In the convectively unstable region frozen states are rather sensitive to small noise [128].

In the core-unstable region (large $|b|$, see above) and close to it one has an even richer richer scenario [118]. In the convectively unstable region an interesting type of spatio-temporal intermittency arizes. An example is shown in Fig. 10 for a fairly small system. In (a) one spiral starts to organize its neighborhood by pushing away STC structures.

In (b) the spiral has organized the whole cell, but the core is accelerating on a spiral trajectory. Then rather suddenly the coherence is destroyed and

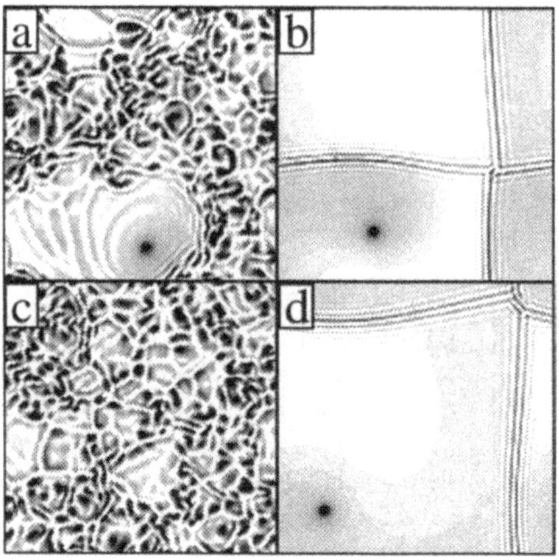

Fig. 10. Snapshots of $|A(x,y)|$ in the intermittency range for four consecutive times. $b = 40$, $c = -0.5$, $\Delta t \approx 800$.

STC spreads over the cell. In (d) again a spiral has succeded to organize the cell. In a large system one expects the region of influence of a spiral to remain localized.

The Coupled CGLE. The coupled CGLE (62) must be considered in particular when left- and right-traveling waves superpose to form *standing waves* (or, in 2D, more complicated patterns). We mention some results on the stability of standing waves

$$A = Fe^{-i(\Omega t + Qx)}, \qquad B = Fe^{-i(\Omega t - Qx)}, \tag{96}$$

$$F^2 = (1 - Q^2)/(1 + h), \; \Omega = uQ - b_2 Q^2 - (c + hd)F^2 \tag{97}$$

with respect to 1D fluctuations. (As in the last section we replace X, T by x, t and consider only the supercritical region.) One needs $h^2 < 1$ for standing waves to be amplitude stable. The modulational stability analysis now leads to a quartic polynomial in the complex growth rate $\lambda(k)$ and a full discussion is tedious. The long-wave limit is of the form (94) with more complicated expressions for V_g and $D_{||}$ [133]. Here are some features:

i) V_g can be real or imaginary, so instability can arise already in this order. The criterion is independent of Q, so the stability limits are vertical lines in the $\epsilon - Q$ plane. For $u = 0$ they collapse to the line $Q = 0$, so that either all states are stable (to this order) or unstable. For $u \neq 0$ the band center always remains stable (to this order)

ii) The Eckhaus parabola obtained at next order is centered around $Q = 0$ only if $u = 0$ and shifted otherwise. It can be wider or narrower than the neutral curve (or even turned upside down) and the states can be stable either inside or outside.

iii) As a result one has almost always states on the neutral curve that are stable according to the long-wave analysis and then short-wave instabilities arise (states on the neutral curve may be stable only at the minimum).

vi) Remembering the rescaling (61) used one may consider the limit $\epsilon \longrightarrow 0$ while keeping the physical system length fixed. This eliminates many of the previous long-wave instabilities and shifts previous short-wave instabilities to long wavelength. Then the dispersion relation for $\lambda(k)$ goes over into (93) with F replaced by the expression for standing waves (see (96)). Alternatively, this result can be obtained by taking the above limit from the beginning on in the derivation of the coupled CGLE, which leads to nonlocal equations [134]. The claim raised sometimes that this is the only valid expansion scheme is in our view not justified. The local equations include the global ones as a limiting case and are applicable as long as higher-order terms have a small influence. Nevertheless, the globally coupled limit is a useful further reduction.

The coupled CGLE is also relevant when left- and right-traveling waves suppress each other ($h^2 > 1$), because under certain conditions both occur separated in space, see e.g. [135]. Above the threshold for absolute stability $\epsilon_{abs} = \frac{\tau^2 v_{gr}^2}{4\xi^2(1+b^2)}$ stationary source solutions separating the two wavetrains are often stable in 1D. These coherent structures select a unique frequency and are therefore reminiscent of spirals in 2D. In addition one can have source-induced chaotic states.

Generalizations. As discussed before the *anisotropic* version of (88) (the ACGLE), where $(1 + ib)\Delta$ is replaced by $(1 + ib_x)\partial_x^2 + (1 + ib_y)\partial_y^2$, is important in particular for finite-k instabilities. This equation has been considered recently [106]. Some highlights are the occurrence of stationary or drifting periodic defect chains and two different versions of PC. Also, one finds vortex glass solutions that drift very slowly with constant velocity. An example is shown in Fig. 9.

The CGLE with *broken chiral symmetry* without breaking other symmetries was considered in [129].

We make some remarks on *the 3D case* where the spiral defects become vortex lines (scroll waves). Until recently it appeared that scroll rings would always contract and disappear in finite time, so there would be no analog of DC [130]. However, in the analysis the analog of the acceleration instability was ignored, which becomes operative in a sizable parameter range. In 3D there is the additional possibility for the vortex line to bend, and this increases the unstable region considerably ($|b|$ larger than about 2 for small $|c|$) [131]. The instability in fact occurs at a finite bending wavenumber and turns out to be supercritical (as in 1D). In a narrow parameter range one finds stable, traveling helices. Otherwise one has spontaneous stretching and entanglement of vortices leading to a complex state of spatio-temporal chaos that may have features in common with turbulence in the Navier-Stokes equation. Also PC was found in 3D [121]. Away from the acceleration instabilty PC may here represent the global attractor.

Finally we remark on the *subcritical bifurcation* case, where the nonlinear saturation term $-|A|^2A$ in eq.(88) is replaced by $+|A|^2A$. Then there is no saturation for plane-wave solutions (plane-wave initial conditions diverge in finite time). Interestingly, in 1D generic initial conditions do not necessarily lead to such explosion (often called collapse). In fact there is a well defined, analytically calculable, region in parameter space around the axis $b = 0$ limited towards small $|c|$ and opening up for increasing $|c|$, where the relevant attractors are bounded. Thus the relevant attractors in fact bifurcate supercritically. This has actually been known for a long time, but recent work has clarified the situation [132].

8 Beyond GL Equations

One can distinguish different directions that lead beyond the range of applicability of the (lowest-order) GL approach presented above (within the framework of a modulational description using PDEs). We will first enumerate some items and then briefly discuss the in some sense complementary approach of phase equations.

8.1 Complications and Generalisations

1. *Systematic expansion to higher order,* with the amplitude as well as the time and space derivatives of modulations being small. This has been done mainly in 1D and was applied in particular to accommodate the effect of distant side walls [136]. There also exists an extensive literature on the equation including quintic nonlinearities in order to capture weakly subcritical bifurcations, see [137, 107, 138]. A particularly interesting feature is the occurence of stable pulse solutions over a finite range of parameters.

2. *Order parameter approach, singular mean flow.* Whereas the restriction to cubic order in the amplitude of the pattern is often reasonable over a meaningful range of ϵ (for a supercritical bifurcation), the restriction to the lowest space derivatives is often not. Dropping restrictions on the (formal) scale of spatial variations and formulating the problem systematically in Fourier space is sometimes called the "order parameter approach". In hydrodynamic quasi-2D systems an expansion around q_c (local approximation) may not be possible due to singular behavior of the kernel even in the situation with rolls modulated slowly around a fixed direction. Then one is left with nonlocal terms, which can be incorporated by coupling the pattern amplitude(s) advectively to an additional real field decribing the strength of the slowly varying vertical vorticity and satifying a Poisson-type equation (for an example see [1]), see e.g. [5, 78, 140]. This "singular mean flow" can be traced back to the incompressibility condition. Its proper inclusion is of great importance leading to the well-known skewed-varicose instability [141]. Mean-flow contributions arise at second order in the amplitude expansion at wavenumbers near zero and thereby contribute at cubic order. Now introducing appropriate spatial scaling it turns that for realistic no-slip boundary

conditions at the bounding plates and a reflection symmetric basic state the mean-flow contributions vanish for $\epsilon \to 0$ faster than the GL contributions. This is different for free-slip boundary conditions, which can be included into the more general class of systems exhibiting an additional symmetry-related neutral mode (see next item). Examples with a non-reflection symmetric basic state, where mean flow comes in already at lowest order, are encountered in rotating Rayleigh-Bénard convection, see Ponty *et al.* [92], and in the destabilisation of plane Poiseuille flow [142].

3. *Degenerate basic state, additional broken symmetry.* If the system has an additional neutral mode, an equation for the amplitude of this mode has to be included into the decription. The relevant cases are related to a spontaneously broken continuous symmetry of the system, and the mode is the corresponding Goldstone mode. Sometimes the slow mode can be eliminated adiabatically (in particular in 1D) and then the remaining equations exibit an additional symmetry (for an example see [148]). Important cases relate to the Galilean invariance present in the case of free-slip boundary conditions in fluid layers [149] and rotational invariance in the case of a spontaneously broken isotropy as in electroconvection in homeotropically aligned nematics with negative dielectric anisotropy, see [1]. In these cases one can have STC at threshold for a stationary bifurcation. One has this situation also in secondary instabilities of a periodic pattern, where the Goldstone mode is the phase of the pattern.

4. *GL equation coupled to a slow mode.* More generally, in a number of systems one has at threshold for pattern formation a weakly damped ("slow") mode which is often related to some material diffusion. The lowest-order description can then be extended by including a dynamic equation for this mode. The study of such coupled amplitude equations is a very active field [143].

5. *Global coupling.* In some cases there is a very rapid diffusive process involved that spreads out over the whole system on time scales that are short compared to those of the actual patterning processes. It is then useful to treat the rapid process as instantaneous, which produces a global coupling in the equations. Examples are pattern formation in catalytic suface reactions [144] and in gas discharge [145]. Actually global coupling can be introduced into any electrically driven system by approriate externeal circuitry. The CGLE with global coupling has been studied extensively [146].

6. *Nonadiabatic effects.* Inherent to the GL approach for instabilities with $k_c \neq 0$ is the fact that correlations between the "fast" space scale (wavelength of the pattern) and the "slow" space scale (the modulations) are discarded. This amounts to an adiabatic elimination of the fast scale. Clearly *"nonadiabatic"* (correlation) effects exist (and can be observed in experiments). They have to be captured by nonperturbative methods. In particular, nonadiabatic effects can lead to the stabilisation of localized structures [150].

7. *Symmetry breaking perturbations.* Interesting effects result from breaking the space or time translation invariance of the underlying system by applying an appropriate periodic perturbation, which can be done elegantly within the GL scheme, see e.g. [90, 151].

8. *Noise.* The inclusion of small, additive, spatially and temporally δ-correlated noise, like thermal noise, into the GL scheme is possible, see e.g. [152]. It leads to a corresponding noise term in the GL equation, which thereby becomes a Langevin equation. The influence of noise on frozen states has been studied in [128]. Also, spiral growth in crystalline surfaces has been described by a noisy CGLE with periodic forcing [147]. The situation with multiplicative (state-dependent, parametric) noise, which does not break the bifurcation is more complicated. Here no general scheme for the reduction of the original (hydrodynamic) equations to GL equations appears to exist. The situation for spatially constant and spatially distributed noise is very different, see e.g. [1, 153].

8.2 Phase Equations

An approach that discards the amplitude expansion altogether but keeps the slow variation approximation for modulations of the periodic pattern makes use of phase equations (PE). It may be viewed as the analog of a hydrodynamic approximation and is as such valid also far from threshold [78]. Our starting point is a situation where one has a family of space- and/or time-periodic solutions described by 2π-periodic functions $U_{\mathbf{q}}(\mathbf{qx} - \omega_0(\mathbf{q})t + const.)$. We assume that slow modulations can be described by introducing a phase $\varphi(\mathbf{x}, t)$ such that the local wavevector $\nabla\varphi(\mathbf{x}, t) = \mathbf{q}(\mathbf{x}, t)$ and $\partial_t\varphi$ are slowly varying functions and $u(x,t) = U_{\mathbf{q}}(\varphi) + correction$. Let us first consider a 1D situation with phase $\varphi(x,t)$. Assuming the applicability of a gradient expansion we may write

$$\partial_t\varphi = \nu + v\partial_x\varphi + D_{\|}\partial_x^2\varphi - \Omega\partial_x^3\varphi - D_4\partial_x^4\varphi + hot \qquad (98)$$

Note the absence of a term proportional to φ which would violate translation invariance. The coefficients depend on q. By choosing $\varphi = qx$ one sees that $\nu(q) + v(q)q = \omega_0(q)$ holds. For stationary patterns one has reflection symmetry, so that $v = \Omega = 0$, and also $\nu = 0$.

Through the q dependence of the coefficients eq. (98) is actually nonlinear. We can capture the Eckhaus bifurcation by expanding around some suitably chosen wavenumber q_0. Thus we write $\varphi = q_0x - \omega_0(q_0)t + \phi(x,t)$ where $\phi(x,t)$ is now small (and slowly varying). Expanding $v(q)$ and $D_{\|}(q)$ one generates nonlinear terms, so that eq. (98) goes over into

$$(\partial_t - v\partial_x)\phi = D_{\|}\partial_x^2\phi - D_4\partial_x^4\phi - g_1(\partial_x\phi)^2 - \Omega\partial_x^3\phi - \frac{1}{2}g_2\partial_x(\partial_x\phi)^2 + hot \qquad (99)$$

The coefficients are now to be taken at q_0. There is a consistency relation that connects the nonlinear coefficients with linear ones, which derives from the fact that there must be invariance under the transformation $q_0 \rightarrow q_0 - \delta q$, $\partial_x \phi \rightarrow \partial_x \phi + \delta q$ [154]. One finds $g_1 = -\frac{1}{2}\partial_q v(q)|_{q_0}$ and $g_2 = -\partial_q D_{||}(q)|_{q_0}$. (99) is a very general nonlinear wave equation which applies also to primary long-wave instabilities like e.g. waves on a viscous fluid film flowing down an inclined plane [155]. Phase equations can in particular be considered near threshold where a weakly nonlinear description (here the CGLE) holds [156], and there the coefficients of (99) can be calculated from those of the CGLE [105]. Thus $D_{||}$ is given by (94) and $v = V_g$. In the mathematical literature this approach is sometimes termed the modulated modulation approach [157].

Clearly $D_{||} > 0$ is needed for stability of the solution. When this is satisfied one can discard the last four terms in (99) to describe the relaxation of long-wave fluctuations. The (generalized) Eckhaus instability is signalized by the phase diffusion constant $D_{||}$ becoming negative, and then further terms are needed (we assume $D_4 > 0$). Rigorous calculations of $D_{||}$ for (stationary) Taylor-Vortex Flow [158] gave excellent agreement with experiments [159]. For stationary patterns one has $v = g_1 = \Omega = 0$ and then eq.(99) describes the subcritical Eckhaus bifurcation [96]. Slow spatial variations of the control parameter(s) can be incorporated into the phase equation for stationary patterns and lead to drift terms. This method was introduced to understand the wavenumber selection process occuring in the presence of a control parameter ramps that render the system subcritical at one end (or both ends) [160]. Experiments in Taylor-Vortex flow support this concept quantitatively [159, 158].

In general, by going into a moving frame and rescaling length and time the number of parameters in Eq. (99) can be reduced to two. If there is reflection symmetry one has $\Omega = g_2 = 0$ and then the equation describes the supercritical BF instability as $D_{||}$ changes sign. In fact, for $D_{||} < 0$ one has the Kuramoto-Sivashinski equation, which gives the simplest description for phase chaos. The full equation describes the crossover from a subcritical to a supercritical Eckhaus bifurcation [105]. Dropping the first three terms on the rhs of (99) one is left with the conservative and completely integrable Korteweg-de Vries equation. Equation (99) has been studied fairly extensively [105, 161, 86].

The simplest generalization to 2D is obtained for axially anisotropic systems with the waves traveling in the x direction. If no instability arises in the y direction it suffices to include a diffusive term $D_\perp \partial_y^2 \varphi$ in (98) (or in (99)). The zigzag instability is indicated by $D_\perp$ going through zero. The nonlinear aspects can again be incorporated by including higher derivative terms and expanding the coefficients of the low-order terms around $\mathbf{q}_0 = (q_0, 0)$. In the situation considered one has reflection symmetry in the y direction, so only terms with an even number of derivatives arise. In the stationary case the resulting equation exhibits undulated solutions [162].

For isotropic systems the lowest-order phase equation for slow modulations of stationary, periodic patterns is of the form [163]

$$\tau(q)\partial_t\varphi = -\nabla\cdot[B(q)\nabla\varphi]. \tag{100}$$

Writing $\varphi = \mathbf{q_0}\cdot\mathbf{x}+\phi(\mathbf{x})$ (as before) and expanding in terms of ϕ one finds a 2D phase equation with $D_{||} = -\tau(q)^{-1}\partial_q[qB()q)]_{q_0}$ and $D_\perp = -\tau(q)^{-1}B(q_0)$. Again the Eckhaus instability is indicated by $D_{||}$ going through zero and the zigzag instability by $D_\perp$ changing sign

More complicated situations may arise in analogy to the items discussed in Subsec. 8.1 [5]. In particular, the smoothness assumptions needed for the applicability of the gradient expansion leading to the phase equations often breaks down in 2D hydrodynamic systems due to singular mean-flow effects. In simple fluids with a reflection symmetric basic state this happens for not too large Prandtl numbers and at finite distance from threshold. In analogy to the situation in (generalized) amplitude equations one then has to couple the phase equation advectively to the mean-flow field, which satisfies a Poisson-type equation. One then can incorporate the skewed-varicose instability. For Rayleigh-Bénard convection in simple fluids this program has been carried out [78, 164]. The resulting equations reproduce the Busse balloon. The equations have also been used to capture some aspects of spiral defect chaos in Rayleigh-Bénard convection [165].

8.3 Final Remarks

The material presented shows that a classification of the primary pattern-forming instabilities in extened media is quite generally possible, and there is a large degree of universality in their description near threshold. The analogous description of secondary instabilities that destabilize a spatial and/or temporal pattern is much more involved. Whereas long-wave secondary instabilities can be classified and described by phase equations, this is not possible for short-wave instabilities. There one has to introduce an appropriate new order parameter (amplitude) for the transition and couple its dynamical equation to the phase of the underlying pattern, which remains a neutral mode. For quasi-1D systems a classification has been presented [166]. Various cases have been studied in detail.

References

1 W. Pesch and U. Behn, this volume.
2 W. Zimmermann, B. Painter, and R. Behringer, this volume.
3 M. Lücke, W. Barten, P. Büchel, C. Fütterer, St. Hollinger, and Ch. Jung, this volume.
4 E. Schöll, F.J. Niedernostheide, J. Parisi, W. Prettl, and H.G. Purwins, this volume.
5 M.C. Cross and P.C. Hohenberg, Rev. Mod. Phys. **65** 851 - 1112 (1993).
6 R. Lefever and I. Prigogine, J. Chem. Phys. **48**, 1695-700 (1968). P.C. Five, J. Stat. Phys. **39**, 687-703 (1985).
7 S. Mizohata, *The Theory of Partial Differential Equations*, Cambridge University Press, 1973.
8 F. Takens, Publ. Math. IHES **43**, 47-100 (1974). B.I. Bogdanov, Funct. Anal. and its Appls. **9**, 144-5 (1975).
9 W. Miller, *Symmetry Groups and Their Applications*, Academic Press, 1972.
10 M. Golubitsky, I.N. Stewart and D. Schaeffer, *Singularities and Groups in Bifurcation Theory, Vol. II*, Springer, 1988.
11 J. Guckenheimer and P. Holmes, *Nonlinear Oscillations, Dynamical Systems and Bifurcations of Vector Fields*, Springer, 1990 (3rd edition). S. Wiggins, *Introduction to Applied Nonlinear Dynamical Systems and Chaos*, Springer, 1990.
12 C. Elphick, E. Tirapegui, M.E. Brachet, P. Coullet and G. Iooss, Physica D **29**, 95-127 (1987).
13 M. Golubitsky and I.N. Stewart, Arch. Rat. Mech. Anal. **87**, 107-65 (1985).
14 M. Golubitsky, E. Knobloch and I.N. Stewart, in preparation.
15 M. Golubitsky and I.N. Stewart, in *Multiparameter Bifurcation Theory*, M. Golubitsky and J. Guckenheimer (eds.), Contemporary Mathematics **56**, A.M.S., Providence, 1986.
16 J.W. Swift, Nonlinearity **1**, 333-77 (1988).
17 M. Krupa, SIAM J. Math. Anal. **21**, 1453-86 (1990).
18 E. Knobloch, J. Hettel and G. Dangelmayr, Phys. Rev. Lett. **74**, 4839-42 (1995). G. Dangelmayr, J. Hettel and E. Knobloch, Nonlinearity **10**, 1093-114 (1997).
19 M. Golubitsky, V.G. LeBlanc and I. Melbourne, *Meandering of the Spiral Tip*, preprint, 1997.
20 M. Field, Trans. Amer. Math. Soc. **259**, 185-205 (1980). F.H. Busse and K.E. Heikes, Science **208**, 173-75 (1980). J. Guckenheimer and P. Holmes, Math. Proc. Cambr. Phil. Soc. **103**, 189-92 (1988).
21 G. Cicogna, Lettere al Nuovo Cimento **31**, 600-2 (1981). A. Vanderbauwhede, *Local Bifurcation and Symmetry*, Pitman, 1982.
22 B. Hassard, N. Kazarinoff and Y-H. Wan, *Theory and Applications of the Hopf Bifurcation*, Cambridge University Press, 1980.
23 J. Carr, *Applications of Centre Manifold Theory*, Springer, 1981. A. Vanderbauwhede and G. Iooss, Dynam. Report. **1**, 125-63 (1992).
24 M. Silber, H. Riecke and L. Kramer, Physica D **61**, 260-78 (1992).
25 M. Wegelin, Doctoral Dissertation, Tübingen, 1993.
26 B. Dionne, Ph.D. thesis, Houston, 1990. B. Dionne and M. Golubitsky, Z. Angew. Math. Phys. **43**, 36-62 (1992).
27 M. Silber and E. Knobloch, Nonlinearity **4**, 1063-106 (1991).
28 B. Dionne, M. Silber and A.C. Skeldon, Nonlinearity **10**, 321-53 (1997).

29 B. Dionne, M. Golubitsky, M. Silber and I.N. Stewart, Phil. Trans. R. Soc. Lond. A **352**, 125-68 (1995).

30 K. Kirchgässner, Math. Meth. in the Appl. Sci. **1**, 453-67 (1979).

31 E. Buzano and M. Golubitsky, Phil. Trans. R. Soc. Lond. A **308**, 617-67 (1983).

32 M. Golubitsky, J.W. Swift and E. Knobloch, Physica D **10**, 249-76 (1984).

33 M. Roberts, J.W. Swift and D.H. Wagner, in *Multiparameter Bifurcation Theory*, M. Golubitsky and J. Guckenheimer (eds.), Contemporary Mathematics **56**, A.M.S., Providence, 1986.

34 T. Clune and E. Knobloch, Physica D **74**, 151-76 (1994).

35 E.A. Spiegel and J.P. Zahn, (eds.), *Problems of Stellar Convection*, Lecture Notes in Physics **71**, Springer, 1977. M. Ghil and S, Childress, *Topics in Geophysical Fluid Dynamics: Atmospheric Dynamics, Dynamo Theory and Climate Dynamics*, Springer, 1987.

36 X. Le Pichon and P. Huchon, Earth Planet Sci. Lett. **67**, 123-35 (1984).

37 M. Michel, in *Statistical Mechanics and Field Theory*, R.N. Sen and C. Weil (eds.), Israel Univ. Press, 1972. D.H. Sattinger, J. Math. Phys. **19**, 1720-32 (1978) and *Group Theoretic Methods in Bifurcation Theory*, Lecture Notes in Math **762**, Springer, 1979.

38 V.G. Babskii and I.L. Sklovskaia, Prikl. Mat. Mekh. **35**, 951-67 (1971). F.H. Busse, J. Fluid Mech. **72**, 65-85 (1975).

39 G.H. Knightly and D. Sather, Arch. Rat. Mech. Anal. **72**, 315-80 (1980).

40 E. Ihrig and M. Golubitsky, Physica D **13**, 1-33 (1984).

41 M. Golubitsky and D. Schaeffer, Comm. Pure Appl. Math. **35**, 81-111 (1982).

42 C. Geiger, J.D. Rodriguez and G. Dangelmayr, Physica D (1998, in press).

43 C. Geiger, G. Dangelmayr, J.D. Rodriguez and W. Güttinger, Fields Inst. Comm. **46**, 225-37 (1996). J.D. Rodriguez, C. Geiger, G. Dangelmayr and W. Güttinger, Fields Inst. Comm. **46**, 239-53 (1996).

44 P. Chossat, R. Lauterbach and I. Melbourne, Arch. Rat. Mech. Anal. **113**, 313-76 (1990). C. Geiger, Doctoral Dissertation, Tübingen, 1994.

45 G. Iooss and M. Rossi, SIAM J. Math. Anal. **20**, 511-32 (1989).

46 E. Knobloch and J. Guckenheimer, Phys. Rev. A **27**, 408-16 (1983). M. Neveling and G. Dangelmayr, Phys. Rev. A **38**, 2536-43 (1988).

47 J. Guckenheimer, SIAM J. Math. Anal. **15**, 1-49 (1984). P. Holmes, Physica D **2**, 449-81 (1981) and J. Fluid Mech. **162**, 365-88 (1986).

48 S.E. Newhouse, D. Ruelle and F. Takens, Comm. Math. Phys. **64**, 35-40 (1978).

49 G. Dangelmayr, Dyn. Stab. Syst. **1**, 159-85 (1986). A. Russo and E. Buzano, Annali di Matematica Pura et Applicata **146**, 217-62 (1987).

50 G. Dangelmayr and D. Armbruster, in *Multiparameter Bifurcation Theory*, M. Golubitsky and J. Guckenheimer (eds.), Contemporary Mathematics **56**, A.M.S., Providence, 1986.

51 D. Armbruster, J. Guckenheimer and P. Holmes, Physica D **29**, 257-82 (1988). M.R.E. Proctor and C.A. Jones, J. Fluid mech. **188**, 301-35 (1988).

52 D. Armbruster, J. Guckenheimer and P. Holmes, SIAM J. Appl. Math. **49**, 676 (1989).

53 F.H. Busse and A.C. Or, Zeitschrift Angew. Math. und Phys. **37**, 608-23 (1986). D. Armbruster, Physica D **47**, 433-39 (1987).

54 H. Riecke and H.G. Paap, Phys. Rev. A **33**, 547-53 (1986).

55 K. Fujimura, Phil. Trans. R. Soc. Lond. A **340**, 95-130 (1992). K. Fujimura and M. Nagata, Physica D **115**, 377-400 (1998).
56 R. Friedrich and H. Haken, Phys. Rev. A **34**, 2100-20 (1986).
57 P. Chossat and D. Armbruster, in *Singularity Theory and its Applications*, M. Roberts and I.N. Stewart (eds.), Lecture Notes in Math. **1463**, Springer, 1990. D. Armbruster and P. Chossat, Physica D **50**, 155 (1991).
58 P. Chossat and F. Guyard, preprint, 1997.
59 P. Chossat and G. Iooss, Japan Journal of Appl. Math. **2**, 36-68 (1985). M. Golubitsky and I. Stewart, SIAM Journal on Math. Anal. **17**, 249-88 (1986). A.S. Hill and I.N. Stewart, Dyn. and Stab. of Syst. **6**, 149-71 (1991).
60 M. Golubitsky and W.F. Langford, Physica D **32**, 362-92 (1988).
61 W.F. Langford, R. Tagg, E. Kostelich, H.L. Swinney and M. Golubitsky, Phys. Fluids **31**, 776-85 (1988).
62 A. Hill and I.N. Stewart, Dyn. and Stab. of Syst. **6**, 267-339 (1991).
63 P. Chossat, M. Golubitsky and B.L. Keyfitz, Dyn. and Stab. of Syst. **1**, 256-91 (1986). P. Chossat, Y. Demay and G. Iooss, Arch. Rat. Mech. Anal. **99**, 213-48 (1987).
64 E. Knobloch and M.R.E. Proctor, J. Fluid Mech. **108**, 291-316 (1981).
65 G. Dangelmayr and E. Knobloch, Phil. Trans. Roy. Soc. Lond. A **322**, 243-79 (1987).
66 G. Dangelmayr and E. Knobloch, Physics Letters A **117**, 394-8 (1986). G. Dangelmayr and M. Neveling, J. Phys. A **22**, 1291-301 (1989).
67 D. Armbruster and G. Dangelmayr, Math. Proc. Cambr. Phil. Soc, **101**, 167-92 (1987) and Dyn. and Stab. of Syst. **1**, 187-200 (1986).
68 J.D. Crawford, M. Golubitsky, M.G.M. Gomes, E. Knobloch and I.N. Stewart, in *Singularity Theory and its Applications*, M. Roberts and I.N. Stewart (eds.), Lecture Notes in Math. **1463**, Springer, 1990. D. Armbruster and P. Chossat, Physica D **50**, 155 (1991).
69 M.G.M. Gomes, Ph.D. thesis, Warwick, 1992. M.G.M. Gomes and I.N. Stewart, in *Dynamics, Bifurcation and Symmetry*, P. Chossat (ed.), NATO ASI Series C **437**, Kluwer, 1994.
70 M.J. Field and M. Golubitsky, J. Nonlinear Sci. **1**, 119-64 (1991).
71 J.D. Crawford, Phys. Rev. Lett. **67**, 441-4 (1991) and Physica D **52**, 429-57 (1991).
72 J.D. Crawford, J.P. Gollub and D. Lane, Nonlinearity **6**, 119-64 (1993).
73 G. Dangelmayr and E. Knobloch, Nonlinearity **4**, 339-427 (1991). G. Dangelmayr, E. Knobloch and M. Wegelin, Europhys. Lett. **16**, 723-9 (1991) and Eur. J. Mech. B/Fluids **10**, 125-30 (1991).
74 J. Fineberg, E. Moses and V. Steinberg, Phys. Rev. Lett. **61**, 838 (1988). P. Kolodner and C.M. Surko, Phys. Rev. Lett. **61**, 842 (1988).
75 G. Dangelmayr, J.D. Rodriguez and W. Güttinger, AMS Lectures in Appl. Math **29**, 145-61 (1993). G. Dangelmayr and J.D. Rodriguez, in *Structure and Dynamics of Nonlinear Waves in Fluids*, A. Mielke and K. Kirchgässner (eds.), World Scientific, 1995.
76 P. Chossat and M. Golubitsky, Physica D **32**, 423-36 (1988). P. Ashwin, J. Buescu and I.N. Stewart, Nonlinearity **9**, 703 (1996). P. Ashwin and A.M. Rucklidge, Physica D (1998, in press). P. Ashwin and P. Aston, Phys. Lett. A (1998, in press). J. Moehlis and E. Knobloch, Phys. Rev. Lett. **80**, 5329 (1998).

77 P. Manneville, *Dissipative Structures and Weak Turbulence* (Academic, San Diego, 1990); G. Nicolis, *Introduction to Nonlinear Science* (Cambridge University Press, 1995).

78 A.C. Newell, T. Passot, and J. Lega, Annu. Rev. Fluid Mech. **25**, 399-453 (1993).

79 G. Dangelmayr, B. Fiedler, K. Kirchgaessner, and A. Mielke, *Dynamics of nonlinear waves in dissipative systems: reduction, bifurcation, and stability* (Addison Wesley Longman, Harlow, 1996).

80 A.C. Newell and J.A. Whitehead, J. Fluid. Mech. **38**, 279-303 (1969); L.A. Segel, J. Fluid Mech. **39**, 203-224 (1969).

81 L.Y. Chen, N. Goldenfeld, and Y. Ono, Phys. Rev. Lett. **73** 1311-1314; Phys. Rev. E**54** 376-394 (1996).

82 S. Sasa, Physica D**108** 45-59 (1997).

83 V.L. Ginzburg and L.D. Landau, JETP **20**, 1064 (1950).

84 A. Hasegawa, *Optical Solitons in Fibres*, Springer Tracts in Modern Physics Vol. **116** (Springer-Verlag, Berlin, 1989).

85 A.C. Newell, Solitons ins Mathematics an Physics (SIAM Philadelphia, 1985). P.G. Drazin and R.S. Johnson, *Solitons: an Introduction* (Cambridge University Press, 1989).

86 A.A. Nepomnyashchy, Physica D**86**, 90-95 1995; and to be published.

87 P. Collet and J.P. Eckmann, Comm. Math. Physics **132**, 139 (1990); A. van Harten, J. Nonlin. Science **1**, 397 (1991); P. Kirrmann, G. Schneider, and A. Mielke, Proc. Roy. Soc. Edinburgh **A122**, 85 (1992); G. Schneider, J. Nonlin. Science **4**, 23 (1994); ZAMP **45**, 433 (1994);

88 T. Leweke and M. Provansal, Phys. Rev. Lett. **72**, 3174-3177 (1994); J. Fluid Mech. **288**, 265 (1995) K. Roussopoulos, P.A. Monkewitz, Physica D**97**, 264 (1996).

89 D. Roth, P. Büchel, M. Lücke, H.W. Müller, M. Kamps, and R. Schmitz, Physica D**97**, 253 (1996); M. Lücke et al., this volume.

90 A. de Wit, D. Lima, G. Dewel, and P. Borckmans, Phys. Rev. E**54**, 261 (1996); D. Walgraef *Spatio-Temporal Pattern Formation* (Springer, New York, 1996); M. Meixner, A. De Wit, S. Bose, and E. Schöll, Phys. Rev. E **55**, 6690 (1997).

91 H.R. Brand, P.S. Lomdahl, and A.C. Newell, Physica D**23**, 345-361 (1986).

92 E. Knobloch, in **Lectures on Solar and Planetary Dynamos**, Editors M.R.E. Proctor and A.D. Gilbert (Cambridge University Press, 1994); Y. Ponty, T. Passot, and P.L. Sulem, Phys. Rev. Lett. **79**, 71 (1997); Phys. Rev. E**56**, 4162 (1997).

93 W. Zimmermann, in *Nematics: Mathematical and Physical Aspects*, NATO ASI Series C 332, edited by J.M. Coron et al. (Kluwer, dordrecht,1991), 401.

94 U. Ebert and W. van Saarloos, Phys. Rev. Lett. **80**, 1650 (1998). W. van Saarloos, Physics Reports, in press (1998). S.J. Di Bartolo and A.T. Dorsey, Phys. Rev. Lett. **77** 4442-4445 (1996);

95 P. Ortoleva and J. Ross, J. Chem. Phys. **63**, 3398 (1975); L. Kramer, Z. Phys. B **41**, 357-363 (1981); **45**, 167-171 (1981).

96 L. Kramer and W. Zimmermann, Physica **16D**, 221-232 (1985); E. Bodenschatz and L. Kramer, Physica **27D**, 249-259 (1987); L. Kramer, H. R. Schober, and W. Zimmermann, Physica D**31**, 212-226 (1988).

97 E. Bodenschatz. Doctoral Dissertation, Bayreuth, 1989; A. Weber, E. Bodenschatz, and L. Kramer, Adv. Mater. **3**, 191 (1991).

98 E. Bodenschatz, W.Pesch, and L. Kramer, Physica D**32**, 135 - 145 (1988); L. Kramer, E. Bodenschatz, and W.Pesch, Phys. Rev. Lett **64**, 2588 (1990); E. Bodenschatz, A. Weber, and L. Kramer, J. Stat. Phys. **64**, 1007 - 1015 (1991).

99 L. Pismen, *Defects in Continuous Media* (Oxford University Press, 1998).

100 W. Pesch and L. Kramer, Z. Phys. B**63**, 121 (1986); L. Kramer, E. Bodenschatz, W. Pesch, and W. Zimmermann, in *The Physics of Structure Formation*, Editors W. Güttinger and G. Dangelmayr (Springer-Verlag, Heidelberg, 1987); E. Bodenschatz, M. Kaiser, L. Kramer, W. Pesch, W. Weber, and W. Zimmermann, in *New trends in Nonlinear Dynamics and Pattern-Forming Phenomena*, Editors Pierre Coullet and P. Huerre, NATO ASI Series B **237** (Plenum Press, 1990).

101 F. H. Busse and M. Auer, Phys. Rev. Lett. **72**, 3178 (1994). M.A. Zaks, M. Auer, and F.H. Busse, Phys. Rev. E**53**, 4807 (1996).

102 P.S.Hagan, SIAM J.Appl.Math. **42**, 762 (1982).

103 S. Popp, Doctoral Dissertation, Bayreuth, 1996. E. Bodenschatz, A. Weber und L. Kramer, in *Nonlinear Wave Processes in Excitable Media*, Editors A.V. Holden, M.Markus and H.G. Othmer, NATO ASI Series B **244** (Plenum Press, 1990).

104 I. S. Aranson, L. Aranson, L. Kramer, and A. Weber, Phys. Rev. A**46**, R2992-R2995 (1992). A. Weber, L. Kramer, I. S. Aranson, and L. B. Aranson, Physica D**61**, 279 - 283 (1992).

105 B. Janiaud, A. Pumir, D. Bensimon, V. Croquette, H. Richter, and L. Kramer, Physica D**55**, 269 (1992). G. Hager and L, Kramer, unpublished (1996).

106 R. Faller and L. Kramer, Phys. Rev. E, in press; Chaos, *Solitons, and Fractals*, in press; and to be published.

107 W. van Saarloos and P.C. Hohenberg, Physica **56D**, 303 (1992).

108 N.N. Akhmediev and V.V. Afanasjew, and J.M. Soto-Crespo, Phys. Rev. E**53**, 1190-1201 (1995).

109 P. Manneville and H. Chaté, Physics Letters **A 171**, 183-188 (1992); S. Sasa and T. Iwamoto, Physics Letters bf A 175, 289 (1992).

110 S. Popp, O. Stiller, I. Aranson, L. Kramer und A. Weber, Phys. Rev. Letters **70**, 3880 - 3883 (1993); S.Popp, O.Stiller, I.S.Aranson und L.Kramer, Physica D**87**, 398-423 (1995); O.Stiller, S.Popp und L.Kramer, Physica D**87**, 424-436 (1995).

111 M. van Hecke, Phys. Rev. Lett. **80**, 1896 (1998).

112 H. Sakaguchi, Prog. Theor. Phys. **84**, 792 (1990); B.I. Shraiman, A. Pumir, W. v. Saarloos, H. Chaté, and M. Holen, Physica D **57**, 241 (1992).

113 H. Chaté, Nonlinearity **7**, 185 (1994).

114 R. Montagne, E. Hernández-Garcia, and M. San Miguel, Phys. Rev. Lett. **77**, 267 (1997); R. Montagne, E. Hernández-García, A. Amengual, and M. San Miguel, Phys. Rev. E**56**, 151 (1997).

115 A. Torcini, Phys. Rev. Lett. **77**, 1047 (1996); A. Torcini, H. Frauenkron, and P. Grassberger, Phys, Rev. E**55**, 5073 (1997).

116 T. Yamada and Y. Kuramoto, Prog. Theor. Phys. **55**, 2035 (1976).

117 Y. Kuramoto, *Chemical Oscillations, Waves, and Turbulence* (Springer-Verlag, Berlin, 1984);

118 I. Aranson, L. Kramer, and A.Weber, Phys. Rev. Lett. **72**, 2316-2319 (1994).

119 D. Barclay, Phys. Rev. Lett. **72**, 164 (1994); D. Barclay and I. G. Kevrekidis, Chaos **4**, 453-460 (1994).
120 I. Aranson, L. Kramer, and A. Weber, Phys.Rev. E**47**, 3231 - 3241 (1993); I. Aranson, L. Kramer, and A. Weber Phys. Rev. E**48**, R9 - R12 (1993).
121 P. Manneville and H. Chaté, Physica D**96**, 30 (1995).
122 Y. Kuramoto and S. Koga, Prog. Theor. Phys. **66**, 1081 (1981); P. Coullet, J. Lega, and L. Gil, Phys. Rev. Lett. **62**, 1619 (1989).
123 L. Gil, J. Lega, and J.L. Meunier, Phys. Rev. **A 41**, 1138 (1990).
124 T. Bohr, A.W. Pedersen and M.H. Jensen, Phys. Rev. **A 42**, 3626 (1991); G. Huber, P. Alstrøm and T. Bohr, Phys. Rev. Lett. **69**, 2380,(1992).
125 H. Chaté and P. Manneville, Physica A**224**, 348 (1996).
126 R. Braun and F. Feudel, Phys. Rev. E**53**, 6562 (1996).
127 T. Bohr, G. Huber, and E. Ott, Physica D**106**, 95 (1997).
128 I. Aranson, H. Chaté, and L.H. Tang, Phys. Rev. Lett **80**, 1770 (1998).
129 K. Nam, E. Ott, M. Gabbay, and P.N. Guzdar, Physica D**118**, 69 (1998).
130 M. Gabbay, E. Ott, and P. Guzdar, Phys. Rev. Lett. **78**, 2012 (1997); Physica D**118**, 371 (1998).
131 I.S. Aranson and A.R. Bishop, Phys. Rev. Lett. **79**, 4174 (1997); I.S. Aranson, A.R. Bishop, and L. Kramer, Phys. Rev. E**57**, 5276 (1998).
132 W. Schöpf und L. Kramer, Phys. Rev. Lett. **66**, 2316 - 2319 (1991); J.A. Powell and P.K. Jacobsen, Physica D**64**, 132 (1993); S. Popp, O. Stiller, E. Kuznetsov, and L. Kramer, Physica D**114**, 81-107 (1998).
133 H. Riecke and L. Kramer, preprint patt-sol/9805005 (1998).
134 E. Knobloch and J.D. Luca, Nonlinearity **3**, 975 (1990); E. Knobloch, in *Pattern Formation in Complex Dissipative Systems*, edited by S. Kai (World Scientific, Singapore, 1992).
135 P. Coullet, T. Frisch, and F. Plaza, Physica D**59**, 73 (1993); I. Aranson and L. Tsimring, Phys. Rev. Lett. **75**, 3273 (1995); M. van Hecke, C. Storm, and W. van Saarloos, Physica D (submitted).
136 M.C. Cross, P.G. Daniels, P.C. Hohenberg, and E.D. Siggia, Phys. Rev. Lett. **45**, 898 (1980); J. Fluid Mech. **127**, 155 (1983); M. Linek and G. Ahlers, preprint 1998.
137 O. Thual and S. Fauve, J. Phys. (France) **49**, 1829 (1988).
138 R.J. Deissler and H.R. Brand, Phys. Rev. Lett **74**, 4847 (1995).
139 M. van Hecke, E. de Wit, and W. van Saarloos, Phys. Rev. Lett. **75**, 3830 (1995).
140 W. Decker and W. Pesch, J. Phys. II (France) **4**, 419 (1994);
141 F.H. Busse, Rep. Prog. Phys. **41**, 1929 (1978).
142 A. Davey and L.M. Hocking, J. Fluid Mech. **63**, 529 (1974).
143 H. Riecke and W.-J. Rappel, Phys. Rev. Lett. **75**, 4035 (1995); H. Herrero and H. Riecke, Phys. Lett. A**235**, 493 (1997); H. Riecke and Glen D. Granzow, Phys. Rev. Lett. (in press). M. Ipsen, Doctoral Dissertation, Copenhagen, 1997; M. Ipsen, F. Hynne, and P. G. Sœrensen, *Amplitude Equations for Reaction-Diffusion Systems with a Hopf bifurcation and Slow Real Modes*, preprint (1998).
144 R. Imbihl, H. Engel, and M. Eiswirth, this volume.
145 E. Ammelt, Yu. A. Astrov, and H.-G. Purwins, Phys. Rev. E**55**, 6731 (1997).
146 M. Bär, M. Hildebrand, M. Eiswirth, M. Falcke, H. Engel, and M. Neufeld, Chaos **4**, 499 (1994).

147 I.S. Aranson, A. Bishop, I. Daruka, and V.M. Vinokur, Phys. Rev. Lett. **80**, 1770 (1998).
148 M.I. Tribelsky and K. Tsuboi, Phys. Rev. Lett. **76**, 1631 (1996) M.I. Tribelsky and M. Velarde, Phys. Rev. E**54**, 4973-4981 (1996).
149 A.J. Bernoff, Euro. Jour. Appl. Math. **5**, 267 (1994).
150 C. Crawford and H. Riecke, Physica D (submitted).
151 J. Lega and J.M. Vince, J. Phys. I France **6**, 1417 (1996).
152 M. Treiber, in *Pattern Formation in Liquid Crystals*, Editors L. Kramer and A. Buka (Springer-Verlag, New York, 1996).
153 J. Garcia-Ojalvo, J.M.R. Parrondo, J.M. Sancho, and C. Van den Broek, Phys. Rev. E**54**, 6918 (1996); A. Becker and L. Kramer, Physica D**90**, 408 (1996).
154 Y. Kuramoto, Prog. Theor. Phys. **71**, 1182 (1984).
155 A. A. Nepomnyashchy, Proc. Perm State Univ. **362**, 114 (1976) (in Russian); A. A. Nepomnyashchy, to be published.
156 Y. Pomeau and P. Manneville, J. Phys. Lett. **40**, L-609 (1979); P. Manneville *Dissipative Structures and and Weak Turbulence* (Academic, San Diego, 1990).
157 A. Van Harten, in *Structure and Dynamics of Nonlinear Waves in Fluids (Advanced Series in Nonlinear Dynamics 7)*, ed. A. Mielke and K. Kirchgaessner (World Scientific, Singapore, 1995) p. 117.
158 H. Riecke and H.G. Paap, Phys. Rev. A**33**, 547 (1986); H. Riecke and H.G. Paap, Phys. Rev. Lett. **59**, 2570 (1987); H. Riecke and H.G. Paap, Europhys. Lett. **14**, 433 (1991).
159 G. Ahlers, D.S. Cannel, M.A. Dominguez-Lerma, and R. Heinrichs, Physica D**23**, 202 (1986).
160 L. Kramer, E. Ben-Jacob, H. Brand, and M. C. Cross, Phys. Rev. Lett. **49**, 1891 (1982); L.Kramer and H. Riecke, Z. Phys. **B59**, 245 (1985); I. Rehberg and H. Riecke, in *The Physics of Structure Formation*, W. Güttinger and G. Dangelmayr (Springer-Verlag, Heidelberg, 1987).
161 H. Aspe and M.C. Depassier, Phys. Rev. A**41**, 3125 (1990); S. Lou, G. Huang, and H. Ruan, J. Phys. A: Math. Gen. **24**, L587 (1991); D.E. Bar and A. A. Nepomnyashchy, Physica D**86**, 586 (1995); D. E. Bar, PhD thesis, Technion, Haifa (1996); A.A. Nepomnyashchy and M. Velarde, Phys. Fluids **6**, 187 (1994).
162 E. Bodenschatz, M. Kaiser, L. Kramer, W. Pesch, A. Weber, and W. Zimmermann, in *New Trends in Nonlinear Dynamics and Pattern Formation Phenomena: The Geometry of Nonequilibrium*, Eds. P. Coullet and P. Huerre, NATO ASI Series, Plenum Press, 1990, p. 111 - 124.
163 M.C. Cross and A.C. Newell, Physica D**10**, 299 (1984).
164 A.C. Newell, T. Passot, C. Bowman, N. Ercolani, and R. Indik, Physica D**97**, 185 (1996).
165 M.C. Cross, Physica D**97**, 65 (1996).
166 P. Coullet and G. Iooss, Phys. Rev. Lett. **64**, 866 (1990).

Formation of Dynamical Structures in Axisymmetric Fluid Systems

F.H. Busse[1], G. Pfister[2] and D. Schwabe[3]

[1] Institute of Physics, University of Bayreuth, D-95440 Bayreuth, Germany
[2] Institute of Applied Physics, University of Kiel, D-24098 Kiel, Germany
[3] Institute of Physics, University of Giessen, D-35392 Giessen, Germany

Abstract. Axisymmetric fluid systems are of special interest because the homogeneity of conditions in the azimuthal dimension can be realized easily in experiments. Numerous mechanisms of instability driven by shear, buoyancy or other effects can be studied in axisymmetric configurations. In this chapter a few examples are presented. The Taylor-Couette system has long been a preferred subject of nonlinear fluid dynamics and several recently discovered dynamical phenomena will be described. Convection driven by centrifugal buoyancy has become another intensively studied example of hydrodynamic instability because of its relevance for the dynamics of stellar and planetary atmospheres. Convection in superimposed layers of immiscible liquids can also profitably be studied in a ring channel. Finally the problem of the instability of thermocapillary flow is investigated in axisymmetric settings.

1 Introduction

Axisymmetric fluid systems are a favourite tool for the investigation of the spontaneous formation of structures in hydrodynamics because the homogeneity of the external conditions is realized in the azimuthal dimension. Homogeneity in the axial direction can often be achieved approximately through the use of long cylinders. Thus the problem remains non-uniform only in the radial direction which represents the minimal inhomogeneity of a non-equilibrium system under stationary conditions.

Apart from these basic considerations researchers have been led to axisymmetric systems because of their important applications in engineering problems and in astro- and geophysics. Fluid flows in circular pipes or between coaxial cylinders or in other types of fluid domains in rotating machinery provide numerous challenges for the understanding of the mechanisms of the transfers of heat, mass and momentum in these systems. Rotating celestial bodies are usually axisymmetric to a high degree of approximation. Non-axisymmetric patterns in their atmospheres are thus typical manifestations of the spontaneous formation of dynamical structures in fluids far from equilibrium. Even in the case of the Earth's atmosphere which is influenced by the non-axisymmetric distribution of oceans and continents a regular pattern of cyclones and anticyclones at mid latitudes can often be seen in the global weather maps.

Nearly all mechanisms of hydrodynamic instability can be studied in axisymmetric configurations. In this chapter we shall focus on some systems for which axisymmetry is essential and which are readily accessible for a comparison between theory and experimental observation. Foremost among the axisymmetric fluid systems is the Taylor-Couette system of flow between differentially rotating cylinders which has become a principal example for pattern formation in nonlinear fluid dynamics. Instabilities driven by thermal buoyancy can be studied in the rotating cylindrical annulus almost as easily as in the horizontal Rayleigh-Bénard layer heated from below. The centrifugal force has become a useful tool for the experimental study of convection in configurations in which the directions of gravity and of the axis of rotation are not parallel such as those of deep planetary or stellar atmospheres or the liquid core of the Earth. A third section of this chapter will be devoted to wave like processes in annular channels. Of interest is the effect of small deviations from axisymmetry introduced through sinusoidal modulations of the boundary conditions. The interaction of two spatially periodic processes, one forced, the other spontaneously generated, gives rise to new dynamical phenomena in this case. Finally we shall consider the wave like dynamical patterns of surface tension driven (thermocapillary) flows in cylindrical liquid bridges with free surface under axial heat flux and in annular gaps with free upper surface under radial heat flux. These are two important configurations used in crystal growth from melts.

2 The Taylor-Couette System

2.1 Historical Remarks

Since the landmark paper of G. I. Taylor in 1923 [1] the Taylor-Couette problem has been an intensively studied area of hydrodynamic stability. The Taylor-Couette system corresponds to the flow between independently rotating cylinders and was originally designed by M. M. Couette [2] and A. Mallock [3] for the measurement of viscosity. This is a closed flow configuration unlike the flow motion through a pipe, which is an open flow, where disturbances may grow as they are carried off downstream. It can be controlled very precisely in experiments and its dynamics can be followed very easily because typical time constants are of the order of seconds. The Taylor-Couette system is characterized by a rich variety of hydrodynamic instabilities and it exhibits almost all well known low-dimensional scenarios of transitions to chaotic states for proper boundary conditions. Therefore it serves as a test system for rotating fluid motion and, more importantly, as a principal example for the study of basic properties of nonlinear hydrodynamics such as pattern formation, bifurcation structures, and transitions to turbulence.From the experimental point of view it is an invaluable advantage to have the possibility to study all these phenomena with a single experimental apparatus.

High precision experiments which allow quantitative comparisons with theories require enormous efforts and the ability to use the same apparatus for realizing well known fluid flows as well as new phenomena is essential for the interpretation of the latter.The aim of this contribution is not a complete review of Taylor-Couette flow but a summary of recent results obtained in the context of pattern formation in dynamical systems. Several reviews concerned with the Taylor-Couette system have appeared recently: R. Tagg [4], P. Chossat and G. Iooss [5], L. Koschmieder [6], T. Mullin [7], R. C.DiPrima and H. L. Swinney [8] and G. Pfister and J. v. Stamm [9]. A comprehensive bibliography has been published by R. Tagg [10].

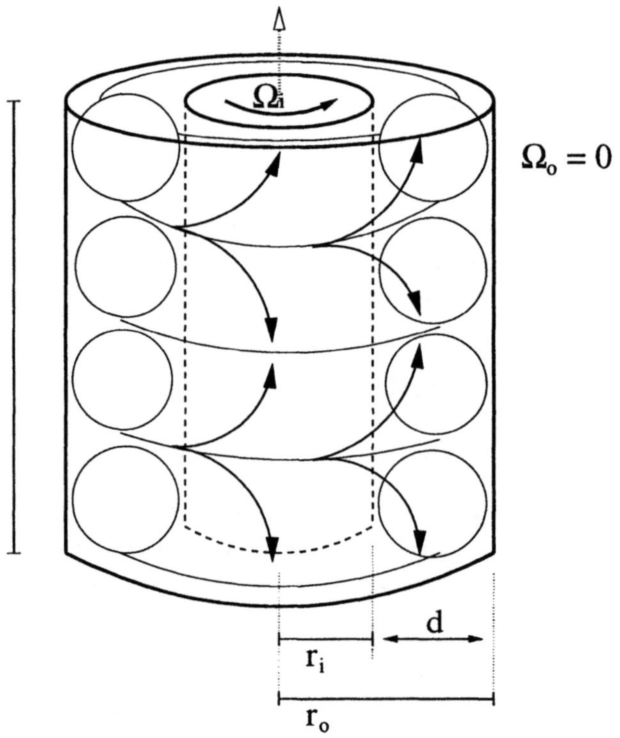

Fig. 1. The Taylor-Couette geometry

2.2 Geometry of the flow

A schematic sketch of the Taylor-Couette system is shown in fig. 1.Various geometry parameters define the details of the flow: r_i and r_o are the inner and outer cylinder radius and $\eta = r_i/r_o$ is the radius ratio; $\Gamma = l/d$

the aspect ratio, i.e. the gap length l to gap width d ratio.In most experiments top and bottom plates are held at rest to achieve well defined boundary conditions. The flow is then determined by the rotation rates Ω_i and Ω_o of the inner and outer cylinder respectively. The Taylor number T, $T = 4\Omega_i d^2(\Omega_i r_i^2 - \Omega_o r_o^2)/\nu^2(r_o^2 - r_i^2)$, has been introduced as a dimensionless measure of the shear between the cylinders where ν denotes the kinematic viscosity. Below the critical value T_c of T a unique solution exists in the form of an axisymmetric basic state which is also approximately independent of the axial coordinate except for the so-called Ekman vortices near the end plates. When the Taylor number exceeds the critical value, other states of flow characterized axisymmetric or non-axisymmetric vortices can be realized in experiments.Because of the multiplicity of solutions in most cases the actual flow state depends on the history of the experimental system. This was first clearly stated by Coles [11] in 1965 in a seminal paper concerning wavy vortex flow. Because of the end effects the transition to axisymmetric vortices corresponds typically to a disconnection of a pitchfork bifurcation [7]. But an exact symmetry breaking bifurcation occurs in the transition to wavy vortices.

The rich variety of flow states in an apparatus with independently rotating cylinders is demonstrated in the experiments by Andereck et al. [12] in which the rotation of the outer cylinder is kept at a fixed rate and the rotation rate of the inner cylinder is slowly increased An apparatus with a radius ratio $\eta = 0.883$, corresponding to the narrow gap case, and an aspect ratio of $\Gamma = 30$ has been used. The theoretical understanding of this general case is not yet as well developed as that of the standard Taylor-Couette case with resting outer cylinder. We will concentrate on this latter case, where top and bottom plate are held at rest and the aspect ratio Γ is varied only for the adjustment of the number and the size of the Taylor vortices. If not stated otherwise, all the results reported here are obtained with a radius ratio of $\eta = 0.5$, corresponding to a wide gap case.

Depending on the location in the control parameter space (Re, Γ) where $Re = \frac{\Omega_i r_i \cdot d}{\nu}$ and on its path, Taylor-Couette flow shows a rich variety of flow states which can easily be studied under controlled conditions. We will focus on some topics such as localised oscillations, frequency locking, homoclinic bifurcations, period doubling and phase dynamics.

2.3 Wavy Taylor vortex flow

One of the advantageous properties of the Taylor-Couette system is the possibility to vary three control parameters.These are the radius ratio η, the aspect ratio Γ and the Reynolds number defined as $Re = (\Omega * r_i * d)/\nu$. In a given experiment the radius ratio is fixed, so the two other parameters may be used to define a control parameter plane.In fig.2 an example of a stability diagram is shown for a 10-vortex state. Here in the $(Re, \Gamma/N)$-plane the first appearances of time dependent flow modes are depicted. N is the number

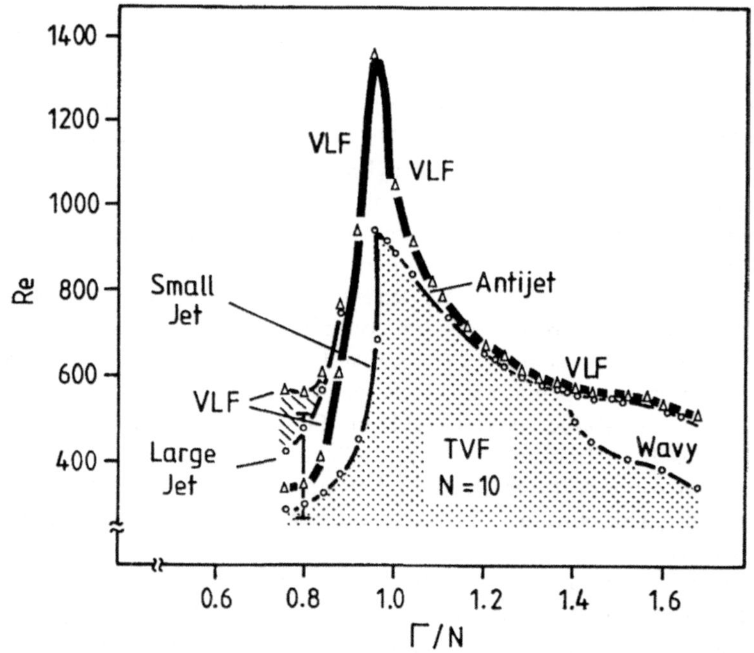

Fig. 2. Stability diagram for the onset of primary and higher oscillatory modes in the ten-vortex flow in the Taylor-Couette system with $\eta = 0.5$. The dotted section marks the stationary10-vortex flow.

of vortices such that Γ/N is the average width of a vortex.Below the first temporal instabilities the flow is stationary. The most stable stationary state is that with a vertical vortex size equal to d, i.e. $\Gamma/N \approx 1$. For all vortex sizes the flow becomes unstable with respect to time periodic flow modes with increasing Re. The time dependent states occurring in Taylor-Couette flow depend strongly on the wavelength λ of the underlying vortices [13], [14].One can find at least eight different kinds of time periodic flow modes which are physically clearly distinguishable [15]. The stability lines for five of them are drawn in the diagram. By controlling the Reynolds number, i.e. the rotation rate and the temperature dependent viscosity of the working fluid, and the geometry of the boundaries, one can perform hydrodynamic precision experiments.

2.4 Localised oscillations

One of the temporal instabilities found in Taylor-Couette flow is the large-jet mode which is characterized by a wavy outward flow while inward flow is still nearly axisymmetric. For proper parameter values there exist besides the global oscillations of all outward flows those where only one or a few vortices oscillate. An example is shown in fig.3 where in addition to a schematic sketch of a 12-vortex flow state ten velocity profiles, taken from Laser-Doppler-Velocimetry (LDV) measurements, are depicted. These velocity profiles were taken while the system exhibits time dependent behaviour. Regions of oscillations thus correspond to a broad line, while regions of stationary flow manifest themselves as small lines. On the far right side one can see the global oscillation of all vortices except for the end vortices which are damped by the boundaries as we know from most of the oscillatory states in this system.

Besides this global mode, local oscillations can occur as the other nine profiles demonstrate. All states are obtained at similar Reynolds numbers with sudden starts. This phenomenon coincides with a change in the vortex sizes, where for this case the oscillating vortices increase in size [16]. The interaction between the underlying vortex structure and its oscillatory modes thus leads to these localised states. They have been found in Taylor-Couette systems with vortex numbers larger than eight.

Localised oscillations also occur with the core mode (the location of the vortex center oscillates) and with a modulated oscillatory state and show that this phenomenon is not a rare effect [17]. It seems that these modes are "self-trapped" states which stabilize themselves by the interaction with the underlying vortex structure.

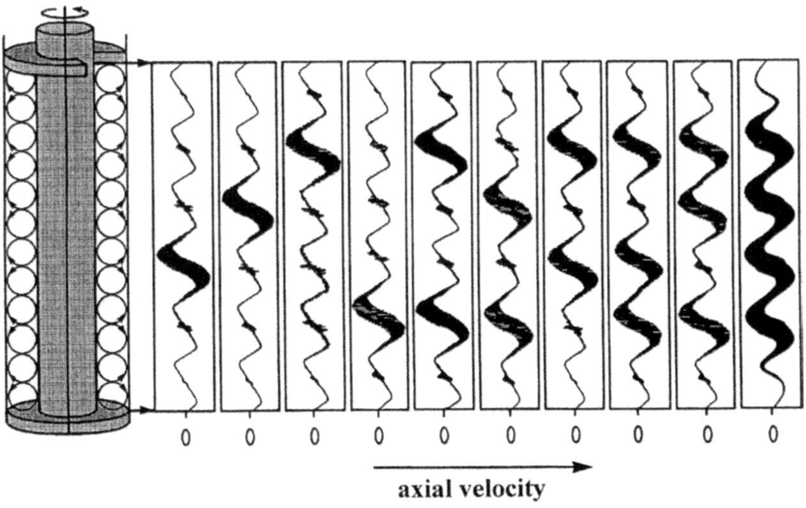

Fig. 3. Axial velocity profiles of localised Large-Jet oscillations in the twelve vortex state.

2.5 Mode coupling

The interaction between the basic flow pattern and its oscillatory excitations leads to another effect found in Taylor-Couette flow. When the Reynolds number is increased, the flow undergoes a series of bifurcations on the route to chaos. One of them is the breaking up of a two-torus after two Hopf bifurcations [18]. According to theoretical considerations [19] based on symmetry group arguments, frequency locking of a doubly periodic flow should not occur in axisymmetric systems. In contradiction to this claim at least two examples of frequency locking have been found in Taylor-Couette flow [20], [17].

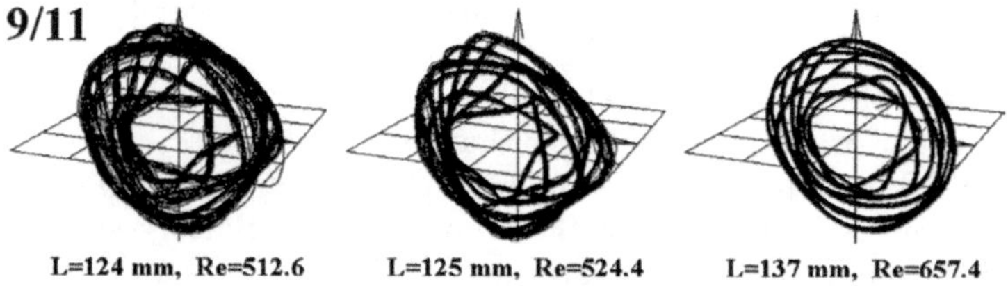

Fig. 4. Reconstructions of coupled states with a frequency ratio of 9/11 for different cylinder length L and Reynolds number. All three states belong to the same Arnol'd tongue.

In the first case two different jet modes, which can be present in the flow as isolated limit cycles for proper boundary conditions or as a torus if both are present, can show frequency locking corresponding to a Farey tree order in a preturbulent regime in the frequency range from $\Omega_1/\Omega_2 = 4/5$ to $\Omega_1/\Omega_2 = 5/6$. The highest resolved frequency ratio was 62/77.The locking is present in a finite Reynolds number range, corresponding to finite width of the Arnold's tongues. Experiments in which the axial symmetry had been slightly perturbed confirm, that no pinning effect of an imperfect geometry is responsible for the locking. Surprisingly the "noise-level" of the complicated limit cycles depends strongly on the widths of the Arnold's tongues as can be seen in fig.4. The question remains open whether this is an amplification of apparatus noise due to a weak coupling strength of the two modes or if this is an unresolved dynamical effect.It seems that whenever two oscillatory flow modes have similar dynamical structures these modes can show frequency locking. In a second example, where two ($m = 1$)-core modes show frequency coupling, it can be seen, that the frequencies of the two modes can be quite different. This is shown in fig.5 and fig.6 in a sequence of frequency locked states for slightly different Reynolds numbers. Starting from the locked state with the frequency ratio 1/6 the two coupled modes become chaotic if the

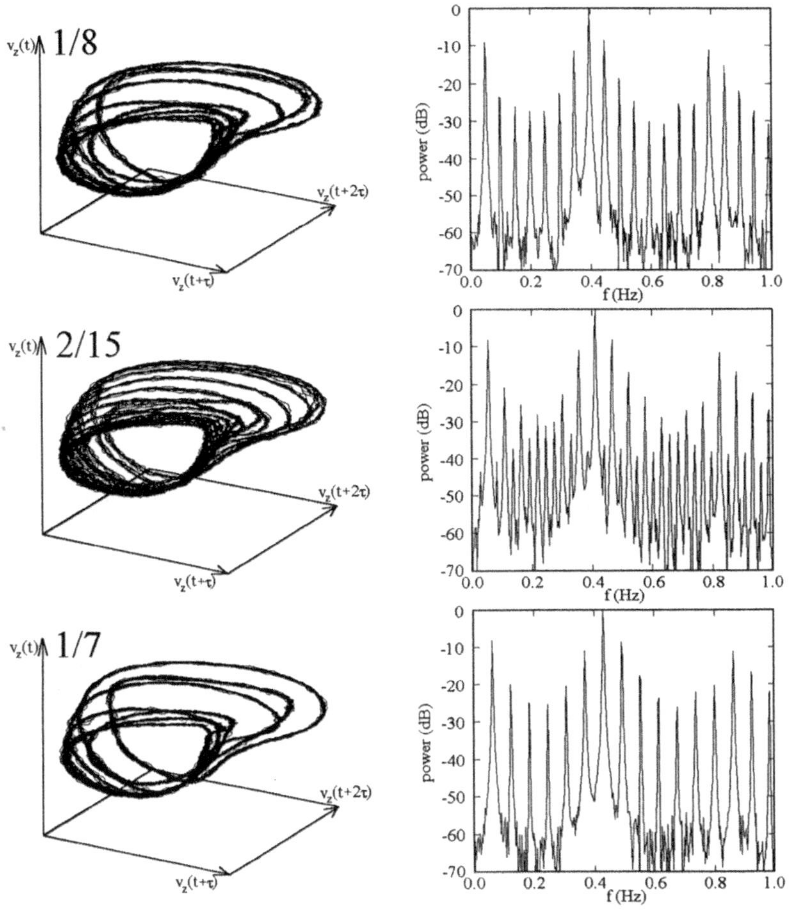

Fig. 5. Phase space reconstructions and spectra of frequency coupled states for $\Gamma = 9.92$ in an eight vortex state. Different frequency ratios ranging from 1/8 up to 1/7are shown.

Reynolds number is increased. The physical origin of the coupling is still unclear and the contradiction to Rand's theory [19] has not been resolved.

2.6 Period doubling scenarios

Experimentally, the period doubling route to chaos was first found in a hydrodynamic system by Libchaber andMaurer [21]. They observed subharmonics up to $f/16$ in a Rayleigh-Bénard convection box filled with liquid Helium and enclosing two convection rolls. In Taylor-Couette flow period doubling was not expected, because a prediction of Swift and Wiesenfeld [22] stated, that this phenomenon should not occur in systems with rotational symmetry.In

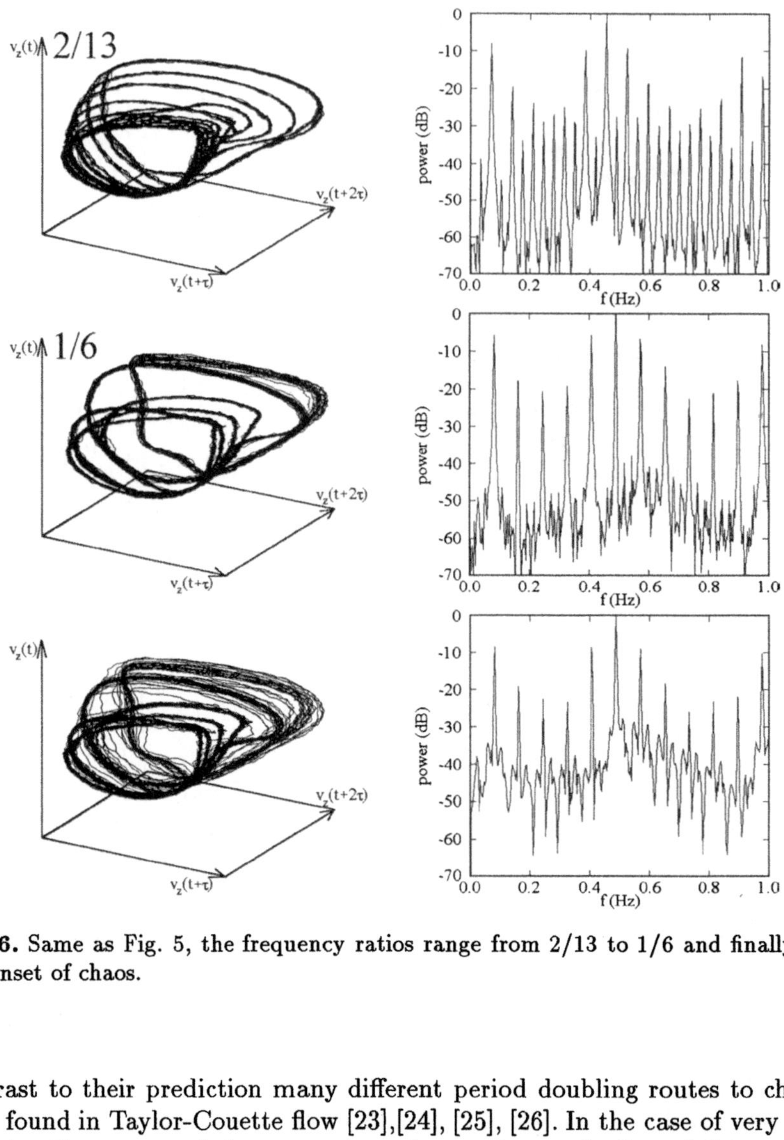

Fig. 6. Same as Fig. 5, the frequency ratios range from 2/13 to 1/6 and finally to the onset of chaos.

contrast to their prediction many different period doubling routes to chaos were found in Taylor-Couette flow [23],[24], [25], [26]. In the case of very low aspect ratios one can induce an extremely small perturbation of the perfect rotational symmetry by inclining the top plate by a small amount. It was found, that the routes to chaos are sensitive to even such small changes in the boundary conditions. But in other scenarios period doubling has been observed with the best rotational symmetry experimentally available, so that an open question remains to be solved here. An example for a more complex route to chaos is the period doubling on the torus [26]. Fig. 7 displays on the left side projections of a three-dimensional reconstruction of the measured attractors, on the right side the corresponding Poincaré sections. The

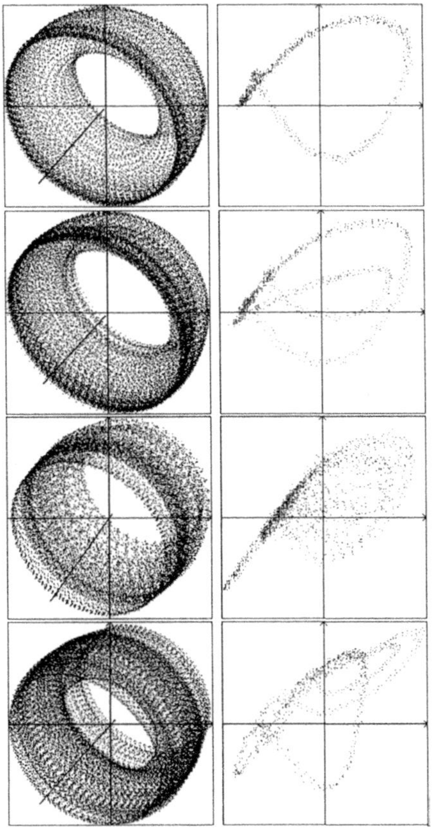

Fig. 7. A period-doubling-on-the-torus scenario in 14-vortex flow at an aspect ratioΓ = 11.76. Left column: Projections of three-dimensional reconstructions of the attractors in phase space, right column: The corresponding Poincaré-sections. From the top to the bottom the simple torus at $Re = 410$, the doubled torus at $Re = 418$, a chaotic attractor at $Re = 422$ and an attractor recorded in a period-3 window at $Re = 423$.

sequence shows a torus, a doubled torus, chaos and a period-three window for increasing Reynolds number. The corresponding bifurcation diagram is depicted in fig. 8.

2.7 Homoclinic bifurcations and chaos

Homoclinic orbits, i.e. the connection of the stable and unstable manifold of a fixed point in phase space, play an important role in nonlinear dynamics. In parameter-dependent dynamical systems a homoclinic orbit can emerge through a collision of a limit cycle with the fixed point leading to a global bifurcation. Depending on the dimension of the phase space and the eigenvalues

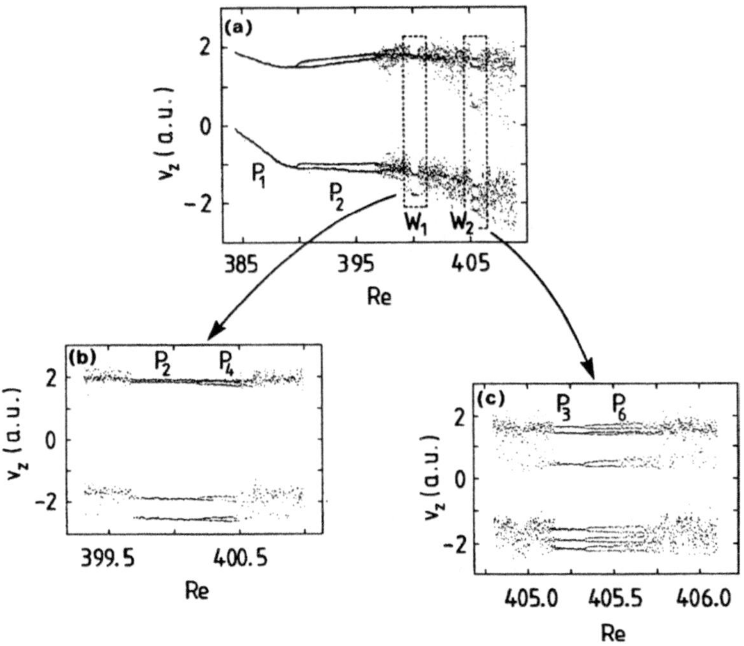

Fig. 8. The bifurcation diagram for the 20-vortex flow for $\Gamma = 16.8$. The Reynolds number was increased quasistatically within $\Delta Re/\Delta t = 1.63 * 10^{-4} s^{-1}$. The regime around the two visible periodic windows is extended: (b) reveals a doubling from period-2 to period-4 and in (c) a doubling from period-3 to period-6 becomes visible.

of the fixed point different kinds of homoclinic orbits can be formed corresponding to different bifurcation scenarios. In the Taylor-Couette system several situations, where homoclinic bifurcations are essential, have been identified.

A SNIPER (Saddle Node Infinite PERiod) bifurcation was found to be responsible for the disappearence and the onset of a time-periodic (m=2)-mode in an asymmetric one-vortex state. The frequency of this oscillation tends to zero as it approaches homoclinicity and increases again after a small, but finite Reynolds number interval of steady fluid motion.The same scenario can be observed at the onset of the time-periodic VLF mode, to be described below, but the homoclinic bifurcation involved is of a different type [26].

A more complex scenario can arise from a collision of a limit cycle with a saddle-focus in a three dimensional phase space. For a proper ratio of the eigenvalues of the saddle-focus this so-called Shil'nikov bifurcation is capable of producing homoclinic chaos. A Shil'nikov bifurcation can be generated e.g. by a codimension-2 bifurcation of the fold-Hopf type. Mullin and Price [27] have investigated a scenario induced by the interaction of a steady and a time-dependent flow in the 4-vortex state of a Taylor-system, where the top and bottom plates were attached to the inner cylinder. The dynamics at proper parameter values seems to be closely related to the Shil'nikov mechanism. Their experimental observations are the first of such a kind in a flow system [27],[28].

Recently the relevance of the Shil'nikov mechanism for the onset of chaos in the VLF-regime has been identified [29]. The physical origin of the dynamics of the VLF mode as well as the bifurcation structure differs substantially from that investigated by Mullin and Price. These results underline the importance of homoclinic bifurcations for the dynamics of the Taylor-Couette system.

2.8 Destabilisation of the vortex wave vectors: VLF mode

Depending on the radius ratio, the first time periodic instability is a travelling wave with azimuthal wave numbers $m = 1, 2, 3, \ldots$. For the wide gap case, with radius ratios $\eta \leq 0.5$, the first wave number is $m = 1$ in most cases. The axial wave number of the underlying vortex structure may show a slight spatial dependence which is stationary. With increasing Reynolds number a very-low-frequency (VLF) mode appears in all cases as a secondary instability corresponding to a very slow time-periodic oscillation with an azimuthal wave number $m = 0$ [30]. The observed periods extend as far as hours, i.e. 10^5 times the rotation period of the inner cylinder. The VLF mode differs qualitatively and quantitatively from all other known time-periodic flows in the Taylor-Couette experiment. This mode exists in the entire wave number range and manifests itself as a modulation of the periodic vortex structure. The modulation appears to be an universal phenomenon since its appearance does not depend on the type of oscillation which is present. This is shown in fig. 2, where the stability diagramm for the onset of primary and higher time-dependent modes in the 10-vortex flow in a Taylor-Couette system with $\eta = 0.5$ is shown. Because the VLF-mode never exists alone, we conclude that the underlying oscillation drives the VLF-mode. The mechanism of this interaction has not yet been clarified and it thus is a problem of current research. One step has already been done through the introduction of external perturbations on the wave vector [30],[31], [15], [18], [32].

2.9 Phase dynamics

Phase dynamics has been proved to be very successful concept for the understanding of the stability and of long-wave modulations of periodic roll patterns in dissipative systems. It has been discovered that under some conditions the evolution of a periodic pattern is governed by a linear diffusion equation for the phase variable [33]. Later a conceptional enlargement to nonlinear phase equations and coupled phase dynamics has been worked out (for a review see [34]). In addition to the description of the dynamical properties

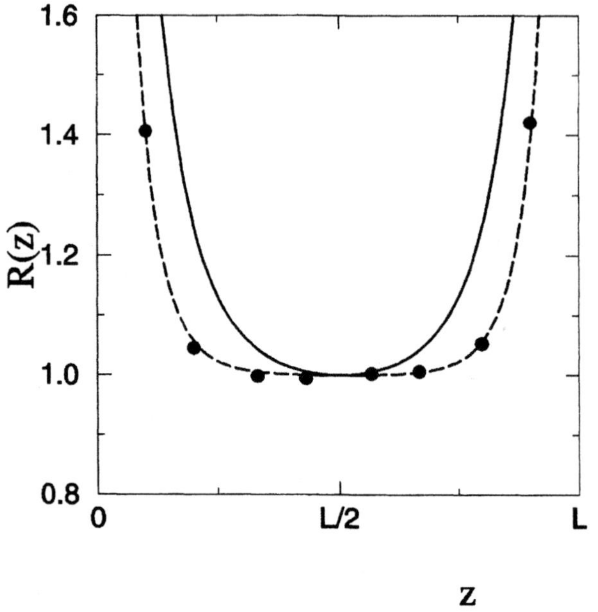

Fig. 9. Envelope $R(z)$ vs. cylinder heigth z: Solution of the Ginzburg-Landau equation (—) and Fit of Jacobian-elliptic function (−−) to measured data (•) from an 8-vortex state

of a periodic pattern, the phase dynamics approach has made important contributions to the understanding of pattern selection problem (see e.g. [35]).

The diffusion coefficient D is the essential quantity of the theory governing the dynamics of the phase. Gerdts [30] has performed measurements of the diffusion coefficient in a wide range of wave numbers k and Reynolds numbers Re. He applied external perturbations to the vortex column and determined the relaxation rates. In his experiments the general applicability of phase diffusion theory to the Taylor experiment has been shown as well.

The agreement between theoretical and experimental results has remained unsatisfactory, however, as long as essential properties of the experiment had not found their correspondence in theory. The phase diffusion equation described in [33] did not cope with the influence of boundary conditions in the way they are realized in experiments. A conceptional framework for the incorporation of boundaries was recently given by Roth et al.[36], who developed a generalised phase diffusion theory.

In Taylor-vortex flow the end plates cause the generation of Ekman vortices, which differ from the bulk vortices with respect to wave number and amplitude. The variation of the amplitude profile can be described by Jacobian-elliptic functions (JEF). The theory of Roth et al. integrates this deviation from a constant amplitude profile within the phase equation, because the latter profile is suitable only for the description of an infinitely extended periodic structure. The phase equation is derived from a Ginzburg-Landau equation (GLE):

$$\tau_0 \partial_t A(z,t) = \left(\varepsilon + \xi_0 \partial_z^2 + g|A(z,t)|^2\right) A(z,t) \tag{1}$$

for the amplitude $A(z,t) = R(z,t)exp(i\Phi(z,t))$.

Equation (1) can be separated into two coupled equations for the modulus $R(z,t)$ and the phase $\Phi(z,t)$. As long as the assumption

$$\beta := \epsilon^{1/2} L \xi_0^{-1} \gg \pi/\sqrt{2} \text{ with } \quad \epsilon = \frac{Re - Re_{crit}}{Re_{crit}} \tag{2}$$

holds where L is the system length and ξ_0 is the coherence length, the phase $\Phi(z,t)$ and the modulus $R(z,t)$ evolve on different time scales and the equations can be separated. The final profile of the modulus $R(z) = R(z, t = \infty)$ is thus sufficient for the phase equation, which reads:

$$\partial_t \Phi(z,t) = D_0 \left\{ \partial_z^2 + \frac{2\partial_z R(z)}{R(z)} \partial_z \right\} \Phi(z,t) \tag{3}$$

This is an equation for the deviation $\Phi(z,t)$ from the final phase $\Phi_\infty(z)$ under the assumption $\partial_z \Phi_\infty(z) \approx 0$, i.e. $k \approx k_{crit}$. The diffusion coefficient is $D_0 = \xi_0^2/\tau_0$. The equation is linear and can be solved by separation. This leads to an eigenvalue-problem of the form of a stationary Schrödinger equation. The time evolution is governed by the smallest eigenvalue λ, which involves the definition of the effective diffusion coefficient defined by $D/D_0 := \min_{\lambda \neq 0}\{\lambda\}$. The eigenvalues λ can be calculated, if the modulus $R(z)$ is known. Following the approach of [36] $R(z)$ is found as a solution of the GLE. If we take into consideration that $R(z) \approx \frac{1}{z}$ and $R(L-z) \approx \frac{1}{L-z}$ near the boundaries, these solution are the Jacobian-elliptic functions ns and ds. From these the effective diffusion coefficient D/D_0 as a function of β can be calculated numerically. Experimentally the diffusion coefficients D were obtained by applying small perturbations on the vortex column and measuring the relaxation rates, as stated above. The details of the measurement procedure as well as the experimental diffusion coefficients can be found in [30]. For comparisons with the theory the experimental coefficients D must be normalised with D_0, the latter being the diffusion coefficient of an infinitely extended system. The coefficient D_0 as function of the parameters of the problem has been calculated numerically from the Navier-Stokes equation with the algorithm of Paap and Riecke [35]. In comparison with the diffusion coefficients calculated for an infinite system, the new approach leads to a significant improvement of the theoretical predictions within the wave number interval $2.8 \leq k \leq 3.57$. As an example fig. 10 shows the normalised diffusion coefficient D/D_0 obtained from the theory (solid line, labeled JEFs) and from experiment (the symbols correspond to flow states having N vortices) for $k \approx 2.8$.The systematic deviation is due to the fact, that the solution of the GLE is not suitable for a sufficiently accurate description of the amplitude profile of the experimental Taylor-vortex flow. Moreover, for larger wave vectors $k > 3.57$ only a qualitative agreement between theory and experiment is obtained. For smaller wave numbers $k < 2.8$ the theory does not work at all. The systematic error can be avoided within the interval $2.8 \leq k \leq 3.57$, if measured amplitude profiles are used instead of calculated profiles. The data of a profile obtained experimentally from an 8-vortex state at $Re = 74$ is plotted together with a fit of a JEF in fig.9. The numerical solution of the GLE for the same parameters, which is also a JEF, is depicted as well. The measurements were performed using low-pass filtered axial velocity-profiles.After these function had been inserted into the Schrödinger equation the solutions for the effective diffusion coefficient D/D_0 were calculated. The result of this procedure for $k \approx 2.8$ is also plotted in fig.10

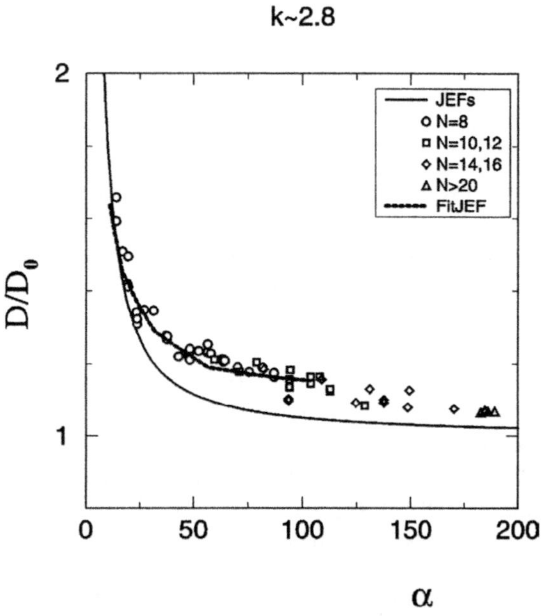

Fig. 10. Effective diffusion coefficient D/D_0 vs. α: numerical calculations from equation 3 with JEFs (—), experimental data measured in an 8- to 20-vortex state with axial wave vector $k \sim 2.8$. In addition calculations with a JEF fit curve (...) is shown.

(dashed line, labeled FitJEF). The profiles for the calculations have been measured in an 8-vortex state. An excellent agreement between theory and experiment is obtained and this quantitative agreement is found in the whole interval $2.8 \leq k \leq 3.57$. Therefore, the generalized phase diffusion theory combined with experimental amplitude profiles gives a sufficient explanation of the influence of boundary conditions on the phase dynamics of stationary

Taylor-vortex flow for wave numbers in the interval $2.8 \le k \le 3.57$ and for a wide range of Reynolds numbers Re [37].The generalized phase diffusion description of stationary Taylor-vortex flow thus serves as an example for the way in which hydrodynamic problems can be successfully treated by simple model equations describing the properties of the flow not only qualitatively, but quantitatively as well.

2.10 Summary

Experiments on closed hydrodynamic flow systems, such as Taylor-Couette flow and its close relative, the flow between rotating spheres [38], cover a wide range of nonlinear phenomena which can be studied in precison hydrodynamic experiments. It has been demonstrated, that the notion of low dimensional chaos is relevant to the understanding of the transition to weakly irregular states in carefully controlled fluid flow systems [39], [40], [41], [42]. The dynamics and pattern formations [20] are caused by the underlying bifurcation structure, such that the global behaviour can be understood as a consequence of local critical events in the control parameter space. The connection of low dimensional chaos to fluid flow turbulence, however, is not yet well understood. The difficulties lie in the fact, that turbulence is an infinite-dimensional phenomenon and not understandable directly in terms of the low dimensional models applied in dynamical systems.The experiments described in this brief review deal with well defined spatial states and some weakly irregular dynamics around them. While the knowledge obtained from these measurements has a fundamental character, it may also be relevant for the understanding of large scale coherent motions such as atmospheric waves and large eddy flows in oceans and in shear layers. From a basic point of view the multiplicity of states, the routes to chaos and the large time scale dynamics found in the Taylor vortex case serve as examples for the understanding of phenomena seen in more complex nonlinear systems in nature as well as in engineering applications.

3 Convection Driven by Centrifugal Buoyancy

3.1 Theoretical Concepts

Since the centrifugal force in a rotating system can be written as the gradient of the potential, it can be employed for the generation of convection in a cylindrical layer cooled from the inside in the same way as gravity is used to drive convection in a Rayleigh-Bénard layer. The main difference is that in a rotating system the Coriolis force operates as well. Only in the limit of large Prandtl numbers does the influence of the Coriolis force vanish for a fixed value of the Rayleigh number in the definition of which the centrifugal force assumes the role of gravity [43]. The Coriolis force acting in the fluid

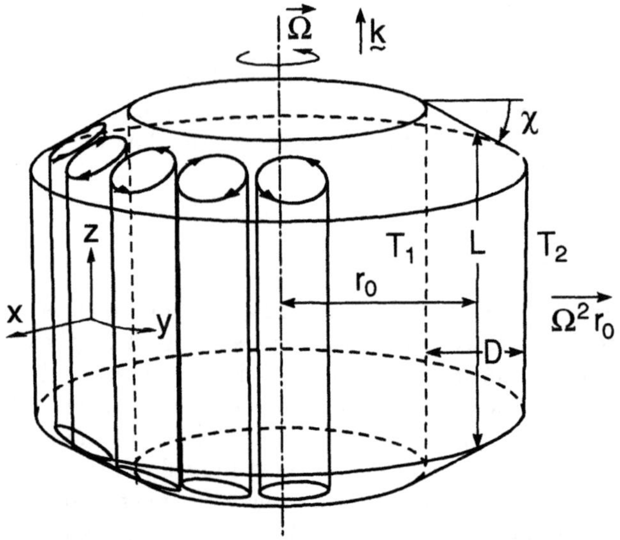

Fig. 11. Geometrical configuration of the rotating cylindrical annulus. The Cartesian coordinate system (x, y, z) used in the small gap limit has been indicated and the convection rolls aligned with the axis of rotation have been sketched. r_0 is the mean radius of the annulus while L is its mean height.

in a cylindrical annulus rotating about a vertical axis as shown in figure 11 offers the advantage that the convection motions simulate the convection flows in the fluid interiors of rotating celestial bodies, such as the gaseous outer planets, rotating stars, or the liquid core of the Earth [44]. For a recent review of the convection driven zonal flows in the major planets and their relationship to laboratory experiments we refer to [45].

Here we want to draw attention to centrifugally driven convection in the rotating annulus as a system in which several dynamical phenomena can be observed in their simplest manifestation. The property that the onset of convection occurs in the form of traveling waves when the end boundaries are inclined as shown in figure 11, the transition to mean flow convection which is associated with strong zonal flows and the occurrence of vacillations are examples of these phenomena which have no counterpart in ordinary Rayleigh-Bénard convection. The theoretical description of such phenomena remains simple since the dependence on the z-coordinate along the axis of rotation can be neglected. Because of the Proudman-Taylor theorem stating that weakly time dependent motions of small amplitude must be independent of the z-coordinate in a rotating system when viscous friction can be neglected (see, for example [46]), the velocity field of convection must be of the form

$$\mathbf{v} = \nabla\psi(x, y, t) \times \mathbf{k} + \ldots \tag{4}$$

in first approximation, where $\mathbf{k}$ is the unit vector parallel to the axis of rotation and where we have introduced the small gap approximation such that a Cartesian system of coordinates can be used as indicated in figure 11. Any velocity field of the form (4) with vanishing time dependence satisfies the inviscid equation of motion in a rotating system. The velocity field (4) can not satisfy, however, the condition of vanishing normal component at the end boundaries

$$v_z \pm v_x \tan\chi = 0 \quad \text{at} \quad z = \pm\frac{L}{2D} \tag{5}$$

But for small angles χ this condition can be taken into account as a perturbation. The solvability condition for the perturbation equation in which viscous friction and thermal buoyancy are taken into account gives rise to an equation for ψ of the form

$$\left(\frac{\partial}{\partial t} + \frac{\partial}{\partial y}\psi\frac{\partial}{\partial x} - \frac{\partial}{\partial x}\psi\frac{\partial}{\partial y} - \Delta_2\right)\Delta_2\psi - \eta^*\frac{\partial}{\partial y}\psi + R\frac{\partial}{\partial y}\Theta = 0 \tag{6a}$$

which must be solved together with the heat equation for the deviation Θ from the static temperature distribution.

$$P\left(\frac{\partial}{\partial t} + \frac{\partial}{\partial y}\psi\frac{\partial}{\partial x} - \frac{\partial}{\partial x}\psi\frac{\partial}{\partial y}\right)\Theta - \Delta_2\Theta + \frac{\partial}{\partial y}\psi = 0 \tag{6b}$$

where dimensionless variables have been used based on the gap width D between the cylinders as length scale, on D^2/ν as time scale where ν is the kinematic viscosity and on $(T_2 - T_1)P$ as temperature scale where T_1 and T_2 are the temperatures on the inner and on the outer cylinder. The Prandtl number P, the Rayleigh number R and the rotation parameter η^* are defined by

$$P = \frac{\nu}{\kappa}, \quad R = \frac{\gamma(T_2 - T_1)\Omega^2 r_0 D^3}{\kappa\nu}, \qquad \eta^* = \frac{4\Omega D^3 \tan\chi}{\nu L} \tag{7}$$

where γ is the thermal expansivity, κ is the thermal diffusivity and r_0 and L are the mean radius and the height of the annular gap. Δ_2 denotes the two-dimensional Laplacian, $\Delta_2 = \partial^2/\partial x^2 + \partial^2/\partial y^2$. For details of the derivation of equations (6) we refer to [47]. The important point is that η^* is a quantity of the same order of magnitude as R even though $\tan\chi$ has been introduced as a small parameter. But this latter property is compensated by the fact that we are interested in the limit of high rotation rates, $\Omega D^2/\nu \gg 1$.

3.2 Experimental Results

In the experimental realisations of the rotating cylindrical annulus it is not easy to achieve high values of η^* without a major effort. With rotation rates of a few hundred r.p.m. values of η^* of several 10^2 can be reached with water or air as fluids. In order to maximize η^* an angle $\chi = 45°$ has been used in the experiments to be discussed in the following. This angle certainly does not satisfy the assumption $\tan\chi \ll 1$ made above. But, as has been shown in [47], higher order terms in $\tan\chi$ have little influence on the theoretical results.

A demonstration of the thermal Rossby waves can be seen in figure 12 which actually shows two experiments in that $\chi = 0$ is used in the upper part while $\chi = 45°$ is used in the lower part. This apparatus was constructed in the connection with the earlier experiments [48] which did allow for accurate determination of the frequency ω. This measurement was accomplished only in recent years [49] with the use of thermistor probes.

Fig. 12. Laboratory realisation of convection driven by centrifugal buoyancy in the rotating cylindrical annulus. The critical Rayleigh number for onset of convection columns in the upper annulus with parallel top and bottom boundaries is lower than in the lower annular region with conical top and bottom boundaries. Accordingly the amplitude of convection and its wavelength are smaller in the lower region. For details see [48].

In figure 13a results are shown for the frequency of thermal Rossby waves as function of Rayleigh number R. In these experiments the temperature has been kept constant such that R increases in proportion to Ω^2. For comparison some numerical computations are shown in table 1. These computations have been carried out in extension of those described in [50]. Rigid cylindrical boundaries are assumed, but the Ekman layer friction at the conical

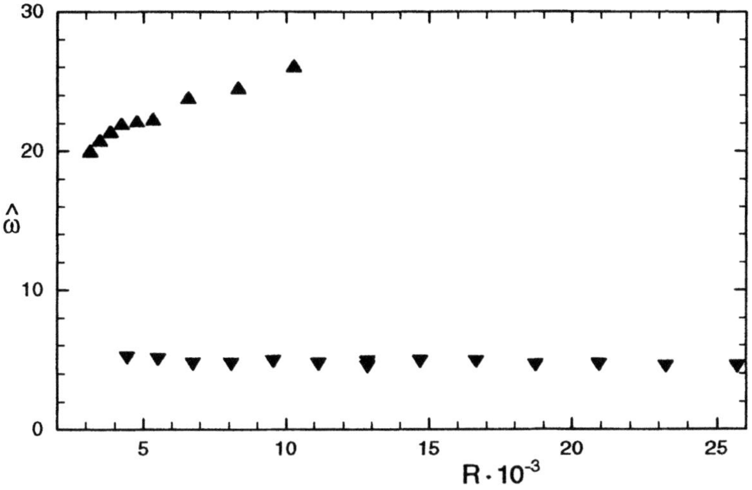

Fig. 13. a) Frequencies $\hat{\omega}$ (as fraction of the rotation rate) of thermal Rossby waves in the cylindrical annulus experiment as function of the Rayleigh number for air (▲) and for water (▼). Experimental parameter in the case of air (water) are $r_0 = 7.5cm, D = 1.0cm(0.5cm), \chi = 45°, L = 6.7cm(4.6cm), T_2 - T_1 = 0.7°C(5°C)$. The value of η^* at the data point corresponding to the lowest Rayleigh number is 109.2 (230.4). Since $T_2 - T_1$ is kept constant during the experiment, η^* increases in proportion to $\sqrt{R}$.

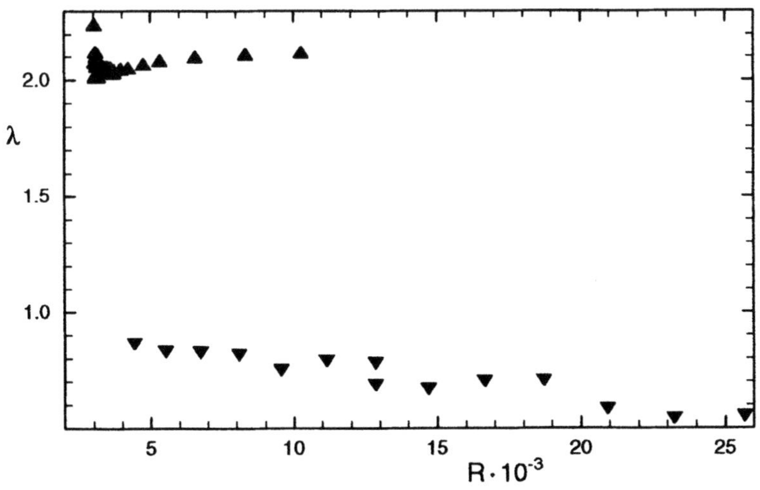

Fig. 13. b) The wavelength λ in cm for the same data as shown in figure 13a.

Table 1. Properties of thermal Rossby waves in the rotating cylindrical annulus for the Prandtl numbers $P = 0.7$ (air) and $P = 7.0$ (water) as function of the Rayleigh number R, the rotation parameter η^*. The wavenumber α has been chosen close to its critical value. $\hat{\omega}$ denotes the ratio ω/Ω. The Nusselt number Nu describes the ratio between the heat transports with and without convection.

P	η^*	α	$R \cdot 10^{-3}$	Nu	ω	$\hat{\omega} \cdot 10^3$
0.7	50	3.12	1.8	1.06	3.03	36.2
0.7	100	3.12	7.2	2.43	5.78	34.5
0.7	141.4	3.12	14.4	3.02	8.37	35.4
0.7	100	3.2	1.8	1.009	6.17	3.68
0.7	141.4	3.2	3.6	1.77	8.47	35.9
0.7	200	3.2	7.2	2.34	12.09	3.61
7.0	50	3.12	1.8	1.04	0.490	4.26
7.0	100	3.12	7.2	2.36	0.955	4.19
7.0	141.4	3.12	14.4	2.98	1.326	4.08

end boundaries has been neglected. The results clearly demonstrate that the ratio $\hat{\omega}$ between frequency and rotation rate is nearly independent of the Rayleigh number and of η^* for the relatively small values of η^* realized in the experiment. Theoretically predicted and experimentally measured values for $\hat{\omega}$ agree very well in the case of water. In the case of air the neglected Ekman layers exert a stronger influence. But it is not clear whether this effect can account for the discrepancy. At least the tendency to approach the theoretical value for larger values of η^* for which the influence of the Ekman layer declines can be noticed. The measured wavelengths shown in figure 13b are close to the theoretically expected value of slightly less than twice the gap width.

3.3 The Influence of Azimuthal Modulations

When the axisymmetry of the cylindrical annulus is broken through sinusoidal variations in the height of the annulus, the dynamics of the propagating thermal Rossby waves is changed in similar ways as in the case of the nonlinear interaction of two different modes. For this reason this problem is considered in the present chapter even though the configuration is not axisymmetric. The problem of convection in the cylindrical rotating annulus with modulated height as shown in figure 14 is of particular interest since the basic static state is not affected by the modulations and the onset of convection still occurs in the form of a perfect bifurcation, at least as long as the effects of laboratory gravity can be neglected. For the theoretical treatment of the problem the term $\eta^* \partial\psi/\partial y$ in equation (6a) must be replaced by

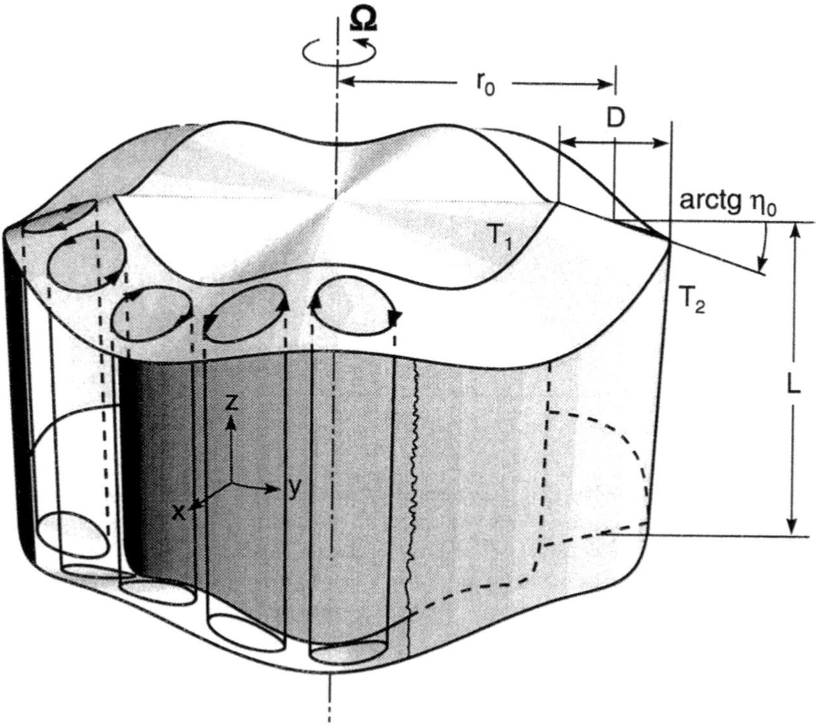

Fig. 14. Geometrical configuration of the modulated rotating cylindrical annulus.

$$\eta^*(1 + \delta \cos \alpha_M y)\frac{\partial \psi}{\partial y} - \varepsilon \sin \alpha_M y \frac{\partial}{\partial x}\psi \tag{8}$$

The corresponding mathematical problem has been solved numerically in the special case $\delta = 0$ in [51]. The linear analysis of the problem indicates that the onset of convection occurs in the form of modulated Rossby waves for values of ε small compared to $\alpha_M \eta$. In this case only a small fraction of the constant height contours in the annular gap are cut by the cylindrical boundaries and the dynamics of convection is not changed much from the unmodulated case. When ε becomes of the same order as $\alpha_M \eta$ the appearance of a steady "locked" mode of convection leads to a substantial decrease of the critical value R_c of the Rayleigh number R as shown in figure 15. This destabilisation is caused by the property that the constraining influence of the change in height in the radial direction is counteracted by the change in height in the y-direction. The maximum reduction of R_c is achieved for $\eta \alpha_M \approx \frac{1}{2}\varepsilon$, i.e. when azimuthally closed contours of constant height cease to exist in the annular gap between the cylinders.

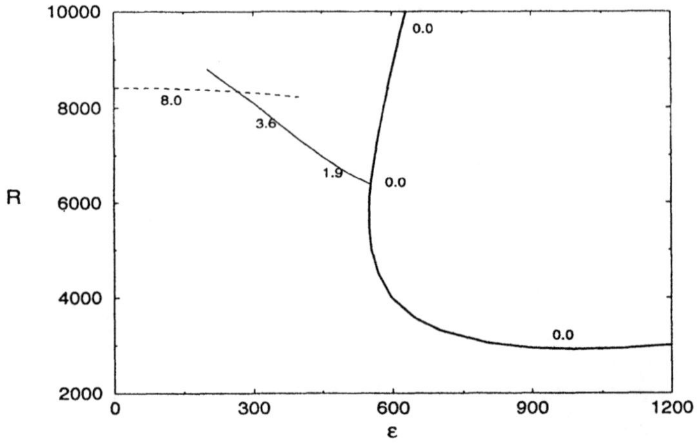

Fig. 15. Rayleigh number R for the onset of convection as a function of the amplitude ε of modulation for $\eta^* = 600, P = 1, \alpha_M = 3.0$. The onset in form of the modified thermal Rossby wave is indicated by the dashed line. The onsets in the form of standing oscillations and in the form of steady convection are indicated by the thin and thick solid lines respectively. The numbers along the curves indicate the imaginary part of the growthrate.

At finite amplitudes subcritical convection may occur and complex spatio-temporal structures are found in numerical simulations [52]. An example is shown in figure 16. The experimental realisation of convection in the modulated rotating cylindrical annulus also exhibits at least two frequencies as indicated in the thermistor record shown in figure 17. Because thermistor records as that shown in the figure are available only in a few places the complex spatio-temporal of convection as shown in figure 16 can not yet be captured by the experimental measurements. Further experimental studies will be required to permit a detailed comparison with the theoretical predictions.

4 Time Dependent Convection in Two Superimposed Fluid Layers

The numerous possibilities for dynamical interactions between convection rolls in superimposed layers of two immiscible liquids have long attracted the attention of theoreticians. Surveys of the literature on the subject can be found in the monographs by Joseph and Renardy [53] and by Simanovskii and Nepomnyashchy [54]. Experimental realisations have been hampered by a lack of truly immiscible liquids and by the limited range of material properties for which nearly immiscible liquids can be found. The onset of convection in superimposed layers heated from below occurs usually in the form of steady

Fig. 16. A time sequence of streamline plots over the interval $0 \leq y \leq 2\pi/\alpha_M$ (abszissa), $-0.5 \leq x \leq 0.5$ (ordinate) for the times $t = n\tau/5$ for $n = 0, \ldots, 5$ with the period $\tau = 0.0983$ in the case $\eta^* = 600, \varepsilon = 400, \alpha_M = 1.0, R = 16000$.

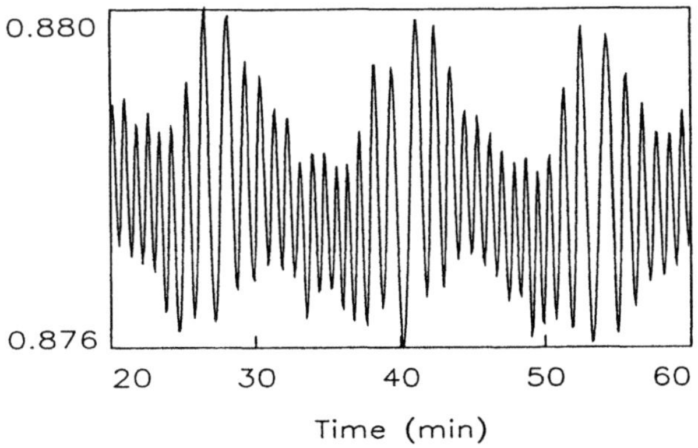

Fig. 17. Thermistor signal indicating temperature variations as a function of time in the case of convection driven by centrifugal buoyancy in the case of the modulated rotating cylindrical annulus as shown in figure 14.

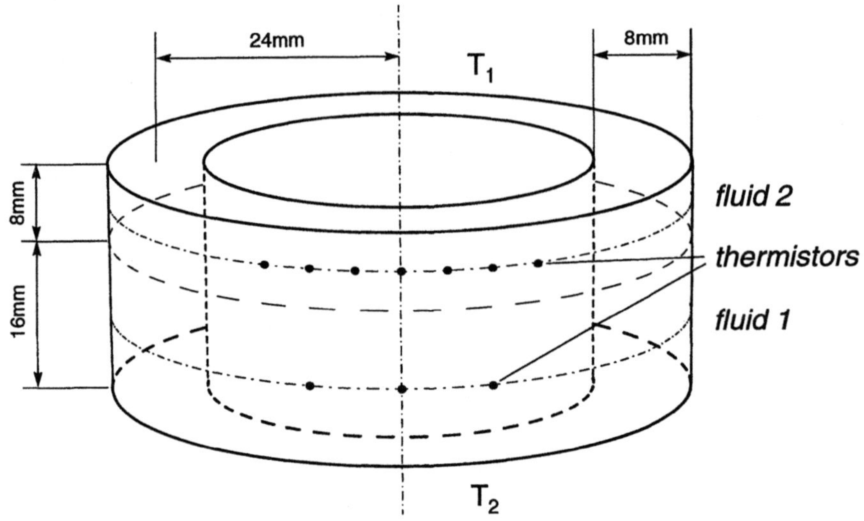

Fig. 18. Sketch of the experimental apparatus in the form of an annular channel.

motions. In the case of primary interest for which the Rayleigh numbers for the separate layers are of the same order of magnitude and the layer heights are not too different, two kinds of linear couplings can be distinguished. Either the convective motions in the two layers are viscously coupled with rolls of opposite vorticity on top of each other, or they are thermally coupled in that the deviation θ of the temperature from the static distribution assumes the same sign in both layers such that the rolls on top of each other rotate in the same sense. Even without taking into account the distortion of the interface and neglecting effects of surface tension, one can find an oscillating onset of convection [55] in addition to the steady onsets. The oscillatory mode corresponds to a vacillation between thermal and viscous coupling [56].

Besides the time dependence of convection in superimposed layers predicted by linear theory, there are numerous possibilities for time dependence caused by nonlinear interactions. An important case is the 1:2 resonance when the heights of the two layers approach a 1:2 ratio such that convection modes with wavelength differing by the factor two are realized in the two layers. This situation has been analyzed by Proctor and Jones [57] who assumed the special case when a thin plate separates the two fluid layers and permits a nonlinear interaction between the two convecting systems solely through the temperature field. Nevertheless Proctor and Jones found a rich variety of time dependent processes involving convection in the form of traveling waves, modulated traveling waves, homoclinic orbits and others.

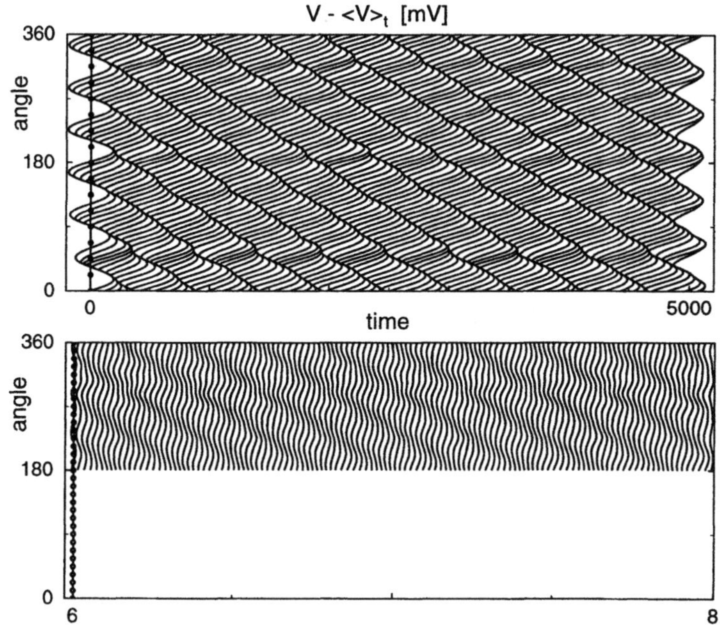

Fig. 19. Time sequence of temperature around the annular channel. There are two records per minute each of which is shifted by 2.083 mV towards the right from the preceding one. The thermistor resistance minus its time average is shown for the upper and the lower convection layers. For the upper layer only records for half of the circumference are shown.

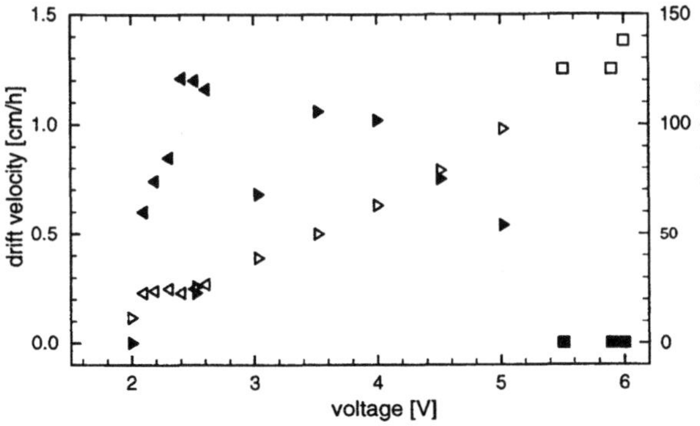

Fig. 20. The phase speed (filled symbols) and rms values (open symbols in 100mV) of traveling wave convection as a function of the heater voltage. ∇ (Δ) symbols indicate measurements in a sequence when the voltage was increased (decreased). The square symbols indicate steady convection.

Since theory of nonlinear convection in superimposed layers has been restricted almost entirely to the two-dimensional case, it appears that convection in a relatively narrow horizontal channel offers the best hope for an approximate realisation of the theoretical models. In order to avoid end effects which would inhibit traveling wave modes, the experimental configuration sketched in figure 18 has been constructed for the purpose of observing time dependent convection [58], [59]. Through thermistors imbedded in the outer cylindrical wall made of Plexiglas, the time dependence of the temperature field in the upper and lower fluid can be monitored. 16 thermistors arranged around the circumference are sufficient for the registration of convection in the lower layer, while 32 thermistors are needed for the shallower upper layer with its smaller convection cells. A typical thermistor record of traveling wave convection is shown in figure 19. This particular experiment was done with a layer of 10 centistokes silicone oil placed on top of a layer of fluorinert F70. The phase speed of the traveling wave as a function of the rate of heating is shown in figure 20 together with a measure of the strength of convection.

5 Thermocapillary Flow Structures

5.1 Introduction

The surface tension σ of a fluid normally decreases approximately linearly with temperature; $\sigma = \sigma_0 - \gamma(T - T_0)$. A temperature gradient along the free surface S induces a surface tension gradient on S giving rise to a shear stress on S that results in fluid motion for any temperature difference ΔT along S. It should be noted that this situation is different from one with ΔT perpendicular to S giving rise to hexagonal Bénard cells when a certain critical ΔT^c is exceeded [60, 61].

For normal liquids with $\gamma > 0$ the flow along the free surface is from hot to cold sites and induces motion in the bulk of the fluid. In bounded fluid domains a return flow exists for reasons of continuity.

The study of axisymmetric thermocapillary systems has focussed on two systems: floating zones (circular liquid bridges) and thin liquid layers, e. g. annular liquid gaps (fig. 21). These two geometries are related to the crystal growth techniques of "floating zone" and of "Czochralski", in which thermocapillarity plays an important role.

To investigate pure thermocapillary systems and to reduce the contribution of buoyancy forces the length l of the zone and the layer depth d must be chosen sufficiently small, i.e. of the order of mm. Figure 21 indicates some of the difficulties and defects of such experiments:

1. The shape of the liquid zone is hydrostatically distorted.
2. The length of stable zones is limited to some mm by gravity.
3. Under zero gravity rectcylindrical zones are possible but the length $l \leq 2\pi r$ (Rayleigh-limit) is limited as well.

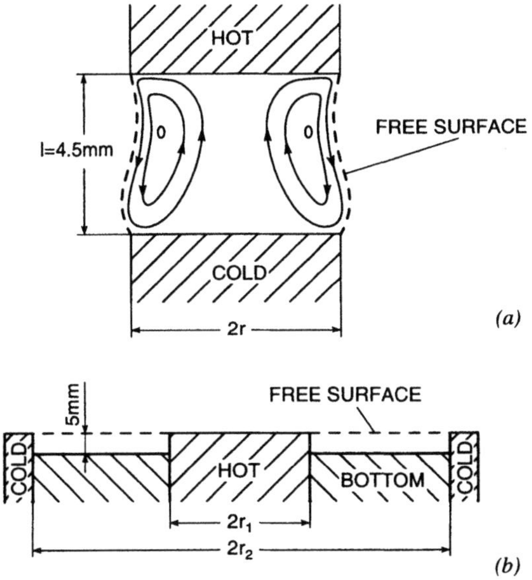

Fig. 21. (a): Sketch of a floating zone (cylindrical liquid bridge) heated from above with streamlines of thermocapillary basic flow and (b): A thin annular layer heated from the inner side (free liquid surface indicated by a broken line).

4. The meniscus-problem of liquid layers can be overcome; entirely flat free surfaces are possible, but only for vanishing temperature gradients.
5. It is virtually impossible to realize adiabatic boundary conditions at the free surface.
6. The time-dependent flow states of thermocapillary flow arise in fluids with Prandtl number $P \leq 100$ only and for large temperature gradients ($|\operatorname{grad}_{\|S} T| \geq 10K/cm$). All suitable room-temperature liquids (ethanol, silicone oils with viscosity 10 cSt) show considerable evaporation at the hot side, limiting $\operatorname{grad}_{\|S} T$.
7. Liquid metals ($P \approx 0.05$) are very difficult to handle; oxide skins can easily suppress the thermocapillary shear stress or alter γ.
8. The extension of the experiments towards higher temperature gradients is technically limited by evaporation/boiling-problems mentioned under 6. We have to deal with "end effects" (turn around regions and thermal boundary layers at the hot and cold walls).
9. On the other hand, analytical theories of dynamic thermocapillary flow structures are only available for unbounded free surfaces. Only numerical work exists for systems with sidewalls as those used in experiments.
10. The temperature distribution in the layer and dynamical effects distort the free surface.

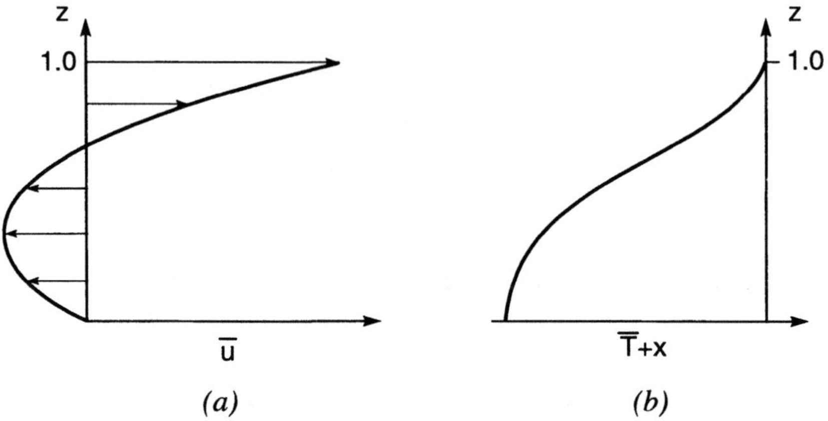

Fig. 22. The return flow basic state. (a) The velocity profile and (b) the flow-induced vertical temperature profile (after Smith [68]).

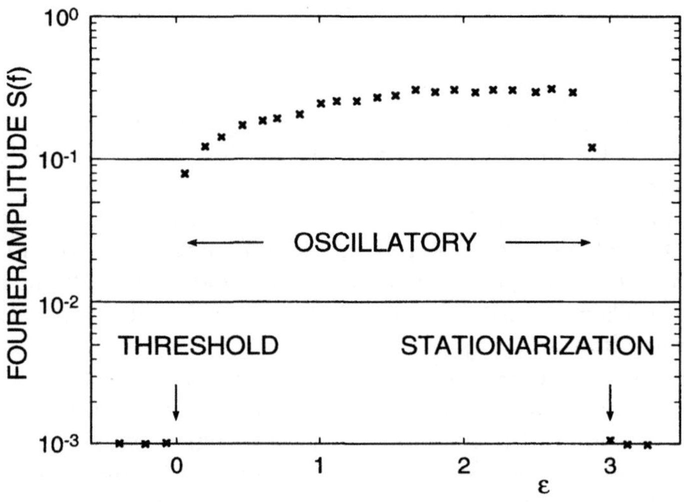

Fig. 23. Restationarisation of time-dependent thermocapillary flow in a $C_{24}H_{50}$-zone with $\phi = 6.00mm, l = 3.89mm$ heated from above. The Fourieramplitude $S(f)$ of the main frequency $f \approx 0.7$ Hz over the normalized distance ϵ from the treshhold for the onset of oscillations is shown; relaminarization at $\varepsilon = 3.0$.

5.2 Theoretical and numerical work

After some first experimental observations of time-dependent thermocapillary flow in floating zones [62, 63, 64, 65, 66] a linear theory for unbounded thermocapillary layers by Smith and Davis [67, 68, 69] gave insight into the physical mechanism and the spatio-temporal structure of oscillatory thermocapillary flow. Besides the flow with constant shear the authors consider the stability of surface flow with return flow in a thin layer subject to a linear temperature gradient $||S$ away from the endwalls. This flow induces a vertical temperature field which, on a static basis, is stabilizing (fig. 22), but not on a dynamic one. They find so called "hydrothermal waves" which are convective instabilities that do not depend strongly on the deflection of S and "surface waves", that depend intrinsically on the interaction of the flow and surface deflection. We must mention that the basic flow state (fig. 22) analyzed by Smith and Davis [67] is not found to precede the thermocapillary wave-instabilities in experiment when the Marangoni number Ma is increased from zero (Ma is a dimensionless measure of grad T). The basic flow state shows first a transition to a stationary multiroll-state (m-state) with transverse convection rolls, all with the same sense of rotation, embedded into the main basic flow [70, 71]. Smith and Davis calculate the critical value Ma^c of the Marangoni number for the hydrothermal wave instability;

$$Ma^c = \gamma \frac{|\,\mathrm{grad}_{||} T\,|\, d^2}{\eta \chi} \tag{9}$$

where d is the height of the layer, η is the dynamic viscosity, and χ is the thermal diffusivity. Ma^c is approximately 10^2 for $P = 1$. Ma^c decreases for $P \to 0$ and increases for $P \to \infty$. The hydrothermal waves are transversely travelling longitudinal rolls for $P \leq 10^{-1}$, obliquely travelling rolls for $P \geq 10^{-1}$ with the axis turning more and more perpendicular to $\mathrm{grad}_{||S} T$ for $P \to \infty$. A simplified description of the mechanism in the wording of Smith [68] proceeds as follows for small P: consider a disturbance in the form of a hot spot. It induces beneath it an upward flow as a result of thermocapillarity and continuity just as in the classical Marangoni mechanism. This upflow brings fluid with a lower basic-state velocity toward the surface, creating an inertial force in the upstream direction that drives an upstream velocity perturbation. This upstream flow cools the hot spot because it brings cooler fluid from downstream toward the spot. (The imposed surface temperature decreases downstream.) The cooling reduces the temperature of the spot and, consequently, the magnitude of the upflow. However, as long as there is an upflow, there will be an upstream inertial force and thus the upstream velocity perturbation increases. The cooling of this flow continues to decrease the temperature until the hot spot has become a cold spot. As a result of thermocapillarity the cold spot causes a surface flow towards it and a downflow beneath it because of continuity. The downflow produces an inertial force in the downstream direction slowing down the upstream flow until it is eliminated. Now, in exactly the reverse process, the temperature of the cold spot increases, overshoots the zero point, and a new hot spot is formed. If the imposed interfacial temperature gradient is high enough, the energy input to the spot associated with streamwise convection overcomes the energy loss as a result of conduction and the temperature of his new hot spot will be greater than the previous one. Thus, exponentially growing oscillations must be expected.

The instability mechanism is different for large P. The hydrothermal waves gain their energy from the inertially driven streamwise flow for small P and from the vertical basic-state temperature field for large P.

Surface-wave instabilities [67] couple the interfacial deflection to the underlying shear flow. They are directly related to the two-dimensional waves on an isothermal layer subject to wind shear as described by Miles [72] and Smith and Davies [73]. Hence the surface-tension gradients are important only in that they drive the basic shear flow. The critical Marangoni-numbers for the occurrence of surface waves and hydrothermal waves are of comparable magnitude. For $P < 0.15$, surface waves are preferred.

Smith [74] has studied the non-linear behaviour of hydrothermal waves. He finds that among the two equivalent obliquely travelling waves one is selected for a nearly adiabatic free surface (small Biot-number Bi) and a combination of the two waves for small P and Bi near one. Both wave-forms exhibit the Eckhaus and Benjamin-Feir sideband instability and a corresponding phase instability.

In the case of combined action of thermocapillarity and buoyancy, a transition from oscillatory modes to stationary longitudinal rolls occurs when buoyancy is increased as, for example, through an increase of d [75, 76]. From the large number of numerical papers on thermocapillary or buoyant thermocapillary instabilities in bounded configurations we cite a selection because they are important to interpret experimental results [77, 78, 79].

For unbounded cylindrical columns Xu and Davis [80] have applied the same linear stability analysis as Smith and Davis [67] and found azimuthally travelling hydrothermal waves as most dangerous mode for $P \leq 100$, and an axissymmetric oscillatory instability near the threshold for liquids with $P \geq 100$. For bounded liquid columns Neitzel et al. [81] found critical Marangoni numbers and azimuthal wavenumbers m in good agreement with experiments [66]. The critical Marangoni number increases significantly with surface cooling. For small P a transition from axisymmetric thermocapillary flow to a non-axisymmetric steady state (e. g. with $m = 2$) has been found [82, 83]. Only at larger Ma the oscillatory state appears. For larger P nearly azimuthally travelling hydrothermal waves appear driven by the same instability mechanisms as the hydrothermal waves in the presence of return flows in plane thermocapillary layers.

5.3 Experimental observations of thermocapillary flow structures in cylindrical liquid bridges

We report only on liquid bridges held between solid supports kept at different temperatures; $T_{upper} - T_{lower} = \Delta T$ as sketched in fig. 21 and not on the side-heated configuration used in floating-zone crystal growth described e. g. in [84]. Early work [63, 64, 65, 66, 85, 86] on transparent fluids with $8 \leq P \leq 64$ gave a critical Marangoni number $Ma^c \approx 1 \cdot 10^4$ for the onset of oscillations (increasing with P). Wavelike disturbances (hydrothermal waves) with wavelength λ organize themselves so that an integer number m of waves fits onto the circumference: $m \cdot \lambda = 2\pi r$. $m \cdot A = \text{const} = 2.2$ has been found for the aspect ratio $A = l \cdot r^{-1}$ for all zones [66]. λ scales with $l(\lambda = 2\pi/2.2 \cdot l)$. The oscillation frequencies are of the order of $f \cong 1Hz$. For $P = 8$ ($NaNO_3$-melt) the phase-speed of the waves is of the order of $0.18cm \cdot s^{-1}$ which is about one tenth of the basic surface flow speed [66]. The measured critical values of Ma are more than one order of magnitude larger than theory predicts for infinitely long zones [80]. End effects and the selection of smaller wavelengths than the one corresponding to the minimum of the critical curve may be the reasons, but their measurement compares favourably with theory on short zones [81].

In a series of microgravity experiments and for comparison under normal gravity this time dependent flow state has been investigated by Schwabe et al. [65, 87, 88] in small liquid columns with $P = 8$. The critical value Ma^c is smaller under microgravity than under the stabilizing action of gravity (heating from above). One possible reason is that the surface cooling (Biot number) is larger under gravity owing to convection in air than under microgravity. Another reason may be a change of the basic flow velocity field by gravity. The oscillation frequencies and temperature amplitudes are smaller under microgravity. Complex time behaviour such as the occurrence of beat frequencies and frequency jumps have been observed [87] when Ma was increased. A transition to chaos also occurs under microgravity as is indicated by figs. 9 and 11 in [65]. Transition to chaos has later been shown by Velten et al. [89, 90] to proceed either on the period doubling route or preferentially via the Ruelle-Takens scenario.

Systematic experiments with liquids of $P = 1$, 7 and 49 for various aspect ratios A and heating from above or below have revealed different periodic and nonperiodic flow structures [91]. Azimuthally travelling waves with $m \geq 1$ and a wave travelling along the axis ($m = 0$) have been found, the latter only near the threshold in the case $P = 49$. Transitions to nonperiodic flow and chaos occur closer to the threshold as A is increased. Flow states in zones with a value of A that fits a particular mode, say $m = 2$, are much more stable against a transition to chaos than zones with intermediate values of A. Interestingly the zone with $P = 49$ became aperiodic near $1.5 \cdot Ma^c$ (except for $A = 0.9$ with $m = 2$) under heating from above for the whole range

$0.5 \leq A \leq 1.3$ but showed only periodic states up to $7 \cdot Ma^c$ when heated from below.

Frank and Schwabe [92] report on experiments to improve the characterization of the temporal structures of time-dependent thermocapillary flow in floating zones and to correlate them with spatial observations. They give an example of quasiperiodic states (QP) with two frequencies (QP_2), locked states with ratio 4 : 1 and 13 : 4, a period doubling state and the chaotic state. Further they observe splitting of the essential frequency peak, appearance of exclusively odd harmonics and frequency skips (side band instability). Their observations of spatial structures are of special interest. They find e.g. a asymmetric pulsating state, consisting of the superposition of rotating $m = 1$ and $m = 2$ mode. Connected with these modes they observe accumulation structures of tracer particles showing the symmetry of the observed mode, eventually becoming asymmetric. Such a sudden occurrence of asymmetry might explain e. g. period doubling. QP behaviour can be understood by the combination of e. g. pulsating and rotating modes.

The inclination of the wave front of hydrothermal waves in the floating zone is under debate [92, 93, 94] as well as the question whether the transition to the oscillatory state is super-or subcritical [95, 87]. Now we mention a very strange phenomenon observed a few times [89, 95], called relaminarization (Fig. 23). This is the transition from a quasiperiodic flow state or the chaotic state to a time independent state when Ma is increased only insignificantly. This transition was observed for $P = 49$ only and is accompanied by a change of the basic flow structure, - the return flow shifts out of the bulk towards the free surface flow.

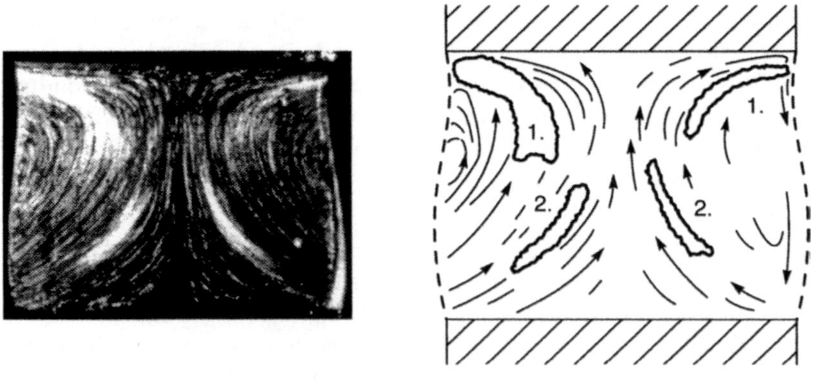

Fig. 24. Particle accummulation structure (PAS) in a $NaNO_3$-zone with 6.0 mm ø, l = 4.5 mm, oscillatory flow, heated from the top with $\Delta T = 32K$, $3\mu m\phi Al_2O_3$-tracers. The inner particle bunches 1 circulate with twice the frequency as particle bunches 2, $f_1 = 1.10$ Hz.

Finally, Schwabe et al. [96] observed in $P = 7$ zones with $A \approx 0.3(R = 10mm)$ an azimuthal net-flow (rectification) of e.g. $3^\circ \cdot s^{-1}$ at $Ma = 2Ma^c$ due to the nonlinear character of the hydrothermal wave and a direct visualization of the structure of the hydrothermal waves (λ and c_p) by particle accumulation structures (PAS) (Fig. 24). PAS is probably a resonant Bernoulli- Effect of the tracers with higher density than the fluid in the very noncircular, locally accelerated thermocapillary vortex.

5.4 Thermocapillary flow structures in rotationally symmetric liquid layers - annular layers

Layers have the advantage over floating zones to show no hydrostatic deformation and to allow easy observation, e. g. of travelling hydrothermal waves by observation of distortions of the flat surface or, by infrared images, interferometry or shadowgraphic techniques. On the other hand, the underlying flow structures are more easily visualized in floating zones by a vertical lightsheet. Annular layers have no sidewall as compared to rectangular layers, but the latter allow an easy observation of the flow structures through transparent plane sidewalls. Before discussing our main topic, we mention that important insights on the structure of hydrothermal waves have been gained also with rectangular geometries by J.M. Vince and Dubois, A.B. Ezersky et al., Daviaud and Vince, De Saedeler et al., Riley and Neitzel, Garcimatin et al., and Mukolobwiez during 1992 - 1998.

We now discuss the findings of thermocapillary flow instabilities in annular thin liquid layers. Schwabe et al. [71] investigated a $P = 17$ fluid in a $20mm$ wide annular gap with an isothermal cold bottom. They found a transition from monocellular to the stationary m-roll state to occur at $Ma = 6 \cdot 10^2$ and from this the transition to wavelike time dependent states near $Ma \approx 3 \cdot 10^3$. For $d < 1.4mm$ they found "short wavelength waves" with a typical phase speed v of $6mm \cdot s^{-1}$ with the features of Smith's [67] hydrothermal waves and for $1.4mm \leq d \leq 3mm$ "long wavelength waves" with $v \approx 20mm \cdot s^{-1}$ with the features of surface waves [67]. They measure the stability diagram, wavelength, frequency, and phase speed for $0.5mm \leq d \leq 3mm$. This experiment has been repeated by Schneider [97] using a modified apparatus with adiabatic bottom, - a situation approaching the theoretical one. His results are similar to the ones by Schwabe et al. [71] with experimental values fitting better the theory. He reports further interesting findings: phase-instabilities (side-band instability) of hydrothermal waves, transition to hydrothermal waves as Hopf-bifurcation with a critical parameter $= 1.15 \pm 0.08$, travelling cells (two rings of pentagonal Bénard-like cells filling the annular gap and travelling azimuthally with e. g. $4mm \cdot s^{-1}$ for $d = 1.5mm$ at $Ma^c = 3 \cdot 10^4$), quasiperiodicity and intermittency for the surface wave and, solitonic behaviour of hydrothermal waves. Examples for wave-structures in the annular gap are shown in fig. 25.

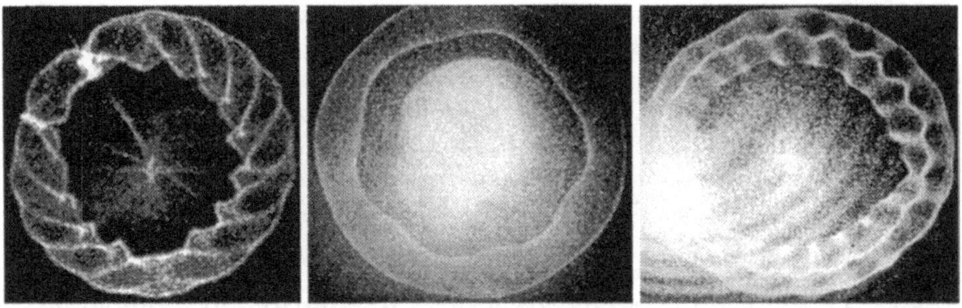

Fig. 25. Shadowgraphic pictures (reflection from the free surface) of azimuthally travelling flow structures in an annular gap as sketched in Fig. 21(b) filled with ethanol ($P = 17$) [97]. a) Hydrothermal waves with source at 11 o'clock and sink at 5 o'clock (number of wavetrains $m = 15, d = 1.2mm, \Delta T = 24K$). b) Surface wave with $m = 6(d = 3.1mm, \Delta T = 24K)$. c) Pentagonal travelling cells (phase speed = $2mms^{-1}, m = 22, d = 2.5mm, \Delta T = 30K$).

Some of the properties of surface waves in Schneider's experiment [97], such as the critical conditions for various d, intermittent behaviour of type I, occurrence of 2D azimuthally standing waves, and 3D travelling waves with source and sink are reported in [97]. The critical conditions found experimentally compare well with Smith's theory. Observations of surface waves, probably because configurations were too small to allow the realisation of "long" wavelengths, e. g. $\lambda^c = 25mm$ for $d = 1.9mm(\Delta T^c = 20K)$, have not been reported previously. Schneider has also measured the shape of the surface wave which becomes more and more non-harmonic for increasing distance ε from the threshold [97]. Ezersky et al. [99] use a circular heater submerged under the 5 cSt silicone oil layer. In the circular gap outside the heater they observe much the same phenomena as reported in [71, 97]. Kamotani et al. [100] observe azimuthally travelling waves in a rather deep ($d \approx 5mm$) cylindrical container (with $d \approx r$) heated centrally by a rod.

E. Favre et al. [101] conduct a related experiment but on a circular extended layer of $P = 10$ silicone oil in a cooled vessel (sapphire-bottom) centrally heated by a rod placed on the free surface. They have a meniscus at both the hot and the cold wall and problems with evaporative loss of fluid. For small $d < 2.8mm$ they find for a sufficiently large ΔT the stationary m-roll state as reported in [71]. For larger depth, $3.2 < d < 5mm$, a transition to 3D-flow becomes visible as "petals" near the hot rod in a shadowgraphic image, and for $d = 10mm$ a ray-like cellular pattern can be seen with the rays extending fully from the hot to the cold wall. The authors do not fully describe the underlying 3D-flows of these flow-states but it can be argued that they are of a buoyant-thermocapillary type like the one found numerically [77]. At a higher ΔT (second threshold) all these patterns become time dependent, either rotating, radially pulsating or oscillating azimuthally. A further increase leads to spatio-temporal chaos. These features resemble those found in [71, 97, 98]. In a similar experiment but with a less conductive glass bottom, Chiffaudel et al. [102] report on a supercritical Eckhaus instability of the travelling waves before they become turbulent in a 1.7mm thick centrally heated layer (10mm wide annulus).

5.5 Conclusions

Thermocapillary flows in liquid bridges and thin layers show a variety of structures and time-dependent states (travelling waves). Sequences of transitions leading to chaos and turbulence can be studied. The subject has not yet been fully studied, neither experimentally for small P nor theoretically for surface deformations and endwall effects. The coupling of the forces of buoyancy and thermocapillarity opens another rich field of interesting dynamical phenomena.

References

[1] G. I. Taylor, "Stability of a viscous liquid contained between two rotating cylinders", Phil.Trans Roy. Soc. London, Ser. A **223**, pp. 289-343 (1923)

[2] M. M. Couette, "Sur un nouvel appareil pour l'étude du frottement des fluids", Comptes Rendus 107, pp. 388-390 (1888)

[3] A. Mallock, "Determination of the viscosity of water", Proc. Roy. Soc. London, Ser. A **45**, pp. 126-132 (1888)

[4] R. Tagg, "The Taylor-Couette Problem", Nonlinear Science **4**, pp. 1-25 (1994)

[5] P. Chossat and G. Iooss, "The Taylor-Couette Problem", Springer Verlag (1994)

[6] L. Koschmieder, "Bénard Cells and Taylor Vortices", Cambridge University Press (1993)

[7] T. Mullin, "Finite-dimensional dynamics and chaos in fluid flows", in *Introduction to Plasma Physics*, Edt. R.O.Dendy , Cambridge University Press, p.129 (1993)

[8] R.C. DiPrima and H.L. Swinney, "Instabilities and transition in flow between concentric rotating cylinders", in *Hydrodynamic Instabilities and the Transition to Turbulence*, Eds. H.L.Swinney and J.P.Gollub, Springer-Verlag ,pp. 139-180 (1985)

[9] G.Pfister and J. v.Stamm, "Routes to Chaos in Rotational Tayler-Couette Flow", Proc. of the 3rd Experimental Chaos Conference, Edinburgh, UK, p.172 (1995)

[10] R. Tagg, "A guide to literature related to the Taylor-Couette problem", in *Ordered and Turbulent Patterns in Taylor-Couette Flow*, Edt. C.D. Andereck and F.Hayot, Plenum Press, p.303 (1992)

[11] D. Coles, "Transition to circular Couette flow", J. Fluid Mech. **21**, pp. 385-425 (1965)

[12] C.D.Andereck, S.S.Liu and H.L.Swinney, "Flow regimes in a circular Couette system with independently rotating cylinders", J.Fluid Mech. **164**, pp. 155-183 (1986)

[13] T. Mullin, A. Cliffe and G. Pfister, "Unusual Time-Dependent Phenomena in Taylor-Couette Flow at Moderately low Reynolds Numbers", Phys. Rev. Letters **58**, p. 2212 (1987)

[14] G. Pfister, H. Schmidt, A. Cliffe and T. Mullin, "Bifurcation Phenomena in Taylor-Couette Flow in a very short Annulus", J. Fluid Mech. **191**, p. 1 (1988)

[15] U. Gerdts, J. von Stamm, Th. Buzug and G. Pfister, "An axisymmetric time-dependent flow in Taylor-Couette system", Phys. Rev. E **49**, p.4019 (1994).

[16] J. Abshagen: Diploma thesis, University of Kiel (1996)

[17] A. Schulz: Diploma thesis, University of Kiel (1996)

[18] J. von Stamm, Th. Buzug and G. Pfister, "Instabilities and Chaos in Taylor-Couette flow", in *Chaos and Nonlinear Mechanics*, Proceedings of Euromech Colloquium 308 "Chaos and Noise in Dynamical Systems", World Scientific, p.182 (1994).

[19] D. Rand, "Predictions for modulated waves in rotating fluids", Arch. Rational Mech. Annu. **51**, p.1 (1973)

[20] J. von Stamm, Th. Buzug and G. Pfister, "Frequency locking in a rotation symmetric hydrodynamic flow", Phys. Lett. A **194**, p.173 (1994).

[21] A. Libchaber, J.Maurer, "Une experience de Rayleigh-Bénard de géometrie reduite; Multiplication, accrochage et démulitiplication de fréquences", J. Phys. (Paris) Coll. **41**, p. 3-51 (1980)

[22] J.W. Swift, K. Wiesenfeld, "Suppression of period-doubling in symmetric systems", Phys. Rev. Lett. **52**, p. 705 (1984)

[23] G. Pfister, "Period Doubling in Rotational Taylor-Couette Flow", in *Selected Papers from the Second International Symposium on Applications of Laser-Doppler Anemometry to Fluid Mechanics*, Edt.: D. F. G. Durao, Lisbon, p. 159 (1985)

[24] G. Pfister, "Deterministic Chaos in Rotational Taylor Couette flow", in *Flow of Real Fluids*, Lecture Notes in Physics,Vol.**235**, Springer Verlag, p. 199 (1985)

[25] G. Pfister, F. Schulz and G. Geister, "Period Doubling and Deterministic Chaos in Rotational Taylor-Couette Flow", in *Optical Measurements in Fluid Mechanics*, Edt.: P. Richards, Conference Series Vol. **77**, Adam Hilger Press, p. 57 (1985)

[26] J. von Stamm, U.Gerdts, Th. Buzug and G.Pfister, "Symmetry breaking and period doubling on a torus in the VLF regime in Taylor-Couette flow", Phys.Rev. E **54**, p. 4938 (1996)

[27] T. Mullin, T.J. Price, "An experimental observation of chaos arising from the interaction of steady and time dependent flows", Nature, Vol. **340**, No.6321, pp.294-296 (1989)

[28] T. Mullin, J.J. Kobine, "Organized chaos in fluid dynamics", in *Nonlinear Mathematics and its applications*, Ed. P.T.Aston, Cambridge UP, pp.147-177 (1996)

[29] J. Abshagen, G. Pfister, "Chaotic homoclinic dynamics in the VLF-regime of the Taylor-Couette flow", in preparation

[30] U. Gerdts, PhD thesis, University of Kiel, (1984)

[31] G. Pfister, Th. Buzug and N. Enge, "Characterization of Experimental Time Series from Taylor-Couette Flow", Physica D **58**, p. 441 (1992)

[32] J. Abshagen, A. Schulz and G. Pfister, "Taylor-Couette Flow - a paradigmatic system for instabilities, pattern formation and routes to chaos", in *Nonlinear*

Physics of Complex Systems - Current Status and Future Trends, Edt.: J.Parisi, S.C.Müller and W. Zimmermann (Springer, Berlin) ,p.63 (1996)

[33] Y.Pomeau, P.Manneville, "Stability and fluctuations of a spatially periodic convective flow", J.Phys.Lett. **40**, p. 609 (1979)

[34] H.R.Brand, "Phase dynamics - a review and a perspective", in *Propagation in systems far from equilibrium*, Ed. J.Wesfreid, H.R.Brand, P.Manneville, G.Albinet, N.Boccara, Springer, p.206 (1988)

[35] H.G.Paap, H. Riecke, "Drifting vortices in ramped Taylor vortex flow. Quantitative results from phase equation", Phys. Fluids A **36**, p.1519 (1991)

[36] D. Roth, M.Lücke, K. Kamps and R. Schmitz,"Phase dynamics of patterns: The effect of boundary-induced amplitude variations", Phys. Rev. E **50**, p. 2756 (1994)

[37] J.von Stamm, D.Roth, J.Abshagen, U.Gerdts, G.Pfister, "Phase dynamics in Taylor-Couette flow", in preparation

[38] C. Egbers, H.J. Rath "The existence of Taylor vortices and wide-gap instabilities in spherical Couette flow", Acta Mechanica, vol. **111**, pp. 125-140 (1995)

[39] A. Chennaoui, K. Pawelzik, W. Liebert, H. G. Schuster and G. Pfister, "Attractor reconstruction from filtered chaotic time series" Phys. Rev. A **41**, p. 4051 (1990)

[40] Th. Buzug, T. Reimers and G. Pfister, "Dimensions and Lyapunov Spectra from Measured Time Series of Taylor-Couette Flow", in *Nonlinear Evolution of Spatio-Temporal Structures in Dissipative Continuous Systems*, Eds. F. H. Busse and L. Kramer, Plenum Press, NY, p. 7 (1990)

[41] H. Kantz, T. Schreiber, I. Hoffmann, Th. Buzug, G. Pfister, L. Flepp, J. Simonet, R. Badii and E. Brun, "Nonlinear noise reduction: A case study on experimental data", Phys. Rev. E **48** (1994).

[42] Th. Buzug, K. Pawelzik, J. von Stamm and G. Pfister, "Mutual information and global strange attractors in Taylor-Couette flow", Physica D **72**, p.343 (1994).

[43] M. Auer, F.H. Busse and R.M. Clever, Three-dimensional convection driven by centrifugal buoyancy, J. Fluid Mech. **301**, 371-382 (1995).

[44] F.H. Busse and C.R. Carrigan, Laboratory simulation of thermal convection in rotating planets and stars, SCIENCE **191**, 81-83 (1976).

[45] F.H. Busse, Convection driven zonal flows and vortices in the major planets, CHAOS **4**, 123-134, (1994)

[46] H.P. Greenspan, *The theory of rotating fluids*, Cambridge University Press, (1968).

[47] F.H. Busse, Asymptotic theory of convection in a rotating, cylindrical annulus, J. Fluid Mech. **173**, 545-556 (1986).

[48] F.H. Busse and C.R. Carrigan, Convection induced by centrifugal buoyancy, J. Fluid Mech. **62**, 579-592 (1974).

[49] F.H. Busse, G. Hartung, M. Jaletzky and G. Sommermann, Thermal Convection Driven by Centrifugal Buoyancy in the Laboratory and its Applications to Planetary Problems, Dyn. Atmospheres Oceans, **27**, 161-174 (1997).

[50] M. Schnaubelt and F.H. Busse, Convection in a rotating cylindrical annulus, part 3: vacillating and spatially modulated flows, J. Fluid Mech. **245**, 155-173 (1992).

[51] Herrmann, J., and Busse, F.H., Stationary and time dependent convection in the rotating cylindrical annulus with modulated height, Phys. Fluids **10**, 1611-1620 (1998).

[52] Herrmann, J., Untersuchungen zur Konvektion im rotierenden Annulus bei kleinen Prandtlzahlen und bei modulierten Rändern, Dissertation, University of Bayreuth, 1996

[53] D.D. Joseph and Y. Renardy, *Fundamentals of Two-Fluid Dynamics: Mathematical Theory*, Interdisciplinary Applied Mathematics Series, Springer Verlag, New York (1992).

[54] I.B. Simanovskii and A.A. Nepomnyashchy, *Convective instabilities in systems with interface*, Gordon and Breach (1993).

[55] G.Z. Gershuni and E.M. Zhukovitskii, Monotonic and oscillatory instabilities of a two-layer system of immiscible liquids heated from below, Sov. Phys. Dokl. **27**, 531-533 (1983).

[56] S. Rasenat, F.H. Busse and I. Rehberg, A theoretical and experimental study of double-layer convection, J. Fluid Mech. **199**, 519-540 (1989).

[57] M.R.E. Proctor and C.A. Jones, The interaction of two spatially resonant patterns in thermal convection I. Exact 1:2 resonance, J. Fluid Mech. **188**, 301-335 (1988).

[58] F.H. Busse and G. Sommermann, Double-Layer Convection: A Brief Review and Some Recent Experimental Results, pp. 33-41 in *Advance in Multi-Fluid Flows*, ed. Y.Y. Renardy, A.V. Coward, D.T. Papageorgiou, S.-M. Sun, SIAM (1996).

[59] G. Sommermann and F.H. Busse, Time Dependent Convection in Superimposed Layers, Zeitschr. Angew. Math. Mech. (ZAMM) (1998).

[60] H. Bénard, Les tourbillions cellulaires dans une nappe liquide, Rev. Gén. Sci. Pures Appl. **11**, 1261-1268 (1900).

[61] J.R.A. Pearson, On convection cells induced by surface tension, J. Fluid Mech. **4**, 489-500 (1958).

[62] D. Schwabe, A. Scharmann, F. Preisser and R. Oeder, Experiments on surface tension driven flow in floating zone melting, J. Crystal Growth **43**, 305-312 (1978).

[63] Ch.-H. Chun and W. Wuest, Experiments on the transition from steady to oscillatory Marangoni-convection of a Floating Zone under reduced gravity effect, Acta Astronautica **6**, 1073-1082 (1979).

[64] D. Schwabe and A. Scharmann, Some evidence for the existence and magnitude of a critical Marangoni-number for the onset of oscillatory flow in crystal growth melts, J. Crystal Growth **46**, 125-131 (1979).

[65] D. Schwabe, F. Preisser and A. Scharmann, Verification of the oscillatory state of thermocapillary convection in a floating zone under low gravity, Acta Astronautica **9**, 265-273 (1982).

[66] F. Preisser, D. Schwabe and A. Scharmann, Steady and oscillatory thermocapillary convection in liquid columns with free cylindrical surface, J. Fluid Mech. 126, 545 567 (1983).

[67] M.K. Smith and S.H. Davis, Instabilities of dynamic thermocapillary liquid layers. Part 1. Convective instabilities, J. Fluid Mech. **132**, 119-144 (1983); Part 2: Surface-wave instabilities, ibid. 145-162.

[68] M.K. Smith, Instability mechanisms in dynamic thermocapillary liquid layers, Phys. Fluids **29**, 3182-3186 (1986).

[69] S.H. Davis, Thermocapillary instabilities, Ann. Rev. Fluid Mech. **19**, 403-435 (1987).

[70] D. Villers and J.K. Platten, Coupled buoyancy and Marangoni convection in acetone: experiments and comparison with numerical simulation, J. Fluid Mech. **234**, 487-510 (1992).

[71] D. Schwabe, U. Möller, J. Schneider and A. Scharmann, Instabilities of shallow dynamic thermocapillary liquid layers, Phys. Fluids A **4**, 2368-2381 (1992).

[72] J.W. Miles, J. Fluid Mech. **8**, 593 (1960).

[73] M.K. Smith and S.H. Davis, The instabilities of sheared liquid layers, J. Fluid Mech. **121**, 187-206 (1982).

[74] M.K. Smith, The nonlinear stability of dynamic thermocapillary liquid layers, J. Fluid Mech. **194**, 391-415 (1988).

[75] P.M. Parmentier, V.C. Regnier and G. Lebon, Buoyant-thermocapillary instabilities in medium-Prandtl-number fluid layers subject to a horizontal temperature gradient, Int. J. Heat Mass Transfer **36**, 2417-2427 (1993).

[76] J.F. Mercier and C. Normand, Buoyant-thermocapillary instabilities of differentially heated liquid layers, Phys. Fluids **8**, 1433-1445 (1996).

[77] M. Mundrane and A. Zebib, Two and three-dimensional buoyant thermocapillary convection, Phys. Fluids A **5**, 810-818 (1993).

[78] J. Priede and G. Gerbeth, Convective, absolute, and global instabilities of thermocapillary-buoyant convection in extended layers. Phys. Rev. E **56**, 1-13 (1997).

[79] R. Birikh, V. Briskman, V. Chernatinsky, J. Schneider, S. Benz and D. Schwabe, Axisymmetrical Marangoni convection in toroidal region, Proceed. of the joint 10th European and 6th Russian Symposium on "Physical Sciences in Microgravity", St. Petersburg, Russia, 1997 (Publ.: Inst. for Problems in Mechanics RAS, Moscow, Russia (1997)). Vol. 1, 91-98.

[80] J.-J. Xu and S.H. Davis, Convective thermocapillary instabilities in liquid bridges, Phys. Fluids **27**, 1102-1107 (1984).

[81] G.P. Neitzel, K.T. Chang, D.F. Jankowski and H.D. Mittelmann, Linear-stability theory of thermocapillary convection in a model of the float zone crystal-growth process, Phys. Fluids A **5**, 108-114 (1993).

[82] M. Levenstam and G. Amberg, Hydrodynamical instabilities of thermocapillary flow in a half-zone, J. Fluid Mech. **297**, 357-372 (1995).

[83] M. Wanschura, V.M. Shevtsova, H.C. Kuhlmann and H.J. Rath, Convective instability mechanisms in thermocapillary liquid bridges, Phys. Fluids **7**, 912-925 (1995).

[84] D. Schwabe, Surface-tension-driven flow in crystal growth melts, Crystals **11**, 75-112, Springer-Verlag, Berlin-Heidelberg (1988).

[85] Ch.-H. Chun, Experiments on steady and oscillatory temperature distribution in a floating zone due to the Marangoni convection, Acta Astronautica **7**, 479-488 (1980).

[86] Y. Kamotani, S. Ostrach and M. Vargas, Oscillatory thermocapillary convection in a simulated floating-zone configuration, J. Crystal Growth **66**, 83-90 (1984).

[87] D. Schwabe and A. Scharmann, Measurements of the critical Marangoni number of the laminar oscillatory transition of thermocapillary convection in floating zones, ESA SP **222**, 281-289 (1984).

[88] D. Schwabe and A. Scharmann, Messung der kritischen Marangoni-Zahl für den Übergang von stationärer zu oszillatorischer thermokapillarer Konvektion unter Mikrogravitation: Ergebnisse der Experimente in den ballistischen Raketen TEXUS 5 und TEXUS 8, Z. Flugwiss. Weltraumforsch. **9**, 21-28 (1985).

[89] R. Velten, Strukturen der in stationären thermokapillaren Konvektion in zylindrischen Schmelzzonen, PhD thesis, University of Giessen (1988).

[90] R. Velten, D. Schwabe and A. Scharmann, Gravity-dependence of the instability of surface-tension-driven flow in floating zones, ESA SP **295**, 271-278 (1990).

[91] R. Velten, D. Schwabe and A. Scharmann, The periodic instability of thermocapillary convection in cylindrical liquid bridges, Phys. Fluids A **3**, 267-279 (1991).

[92] S. Frank and D. Schwabe, Temporal and spatial elements of thermocapillary convection in floating zones, Experiments in Fluids **23**, 234-251 (1997).

[93] K.A. Muehlner, M.F. Schatz, V. Petrov, W.D. McCormick, J.B. Swift and H.L. Swinney, Observation of helical travelling-wave convection in a liquid bridge, Phys. Fluids **9**, 1850-1852 (1997).

[94] L. Carotenuto, D. Castagnolo, C. Albanese and R. Monti, Instability of thermocapillary convection in liquid bridges, Phys. Fluids **10**, 555-565 (1998).

[95] S. Frank, Zeitliche und räumliche Elemente thermokapillarer Konvektion in zylindrischen Schmelzzonen, PhD-thesis, University of Giessen (1995).

[96] D. Schwabe, P. Hintz and S. Frank, New features of thermocapillary convection in floating zones revealed by tracer particle accumulation structures (PAS), Microgravity sci. technol. IX/3, 163-183 (1996).

[97] J. Schneider, Strukturen thermokapillarer Konvektion in einem Ringspalt, PhD Thesis, University of Giessen (1995).

[98] J. Schneider, D. Schwabe and A. Scharmann, Experiments on surface waves in dynamic thermocapillary liquid layers, Microgravity sci. technol. IX/2, 86-94 (1996).

[99] A.B. Ezersky, A. Garcimartin, J. Burguete, H.L. Mancini and C. Pérez-Garcia, Hydrothermal waves in Marangoni convection in a cylindrical container, Phys. Rev. E **47**, 1126-1131 (1993).

[100] Y. Kamotani, J.H. Lee and S. Ostrach, An experimental study of oscillatory thermocapillary convection in cylindrical containers, Phys. Fluids A **4**, 955-962 (1992).

[101] E. Favre, L. Blumenfeld and F. Daviaud, Instabilities of a liquid layer locally heated on its surface, Phys. Fluids **9**, 1473-1475 (1997).

[102] A. Chiffaudel, M. Mukolobwiez and F. Daviaud, Supercritical Eckhaus instability for travelling waves on their way to turbulence, to appear in Phys. Rev. Lett. **80**, 4661-4664 (1998).

Pattern Formation in Binary Fluid Convection and in Systems with Throughflow

M. Lücke[1], W. Barten[2], P. Büchel[1], C. Fütterer[1], St. Hollinger[1], and Ch. Jung[1]

[1] Institut für Theoretische Physik,
Universität des Saarlandes, D-66041 Saarbrücken, Germany
[2] Paul Scherrer Institut,
Würenlingen und Villigen, CH-5232 Villigen PSI, Switzerland.

Abstract. Linear, weakly nonlinear, and strongly nonlinear pattern forming properties of binary fluid convection and of systems with throughflow are elucidated with a broad spectrum of numerical and analytical methods and reviewed in comparison with experimental and other theoretical work. Growth dynamics, spatiotemporal behavior, stability, scaling properties, and bifurcation behavior of spatially extended stationary, traveling wave, and standing wave convective roll structures are presented. Also pulselike spatially localized traveling wave convection is investigated. The Dufour effect in gas mixtures is reviewed. 3D square and crossroll patterns and oscillations between them which occur in liquid mixtures are analyzed. In the second part the influence of an externally imposed shear flow on structure formation is investigated. The unique selection of a self sustained forwards bifurcating pattern in the absolutely unstable parameter regime is shown to be a nonlinear eigenvalue problem. Noise sustained pattern growth in the convectively unstable regime is analyzed as a result of bulk thermal fluctuations and of perturbations that are swept by the throughflow into the system via the inlet. Finally, the influence of a lateral flow on linear and nonlinear properties of binary mixture convection is studied with special emphasis on the symmetry breaking of Hopf degenerate traveling waves and their subcritical bifurcation topology.

Contents

1 Introduction

Convection in binary miscible fluids like ethanol-water, $^3\mathrm{He}-^4\mathrm{He}$, or various gas mixtures is an example of systems that show rich and interesting pattern formation behavior when driven away from equilibrium. It is paradigmatic for problems related to instabilities, bifurcations, selforganisation, complex spatiotemporal behavior, and turbulence and it continues to attract research activities — the setup is sufficiently simple for controlled experiments and the governing field equations are well known, thus allowing quantitative comparisons of theory and experiment.

Compared to convection in one-component fluids the spatiotemporal properties are far more complex due to the influence of Soret sustained concentration gradients. The strength of the Soret coupling is measured by the dimensionless separation ratio ψ. The evolution of the concentration field is governed by the interplay of typically strong nonlinear convective transport and mixing, weak dissipative solutal diffusion, and the Soret effect. The latter is a source of concentration fluctuations. It generates concentration gradients in response to the applied temperature difference and to local temperature gradients. Without Soret coupling, $\psi = 0$, initial concentration deviations from the mean diffuse away. For $\psi \neq 0$, however, the externally imposed temperature stress sustains concentration variations against the mixing action of convection via the Soret effect. The concentration field changes the convective dynamics via solutal buoyancy forces entering into the momentum balance. In this way concentration gradients directly influence the flow which in turn changes and mixes the concentration. In binary *liquids*, this strongly nonlinear feed back is only weakly damped by small diffusive homogenization so that the concentration field distribution shows significant anharmonic and boundary layer structures.

It is this feed-back loop that causes convection in binary mixtures to exhibit such a rich variety of patterns arising from stationary and oscillatory [1] instabilities: Depending on the parameters the hydrodynamic balance equations show convective solutions that bifurcate out of the quiescent conductive basic state in the form of (*i*) straight, stationary, parallel rolls that are called for historical reasons states of stationary overturning convection (SOC), (*ii*) traveling waves (TWs) of roll structures, (*iii*) standing wave (SW) oscillations, and (*iv*) stationary squares. Besides these primary states, there are close to onset of convection (*v*) pulse like, spatially localized traveling wave (LTW) states consisting of only few TW rolls, (*vi*) oscillating square patterns, (*vii*) crossroll structures, and (*viii*) also (spiral) defect chaos. Note that patterns (*ii*) - (*vii*) are concentration field induced — they disappear in the pure fluid limit, $\psi \to 0$, upon switching off the Soret coupling to the concentration field [2]. Moreover, they are typically strongly nonlinear involving boundary layers and concentration plumes. Here we elucidate and discuss in Secs. 2–5 the convection structures (*i*) - (*vii*).

In Sec. 6 we review the response of supercritically bifurcating structures to an externally imposed throughflow parallel to the wavevector of the pattern. In this way one can investigate pattern formation and instabilities in well controlled open-flow systems with experimentally well defined and accessible setups like (Taylor) vortices in the annulus between concentric rotating cylinders or convection vortices in the form of straight rolls in fluid layers heated from below. Concepts, e.g., of absolute versus convective instability that were developed originally for other open-flow problems have been applied only recently to these simple and paradigmatic hydrodynamic instabilities and provide new and interesting insights:

In the absolutely unstable parameter regime at large driving and/or small throughflow the pattern can expand against the throughflow in upstream direction towards the inlet. The final structure is self-sustained and uniquely selected, i.e., independent of parameter history, initial conditions, and system length provided the latter is large enough to allow for a saturated bulk region. The selected frequency of the *globally* oscillating phase of the throughflow-induced propagating pattern can be understood as an eigenvalue of a nonlinear eigenvalue problem with the time periodic fields defining the eigenfunction. Different inlet boundary conditions entail different frequencies and structures.

In the convectively unstable regime at larger throughflow initial perturbations are blown out of the system. A downstream pattern growth can be sustained by persistently injecting those perturbations – via the inlet or in the bulk – that are amplified by the supercritical, deterministic, macroscopic pattern forming instability. The spatiotemporal behavior of the resulting "noise-sustained" structures reflect properties of their sources. Thus, one can investigate whether instability-driven amplification of thermal noise is relevant and observable in these systems, e. g., in comparison with other perturbations that might enter with the flow into the system via the inlet.

In Sec. 7 we investigate the influence of lateral flow on roll structures in binary fluid convection. Here one has to distinguish between the response of SOC and left and right propagating TWs whose symmetry degeneracy is broken – the throughflow changes the spatiotemporal structure and bifurcation behavior of these three states in significantly different ways. Furthermore, the subcritical nature of the transition between the homogeneous and the structured states requires to go beyond a linear concept of convective versus absolute instability in order to decide when the pattern in question can expand to fill the whole system.

To obtain our results we used various methods: Linear stability properties were determined analytically, with shooting methods, and with Galerkin techniques. The linear Ginzburg Landau equation (GLE) with additive stochastic forcing was solved in semiinfinite systems. Weakly nonlinear analyses where performed in quantitative manner within the nonlinear GLE approach where this approximation can be applied. Finite differences simulations of the full nonlinear 2D and 3D hydrodynamic field equations were done with Marker and Cell (MAC) type codes — with different time stepping procedures, with standard pressure relaxation methods and with multigrid techniques, in the laboratory system, in comoving frames of references, and with control processes that allowed evaluation of unstable states. In addition various multimode Galerkin (MMG) expansions including up to about 800 modes were used. Last, not least analytically manageable few mode Galerkin models were derived. The combination of all these methods has contributed in comparison with experiments and other work to a picture and understanding of pattern formation in binary fluid convection and in systems with throughflow that is becoming more and more coherent and also quantitative while simultaneously allowing for a better characterization of the multitude of open problems.

2 Extended 2D convection structures in fluid mixtures

Here we review convection structures for a stabilizing, negative Soret coupling [3–10]. For a recent overview of the experimental and theoretical work see [6, 11]. We discuss stable and unstable TW and SOC structures and also unstable SWs in the experimentally accessible parameter range for ethanol-water mixtures. Experiments in narrow, straight, rectangular containers or in annular channels [12–51] are best suited to compare with our results. They agree well with experimental bifurcation behavior [12–14] and measured top-view [15] and side-view shadowgraph intensity distributions [16]. The analysis of these extended states is useful and necessary to understand more complex structures: it is most likely that the strongly non-linear behavior of the former plays an important role, e.g., for the 3D convection structures investigated recently in cylindrical cells [52, 53].

After describing the system we discuss in Sec. 2.2 typical bifurcation diagrams. The structure, especially of the important concentration field of stable and unstable TW and SOC states is elucidated in Sec. 2.3 together with the changes along the respective bifurcation branches. For stronger Soret coupling in Sec. 2.4 we predict a second stable TW branch with smaller convection amplitude. Bifurcation behavior and concentration structure provide quantitative criteria (Sec. 2.5) to distinguish between weakly nonlinear unstable TW states close to onset and strongly nonlinear TW states in the dominant part of the bifurcation diagram. For wide parameter ranges the TW frequency and the structure of the concentration field are universally related to each other and to a properly defined "distance" from pure fluid convection as described in Sec. 2.6. The growth of convection out of the conductive state with large concentration gradients into a well mixed, strongly nonlinear final state is nonmonotonous and very complex. It generically involves SW transients as shown in Sec. 2.7. In Sec. 2.8 we present a few-mode model that is based on the detailed knowledge especially of the concentration field in TWs. The resulting TW and SOC states agree almost quantitatively with experimentally and numerically determined states. The model yields also unstable SW states.

2.1 The system

We consider a horizontal layer of a binary fluid mixture under a homogeneous gravitational field, $\mathbf{g} = -g\,\mathbf{e}_z$, that is directed downwards. A positive temperature difference ΔT between the lower and upper confining boundaries is imposed, e.g., via high-thermal-conductivity plates in experiments. Here we deal with convection in the form of straight, parallel rolls, as seen in many experiments. Ignoring variations along the roll axes, we investigate 2D convection in an $x-z$ plane perpendicular to the axes described by the balance equations for mass, heat, concentration, and momentum in Oberbeck-Boussinesq approximation [54–57]

$$0 = -\boldsymbol{\nabla}\cdot\mathbf{u}\,, \tag{2.1}$$

$$\partial_t\,\delta T = -\boldsymbol{\nabla}\cdot\mathbf{Q}\ ;\ \mathbf{Q} = \mathbf{u}\,\delta T - \boldsymbol{\nabla}\,\delta T\,, \tag{2.2}$$

$$\partial_t\,\delta C = -\boldsymbol{\nabla}\cdot\mathbf{J}\ ;\ \mathbf{J} = \mathbf{u}\,\delta C - L\boldsymbol{\nabla}\,(\delta C - \psi\,\delta T)\,, \tag{2.3}$$

$$\partial_t\,\mathbf{u} = -\boldsymbol{\nabla}\,(\mathbf{u}:\mathbf{u} + p - \sigma\,\boldsymbol{\nabla}:\mathbf{u}) + \mathbf{B}\ ;\ \mathbf{B} = \sigma\,R\,(\delta T + \delta C)\mathbf{e}_z\,. \tag{2.4}$$

Taking the divergence of (2.4), one gets a Poisson equation for the pressure p. It substitutes the continuity equation and builds together with (2.2–2.4) a complete set of equations for the fields $\mathbf{u}$, δT, δC, and p.

Lengths are scaled with the height d of the layer, time with the vertical thermal diffusion time d^2/κ, and the velocity field $\mathbf{u} = (u,\,0,\,w)$ by κ/d. Here, κ is the thermal diffusivity of the mixture. δT denotes the deviation of the temperature from its mean in the fluid, scaled by ΔT. The field δC is the deviation of the concentration from its mean scaled by $\Delta T\alpha/\beta$. The thermal expansion coefficient α and the solutal expansion coefficient β come from a linear equation of state of the mass density ρ for small deviations from the mean temperature and concentration. $\mathbf{Q}$ denotes the reduced heat current. The reduced concentration current $\mathbf{J}$ consists of the convective part $\mathbf{u}\,\delta C$ and the diffusive part $-L\,\boldsymbol{\nabla}\,(\delta C - \psi\,\delta T)$. The Lewis number $L = D/\kappa$ is the ratio of the concentration diffusivity D to the thermal diffusivity κ, and the Prandtl number $\sigma = \nu/\kappa$ is the ratio of the momentum diffusivity ν and κ. For room temperatures the Prandtl number of ethanol-water mixtures lies between 5 and 20 [17]. The Lewis numbers are around 0.01. For most of our calculations we use $L = 0.01$ and $\sigma = 10$ as typical parameters.

The Rayleigh number $R = \frac{\alpha g d^3}{\kappa\nu}\Delta T$ and the separation ratio $\psi = -\frac{\beta}{\alpha}\frac{k_T}{T}$ [11, 54, 55] are control parameters. R measures the externally imposed thermal stress. The Soret coupling ψ between temperature and concentration, into which enters the thermo diffusion ratio k_T of the mixture, reflects the influence of temperature gradients on the concentration field. For room temperature ethanol-water mixtures ψ-values between about -0.5 and $+0.2$ can be easily realized experimentally [17, 18]. The buoyancy force $\delta\rho\mathbf{g}$ is the driving mechanism for convective motion. It enters into the momentum balance (2.4) via the buoyancy term $\mathbf{B}$.

The horizontal boundaries of the layer, the plates, are at $z = 0, 1$. The lateral boundaries are at $x = 0,\ \Gamma$. The plates are rigid so that $u = w = 0$ at $z = 0,\ 1$. They are perfect heat conductors, so that the temperature of the fluid at $z = 0,\ 1$ is constant and laterally homogeneous, $\delta T = 1/2$ at $z = 0$, and $\delta T = -1/2$ at $z = 1$. Impermeability of the plates implies that there is no concentration current through the plates:

$$\mathbf{J}\cdot\mathbf{e}_z = \mathbf{0} \quad \text{or} \quad \partial_z\,\delta C = \psi\,\partial_z\,\delta T \qquad \text{at } z = 0,\ 1\,. \tag{2.5}$$

These NSI (no-slip impermeable) boundary conditions have to be contrasted with the idealized FSP (free-slip permeable) boundary conditions that are often used in theoretical approaches. However, since the latter allow vertical concentration transport through the plates, they misrepresent the delicate concentration balance [58] that is involved, e.g., in explaining structure, dynamics, and stability of LTW states. In lateral direction, all fields are periodic, with a given lateral periodicity length Γ.

The conductive state $\mathbf{u}_{cond} = \mathbf{0}\ ,\delta\,T_{cond} = \frac{1}{2} - z\ ,\delta C_{cond} = \psi\left(\frac{1}{2} - z\right)$, is globally stable for small thermal stress. It describes vertical heat diffusion through the layer without convective motion. The heat current is $\mathbf{Q}_{cond} = \mathbf{e}_z$. The constant temperature gradient induces a vertical concentration gradient given by $-\psi$ which enforces the concentration current to vanish: $\mathbf{J}_{cond} = \mathbf{0}$. The concentration stratification causes in the buoyancy $\mathbf{B}_{cond} = \sigma\,R\,(1+\psi)\left(\frac{1}{2} - z\right)\mathbf{e}_z$ a modification by the factor $(1+\psi)$ relative to the thermal part. So, depending on the sign of ψ, the Soret effect enhances or depresses the buoyancy.

To describe convection, we use the velocity field, in particular the vertical component w and also the deviations

$$\theta = \delta T - \delta T_{cond}\,, \; c = \delta C - \delta C_{cond}\,, \; (\mathbf{B} - \mathbf{B}_{cond}) \cdot \mathbf{e}_z = b = \theta + c \tag{2.6}$$

from the conductive state. To characterize the convective solutions we use different order parameters. (i) The maximal vertical velocity w_{max} directly measures the convective amplitude. (ii) The Nusselt number N is the total vertical heat current through the fluid layer reduced by the conductive part. (iii) Since w_{max} and the reduced vertical heat current carried by convection alone, $N-1$, do not characterize the concentration field appropriately, we introduce a quantity that characterizes the magnitude of concentration variations. The mixing parameter

$$M = \sqrt{\overline{\langle \delta C^2 \rangle} \Big/ \overline{\delta C^2_{cond}}} \tag{2.7}$$

is the variance of the concentration field reduced by its value in the conductive state, thus $M_{cond} = 1$. Here brackets and overbar denoting lateral and vertical averages, respectively, taken together imply a spatial average over the whole fluid volume. M would be zero in a perfectly mixed mixture with $\delta C = 0$. (iv) TW states with wave number k are characterized by the oscillation frequency ω or phase velocity $v = \omega/k$.

The stability properties of the conductive state against infinitesimal convective perturbations [59–62] are summarized for ethanol-water parameters, $L = 0.01$ and $\sigma = 10$, in [6, Fig. 1] as a function of ψ for the experimentally-accessible ψ-range. We use the scaled Rayleigh number $r = R/R_c^0$ that is reduced by the critical Rayleigh number R_c^0 for onset of convection in a pure fluid with the critical wave number k_c^0. The analytical values are $R_c^0 = 1707.762$ and $k_c^0 = 3.11632$. To compare our finite differences numerical results with analytical or numerical ones we scale R by the threshold R_c^0 appropriate for the respective code [63, 64]. For example $R_c^0 = 1685.8$ for an uniform spatial resolution of $\Delta = 1/20$ [65]. All MAC results of this paper were obtained with this resolution except for Fig. 2.1 where $\Delta = 1/40$.

2.2 Bifurcation behavior

Fig. 2.1 displays in a representative way the subcritical bifurcation topology that is typical for moderately negative Soret coupling ψ [6] considered in most experiments. The conductive state loses its stability at r_{osc}. Just above r_{osc}, the system does not saturate in a state with small convective amplitude as in a pure fluid; rather, a nonlinear TW state of large amplitude is stable. This TW state has a frequency which is much smaller than the Hopf frequency ω_H. The transition behavior between conductive state and final stationary TW is discussed in Sec. 2.7. The arrows in Fig. 2.1d indicate the hysteretic character of the bifurcation.

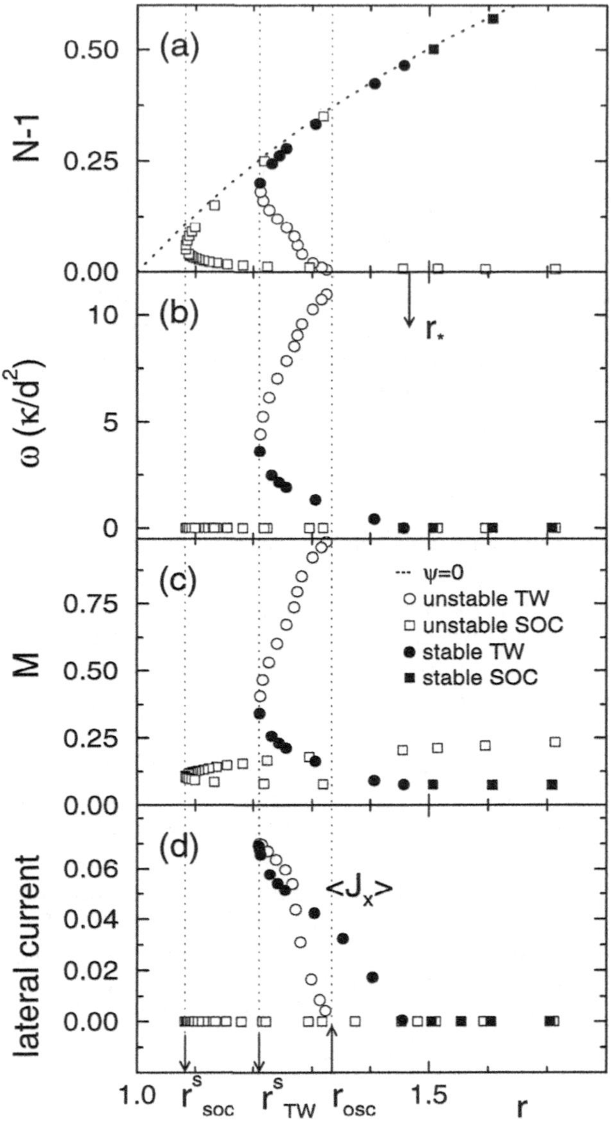

Fig. 2.1. Bifurcation diagrams of (a) Nusselt number $N-1$, (b) Frequency ω, (c) mixing parameter M, (d) maximal value of lateral concentration current $\langle J_x \rangle$ of extended convective states with wavelength $\lambda = 2$ for $\psi = -0.25, L = 0.01, \sigma = 10$. The TW solution branch bifurcates backward at r_{osc} with frequency $\omega_H \simeq 11.2$ [61], becomes stable at r^s_{TW}, and ends at r^*. SOCs are stable only above r^*. Unstable SOC states [lower branch in (a), upper branch in (c)] are disconnected from the conductive solution and lie almost parallel to the r-axis. The dotted line in (a) shows N in a pure fluid.

The reason for the associated discontinuous onset of convection is a nonlinear feedback: For $\psi < 0$, the Soret-induced conductive concentration distribution weakens the buoyancy. Convection, on the other hand, reduces the strength of the concentration field and thus increases the buoyancy much more strongly than the flattening of the vertical temperature profile in the bulk decreases the buoyancy. This increased buoyancy in turn strengthens convection, which again amplifies the buoyancy.

With increasing r, the TW frequency decreases until there is a continuous transition, at r^*, to a branch of SOC states. There, ω decreases as $(r^* - r)^{\frac{1}{2}}$. From a fit $\omega(r) = a\,(r^* - r)^b$ with parameters a, r^*, and b, we got from our data $b = 0.5(\pm 10\%)$. Ohlsen et al.[14] obtained a slightly different exponent $b = 0.58$ from their experiments. Bensimon et al. [66] got $b = 0.5$ from their concentration-boundary-layer model. The transition at r^* is smooth and nonhysteretic for our virtually infinite system, while it is hysteretic with a jump in the frequency in finite experimental cells with rigid lateral boundaries [19–21].

It is possible to stabilize the otherwise unstable SOC branch below r^* down to the SOC saddle r^s_{SOC} with the additional phase pinning condition $u = 0$ at $x = 0, \Gamma$ which forbids lateral convective transport through the line $x = 0$. So, in a short system of $\Gamma = 2$, the roll pair cannot travel. In somewhat longer systems, this boundary condition is not sufficient to prevent TWs in the bulk, as has been seen in experiments [21–24] and in simulations of a system of length $\Gamma = 10$ with $\psi = -0.6$ [64, Sec. 4.3]. If we start from a phase pinned SOC state below r^* as initial condition and then release the phase pinning, the SOC state becomes unstable. The rolls start to travel and v increases as N and $w_{\max}$ decrease. Finally, for $r^s_{TW} < r < r^*$, the system ends in the stable TW state. For $r < r^s_{TW}$, on the other hand, there is no stable convective state. Below r^s_{TW}, the convection amplitude decreases, while v increases, until finally the system approaches the conductive state as a linear TW of exponentially-decreasing amplitude.

The Nusselt numbers of stable TW and SOC states typically lie just below the values for a pure fluid when the fluid is well mixed so that the concentration field does not influence the vertical heat transport significantly. Only near the saddles there are stronger deviations. That our stable SOC and TW states are closer to SOC states in a pure fluid than to the conductive state of the mixture is reflected not only in the Nusselt number, but also in the mixing parameter M (cf. Fig. 2.1c). Convective mixing dramatically reduces the *concentration variation* from $M_{cond} = 1$ in the conductive state to $M \approx 0.1$ for the SOCs on the upper branch while the similarly-defined *temperature variation* is reduced by only 10% in the strongest stable convective state in Fig. 2.1 indicating that the temperature field remains much closer to the conductive temperature profile. M slightly increases for the unstable SOCs on the "lower branch" with weaker velocity field. All stable and unstable SOC states shown in Fig. 2.1 are strongly nonlinear with $M \ll 1$ and thus cannot be described adequately with weakly nonlinear models. The linear relation between M and TW frequency ω is discussed in [7] and in Sec. 2.6 for a broad range of parameters.

The bifurcation diagrams in Fig. 2.1 were obtained in a system with periodicity length $\Gamma = \lambda = 2$. Thus, e.g., the Eckhaus instability [27, 28] and more complicated dynamical behavior [5, 11, 29–31, 58, 64, and references cited therein] occurring in large systems are suppressed here.

2.3 Structure of TW and SOC states

To discuss how the nonlinear convective states change along the upper bifurcation branch we display in Fig. 2.2 a typical stable TW and a stable SOC state. The TW travels with phase velocity $v = \omega/\pi$ to the right. Close to the maximal upflow at $x = 0$, the isotherms are bent upwards, because warm fluid is transported upwards from the warmer lower plate there. Correspondingly, the isotherms are bent downwards close to the maximal downflow at $x = \pm 1$. Both velocity and temperature fields look like those in a pure fluid. But here in the mixture, the temperature wave δT (thin line with triangles in the fourth row) lags by a slight phase shift behind the wave of w (thin line). While the velocity and temperature fields are laterally nearly harmonic, the lateral concentration profile has a trapezoidal, strongly nonharmonic shape (thick line with squares). For a TW propagating to the right, δC has a positive plateau with a concentration surplus (dark area) in the region of the right turning convection rolls and correspondingly a negative plateau with a concentration deficit (light area) in the region of the left turning rolls.

To explain this structure of the TW's concentration field self-consistently, we view the fluid layer from the frame of reference $\widetilde{\Sigma}$ that comoves with the phase velocity $v\,\mathbf{e}_x$ of the TW so that $\widetilde{x} = x - v\,t$. In this frame, all fields $\delta\widetilde{F}$ of the stationary TW are time-independent so that $\left(\widetilde{\mathbf{u}}\cdot\widetilde{\boldsymbol{\nabla}}\right)\delta\widetilde{C} = L\,\widetilde{\nabla}^2\left(\delta\widetilde{C} - \psi\,\delta\widetilde{T}\right)$. Since L is small, we can set the diffusive term on the r.h.s. of this equation to zero for the present discussion. Thus, there is no gradient of $\delta\widetilde{C}$ parallel to $\widetilde{\mathbf{u}}$, i.e., the isolines of $\delta\widetilde{C}$ lie practically parallel to $\widetilde{\mathbf{u}}$. So, if we define the velocity streamfunction $\widetilde{\Phi}$ in $\widetilde{\Sigma}$ via

$$(-\partial_z\widetilde{\Phi}, 0, \widetilde{\partial}_x\widetilde{\Phi}) = (\widetilde{u}, 0, \widetilde{w}), \tag{2.8}$$

then the isolines of $\widetilde{\Phi}$, i.e., the streamlines of $\widetilde{\mathbf{u}}$ along which the fluid moves in $\widetilde{\Sigma}$, are almost identical with the isolines of the concentration. That this holds quite well can be seen in Fig. 2.2 by comparing the grey-scale coded concentration distributions (fifth row) with the streamlines of $\widetilde{\mathbf{u}}$ (sixth row).

2.3.1 Concentration distribution and streamfunction

In order to understand the structure of the concentration field, we first investigate the form of the streamlines. The latter can be most easily understood within a single–mode representation [8–10, 67],

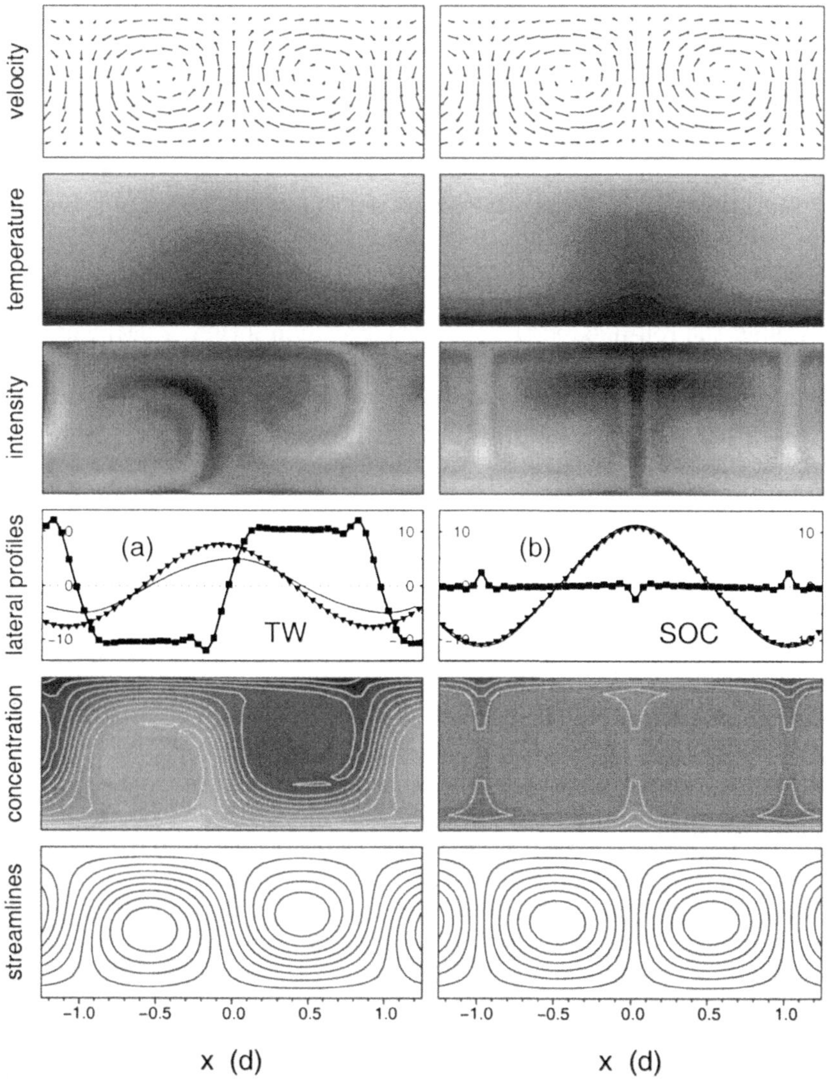

Fig. 2.2. Snapshots of a representative stable TW close to the saddle traveling to the right with $\omega \approx 0.3\omega_H$ and a stable SOC state for $L = 0.01$, $\sigma = 10$, $\psi = -0.25$. The wavelength is $\lambda = 2$, and $x = 0$ denotes the position of maximal upflow. The vector field of the velocity $\widetilde{\mathbf{u}}$ is shown in the first row. The temperature field δT is grey-scale coded in the second row where dark (light) areas denote high (low) temperature. The third row shows the side-view shadowgraph intensity $I(x, z)$ and the fifth row shows the concentration field in grey-scale together with white iso-concentration lines. Dark (light) areas denote high (low) concentration. The structure of the concentration field can be related to the streamlines of $\widetilde{\mathbf{u}}$ in the comoving frame (last row). The fourth row shows lateral profiles of w (thin line), $40\,\delta T$ (triangles), and $400\,\delta C$ (squares) at midheight, $z = 0.5$.

$$\widetilde{\Phi}(\widetilde{x}, z) = \frac{w_{\max}}{k}\, \hat{w}(z)\, \sin(k\widetilde{x}) + v\, z\,. \tag{2.9}$$

This streamfunction captures the characteristic properties of the velocity field being basically harmonic in lateral direction. The vertical profile $\hat{w}(z)$ might be given by the first Chandrasekhar function with maximum normalized to 1. Then $w_{\max}$ is the maximal vertical velocity. The second contribution, $v\, z$, to (2.9), that arises in $\widetilde{\Sigma}$ from the phase velocity of the TW, has two effects: it shifts the closed isolines of the ordinary first contribution to (2.9) – corresponding to stationary roll convection – alternately towards the top and bottom plates, and it generates open streamlines. These meander between and around the roll-like regions of closed streamlines (Fig. 2.2a), whereas in an SOC with $v = 0$ (Fig. 2.2b), the only open streamlines are the vertical separation lines between the oppositely turning rolls joining top and bottom plate. Consequently, in a right-propagating TW, (*i*), the dark, concentration rich (light, concentration poor) regions of closed streamlines for the right (left) turning fluid domains in $\widetilde{\Sigma}$ are displaced towards the upper cold (lower warm) plate, where the Soret effect has caused a concentration surplus (deficiency) for $\psi < 0$. In addition, (*ii*), open streamlines separate the dark (light) regions from the bottom low-concentration (top high-concentration) boundary layer. Therefore, the dark (light) boundary layers feed high concentration (low concentration) only into the respective rolls. Within the regions of closed streamlines, the concentration field behaves like a passive scalar, mixing up diffusively [68] and producing the concentration plateaus discussed above as the fluid is advected around in $\widetilde{\Sigma}$ along the closed streamlines. So the concentration wave moves about 1/4 wavelength ahead of the w-wave.

This explanation of the concentration field structure is based on the smallness of L and on the fact that δC shows some of the characteristic behavior of a passively convected field. Within this picture it is straightforward to relate the changes in the concentration distribution when moving along the TW bifurcation branch to the structural changes of the streamfunction $\widetilde{\Phi}$ (2.9). To that end we show in Fig. 2.3 bifurcation diagrams of $w_{\max}$ and v and in Fig. 2.4 concentration field profiles obtained with a MMG expansion. The first occurrence of local extrema in $\widetilde{\Phi}$ gives rise to the first appearance of areas of closed streamlines. That happens at $u_{\max}/v = 1$ where the velocity ratio $\chi = w_{\max}/v \simeq 1$ for $k = \pi$. Thus, for $\chi \lesssim 1$, i. e., in all TW states between the Hopf bifurcation threshold r_{osc} and state A the streamlines are all open. Increasing the convective amplitude $w_{\max}$ yields a decreasing phase velocity v and increasing values of χ so that areas of closed streamlines appear that grow on cost of those occupied by open streamlines. All streamlines in the bulk are closed for $v = 0$, i. e., at the point D where the TW branch ends on the SOC branch.

Within closed streamlines a weakly diffusing scalar is homogenized [68, 69]. This behavior is elucidated in Fig. 2.4. In the states B, C, and D with $\chi > 1$ where areas of open streamlines exist we see a plateau characteristic of both profiles. However, the state A with only open streamlines shows no such

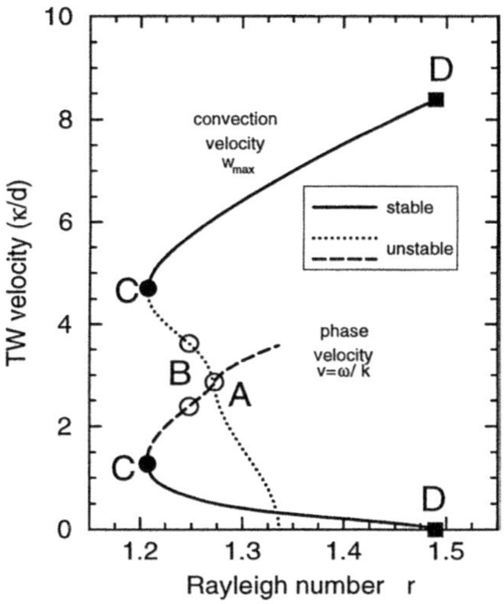

Fig. 2.3. TW bifurcation diagram of $w_{\max}$ and v for $L = 0.01$, $\sigma = 10$, $\psi = -0.25$, and $k = \pi$. Stable (unstable) branches are shown by solid lines (dotted for $w_{\max}$ and dashed for v).

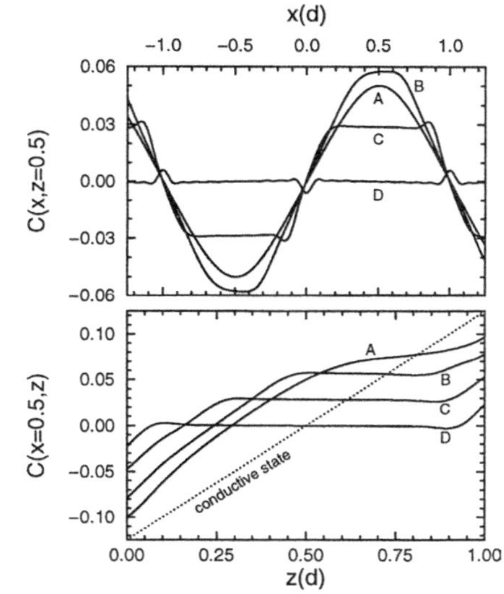

Fig. 2.4. Lateral and vertical concentration profiles of the four TW states identified in Fig. 2.3. Maximal upflow is at $x = 0$.

plateaus. Note that the lateral concentration wave profiles between the Hopf bifurcation and the state A are basically harmonic and that their amplitudes increase first with increasing convective amplitude. This amplitude growth of the harmonic concentration wave occurs as long as there are not yet closed streamlines, i.e., up to a limiting point in the vicinity of state A (the wave crest of B in Fig. 2.4 is larger than that of A since the lateral profiles are not taken at the center of the closed streamlines but in the center of the convection cell). Beyond this maximum amplitude the concentration wave crests are "cut off" and a plateau develops with the appearance of closed streamlines. The plateau extension, i.e., the region of constant concentration within the areas of closed streamlines broadens with increasing χ while simultaneously the plateau value of the concentration wave reduces.

2.3.2 Concentration plumes

The picture of mixing of the TW concentration field within a region of closed streamlines of $\tilde{\mathbf{u}}$ is based on the dominance of advection in the concentration balance. For small phase velocities, and especially for SOC states, this picture has to be extended, since these closed streamlines cover an entire convection roll. Since the velocity field vanishes at the plates, concentration diffusion has "long enough time" to work before a fluid volume is transported away from the plates by the velocity field, and a Soret-induced concentration gradient develops at the plates. The turnover time of a fluid element that starts very close to a plate on a closed streamline is indeed much bigger than the vertical concentration diffusion time $1/L$. Since streamlines through the top layer also pass through the bottom boundary layer, a fluid element on such a streamline experiences a strong change of its concentration content.

However, since diffusion is slow compared to vertical advection the fluid element retains positive concentration during its vertical motion down to the vicinity of the lower plate. This causes the little spikes (dips) in the lateral profile of δC in Figs. 2.2 and 2.4 at the positions $x = \pm 1$ ($x = 0$) of maximal downflow (upflow). In the vicinity of the lower plate, the volume element is transported very slowly further to the lower plate and then sideways. Surplus concentration is now transported diffusively – mostly upwards – out of the volume element, and the concentration content of the fluid volume becomes negative. All these effects together generate the typical plume structure of the SOC concentration field with the shafts of the plumes centered at the positions of maximal up- and downflow. The advective injection of concentration into the regions of closed streamlines takes – with diffusion being small compared to advection – the form of an inwards spiralling concentration jet [70] that can be seen in [6, Fig. 3] and identified in experimental side-view shadowgraphs [16, Fig. 13].

However, the plumes are rather narrow and modify the concentration field in the interior of the rolls only very slightly. In the TW of Fig. 2.2a, the weak plume structure of the concentration field is superimposed on the stronger plateau structure. Since the turnover time also diverges [67, 71] on the separatrix of the closed TW streamlines, this plume fine structure on top of the smooth background of the interior of the rolls has the same origin in TWs as in SOCs. But while the plumes, or concentration jets, are mirror symmetric in the SOC, this is not the case in TWs.

2.3.3 Mirror–glide symmetry

SOC and TW states discussed here are stationary, periodic structures that show a mirror-glide symmetry [6] $\delta F(x,z;t) = \pm\delta F\,(x+\lambda/2, 1-z;t)$ with $+$ for u, p and $-$ for w, δT, δC and B. This combination of a translation by half a wavelength in the x direction and a reflection through the $z = 1/2$ plane can easily be seen in the light intensity distribution $I(x,z) \propto r\,\nabla^2\,[\delta C(x,z) + b\,\delta T(x,z)]$ that one can observe in side-view shadowgraphs obtained with horizontal light through the fluid layer along the roll axes [15–17, 72, 73]. For our parameters $b = -0.919$ [15, 17, 65]. Fig. 2.2 shows this light intensity distribution in grey-scale. Its characteristic structures are dominated by the large Laplacians of the concentration field. Eaton et al. [15] using top-view shadowgraphs and then Winkler and Kolodner [16] using side-view shadowgraphs were the first to establish experimentally that the shadowgraph technique can give quantitative information also on the concentration field.

2.3.4 Large-scale lateral concentration current

The phase shift $\varphi_{C_1} - \varphi_{w_1}$ between the concentration and velocity wave in extended TW states drives a z-dependent mean *lateral* concentration current [3, 67, 74]

$$\langle J_x \rangle (z) = \langle \delta C\, u \rangle \simeq \frac{-1}{2k} \left| \delta \widehat{C}_1 \right| \left(\partial_z \, |\widehat{w}_1| \right) \sin\left(\varphi_{C_1} - \varphi_{w_1} \right) . \qquad (2.10)$$

For stable TWs with the velocity field changing only slightly, the concentration current is roughly proportional to the frequency – compare the full dots in Fig. 2.1d representing the maxima of $\langle J_x \rangle$ with the TW frequency in Fig. 2.1b. The ω dependence arises from the first lateral Fourier mode $\delta \widehat{C}_1$ and not from the phase difference: the concentration wave is about $\lambda/4$ ahead of the w-wave. However, on the unstable TW branch (open dots in Fig. 2.1d) the concentration current decreases with decreasing velocity field and reaches zero at r_{osc}. $\langle J_x \rangle$ is in the upper half of the fluid layer parallel and in the lower antiparallel to the phase velocity and integrates vertically to zero [3, 4, 6]. This large-scale concentration current plays an important role in the dynamics of LTWs, as discussed below.

2.4 Appearance of bistable TWs with increasing Soret coupling

Structure, dynamics, and bifurcation behavior of stable TWs on an upper branch observed in many experiments are well established today, see [6, 11] for an overview. However, the complete bifurcation behavior of TWs with unstable and stable TWs and the transition from weakly to strongly nonlinear stationary TWs has been elucidated only very recently [7, 70] by studying the changes in the bifurcation topology with increasing the strength of the negative Soret coupling. With increasing the stabilizing effect of the concentration field on the thermally unstable layering there opens up the possibility of an interesting competition between a strong Soret-induced concentration separation and the tendency to mix the fluid. This mixing is mainly driven by convection whose amplitude increases with increasing the stabilizing Soret coupling since the convection starts at higher Rayleigh numbers. At these heating rates the pure fluid – and so the well mixed mixture – show large velocity amplitudes. The results in Fig. 2.5 show how an increasing Soret coupling strength changes the bifurcation diagrams of $w_{\max}^2$ and ω. Starting at the respective r_{osc} with a large Hopf frequency ω_H and $w_{\max} = 0$ the frequency monotonically decreases along each TW solution branch while the convection amplitude $w_{\max}$ grows. States with $\chi = w_{\max}/v < 1$ (shaded region in Fig. 2.5a) are weakly nonlinear, while those with larger χ are strongly nonlinear. In the former there are only open streamlines in the frame comoving with the TW and the concentration wave is basically harmonic. On the other hand, for $\chi > 1$ there are also regions of closed streamlines. With growing χ their size and with it the anharmonicity of the concentration wave increases since within regions of closed streamlines concentration is diffusively mixed to alternatingly high and low plateau levels – cf. Fig. 2.4 and Sec. 2.3.1.

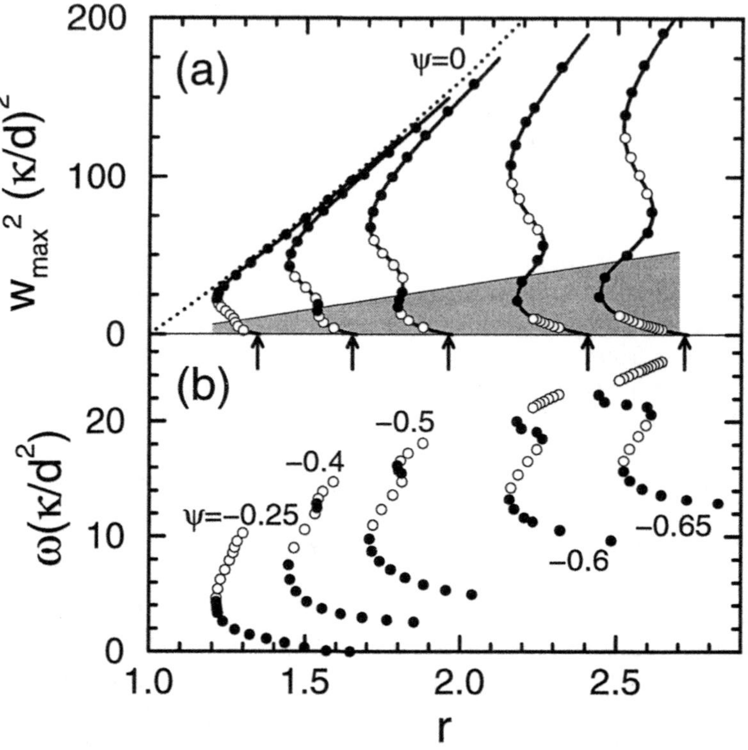

Fig. 2.5. Evolution of TW bifurcation diagrams with Soret coupling strength ψ: (a) squared maximal vertical flow $w_{\max}^2$, (b) frequency ω vs. r. Stable (unstable) TW states are marked by filled (open) symbols. Arrows mark Hopf thresholds r_{osc}. The $\psi = 0$ pure fluid limit is included in (a) by the dotted line. Full lines through the data points of (a) represent the fit discussed in the text. Only states in the shaded region of (a) are weakly nonlinear (cf. text). Parameters are $L = 0.01, \sigma = 10$.

With increasing $|\psi|$ the TW solution branches in Fig. 2.5 become more and more contorted. Already at $\psi \simeq -0.001$ the lower unstable branches of $w_{\max}^2$ have two inflection points with a dent in between that can clearly be seen in Fig. 2.5 at $\psi = -0.25$. The significance of such a dent on the lower unstable branch of solutions being unaccessible to experiments [14–16, 18, 22, 25, 32–34, 52] as well as to earlier numerical simulations [3, 6] has not been appreciated appropriately although it can be seen in the paper of Bensimon et al. [66] in the limit of small Soret coupling and $\sigma \to \infty$. With a modified finite differences method using a control process as well as by means of a MMG scheme it became possible to evaluate also unstable TWs. With these methods the complete solution branch can be determined and how its bifurcation topology changes with ψ. The above mentioned dent develops at $\psi \simeq -0.4$ into a forward bending arc when the first inflection point emits two new saddles. So, for $\psi \lesssim -0.4$ three stable states occur: Two different stable TWs coexist – one fast, the other one slow – that compete with each other and with the stable quiescent basic state. Furthermore, a new scenario appears for $\psi \lesssim -0.61$. There, the Rayleigh number of the saddle with small w_{max} and large ω drops below the saddle with large w_{max} and small ω so that a two-step hysteresis opens up between basic state, slow TWs, and fast TWs [75].

2.5 Application range of amplitude expansions – weakly versus strongly nonlinear TW states

The first inflection point of the TW bifurcation branch coincides closely with the boundary $\chi = 1$ between weakly and strongly nonlinear states – the change in curvature of the bifurcation branch reflects the change in the TW structure. Even for ψ as small as about -0.001 amplitude equations of quintic order might possibly reproduce realistically only those TW states that lie between r_{osc} and the first inflection point. The fact that $\chi \simeq 1$ is an upper limit to amplitude expansions around the Hopf bifurcation threshold follows also directly from the behavior of the multivalued functional dependence of the flow intensity $w^2_{\max}$ on r (Fig. 2.5a). Its singlevalued inverse $r(w^2_{\max})$ – to see it just turn Fig. 2.5a by 90° – can be represented extremely well by the product $r(w^2_{\max}) = r_0(w^2_{\max})\,\mathcal{R}(w^2_{\max})$ cf. the full lines in Fig. 2.5a. Here $r_0(w^2_{\max})$ is the $\psi = 0$ border curve for the pure fluid (dotted line in Fig. 2.5a) that starts at $r_0(w^2_{\max} = 0) = 1$ and grows for large $w^2_{\max}$ slightly sublinearly. The rational function

$$\mathcal{R}(w^2_{\max}) = \frac{a_0 + a_1 w^2_{\max} + a_2 w^4_{\max} + a_3 w^6_{\max}}{1 + b_1 w^2_{\max} + b_2 w^4_{\max} + b_3 w^6_{\max}} \tag{2.11}$$

ensures with $a_0 = r_{\text{osc}}$ that $r(w^2_{\max} = 0) = r_{\text{osc}}$ and with finite a_3/b_3 that $r - r_0$ does not diverge for large $w^2_{\max}$. With the third order polynomials in $\mathcal{R}$ four different $w^2_{\max}$ can have the same r. In that sense the above $r(w^2_{\max})$ is a minimal representation of the bifurcation diagrams of Fig. 2.5a. Now, the radius of convergence $w^2_{\max,c}$ of an amplitude expansion of $\mathcal{R}$ in a power series in $w^2_{\max}$ is given by the absolute value of that pole of $\mathcal{R}$ closest to $w^2_{\max} = 0$ in the complex $w^2_{\max}$–plane. Our fits for $\mathcal{R}(w^2_{\max})$ show that this pole lies on the negative real $w^2_{\max}$–axis with $w^2_{\max,c}$ being close to v^2 in each ψ–case so that $\chi = w_{max}/v \simeq 1$ is indeed an *upper limit* for the applicability of amplitude expansions. For further discussion see Sec. 2.8.

2.6 Universal scaling of TW frequencies and mixing properties

TW frequencies are universally related to the "distance" $r(w^2_{\max}) - r_0(w^2_{\max})$ of the TW states from pure fluid convection ($\psi = 0$). For any TW with a particular $w^2_{\max}$ this distance parallel to the r-axis can be read off in Fig. 2.5a. It measures also the "vertical distance" in flow intensity between the $\psi = 0$ state and the TW state in question. We confirm this prediction of the model [8] of

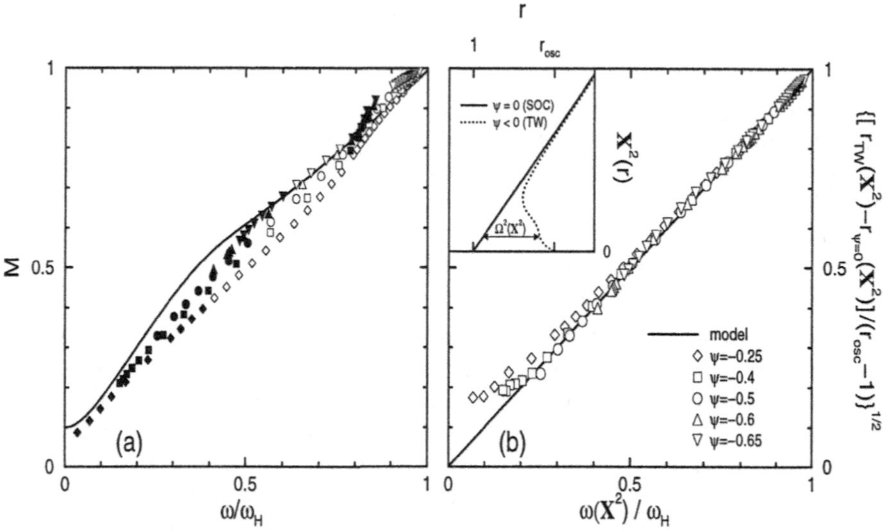

Fig. 2.6. Universal scaling relations between M and ω (a) and convective velocity amplitude $\mathbf{X} \propto w_{\max}$ and ω (b). Solid lines denote predictions of the model of Sec. 2.8. Numerical, finite differences data are shown by symbols. In (a) stable (unstable) states are shown by filled (open) symbols. In (b) all states stable or unstable are shown with open symbols.The inset explains the velocity scaling relation. L and σ were fixed to values of 0.01 and 10, respectively. But the scaling relations should not depend on them as long as $\sigma \gtrsim 1$ and $L \ll 1$.

Sec. 2.8 in Fig. 2.6b. It shows that the scaled frequencies ω/ω_H are indeed linearly related to $\sqrt{(r_{\mathrm{TW}} - r_{\mathrm{O}})/(r_{\mathrm{osc}} - 1)}$ in an universal, ψ-independent manner that holds for stable as well as for unstable TWs.

Furthermore, according to Fig. 2.6a the mixing parameter M varies linearly with ω/ω_H and thus also M is roughly linearly related to the expression $\sqrt{(r_{\mathrm{TW}} - r_{\mathrm{O}})/(r_{\mathrm{osc}} - 1)}$. Hence slow TWs with ω small compared to the frequency ω_H at onset imply good mixing, i.e., small concentration variations M — the coupled effects of increasing the mixing and decreasing the TW speed are a consequence of the monotonous growth of flow intensity along the TW bifurcation branches.

2.7 Growth dynamics

While final state patterns have been investigated intensively the transient growth dynamics has attracted much less attention [16] despite the fact that it is very interesting yet also complex, e.g., including the transition from weakly to strongly nonlinear behavior. We consider here the growth of convection out of initially small perturbations of the homogeneous conductive state after a sudden increase in the heating rate shortly above r_{osc}.

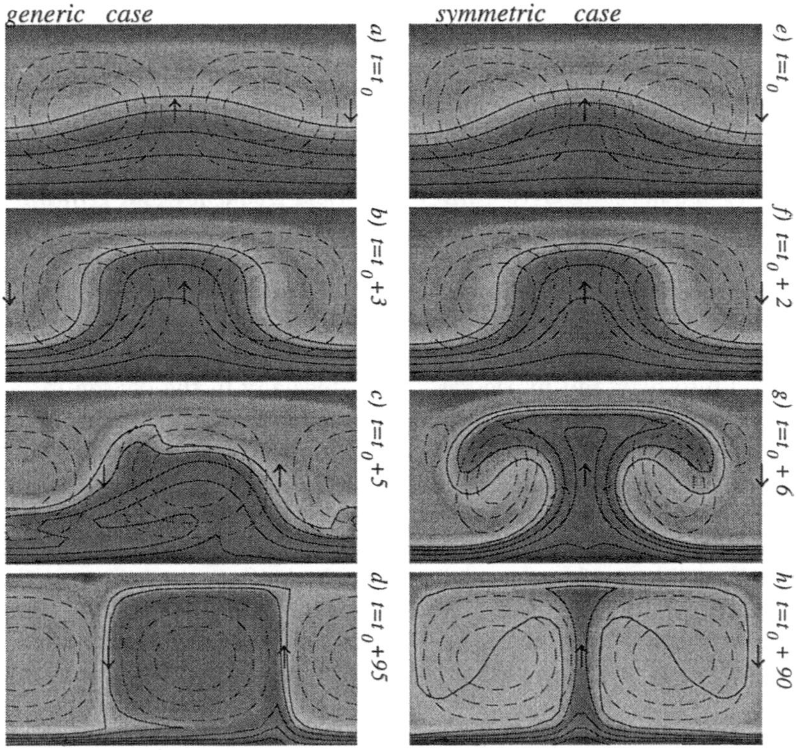

Fig. 2.7. Growth dynamics of the concentration field. Full lines are iso-concentration lines. Dashed lines are contour lines of the instantaneous stream-function. In the generic case the initial condition at $t_0 - 8.3$ is the conductive state plus small noise, and the final state is a slow nonlinear TW propagating to the right. In the right column the perfect mirror symmetry of the initial state is not broken by the MAC algorithm and the final state is an unstable symmetric SOC state. Parameters are $r = 1.42$, $\psi = -0.25$, $L = 0.01$, $\sigma = 10$.

2.7.1 Concentration redistribution.

The initial state is characterized by a strong vertical concentration contrast with *rapidly* oscillating structures growing out of perturbations while in the final nonlinear convective state, say, a *slow* TW the fluid is well mixed up with the peculiar vertical and lateral concentration profiles discussed in Sec. 2.3. The transition between these two field configurations with dramatically different concentration structures is highly complex and non smooth. An analysis revealed that it consists of four different parts: (*i*) the initial value dominated very beginning that is not discussed here, (*ii*) an exponentially growing part where the GLE approach works properly, (*iii*) an interaction period, and (*iv*) a relaxation regime into the final pattern.

Part (*ii*) is characterized by the exponential growth of linear superpositions of two independent critical modes of left- and right-moving TWs. Note that under the linear dynamics that applies as long as the perturbation fields are small enough all the other modes are damped. ***Generically***, the initial condition, e.g., noisy or other unbiased perturbations of the conductive state will contain the two critical TW modes with almost the same weight. So in part (*ii*) one observes SWs with near critical frequency and growing amplitude. See Fig. 2.7a and (e) for snapshots of the concentration field in part (*ii*). Then, in Fig. 2.7b and (f) the amplitudes increase further and the concentration field becomes slightly anharmonic. In addition a phase shift between w and δC appears in (b) reflecting the mirror-symmetry breaking deformation of the concentration field. After that, in (c) and (g) nonlinearities come strongly into play and particularly and especially in the concentration field higher modes are excited via the nonlinear terms in the balance equations. The concentration wave breaks and tilts over to the left in (c) forming a phrygian-cap like structure. Then in (d) concentration is advectively caught in the closed streamline area of the then well established TW that is propagating to the right. Its concentration contrast between left and right turning rolls monotonically decreases together with the frequency towards the final TW state values.

In the right column of Fig. 2.7 laterally symmetric initial conditions are imposed. Since our formulation of the MAC algorithm does not break this mirror symmetry [76] the pattern cannot move. This is similar to the stabilization of SOCs by phase pinning discussed in Sec. 2.2. Hence the concentration distribution becomes mushroom and plume like.

2.7.2 Time evolution of the first lateral Fourier modes

It is instructive to compare the transient velocity and concentration dynamics, say, at midheight $z = 1/2$ with the help of a lateral Fourier decomposition, e.g.,

$$w(x, z = 1/2; t) = \sum_n \hat{w}_n(t) e^{-inkx} + c.c. \quad . \tag{2.12}$$

To that end we show in Fig. 2.8 the dominating first lateral Fourier coefficients $\hat{w}_1$ and $\hat{C}_1$ — $\hat{T}_1$ behaves quite similar to $\hat{w}_1$. The initial behavior is well described by

$$\hat{w}_1(t) = \left(A\, e^{i\,\omega\,t} + B\, e^{-i\,\omega\,t}\right) e^{\gamma\,t} \tag{2.13}$$

i.e., a superposition of an anticlockwise (A) and a clockwise (B) rotating and exponentially growing complex amplitude. It gives rise to right (A) and left (B) propagating TWs with initial frequency ω that is close to the Hopf frequency ω_H. For generic conditions, $|A| \approx |B|$, as used in Fig. 2.8, this superposition amounts to an initial SW behavior — $\hat{w}_1$, $\hat{C}_1$, and $\hat{T}_1$ oscillate with frequency ω along a line through the origin in the complex plane of Fig. 2.8a and (c).

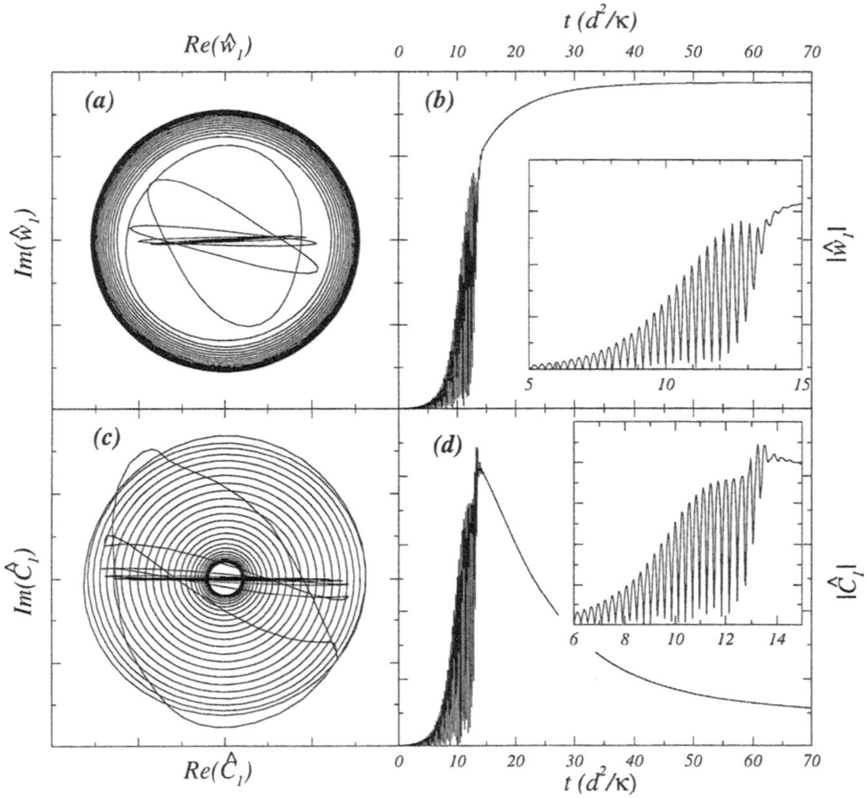

Fig. 2.8. Time evolution of the first lateral Fourier coefficients $\hat{w}_1$ and $\hat{C}_1$ of $w(x, z = 1/2; t)$ and of $\delta C(x, z = 1/2; t)$, respectively, starting from generic initial conditions. Parameters as in Fig. 2.7.

During this SW period the moduli of these modes oscillate with 2ω as indicated in (b) and (d) and in the insets. The ratio $\eta = (|A|-|B|)/(|A|+|B|)$ measures the deviation from a pure SW — $\eta = 0$ implies a pure SW while $\eta = +1(-1)$ represents a pure right (left) propagating TW.

Suppose that initially η is very small and positive, i.e., $|A| > |B|$. With increasing time this ratio grows since the larger mode becomes more and more dominant due to nonlinear interactions and the initial SW transforms into a TW. For the velocity field this evolution occurs in the complex plane of Fig. 2.8 from a straight line via a widening and slightly rotating ellipse to a circle in a monotonous manner — cf. the growth of the phase in Fig. 2.8. Thus, the w-field transformation from SW to TW has ended at $t \simeq 14$ in Fig. 2.8a and (b) when the amplitude B of the left TW has died. After that the frequency of the right TW slows down to its final value while $|\hat{w}_1|$ still grows — cf. the outwards spiralling decelerating motion of $\hat{w}_1(t)$ in Fig. 2.8a.

On the other hand, the evolution of the concentration mode $\hat{C}_1$ in Fig. 2.8c and (d) is more complex except for its initial, linear SW behavior that is similar to that of $\hat{w}_1$ and $\hat{T}_1$. During the transition from SW to TW there are loops with a backward moving phase and a strong overshooting of the modulus. This overshoot at $t \simeq 13$ in Fig. 2.8d surpasses the final value by almost a factor of 10.

At this overshoot the concentration field of the TW not only has a huge amplitude but it still has almost the original large frequency $\omega \simeq \omega_H$ and thus the concentration contrast and M is large, too. Then, because of the smallness of the Lewis number, the concentration is very slowly degraded accompanied by a slow decrease of ω, $|\hat{C}_1|$, and M towards the small final state values of a slow TW. This is reflected in Fig. 2.8c by an inwards spiralling $\hat{C}_1$ with decreasing phase velocity.

In the interaction part (*ii*) the higher Fourier modes play also an important rôle. Due to the mirror-glide symmetry even modes vanish at the midplane but the higher odd modes grow to a considerable magnitude. This is not surprising since the bending and mixing process in Fig. 2.7c and (g) creates complicated nonharmonic shapes.

2.8 Models

It has always been a theoretical challenge to derive a minimal, analytically manageable model which shows the essentials of extended states in binary liquids at negative ψ: Backward SOC and TW bifurcations, saddle node transitions to stable strongly nonlinear states, merging of TWs with stationary states, flow and mixing induced changes in the concentration field, and the fact that the frequency decreases along the TW solution branch with increasing flow intensity. The most severe problem is the weak diffusion of concentration leading in combination with the strong advection to Péclet numbers of $w_{\max}/L = \mathcal{O}(1000)$. Hence, convection in mixtures is a strongly nonlinear problem already close to onset in contrast to pure fluid convection. This is the reason why amplitude equations – also with various ad hoc quintic supplementations – are not as successful as they are for pure fluids: They do not incorporate the weakly diffusing but strongly advected concentration field adequately and hence they do not reproduce the relations between ω, M, $w^2_{\max}$ or N, and r. Typically, already the simpler relation between $w^2_{\max}$ or N and r is wrong on the upper TW branch – not to mention the more sensitive relations between ω, M, and r.

Also the results for LTWs based on this approach [77–80] have to be questioned: The spatiotemporal field properties under the LTW envelope being closely related to those in extended TWs are not captured correctly. Riecke [81, 82] has proposed an extension of these complex equations by an additional degree of freedom for the concentration taking its weak diffusion into account by scaling L to be of the same order as the distance from onset. In his investigation of pulses and other typical localized structures he uses however parameter combinations which do *not* reproduce the bifurcation topology of extended states with the aforementioned relations between ω, M, N, and r and which are also unrelated to the basic equations.

Another modeling approach for TWs was given by Bensimon et al. [66] using an expansion around the pure fluid for small negative ψ. This was the first theory which has produced a backward Hopf bifurcation, a saddle node, and the merging with the stationary branch via a boundary layer instability. However, it required a complete *numerical* determination of the concentration field and some integrals of it in order to compute the bifurcation branch. Additionally, this approach is limited to weak ψ and to regimes where phase and convective velocities which are $\propto \sqrt{-\psi}$ are much faster than diffusive processes ($L \ll 1$).

Finally, there have been *simple* extensions of the classical Lorenz model [83] which reproduces convection in pure fluids close to onset in a satisfying way. Based upon an earlier few–mode Galerkin model [84], Cross [85] and also Ahlers and Lücke [86] investigated Soret–driven convection with permeable boundary conditions. They found TW states only locally at the onset of convection since for permeable concentration boundary conditions the onset of TW convection is tricritical. Linz et al. [67] implemented impermeable conditions yielding a subcritical primary bifurcation to TWs. This observation is in agreement with small amplitude computations of Schöpf and Zimmermann [62]. Both approaches, however, did not explain the stabilization of the TW branch via a saddle node bifurcation. Also extensions with more modes [87] in the combined field $\delta\zeta = \delta C - \psi\delta T$ which has been introduced in order to fulfill the impermeable concentration boundary condition *exactly* were not successful: When using $\delta\zeta$ also δT has to be expanded up to similar order if L is small [70].

Recently, this theoretical dilemma was circumnavigated by a *direct* expansion of the concentration field being the primary field rather than ζ. The mode selection relied on a systematic analysis of the concentration balance equations: the structure of the field components occurring in a symmetry decomposition was investigated [10], a separation into lateral mean fields and deviations thereof was used, and the effect of the Soret coupling in the bulk of the fluid layer was quantitatively assessed to be small compared to that at the plates [9]. With advection being strong and L being small the concentration source or sink, i. e., the Soret effect can be omitted in the bulk. However, in regions where advection needs to be balanced by another transport mechanism strong concentration gradients occur while δT shows no such boundary layers. Thus, close to the impermeable boundaries, the Soret effect leads to a sizeable nonvanishing mean concentration gradient which cannot be ignored. With the field representation [8]

$$w(x,z;t) = \left[\mathbf{w}_{11}(t)e^{-ikx} + \text{c.c.}\right]\sin^2\pi z \tag{2.14a}$$

$$\delta T(x,z;t) = \frac{1}{2} - z + T_{02}(t)\sqrt{2}\sin 2\pi z + \left[\mathbf{T}_{11}(t)e^{-ikx} + \text{c.c.}\right]\sqrt{2}\sin\pi z \tag{2.14b}$$

$$\begin{aligned}\frac{\delta C(x,z,t)}{-\psi} = & \left\{1 - 2\pi\sqrt{2}\left[c_{02}(t) + 2c_{04}(t)\right]\right\}\left(z - \frac{1}{2}\right) \\ & + c_{02}(t)\sqrt{2}\sin 2\pi z + c_{04}(t)\sqrt{2}\sin 4\pi z \\ & + \left[\mathbf{c}_{10}(t)e^{-ikx} + \text{c.c.}\right]\sqrt{2} + \left[\mathbf{c}_{12}(t)e^{-ikx} + \text{c.c.}\right]\sqrt{2}\cos 2\pi z\end{aligned} \tag{2.14c}$$

impermeability of the plates in the lateral mean and with it the Soret forcing near the plates is guaranteed approximately but in an adequate manner [8].

The field representations (2.14a)–(2.14c) are minimal for convection in binary mixtures. They contain the Lorenz model modes, $\mathbf{w}_{11}$, $\mathbf{T}_{11}$, and T_{02}, i. e., the simplest truncation for the pure fluid and a minimal extension for the concentration field with two linear complex modes, $\mathbf{c}_{10}$ and $\mathbf{c}_{12}$, and two nonlinear real modes c_{02} and c_{04}. Previous extensions [67,85–87] with only one linear complex mode and one nonlinear real mode, partially in terms of the ζ–field, were too simple and failed to reproduce, e. g., the TW saddle. In fact, we have found that any truncation of the concentration field with only one linear and one nonlinear mode produces TW solutions with *linear* relations between all pairs of the three quantities v^2, w^2_{max}, and r.

Inserting the fields (2.14a)–(2.14c) into the balance equations without bulk Soret effect one obtains a system of seven ordinary differential equations for the mode amplitudes after projecting onto the bases used in the truncations [8]. In the framework of this model SOC and TW convection can be investigated by analytical expressions as a function of the control and fluid parameters.

2.8.1 Stationary states

For SOC, the reduced Rayleigh number r_{SOC} for a state with flow intensity $\mathbf{X}^2 = \left(\frac{8}{5\pi}\right)^2 \mathbf{w}^2_{11} = \left(\frac{4}{5\pi}\right)^2 w^2_{max}$ is calculated as

$$r_{SOC}(\mathbf{X}^2) = (1+\mathbf{X}^2)\left\{1 + \frac{11}{5}a\frac{\psi}{L}\frac{(1+\mathbf{X}^2)\left[1+\frac{25}{99}a\left(\frac{\mathbf{X}}{L}\right)^2\right]}{1+\frac{185}{36}a\left(\frac{\mathbf{X}}{L}\right)^2 + \frac{625}{648}a^2\left(\frac{\mathbf{X}}{L}\right)^4}\right\}^{-1} \tag{2.15}$$

with $a = 9\pi^2/128 \simeq 0.69$. The saddle $r^s_{SOC} \simeq 1 + 1.52\sqrt{-L\psi} + \mathcal{O}(L\psi)$ where $\partial r/\partial \mathbf{X}^2 = 0$, the tricritical value ψ^t_{SOC} where the saddle lies at $\mathbf{X} \equiv 0$, and the value ψ^∞_{SOC} where the SOC branch becomes disconnected from the basic state all agree well with the full numerical MAC and MMG results. This applies also to the whole bifurcation diagrams of Nusselt number N_{SOC} and mixing number M_{SOC} [8, Fig. 2]. Fig. 2.9 shows N_{SOC} for a representative parameter combination.

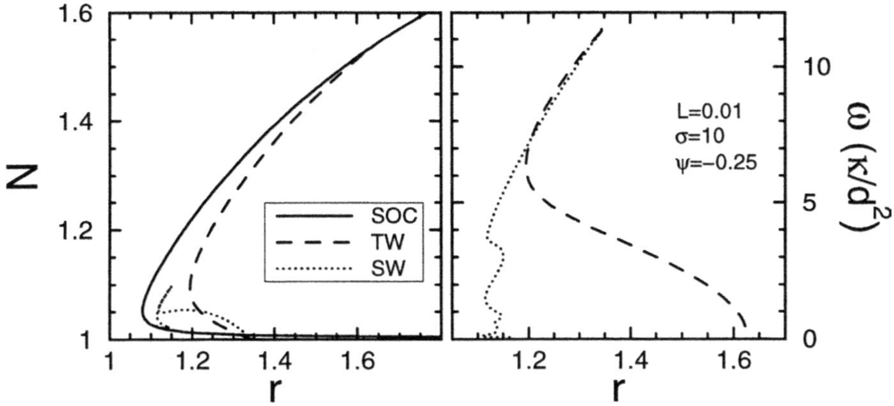

Fig. 2.9. Bifurcation diagram of the Nusselt number (left column, time averaged for SWs) of all types of convection connected with the ground state by a primary bifurcation: SOC, TW, and SW. The right column shows the frequency bifurcation diagrams of TWs and SWs.

M_{SOC} changes from 1 at onset ($\mathbf{X} = 0$) to about 0.1 for large amplitudes implying a strong reduction of concentration gradients throughout the system. Furthermore, r_{SOC} (2.15) deviates for convective amplitudes $\mathbf{X} \gg L$ from the pure fluid branch $r_{\psi=0} = 1 + \mathbf{X}^2$ only by terms $\propto L\psi$. Hence, for $\mathbf{X}^2 \gg L|\psi|$ the SOC bifurcation diagrams of a mixture are the same as for a pure fluid. This equality reflects the fact that strong convective mixing combined with diffusion equilibrates the concentration over wide areas so that it does not influence the bifurcation behavior any more: The stronger the mixing the less the deviations from the pure fluid.

2.8.2 Traveling waves

The more problematic type of convection as far as the description by a model of few degrees of freedom is concerned are TWs for the reasons discussed in Sec. 2.3 and in the introduction of this Subsection.

The TW solutions of our model with a frequency ω have time dependences $\propto e^{i\omega t}$ of the complex modes $\mathbf{w}_{11}$, $\mathbf{T}_{11}$, $\mathbf{c}_{10}$, and $\mathbf{c}_{12}$ – e. g., $\mathbf{w}_{11}(t) = \frac{5\pi^2}{8}\mathbf{X}e^{i\omega t}$ – because they are the amplitudes of a lateral variation $\propto e^{-ikx}$. So, $\omega > 0$ corresponds for $k > 0$ to TWs traveling to the right. The zeroth lateral modes T_{02}, c_{02}, and c_{04} are time independent and real in TWs. We can also take $\mathbf{X} \in \mathbb{R}$ by choosing the temporal phase. In order $\mathcal{O}(L^0)$ this leads [8] to

$$-\frac{1}{\psi} = \frac{3}{2}a\frac{\tilde{\sigma}}{\tilde{\sigma}+1}\frac{1+\Omega^2+\mathbf{X}^2}{\Omega^2}\frac{1+\frac{175}{108}a\left(\frac{\mathbf{X}}{\Omega}\right)^2}{1+\frac{1025}{144}a\left(\frac{\mathbf{X}}{\Omega}\right)^2+\frac{15625}{10368}a^2\left(\frac{\mathbf{X}}{\Omega}\right)^4} \quad (2.16a)$$

$$r_{\mathrm{TW}} = 1 + \Omega^2 + \mathbf{X}^2 \quad . \quad (2.16b)$$

Here, we have introduced $\Omega = \omega\tau$ with $\tau = 1/2\pi^2$ being the intrinsic time scale of the model and $\tilde{\sigma} = 27/14\,\sigma \simeq 1.93\,\sigma$. The neglection of terms $\mathcal{O}(L)$ causes the TW to merge at $\omega = 0$ with the pure fluid SOC instead of with the SOC of a binary mixture. Cancelling $\mathcal{O}(L^2)$ is allowed for all states with $L \ll \Omega \vee L \ll \mathbf{X}$ which is fulfilled for all separation ratios away from the codimension-two (CT) point where none of the relations $L \ll \Omega \vee L \ll \mathbf{X}$ can be fulfilled, since at the CT point the Hopf bifurcation *and* the merging of TWs and SOCs coincide.

According to (2.16a) TWs exist only for $\psi < 0$. This is in line with the absence of TWs for $\psi > 0$ in numerical simulations – the TWs seen in the model of Ref. [67] for $\psi > 0$ result from the low–order truncation. Relation (2.16b) allows to determine the frequency Ω of a TW with velocity amplitude $\mathbf{X}$:

$$r_{\mathrm{TW}}(\mathbf{X}^2) = 1 + \mathbf{X}^2 + \Omega^2(\mathbf{X}^2) \;=\; r_{\psi=0}(\mathbf{X}^2) + \Omega^2(\mathbf{X}^2) \quad (2.17a)$$

or

$$\frac{\omega^2(\mathbf{X}^2)}{\omega_{\mathrm{H}}^2} = \frac{r_{\mathrm{TW}}(\mathbf{X}^2) - r_{\psi=0}(\mathbf{X}^2)}{r_{\mathrm{osc}} - 1} \quad . \quad (2.17b)$$

Thus, the model predicts that the square of the frequency Ω of a TW state is given by the distance in the control parameter r between the TW under consideration and the state of the pure fluid with the same velocity amplitude $\mathbf{X}$. Another equivalent interpretation of (2.16b) is that for a given fixed r the squared frequency of a TW with velocity amplitude $\mathbf{X}_{\mathrm{TW}}$,

$$\Omega^2 = \mathbf{X}^2_{\psi=0} - \mathbf{X}^2_{\mathrm{TW}} \quad , \quad (2.18)$$

is given by the difference between the pure fluid flow intensity, $\mathbf{X}^2_{\psi=0} = r - 1$, and the flow intensity $\mathbf{X}^2_{\mathrm{TW}}$ of the TW in question. Thus, Ω^2 measures also the "vertical" distance in the bifurcation diagrams of $\mathbf{X}^2$ vs. r between the $\psi = 0$ pure fluid SOC solution and the TW solution in the mixture. Hence, the TW frequency is a direct measure of the distance of the system from the pure fluid, i. e., the influence of the concentration.

Eq. (2.17a) has an explicit dependence on the Soret coupling strength ψ since Ω varies between 0 and the scaled Hopf frequency Ω_{H}. This ψ–dependence is cancelled by scaling (2.17a) with the Hopf frequency so that the left hand side of (2.17b) varies for all ψ between 0 and 1. Thus, Eq. (2.17b) is an universal scaling relation for TW frequencies resulting from our model for small L. In Fig. 2.6 this prediction of the model is compared with MAC results for $\psi \in [-0.65, -0.25]$ [7]. For all these Soret couplings the scaling relation is confirmed. Only for small ω there are deviations. They are due to the fact that TWs with small frequencies do not approach the SOC states of the pure fluid as implied by (2.16b) but rather the SOC states of the mixture. These two stationary states differ in r by $\mathcal{O}(L)$ according to (2.15) so that deviations of $\mathcal{O}(\sqrt{L}) \simeq 0.1$ can be expected on the ordinate of Fig. 2.6b. Fig. 2.6a shows a second universal scaling relation between the reduced variance M of the concentration distribution and the reduced TW frequency. Both scaling relations hold for stable as well as for unstable TWs as an outcome of the model and are confirmed by the numerical results [7].

The Nusselt number of TWs is plotted in Fig. 2.9. The three main postulates for the bifurcation we have formulated above are fulfilled: backward bifurcation, saddle node transition, and merging with the stationary solution branch. The fourth one concerning the variation of concentration distribution and mixing with TW frequency is fulfilled because of the scaling relation between M and ω. However, the model does not contain enough degrees of freedom to reproduce also the bistability of fast and slow TWs at larger Soret coupling discussed in Sec. 2.4.

To generate TW bifurcation diagrams for the model we rewrite (2.16a) and (2.16b) in the form

$$r_{\mathrm{TW}}(\chi^2) = \frac{1}{1 + \frac{27\pi^2}{256}\psi \frac{\tilde{\sigma}}{1+\tilde{\sigma}} \frac{\left(1+\frac{64}{25\pi^2}\chi^2\right)\left(1+\frac{7}{24}\chi^2\right)}{1+\frac{41}{32}\chi^2+\frac{25}{512}\chi^4}} \tag{2.19}$$

with $\chi = w_{\max}/v = \frac{5\pi}{8}\frac{|\mathbf{X}|}{\Omega}$ being the ratio of convective and phase velocity. The characteristic points $r_{\mathrm{osc}} = r_{\mathrm{TW}}(\chi^2 = 0)$, $r^s_{\mathrm{TW}} = \min_{\chi^2 \in \mathbb{R}^+} r_{\mathrm{TW}}(\chi^2)$, and $r^* = \lim_{\chi \to \infty} r_{\mathrm{TW}}(\chi^2)$ in the bifurcation diagram can easily be read off Eq. (2.19). Quantitative comparisons of model and numerical results are given in Ref. [8].

It is worth to be noticed that also the model which describes TW convection semiquantitatively has a critical velocity ratio χ_c limiting small amplitude expansions of (2.19) in a series in χ^2. The radius of convergence of such an expansion is $\chi_c \simeq 0.8973$ [8] for all ψ and σ. This ratio confirms the result of Sec. 2.4 where $\chi \simeq 1$ was found as the transition from weakly to strongly nonlinear states and the limiting point of small amplitude theories.

2.8.3 Standing waves

The third type of primary extended states at $\psi < 0$ are SWs. They occur at the Hopf bifurcation as a superposition of a left and a right traveling wave due to the left/right symmetry of the system. In SWs, the phases of complex field modes are time independent so that all modes may be chosen real without restriction of generality. In a temporal Fourier series of the linear modes $\mathbf{w}_{11}$, $\mathbf{T}_{11}$, $\mathbf{c}_{10}$, and $\mathbf{c}_{12}$ only odd multiples of the basic frequency ω of the SW occur whereas in the nonlinear modes only even multiples exist. Inserting these temporal Fourier series one obtains algebraic equations for its coefficients which can be solved numerically.

The resulting bifurcation diagram for the time averaged Nusselt number [8] is shown in Fig. 2.9. TW and SW bifurcate both subcritically so that according to Knobloch [88] both are unstable. This is consistent with numerically determined weakly nonlinear results [62]. The contorted bifurcation behavior of ω_{SW} and N_{SW} in Fig. 2.9 below $\omega = 4$ is genuine to our few-mode truncation and is not an artifact of the numerical evaluation of the time-dependence of the modes. The values of N and r for a given ω agreed in calculations within less than 1% taking temporal modes into account up to 10ω and 20ω. Only for $\omega \leq 0.4$, the time dependence of the modes requires a higher resolution than that used for the calculation of the SWs in Fig. 2.9.

This is the first investigation of strongly nonlinear SWs for experimentally relevant boundary conditions. A numerical determination of SWs using the full field equations and also the experimental preparation of unstable SWs remain a challenge.

3 Localized traveling convection waves in mixtures

During the last decade much theoretical and experimental research activities were devoted to LTWs – for an overview see [11, 58]. They are spatially confined, strongly nonlinear convective structures surrounded by quiescent fluid. Like extended TWs they are observed for negative ψ only. The convectively active region consists of a few roll vortices traveling under an intensity envelope which itself drifts, albeit very slowly compared to the phase velocity of the rolls. The LTW rolls show many features of those in extended TWs. Structure and dynamics of the LTWs are governed by the concentration field: a large-scale concentration current stabilizes the LTW and controls its drift velocity into the quiescent fluid.

Here we review properties of stable LTWs [5, 58, 64, 89] at $\psi = -0.08$ and $\psi = -0.25$ for which experiments [12, 13, 23, 90–99] have been performed. Experiments seem to suggest that the proximity of walls in convection channels which orient the axes of the rolls perpendicular to the walls – thus enforcing a 2D configuration we are simulating – is necessary to stabilize LTW convection rolls [100]. The experimental top-view shadowgraph intensity distributions of LTWs [91–93] show structural properties that agree well with our numerical predictions. Also side-view shadowgraph intensity distributions of LTWs situated close to the lateral walls of a finite container [16] reveal, partly in detail, the predicted field structure quite well.

In Sec. 3.1 we present LTW and extended states in a common bifurcation diagram. The LTWs spatiotemporal structure is analyzed in Sec. 3.2 and compared for different Soret couplings. The close interplay of concentration currents, field structure, buoyancy force profile, and the effect on the propagation of LTWs is elucidated. In Sec. 3.3 we relate our findings to experiments and other theoretical investigations.

3.1 Bifurcation behavior

In the bifurcation diagrams of Fig. 3.1, we show our LTW attractors by stars in comparison with stable and unstable extended states. For $\psi = -0.25$ the LTWs are subcritical with respect to r_{osc} and they compete with extended TWs and the stable quiescent conductive state. For the weaker cou-

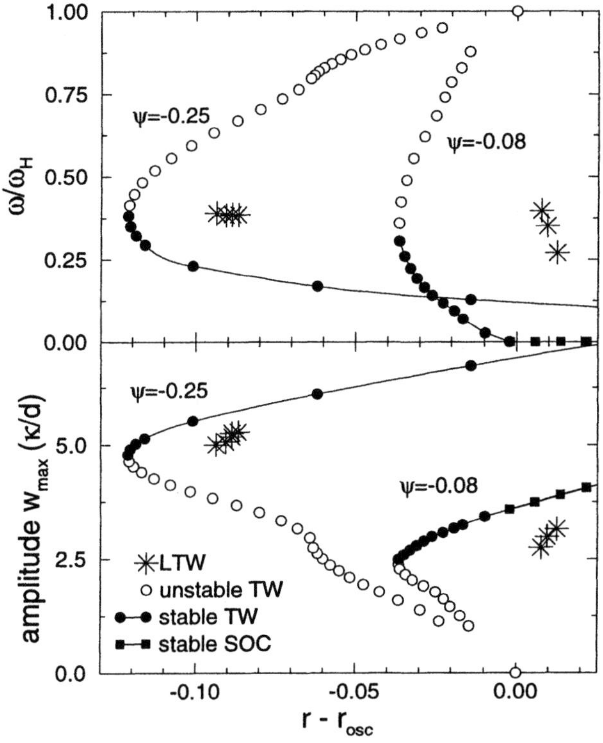

Fig. 3.1. Bifurcation diagrams of ω and of $w_{\max}$ for localized and extended states obtained for $L = 0.01, \sigma = 10$ and two different ψ.

pling $\psi = -0.08$, LTWs lie above r_{osc} in a regime where the conductive state is only convectively unstable [60, 62] but not yet absolutely unstable. Here, the competing extended state attractor is SOC convection – for $\psi = -0.08$, the band of stable LTWs lies above r^* [58, 89]. To explain why stable LTWs exist above r_{osc}, it has been proposed [5, 89, 101], experimentally seen [94] and analyzed [92, 102, 103] that with laterally periodic boundaries, initially *localized* disturbances of the conductive state move with the fast linear group velocity into the LTW and get absorbed by the LTW state if they do not have enough time/space to grow to an extended state.

3.2 Spatiotemporal structure

The LTWs of Fig. 3.1 display similar physical properties. After transients have died out, they drift with a finite but small group velocity v_g forward, i.e., into the propagation direction of the phase of its TW constituents, while the width ℓ remains constant. For other ψ, however, LTWs change their drift direction as a function of r [92, 93, 96–98]. In the coordinate system $\breve{\Sigma}$ comoving with v_g, the LTW fields are periodic in time, with $\breve{\tau}$ being the oscillation period of the LTW in its rest frame $\breve{\Sigma}$. Remarkably enough, our LTW states display an additional nontrivial spatiotemporal symmetry, $\delta\breve{F}(\breve{x}, z; t) = \pm\delta\breve{F}(\breve{x}, 1-z; t+\breve{\tau}/2)$, with + for u, p and − for w, δT, δC, and B which is the analog of the mirror-glide symmetry of extended TWs.

While LTWs differ in spatial extent and somewhat less in frequency, group velocity, and convection intensity, they exhibit similar field structure and behavior. LTWs consist of three characteristic parts: a leading front, a central part, and a trailing front, relative to the propagation direction of the TW phase. The fields in the center of LTW states are similar to each other and moreover have a structure similar to that in extended TW states of similar frequency. This includes the mixing and boundary-layer behavior of the concentration field with its characteristic plumes and trapezoidal lateral profiles. Structural properties at the leading front seem to be universal. The trailing fronts of the LTW states also resemble each other, but the field structure there is distinctly different from the leading front. These features can be seen in Figures 3.2 and 3.3. The structural properties at the leading and trailing fronts, respectively, of the two different LTWs agree with each other even in detail. Because the phase velocity of the LTWs is bigger than of the respective TWs, the concentration contrast between adjacent rolls is significantly higher in LTWs. Thus the mean lateral concentration current is also increased. The local LTW wavelength (Fig. 3.2a) varies laterally [29, 89, 91–93]. It increases towards the leading front and with it the local phase velocity. In the center part near $x = 0$ the wavelength is slightly below $\lambda = 2$, the value of an extended state.

3.2.1 Buoyancy force profiles

Analysing the balance of buoyancy forces that drive convection is central to understanding the dynamics of LTWs. The distribution of the mean concentration, $\langle\delta C\rangle = \frac{1}{\tau}\int_0^\tau \delta C(x, z; t')\, dt'$, averaged over one oscillation period $\tau = 2\pi/\omega$ of the LTW is associated with a large-scale, closed, mean concentration current circulating over the whole LTW state and influences the local effective driving force for convection. In the center part of LTWs, as in an extended TW, convection reduces the time averaged vertical temperature and the concentration gradient in the bulk of the fluid layer. Fig. 3.2d presents evidence for this convectively induced flattening of the vertical field profiles in the pulse centers: in the lower warm and alcohol-poor region of the layer, say, at $z = 0.25$ one has $\langle\theta\rangle < 0$ and $\langle c\rangle > 0$, while in the upper cold and alcohol-rich region of the layer, the reverse holds. This convection-induced vertical concentration redistribution enhances the mean buoyancy force $\langle b\rangle = \langle c + \theta\rangle$ (Fig. 3.2d) relative to that in the conductive region, while the convective temperature equilibration in the bulk of the layer has the opposite effect of weakening $\langle b\rangle$. Note that a positive (negative) $\langle b\rangle$ in the lower (upper) half of the fluid layer implies an enhancement of the force $\langle b\rangle\, \mathbf{e_z}$ in upwards (downwards) direction.

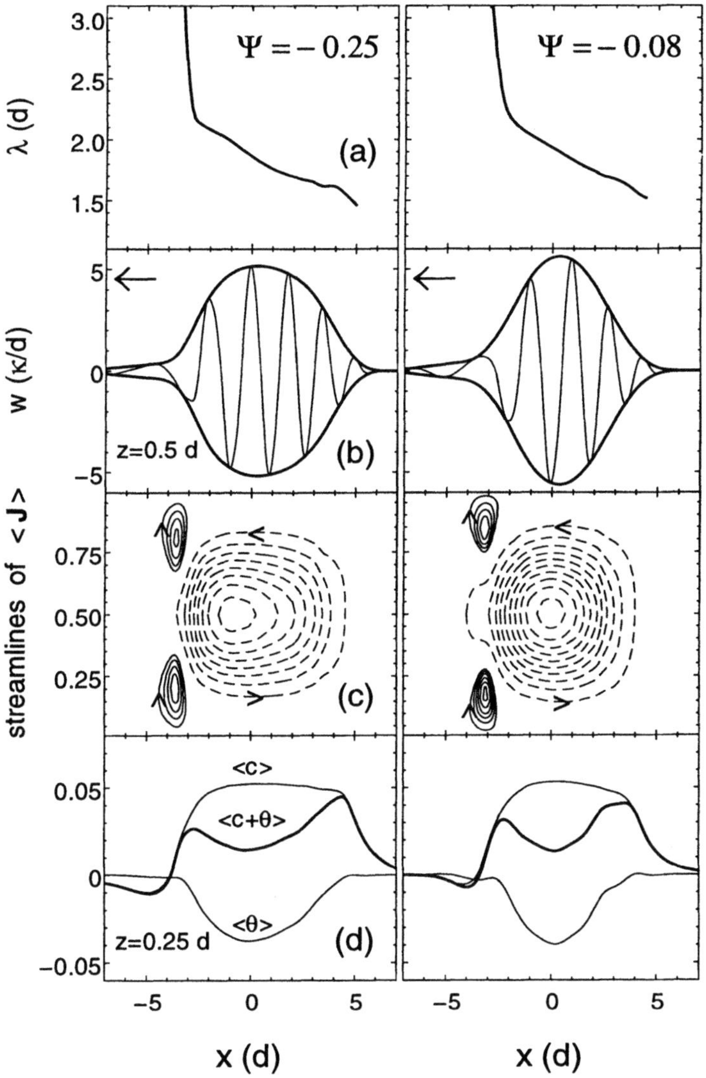

Fig. 3.2. LTWs at $r = 1.246$, $\psi = -0.25$ (left column) and at $r = 1.104$, $\psi = -0.08$ (right column) for $L = 0.01$, $\sigma = 10$. Both are presented as traveling to the left. (a) The wavelength $\lambda(x) = \breve{v}(x)2\pi/\breve{\omega}$ in the center part with sufficient convective intensity was determined from the frequency $\breve{\omega}$ and the phase velocity $\breve{v}(x)$ of the nodes of $w(x, z = 0.5)$ in the frame $\breve{\Sigma}$ and was checked against the node distances. (b) Snapshot (thin line) and envelope (thick line) of vertical velocity field $w(x, z = 0.5)$. (c) Streamlines of the time-averaged concentration current $\langle \mathbf{J} \rangle$. (d) Lateral profile of the time-averaged temperature field $\langle \theta \rangle$, concentration field $\langle c \rangle$, and the buoyancy force $\langle b \rangle = \langle c + \theta \rangle$ at $z = 0.25$. Profiles at $z = 0.75$ are the inverse of those at $z = 0.25$. For better resolution, the right column for $\psi = -0.08$ shows $2w$ in (b) and $3\langle \theta \rangle$, $3\langle c \rangle$, and $3\langle c + \theta \rangle$ in (d).

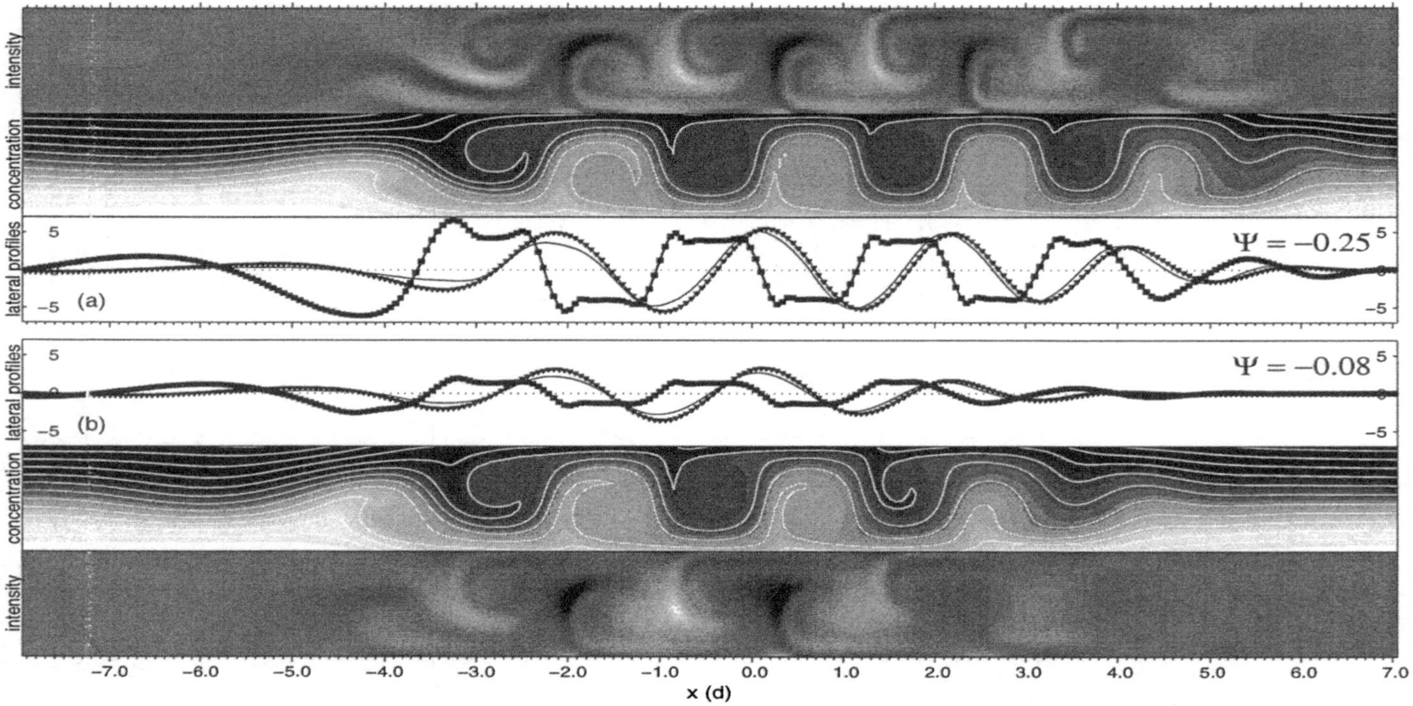

Fig. 3.3. Structure of the LTWs of Fig. 3.2, however, presented at a different time, so that, e.g., the node positions of w are different. In each case we show the side-view shadowgraph intensity $I(x, z)$, the concentration field $\delta C(x, z)$ in grey-scale together with white iso-concentration lines, and lateral profiles of w (thin line), $30\,\delta T$ (triangles), and $150\,\delta C$ (squares) at midheight, $z = 0.5$. The concentration grey-code is chosen such that the conductive profiles for both ψ look the same.

The mean buoyancy profile $\langle b \rangle = \langle c + \theta \rangle$ is dominated by the concentration contribution, especially far outside the center part of the LTW where the temperature contribution vanishes. Right under the two fronts where $\langle b \rangle$ is largest the LTW stabilizes itself against an invasion of the surrounding conductive state. On the other hand, ahead of the leading front, the concentration distribution is such as to weaken the buoyancy relative to the conductive state. By symmetry, $\langle c \rangle$ and $\langle b \rangle$ are positive in the upper half, thus depressing the *downwards* buoyancy. The concentration depression in the lower half of the layer ahead of the front at $x \approx -4.5$ can also be seen in the snapshots of the LTW concentration fields Fig. 3.3. The large, upwards-bending bulge at $x \approx -4.5$ reflects a concentration-depressed region that contains less alcohol than in the conductive state. This peculiar concentration distribution ahead of the leading LTW front acts like a "barrier". The time-averaged concentration "barrier" shows a surplus, $\langle c \rangle > 0$, in the upper half of the layer and symmetrically a depletion, $\langle c \rangle < 0$, in the lower half relative to the conductive concentration distribution. So the enhanced vertical concentration gradient in the "barrier" reduces the local buoyancy force there and stabilizes the conductive region ahead of it against invasion of convection located behind it.

3.2.2 Large-scale concentration redistribution

The concentration "barrier" ahead of the leading front is related to the existence of a strong mean circulating concentration current $\langle \mathbf{J} \rangle$. Its streamlines are shown in Fig. 3.2c. The origin of this current is the phase shift between δC and w in the center part of the LTW. This shift leads, as in extended TWs to a strong time-averaged concentration current in the center of the LTW. In the upper half of the layer, $\langle \mathbf{J} \rangle$ flows parallel to the phase velocity of the LTW, i.e., here to the left, and vice versa in the lower half of the layer. So $\langle \mathbf{J} \rangle$ has nearly horizontal streamlines in the central part of the LTW. On the other hand, there is almost no concentration current in the conductive area surrounding the LTW. Hence, the two currents that flow oppositely to each other in the center part of the LTW bend over vertically, thus forming a large-scale, closed circulation loop (dashed lines in Fig. 3.2c) extending over the whole LTW.

In addition to the large-scale concentration current that circulates counterclockwise along the dashed streamlines in Fig. 3.2c, there are two small, secondary loops with clockwise circulation under the leading front. They are caused by the strong convective reduction (increase) of the mean concentration $\langle \delta C \rangle$ in the center part of the LTW in the upper (lower) half of the fluid layer relative to the surrounding conductive state. Near the plates, where diffusion dominates, the resulting lateral concentration gradients, $\partial_x \langle \delta C \rangle$, drive a lateral diffusive current, which transports concentration out of the conductive region into the LTW at the upper plate and vice versa at the lower plate. This increases slightly the primary concentration circulation at the trailing front; see the small dents in the outer concentration streamlines.

However, at the leading front this diffusive effect counteracts the primary concentration current and generates the two small secondary countercirculating current loops (solid lines in Fig. 3.2c) under the leading front.

All in all, the time-averaged concentration current sustains a small concentration surplus in the upper half of the layer ahead of the leading front and a small concentration deficiency in the lower half of the layer, as compared to the conductive state, which reduces the buoyancy force there. This current-induced "barrier" seems to be the reason why the LTW drifts with group velocity much smaller than its phase velocity. This explanation is supported by the experimental and numerical fact that v_g increases as v, i.e., ω decreases with growing r: As ω decreases, so does the mean concentration current, cf. Fig. 2.1. Thus the concentration redistribution forming the "barrier" and its "braking" effect becomes weaker. Also, the experimentally-observed [92, 93, 96–98] reduction of v_g below zero for smaller r, i.e., larger ω, fits into this picture.

3.3 Comparison with experiment and theory

The calculated LTW bifurcation diagrams are discussed in detail and compared to experiments in [58]. For $\psi = -0.08$, we found stable LTWs of uniquely selected narrow width as in experiments [92, 93, 95, 99]. The $r-$variation of various properties of the pulses within the stable band is in good agreement with experiments by Kolodner [92, 93] and Niemela et al. [99], if one ignores that the latter pulses [99] have zero drift velocity – presumably due [92, 96] to some experimental pinning inhomogeneities. At $\psi = -0.25$, LTWs show richer and more complex bifurcation properties than at $\psi = -0.08$. Again, there is a finite, narrow band of r-values in which stable LTWs exist. Here, the band lies below r_{osc}, and the attractors of LTW, extended TW, and conductive state coexist therein with each other.

While several theoretical models have been investigated [77–79, 81, 82, 104–109] a quantitative model description for LTWs is lacking. Standard amplitude equations fail to account for and to incorporate the complex spatiotemporal structure of the concentration field [77–79, 104–108]. Yet it is this field structure that causes the existence of LTWs. The presently most advanced model for LTWs developed by Riecke and coworkers [81, 82, 109–112] is an extension of the complex GLE taking some aspects of the slow dynamics of the time-averaged concentration field into account. While it describes qualitative features of LTW convection with more success than ordinary complex GLE, in its present form its quantitative relevance for describing real LTWs is still an open question [58, 70], see also Sec. 2.8.

4 Dufour effect in gas mixtures

Besides liquid mixtures, the investigation of convection in pure gases and gas mixtures has obtained increasing attention in the last few years. There are several reasons for this interest: In gases, Prandtl numbers of $\mathcal{O}(1)$ or even smaller down to 0.16 in a H_2–Xe mixture [113] can be realized. In this regime, spiral defect chaos (SDC) has been discovered in cells of large aspect ratio Γ [114] and has been confirmed by numerical simulations of the proper Oberbeck–Boussinesq field equations [115]. Recent experiments [116] showed that with decreasing σ the appearance of SDC out of the weakly stable rolls shifts to lower heating rates. An advantage of gases compared to normal liquid He mixtures where also σ of about 0.6 can be realized is that flow visualization is easy with room temperature gases.

The values of ν, κ, and thermal expansion coefficient α are another important point. Their size allows Rayleigh numbers in the neighbourhood of the critical one to be realized at Boussinesq uncritical temperature differences of about 5°C in systems of $d = 0.5$mm height (in liquids usually 5mm). This, in turn, reduces the characteristic time scale $\propto d^2$ from 1–5 min to 1s. Furthermore, with thin layers large Γ being necessary for SDC are realizable with manageable experimental setups. Also, one can wait for the healing of transients on horizontal diffusion times $\propto \Gamma^2$.

Giacobbe [117] showed that σ can be drastically reduced by switching from pure gases to mixtures. Liu and Ahlers [113] used this fact and investigated SDC in the Soret regime at positive ψ, i. e., at Rayleigh numbers where in pure fluids no convection occurs. These studies required theoretical background concerning the stability analysis and weakly nonlinear convection properties including the coupling of concentration gradients into the heat current. This so–called Dufour effect causes the diffusive heat current to be changed from $-\nabla T$ to $-(1 + QL\psi^2)\nabla T + QL\psi\nabla C$ with the positive Dufour number Q [54, 118] measuring the strength of the diagonal and offdiagonal couplings in the above current. In typical *liquid* mixtures Q is so small that the Dufour effect is commonly ignored.

Stability analyses in gas mixtures started with Gutkowicz-Krusin et al. [119]. However, ignoring the diagonal correction $\propto QL\psi^2$ in the heat current they incorporated the Dufour effect only incompletely. The first complete analysis was performed by Lee et al. [46] albeit in a somewhat unusual notation. They also discussed the Dufour effect as essential in He mixtures near the λ–point. In the commonly used notation the results are compiled, reviewed, and extended in Refs. [118, 120]. Nonlinear results have been obtained in the framework of an eight-mode Galerkin model [120]. These predictions have been confirmed recently by a number of experiments in several different gas mixtures [113].

4.1 Stability analysis of the basic state

The stationary stability threshold of the quiescent heat conducting state in binary gas or liquid mixtures shows an interesting scaling behavior. It does not depend on L, Q, and ψ separately (and not at all on σ), but only on a particular combination of them which we call $H = \psi/[L(1 + Q\psi^2)(1 + \psi)]$ following [46] and on ψ/L. The marginal stability curve is given by

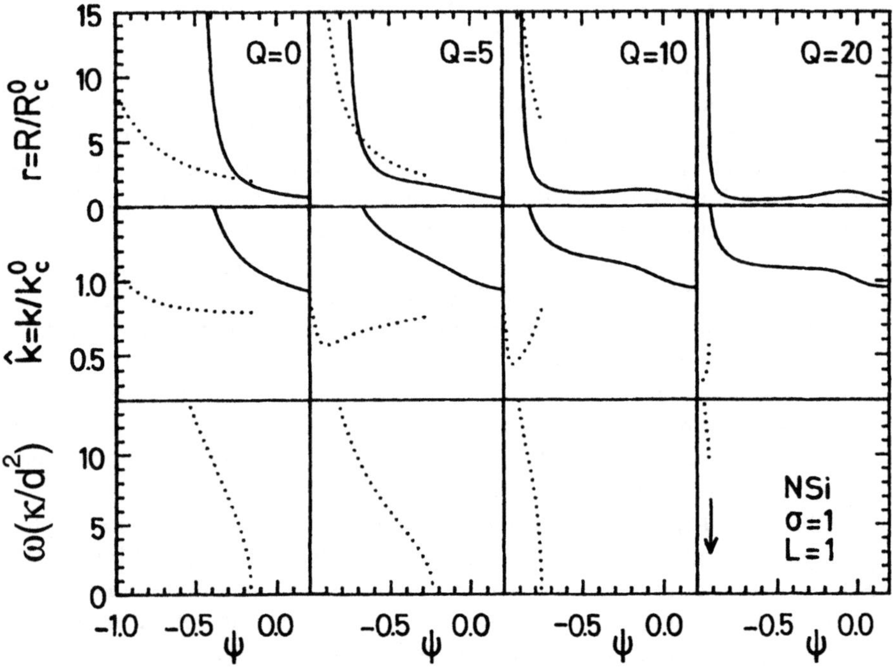

Fig. 4.1. Stability properties of gas mixtures ($L = 1$, $\sigma = 1$). The stationary (solid line) and oscillatory (dashed line) stability thresholds r^c_{stat} and r^c_{osc}, the corresponding reduced wave numbers $\hat{k}^c_{\text{stat}}$ and $\hat{k}^c_{\text{osc}}$, and the critical Hopf frequency ω_{H} result from a NSI shooting analysis [118] for which $R^0_c = 1707.762$ and $k^0_c = 3.116$.

$$\frac{\psi}{L} R_{stab}(k; L, Q, \psi) = \frac{H}{1+H} \widetilde{R}_{stab}(k; H) \tag{4.1}$$

so that the critical wave number k^c_{stat} depends only on H. It vanishes at $H_0 = 131/34$ [120, 121]. For $H \geq H_0$ large-scale stationary convection occurs. The marginal curve can then be expanded

$$R_{\text{stab}} \frac{\psi}{L} = 720 + \frac{2040}{77H} \left(H - \frac{131}{34} \right) k^2 + \mathcal{O}(k^4) \ . \tag{4.2}$$

As an aside we mention that (4.2) also holds for $\psi < 0$ and heating from above, $R < 0$, in which case $k^c_{stat} = 0$. All this scaling allows the influence of the Dufour effect on stationary stability properties to be mapped to a mixture without Dufour effect but rescaled L, ψ. The resulting change in L, ψ depend only on the product $Q\psi^2$ which is quite small according to experiments and numerical kinetic transport theory [113] or a Lorentz gas model [122]. Note however, that the oscillatory stability threshold and the respective critical wave number do not show such a scaling behavior.

Fig. 4.1 shows the main results of a stability analysis: The Dufour effect destabilizes (stabilizes) the conductive state against the growth of stationary

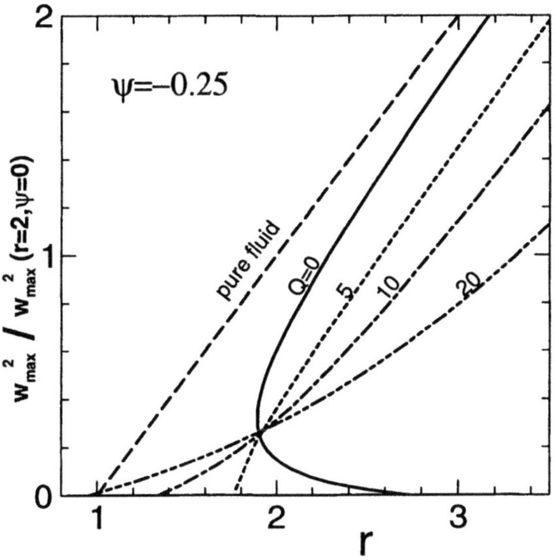

Fig. 4.2. Square of the vertical velocity of stationary convection reduced by the pure fluid value at $r = 2$ vs. r. Parameters are $\psi = -0.25$, $L = 1$, $k = k_c^0$, and arbitrary σ.

(oscillatory) convection and it shifts the critical curves $r_{\text{stat}}^c(\psi)$ and $r_{\text{osc}}^c(\psi)$ towards smaller ψ. The critical stationary wave number $k_{\text{stat}}^c(\psi)$ forms a saddle for $Q = 27$ and $\psi = -1/3$ as can be calculated from $k_{\text{stat}}^c = k_{\text{stat}}(H)$. The wave number $k_{\text{osc}}^c(\psi)$ of critical oscillatory patterns decreases and the difference, $k_{\text{stat}}^c - k_{\text{osc}}^c$, increases with increasing Dufour effect. The Hopf frequency decreases with growing Q.

4.2 Nonlinear SOC and TW properties

Nonlinear results concerning convection in gas mixtures were obtained by a Galerkin model [120], by numerical simulations of the full field equations [120], and by recent experiments [113]. Here the influence of the Dufour effect in particular on the bifurcation behavior of flow intensity, convective heat current $N-1$, and convective mixing of the concentration field M as functions of r is reviewed for several Q.

4.2.1 Stationary convection

Fig. 4.2 shows SOC bifurcation diagrams resulting from an eight–mode Galerkin model [120] for the squared velocity amplitude reduced by the amplitude square in a pure fluid at $2R_c^0$. This ratio should eliminate scale differences in the data from model, simulations, and experiments. The main results are: (1) The Dufour–induced destabilization of the conductive state against stationary perturbations shifts the onset of SOC to smaller r. (2) The range of subcritical SOC bifurcations in the L–ψ parameter plane shrinks upon increasing Q. (3) With growing Dufour coupling there is a gradual change from a strongly backward SOC bifurcation via a tricritical one to a forward one and the initial slope of a supercritical bifurcation curve decreases. (4) In addition, the Dufour effect reduces the convective intensity and with it the mixing of the concentration field at larger r so that the bifurcation curves of $N-1$, w^2_{max}, and M as functions of r become flatter. (5) Furthermore, structural changes in the convective fields resulting from either increasing Q or decreasing r are the same. (6) Comparison with numerical simulations of the full OBE's shows that the Galerkin approximation provides good results for SOC states well above onset not only for N but also for M. The reason is that the field expansion in trigonometric functions is quite effective since with $L = \mathcal{O}(1)$ the problem of concentration boundary layers is less severe than in liquids.

4.2.2 Traveling wave convection

The main TW results [120] are: (1) The ψ-range in which TW solutions – stable or unstable – are present is shifted to more negative values when the Dufour coupling increases. (2) Forward bifurcating TWs disappear already for smaller Q while backward bifurcating ones still exist for larger Q when ψ is sufficiently negative. (3) However, for $L = \sigma = 1$ these subcritically bifurcated TW solution branches do not develop a saddle before they merge with zero frequency with the SOC solution branch. Thus, ***stable*** TW states on an ***upper*** TW solution branch beyond a saddle that can be seen in liquid mixtures for $L = \mathcal{O}(10^{-2})$ and $\sigma = \mathcal{O}(10)$ were not found, neither in the model nor in the numerical simulations of the full field OBE's. The reason for their absence is the large Lewis number $L = \mathcal{O}(1)$ of gas mixtures.

As an outlook we mention that up to now only gas mixtures with positive ψ have been investigated. Negative ψ might be realizable at pressures higher than the present upper limit of 30 bars. It remains an experimental challenge to prepare mixtures with the right $L-\sigma-Q-\psi$ parameter combinations that allow to study TWs on the fast time scale of gases and to compare with our predictions.

5 3D convection structures in fluid mixtures: squares, crossrolls, and a subharmonic bifurcation cascade

For ***positive*** separation ratios, $\psi > 0$, the heavier component, say water in an ethanol-water mixture, tends to accumulate at the upper cool plate. This Soret induced additional destabilization of the conductive state causes a substantial depression of the convective onset r^c_{stat}. Furthermore, for $\psi > \psi_0 = L/(\frac{34}{131} - L)$ the critical wave number k_c vanishes and $R_c = 720L/\psi$, cf. Sec. 4.1 and Refs. [120, 121]. For ethanol-water mixtures L is so small that $\psi > \psi_0$ can easily be realized experimentally [35, 41, 43, 123]. Hence, one can prepare a wide so-called Soret driving regime, $1 \gtrsim r \gtrsim r^c_{stat} \approx 0.42L/\psi$, in which convection is primarily driven by solutal buoyancy, i.e., by concentration gradients. The resulting convection structures show spatiotemporal characteristics that differ from those in the so-called Rayleigh regime, $r > 1$, where thermal driving dominates.

Experiments done in the Soret regime typically revealed 3D structures, usually square patterns [35, 41, 43, 123]. Clune and Knobloch [124] showed that squares are indeed stable slightly above onset starting at small positive ψ up to ψ_0. But for $\psi > \psi_0$ there is no rigorous theory establishing the stability of squares or understanding the wave number selection. Indeed, for these separation ratios large-scale structures like a spokelike pattern [35] or a single large convection roll [42] were reported.

In the crossover driving range $r \simeq 1$ between the Soret and the Rayleigh regimes experiments showed an oscillatory competition between squares and rolls [35, 43, 123]. It reflects the presence of both driving mechanisms and the fact that with increasing heating and flow amplitudes concentration gradients are more and more reduced. Finally, at larger r, where the fluid is well mixed, thermal driving dominates and stationary 2D roll convection becomes stable as in one-component fluids.

On the theoretical side there was first a weakly nonlinear, few mode Galerkin approximation [125] aimed at describing squares, oscillations, and rolls at larger r. However, it required the ad hoc inclusion of symmetry breaking side wall forces. Later is was shown [121, 124] that such forces are not necessary, at least for stabilizing squares. Other theoretical work provided more qualitative insights then detailed quantitative descriptions of this phenomena [126–129]. Recently we could confirm not only the experimentally observed oscillations but we also found a new phenomena: a subharmonic bifurcation cascade (SC) of pattern oscillations caused by an entrainment by a stationary crossroll state [65, 130].

5.1 Full-scale 3D numerical simulations

In view of the spatiotemporal complexity of these 3D convection structures in the Soret and crossover regimes we performed finite differences MAC simulations with an uniform spatial resolution $\Delta = 0.05$ and a second order DuFort-Frankel time step of $\Delta t = 10^{-3}$. The pressure field was iteratively calculated with a fast multigrid method. This method uses the fact that, e.g., SOR or Jacobi iterative solvers reduce the high-k components of the defect of the solution very fast while the low-k components of the defect are eliminated ineffectively. Thus, several iterations on the finest grid yield a smooth, but not necessary a small defect. This smooth defect can then be 'restricted' to a coarser grid and the solution process can be continued on this grid. Having solved the problem on the coarser grid, the defect is interpolated back to the finest grid to improve the approximation first obtained there. We have analyzed the multigrid method intensively [65] and found a speedup by factors of 10 and larger for 2D TWs compared with previous calculations. Furthermore, the convergence properties of our multigrid method are independent of the fineness of the grid in contrast to classical methods. This fact is useful, because fine grids are necessary to resolve the boundary layers for small L.

Here we present results for 2×2 systems with laterally periodic boundary conditions. Thus, the wave number is fixed to $k = \pi$ as suggested by experimental measurements [35, 43, 123] for $r \simeq 1$. We have checked that the observed squares, oscillations, and crossrolls appear also in 16×16 systems with rigid insulating sidewalls. The parameters $L = 0.0045, \sigma = 27$, and $\psi = 0.23$ considered here are typical for ethanol-water mixtures.

5.2 Stationary squares

Above the bifurcation threshold $r_{stat}(k = \pi) \approx 0.0085$ stable stationary square patterns appear in our simulations. In Fig. 5.1a we show the horizontal intensity distribution $I(x, y)$ of a top-view shadowgraph image of a square pattern. It is determined by the Laplacian of the vertically averaged temperature and concentration fields $I \propto r\nabla^2 \left(\overline{\delta T} - 0.431\,\overline{\delta C}\right)$. For the parameters of Fig. 5.1 $\overline{\delta C}$ enters with relative weight -0.431. Fig. 5.1b shows the variation along the line $y = 0.45$ marked by the arrow in Fig. 5.1a. In (c) we show in a vertical crossection at $y = 0.45$ the grey-scale coded concentration field together with isotherms. Their downwards bending near the maximal downflow at $(x, y) \approx (1.4, 0.45)$ causes a temperature induced light intensity enhancement [cf. triangles in Fig. 5.1b]. On the other hand, δC is laterally strongly anharmonic. For example, the downwards directed concentration plume at $(x, y) \approx (1.4, 0.45)$ gives rise in I to a contribution (cf. squares in Fig. 5.1b) that causes a dark spot there. The shadowgraph intensity profiles of Fig. 5.1b with the fine structure resulting from narrow concentration plumes have been observed for roll convection also [6, Sec. V]. There, however, the plumes are narrower as can be seen by comparing Figs. 5.1b and (c).

5.3 Oscillations between x–rolls, squares, and y–rolls

Roughly one can consider the square pattern to consist of two orthogonal roll structures. Then the lateral Fourier mode I_{10} (I_{01}) of the shadowgraph intensity is a measure of the amplitude of x–rolls (y–rolls) with wavevector in x (y) direction. For the parameters of Figs. 5.1, 5.2 the stationary square pattern ($I_{10} = I_{01}$) becomes unstable at $r^{OS} \approx 1.063$ and I_{10} and I_{01} begin to oscillate with opposite phase about a common mean value. The oscillation frequency varies roughly linearly with r — in agreement with experiments [35, 43] — such that $\omega(r^{OS} = 1.063) = 0.251$ and $\omega(r = 1.173) = 0.0586$ for the parameters of Fig. 5.2. First the oscillation is harmonic with small amplitude. Then, with increasing r it becomes more and more anharmonic and relaxational – cf. Fig. 5.2b.

Note that I_{10} and I_{01} always remain positive so that x– and y–rolls never disappear and that they always maintain their turning direction. So, during a cycle first x–rolls, say, dominate and half a period later y–rolls as indicated in Fig. 5.2a with squares being realized whenever $I_{10} = I_{01}$. The Nusselt number

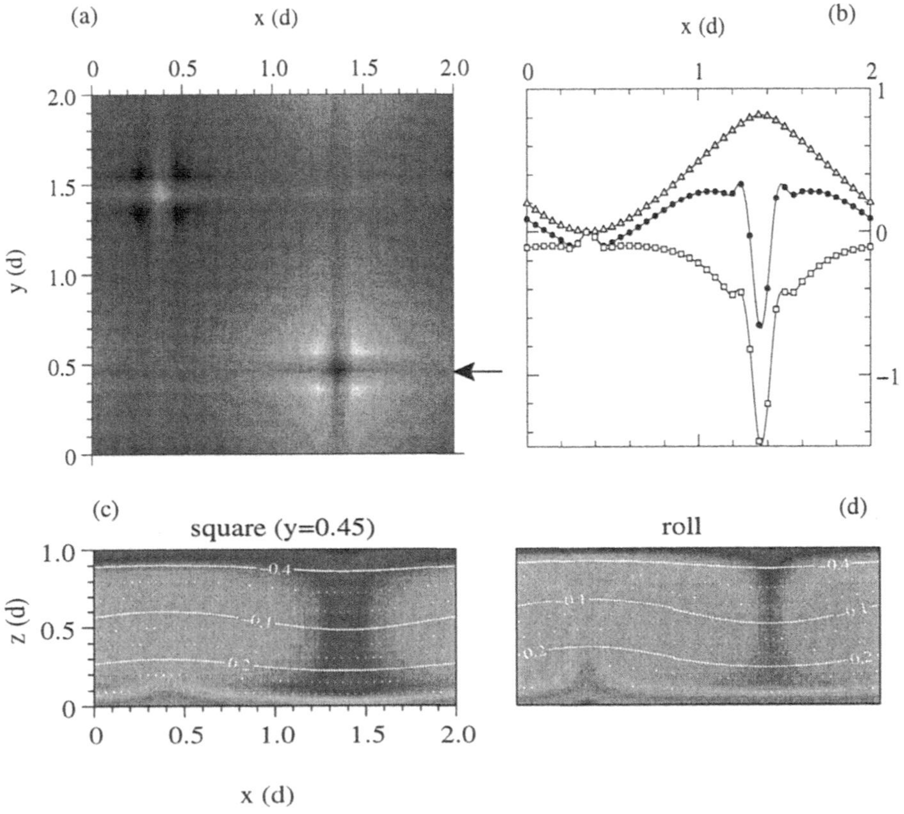

Fig. 5.1. Structure of stationary convection squares at $r = 1$ and $\psi = 0.23$. (a) Intensity $I(x, y)$ of top-view shadowgraph. (b) Profiles of I (dots), $\nabla^2 \overline{\delta T}$ (triangles), and $-0.431\nabla^2 \overline{\delta C}$ (squares) along the line $y = 0.45$ marked by an arrow in (a). (c) Grey-scale coded concentration field and isotherms in $x - z$ crossection along the line $y = 0.45$. (d) The same for an x−roll pattern. Parameters $L = 0.0045, \sigma = 27$ are typical for ethanol-water mixtures.

$N(t)$ oscillates with twice the pattern oscillation frequency since every field structure in the pattern oscillation cycle appears half a cycle later in a form that is rotated by 90° in the $x - y$−plane. As an interesting aside we mention that throughout the cycle the mirror glide symmetry $\mathcal{F}(x, y, z) = -\mathcal{F}(x + 1, y + 1, 1 - z)$ for $\mathcal{F} = w, \delta T$, and δC is fulfilled at all times.

In particular at larger r the intermediate square state — in Fig. 5.2b at $t \simeq 16$ and $t \simeq 40$ — is quickly swept through while the system spends more and more time in a crossroll configuration, e.g., $I_{10} \gg I_{01} > 0$ where the x−roll amplitude is large and the y−roll amplitude is small but nonzero or vice versa.

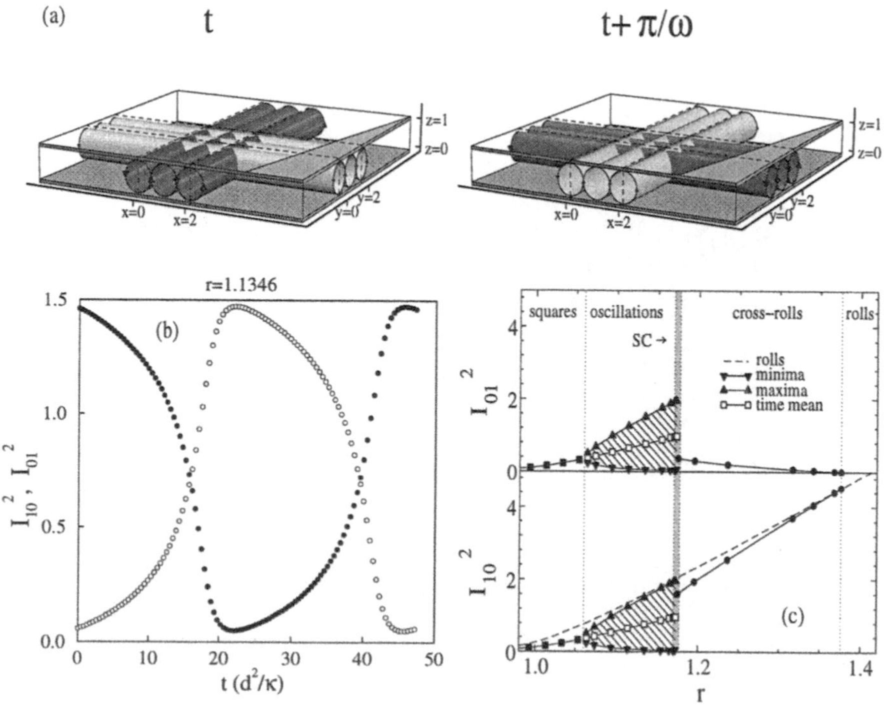

Fig. 5.2. Oscillation and bifurcation behavior. (a) The system oscillates from an $x-$roll dominated crossroll state (left part) to a $y-$roll dominated crossroll state half a period later (left part) and then back. (b) Counterphase oscillation of the lateral shadowgraph intensity modes I_{10} and I_{01} measuring the amplitudes of $x-$roll and $y-$rolls, respectively. (c) Bifurcation diagram. Parameters are $\psi = 0.23, L = 0.0045, \sigma = 27$.

5.4 Subharmonic bifurcation cascade to stationary crossrolls

Before the above described oscillations eventually get locked into one of the symmetry degenerate crossroll states we have found evidence for a subharmonic bifurcation cascade caused by an entrainment in a very narrow $r-$interval that is shaded and marked SC in Fig. 5.2c. The transition from SC regime to crossrolls is complicated and slightly hysteretic [65, 130].

Beyond the SC regime there is in Fig. 5.2c a relative wide $r-$range of stable crossrolls which was theoretically not known to exist but for which some experimental evidence has been reported [41]. At the upper end of the crossroll regime the weaker roll amplitude, say I_{01} in Fig. 5.2c, drops continuously to zero marking at r^R the transition to a pure $x-$roll state.

These squares, oscillations, and crossrolls have also been obtained with a MMG expansion [130] in agreement with our MAC results. A more extensive investigation of the behavior as function of the fluid parameters seems worthwhile.

6 The effect of throughflow on supercritically bifurcating patterns

Here we review how a throughflow changes the spatiotemporal pattern forming properties [131] of systems that in the absence of throughflow undergo a continuous, hysteresis free transition from a spatially homogeneous state to a stationary spatially structured one. Without throughflow a continuous or discrete family of patterns with different wave numbers is often stable for a fixed configuration of parameters and boundary conditions (BCs). The wave number band of these stable states is usually limited by secondary instabilities. The stable structures within such a band can be generated by appropriately engineered time histories of the control parameters and/or by properly changing the BCs, e. g., the system size. The most intensively investigated examples in this respect are Taylor vortices [11, 132–137] in an annulus between concentric cylinders of which the inner one rotates and convective roll vortices in horizontal layers of one-component fluids heated from below [11, 135, 138–141].

This multiplicity of solutions of the underlying field equations that stably coexist for a fixed configuration of parameters and BCs seems to disappear in open-flow systems: Recent numerical simulations of the full field equations – called Navier Stokes equations (NSE) for short – for convective roll vortices traveling downstream in an imposed horizontal Poiseuille flow [142–145] and of Taylor vortices propagating downstream in an axial throughflow [146] showed that in the absolutely unstable regime the spatiotemporal structure of these propagating vortex (PV) flows is uniquely selected, i.e., independent of parameter history, initial conditions, and system size. The pattern shows a characteristic streamwise profile of the amplitude growing with increasing distance from the inlet and of the wave number, and a characteristic global, i.e., spatially constant oscillation frequency associated with the downstream motion of the pattern. These properties depend, however, on the BCs. On the other hand, in the convectively unstable regime [147] initial perturbations are blown out of the system. Therefore it requires a persistent perturbation source like, e. g., noise to sustain a vortex pattern in the convectively unstable regime [148]. These noise sustained patterns are sensitive to initial conditions and perturbations.

While most of the results reviewed here should pertain to a broad class of systems we use for definiteness the notation and terminology of the Taylor-Couette or Rayleigh-Bénard system with an externally imposed mean flow – measured by its Reynolds number Re – through an annulus between concentric cylinders or a not too wide rectangular convective channel [131, 149], respectively, such that the vortices are oriented tranversally to the throughflow. Then the deterministic field equations show for $Re > 0$ an *oscillatory* instability: The translationally invariant stationary basic flow state is stable for Taylor numbers T (Rayleigh numbers Ra) below a critical threshold $T_c(Re)$ ($Ra_c(Re)$). At the critical threshold a spatially extended, temporally oscillatory solution of downstream propagating vortices branches supercritically off the basic state. We use the relative control parameters

$$\mu = \frac{T}{T_c\,(Re)} - 1 \quad \text{and} \quad \epsilon = \frac{T}{T_c\,(Re=0)} - 1 \tag{6.1}$$

with $\mu_c = 0$, i.e., $\epsilon_c\,(Re) = \frac{T_c(Re)}{T_c(Re=0)} - 1$ denoting the critical reduced threshold for onset of PV flow. The same formulas with T replaced by Ra hold for the Rayleigh-Bénard system. The shear forces of the throughflow slightly stabilize the homogeneous basic state, so $\epsilon_c\,(Re)$ slightly increases with Re [131, 143, 144, 150–156].

For small supercritical control parameters, $0 < \mu \ll 1$, the vortex flow is governed by a narrow band of near critical modes. This allows to represent the hydrodynamic fields within a GLE approximation [11] by a slowly varying complex amplitude $A(x,t)$ multiplied with the critical mode

$$w\,(x, \mathbf{r}_\perp; t) = A\,(x,t)\, e^{i(k_c x - \omega_c t)}\, \hat{w}\,(\mathbf{r}_\perp) + c.c.\,. \tag{6.2}$$

Here w is, for example a component of the vortex velocity field. The critical wave number k_c, frequency ω_c [131, 143, 144, 150–152, 154–156], and eigenfunction $\hat{w}(\mathbf{r}_\perp)$ [151] depending on the coordinates $\mathbf{r}_\perp$ perpendicular to the critical wavevector of the PV structures and the throughflow direction have been obtained from a linear stability analysis. A is given by the solution of the complex GLE

$$\tau_0 \left(\dot{A} + v_g A'\right) = \mu\,(1 + ic_0)\,A + \xi_0^2\,(1 + ic_1)\,A'' - \gamma\,(1 + ic_2)\,|A|^2\,A. \tag{6.3}$$

Dot and primes denote temporal and spatial derivatives, respectively. The coefficients τ_0, ξ_0^2, γ are even in Re while the group velocity v_g and the imaginary parts c_0, c_1, c_2 are odd in Re [144, 150].

6.1 Absolute and convective instability

For small μ and Re the control parameter plane is divided into three stability regimes - cf. Fig. 6.1 - characterized by different growth behavior of *linear* perturbations of the basic flow state. In the presence of throughflow one has to distinguish [157] between the spatiotemporal growth behavior of spatially localized perturbations and of spatially extended ones. Below the critical threshold $\mu = 0$, i.e., $\epsilon_c\,(Re)$ for onset of PV flow (dashed line in Fig. 6.1) any perturbation, spatially localized as well as extended, decays. This is the parameter regime of absolute stability of the basic state.

For $\mu > 0$ extended perturbations can grow. However, a spatially localized perturbation, i.e., a wave packet of plane wave perturbations is advected in the so-called convectively unstable parameter regime faster downstream than

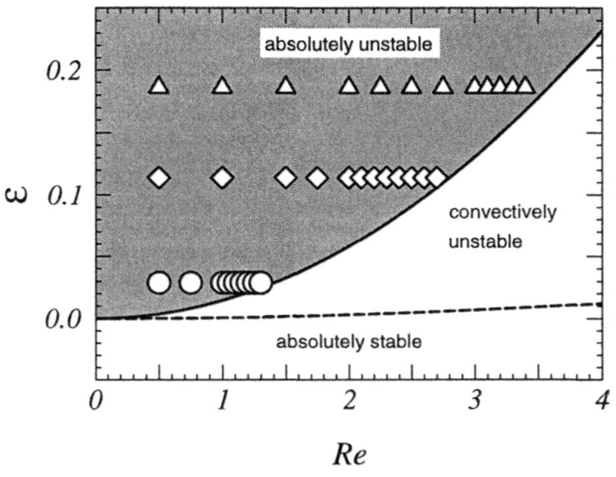

Fig. 6.1. Stability domains of the basic state. Numerical simulations of the Taylor-Couette system have been performed for the parameters marked by symbols. Dashed line is the critical threshold for onset of extended PV flow and full line the boundary (6.4) between absolute and convective instability.

it grows – while growing in the comoving frame it moves out of the system [147, 148, 157, 158]. Thus, both the downstream as well as the upstream facing intensity front of the vortex packet move downstream. A spatially localized perturbation is blown out of any system of finite length and the basic state is reestablished. It therefore requires a persistent perturbation source like, e. g., noise operating in the bulk part of the system or fluctuations entering the system via the inlet to sustain a vortex pattern in the convectively unstable regime [148] – c.f. Sec. 6.3.

In the absolutely unstable regime (shaded region in Fig. 6.1) a localized perturbation grows not only in downstream direction but it grows and spatially expands also in upstream direction until the upstream propagating front encounters the inlet in a finite system. The final pattern resulting in such a situation is self-sustained and stable. For sufficiently large amplitudes this structure is dominated by the deterministic contributions in the relevant balances of momentum and heat and thus it is insensitive to, say, thermal noise. This pattern shows in the downstream direction a characteristic intensity profile under which the PV flow develops with increasing distance from the inlet. Approaching the boundary between absolute and convective instability the growth length of the PV structure diverges in the absence of any perturbation source and the PV pattern is blown out of the system. However, in the presence of noise there is a transition to a noise sustained structure with a characteristic finite growth length depending on noise properties and control parameters.

Within the GLE approximation the boundary (full line in Fig. 6.1) between absolute and convective instability is given by [148]

$$\mu^c_{conv} = \frac{\tau_0^2 v_g^2}{4\xi_0^2\,(1 + c_1^2)} \tag{6.4}$$

corresponding to $\epsilon^c_{conv} = \epsilon_c + (1+\epsilon_c)\mu^c_{conv}$. It agrees very well with the boundary resulting from the full NSE for the small Reynolds numbers considered here [143–145, 156].

6.2 Pattern selection in the absolutely unstable regime: a nonlinear eigenvalue problem

Here we review how the unique spatiotemporal selection of PV patterns in the absolutely unstable regime can be understood as a nonlinear eigenvalue problem with the oscillation frequency being the eigenvalue and the profiles of pattern intensity and wave number determining the corresponding eigenfunction. Although the selected PV structures resulting from the NSE and from the GLE approximation differ in a characteristic way the selection mechanism looks similar. We also show how the pattern selection process is related to the one occurring behind a front that spatially separates an unstable, homogeneous state from a stable, structured one.

We have performed numerical simulations of the full 2D NSE for rotationally symmetric Taylor vortices and for straight convective roll vortices in systems of finite length [142–146]. They revealed in the absolute unstable regime for various end conditions patterns of vortices propagating downstream under a stationary intensity envelope after transients have died out. The oscillation frequency ω of the PV flows is spatially constant while the local wave number k and the vortex flow intensity varies. The frequency and the spatial variation of the PV pattern depends on the control parameters and on the BCs but not on parameter history, initial conditions, and system length provided the latter is large enough to allow for a saturated bulk region.

To illustrate the global properties of the flow patterns we present in Fig. 6.2 a hidden-line plot of the axial velocity field w in the Taylor-Couette system. The intensity variation of PV flow at the fronts of the envelope causes there a spatial variation in the local wavelength $\lambda(x)$. However, the oscillation frequency ω of PV flow is globally constant [142, 146]. Sufficiently away from the inlet and outlet we observe a bulk region of nonlinear saturated PV flow with spatially uniform amplitude and wavelength. With increasing *Re* this bulk pattern is pushed further and further downstream. The growth length l from the inlet depends on μ and *Re*. It increases and finally diverges when the control parameters μ, *Re* approach the absolute-convective instability border. The *scaled* growth length $L = \sqrt{\mu}\, l/\xi_0$ computed from the GLE agrees quite well with results from the NSE [146]. Due to the scaling property of the GLE [143–145] keeping in mind the smallness of the imaginary parts c_i all GLE values for L fall onto one curve for different control parameters.

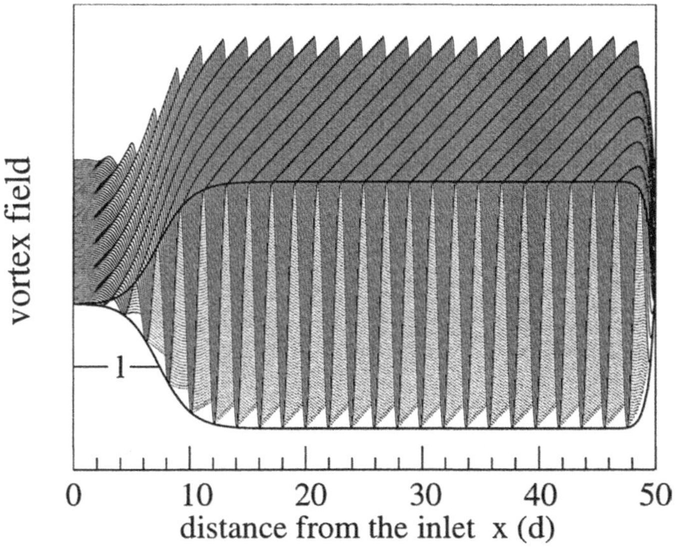

Fig. 6.2. Spatiotemporal structure of PV flow in the Taylor-Couette system. Thin lines show vertically displaced snapshots of the vortex velocity field at successive, equidistantly spaced times. Thick lines show the stationary envelope. The BC at the inlet (left) and outlet (right) suppresses any vortex flow there.

6.2.1 Sensitivity to boundary conditions

The deviations between GLE and NSE can be large for inlet BCs which do not enforce the basic state's field values for all fields [142]. This is demonstrated in Fig. 6.3 for the Rayleigh-Bénard system with various BCs at inlet and outlet: no-slip (N) or free-slip (F) for the vertical velocity field w and the vertically linear conductive profile (T_{cond}) or the mean temperature (T_0) or thermal insulation $\partial_x T = 0$ (Q_0) for the temperature field. These BCs have different phase pinning properties for the PV structure and therefore cause different frequencies. For example, for the combination $F - Q_0$ the phase pinning is strongest and consequently ω is the smallest, because the two von Neumann BCs $\partial_x w = 0$ and $\partial_x T = 0$ yield nonvanishing field amplitudes and leave a free-phase condition only for the lateral component of the vortex velocity field.

The conditions $N - T_0$ and $F - T_0$ induce a large stationary boundary vortex to be seen in the lateral profiles of the zeroth temporal modes of the velocity field. The boundary vortex is similar to the stationary Ekman vortex in the Taylor-Couette system with throughflow when the azimuthal velocity is enforced to vanish at the end plates. In addition there are stationary "flank" vortices in the region of largest lateral growth of $|u_1|$ that are driven by the lateral gradients of $|u_1|$ and which also feed back on the lateral variation of $|u_1|$. With increasing distance from inlet/outlet all these additional structures

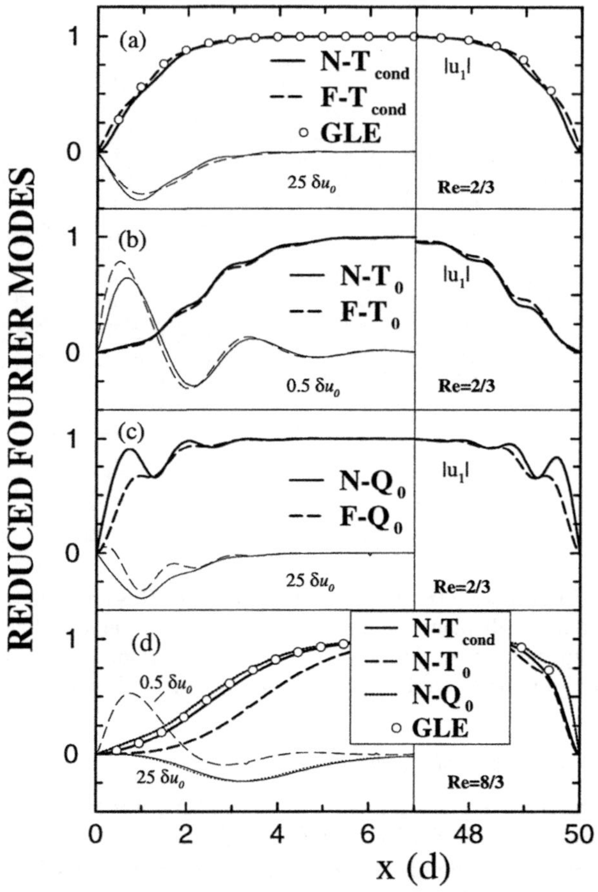

Fig. 6.3. Lateral profiles of convection structures for different BCs. Temporal Fouriermodes $|u_1|$ and δu_0 reduced by the bulk value of $|u_1|$ are shown at $z = 0.225$ close to the inlet ($x = 0$) and outlet ($x = 50$) boundaries. For clarity δu_0 is suppressed in the outlet region $47 < x < 50$. Lines refer to NSE results and open circles to the GLE subject to $N\text{-}T_{cond}$ conditions. Parameters are $\epsilon = 0.215$ and $\sigma = 1$.

smooth out and a saturated bulk appears with constant wave number and field amplitudes – provided that the system is long enough.

6.2.2 Pattern selection within the GLE

Our investigations [142–146] have shown that the throughflow causes an unique pattern selection: Possible wave numbers within the Eckhaus-stable band of stationary patterns for $Re = 0$ collapse to only one uniquely selected PV structure for $Re > 0$. Here we elucidate how this pattern selection mechanism can be understood within the GLE approach as a nonlinear eigenvalue

value problem where the frequency of the PV pattern is the eigenvalue and the complex amplitude is the eigenfunction.

We look for solutions of the GLE of the form

$$A(x,t) = a(x)e^{-i\Omega t} = R(x)e^{i[\varphi(x)-\Omega t]}, \tag{6.5}$$

separating spatial and temporal variations with stationary envelope $R(x)$, stationary wave number $q(x) = k(x) - k_c = \varphi'(x)$, and constant frequency $\Omega = \omega - \omega_c$. We have numerically solved the GLE and verified that its long-time solution is indeed of the form (6.5). Inserting this solution ansatz into the GLE one obtains the nonlinear eigenvalue problem

$$\tau_0 \left(-i\Omega a + v_g a'\right) = \mu \left(1 + ic_0\right) a + \xi_0^2 \left(1 + ic_1\right) a'' - \gamma \left(1 + ic_2\right) |a|^2 a \tag{6.6}$$

with Ω and $a(x)$ being the eigenvalue and associated eigenfunction, respectively. We are interested in solutions for which the variation of $R(x)$ and in particular of $q(x)$ is "as small as possible". Such a solution type seems to be connected with the eigenvalue Ω that is closest to zero. While we cannot make statements about the eigenvalue spectrum our numerical solutions indicate that initial conditions always evolve into a pattern with smooth amplitude $a(x)$ and small Ω. In that sense the eigenvalue problem (6.6) resembles ground state problems of the Schrödinger equation – increasing the spatial variation of the wave function (amplitude a) implies higher kinetic energy (frequency Ω).

The eigenfrequency Ω is determined by *global* properties and not by the *local* variation of the eigenfunctions $R(x)$ and $q(x)$, say, at the inlet or in the bulk. From (6.6) one finds that with $R = 0$ enforcing the basic state at the inlet and outlet the wave number is always fixed there to the value $q_{in} = q_{out}$,

$$\xi_0 q_{in,out} = -\frac{1}{2} c_1 \frac{\tau_0 v_g}{\xi_0(1+c_1^2)} = -\frac{c_1}{\sqrt{1+c_1^2}} \sqrt{\mu_{conv}^c} \tag{6.7}$$

whenever the modulus grows with finite slope, $R'_{in,out} \neq 0$. For $Re = 0$ one obtains $q_{in,out} = 0$ [159]. Furthermore, one can derive the relation $\tau_0 \Omega = (c_1 - c_0)\mu - 2c_1\mu_{conv}^c - 3(1+c_1^2)\xi_0^2 q'_{in,out}$ between the eigenfrequency Ω and the slope $q'_{in,out}$ at the inlet and outlet [146]. On the other hand, in the bulk part with homogeneous modulus R_b and wave number q_b, i. e., where $R'_b = R''_b = q'_b = 0$, the dispersion relation

$$\tau_0 \Omega = (c_2 - c_0)\mu + \tau_0 v_g q_b + (c_1 - c_2)\xi_0^2 q_b^2 \tag{6.8}$$

provides a relation between Ω and the *bulk* wave number q_b. Approaching the convective instability boundary (6.4) we found that $q'_{in} \to 0$ so that $\Omega \to \Omega_{conv}^c = -(c_0 + c_1)\mu_{conv}^c/\tau_0$ follows from the first relation, and q_b approaches a limiting value that is determined by (6.8).

6.2.3 Comparison with front propagation

It is suggestive to compare PV patterns like the one in Fig. 6.2 with those behind an upstream facing front that connects in an infinite system the basic state at $x = -\infty$ with the developed PV state at $x = +\infty$. A *linear* growth analysis of the GLE along the lines of [Sec. VI B 3;11] shows that the far tail of a *linear* front of PV perturbations moves with the front velocity $v_F = v_g - 2\frac{\xi_0}{\tau_0}\sqrt{\mu(1+c_1^2)}$. At the boundary $\mu = \mu_{conv}^c$ the front is stationary that is $v_F = 0$. We compare the properties of such a stationary front with those of the solution $A(x,t)$ (6.5) with stationary modulus in a semi-infinite system, $0 \leq x < \infty$. The reason for restricting the comparison to μ_{conv}^c is that an upstream moving front in the absolutely unstable regime will be pushed against the inlet whence the spatiotemporal structure of the free front gets modified by the inlet.

In the linear part of the stationary front the flow amplitude $A(x,t) \sim e^{i[Q_s x - \zeta(Q_s)t]}$ varies with a wave number $q_F = \Re Q_s$ and an oscillation frequency $\Omega_F = \Re\zeta(Q_s)$. Here Q_s is the saddle position of the complex dispersion relation of the linear GLE in the complex Q-plane. So Ω_{conv}^c (6.8) coincides with Ω_F, and q_{in} (6.7) agrees with q_F. Thus, for $\mu = \mu_{conv}^c$ the frequency eigenvalue of the PV pattern growing in the downstream direction from the inlet at $x = 0$ agrees with the oscillation frequency selected by the stationary, *linear* front. But the PV front pattern develops from the basic state with an exponential growth rate κ of the modulus. This leads for $R \to 0$ to $R' = \kappa R, R'' = \kappa^2 R, ... \to 0$ for $x \to -\infty$. On the other hand, the PV flow intensity in the semi-infinite system in general drops to zero at the inlet with finite R' and R''. The difference disappears only at the border line $\mu = \mu_{conv}^c$ since there $R', R'', ... \to 0$ for $x = 0$.

For $\mu > \mu_{conv}^c$, however, $\Omega > \Omega_{conv}^c$ – cf. the half-tone symbols in Fig. 6.4. Therein we compare the eigenfrequencies of the *nonlinear* NSE (open symbols) with those (solid line) that would be selected by a *linear* front that is stationary at μ_{conv}^c. The solid line in Fig. 6.4 represents the frequency Ω_F selected by the stationary front of the *linear* NSE. This frequency is determined by the saddle Q_s of the linear dispersion relation $\zeta(Q)$ of the NSE. We have evaluated the latter with a shooting method as described in [156]. With increasing *Re* the eigenfrequency Ω of the *nonlinear* NSE approaches the frequency Ω_F selected by the *linear* stationary front at the border μ_{conv}^c of the absolutely unstable regime as indicated for the three ϵ-values shown in Fig. 6.4. Thus there, the eigenvalue problem becomes effectively linear and the pattern selection mechanism becomes that one of linear front propagation.

This NSE behavior is very similar to the GLE behavior. Note, however, that the linear dispersion relations of NSE and GLE are different: the front frequency $\Re\zeta_{NSE}(Q_s)$ is positive and increases with throughflow while $\Re\zeta_{GLE}(Q_s)$ is negative and decreases with throughflow. This difference in

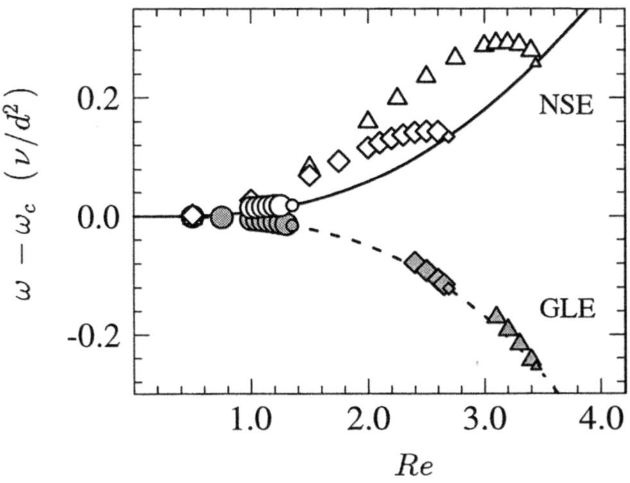

Fig. 6.4. Selected oscillation frequency of PV flow in the Taylor-Couette system obtained from the *nonlinear* NSE (open symbols) and GLE (half-tone symbols). The full (dashed) line is the front frequency Ω_F of the *linearized* NSE (GLE) at μ^c_{conv}. Parameters are $\epsilon = 0.0288$ (circles), $\epsilon = 0.114$ (diamonds), and $\epsilon = 0.186$ (triangles). The small symbols marking Ω_F for these three ϵ values show that the eigenfrequencies of the nonlinear equations in the absolutely unstable regime approach the Ω_F limit curve at the right places: $\Omega \to \Omega^c_{conv} = \Omega_F$.

the dispersion relations seems to be the major cause for the differences in the selected patterns.

6.3 Noise sustained pattern growth in the convectively unstable regime: bulk versus boundary effects

Nonequilibrium extended systems that undergo a pattern forming instability with nonzero group velocity [11] – whether the latter arises in open-flow configurations [131, 160] via an imposed throughflow [142–144, 146, 154–156, 161–165] or internally from a bifurcation to TWs [166–168] – show a peculiar sensitivity to perturbations in the convectively unstable regime [147, 157] where initially spatially localized perturbations are advected downstream by the group velocity while being amplified selectively [147, 148] according to the deterministic amplification mechanism that occurs near the pattern forming bifurcation. Thus, advection spatially deconvolutes the pattern growth from its cause – the perturbation source. But the latter leaves its signature via the bifurcation-induced, deterministic amplification process on the spatiotemporal properties of the downstream growing macroscopic structures. The statistical dynamics of fluctuations of these structures is easily detectable [144, 154–156, 162–164, 167, 168] thus providing information about the small perturbations that have triggered their growth. The feasibility to distinguish the influence of different perturbation sources – e.g., those that are external to the system under study versus internal thermal noise – on the macroscopic pattern growth in a very direct way has attracted much research activity recently [154–156, 162–164, 167, 168].

The deterministic dynamics of the macroscopic structure forming instability determines which of the modes that a perturbation source emits is amplified and which one decays in downstream direction. And as long as the mode amplitudes are still small each of the perturbation modes evolves separately and independently: the bifurcation dynamics is initially linear with saturating nonlinearities being irrelevant thus producing exponential growth with time and downstream distance. Hence, in a system with group velocity in positive x direction one expects that fluctuations generated near the inlet boundary at $x = 0$ and/or perturbations that might enter the system via the inlet are most important. They have the longest advection/amplification time to grow in the flow before they reach a particular observation point at a downstream location $x > 0$. Therefore a careful analysis of the influence of the inlet boundary condition relative to the effect of bulk (fluctuating) forces is necessary in particular in view of the fact that both sources cause exponential growth. The results of such an analysis [169] sheds also light on the problem whether in bounded experimental systems it is internal bulk generated thermal noise that triggers and sustains the growth of macroscopic vortex structures in the convectively unstable regime or other kind of environment-produced perturbations that might, e. g., be advected from the inlet boundary into the bulk.

6.3.1 Linear stochastic GLE in semiinfinite systems – solution via spatial Laplace transformation

The starting point of most theoretical approaches for describing the effect of thermal noise on macroscopic hydrodynamic pattern growth is the linear stochastic GLE [62, 170–173]

$$\left[\tau_0(\partial_t + v_g\partial_x) - (1 + ic_0)\mu - (1 + ic_1)\xi_0^2\partial_x^2\right] A(x,t) = f(x,t) \tag{6.9}$$

with Gaussian distributed complex forces $f(x,t)$ of zero mean that are δ-correlated in space and time

$$< f(x,t)\,[f(x',t')]^* >= F\delta(x - x')\delta(t - t'). \tag{6.10}$$

Here the star implies complex conjugation. The noise strength F has been evaluated [62, 170–174] for the RB and TC systems without throughflow and with different boundary conditions. We mention that the value of F characterizing a particular system depends on the arbitrary choice of the normalization condition for the linear critical mode $\hat{w}\,(\mathbf{r}_\perp)$ that enters into Eq. (6.2) and that establishes the relation between the physically observable order parameter field $w\,(x, \mathbf{r}_\perp; t)$ and the complex amplitude A of the critical mode.

We consider a semiinfinite system with the throughflow entering the system at $x = 0$. This flow transports fluctuations into the system that we specify by imposing an inlet boundary condition $A\,(x = 0, t)$ on the amplitude A. We consider situations in which $A\,(x = 0, t)$ is the result of a statistically stationary process. Thus, we restrict our investigations to the long-time statistically stationary fluctuations of $A(x, t)$ – initial transients occurring in experiments after, e.g., changing control parameters have died out. Under these circumstances the theoretical analysis is most conveniently done in frequency space. Then the stochastic GLE becomes an ordinary, second order, inhomogeneous differential equation [169]

$$\left[\tau_0(-i\omega + v_g\partial_x) - (1 + ic_0)\mu - (1 + ic_1)\xi_0^2\partial_x^2\right] A(x, \omega) = f(x, \omega) \tag{6.11}$$

that is to be solved subject to the inlet boundary condition $A(x = 0, \omega)$. A condition on the derivative of A at the inlet or, equivalently, on the behavior of A for $x \to +\infty$ follows from a physical requirement discussed below.

In the solution of Eq. (6.11) there appear contributions of the form e^{iK_1x} and e^{iK_2x} where $K_{1,2}$ are the complex ω-dependent characteristic exponents, i.e., spatial eigenvalues of Eq. (6.11)

$$\xi_0\sqrt{1 + ic_1}K_{\frac{1}{2}} = \pm i\sqrt{(1 - ic_1)\mu_{conv}^c - (1 + ic_0)\mu - i\omega\tau_0} - i\sqrt{(1 - ic_1)\mu_{conv}^c}. \tag{6.12}$$

The physically relevant characteristic exponent that controls the spatial growth of vortices in downstream direction is K_1. For $\mu < 0$, the spatial growth rate $-\Im K_1$ is negative implying spatial decay of each frequency mode. For $\mu > 0$ selective amplification of frequency modes out of a finite band $\omega_- < \omega < \omega_+$ occurs with band limits $\tau_0\omega_\pm = \pm 2\sqrt{(1 + c_1^2)\mu\mu_{conv}^c} + (c_1 - c_0)\mu$. For plots of $-\Im K_1$ versus ω see, e.g., [175, Fig. 1] and [156, Fig. 14]. The relations to our exponent K_1 are $iK_1 = \kappa[175] = \beta[156]$.

The GLE (6.11) was solved with the method of Green's functions and, equivalently, using a spatial Laplace transformation [169]. The latter method most clearly displays the influence of the inlet boundary properties, e.g., on the behavior at large $x \to +\infty$. In this way one can easily impose the "physical" condition that forbids exponential growth $\sim e^{iK_2x}$ for $x \to +\infty$. This condition is related to selecting the proper Green's function which ensures that (*i*) for $\mu < 0$ the response to a perturbation pulse $\sim \delta(x - x_0)\delta(t - t_0)$ decays and that (*ii*) in the convectively unstable regime, $0 < \mu < \mu_{conv}^c$, longtime growth of this response occurs only in the frame comoving with the group velocity v_g while at any fixed x one has decay for $t \to +\infty$. In the framework of the Laplace transformed solution this restriction amounts to the suppression of a pole at K_2 which leads to

$$i\,(\partial_x - iK_1)\,A(x, \omega)\,|_{x=0} + \frac{f(K_2, \omega)}{(1 + ic_1)\xi_0^2} = 0. \tag{6.13}$$

Hence, for a given forcing $f(K_2, \omega) = -i \int_0^\infty dx e^{-iK_2 x} f(x, \omega)$ and a given inlet condition $A(0, \omega)$ the slope $A'(0, \omega)$ is fixed by (6.13) such as to guarantee the above discussed limiting behavior.

The physical solution of the stochastic GLE that obeys the above restriction consists of the sum of two contributions, $A = A_b + A_{in}$[169]. The bulk part $A_b(x, \omega)$ with BC $A_b(x = 0, \omega) = 0$ contains amplitude fluctuations that are caused and sustained by bulk forces $f(x, \omega)$ whereas $A_{in}(x, \omega) = A(x = 0, \omega) e^{iK_1(\omega)x}$ describes how the (imposed) inlet fluctuations $A(x = 0, \omega)$ evolve downstream.

6.3.2 Frequency spectra of amplitude fluctuations

Due to the linear dynamics the two types of amplitude fluctuations, A_{in} and A_b, evolve independently from each other and since fluctuations can not propagate upstream in the convectively unstable regime, A_{in} and A_b are always uncorrelated. Hence, the frequency spectrum of the time-displaced autocorrelation function $C(x, \mid t - t' \mid) =< A(x, t) [A(x, t')]^* >$ of amplitude fluctuations at the same downstream position x is given by the sum $C(x, \omega) = C_b(x, \omega) + C_{in}(x, \omega)$ of a bulk contribution C_b and of an inlet contribution $C_{in} = e^{-2\Im K_1 x} D(\omega)$. The latter reflects how the spectrum $D(\omega) = C_{in}(x = 0, \omega) = C(x = 0, \omega)$ of inlet fluctuations evolves downstream under the GLE dynamics. Since the semiinfinite system with throughflow is spatially not translational invariant the amplitude correlations depend explicitly on the position x. On the other hand, the statistical stationarity of the forcing processes in the bulk and at the inlet causes a temporal dependence only via $\mid t - t' \mid$.

The exponential growth of the fluctuation spectrum that occurs for $\omega_- < \omega < \omega_+$ in the convectively unstable regime $0 < \mu < \mu^c_{conv}$ is given for large x by

$$C(x \to \infty, \omega) \to \left[D(\omega) - \frac{F \Im(K_1 + K_2)}{2(1 + c_1^2) \xi_0^4 \mid K_1^* - K_2 \mid^2} \right] e^{-2\Im K_1 x}. \tag{6.14}$$

To make quantitative estimates of the relative contributions to $C(x, \omega)$ a simple model for the frequency spectrum $D(\omega)$ of inlet fluctuations was used [169]. It captures an experimental situation with an upstream section $-\underline{L} < x < 0$ to the left of the inlet at $x = 0$ in which the fluid is equilibrated at a finite subcritical driving $\underline{\mu} < 0$. Here we mark quantities characterizing such a subcritical upstream part of the system by an underbar.

Babcock et al [156] have numerically solved the stochastic GLE for such a setup with $\underline{L} = 0.1, \underline{\mu} = -0.1$, and boundary condition $A(x = -\underline{L}, t) = 0$. We consider a situation where $\underline{L}$ is sufficiently large. Then perturbations, e.g., from the pumping machinery that are swept at $x = -\underline{L}$ into the subcritical upstream part of the apparatus have decayed by the time the fluid reaches the inlet, $x = 0$, of the supercritical part. Thus, one ideally has at $x = 0$ only fluctuations in the hydrodynamic fields that are sustained by thermally activated fluctuating forces. The most important field fluctuations are those with wave numbers close to the critical one and with spatial variation in the directions perpendicular to the throughflow similar to the critical mode. We approximately describe the dynamics of the amplitude of these modes with the linear stochastic GLE (6.9), now however for subcritical driving $\underline{\mu} < 0$. Therein, the spectrum of amplitude fluctuations at $x = 0$ are generated at subcritical driving by thermal forcing $f(x \leq 0, t)$ alone since $\underline{L}$ is large.

The total spectral power of these noise generated fluctuations of the amplitude $A(x = 0)$ of the critical mode at the inlet increases with increasing driving in the subcritical upstream section and diverges in the absence of nonlinearities in (6.9) at the critical driving $\underline{\mu} = 0$. Thus, the model is only realistic for a situation where the noise generated amplitudes in the upstream part are small enough to neglect the nonlinear contributions in the hydrodynamic field equations, i.e., for $\underline{\mu}$ not too close to zero. If that is the case then a variation $\propto 1/\sqrt{-\underline{\mu}}$ in the inlet contribution should be experimentally observable.

6.3.3 Comparison of inlet and bulk generated fluctuations

In Fig. 6.5 we show the downstream growth of the spectral peak $C_b(x, \omega = 0)$ in comparison with the peak height $C_{in}(x, \omega = 0)$ of the inlet contribution – for our parameters the spectral maxima are located practically at $\omega = 0$. The contributions are reduced by F/ξ_0 measuring the forcing strength. C_{in} was evaluated with the above described model for several subcritical upstream driving values of $\underline{\mu}$ as indicated. The dashed line labelled C_{SBH} is the result of Swift et al [172]. The latter was obtained with inlet fluctuations which are correlated with the bulk forcing. This enhances the amplitude fluctuations of the growing modes and thereby tends to overestimate the effect of thermal bulk forcing.

For the values of $\underline{\mu}$ in Fig. 6.5 the bulk contribution rapidly outgrows the inlet contribution. A similar observation was made also in the numerical treatment of Babcock et al [156]. Thus, *typically* the large (small)-x behavior is governed by bulk (inlet) fluctuations. However, the upstream model clearly shows that in appropriately designed experimental upstream sections – with $\underline{L}$ being large enough to avoid influence of the pumping machinery operating at $x = -\underline{L}$ – one can tune and enhance the inlet noise at $x = 0$ almost arbitrarily by selecting the subcritical upstream driving $\underline{\mu}$ accordingly.

Finally we should like to stress that the model for inlet fluctuations contains only internal thermal fluctuations generated in the subcritical upstream part. It does not contain perturbations, e.g., from the pumping machinery that might be advected towards the inlet in an experimental setup. Their contribution to the inlet fluctuation spectrum would have to be added to $D(\omega)$.

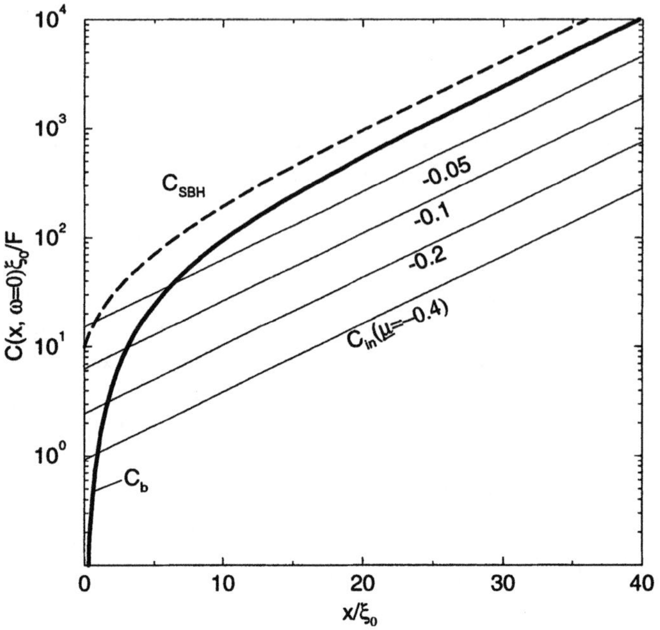

Fig. 6.5. Downstream growth of spectral peaks. They are located very close to $\omega = 0$ for our parameters Re=2, $\mu = \mu^c_{conv}/2 = 0.03$. Thin lines show $C_{in}(x, \omega = 0)\xi_0/F$ coming from model inlet fluctuations obtained for different subcritical upstream driving $\underline{\mu}$. For $C_{SBH}(x, \omega = 0)\xi_0/F$ [172] (thick dashed line) the inlet fluctuations are correlated with the bulk forcing and fixed by construction via μ, Re.

7 Influence of throughflow on binary fluid convection

We should like to draw attention to the fact that the spatio temporal pattern forming phenomena that occur when a mean flow is coupled to convection in binary mixtures are much richer than those, say, in pure fluid convection or Taylor vortex flow. Yet, while there are several experimental results for the latter systems we are not aware of experiments investigating the influence of throughflow on convective structures in binary mixtures although they seem to be realizable and in addition highly rewarding. So we strongly recommend such experiments.

In this section we first describe how a lateral throughflow changes structure and dynamics of convection fields at onset. The relevant characteristic exponents, the spatial structure of their associated eigenfunctions for different perturbations of the conductive basic state, and the throughflow induced changes of the bifurcation thresholds for different types of convective solutions have been determined in quantitative detail for various parameters

[176]. These thresholds are defined by vanishing real parts of the characteristic exponents. They mark stability boundaries of the conductive state against different convective perturbations in the control parameter space spanned by the Rayleigh number Ra, the Soret coupling ψ, and the throughflow Reynolds number Re. The full linear field equations subject to realistic boundary conditions were solved using a shooting method for perturbations with wavevectors parallel to the throughflow. The solutions for other wavevectors can be obtained from the former by a straightforward symmetry transformation.

For $\psi < 0$ the most important result of this linear analysis is that the throughflow lifts the Hopf symmetry degeneracy of left and right TWs at $Re = 0$: frequencies, bifurcation thresholds, and structural properties of the two waves are changed dramatically. Mixtures with more negative ψ require a larger Re for the changes to reach a comparable relative size. For sufficiently large Re the lowest relevant bifurcation threshold of binary mixtures with any ψ asymptotically approaches the critical Rayleigh number $Ra_c(Re, \psi = 0)$ of a pure fluid with imposed throughflow. Then the externally imposed shear flow eliminates the Soret-induced coupling effects between the convective concentration field and the other fields by suppressing vertical convective transport of Soret driven concentration perturbations. In the final subsections we describe problems associated with the pattern formation of subcritically bifurcating *nonlinear* TW solutions.

7.1 Linear properties

For small Ra, Re a laterally homogeneous conductive state without vertical convective flow is stable. It is a combination of vertically linear temperature and concentration fields and a plane horizontal Poiseuille flow. The linearized field equations for the deviations w, $\theta = T - T_{cond}$, $c = C - C_{cond}$ from the conductive state are solved with a plane-wave solution ansatz $\Phi(\mathbf{r}, t) = \widehat{\Phi}(z)\; e^{i(k_x x + k_y y)}\; e^{st}$ for the perturbations $\Phi = (w, \theta, \zeta)$ with $\zeta = c - \psi\theta$. Here $s = \Re s + i\,\Im s = \gamma - i\omega$ is a complex characteristic exponent. $\widehat{\Phi} = (\widehat{w}, \widehat{\theta}, \widehat{\zeta})$ denote complex z-dependent amplitude functions and $\mathbf{k} = k_x \mathbf{e}_x + k_y \mathbf{e}_y$. Inserting this ansatz into the linearized field equations yields a 3×3 linear eigenvalue problem for the eigenvalues s and eigenvectors $\widehat{\Phi}$ [176]. Due to the boundary conditions the eigenvalue spectrum is discrete. We are interested in the three characteristic exponents s_j $(j = 1, 2, 3)$ whose growth rates γ_j are closest to zero and whose eigenfunctions $\widehat{\Phi}_j(z)$ have no nodes other than at the horizontal boundaries $z = 0, 1$.

Since $\mathbf{k}$ and Re enter only as k^2 and $k_x Re$ the dependence of the functions $f = s$, $\widehat{\Phi}$ on $\mathbf{k}$ and Re is $f = f\left(k^2,\; k_x Re\right)$. Using this behavior the Squire transformation [177]

$$f\left(k_x^2 + k_y^2,\; k_x Re\right) = f\left(\widetilde{k}_x^2, \widetilde{k}_x \widetilde{Re}\right)\; ; \; \widetilde{k}_x^2 = k_x^2 + k_y^2 ; \; \widetilde{Re} = \frac{k_x}{\widetilde{k}_x}\, Re \qquad (7.1)$$

relates the functions f for a wavevector $\mathbf{k}$ to the functions that have been determined for $\widetilde{\mathbf{k}} = \widetilde{k}_x \mathbf{e}_x$ in throughflow direction and Reynolds numbers $\widetilde{Re}$. Therefore, it suffices to consider only $k_y = 0$ perturbations with $\mathbf{k} = k_x \mathbf{e}_x$ parallel or antiparallel to the throughflow. For $Re = 0$ the Squire relations reflect the horizontal rotational symmetry of the system in the absence of throughflow.

Upon flow reversion, $Re \to -Re$, the set $\{s_j,\ \widehat{\Phi}_j\}$ of eigenvalues and eigenfunctions transforms into each other since the balance equations are invariant under the parity operation $(x, u) \to -(x, u)$. The transformation behavior of $\{s_j,\ \widehat{\Phi}_j\}$ follows explicitly from the fact that the linear operator $\mathcal{L}$ entering the eigenvalue equation transforms as

$$\mathcal{L}(-q_x) = \mathcal{L}^*(q_x) \quad ; \quad q_x = k_x Re \tag{7.2}$$

under $Re \to -Re$ with the star denoting complex conjugation. From this relation one finds [176] that one eigenvalue, say $j = 3$, does not change so that $\gamma_3(-q_x) = \gamma_3(q_x)$, $-\omega_3(-q_x) = \omega_3(q_x)$, and $\widehat{\Phi}_3^*(-q_x) = \widehat{\Phi}_3(q_x)$. The other two eigenvalues and eigenfunctions are crossrelated to each other so that $\gamma_1(-q_x) = \gamma_2(q_x)$, $-\omega_1(-q_x) = \omega_2(q_x)$, and $\widehat{\Phi}_1^*(-q_x) = \widehat{\Phi}_2(q_x)$. In the absence of throughflow (i) the solutions depend on k_x^2 only, (ii) one eigenvalue is real, say $\omega_3 = 0$, with real eigenfunction $\widehat{\Phi}_3 = \widehat{\Phi}_3^*$, and ($iii$) the other two are complex conjugate – $\gamma_1 = \gamma_2$, $\omega_1 = -\omega_2$ with $\widehat{\Phi}_1^* = \widehat{\Phi}_2$ – describing symmetry degenerate left and right TWs for $Re = 0$.

In the remainder of this paper we mark critical quantities for which the growth rates γ_j first pass through zero when increasing r by a subscript c. In addition we replace the running index j for the three eigenvalues by a superscript S, U, or D. These superscripts identify the critical solution behavior, $e^{i\,(k_c\,x - \omega_c\,t)}$, in the limit $Re \to 0$: Eigenvalues for which $\omega_c(Re \to 0) = 0$ are marked by S since these perturbations are stationary for $Re = 0$. Eigenvalues for which $\omega_c(Re \to 0)$ is positive (negative) carry the superscript D (U) since they characterize for $Re \to 0$ perturbations which propagate in downstream (upstream) direction. So our classification of the different convective solutions is based on their behavior for $Re \to 0$.

As a representative example for a moderately negative Soret coupling we consider a mixture like ethanol-water with $\psi = -0.25$, $\sigma = 10$, $L = 0.01$. For this small Lewis number the threshold for stationary S perturbations has disappeared already at much weaker Soret coupling. See Ref. [176] for a detailed discussion and results for other parameter combinations. For convenience we scale Rayleigh numbers and wave numbers $r = Ra/Ra_c(Re = 0, \psi = 0)$ and $\widehat{k} = k/k_c(Re = 0, \psi = 0)$, respectively, by the critical ones of a pure fluid ($\psi = 0$) in the absence of throughflow.

7.1.1 Critical behavior

Thin lines in Fig. 7.1 show critical thresholds, wave numbers, and frequencies versus Re for waves traveling upstream (TW^U) and downstream (TW^D).

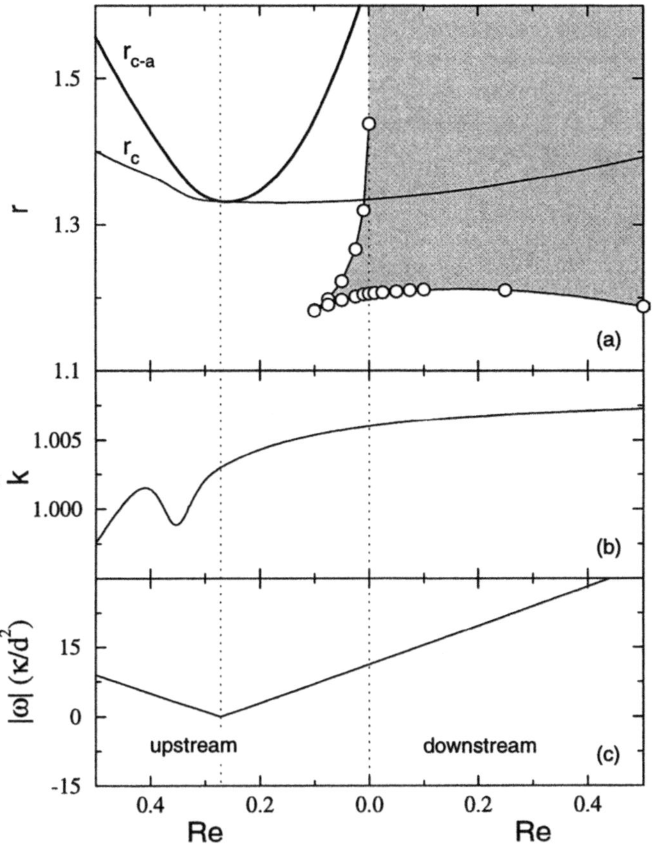

Fig. 7.1. Linear and nonlinear properties of TW^U (left column) and TW^D (right column) convection vs. *Re*. Note that *Re* increases from right to left in the left column. Critical thresholds (a), wave numbers (b), and frequencies (c) are shown by thin lines. Thick line is the border between the absolutely and convectively unstable parameter regime. The shaded area shows the existence range of ***stable, nonlinear*** extended TW^U and TW^D patterns. Circles mark TW saddle node positions taken from bifurcation diagrams like those of Fig. 7.3. Parameters are $L = 0.01$, $\sigma = 10$, $\psi = -0.25$.

For $Re = 0$ the critical values of the symmetry degenerated TWs are $r_{osc} = 1.3348$, $\widehat{k}_{osc} = 1.0060$ and $\omega_H = \pm 11.21$. For small *Re*, the bifurcation threshold r_c^U (r_c^D) is depressed (enhanced) by the throughflow so that the conductive state is destabilized (stabilized) against convective TW^U (TW^D) perturbations. So at very small *Re* it is TW^U convection that grows first when increasing r quasistatically. However, while r_c^D increases monotonically with *Re* the initial linear downwards shift of r_c^U changes at larger *Re* to a precipitous increase and a subsequent flattening; the latter occurs outside the plot range

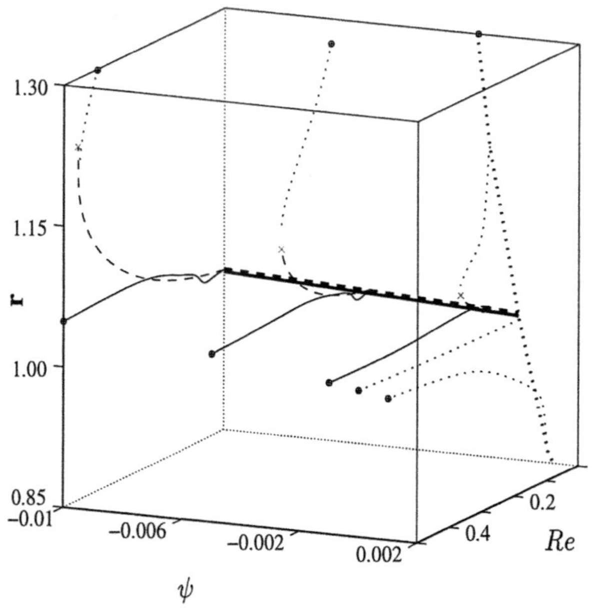

Fig. 7.2. Bifurcation surfaces in $r - Re - \psi$ space: r_c^U (thin full lines), r_c^D (thin dashed lines), and r_c^S (thin dotted lines) together with their ψ-dependence for $Re = 0$ (thick lines). Parameters are $L = 0.01$, $\sigma = 10$.

of Fig. 7.1. The variation of the critical wave numbers k_c^U and k_c^D is very small such that $k_c^U < k_c^D$.

The critical frequencies shown in Fig. 7.1c are practically linear functions of Re. They start at $Re = 0$ with the Hopf values and they can be very well approximated by $\omega_c^D \approx |\omega_H| + 41.9\,Re$ and $\omega_c^U \approx -|\omega_H| + 41.9\,Re$, respectively. Thus ω_c^U becomes positive for $Re > 0.27$. So the identification as "upstream" propagating perturbation made for $Re \to 0$ does not imply that the phase velocity is always opposite to the throughflow direction. The linear TW^U solution with $\omega_c^U = 0$ is stationary in the laboratory frame at $Re \approx |\omega_H|/41.9$. Close to this Reynolds number, r_c^U and $\widehat{k}_c^U$ show a marked variation that is associated with structural changes in the TW^U solution induced by reverting the propagation direction.

7.1.2 Bifurcation surfaces in $r - Re - \psi$ space for TW^U, TW^D, and TW^S solutions

To give an impression of the form of the three critical surfaces in the $r - Re - \psi$ space where TW^U, TW^D, and TW^S convection patterns bifurcate out of the conductive state we combine in Fig. 7.2 in a 3-D plot the Re-dependence of the bifurcation thresholds r_c^U (thin full lines), r_c^D (thin dashed lines), and r_c^S (thin dotted lines) together with their ψ-dependence for $Re = 0$ (thick lines).

The bifurcation surfaces $r_c^D(Re, \psi)$ and $r_c^U(Re, \psi)$ emanate for $\psi < 0$ out of the degenerate Hopf threshold line $r_c^D(Re = 0, \psi) = r_c^U(Re = 0, \psi)$ (thick dashed and full line) and split apart when the throughflow is switched on. Upon increasing Re further r_c^U gets indented slightly. On the other hand, r_c^D curls up and comes very close to the surface $r_c^S(Re, \psi)$ that is strongly bent down by the throughflow for $\psi < 0$. See [176] for a discussion of the further fate of these bifurcation surfaces.

The physically relevant, i.e., lowest lying surface at larger Re is $r_c^U(Re, \psi)$ (thin full lines). For any $\psi < 0$ it asymptotically approaches with increasing throughflow the $\psi = 0$ stability threshold, i.e., $r_c^U(Re, \psi) \to r_c^S(Re, \psi = 0)$. Thus a sufficiently large throughflow effectively eliminates the influence of any Soret coupling between *convective* concentration field and temperature and velocity fields (the diffusively induced concentration stratification in the conductive state, on the other hand, is not altered by the lateral shear flow) and the bifurcation thresholds of the mixture approach that one, $r_c^S(Re, \psi = 0)$, of the pure fluid. This also holds for mixtures with positive Soret coupling – cf. the thin dotted line for $r_c^S(Re, \psi)$ at $\psi > 0$.

7.1.3 Absolute versus convective instability

Since the critical group velocities $v_g = (\partial \omega(k)/\partial k)_c$ of the three different TW^U, TW^D, and TW^S patterns are finite in the presence of throughflow one has to distinguish between spatiotemporal growth behavior of spatially extended and of spatially localized perturbations. Here we deal only with a linear analysis. Thus, we do not investigate whether, e.g., at a subcritical bifurcation the wave packet grows to a stable nonlinear state that then expands back into the system with a larger nonlinear front velocity so that the nonlinear structured state ultimately invades the region occupied by the homogeneous state [79, 107, 178]. So we investigate the spatiotemporal growth behavior of three different packets consisting of superpositions of linear TW^U, TW^D, and TW^S perturbations. To that end the linear complex dispersion relations $s(k)$ of these three waves have been determined over the complex k-plane by solving the full linear field equations with a shooting method [176]. A saddle point analysis [147] of the three $s(k)$ functions then yields three functions r_{c-a}^U, r_{c-a}^D, and r_{c-a}^S that mark the boundary between the convectively and absolutely regime of the basic state against type TW^U, TW^D, and TW^S perturbations. This boundary is marked by parameter combinations for which one of the fronts of the linear wave packet reverts its propagation direction in the laboratory frame. The results of this analysis for $\psi = -0.25$ are shown in Fig. 7.1 by thick lines: r_{c-a}^U first decreases with increasing Re, coincides at its local minimum with r_c^U where $\omega_c^U = 0$, and afterwards increases with Re. The curve for r_{c-a}^D increases monotonically with Re but lies outside the plot range of Fig. 7.1.

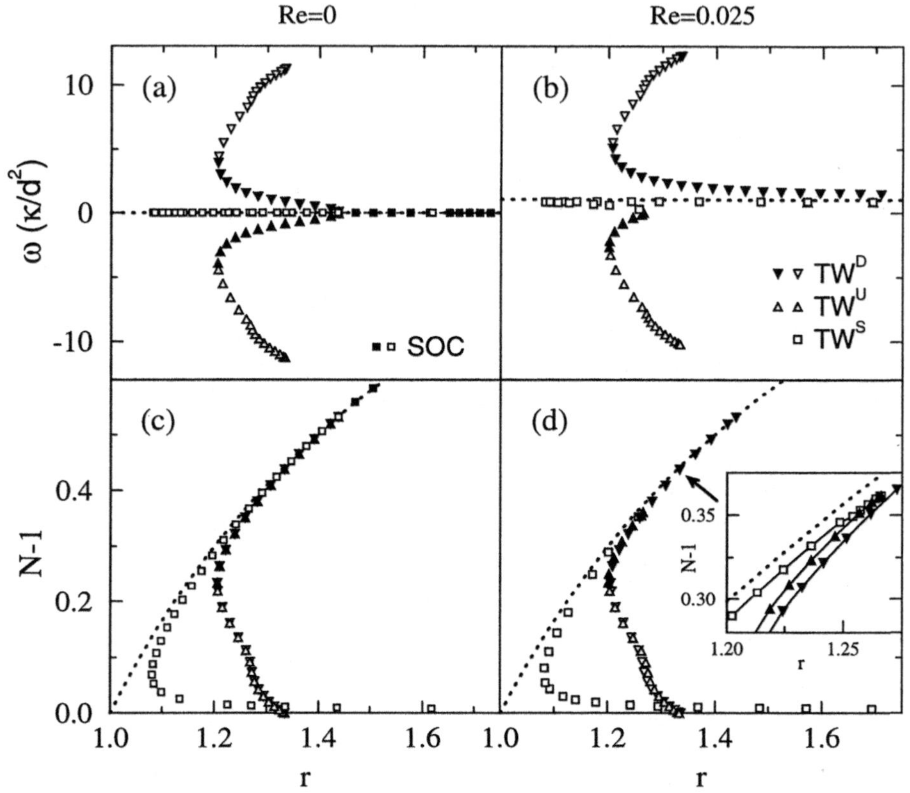

Fig. 7.3. Throughflow induced changes in the bifurcation diagrams of extended TW and SOC states. Open (filled) symbols refer to unstable (stable) states obtained with a MMG expansion using 783 modes. The $Re = 0$ diagrams correspond to Fig. 2.1a and b, however with the symmetry degenerate frequencies for left ($\omega < 0$) and right ($\omega > 0$) TWs being shown explicitely. The inset in (d) elucidates the connection between U and S solution branches of the Nusselt number that is seen more clearly in the frequency bifurcation diagram (b). Dotted lines represent N and ω in a pure fluid, $\psi = 0$. Parameters are $L = 0.01, \sigma = 10, \psi = -0.25$.

7.2 Nonlinear TW^U, TW^D, and TW^S states

7.2.1 Throughflow induced changes of the bifurcation behavior of extended states

Fig. 7.3 shows the r-dependence of frequency and Nusselt number of extended convective states with wave number $k = \pi$ for a very small $Re = 0.025$. The bifurcation thresholds for TW^U and TW^D are slightly split up and the critical oscillation frequencies are slightly shifted by the lateral throughflow.

The bifurcation behavior of the *stable, nonlinear* TW states (▲ and ▼) is more interesting. For $Re = 0$ the two symmetry degenerate stable TW

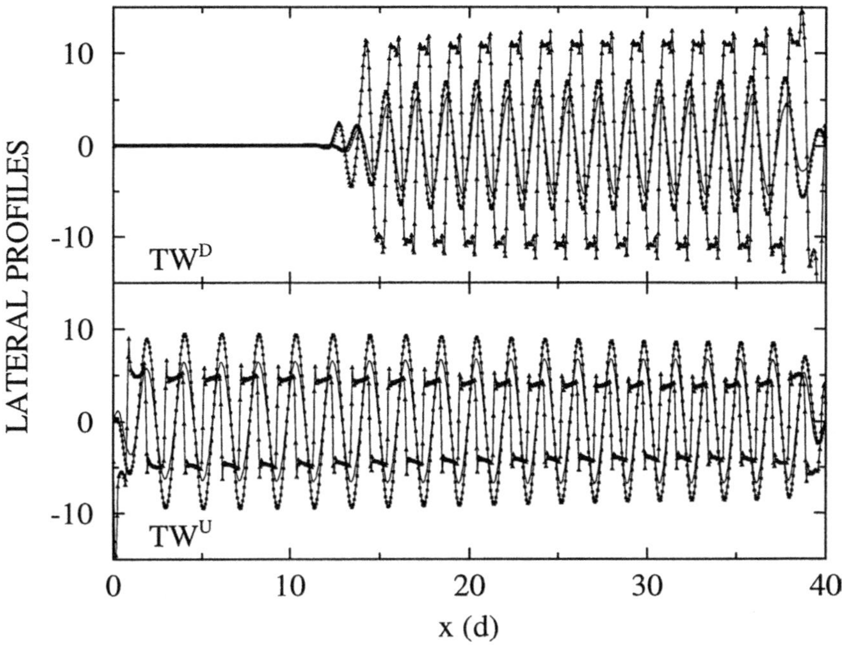

Fig. 7.4. Lateral profiles of TW^D and TW^U patterns at $r \approx 1.3$ and $Re = 0.01$ for w (lines), $20\,\delta T$ (lines with triangles), and $200\,\delta C$ (lines with squares). The basic state is enforced at inlet ($x = 0$) and outlet ($x = 40$). Parameters are $L = 0.01$, $\sigma = 10$, $\psi = -0.25$.

solution branches end by merging at r^* with zero frequency with the SOC branch, thereby changing the stability of the latter – c. f. the associated backward pitchfork bifurcation in (a) at $r^* \approx 1.43$. A small throughflow perturbs and breaks up this bifurcation leading to new branch connections in (b) and (d): The TW^D solution ($\blacktriangledown$) remains stable with growing r and its frequency stays finite. The stable TW^U branch ($\blacktriangle$) connects back and looses its stability via an additional new saddle to the unstable former SOC solution at smaller r which now, for $Re > 0$, becomes a downstream moving pattern TW^S ($\square$) with small frequency. The TW^D saddle and the two TW^U saddles, that delimit the existence range of stable TW^Us depend on Re. In particular the new TW^U saddle (joining the $\blacktriangle$ and $\square$ branch) approaches the old TW^U saddle (joining $\triangle$ and $\blacktriangle$ branches) so that already at $Re \approx 0.1$ the existence range of stable TW^Us is pinched to zero – cf. the open circles in the upstream part of Fig. 7.1.

7.2.2 Stable TW patterns in finite systems

In a system of finite length enforcing zero amplitudes at the lateral inlet and outlet boundaries one obtains patterns with inhomogeneous lateral amplitude profiles if the driving is strong enough. However, the parameter regime for which a ***supercritically bifurcating nonlinear*** self sustained pattern occurs in the bulk of a finite system is not determined by a linear analysis as for forward bifurcating patterns investigated in Sec. 6. There the propagation velocity of the upstream facing front of a ***linear*** packet of plane wave perturbations determined whether the forward bifurcating nonlinear PV structure can move upstream and fill the system or not. Here one has to determine the propagation direction of the ***nonlinear*** upstream facing front that connects the upstream basic conductive state with a downstream fully nonlinear TW^D or TW^U pattern whose extended version bifurcates subcritically. Thus, one has to distinguish whether this front moves upstream and invades the conductive state or whether it recedes downstream so that the structureless basic state invades the whole system. Stationarity of this front can then be used to define the border between nonlinear absolute and convective instability [178] as a generalization of the Maxwell criterium, say, for quintic GLE with real coefficients that are derivable from a potential.

In view of strongly reduced drift velocity of LTWs compared to the group velocity for linear perturbations, one expects that the propagation velocities of the nonlinear front solutions are also small. Therefore, the relative throughflow induced changes are much stronger for nonlinear states than for linear perturbations. This should lead to a much stronger Re dependence of the nonlinear absolutely vs. convectively unstable boundaries compared to the linear ones. Fig. 7.4 shows that the lateral amplitude profiles of TW^U and TW^D patterns differ strongly already for the very small throughflow rate of $Re = 0.01$, which supports the above hypothesis concerning the Re dependence of the propagation velocity of nonlinear fronts.

Acknowledgments

This work was supported by the Deutsche Forschungsgemeinschaft and the Volkswagen-Stiftung. We gratefully acknowledge the contributions of G. Ahlers, K. D. Eaton, W. Hort, M. Kamps, P. Kolodner, S. J. Linz, H. W. Müller, J. Niederländer, D. R. Ohlsen, M. T. Ouazzani, J. K. Platten, A. Recktenwald, D. Roth, R. Schmitz, C. M. Surko, A. Szprynger, and S. Y. Yamamoto as coauthors of different parts of the work reviewed here and discussions with them. Also the discussions with and the diploma theses of B. Huke, M. Junk, D. Papathanassiou, and H. Schmitt have contributed significantly to an understanding of the pattern formation problems discussed here. Last but not least we thank M. Kamps, R. Schmitz, and M. Fücker for their competent and engaged numerical support.

References

[1] Contrary to misleading statements found in the literature the oscillatory instability is not due to concentration diffusion being small: even if the diffusive time scales for momentum, heat, and concentration are the same ($L = \sigma = 1$) there are oscillations. They are caused by a sufficiently strong "off diagonal" Soret coupling between the degrees of freedom of concentration and temperature and their feedback into the buoyancy.

[2] Spiral defect chaos that was first seen in pure fluids has recently been observed also in binary mixtures; see J. Liu and G. Ahlers, Phys. Rev. E **55**, 6950 (1996) and references therein.

[3] W. Barten, M. Lücke, W. Hort, M. Kamps, Phys. Rev. Lett. **63**, 376 (1989).

[4] W. Barten, M. Lücke, M. Kamps, in *Nonlinear Evolution of Spatio-Temporal Structures in Dissipative Continuous Systems*, edited by F. H. Busse, L. Kramer, NATO ASI Series B **225**, (Plenum, New York, 1990), p. 131.

[5] M. Lücke, W. Barten, M. Kamps, Physica D **61**, 183 (1992).

[6] W. Barten, M. Lücke, M. Kamps, R. Schmitz, Phys. Rev. E **51**, 5636 (1995).

[7] St. Hollinger, P. Büchel, M. Lücke, Phys. Rev. Lett. **78**, 235 (1997).

[8] St. Hollinger, M. Lücke, H. W. Müller, Phys. Rev. E **57**, 4250 (1998).

[9] St. Hollinger, M. Lücke, Phys. Rev. E **57**, 4238 (1998).

[10] St. Hollinger, M. Lücke, Z. Phys. B **103**, 531 (1997).

[11] M. C. Cross, P. C. Hohenberg, Rev. Mod. Phys. **65**, 851 (1993).

[12] P. Kolodner, D. Bensimon, C. M. Surko, Phys. Rev. Lett. **60**, 1723 (1988).

[13] D. Bensimon, P. Kolodner, C. M. Surko, H. L. Williams, V. Croquette, J. Fluid Mech. **217**, 441 (1990).

[14] D. R. Ohlsen, S. Y. Yamamoto, C. M. Surko, P. Kolodner, Phys. Rev. Lett. **65**, 1431 (1990).

[15] K. D. Eaton, D. R. Ohlsen, S. Y. Yamamoto, C. M. Surko, W. Barten, M. Lücke, M. Kamps, P. Kolodner, Phys. Rev. A **43**, 7105 (1991).

[16] B. L. Winkler, P. Kolodner, J. Fluid Mech. **240**, 31 (1992).

[17] P. Kolodner, H. L. Williams, C. Moe, J. Chem. Phys. **88**, 6512 (1988).

[18] H. Touiri, J. K. Platten, G. Chavepeyer, Eur. J. Mech. B **15**, 241 (1996).

[19] E. Moses, V. Steinberg, Phys. Rev. A **34**, 693 (1986).

[20] V. Steinberg, E. Moses, in *Patterns, Defects, and Microstructures in Nonequilibrium Systems*, edited by D. Walgraef (NATO ASI Series E **121**, 1987), p. 309.

[21] J. Fineberg, E. Moses, V. Steinberg, Phys. Rev. A **38**, 4939 (1988).

[22] R. W. Walden, P. Kolodner, A. Passner, C. M. Surko, Phys. Rev. Lett. **55**, 496 (1985).

[23] R. Heinrichs, G. Ahlers, D. S. Cannell, Phys. Rev. A **35**, 2761 (1987).

[24] V. Steinberg, J. Fineberg, E. Moses, I. Rehberg, Physica D **37**, 359 (1989).

[25] E. Moses, V. Steinberg, Phys. Rev. Lett. **60**, 2030 (1988).

[26] E. Moses, V. Steinberg, Physica D **37**, 341 (1989).

[27] G. W. Baxter, K. D. Eaton, C. M. Surko, Phys. Rev. A **46**, 1735 (1992).

[28] P. Kolodner, Phys. Rev. A **46**, 1739 (1992); **46**, 6431 (1992).

[29] V. Steinberg, E. Kaplan, in *Spontaneous Formation of Space-Time Structures and Criticality*, edited by T. Riste, D. Sherrington (Kluver Academic Publishers, Boston, 1991), p. 207.

[30] P. Kolodner, Phys. Rev. A **46**, 6452 (1992).

[31] E. Kaplan, V. Steinberg, Phys. Rev. Lett. **71**, 3291 (1993).

[32] G. Ahlers, I. Rehberg, Phys. Rev. Lett. **56**, 1373 (1986).
[33] H. Gao, R. P. Behringer, Phys. Rev. A **34**, 697 (1986).
[34] G. Zimmermann, U. Müller, Int. J. Heat Mass Transfer **35**, 2245 (1992).
[35] E. Moses, V. Steinberg, Phys. Rev. A **43**, 707 (1991).
[36] P. Kolodner, A. Passner, C. M. Surko, R. W. Walden, Phys. Rev. Lett. **56**, 2621 (1986).
[37] M. Liu, J. R. de Bruyn, Canadian J. Phys. **70**, 689 (1992).
[38] V. Steinberg, E. Moses, J. Fineberg, Nucl. Physics B **2** (Proc. Suppl.), 109 (1987).
[39] P. Kolodner, C. M. Surko, H. L. Williams, Physica D **37**, 319 (1989).
[40] G. Zimmermann, U. Müller, S. H. Davis, J. Fluid Mech. **238**, 657 (1992); **254**, 720(E) (1993).
[41] P. Bigazzi, S. Ciliberto, V. Croquette, J. de Phys. **51**, 611 (1990).
[42] O. Lhost, J. K. Platten, Phys. Rev. A **40**, 6415 (1989).
[43] P. Le Gal, A. Pocheau, V. Croquette, Phys. Rev. Lett. **54**, 2501 (1985); P. Le Gal, Ph. D. thesis (Orsay, 1986).
[44] E. Moses, V. Steinberg, Phys. Rev. Lett. **57**, 2018 (1986).
[45] D. T. J. Hurle, E. Jakeman, J. Fluid Mech. **47**, 667 (1971).
[46] G. W. T. Lee, P. Lucas, A. Tyler, J. Fluid Mech. **135**, 235 (1983).
[47] R. P. Behringer, Rev. Mod. Phys. **57**, 657 (1985).
[48] I. Rehberg, G. Ahlers, Phys. Rev. Lett. **55**, 500 (1985); Contemporary Mathematics **56**, 277 (1986).
[49] H. Gao, R. P. Behringer, Phys. Rev. A **35**, 3993 (1987).
[50] T. S. Sullivan, G. Ahlers, Phys. Rev. Lett. **61**, 78 (1988); Phys. Rev. A **38**, 3143 (1988).
[51] T. J. Bloodworth, M. R. Ardron, J. K. Bhattacharjee, P. G. J. Lucas, N. D. Stein, Nonlinearity **3**, 981 (1990).
[52] A. La Porta, K. D. Eaton, C. M. Surko, Phys. Rev. E **53**, 570 (1996).
[53] A. La Porta, C. M. Surko, Phys. Rev. Lett. **77**, 5916 (1996); Phys. Rev. E **55**, R6327 (1997).
[54] L. D. Landau, E. M. Lifshitz, *Fluid Mechanics*, (Pergamon Press, Oxford, 1959).
[55] J. K. Platten, J. C. Legros, *Convection in Liquids*, (Springer, Berlin, 1984).
[56] G. Z. Gershuni, E. M. Zhukhovitskii, *Convective Stability of Incompressible Fluids*, (Keter Press, Jerusalem, 1976).
[57] S. Chandrasekhar, *Hydrodynamic and Hydromagnetic Stability*, (Dover Publications, New York, 1981).
[58] W. Barten, M. Lücke, M. Kamps, R. Schmitz, Phys. Rev. E **51**, 5662 (1995).
[59] E. Knobloch, D. R. Moore, Phys. Rev. A **37**, 860 (1988).
[60] M. C. Cross, K. Kim, Phys. Rev. A **37**, 3909 (1988).
[61] W. Hort, diploma thesis (Saarbrücken, 1990); private communication.
[62] W. Schöpf, W. Zimmermann, Phys. Rev. E **47**, 1739 (1993).
[63] J. Niederländer, M. Lücke, M. Kamps, Z. Phys. B **82**, 135 (1991).
[64] W. Barten, Ph. D. thesis (Saarbrücken, 1992; Dengmerter Heimatverlag, St. Ingbert, 1993).
[65] Ch. Jung, Ph. D. thesis (Saarbrücken, 1997).
[66] D. Bensimon, A. Pumir, B. I. Shraiman, J. de Phys. **50**, 3089 (1989).
[67] S. J. Linz, M. Lücke, H. W. Müller, J. Niederländer, Phys. Rev. A **38**, 5727 (1988).

[68] B. Shraiman, Phys. Rev. A **36**, 261 (1987).
[69] P. B. Rhines, W. R. Young, J. Fluid Mech. **133**, 133 (1983).
[70] St. Hollinger, Ph. D. thesis (Saarbrücken, 1996).
[71] A. Joets, R. Ribotta, Europhys. Lett. **10**, 721 (1989).
[72] A. Pocheau, V. Croquette, P. LeGal, Phys. Rev. Lett. **55**, 1094 (1985).
[73] S. Rasenat, G. Hartung, B. L. Winkler, I. Rehberg, Experiments in Fluids **7**, 412 (1989).
[74] S. J. Linz, M. Lücke, in *Propagation in Systems Far from Equilibrium*, edited by J. E. Wesfreid, H. R. Brand, P. Manneville, G. Albinet, (Springer, Berlin, 1988), p. 292.
[75] The fast stable TWs in Fig. 2.5 are genuine extended states that occur only for sufficiently strong Soret coupling. They are unrelated to the weakly nonlinear fast waves shown schematically in [11] Fig. 53. The latter appear even for small $|\psi|$ but only in finite–length non–annular convection channels with lateral endwalls that have a finite reflectivity for convection perturbations, see also [24].
[76] W. Barten, M. Lücke, M. Kamps, J. Comp. Phys. **91**, 486 (1990).
[77] B. A. Malomed, A. A. Nepomnyashchy, in *Nonlinear Evolution of Spatio-Temporal Structures in Dissipative Continuous Systems*, edited by F. H. Busse, L. Kramer, NATO ASI Series B **225**, (Plenum, New York, 1990), p. 411; Phys. Rev. A **42**, 6009 (1990).
[78] S. Fauve, O. Thual, Phys. Rev. Lett. **64**, 282 (1990).
[79] W. van Saarloos, P. C. Hohenberg, Phys. Rev. Lett. **64**, 749 (1990).
[80] R. J. Deissler, H. R. Brand, Phys. Lett. A **146**, 252 (1990).
[81] H. Riecke, Phys. Rev. Lett. **68**, 301 (1992).
[82] H. Riecke, Physica D **61**, 253 (1992).
[83] E. N. Lorenz, J. Atmospheric Sciences **20**, 130 (1963).
[84] G. Veronis, J. Marine Research **23**, 1 (1965).
[85] M. C. Cross, Phys. Lett. A **119**, 21 (1986).
[86] G. Ahlers, M. Lücke, Phys. Rev. A **35**, 470 (1987).
[87] O. Lhost, Ph. D. thesis (Mons, 1990).
[88] E. Knobloch, Phys. Rev. A **34**, 1538 (1986).
[89] W. Barten, M. Lücke, M. Kamps, Phys. Rev. Lett. **66**, 2621 (1991).
[90] E. Moses, J. Fineberg, V. Steinberg, Phys. Rev. A **35**, 2757 (1987).
[91] C. M. Surko, D. R. Ohlsen, S. Y. Yamamoto, P. Kolodner, Phys. Rev. A **43**, 7101 (1991).
[92] P. Kolodner, Phys. Rev. A **44**, 6448 (1991).
[93] P. Kolodner, Phys. Rev. E **50**, 2731 (1994).
[94] K. E. Anderson, R. P. Behringer, Phys. Lett. A **145**, 323 (1990); Physica D **51**, 444 (1991).
[95] P. Kolodner, Phys. Rev. A **43**, 2827 (1991).
[96] P. Kolodner, Phys. Rev. Lett. **66**, 1165 (1991).
[97] P. Kolodner, Phys. Rev. A **44**, 6466 (1991).
[98] P. Kolodner, Phys. Rev. E **48**, R4187 (1993).
[99] J. J. Niemela, G. Ahlers, D. S. Cannell, Phys. Rev. Lett. **64**, 1365 (1990).
[100] K. Lerman, E. Bodenschatz, D. S. Cannell, G. Ahlers, Phys. Rev. Lett. **70**, 3572 (1993).
[101] G. Ahlers, Physica D **51**, 421 (1991).
[102] P. Kolodner, J. A. Glazier, Phys. Rev. A **42**, 7504 (1990).

[103] J. A. Glazier, P. Kolodner, Phys. Rev. A **43**, 4269 (1991).

[104] O. Thual, S. Fauve, J. de Phys. **49**, 1829 (1988).

[105] V. Hakim, P. Jakobsen, Y. Pomeau, Europhys. Lett. **11**, 19 (1990).

[106] V. Hakim, Y. Pomeau, Eur. J. Mech. B **10** (Suppl.), 19 (1991).

[107] W. van Saarloos, P. C. Hohenberg, Physica D **56**, 303 (1992); **69**, 209(E) (1993).

[108] M. van Hecke, E. Dewit, W. van Saarloos, Phys. Rev. Lett. **75**, 3830 (1995).

[109] H. Riecke, Physica D **92**, 69 (1996).

[110] H. Herrero, H. Riecke, J. Bif. and Chaos **4**, 1343 (1994); Physica D **85**, 79 (1995).

[111] H. Riecke, W. J. Rappel, Phys. Rev. Lett. **75**, 4035 (1995).

[112] H. Riecke, Phys. Rev. E **52**, 5685 (1995).

[113] J. Liu, G. Ahlers, Phys. Rev. E **55**, 6950 (1997).

[114] S. W. Morris, E. Bodenschatz, D. S. Cannell, G. Ahlers, Phys. Rev. Lett. **71**, 2026 (1993).

[115] W. Decker, W. Pesch, A. Weber, Phys. Rev. Lett. **73**, 648 (1994).

[116] J. L. Liu, G. Ahlers, Phys. Rev. Lett. **77**, 3126 (1996).

[117] F. W. Giacobbe, J. Acoust. Soc. Am. **96**, 3568 (1994).

[118] W. Hort, S. J. Linz, M. Lücke, Phys. Rev. A **45**, 3737 (1992).

[119] D. Gutkowicz-Krusin, M. A. Collins, J. Ross, Phys. Fluids **22**, 1443 (1979); **22**, 1451 (1979).

[120] St. Hollinger, M. Lücke, Phys. Rev. E **52**, 642 (1995).

[121] E. Knobloch, Phys. Rev. A **40**, 1549 (1989).

[122] S. J. Linz, Phys. Rev. E **49**, 5881 (1994).

[123] M. A. Dominguez-Lerma, G. Ahlers, D. S. Cannell, Phys. Rev. E **52**, 6159 (1995).

[124] T. Clune, E. Knobloch, Phys. Rev. A **44**, 8084 (1991).

[125] H. W. Müller, M. Lücke, Phys. Rev. A **38**, 2965 (1988).

[126] G. Ahlers, *Lectures in the Sciences of Complexity*, edited by D. Stein (Addison-Wesley Longman, 1989), p. 175.

[127] M. Bestehorn, Physica D **61**, 59 (1992); Phys. Lett. A **174**, 43 (1993).

[128] J. Fernández-Vela, C. Pérez-García, Int. J. Bif. and Chaos **4**, 1333 (1994).

[129] M. Bestehorn, *Strukturbildung durch Selbstorganisation in Flüssigkeiten und in chemischen Systemen*, (Harri Deutsch, Frankfurt am Main, 1995).

[130] Ch. Jung, B. Huke, M. Lücke, submitted to Phys. Rev. Lett. (1998).

[131] For a review see, e. g., R. E. Kelly, Adv. Appl. Mech. **31**, 35 (1994).

[132] G. Ahlers, D. S. Cannell, M. A. Dominguez-Lerma, R. Heinrichs, Physica D **23**, 202 (1986).

[133] G. Ahlers, D. S. Cannell, Phys. Rev. Lett. **50**, 1583 (1983).

[134] R. Heinrichs, G. Ahlers, D. S. Cannell, Phys. Rev. Lett. **56**, 1794 (1986).

[135] M. Lücke, M. Mihelcic, B. Kowalski, K. Wingerath, in *The Physics of Struture Formation*, edited by W. Güttinger and G. Dangelmayr (Springer, Berlin, 1987), p. 97.

[136] M. Lücke, M. Mihelcic, K. Wingerath, Phys. Rev. A **31**, 396 (1985).

[137] H. Riecke, H. G. Paap, Phys. Rev. A **33**, 547 (1986); Phys. Rev. Lett. **59**, 2570 (1987).

[138] V. Croquette, Contemporary Physics **30**, 113 (1989); **30**, 153 (1989).

[139] I. Rehberg, E. Bodenschatz, B. Winkler, F. H. Busse, Phys. Rev. Lett. **59**, 282 (1987).
[140] J. Fineberg, V. Steinberg, Phys. Rev. Lett. **58**, 1332 (1987).
[141] M. Lücke, M. Mihelcic, B. Kowalski, Phys. Rev. A **35**, 4001 (1987).
[142] D. Roth, P. Büchel, M. Lücke, H. W. Müller, M. Kamps, R. Schmitz, Physica D **97**, 253 (1996).
[143] H. W. Müller, M. Lücke, M. Kamps, Europhys. Lett. **10**, 451 (1989).
[144] H. W. Müller, M. Lücke, M. Kamps, Phys. Rev. A **45**, 3714 (1992).
[145] H. W. Müller, M. Lücke, M. Kamps, in *Ordered and Turbulent Patterns in Taylor Couette Flow*, edited by D. Andereck, F. Hayot, Vol. 297 of NATO ASI Series B (Plenum, New York, 1992), p. 187.
[146] P. Büchel, M. Lücke, D. Roth, R. Schmitz, Phys. Rev. E **53**, 4764 (1996); P. Büchel, M. Lücke, in *Nonlinear Physics of Complex Systems*, edited by J. Parisi, St. C. Müller, W. Zimmermann (Springer, Berlin, 1996), p. 35.
[147] P. Huerre, P. A. Monkewitz, Annu. Rev. Fluid Mech. **22**, 473 (1990), J. Fluid Mech. **159**, 151 (1985); P. Huerre, in *Instabilities and Nonequilibrium Structures*, edited by E. Tirapegui, D. Villarroel (Reidel, Dordrecht, 1987), p. 141.
[148] R. J. Deissler, J. Stat. Phys. **40**, 371 (1985); **54**, 1459 (1989); Physica D **25**, 233 (1987); **18**, 467 (1986).
[149] M. T. Ouazzani, J. K. Platten, A. Mojtabi, Int. J. Heat and Mass Transfer **33**, 1417 (1990).
[150] A. Recktenwald, M. Lücke, H. W. Müller, Phys. Rev. E **48**, 4444 (1993).
[151] K. C. Chung, K. N. Astill, J. Fluid Mech. **81**, 641 (1977).
[152] D. J. Takeuchi, D. F. Jankowski, J. Fluid Mech. **102**, 101 (1981).
[153] K. S. Gage, W. H. Reid, J. Fluid Mech. **33**, 21 (1968).
[154] K. L. Babcock, G. Ahlers, D. S. Cannell, Phys. Rev. Lett. **24**, 3388 (1991).
[155] K. L. Babcock, D. S. Cannell, G. Ahlers, Physica D **61**, 40 (1992).
[156] K. L. Babcock, G. Ahlers, D. S. Cannell, Phys. Rev. E **50**, 3670 (1994).
[157] A. Bers, *Basic Plasma Physics I*, edited by A. A. Galeev, R. N. Sudan (North-Holland, New York, 1983); R. J. Briggs, *Electron Stream Interaction with Plasmas* (MIT Press, Cambridge, MA, 1964).
[158] R. Tagg, W. S. Edwards, H. Swinney, Phys. Rev. A **42**, 831 (1990).
[159] M. C. Cross, P. G. Daniels, P. C. Hohenberg, E. D. Siggia, J. Fluid Mech. **127**, 155 (1983).
[160] P. A. Monkewitz, Eur. J. Mech. B **9**, 395 (1990).
[161] A. Tsameret, V. Steinberg, Europhys. Lett. **14**, 331 (1991).
[162] A. Tsameret, V. Steinberg, Phys. Rev. E **49**, 4077 (1994).
[163] A. Tsameret, V. Steinberg, Phys. Rev. Lett. **67**, 3392 (1991).
[164] A. Tsameret, G. Goldner, V. Steinberg, Phys. Rev. E **49**, 1309 (1994).
[165] M. T. Ouazzani, J. K. Platten, H. W. Müller, M. Lücke, Int. J. Heat and Mass Transfer **38**, 875 (1995).
[166] R. J. Deissler, H. R. Brand, Phys. Lett. **130**, 293 (1988).
[167] W. Schöpf, I. Rehberg, Europhys. Lett. **17**, 321 (1992).
[168] I. Rehberg, S. Rasenat, M. la Torre-Juarez, W. Schöpf, F. Hörner, G. Ahlers, H. R. Brand, Phys. Rev. Lett. **67**, 596 (1991).
[169] M. Lücke, A. Szprynger, Phys. Rev. E **55**, 5509 (1997).
[170] R. Graham, Phys. Rev. A **10**, 1762 (1974); **45**, 4198 (E) (1992); **18**, 467 (1986).

[171] P. C. Hohenberg, J. B. Swift, Phys. Rev. A **46**, 4773 (1992).
[172] J. B. Swift, K. L. Babcock, P. C. Hohenberg, Physica A **204**, 625 (1994).
[173] M. Treiber, Phys. Rev. E **53**, 577 (1996).
[174] R. Deissler, Phys. Rev. E **49**, 31 (1994).
[175] M. Lücke, A. Recktenwald, Europhys. Lett. **22**, 559 (1993).
[176] Ch. Jung, M. Lücke, P. Büchel, Phys. Rev. E **54**, 1510 (1996).
[177] H. B. Squire, Proc. Roy. Soc. (London) A **142**, 621 (1933).
[178] J. M. Chomaz, Phys. Rev. Lett. **69**, 1931 (1991).

Dynamical Structures in Open Fluid Systems

Dietmar Rempfer

Sibley School of Mechanical and Aerospace Engineering, Cornell University, Ithaca, NY 14853, USA

Abstract. In this chapter, an introduction into the analysis of structures in open fluid systems and their dynamics is given. In addition to presenting an overview over this area of research, we describe three particularly important approaches to a mathematical definition of the notion of a "structure", and we discuss two examples of the application of these concepts from our own work. We also briefly review a number of important studies by other authors. Due to the vast amount of material that is available and constantly being added to in this field, this introduction necessarily has to be partial, and can only cover a small part of the activity that is centered around the study of flow structures.

1 Introduction

The last few decades of turbulence research have seen a growing interest in the study of coherent structures of turbulent flows (Robinson 1991). What is particularly interesting about this type of research is that the philosophy behind the idea of coherent structures appears — to a certain extent — opposite the philosophy of the statistical theory of turbulence, and thus the coherent structures approach represents a novel direction in the investigation of turbulence.

Historically, because instantaneous flow fields were assumed to be hopelessly complex, the classical statistical theory of turbulence (see, for example, Monin & Yaglom 1973) attempted to circumvent this complexity by considering only averaged quantities of the flow variables. While this approach yielded a lot of important results and is probably the one that, along with dimensional analysis (Tennekes & Lumley 1972), has contributed the most to our current understanding of the phenomenon of turbulence, it is well known that, with the use of averages, the problem of complexity simply changes its name to the 'closure problem'. Furthermore, at least from an intuitive point of view, the idea of completely giving up on understanding the details of the motions of a turbulent fluid seems unsatisfactory.

However, as mentioned above, another view of turbulence, closely connected to the concept of "coherent structures" , gained attention about four decades ago (Townsend 1956). As an aside, it is interesting to note that in fact the presence of structured fluid motion in turbulent flow was actually well-known since the very beginning of what can be called the scientific observation of turbulent flows, see, for example, the well-known drawings of

Fig. 1. Leonardo da Vinci: seated man and water studies. Royal Library, Windsor Castle; Leoni Volume (12579).

Leonardo da Vinci on this very subject (Fig. 1). Indeed, any observer contemplating, say, the large eddies evolving in the flow of a river cannot fail to notice the presence of large, seemingly organized structures that are persistent in space as well as in time. Thus it is actually somewhat surprising that it took so long for this common observation to find the attention that it actually deserves.

The coherent structure approach to turbulence is motivated by the idea that turbulent flows are not just fluids in a state of complete disorder and chaos. Rather, such flows are composed of the superposition of a number of organized motions, the so-called "coherent structures", possibly immersed in some unorganized background of random fluid motion that is not dynamically significant (and thus can be modeled accurately using eddy-viscosity-type models). The relative role that the organized motions play in a flow compared to truly random motions is highly variable in principle, ranging from dominant as in the flows sketched in Fig. 1 to nearly undetectable, depending on the boundary conditions and history of the flow. The central idea then is that turbulent flows — at least some of them — are not as complex as they appear to be at first sight. Rather, such flows are assumed to be composed essentially of a set of organized motions, each possessing a relatively simple spatial structure. The apparent complexity of such turbulent flows is then generated by the superposition in space and time of these organized, so-called 'coherent structures'. An important implicit ingredient of this approach is the assumption — or hope — that there is only a comparably small number of such structures that are relevant to the essential dynamics of the flow. If this is true, then the investigation of the dynamics of these structures would be a feasible task, promising not only deeper insight into the "nature" of turbulent flows, but also the possibility of gaining additional information for modeling purposes.

Since the work by Townsend, literally hundreds of papers have been written on the topic of coherent structures. Instead of attempting a general review of these, we refer the reader to the review articles by Cantwell (1981), Fiedler (1987), and Robinson (1991). Here, we will confine ourselves to pointing out a few of the contributions that are closely related to our subject.

Some of the first detailed investigations of the near-wall structures of turbulent boundary layers on a mainly phenomenological level were reported by Einstein & Li (1956) and Kline & Runstadler (1959). In the 1960's, further important contributions towards an understanding of the structures of turbulent flows followed, represented by, among others, the works of Runstadler, Kline & Reynolds (1963), Kline *et al.* (1967), and Kim, Kline & Reynolds (1971) of the Stanford group. These investigations confirmed the relevance of coherent structures to the dynamics of turbulent flows, particularly because of their important contribution to turbulence production and momentum transport in the turbulent boundary layer. However, it was not until the impressive flow-visualizations by Brown & Roshko (1974) that the idea of coherent structures having a significant role in turbulent flows was finally established.

Important advances in the investigation of coherent structures took place at the end of the 1960's and in the beginning of the 1970's. In particular, progress was made in the formulation of explicit definitions of the term "coherent structure" and the development of corresponding, widely accepted methods to determine these structures. There were essentially two different approaches. One of these approaches can be traced back to investigations by Gupta, Laufer & Kaplan (1971), Wallace, Eckelmann & Brodkey (1972), and Willmarth & Lu (1972). These authors introduced the techniques of *conditional sampling* or *phase averaging* to determine coherent structures. Detailed accounts of the development and results of these methods are given by Hussain (1983, 1986).

The biggest difficulty with these techniques results from the fact that they require the researcher to already have an idea of some characteristic properties of the structure to be determined. Based on that, a criterion for the detection of this structure is developed that allows one to choose, from a set of data or flow visualizations of a flow field, all those realizations that meet the criterion. By an appropriate averaging of the selected flow fields, the prototypical coherent structure can be obtained.

The main drawback of the methods based on conditional sampling is obvious in that the results depend strongly on the choice of criteria for the coherent structure. This choice, in turn, requires that information on the coherent structures be present before they are actually known. Thus, there is actually no fixed, *a priori* definition present at the base of techniques using conditional sampling. Some of the pitfalls of this type of approach are discussed in Lumley (1981), where it is shown that this may lead to the detection of the presence of non-existing objects (see also Lumley (1990) for an extensive discussion giving a good impression of the range of opinions as to the proper definition of the notion of a "coherent structure"). This is, of course, a serious disadvantage, since it appears that the notion of a coherent structure would be of little use as long as it is not possible to give a precise, quantitative description of the role that such structures play in the flow. Such a description, however, requires an exact definition and a mathematical description of the coherent structures.

In this contribution, we will therefore only deal with methods that meet these criteria. In the following section, we will start by briefly discussing the fundamentals of three such methods, namely *wavelet decomposition*, *archetypes*, and *proper orthogonal decomposition (POD)*. These methods share the advantage of completely eliminating the need for any kind of external information or prior knowledge concerning the structures to be identified. In Sects. 3, 4 we will then present some examples for the type of investigations and results that can be achieved with this type of method, concentrating on studies using POD as an example.

Section 4 of this chapter is devoted to the topic of the dynamics of these structures. This field of research has strong ties to another development that has gained importance within the last few decades, which is the study of nonlinear dynamical systems. As in the case of the coherent structures approach to turbulence, the idea of simplicity behind complexity forms a centerpiece of this approach. In the case of dynamical systems theory, it has been shown that very simple systems of ordinary differential equations (ODEs) possessing as few as three degrees of freedom can give rise to extremely complex, 'chaotic' behavior (Lorenz 1963). In connection with the success of dynamical systems theory, it seemed natural to look at turbulence as a chaotic phenomenon in the sense of the theory of ordinary differential equations. In trying to use dynamical systems theory — particularly for the case of open flow systems — one is confronted with the question of what kind of modes of the flow are going to be described by the ODEs.

This is exactly the issue where the concept of coherent structures becomes relevant again: If a turbulent flow could be considered as being composed of the superposition of a possibly small number of coherent structures, the equations describing the evolution of such structures are candidates for a low-dimensional description of turbulence.

Finally, in Sec. 5 we briefly discuss a few important contributions to the analysis of dynamical structures in open fluid systems by other authors.

2 Rigorous Definitions for Coherent Structures: Wavelets, Archetypes, and Karhunen-Loève Eigenfunctions

In this section we will present three alternative approaches to a mathematically precise definition for the concept of a coherent structure. We will only briefly introduce wavelets and archetypes, and spend more time on KL eigenfunctions, as these are used for the research described later in this chapter.

2.1 Wavelets

The method of wavelet transformation can be understood by first looking at its older "sibling", the Fourier transform. In the case of a discrete Fourier transform, a function $f(x)$ is represented as a sum of sine and cosine waves:

$$f(x) = a_0 + \sum_{k=1}^{\infty}(a_k \cos kx + b_k \sin kx) \tag{1}$$

One way to interpret this transformation is by saying that the function f is decomposed into its "wavelength components". For instance, the quantity $\sqrt{(a_k^2 + b_k^2)}$ gives the energy that is contained in the kth Fourier mode. Thus one can say that the representation of the function $f(x)$ in "physical space", giving the value of f at each point x, is replaced by a representation in "wavenumber space", giving a value of $\hat{f}$, the Fourier transform of f at each wavenumber k.

There is one disadvantage of this representation, however: once f has been transformed into Fourier space, any information on local features (local in "physical space") is lost. This is basically due to the fact that the trigonometric functions used for this transform are not localized in the sense that they do not decay in space, and show the same behavior everywhere in an infinite domain.

This is not ideal if one wants to analyze a turbulent flow. What is typical there is that one has structures (or "eddies") that are characterized not only by a definite characteristic scale, which is equivalent to a definite location in wavenumber space, but also by a definite location in physical space. Wavelet analysis takes care of this problem by replacing the sine and cosine waves of the Fourier transform by a new function basis that is defined as follows: One starts out with a "mother wavelet" or "analyzing wavelet" $\Phi(x)$ that has compact support both in physical and in wavenumber space (so that it vanishes outside of a finite domain in x and in k). The complete wavelet basis is then created by rescaling and shifting this mother wavelet as in

$$\phi_{(s,l)}(x) = 2^{s/2}\Phi(2^s x - l), \tag{2}$$

where s and l are integers that give the wavelet's width and position, respectively. We can then take our function $f(x)$, and decompose it into the wavelets $\phi(x)$ as in

$$f(x) = \sum_s \sum_l c_{(s,l)}\phi_{(s,l)}(x). \tag{3}$$

Please note that the above only serves as a very brief introduction into a few of the main ideas of wavelet analysis, which necessarily brushes very lightly over a much larger number of issues involved there. For more in-detail information, the reader is referred to textbooks on the subject, such as Chui (1992), and Kaiser (1994).

The wavelet decomposition as described above appears as a more appropriate tool to investigate the structure of turbulent flows than the traditional Fourier transform. Like the latter, it does give information on the wavenumber content of the turbulent velocity fields, but in addition to that it also retains information on the spatial location of turbulent "eddies".

In this contribution, we cannot go into a detailed description of the applications of wavelet analysis to turbulent flow, for a recent review of this material we refer the reader to Farge (1992). We do however, want to note the one disadvantage of wavelet transforms that we see in the context of the analysis of the structure of turbulent flow. This is due to the fact that the functional form of wavelets is prescribed from purely mathematical reasoning in such a way that certain abstract properties like orthogonality, compactness, etc. are met. What this means is that there is no direct connection between the physical notion of an eddy or a coherent structure, and a specific wavelet function ϕ. This is because there are normally no physical considerations involved in the construction of wavelet bases which would allow them to be particularly tailored to the physics of a turbulent flow. While in many applications of wavelet analysis this is not really a critical drawback, this property is important to mention here, as it is an area where the other two approaches to be discussed below have advantages.

2.2 Archetypes

Archetypal analysis was first introduced by Cutler & Breiman (1994) as a new statistical technique for extracting "archetypal patterns" from data sets. A major difference to the wavelet analysis presented above lies in the fact that archetypal analysis first needs a data set, consisting of an ensemble of velocity fields of the flow to be analyzed, in order to proceed. The advantage of this is that the modes that are used by this technique are in a sense defined and generated by the flow that one is interested in. We notice that archetypal analysis shares this trait with the method of proper orthogonal decomposition to be discussed in the following section.

Archetypal analysis proceeds as follows: We start out from a set of velocity fields $u^i(x) = u(x, t_i)$, where the index i runs from 1 through the number of velocity fields available, say n. In practical applications, the velocity field u will always be discretized, so that we can think of u as an m-vector of velocities. We then need to find p ($p < n$) m-vectors $z_1 \ldots z_p$ that characterize "archetypal patterns" in the data such that it can be represented as mixtures of these archetypal patterns. We define these z_i by requiring this specific set of patterns to minimize in a mean square sense the difference between the actual velocity fields and the representation of these velocity fields in terms of the archetypes,

$$\sum_{i=1}^{n} ||u_i - \hat{u}_i|| \stackrel{!}{=} \text{Min}, \tag{4}$$

where

$$\hat{u}_i = \sum_{k=1}^{p} \alpha_{ik} z_k, \quad \alpha_{ik} \leq 0, \quad \sum_k \alpha_{ik} = 1, \tag{5}$$

$$z_k = \sum_{j=1}^{n} \beta_{kj} x_j, \quad \beta_{kj} \leq 0, \quad \sum_i \beta_{ki} = 1. \tag{6}$$

The above prescription means that for our minimization we only consider representations of the velocity fields in terms of archetypes that result in positive coefficients. Again leaping ahead to the following section we mention that this requirement of positive coefficients constitutes the main difference between archetypal analysis and proper orthogonal decomposition. It is possible to show that this leads to the archetypes lying on the convex hull of the dataset (instead of representing eigendirections in data space as in the case of POD). Finally, we mention that it is possible to extend this type of analysis to "moving archetypes" by adding another optimization step to the above procedure which basically searches for an optimal set of displacements of the fixed archetypal patterns (Cutler & Stone 1997).

Compared to wavelet analysis, archetypal analysis has the advantage that it explicitly includes information on the behavior of the flow in the shape of the eigenmodes that it is using. Thus, in a sense the archetypes are more physical, and definitely contain more physics, than wavelets. Also, due to the requirement of positive expansion coefficients mentioned above, archetypal decomposition is closer to the intuitive notion of turbulent flow as a superposition of certain "eddies" that have a direct physical meaning attached to them, than proper orthogonal decomposition. On the other hand, compared to both wavelet analysis and POD, archetypes have the drawback of not necessarily possessing such nice mathematical properties like orthogonality. In particular, archetypes lack an important type of "universality" in that the shape of archetypes depends on the order of the representation. For example, if one is looking for only four archetypes, say, then these are different from the first four archetypes of a decomposition into eight archetypes.

2.3 Proper Orthogonal Decomposition

The Karhunen-Loève procedure of probability theory (Loève 1955) was first suggested by Lumley (1967) for the purpose of identifying the coherent structures of turbulent flows. This method, which is now widely known as proper orthogonal decomposition (POD) or KL analysis, shares with archetypal analysis the trait that an ensemble of flow fields is required to obtain the eigenmodes of a flow. As this method will be used in the examples presented later in this chapter, we will discuss it in somewhat more detail than the ones above. We will also mention some of the properties of the eigenfunctions of POD that are relevant for the following. For a more complete treatment, the reader is referred to the review paper by Berkooz, Holmes & Lumley (1993).

The method of proper orthogonal decomposition can be motivated by asking for the "characteristic velocity fields" (*viz.* coherent structures) of a given ensemble of flow fields $\boldsymbol{u}^\ell(\boldsymbol{x}) = \boldsymbol{u}(\boldsymbol{x}, t_\ell)$. Here, the boldface symbols $\boldsymbol{u}$ and $\boldsymbol{x}$ denote the vectors of velocity and position, respectively, the individual components of which are u, v, and w (u_1, u_2, u_3), and x, y, and z (x_1, x_2, x_3). If the term "characteristic velocity field" is taken to denote a distribution that is — in some average sense — as similar as possible to the instantaneous flow fields $\boldsymbol{u}^\ell$, then a characteristic velocity field $\boldsymbol{\sigma}(\boldsymbol{x})$ may be defined by the variational expression

$$\frac{\langle(\boldsymbol{u}^\ell, \boldsymbol{\sigma})^2\rangle}{\|\boldsymbol{\sigma}\|^2} \stackrel{!}{=} \text{Max.}, \tag{7}$$

where $(\cdot,\cdot)$ denotes a scalar product in the Hilbert space of square-integrable functions, $\|\cdot\|$ is the corresponding norm, and the brackets $\langle\cdot\rangle$ denote a time average. Using the calculus of variations, the problem can be reduced to the Fredholm integral equation

$$\int_D \boldsymbol{R}(\boldsymbol{x}, \boldsymbol{x}') \cdot \boldsymbol{\sigma}(\boldsymbol{x}')\, d\boldsymbol{x}' = \lambda \boldsymbol{\sigma}(\boldsymbol{x}) \tag{8}$$

representing an eigenvalue problem for the structure $\boldsymbol{\sigma}$ (see Lumley (1970) for details of the derivation). The kernel $\boldsymbol{R}$ of this equation is the autocorrelation function

$$\boldsymbol{R}(\boldsymbol{x}, \boldsymbol{x}') = \langle \boldsymbol{u}(\boldsymbol{x})\boldsymbol{u}(\boldsymbol{x}')\rangle, \tag{9}$$

where the term inside the brackets denotes a dyadic product. Because this kernel has the form of a symmetric, nonnegative definite Hilbert-Schmidt operator, there exists a denumerable infinite set of real solutions $\boldsymbol{\sigma}_j$ and corresponding eigenvalues λ_j.

For the case of three-dimensional flow fields $\boldsymbol{u}^\ell$, the eigenvalue problem (8) soon becomes intractable even at only moderate spatial resolution. It is therefore beneficial to transform (8) by expressing $\boldsymbol{\sigma}$ as a linear combination of the instantaneous flow fields

$$\boldsymbol{\sigma} = \sum_{\ell=1}^{M} q_\ell \boldsymbol{u}^\ell, \tag{10}$$

where M is the number of realizations given. By introducing this ansatz together with the expression for the autocorrelation tensor,

$$\boldsymbol{R}(\boldsymbol{x},\boldsymbol{x}') = \frac{1}{M}\sum_{\ell=1}^{M} \boldsymbol{u}^{\ell}(\boldsymbol{x})\boldsymbol{u}^{\ell}(\boldsymbol{x}'), \tag{11}$$

into (8), we get the algebraic eigenvalue problem

$$\boldsymbol{C}\boldsymbol{q} = \lambda\boldsymbol{q}, \tag{12}$$

where

$$C_{jk} = \frac{1}{M}(\boldsymbol{u}^j,\boldsymbol{u}^k). \tag{13}$$

In this representation, the order of the eigenvalue problem only depends on the number, M, of realizations given and is unaffected by spatial resolution. Therefore this method, referred to by Sirovich (1987) as the *method of strobes*, is used to calculate the eigenfunctions.

The eigenfunctions of proper orthogonal decomposition have a number of interesting properties, of which the ones that are particularly relevant for the study of the structure of turbulent flows are the following:

1. The different eigenfunctions are orthogonal, which is particularly convenient when one is interested in deriving dynamical equations for the behavior of the structures.
2. The set of the eigenfunctions $\boldsymbol{\sigma}_j$ is complete in the sense that the flow fields of the ensemble given can be expanded in the eigenfunctions via

$$\boldsymbol{u}(\boldsymbol{x},t_\ell) = \sum_j \zeta_j(t_\ell)\boldsymbol{\sigma}_j(\boldsymbol{x}), \tag{14}$$

   where

$$\zeta_j(t_\ell) = \frac{(\boldsymbol{u}^\ell,\boldsymbol{\sigma}_j)}{\lambda_j}. \tag{15}$$

   We note that, instead of using (15), the expansion coefficients ζ_j could be calculated directly from an eigenvalue problem that is completely analogous to (8). This eigenvalue problem can be obtained by exchanging time and space in (8) and (9), (Sirovich (1987), see also Aubry, Guyonnet & Lima (1992), where the resulting temporal eigenfunctions — the expansion coefficients $\zeta_j(t)$ in our terminology — are called *chronoi* and our spatial eigenfunctions $\boldsymbol{\sigma}_j(\boldsymbol{x})$ are termed *topoi*).
3. The expansion coefficients are uncorrelated in time, and can be normalized such that

$$\langle\zeta_j\zeta_k\rangle = \delta_{jk}. \tag{16}$$

   Thus, the eigenfunctions $\boldsymbol{\sigma}_j$ represent the root-mean-square values of the flow fields that are induced by the corresponding structure.

4. We define the energy of a flow field as

$$e = \frac{1}{2} \int_D u^2(x)\, dx. \tag{17}$$

Introducing (14) into (17), we find, for the time average of the energy,

$$\langle e \rangle = \frac{1}{2} \sum_j \lambda_j. \tag{18}$$

The eigenvalues of (8) are all nonnegative, and they represent the mean value of the energy that the corresponding coherent structure contributes to the flow. In that sense the eigenvalues λ_j are a measure of the related structure's importance to the flow.

5. By comparing (14) and (7), one of the most significant properties of the system of Karhunen-Loève eigenfunctions can be seen immediately. It becomes obvious that the eigenfunctions have been defined in such a way that the convergence of a representation of the flow field is optimally fast in the mean square. This means that of all possible systems of eigenfunctions, the one described by (7) needs the smallest number of terms to represent the flow fields to a given accuracy. This property is clearly attractive if one is interested in a low-dimensional description of a turbulent flow via a system of ordinary differential equations. Once one has decided that time should be the independent variable of this system of ODEs, which — although not being mandatory — seems to be a natural choice, the POD delivers an optimum system of eigenfunctions for such a description.
Note also that this property is completely analogous to the requirement (5, 6) in archetypal analysis, and that in fact it would have been possible to develop the KL procedure starting from this requirement. The main difference between KL analysis and archetypal analysis consists in the presence of the positivity requirements in (5, 6) which do not exist in KL analysis.

6. Finally, we point out that, as the coherent structures can be represented as a linear combination of the given flow fields, these eigenfunctions inherit all those properties of the instantaneous flow fields that can be expressed in the form of linear homogeneous equations.

KL analysis shares with archetypal analysis the advantage that information on the behavior of the flow is encoded in the shape of the eigenmodes that are used. In contrast to the latter, however, KL eigenfunctions have a number of very favorable mathematical properties that render them especially useful for studies that go beyond phenomenological representations of a flow. In particular, if one is interested in studying the dynamics of the flow and its structures, then Galerkin projections onto KL bases are a very attractive proposition, one of the main reasons being the optimality property of KL eigenfunctions.

3 POD Analysis of Boundary-Layer Transition

3.1 Introduction

In this section we present an application of proper orthogonal decomposition as a tool to study the structure of complex flows for the case of a transitional boundary layer. While investigations of coherent structures of fully turbulent flow in general, and the application of POD analysis for this purpose in particular, have seen a growing interest during the last two decades, such an approach is rather novel in the field of transition research. Nevertheless, the motivation in this case is quite similar to the one that fostered the growth of such studies in turbulence research.

While in the early days of transition research one was happy to be able to prove that linear stability theory can indeed predict the initial stages of the transition process, and, with the advent of secondary instability theory one was even able to go one step beyond that, the problem of getting insight into the very last stages of the transition process has proven a much more formidable task. Here, there seems to be some analogy to the situation in turbulence research, in that traditional tools and notions like waves and Fourier modes just do not seem to be powerful enough to further our knowledge very much beyond its present state. The concept of a coherent structure as defined via eigenfunctions of POD, on the other hand, promises the possibility of an extremely compact description of complex flow phenomena that might make a detailed understanding possible.

The ultimate goal of the work presented in the following was to find general features and mechanisms of transition in wall layers. In a first step that is reported here, a detailed comparison of the transition process in two types of boundary layers, one of them at zero pressure gradient, the other one under a strong adverse pressure gradient, was performed. For this purpose, data from direct numerical simulations have been used. A thorough comparison of these flows is difficult due to the sheer complexity of the transient velocity fields of the flows, particularly at the late stages of transition. It is exactly this type of a situation where KL analysis can simplify the problem considerably, as it allows decomposing the flow into its relevant building blocks, thus reducing the problem to a comparison of a relatively small number of entities that are characteristic for each flow.

For both cases considered here, there is a distinct spatial evolution of the flow, from a laminar to a turbulent state, which takes place in the streamwise direction. To cover the development of these flows in that direction, it is of interest to analyze the flow at different streamwise positions. Consequently, two-dimensional structures within cross-stream planes at different stages of the transition process were analyzed for the cases of a boundary layer under two different pressure gradients.

3.2 Data Base

To reliably calculate coherent structures of a flow, POD requires a data base of spatially highly resolved flow fields at a number of time steps large enough for the statistical averages in time to be meaningful. For the investigations reported here, data from numerical simulations by Rist & Fasel (1995), and by Kloker (1993) were used. In their work, Rist and Kloker performed direct Navier-Stokes simulations of the spatial development of an incompressible flow. The integration domain that was used and the corresponding coordinate system are shown schematically in Fig. 2.

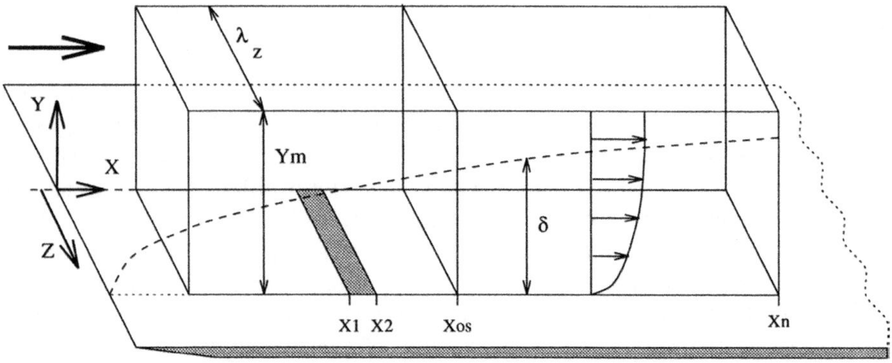

Fig. 2. Integration domain of the direct numerical simulations

The numerical method used employs finite difference approximations for the derivatives in the x- and y-directions, and Fourier spectral modes for the z-coordinate. For the Blasius boundary layer, the discretization consists of a maximum of 4570×121 nodes for the streamwise and wall-normal directions, respectively, and 16 dealiased Fourier harmonics for the spanwise direction. For the simulation of the boundary layer with adverse pressure gradient (Falkner-Skan flow with Hartree parameter $\beta_H = -0.18$), 1850×121 nodes were used for the x- and y-coordinates, and 41 dealiased Fourier modes for the spanwise direction. To induce transition, regular disturbances were introduced locally within a narrow disturbance strip (x_1–x_2 in Fig. 2) that modeled timewise periodic blowing and suction. These disturbances consisted of a monochromatic two-dimensional portion and a three-dimensional portion that is symmetric with respect to $z = 0$, to cause the so-called *fundamental type breakdown* of the laminar flow in both cases.

From the data generated by the simulations above, the values of the velocity vector within 55 evenly spaced cross-sections normal to the bulk flow direction were extracted. The size of these cross-sections is equal to the cross-section of the integration domain, which is 9.6 mm high and 24.5 mm wide for the Blasius boundary layer, and 21.8 mm high and 33.9 mm wide for the pressure gradient case. The streamwise positions of the chosen cross-sections are shown in Table 1. They were selected such that corresponding stages of the transition process between the two cases matched. It turned out that this could be achieved by a linear transformation between the x-coordinates of the two cases (see below).

Table 1. x-positions of the investigated cross-sections.

	x-Position	
Index i	x_{BLASIUS} [mm]	x_{FSKAN} [mm]
1	302	665
10	333	695
20	368	731
30	402	766
40	437	801
50	471	835
55	488	853

The Reynolds number based on laminar displacement thickness is $Re_{\delta_1} = 736$ at the first and $Re_{\delta_1} = 936$ at the last cross-section considered for the Blasius boundary layer. The respective values for the pressure gradient case are $Re_{\delta_1} = 1688$ and $Re_{\delta_1} = 1892$. The regions considered in both cases cover the stages of the transition process from the beginning three-dimensional, nonlinear development of Tollmien-Schlichting waves up to just beyond the spike stages, which is right at the onset of turbulence. The ensemble of flow fields used consisted of 440 and 70 equally spaced time steps within one time period of the prescribed disturbances for the cases without and with pressure gradient, respectively. The two cases are referred to as "BLASIUS" for the Blasius boundary layer, and "FSKAN" for the boundary layer with adverse pressure gradient in the following.

It should be noted that the zero pressure-gradient flow investigated here does not show spanwise homogeneity, but rather has very different characteristics in its "peak position" $z = 0$, compared to its "valley position" $z = \lambda_z/2$. There is much more activity in the peak plane than in the valley plane, which was the reason to create the term "peak plane" (Klebanoff *et al.*, 1962) (because rms-values of disturbance velocities when viewed as a function of the z-coordinate show a pronounced peak at that position); in the adverse pressure gradient case, the term "valley plane" is in fact no longer justified (Kloker, 1993). For the investigations reported below, the integration domain was therefore split up in a peak region $-1/4\lambda_z \leq z \leq 1/4\lambda_z$ and a valley region $1/4\lambda_z \le z \le 3/4\lambda_z$. In the following, only the peak region will be considered.

In addition to the general properties just discussed, the eigenfunctions of POD applied to the case of spatially evolving boundary layers have another characteristic property. This can be seen particularly clearly from Fig. 3 that shows the eigenvalues that are obtained when solving (8) for each of the 55 cross-sections in Table 1. What is observed is that the eigenvalues occur in pairs of almost equal value. It can be shown that this property results from the fact that in our boundary layers the structures in the cross-stream planes show distinct motions at almost constant energy within these planes. This is in complete analogy to the three-dimensional case that is discussed in Rempfer & Fasel (1994a).

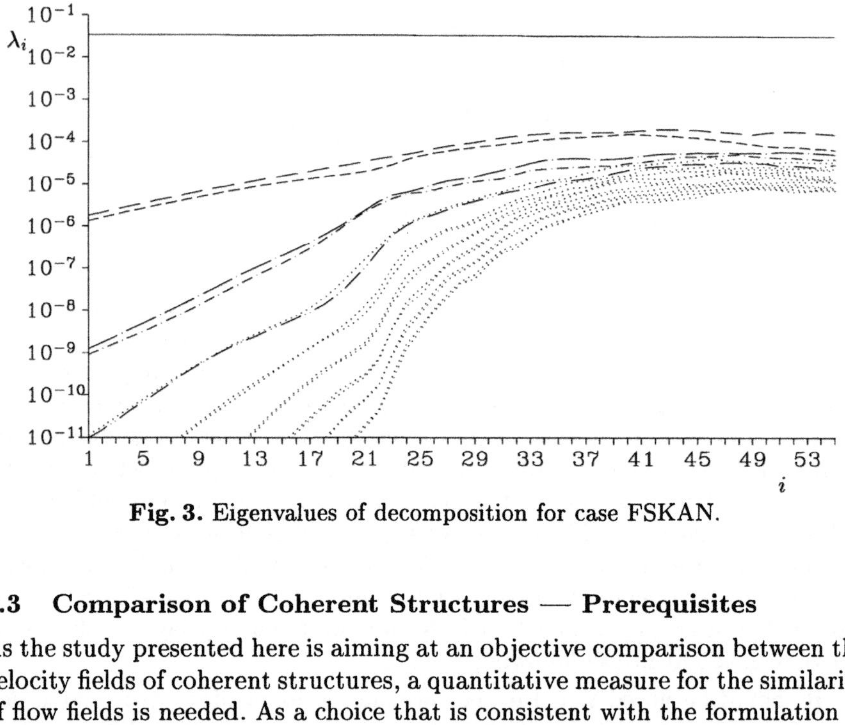

Fig. 3. Eigenvalues of decomposition for case FSKAN.

3.3 Comparison of Coherent Structures — Prerequisites

As the study presented here is aiming at an objective comparison between the velocity fields of coherent structures, a quantitative measure for the similarity of flow fields is needed. As a choice that is consistent with the formulation of the eigenvalue problem that was used to define the coherent structures, the scalar product in the function space spanned by the velocity fields within the cross-stream planes considered here was used. Therefore, the expression

$$\chi^* = \frac{(\boldsymbol{\sigma}_1, \boldsymbol{\sigma}_2)^2}{\|\boldsymbol{\sigma}_1\|^2 \, \|\boldsymbol{\sigma}_2\|^2} \tag{19}$$

serves as a measure for the similarity of $\boldsymbol{\sigma}_1$ and $\boldsymbol{\sigma}_2$. The "similarity measure" χ^* varies between 0 and 1, where a value of 1 means that $\boldsymbol{\sigma}_1$ and $\boldsymbol{\sigma}_2$ are identical, whereas $\chi^* = 0$ means that they are orthogonal.

When σ_1 and σ_2 are structures from the cases BLASIUS and FSKAN, respectively, then the above definition still leaves the problem of actually carrying out the integration involved in the evaluation of the scalar products. The problem there is that σ_1 and σ_2 are defined on different domains, and to be able to carry out the integration, the two domains have to be matched by an appropriate transformation of the wall-normal and spanwise coordinates. For the spanwise coordinate, the periodicity length assumed in the direct numerical simulation is a characteristic length that has to be matched for the two cases. The z-coordinate is therefore stretched in such a way that the transformed periodicity length (principal spanwise wavenumber λ_z) is the same for both cases.

To find a transformation of the wall-normal coordinate the following approach is used. It is assumed that there is an "optimal" transformation that correlates corresponding y-coordinates for the two cases. Under that transformation, the value of χ^* for any two corresponding structures of the two boundary layers should assume a maximum. Thus the factor κ of the transformation $y^* = \kappa y$ has to be chosen such that the value of χ^* is maximized. This has been done for eigenfunctions of different orders to check that the transformation factor obtained does indeed have a physical meaning. This technique yielded consistent values for κ in all cases. Fig. 4 shows the values of the transformation factor κ at different streamwise positions.

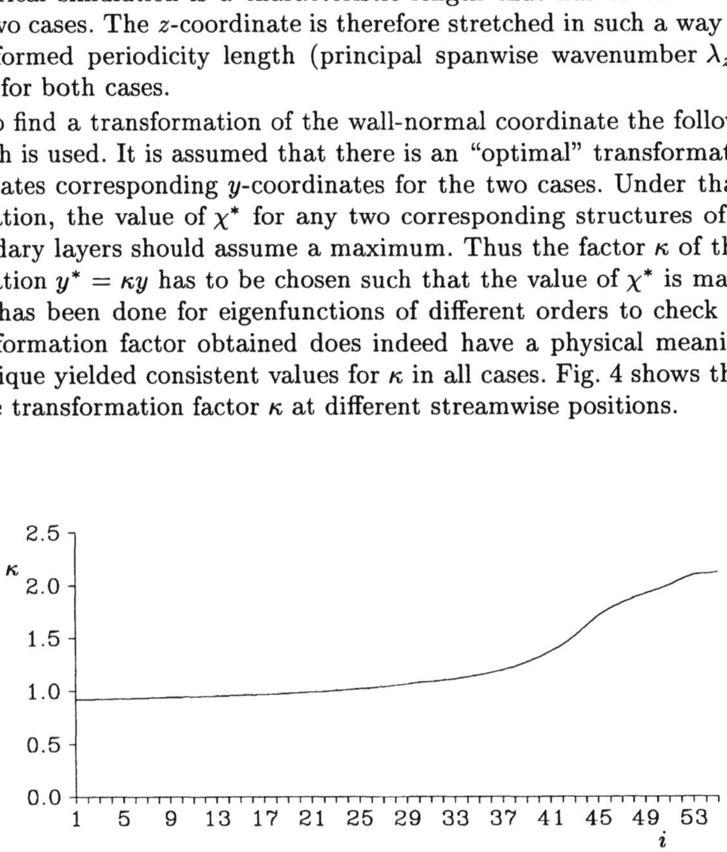

Fig. 4. Factor κ for the stretching of y_{BLASIUS} in the wall-normal coordinate

Finally, the phenomenon of pairs of eigenfunctions mentioned above had to be dealt with. This situation can lead to problems when the eigenfunctions of such pairs exchange their index in a "crossing over" that occurs when the sign of the (small) difference between the eigenvalues of a pair changes (note that the eigenfunctions are numbered strictly according to the size of their eigenvalues). As the eigenfunctions of a pair are orthogonal, this would cause χ^* to drop from a value near 1 to a value near 0. To eliminate this effect, the following approach was chosen. For each of the figures shown in the next section, the value of χ^* for eigenfunctions carrying the same index and belonging to our two cases, respectively, is calculated. In addition, a "conjugate value" of χ^* is computed, using the second "partner" of the pair of eigenfunctions for case FSKAN, and the two values are added. The final similarity measure considered is therefore

$$\chi_i = \frac{(\boldsymbol{\sigma}_i^{\mathrm{BLASIUS}}, \boldsymbol{\sigma}_i^{\mathrm{FSKAN}})^2}{\|\boldsymbol{\sigma}_i^{\mathrm{BLASIUS}}\|^2 \, \|\boldsymbol{\sigma}_i^{\mathrm{FSKAN}}\|^2} + \frac{(\boldsymbol{\sigma}_i^{\mathrm{BLASIUS}}, \boldsymbol{\sigma}_{i+1}^{\mathrm{FSKAN}})^2}{\|\boldsymbol{\sigma}_i^{\mathrm{BLASIUS}}\|^2 \, \|\boldsymbol{\sigma}_{i+1}^{\mathrm{FSKAN}}\|^2}. \tag{20}$$

Due to the fact that both components of a pair of eigenfunctions are taken into account, the "crossing-overs" mentioned above do not affect our data.

3.4 Comparison of Coherent Structures — Results

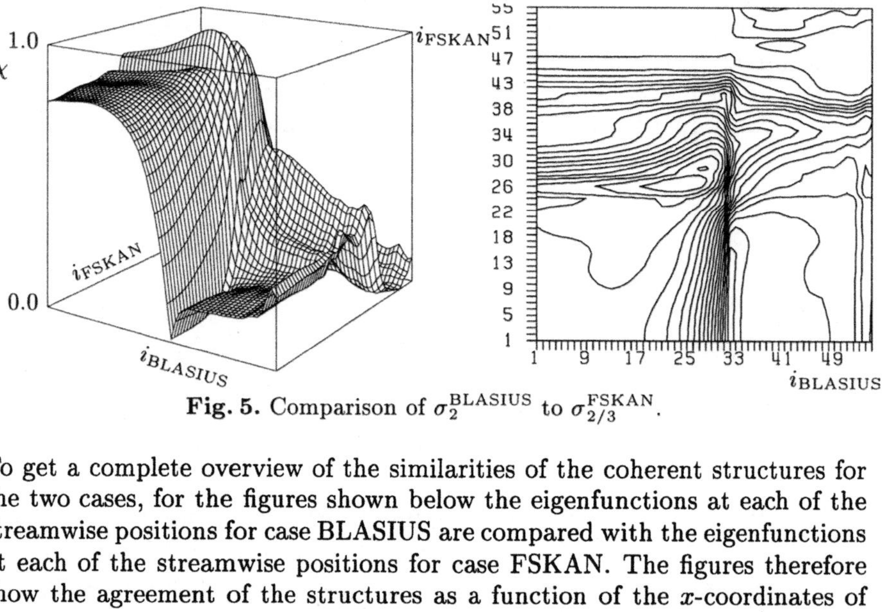

Fig. 5. Comparison of $\sigma_2^{\mathrm{BLASIUS}}$ to $\sigma_{2/3}^{\mathrm{FSKAN}}$.

To get a complete overview of the similarities of the coherent structures for the two cases, for the figures shown below the eigenfunctions at each of the streamwise positions for case BLASIUS are compared with the eigenfunctions at each of the streamwise positions for case FSKAN. The figures therefore show the agreement of the structures as a function of the x-coordinates of both cases. If there was perfect agreement between the structures of these two cases, the values of $\chi(i_{\mathrm{BLASIUS}}, i_{\mathrm{FSKAN}})$ should be exactly equal to unity on the diagonal $i_{\mathrm{BLASIUS}} = i_{\mathrm{FSKAN}}$ and decrease symmetrically away from that diagonal.

The data are shown in a three-dimensional perspective representation as well as in the form of iso-lines. As there are profound differences between the two flows in the valley region, we restrict our attention to the peak region $-1/4\lambda_z \leq z \leq 1/4\lambda_z$. Fig. 5 shows the values of χ for a comparison of the second eigenfunction $\boldsymbol{\sigma}_2$ of case BLASIUS to the eigenfunctions $\boldsymbol{\sigma}_2$ and $\boldsymbol{\sigma}_3$

of case FSKAN. These eigenfunctions represent the most energetic structure of the fluctuating flow $\boldsymbol{u}' = \boldsymbol{u} - \langle \boldsymbol{u} \rangle$. In the upstream portion of the domain considered here, this fluctuating flow is dominated by two-dimensional Tollmien-Schlichting-like waves that are generated in an analogous manner for both cases. It is therefore to be expected that the most energetic structures of the fluctuating flow — that represent just these waves (Rempfer & Fasel, 1994a) — are quite similar during the initial stages of transition, and Fig. 5 indeed shows high values of χ up to about half the length of the domain.

Still, χ is lower than one for two reasons. First, the scaling of the y-axis cannot remove the non-similar Re_{δ_1}-dependence of the u'-amplitude distribution over y. Thus, even for a single base flow, the structures at the different Re_{δ_1} considered here cannot be matched completely. Second, the $u'(y)$-profiles of Tollmien-Schlichting waves in an adverse pressure gradient base flow differ from the ones for Blasius flow by an additional (third) local maximum that is due to an "inviscid" instability of the inflexional base flow profile for boundary layers with adverse pressure gradients.

From the extremely flat plateau in the first quadrant of Fig. 5 it can be seen that the structures change only very slowly during the first stages of transition. However, in sharp contrast to their initial behavior, these structures then obviously change considerably within a very short distance. To find out more about the reason for this sudden change, in Fig. 6 a measure for the "three-dimensionality" of the second and third eigenfunctions for both cases is plotted. The measure that is used is the relative energy of the flow field contained in the non-constant spanwise Fourier modes. The value plotted is thus

$$r = 1 - \frac{\int \left(\frac{\int u\,dz}{\lambda_z}\right)^2 + \left(\frac{\int v\,dz}{\lambda_z}\right)^2 dy}{1/\lambda_z \int u^2 + v^2 + w^2\,dy\,dz}, \tag{21}$$

where λ_z is the principal spanwise wave number. If $r = 0$, there is completely two-dimensional flow, and for $r = 1$ the flow is "purely three-dimensional", meaning that the average over z of the velocities is equal to zero. From Fig. 6 it can be seen that the three-dimensionality of the eigenfunctions for case FSKAN rises sharply at precisely the point where the χ-values drop in Fig. 5, while for case BLASIUS this evolution proceeds much more smoothly. We can thus conclude that the sudden drop in similarity of the eigenfunctions is caused by the structures in the pressure gradient case getting three-dimensional much faster than the ones for case BLASIUS. Also, Fig. 5 shows that even when the eigenfunctions of the BLASIUS case get three-dimensional, they do not resemble the eigenfunctions of case FSKAN as closely as they did during the predominantly two-dimensional phase of their evolution.

Figure 7 shows the results of a comparison of the fourth eigenfunction of case BLASIUS to the eigenfunctions $\boldsymbol{\sigma}_4$ and $\boldsymbol{\sigma}_5$ of case FSKAN. The corresponding structures describe oblique waves that were generated in addition

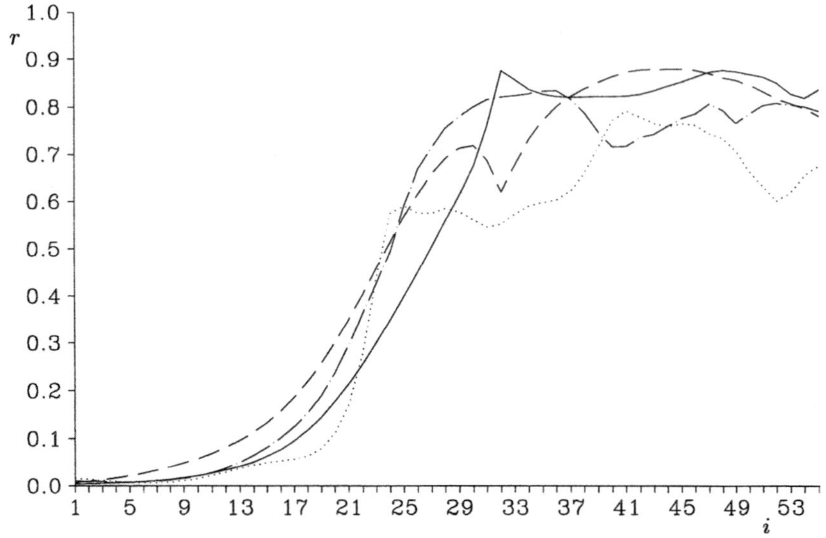

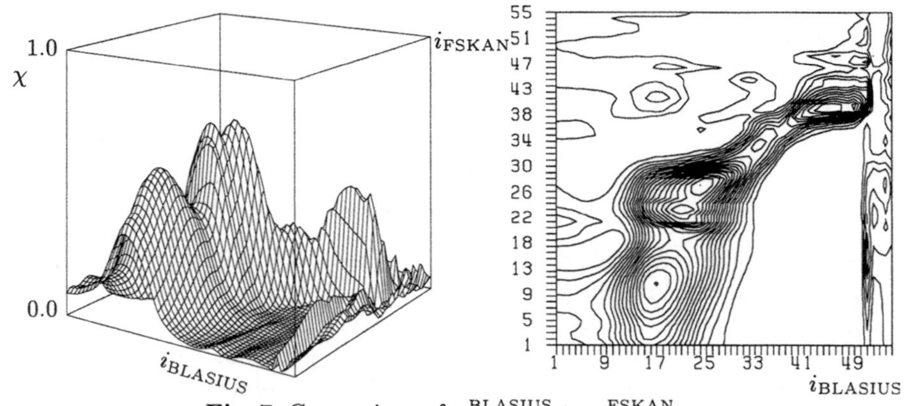

Fig. 7. Comparison of $\sigma_4^{\mathrm{BLASIUS}}$ to $\sigma_{4/5}^{\mathrm{FSKAN}}$.

to the two dimensional Tollmien-Schlichting waves at the disturbance strip (see Fig. 2). Figure 7 exhibits a clear pattern in the distribution of the χ-values. For low values of i_{BLASIUS}, only a very poor agreement between the structures of the two cases can be detected. The reason for this is due to the slightly different methods that were used in the respective numerical simulations to induce transition. For case BLASIUS, in addition to a large-amplitude Tollmien-Schlichting wave, a longitudinal vortex was created. In such a situation the nonlinear terms of the Navier-Stokes equations lead to the generation of a pair of oblique waves. In contrast to this, in the simulation for case FSKAN a three-dimensional disturbance was introduced in the form of oblique waves, so that now the nonlinear terms had to create a longitudinal vortex. However, this requires the Tollmien-Schlichting waves and the oblique waves to synchronize, and to obtain a certain phase relation with respect to each other so that a resonance mechanism can set in. Thus the process of formation of an "active resonant triad" takes longer with a disturbance input like the one used for case FSKAN. The different mechanisms account for the initial lack of similarity of the structures as depicted in Fig. 7.

The high values of χ indicating a good agreement between the structures of the two cases are clearly grouped along a certain line. There are two facts that are particularly remarkable about this line. First, the slope of this line diminishes strongly downstream of $i_{\text{BLASIUS}} = 40$. This confirms that, under adverse pressure gradient, the events of later breakdown stages take place in a much shorter streamwise region compared to the BLASIUS case. Second, this line ends abruptly at a position of about $i_{\text{BLASIUS}} = 50$. Beyond that point, the streaks that appear in the iso-lines seem to be erratic. Considering the fact that the numerical simulation of the BLASIUS case began to show insufficient resolution near the end of our domain, this behavior is most probably a consequence of the degrading quality of the BLASIUS simulation.

In parentheses we would like to note that this behavior also demonstrates the ability of POD to serve as a tool to compare the results of different numerical codes treating the same physical problem. A comparison of the coherent structures delivered by each of the codes in question can serve as an objective test of the ability of such codes to resolve the spatial structures of the flow. As this test is not affected by different initial conditions, it can also be used, for instance, to compare simulations that used "random" excitations, which, under normal circumstances poses the problem that it is often not possible to *exactly* reproduce initial conditions obtained that way.

3.5 Evolution of Spatial Complexity

The results discussed in the previous section already suggest that the flow fields in the case of transition in a boundary layer with adverse pressure gradient are more complex than the ones of the Blasius boundary layer. In this section, a quantitative measure for the concept of spatial complexity is introduced that allows to describe the evolution of this type of complexity in an objective manner.

As discussed above, the decomposition in POD eigenfunctions has the property of being optimal in the sense that there exists no other decomposition that could approximate the velocity fields using fewer terms than the POD to reach a prescribed accuracy. Thus the number of terms d_{KL} that are needed in a POD representation to approximate the flow field to an accuracy of ε in the mean square can be used as a measure for the spatial complexity of the flow. In Fig. 8 this so-called *Karhunen-Loève dimension* d_{KL} is shown for a value of $\varepsilon = 10^{-6}$. This dimension gives the number of eigenfunctions needed to represent the velocity field to an accuracy of 99.9% in the mean square over time. The x-coordinate for the two cases has again been transformed in such a way that corresponding stages of the transition process are situated at the same streamwise position.

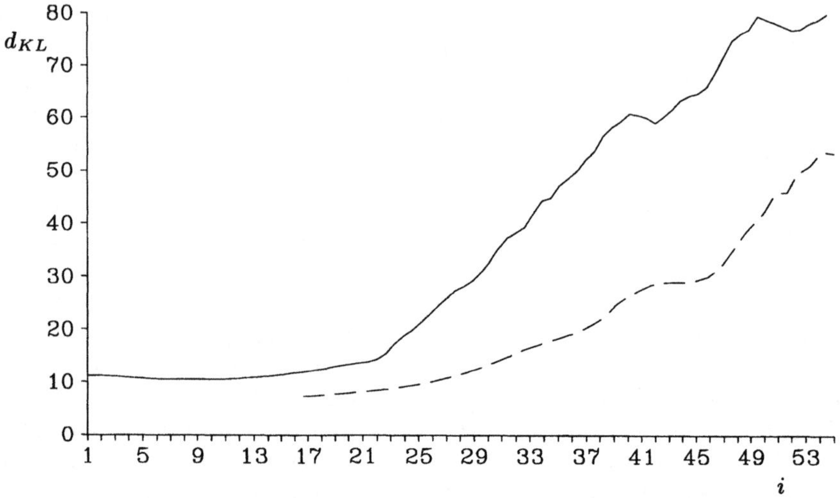

Fig. 8. Karhunen-Loève dimension d_{KL} over x. ————, FSKAN; — — —, BLASIUS.

Figure 8 directly confirms the impression from the above section, in that the Karhunen-Loève dimension is always higher for case FSKAN. Also, it can be observed that the complexity of the BLASIUS boundary layer initially rises more smoothly than that of the pressure gradient case, where there is quite an abrupt change in the streamwise derivative of this property. Interestingly, after the intermediate plateau at $i \approx 43$, the rise in complexity of the Blasius boundary layer seems to occur at the same rate as for the pressure-gradient case. Comparing Fig. 5 and 8 we can see that the sharp drop in similarity of the second order eigenfunctions and the sudden rise in complexity of the pressure gradient boundary layer occur at the same streamwise position. Obviously, this position marks a particularly important stage in the process of transition. In the above section, it has been shown that at this position there is a marked change in the character of the most energetic eigenfunction which can be described as a sudden three-dimensional deformation of Tollmien-Schlichting waves.

3.6 Comparison of Coherent Structures — Conclusions

The investigation presented above started out from the question of which parallels can be found for the transition processes in two different flat-plate boundary layers. The results show, first of all — and not surprisingly — that the disturbance flow field for the two cases is very similar during the initial (almost linear) stages of transition. In addition to that, by using a quantitative measure for the concept of similarity it was also possible to find a certain point in the transition process where the similarity between the two cases diminishes abruptly. It was shown that this sudden change is caused by the structures of the pressure gradient case getting three-dimensional much more rapidly than in the Blasius boundary layer.

An analogous investigation for higher order structures showed evolutions that could not have been observed at all by considering instantaneous flow fields. The results clearly show that initial differences — that are caused by different mechanisms that were used to generate disturbances in the numerical simulations — are diminishing during the course of transition, and it can be concluded that the "active resonant triads" that start the actual nonlinear breakdown of the boundary layer are quite similar for both cases.

The above study of the evolution of spatial complexity confirms the impression of a higher complexity of the flow field in the case of transition under a strong adverse pressure gradient. Nevertheless, in a frame of reference that is adjusted such that corresponding stages of the early transitional phases for the two cases match, the rate by which new structures are formed is very nearly the same for both cases considered, so that at this very general level there seem to exist basic parallels in the formation of turbulence.

4 POD-Galerkin Models for the Dynamics of Coherent Structures

As a second example, we want to discuss the application of structure analysis of flows in conjunction with Galerkin Projection to understand the dynamics of complex flows. While the work described below is again based on KL eigenfunctions, we want to emphasize that in principle, the same type of analysis is also possible based on wavelets, or archetypes.

The data base that was used for the investigations described below was the same as the one described for case BLASIUS in the previous section. While the transitional boundary layer is an interesting research topic *per se*, because of numerous analogies of transitional boundary layers to fully turbulent boundary layers (see, for example, Blackwelder 1983 and Rempfer & Fasel 1994a), the results below may also shed some light on open questions concerning fully turbulent flows.

The motivation for the investigations presented in the following was to gain deeper insight into the mechanisms behind the dynamical behavior of coherent structures in transitional boundary layers. As a particular example, we restrict ourselves to an especially interesting result from work that is presented in more detail in Rempfer & Fasel (1994b).

4.1 The Dynamical Equations

To facilitate a better understanding of the following, we first derive equations for the dynamical behavior and for the energy of coherent structures. This derivation will be based on the properties of the structures as defined by the eigenfunctions of a Karhunen-Loève decomposition (or proper orthogonal decomposition, POD) of the flow.

The coherent structures for the investigations reported here were calculated from the fluctuating flow, $\boldsymbol{u}' = \boldsymbol{u} - \boldsymbol{u}_B$, where $\boldsymbol{u}_B = \langle \boldsymbol{u} \rangle$ denotes the mean flow. We again use the angles $\langle \cdot \rangle$ to describe time averages that, in our case of equally spaced time steps, reduce to ordinary arithmetic mean values. For convenience of notation we also use the symbol $\boldsymbol{\sigma}_0 = \langle \boldsymbol{u} \rangle$ for the base flow.

As in the previous section, see also Rempfer (1994a), the Karhunen-Loève problem for the spatially evolving boundary layer is near degenerate, yielding pairs of eigenfunctions with almost identical eigenvalues (see also Aubry, Guyonnet & Lima 1992). In Rempfer (1994a) it was concluded that the dynamical coherent structures $\varsigma_i(\boldsymbol{x}, t)$ of our transitional boundary layer are represented by pairs of eigenfunctions with almost identical eigenvalues according to

$$\varsigma_i(x, y, z, t) = \zeta_{2i-1}(t)\boldsymbol{\sigma}_{2i-1}(x, y, z) + \zeta_{2i}(t)\boldsymbol{\sigma}_{2i}(x, y, z), \quad i = 1, 2, \ldots \quad (22)$$

Each of these pairs of eigenfunctions describes a coherent structure that is moving in the streamwise direction. In parentheses we note that for the case of a parallel flow where structures of constant shape are traveling downstream at constant speed, the eigenvalue problem of the POD would be exactly degenerate, yielding pairs of identical eigenvalues, and corresponding pairs of eigenfunctions that are exact phase-shifted copies of each other. We will refer to ς_i as the "coherent structure of order i" or the "ith-order coherent structure".

Our goal is the derivation of a low-dimensional model of the flow in the form of a set of ordinary differential equations for the dynamical behavior of the coherent structures. For this model we will introduce the additional assumption that influences from the upstream and downstream boundaries can be neglected. That this assumption is indeed justified was demonstrated in Rempfer & Fasel (1994b). We can therefore start from the precondition that all of the boundary conditions are already met by each of the eigenfunctions $\boldsymbol{\sigma}_i$. In the following, we will derive evolution equations from the vorticity-transport form of the Navier-Stokes equations,

$$\frac{\partial \boldsymbol{\omega}}{\partial t} = \boldsymbol{\omega} \cdot \operatorname{grad} \boldsymbol{u} - \boldsymbol{u} \cdot \operatorname{grad} \boldsymbol{\omega} + \nu \Delta \boldsymbol{\omega}, \quad (23)$$

with the vorticity vector $\boldsymbol{\omega} = \operatorname{curl} \boldsymbol{u}$. To get an evolution equation for the coherent structures, we introduce the expansion (14) of the flow field in the eigenfunctions of the POD. The vorticity required in (23) is given by

$$\omega(x,y,z,t) = \sum_i \zeta_i(t)\boldsymbol{\sigma}_i^\omega(x,y,z), \tag{24}$$

where $\boldsymbol{\sigma}_i^\omega$ is the vorticity field of $\boldsymbol{\sigma}_i$,

$$\boldsymbol{\sigma}_i^\omega = \operatorname{curl}\boldsymbol{\sigma}_i = \sum_\ell q_\ell \boldsymbol{\omega}^\ell. \tag{25}$$

Introducing (14) and (24) in (23) we arrive at

$$\sum_i \frac{d\zeta_i}{dt}\boldsymbol{\sigma}_i^\omega = \sum_{i,j}(\boldsymbol{\sigma}_i^\omega \cdot \operatorname{grad}\boldsymbol{\sigma}_j - \boldsymbol{\sigma}_i \cdot \operatorname{grad}\boldsymbol{\sigma}_j^\omega)\zeta_i\zeta_j + \sum_i \nu\zeta_i\Delta\boldsymbol{\sigma}_i^\omega. \tag{26}$$

By applying Galerkin projection to the equation above, claiming that the equation holds exactly within the function space spanned by the $\boldsymbol{\sigma}_i^\omega$, we get the system of ordinary differential equations

$$\sum_j T_{ij}\frac{d\zeta_j}{dt} = \sum_{j,k}\breve{N}_{ijk}\zeta_j\zeta_k + \sum_j \nu\breve{D}_{ij}\zeta_j, \tag{27}$$

where

$$T_{ij} = (\boldsymbol{\sigma}_i^\omega, \boldsymbol{\sigma}_j^\omega), \tag{28}$$

$$\breve{N}_{ijk} = (\boldsymbol{\sigma}_i^\omega, \boldsymbol{\sigma}_j^\omega \cdot \operatorname{grad}\boldsymbol{\sigma}_k - \boldsymbol{\sigma}_j \cdot \operatorname{grad}\boldsymbol{\sigma}_k^\omega), \tag{29}$$

$$\breve{D}_{ij} = (\boldsymbol{\sigma}_i^\omega, \Delta\boldsymbol{\sigma}_j^\omega). \tag{30}$$

The matrix $\boldsymbol{T}$ on the left-hand side of (27) is a consequence of the fact that the vorticities $\boldsymbol{\sigma}_i^\omega$ of the eigenfunctions $\boldsymbol{\sigma}_i$, in general, will not form an orthogonal system (but they will still form a linear independent system). Multiplying with the inverse $\boldsymbol{T}^{-1}$ we finally get

$$\frac{d\zeta_i}{dt} = \sum_{j,k} N_{ijk}\zeta_j\zeta_k + \nu\sum_j D_{ij}\zeta_j \tag{31}$$

with

$$N_{ijk} = \sum_l T_{il}^{-1}\breve{N}_{ljk}, \tag{32}$$

$$D_{ij} = \sum_l T_{il}^{-1}\breve{D}_{lj}. \tag{33}$$

Equation (31) describes the temporal evolution of the expansion coefficients $\zeta_i(t)$ and thus the dynamical behavior of the coherent structures. The first, quadratic term on the right-hand side of (31) describes nonlinear interactions between the coherent structures, and the second term describes the influence of viscosity on the motion of the structures.

In deriving the energy equations of the coherent structures, we define the energy of the flow by

$$e(t) = \frac{1}{2}\int_D \boldsymbol{u}^2(\boldsymbol{x},t)\,d\boldsymbol{x}. \tag{34}$$

The contribution of an eigenfunction $\boldsymbol{\sigma}_i$ to the energy of the flow is given by

$$e_i(t) = \frac{1}{2}\lambda_i\zeta_i^2(t), \tag{35}$$

so that

$$\frac{de_i}{dt} = \frac{d(\frac{1}{2}\lambda_i\zeta_i^2)}{dt} = \lambda_i\zeta_i\frac{d\zeta_i}{dt}. \tag{36}$$

Using (31), we arrive at

$$\frac{de_i}{dt} = \lambda_i\sum_{j,k} N_{ijk}\zeta_i\zeta_j\zeta_k + \nu\lambda_i\sum_j D_{ij}\zeta_i\zeta_j. \tag{37}$$

Equation (37) describes the variation in time of the energy of the flow field being induced by the eigenfunction with index i depending on the values of all the expansion coefficients. The first term on the right-hand side of (37) describes variations of the energy that are due to the nonlinear terms of the Navier-Stokes equations. The second term on the right-hand side describes the influence of dissipation on the energy of the flow field induced by the eigenfunction $\boldsymbol{\sigma}_i$.

As it has been shown (Rempfer & Fasel 1994a) that the dynamical coherent structures of the flow are formed by the superposition of a pair of eigenfunctions, from a coherent structures point of view it is more interesting to look at the energy flows between the coherent structures than at those between eigenfunctions. Accordingly, we define the energy e_i^ς of a coherent structure by

$$e_i^\varsigma(t) = \frac{1}{2}(\lambda_{2i-1}\zeta_{2i-1}^2(t) + \lambda_{2i}\zeta_{2i}^2(t)), \tag{38}$$

and we will investigate the production terms Π_{ijk}^ς, where

$$\Pi_{ijk}^\varsigma = \breve{\Pi}_{ijk}^\varsigma + \breve{\Pi}_{ikj}^\varsigma, \quad k \neq j, \qquad \Pi_{ijj}^\varsigma = \breve{\Pi}_{ijj}^\varsigma, \tag{39}$$

with

$$\breve{\Pi}_{ijk}^\varsigma = \sum_{l=2i-1}^{2i}\ \sum_{m=\max(0,2j-1)}^{2j}\ \sum_{n=\max(0,2k-1)}^{2k} \Pi_{lmn}, \qquad \begin{array}{c} i = 1,2,\ldots, \\ j,k = 0,1,\ldots \end{array}, \tag{40}$$

and

$$\Pi_{ijk} = \lambda_i N_{ijk}\zeta_i\zeta_j\zeta_k. \tag{41}$$

4.2 Instantaneous Energy Flows

After these preliminaries, we can now explore the behavior of the Π^{ς}_{ijk} using a selected example. Looking at the time behavior of the expansion coefficients corresponding to the third-order structure (see Rempfer & Fasel 1994b), a significant rise in energy of this structure can be observed between time steps 200 and 300. In Rempfer & Fasel (1994a) it has been shown that this behavior is connected to the occurrence of the characteristic "spikes" of the transition stage that we are investigating here. Therefore, it is of particular interest to find the causes of the strong variation in energy of the third-order structure. This variation can only be caused by nonlinear interactions.

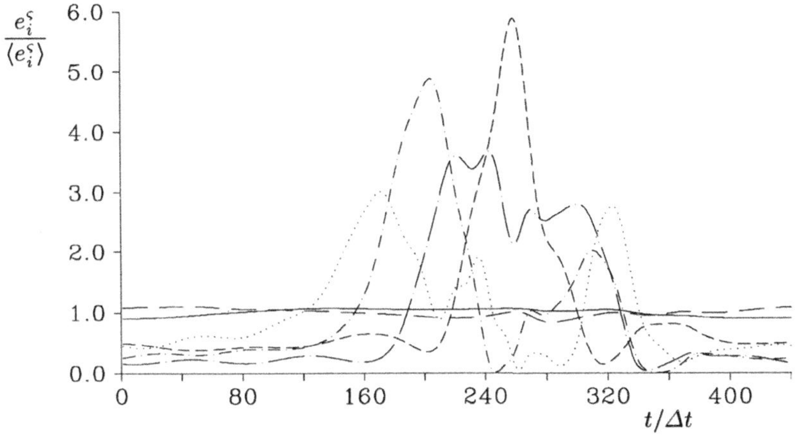

Fig. 9.: Behavior of energy $e^{\varsigma}_i(t)$ of the first six structures. The energies have been divided by their time average. ————, e^{ς}_1; — — —, e^{ς}_2; – – – – – –, e^{ς}_3; ——·——, e^{ς}_4; —·—·—·—, e^{ς}_5; ··············, e^{ς}_6.

In Fig. 9, the development of the energy, $e^{\varsigma}_i(t)$, of the first six coherent structures is shown. The most noticeable behavior is displayed by the energy of the third-order structure. During time step 200 to 260, this energy rises to six times its mean value and more than ten times its average value from the time before. Up to time step 320, the energy falls back again to its previous value. Thus the question arises, "What causes these strong variations in energy?" To explore this issue, the instantaneous values of the matrix $\Pi^{\varsigma}_{3jk}(t)$ at time step 250 (at the moment of steepest raise of energy) and at time step 270 (at the moment of steepest decay) are shown in perspective representation in Fig. 10.

These histograms demonstrate that, in essence, one single interaction determines the behavior of the energy of the third-order structure. The increase as well as the decrease of the energy are caused by the energy gain and loss that the third-order structure suffers in an interaction with the base

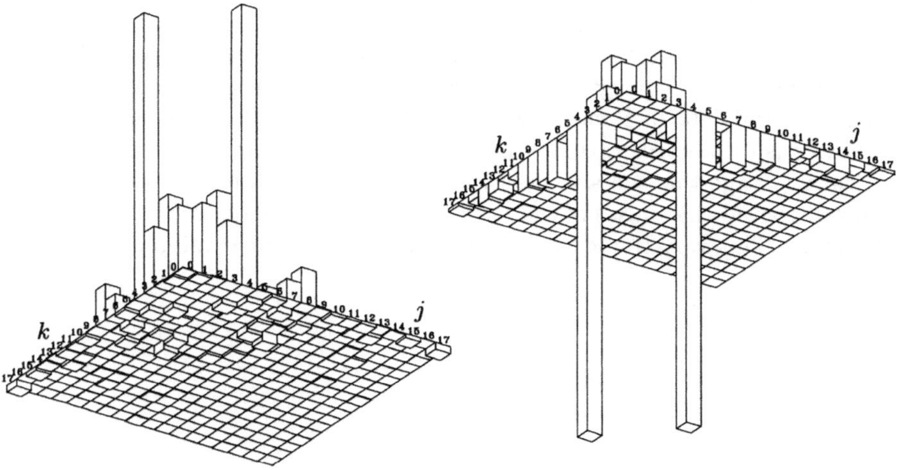

(a) Time step $t/\Delta t = 250$. (b) Time step $t/\Delta t = 270$.

Fig. 10.: Histograms of instantaneous energy flows Π^{ς}_{3jk} of third-order structure ($i = 3$).

flow and the fourth-order structure. This can be seen most clearly when the time derivative of the energy of the third-order structure is compared to the contributions provided by the interaction 3-0-4 (i.e. the interaction between third-order structure, base flow, and fourth-order structure) and dissipation (Fig. 11).

From Fig. 12, which shows instantaneous energy flows of the fourth-order structure in an analogous way to Fig. 10 (both figures are drawn using the same scale factors), one can observe a kind of alternating play: almost exactly the energy that the third-order structure gains in the interaction 3-0-4 is lost by the fourth-order structure in the interaction 4-0-3, and *vice versa*. The fact that the development of the energy of the fourth-order structure is not just a mirror image of the one of the third-order structure is due to additional interactions which contribute to the energy balance of the fourth-order structure, as seen in Fig. 13.

From a coherent structures point of view, the cause of spikes can thus be explained as follows: First, it should be noted that the motion of the

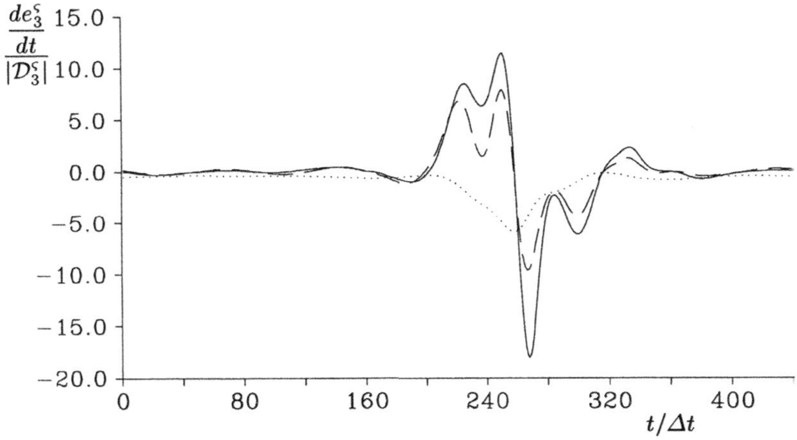

Fig. 11.: Contributions to the derivative $\frac{de_3^\varsigma}{dt}$ of the energy of the third-order structure. Values have been divided by the absolute of the average dissipation $\mathcal{D}_3^\varsigma$. ————, de_3^ς/dt; — — —, Π_{304}^ς; ··········, instantaneous dissipation.

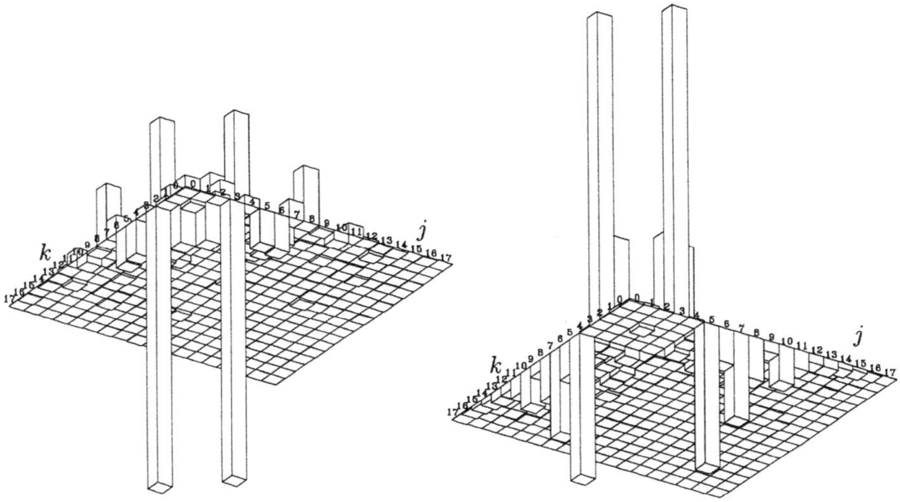

(a) Time step $t/\Delta t = 250$. (b) Time step $t/\Delta t = 270$.

Fig. 12.: Histograms of instantaneous energy flows Π_{4jk}^ς of fourth-order structure ($i = 4$).

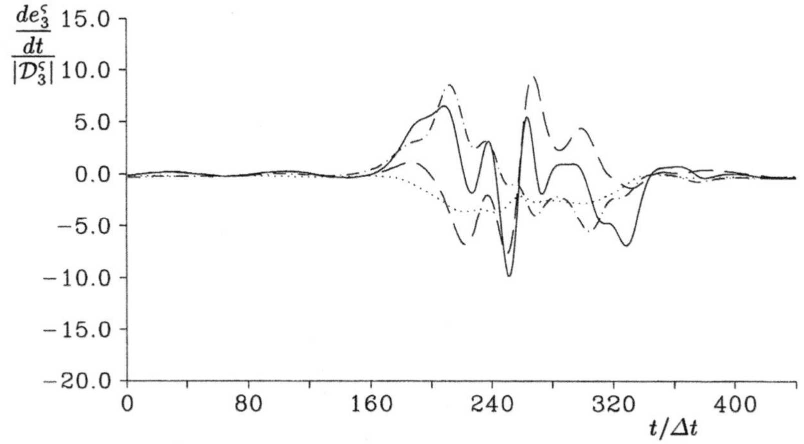

Fig. 13.: Contributions to the derivative $\frac{de_4^\varsigma}{dt}$ of the energy of the third-order structure. Values have been divided by the absolute of the average dissipation $\mathcal{D}_4^\varsigma$. ————, de_4^ς/dt; — — —, Π_{403}^ς; —·—·—·, Π_{405}^ς; ············, instantaneous dissipation.

Λ-vortex (first-order structure) alone induces some negative u' disturbance velocity (see Rempfer & Fasel 1994b). The sharp spike, however, is caused by a separate event. At the time instant of the formation of a spike, a certain coherent structure is "created", the velocity field of which strongly enforces the negative deflection of the u'-signal. Although this third-order structure is actually present all the time, the term 'created' seems to be justified in view of the considerable growth of energy of this structure during the spike.

5 Other Applications and Related Work

Whith the above examples focusing exclusively on work done or directed by this author, at the end of this chapter we also want to briefly discuss a few studies analyzing the dynamics of coherent structures in open flow systems by other authors. Of course, the choice of studies presented in the following is far from complete, for additional information we refer the reader to the review articles cited above, as well as the very recent book by Holmes, Lumley & Berkooz (1996).

Certainly one of the pioneering works on the dynamics of coherent structures was the study of the near-wall layer of a turbulent pipe flow by Aubry *et al.* (1988). In this investigation, KL eigenfunctions obtained from experimental data (Herzog 1986) were used to derive a low-order system of ordinary differential equations describing the dynamics of the coherent structures in the lower part of the turbulent wall layer ($y^+ \leq 40$). Using tools from dynamical systems theory, this system of ODEs was then analyzed in detail. One of the most significant findings was the identification of a characteristic structure in the phase space of the system (a so-called *heteroclinic orbit*), and the discovery that this structure is closely connected to the well-known bursting phenomenon in such a turbulent flow. As the ejection event that is part of the bursting (which consists of a sweep-ejection cycle) is a major contributor to the turbulent drag in a boundary layer, a deeper understanding of this process is of greatest interest. Thus the results of this study may be useful for the design of control strategies aiming at influencing the frequency of the bursting, which could then directly translate into a reduction of turbulent drag.

Another important open flow configuration that was studied in detail by George, Glauser, and co-workers (Glauser, Zheng & George 1992) is the circular jet. One of the goals of their work was to study the mechanisms that govern the dynamics of large-scale structures in this flow. These structures were again defined via eigenfunctions of proper orthogonal decomposition, and, as in the derivation of the eigenfunctions for the pipe flow above, a Fourier decomposition was performed in both the azimuthal and streamwise directions. The data for the correlation tensor that is needed to determine the KL eigenfunctions was obtained using a clever system of hot wire rakes. The authors were finally able to study the behaviors of dynamical systems derived by projection of the momentum equations onto their KL modes. The time histories and phase portraits of these dynamical systems suggest that an intermittent, bursting-like behavior occurin the jet, hinting at unexpected parallels between this flow and the turbulent boundary layer on a very fundamental level.

One more interesting experimental study we want to metion here was the one by Rajaee, Karlsson & Sirovich (1994), who used a data base from experimental hot-wire measurements in a plane, acoustically forced transitional mixing layer. As in the case of the transitional boundary layer above, they obtained KL eigenfunctions that occurred in pairs of almost equal eigenvalues that are due to the approximate streamwise homogeneity of the problem. Based on these eigenfunctions, dynamical systems equations were then derived that consisted of eight and sixteen real modes, respectively. In contrast to the systems in the above two examples, in this case the systems do not appear to have a particular symmetry structure in phase space. In contrast to the two examples above, here the authors were able to directly compare the solutions of their dynamical systems to experimental data. These comparisons display a remarkably good tracking of the experimental time evolution for the most energetic modes, while higher-order modes are less accurate, but still show a generally correct level in the case of the more detailed (16-mode) model. In the lower-order (8-mode) model, the fifth and sixth modes exhibit unphysically growing magnitudes, which the authors interpret as an indication that an energy transfer mechanism (which would normally feed the ninth and higher modes) is blocked in the eight-mode truncation.

An example of a study using KL eigenfunctions in a slightly more complex geometry is the work done by Deane and co-workers (Deane *et al.* 1991) on flows in a grooved channel and in the wake of a circular cylinder. In this case the data bases were generated numerically, by direct numerical simulations using spectral element methods. In both cases the Reynolds numbers were chosen such that the system exhibited a limit cycle behavior so that the temporal dynamics are fairly simple. An important question addressed in these studies concerned the range of validity of models that are based on data at a particular Reynolds number, and what may be done to extend that range. From comparisons of solutions of the low-dimensional models derived by the authors and solutions of their direct simulations, they found good short-term tracking for the Reynolds numbers for which the models were derived. The authors also showed that by using a modified base-flow profile, they could achieve good short-term tracking for extrapolated Reynolds numbers.

Finally, we want to mention a number of interesting studies on the dynamics of the wake of a circular cylinder that have been done by Eckelmann, Noack, and co-workers (Noack & Eckelmann 1992, Noack & Eckelmann 1994, Zhang *et al.* 1995). This work comprises experimental investigations, low-dimensional models, and direct numerical simulations. Here, we want to focus in particular on the low-dimensional models designed by Noack (1992). In contrast to the investigations reported so far, Noack used a special basis of eigenfunctions that was specifically designed to model the cylinder wake. The eigenfunctions that were used were global in the sense that their support included all of the integration domain, so these modes cannot be interpreted as coherent structures of the flow. Nevertheless, through a judicious choice of these eigenmodes, Noack was able to derive an extremely elegant low-dimensional model for the cylinder wake. Using this model, and in conjunction with experimental as well as numerical studies, it was possible to perform a detailed analysis of the bifurcation scenario of the transitional cylinder wake. Among the most significant results of this set of studies is the demonstration of three different generic flow instabilities which are responsible for the phenomena that can be observed in a cylinder wake during transition to turbulence.

6 Discussion

In summary, we would like to point out the considerable advantages of the decomposition of complex flows into structures based on rigorous definitions of this concept. In our view, it is crucial to realize the major importance of a completely unambiguous, precise mathematical definition of what is meant by the term "structure", such as the ones given in Sect. 2 of this chapter. Apart from the purely mathematical attractiveness of this kind of a definition, this property is one of the prime preconditions for such a notion to be meaningful. Only a strict definition actually allows comparison of results obtained by different researchers and in different laboratories. In contrast, such strict definitions are often lacking in the conditional-sampling community, thus making it difficult to formulate statements that go beyond purely qualitative descriptions.

Furthermore, we feel that the concept of a structure is of little use as long as it is not possible to give a precise and comprehensive description of the role that these structures play in the dynamics of a flow. Often, such descriptions have been limited at most to statements with regard to the contribution of a structure to the turbulent energy, or an estimation of its addition to turbulent production. In contrast, by investigating dynamical equations for the structures as described in Sect. 4, a detailed description of the dynamics of coherent structures can be obtained. With this approach, it is possible to uncover the mechanisms behind the behavior of the coherent structures and thus find out, so to speak, "how the flow works".

We also note that, by studying the system of ODEs (31) describing the evolution in time of the structures, it is also possible to apply the well-developed methods of dynamical systems theory to the dynamics of fluid flow. Such investigations have been reported by Aubry *et al.* (1988), Deane *et al.* (1991), Zhou & Sirovich (1992), and Rempfer (1996) and it is hoped that this kind of research will enhance our fundamental understanding of turbulent flows. From such studies, we may expect an answer to the question, "What relevance do the ideas around deterministic chaos have in the case of turbulent flows?"

References

Aubry N., Holmes P., Lumley J. L., Stone E. (1988): The dynamics of coherent structures in the wall region of a turbulent boundary layer. J. Fluid Mech. **192**, 115–173

Aubry N., Guyonnet R., Lima R. (1992): Spatio-temporal symmetries and bifurcations via bi-orthogonal decompositions. J. Nonlin. Sci. **2**, 683–739

Berkooz G., Holmes P., Lumley J. L. (1993): The proper orthogonal decomposition in the analysis of turbulent flows. Ann. Rev. Fluid Mech. **25**, 539–575

Cantwell B. J. (1981): Organized motion in turbulent flow. Ann. Rev. Fluid Mech. **13**, 457–515

Chui C. K. (1992): *An introduction to wavelets.* Academic Press, Boston.

Cutler A., Breiman L. (1994): Archetypal analysis. Technometrics **36**, 338–347

Cutler A., Stone E. (1997): Moving archetypes. Physica D, **107**, 1, 1–16

Deane A. E., Kevrekidis I. G., Karniadakis G. E., Orszag S. A. (1991): Low-dimensional models for complex geometry flows: Application to grooved channels and circular cylinders. Phys. Fluids A **3** (10), 2337–2354

Einstein H. A., Li H. (1956): The viscous sublayer along a smooth boundary. J. Eng. Mech. Div. ASCE **82**(EM2): Pap. No. 945.

Farge M. (1992): Wavelet transforms and their applications to turbulence. Ann. Rev. Fluid Mech. **24**, 395–457

Fiedler H. E. (1987): Coherent structures. In *Advances in Turbulence* (eds. G. Comte-Bellot, J. Mathieu), 320–336, Springer, New York.

Glauser M. N., Zheng X., George W. K. (1992): The streamwise evolution of coherent structures in the axisymmetric jet mixing layer. In *Studies in Turbulence* (eds. Gatski T. B., Sarkar S. & Speziale C. G.), pp. 207–222. Springer Verlag, New York

Gupta A. K., Laufer J., Kaplan R. E. (1971): Spatial structure in the viscous sublayer. J. Fluid Mech. **50**, part 3, 493–512

Herzog S. (1986): *The large scale structure in the near wall region of a turbulent pipe flow.* PhD thesis, Cornell University

Holmes P., Lumley J. L., Berkooz G. (1996): *Turbulence, coherent structures, dynamical systems and symmetry.* Cambridge University Press, Cambridge

Hussain A. K. M. F. (1983): Coherent structures – reality and myth. Phys. Fluids A **26** (10), 2816–2850

Hussain A. K. M. F. (1986): Coherent structures and turbulence. J. Fluid Mech. **173**, 303–356

Kaiser G. (1994): *A friendly guide to wavelets.* Birkhäuser, Boston.

Kim H. T., Kline S. J., Reynolds W. C. (1971): The production of turbulence near a smooth wall in a turbulent boundary layer. J. Fluid Mech. **50**, part 1, 133–160

Kline S. J., Runstadler P. W. (1959): Some preliminary results of visual studies of the flow model of the wall layers of the turbulent boundary layer. Trans. ASME, Ser. E **2**, 166–170

Kline S. J., Reynolds W. C., Schraub F. A., Runstadler P. W. (1967): The structure of turbulent boundary layers. J. Fluid Mech. **30**, 741–773

Kloker M. (1993): *Numerische Untersuchungen zum laminar-turbulenten Strömungsumschlag in verzögerten Grenzschichten.* Dissertation, Universität Stuttgart.

Lorenz E. N. (1963): Deterministic nonperiodic flow. J. Atmos. Sci. **20**, 130–141

Loève M. (1955): *Probability theory.* Van Nostrand, Princeton

Lumley J. L. (1967): The structure of inhomogeneous turbulent flows. In *Atmospheric Turbulence and Radio Wave Propagation* (eds. A. M. Yaglom & V. I. Tatarski), 166–178, Nauka, Moscow

Lumley J. L. (1981): Coherent structures in turbulence. In *Transition and Turbulence* (ed. Meyer R. E.), Academic Press, NY

Lumley J. L. (ed.) (1990): *Whither turbulence? Turbulence at the crossroads.* Volume 357 of *Lecture Notes in Physics*, Springer-Verlag, New York

Noack B. R., Eckelmann H. (1992): On chaos in wakes. Physica D **56**, 151–164.

Noack B. R., Eckelmann H. (1994): A global stability analysis of the steady and periodic cylinder wake. J. Fluid Mech. **270**, 297–330

Rajaee M., Karlsson S. K. F., Sirovich L. (1994): Low-dimensional description of free-shear-flow coherent structures and their dynamical behaviour. J. Fluid Mech. **258**, 1–29

D. Rempfer and H. Fasel (1994a): Evolution of three-dimensional coherent structures in a flat-plate boundary layer. J. Fluid Mech. **260**, 351–375

D. Rempfer and H. Fasel (1994b): Dynamics of three-dimensional coherent structures in a flat-plate boundary layer. J. Fluid Mech. **275**, pp. 257–283

Rempfer D. (1996): Investigations of boundary layer transition via galerkin projections on empirical eigenfunctions. Phys. Fluids **8** (1), 175–188

Rist U., Fasel H. (1995): Numerical simulation of transition in a flat-plate boundary layer. J. Fluid Mech. **298**, 211–248

Robinson S. K. (1991): Coherent motions in the turbulent boundary layer. Ann. Rev. Fluid Mech. **23**, 601–639

Runstadler P. G., Kline S. J., Reynolds W. C. (1963): An experimental investigation of flow structure of the turbulent boundary layer. Rep. No. MD-8 Dep. Mech. Eng., Stanford Univ., Stanford, Calif.

Sirovich L. (1987): Turbulence and the dynamics of coherent structures. Quart. Appl. Math. **45**, 561–590

Tennekes H., Lumley J. L. (1972): *A first course in turbulence* MIT Press, Cambridge

Townsend A. A. (1956): *The structure of turbulent shear flow.* Cambridge University Press, Cambridge.

Wallace J. M., Eckelmann H., Brodkey R. S. (1972): The wall region in turbulent shear flow. J. Fluid Mech. **54**, part 1, 39–48

Willmarth W. W., Lu S. S. (1972): Structure of the reynolds stress near the wall. J. Fluid Mech. **55**, part 1, 65–92

Zhang H. Q., Fey U., Noack B. R., König M., Eckelmann H. (1995): On the transition of the cylinder wake. Phys. Fluids **7**, 779–794.

Zhou X., Sirovich L. (1992): Coherence and chaos in a model of turbulent boundary layer. Phys. Fluids A **4** (12), 2855–2874

Theoretical and Experimental Investigations of the Faraday Instability

H. W. Müller[1], R. Friedrich[2], D. Papathanassiou[1]

[1] Universität des Saarlandes, Institut für Theoretische Physik, Postfach 15 11 50, D-66041 Saarbrücken, Germany
[2] Universität Stuttgart, Institut für Theoretische Physik
Pfaffenwaldring 57
D-70550 Stuttgart, Germany

Abstract. In the Faraday experiment waves on the surface of a fluid layer are excited by a vertical vibration of the vessel. By virtue of its simplicity this system is outstanding among the pattern forming hydrodynamic systems. The variety of observed ordered patterns is extremely rich and ranges from simple lines to complicated quasi-periodic structures. The aim of this review article is to provide an introduction into the field of Faraday waves, as well as to give an account of the developments during the last decade. Within this period three major lines of interest can be identified: (i) the linear stability theory for viscous fluids, (ii) the interaction of a few spatial degrees of freedom in confined geometries giving way to chaos, and (iii) the formation of well ordered surface patterns in extended geometries and their transition towards spatio-temporal disorder. Special emphasis in this article is placed upon a comparison between theory and experiment.

1 Introduction

The generation of surface waves on a fluid via a vertical vibration of its vessel has a long history which dates back to Faraday (1831). He realized that the waves have a frequency equal to one half of that of the excitation. Rayleigh (1883) suspected that this phenomenon was related to a parametric driving mechanism, but the mathematical proof took until the 1950s, when Benjamin and Ursell (1954) demonstrated that the linear behaviour of inviscid surface waves can be understood in terms of a Mathieu oscillator.

Recent studies in nonlinear patterns forming instabilities, bifurcation and chaos theory, and the transition to spatio-temporal disorder has increased our interest in the Faraday system. The system is remarkable for its great variability and richness of the observed patterns. The capability of controlling the wavelength by the external drive frequency via dispersion allows one to switch easily from small to large aspect ratios. In small aspect ratio systems the behaviour is governed by a small number of nonlinearly interacting spatial modes, which nevertheless may generate chaotic dynamics. In the opposite large aspect ratio limit, the effects of lateral boundaries become negligible. As a result of nonlinear wave interaction, a variety of well ordered patterns occurs close above the onset, resembling two-dimensional crystals and even quasi-crystals. Secondary transitions can be observed by a further increase of the drive amplitude.

Besides being easily accessible by experimental means, the Faraday system also has the advantage of a very small dynamical relaxation time. Usually it takes only a few seconds until a stationary pattern has relaxed. This value must be compared with relaxation times of up to hours which apply to thermal convection experiments. Also unlike convection, where the dynamics is governed by dissipative mechanisms (diffusion of heat and momentum), the damping of surface waves can often be treated as a perturbation. This opens the opportunity for a theoretical description in terms of an (almost) Hamiltonian or variational approach. However, this approximation ceases to apply if dissipation increases. Until now, no reasonable nonlinear theory for strongly damped surface waves has been established. Difficulties arise due to the interplay of intricate nonlinear boundary conditions on the free surface and the parametric forcing, which makes the problem non-autonomous. Since the primary pattern at onset is rapidly oscillating, numerical simulations of the full hydrodynamic equations are rather time-consuming.

The amount of literature on parametrically driven surface waves is enormous. The recent review of Miles and Henderson (1990) gives an account of the literature until 1989 but it is far from being complete. The Lagrangian formulation of the problem and several resonance mechanisms are investigated, which give rise to chaotic surface dynamics in a closed vessel experiment.

The present article will review the developments in the field during the last fifteen years. Our intention is not to cover the entire subject, but we will discuss some major issues which, in our opinion, are of special importance. Furthermore this article tries to provide an introduction into the subject of pattern formation in driven surface waves.

During the last decade extensive work has been done especially on the following four issues:

- The development of a stability theory for viscous fluids.
- The investigation of mode interactions in confined geometries.
- The occurence of well-ordered standing wave patterns in large aspect ratio systems.
- The transition towards disordered patterns via secondary instabilities.

The present review follows these main topics.

2 Basic equations

The mathematical description of the surface waves in the Faraday experiment has to take into account the motion of the liquid, the deformation of the interface as well as the properties of the medium above the surface. In the following we shall consider a surface bounded by air, which can be characterized by a constant atmospheric pressure $p_{\rm atm}$. The liquid layer of depth h is considered incompressible with uniform density ρ, kinematic viscosity ν and to be subjected to a periodic vertical vibration with frequency 2Ω (see Fig. 1). In the frame of reference co-moving with the oscillating con-

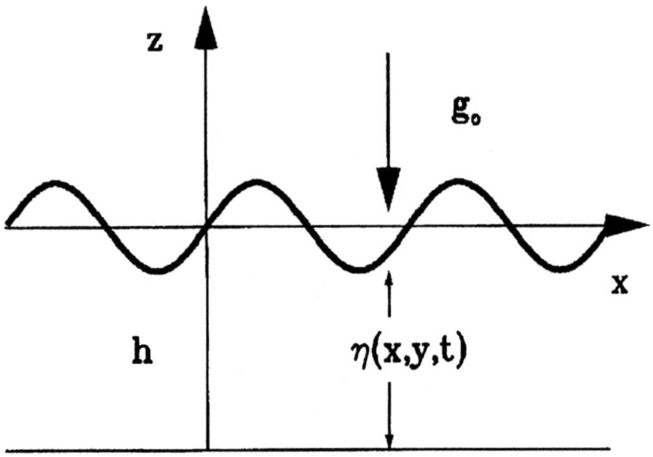

Fig. 1. Setup of the Faraday experiment.

tainer the earth's gravitational acceleration g_0 is modulated according to $\mathbf{g}(t) = (g_0 + a\,W(t))\,\mathbf{e}_z$. If not otherwise specified we study the case of a sinusoidal modulation $W(t) = \cos 2\Omega t$. The time evolution of the velocity field $\mathbf{v} = \mathbf{v}(\mathbf{r},t) = (u,v,w)$ is described by the mass-continuity and the Navier-Stokes equation

$$\nabla \cdot \mathbf{v} = 0 \ , \tag{1}$$

$$\partial_t \mathbf{v} + (\mathbf{v}\cdot\nabla)\,\mathbf{v} = -\frac{\nabla p}{\rho} - g(t)\mathbf{e}_z + \nu\nabla^2\mathbf{v} \ , \tag{2}$$

where $p(\mathbf{r},t)$ denotes the pressure. In what follows we combine the horizontal coordinates into the vector $\mathbf{r}_\perp = (x,y)$. The location of the free surface is determined by the function $z = \eta(\mathbf{r}_\perp,t)$. Its dynamics is governed by the kinematic surface condition

$$\partial_t\eta + (\mathbf{v}_\perp|_\eta \cdot \nabla_\perp)\,\eta = w|_\eta \ , \tag{3}$$

where $|_\eta$ abbreviates evaluation at $z=\eta$. The surface normal vector in the outward direction is $\mathbf{n}(\mathbf{r}_\perp,t) = (-\partial_x\eta, -\partial_y\eta, 1)/\sqrt{1+(\nabla_\perp\eta)^2}$. In order to allow for a unique solution of the Navier-Stokes equation one must specify boundary conditions. At the liquid-air interface the net force per unit area is balanced by the surface tension α

$$\left(\sigma^{\mathrm{air}} - \sigma\right) \cdot \mathbf{n} = \alpha \left(\nabla \cdot \mathbf{n}\right) \mathbf{n} \ . \tag{4}$$

Here $\sigma_{ij} = -p\delta_{ij} + \sigma'_{ij} = -p\delta_{ij} + \rho\nu(\nabla_i v_j + \nabla_j v_i)$ denotes the stress tensor of the liquid. The stress tensor of the air above the liquid layer can be approximated by the constant atmospheric pressure, $\sigma_{ij}^{\mathrm{air}} \simeq -p_{\mathrm{atm}}\delta_{ij}$, because the dynamic viscosity of air $\rho^{\mathrm{air}}\nu^{\mathrm{air}}$ is at least one order of magnitude smaller than that of a liquid. At the bottom of the container, at $z = -h$, one must use the no-slip condition

$$\mathbf{v}\left(\mathbf{r}_{\perp}, z = -h\right) = 0 \ . \tag{5}$$

For sufficiently small drive amplitude a, the fluid remains at rest and the surface is flat. It is therefore convenient to introduce the pressure deviation $\pi(\mathbf{r}, t) = p(\mathbf{r}, t) - p_{\mathrm{atm}} + \rho g(t) z$ from the quiescent state profile. We then get

$$\nabla \cdot \mathbf{v} = 0 \ , \tag{6}$$

$$\partial_t \mathbf{v} + (\mathbf{v} \cdot \nabla)\, \mathbf{v} = -\frac{\nabla \pi}{\rho} + \nu \nabla^2 \mathbf{v} \tag{7}$$

for the bulk of the fluid. At the upper surface the following two conditions must be fulfilled

$$\pi|_{\eta} - \mathbf{n} \cdot \sigma'|_{\eta} \cdot \mathbf{n} = \rho\, g(t)\, \eta + \alpha(\nabla \cdot \mathbf{n}) \ , \tag{8}$$

$$\mathbf{t} \cdot \sigma'|_{\eta} \cdot \mathbf{n} = 0 \ , \tag{9}$$

where $\mathbf{t}$ denotes either of two linearly independent tangential vectors perpendicular to $\mathbf{n}$.

The Faraday experiment in a finite geometry needs analogous conditions to be stated at the lateral walls. However, often it is appropriate to consider an infinite lateral extension or periodic boundary conditions, which should yield accurate results in the case of large aspect ratio systems. Equations (3), (5), (6), (7), (8), and (9) represent the full set of nonlinear hydrodynamic evolution equations for Faraday waves (Kumar and Tuckerman (1994), Beyer and Friedrich (1995)).

In the case of ideal fluids the viscous term $\nabla^2 \mathbf{v}$ in the Navier-Stokes equation (7) and the viscous stress tensor σ' are absent. The resulting potential flow problem with $\mathbf{v} = \nabla\phi$ is governed by the Euler equation

$$\partial_t \phi + \frac{1}{2}\left(\nabla\phi\right)^2 + \pi = 0 \tag{10}$$

and the incompressibility condition

$$\nabla^2 \phi = 0 \ . \tag{11}$$

As usual, a potential flow field cannot fulfill all boundary conditions (5), (8), (9), instead only the normal components of these conditions apply

$$\pi|_\eta = \rho\, g(t)\, \eta + \alpha\, (\nabla \cdot \mathbf{n}) \ , \tag{12}$$

$$\partial_z \phi\, (\mathbf{r}_\perp, z = -h) = 0 \ . \tag{13}$$

For Faraday waves in *weakly viscous* fluids the bulk velocity field remains approximately potential, whereas it is modified along the container walls and beneath the free surface by viscous boundary layers of thickness $\delta = \sqrt{\nu/\Omega}$. These boundary layers carry vorticity. If they are thin compared to the wavelength and to the geometrical dimensions of the vessel, viscous effects can be accounted for by an effective boundary condition upon the velocity potential. This is essentially the quasi-potential approximation (see Lundgren and Mansour (1988), Ruvinsky et al. (1991)). In deep vessels of large lateral aspect ratio only the effect of the surface boundary layer remains. By expanding (3), (7), (8), and (9) in powers of δ and on neglecting nonlinear viscous contributions Zhang and Viñals (1997a) show that the equations of motion can be expressed in terms of variables on the free surface alone. In their approach they obtain two evolution equations for the surface elevation $\eta = \eta(\mathbf{r}_\perp, t)$ and for the potential at the free surface $\Phi = \Phi(\mathbf{r}_\perp, t) = \phi(\mathbf{r}, t)|_\eta$ of the form

$$\begin{aligned} \partial_t \eta = {} & 2\nu \nabla_\perp^2 \eta + \mathcal{D}\Phi - \nabla_\perp \cdot (\eta \nabla_\perp \Phi) + \frac{1}{2}\nabla_\perp^2 \left(\eta^2 \mathcal{D}\Phi\right) \\ & - \mathcal{D}\left(\eta \mathcal{D}\Phi\right) + \mathcal{D}\left[\eta \mathcal{D}\left(\eta \mathcal{D}\Phi\right) + \frac{1}{2}\eta^2 \nabla_\perp^2 \Phi\right] \ , \end{aligned} \tag{14}$$

$$\begin{aligned} \partial_t \Phi = {} & 2\nu \nabla_\perp^2 \Phi + g(t)\eta + \frac{\alpha}{\rho}\nabla_\perp^2 \eta + \frac{1}{2}\left(\mathcal{D}\Phi\right)^2 - \frac{1}{2}\left(\nabla_\perp \Phi\right)^2 \\ & - \left(\mathcal{D}\Phi\right)\left[\eta \nabla_\perp^2 \Phi + \mathcal{D}\left(\eta \mathcal{D}\Phi\right)\right] - \frac{\alpha}{2\rho}\nabla_\perp \cdot \left[\nabla_\perp \eta \left(\nabla_\perp \eta\right)^2\right] \ . \end{aligned} \tag{15}$$

Unlike the original problem, which is 3-dimensional, these equations depend on the two horizontal coordinates $\mathbf{r}_\perp = (x, y)$ only. By virtue of (11) the velocity potential $\phi(\mathbf{r})$ can be eliminated in favour of its values at the surface $\Phi(\mathbf{r}_\perp)$. This gives rise to the non-local operator $\mathcal{D}$ in (14), (15) which implies multiplication by $|\mathbf{k}|$ in the lateral Fourier-space. This operation is easy to perform if spectral codes are used for a numerical integration.

3 Linear stability

3.1 The eigenvalue problem

In order to investigate the stability of the rest state the evolution equations and boundary conditions must be linearized around the quiescent solution $\mathbf{v} = \pi = \eta = 0$. Then, all surface evaluations $|_\eta$ are to be taken at $z = 0$. We expand the lateral dependencies of the hydrodynamic fields into plane waves with wavenumber $\mathbf{k} = (k_x, k_y)$ according to

$$\eta(\mathbf{r}_\perp, t) = \eta_k(t)\, \mathrm{e}^{\mathrm{i}\mathbf{k}\cdot\mathbf{r}_\perp} \ , \tag{16}$$

$$w(\mathbf{r}, t) = w_k(z, t)\, \mathrm{e}^{\mathrm{i}\mathbf{k}\cdot\mathbf{r}_\perp} \ . \tag{17}$$

After elimination of the pressure one arrives at the following linear problem (Kumar and Tuckerman (1994), Kumar (1996)) which determines the onset of the Faraday instability. The equation

$$\left[\partial_t - \nu\left(\partial_{zz} - k^2\right)\right]\left(\partial_{zz} - k^2\right) w_k = 0 \tag{18}$$

holds in the bulk of the fluid, while

$$\left[\partial_t - \nu\left(\partial_{zz} - k^2\right) + 2\nu k^2\right] \partial_z w_k|_0 = -\left(g(t) + \frac{\alpha}{\rho}k^2\right) k^2\, \eta_k \ , \tag{19}$$

$$\nu\left(\partial_{zz} + k^2\right) w_k|_0 = 0 \ , \tag{20}$$

$$\partial_t \eta_k = w_k|_0 \tag{21}$$

apply at the free surface and

$$w_k|_{-h} = 0 \ , \tag{22}$$

$$\partial_z w_k|_{-h} = 0 \tag{23}$$

at the bottom.

3.2 Ideal fluids and the limit of weak dissipation

For inviscid fluids ($\nu = 0$) (18) reduces to the linear Euler equation and the boundary conditions (20) and (23) drop out. Equation (19) can be solved for the velocity field $w_k(z, t)$, which can be eliminated, giving an ordinary differential equation for the surface elevation (Benjamin and Ursell (1954))

$$\ddot{\eta}_k + \left[\omega_0^2(k) + a\, k \tanh(kh) \cos(2\Omega t)\right] \eta_k = 0 \ , \tag{24}$$

with the known dispersion for inviscid free surface waves

$$\omega_0^2(k) = \tanh(kh)\left(\mathrm{g}_0 k + \frac{\alpha}{\rho}k^3\right) \ . \tag{25}$$

At small frequencies (long wavelengths) the dispersion is dominated by gravity (gravity waves). For wavelengths smaller than the capillary length $\sqrt{\alpha/(\rho \mathrm{g}_0)}$ (a few millimeters for most liquid-air interfaces) the restoring force is governed by surface tension and the waves are called capillary ripples. Equation (24) is the evolution equation of a parametrically driven pendulum (Mathieu oscillator). The rest state of the oscillator corresponds to the undeformed flat surface of the hydrodynamic problem. It is well known that the quiescent pendulum can be destabilized by a gravity modulation. By virtue of the dispersion the characteristic resonance tongues of Mathieu's stability chart (see e.g. Nayfeh and Mook (1979)) can be translated into the marginal stability diagram $a(k)$ for the onset of Faraday waves (see Fig. 2). The tongues touch the abscissa whenever the wavevector k fulfills the resonance condition $\omega_0(k) = n\Omega$ (n integer). The curves $a_S(k)$ ($a_H(k)$) labeled by odd (even) n are called subharmonic (harmonic), because the spectrum of $\eta_k(t)$ consists of odd (even) multiples of the basic frequency Ω.

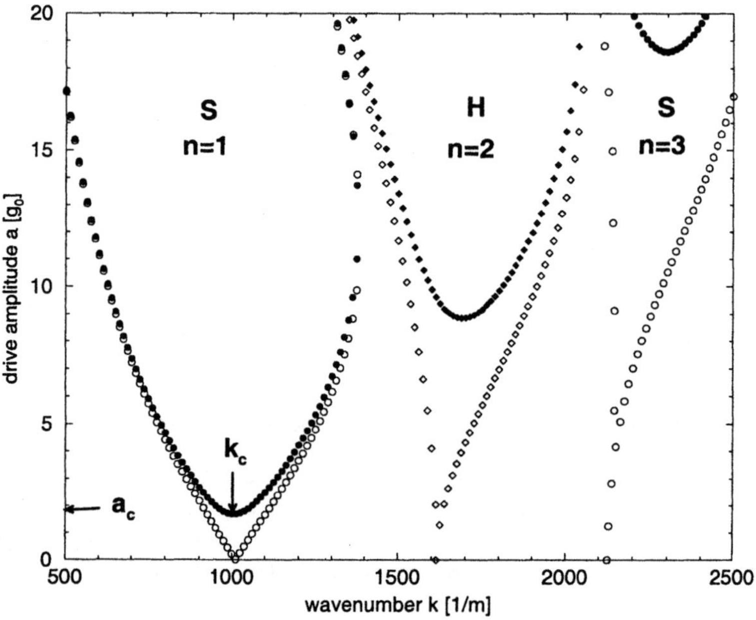

Fig. 2. The neutral stability chart for the onset of Faraday waves. Below (above) the resonance tongues the flat surface is stable (unstable). Tongues labeled by odd (even) n correspond to a subharmonic (harmonic) time-dependence of the surface waves. Open symbols: for an inviscid fluid ($\nu = 0$) resulting from the undamped Mathieu oscillator (24). Solid symbols: for a viscous fluid ($\nu = 10\text{mm}^2/\text{s}$) resulting from a damped Mathieu oscillator with damping of the form $2\gamma\dot{\eta}_k = 4\nu k^2 \dot{\eta}_k$. Parameters are $\rho = 0.934\text{g/cm}^3$, $\alpha = 20.1 \star 10^{-3}\text{N/m}$, $h = 1\text{mm}$, $\Omega/\pi = 50\text{Hz}$.

It is an artifact of the inviscid approximation that all resonances can be excited at drive amplitude $a = 0$. In order to account for viscous dissipation, one can heuristically introduce a linear damping term $2\gamma\dot{\eta}_k$ into (24). The damping coefficient can be estimated (Landau and Lifshitz (1987)) by the ratio $\gamma = |\langle \dot{E} \rangle / 2 \langle E \rangle| = 2\nu k^2$, where E denotes the total mechanical energy in the system and $\dot{E}$ is the rate of viscous energy dissipation. Since E and $\dot{E}$ are evaluated by using an *inviscid* flow field, this phenomenological approach accounts only for the damping within the bulk of the fluid. Dissipation in the viscous boundary layers along the walls, close to the bottom, and just beneath the free surface is ignored. Figure 2 compares the neutral stability curves resulting from the damped Mathieu oscillator (solid symbols) with the inviscid one (open symbols). A systematic perturbation expansion (Nayfeh and Mook (1979)) in powers of the dimensionless damping coefficient $\varepsilon = \nu k^2 / \Omega$ shows that the minima of the succeeding resonance tongues scale with $\varepsilon^{1/n}$. Thus in the weak dissipation limit $\varepsilon \ll 1$ it is always the first subharmonic tongue ($n = 1$) which provides the lowest minimum. This defines the critical threshold a_c and the critical wavenumber k_c for the onset of the Faraday instability. At leading order the perturbation calculation yields for the onset amplitude

$$a_c = 8 \frac{\Omega^2}{k_c} \varepsilon \coth(k_c h) = 8 \nu \Omega k_c \coth(k_c h) \ , \tag{26}$$

where k_c is determined at this stage of the expansion by the *inviscid* dispersion relation $\omega_0(k_c) = \Omega$.

3.3 Rigorous treatment of viscous dissipation

A consistent treatment of viscous dissipation (see Kumar and Tuckerman (1994), Kumar (1996)) requires the consideration of the full hydrodynamic eigenvalue problem. For *free* surface waves (drive amplitude $a = 0$) the surface elevation can be written in the form $\eta_k(t) \propto \bar{\eta}_k(\omega)\, \mathrm{e}^{\mathrm{i}\omega t}$, which leads to the viscous dispersion relation

$$\begin{aligned} 0 &= A(k,\omega) \\ &= \omega_0^2(k) + \nu^2 \frac{q\left(q^4 + 2k^2q^2 + 5k^4\right)\coth(qh) - \left(k^4 + 6k^2q^2 + q^4\right)\tanh(kh)}{q\coth(qh) - k\coth(kh)} \\ &\quad -\nu^2 \frac{4qk^2(q^2+k^2)\tanh(kh)}{[\sinh(kh)\sinh(qh)]\,[q\coth(qh) - k\coth(kh)]} \ , \end{aligned} \tag{27}$$

with $q = \sqrt{k^2 + \mathrm{i}\,\omega/\nu}$ (square root with positive real part). When $a > 0$ the response of the surface is no longer monochromatic as the parametric drive couples Fourier modes of different frequencies ω. The result is an infinite tridiagonal system of the form

$$A(k,\omega)\,\bar{\eta}_k(\omega) + \frac{ak\tanh(kh)}{2}\left[\bar{\eta}_k(\omega + 2\Omega) + \bar{\eta}_k(\omega - 2\Omega)\right] = 0 \ . \tag{28}$$

The Floquet theorem states that the solution at onset of the instability can be represented by a discrete Fourier sum

$$\eta_k(t) = \sum_n \bar{\eta}_{k,n}\, \mathrm{e}^{\mathrm{i}(2n+\beta)\Omega t} \ , \tag{29}$$

where $\beta = 0$ corresponds to the harmonic (H) response and $\beta = 1$ to the subharmonic (S). These two cases are to be investigated separately. The tridiagonal system (28) with ω replaced by $(2n+\beta)\Omega$ defines an infinite dimensional eigenvalue problem for the forcing amplitude a (Hill's infinite determinant). Any desired accuracy can be achieved by an appropriate high-dimensional cut-off. For given parameters $\nu, \alpha, \rho, h, \Omega$ one can follow the eigenvalues a as a function of the wavenumber k. This procedure defines the neutral stability curves $a_{\rm S}(k)$ and $a_{\rm H}(k)$ as shown in Fig. 2. A subsequent minimization with respect to k provides the critical threshold.

As discussed above, for weak dissipation $\varepsilon \ll 1$ it is usually the first subharmonic tongue ($n = 1$) which defines the threshold $a_{\rm c}$. For a typical experimental situation Fig. 3 compares the numerically computed (exact) onset amplitude $a_{\rm c}$ (solid line) with the outcome of the heuristic approach (dotted line according to (26)). Apparently, the phenomenological damping coefficient γ over-estimates the dissipation at higher drive frequencies, while it underestimates at small Ω. The first shortcoming arises because the evaluation $\gamma = |\langle \dot{E} \rangle / 2 \langle E \rangle|$ implicitly assumes that the fluid rests along the integration boundaries. However, since this assumption fails at the free surface, the actual damping is reduced there. At small Ω the heuristic treatment also fails as it predicts the damping coefficient $\gamma = 2\nu k^2$ to vanish in the long wavelength limit. This appears to be unphysical, as the dissipation is expected to re-increase when the wavelength of the pattern compares to the depth of the layer (i.e. $kh = O(1)$). In this case, which we denote as the "shallow water limit", neglecting the viscous boundary layer at the bottom is no longer appropriate.

3.4 Analytic stability analysis in the limit of weak dissipation

In the limit of weak dissipation $\varepsilon \ll 1$ the linear eigenvalue problem can be solved perturbatively (Müller et al. (1997), Müller (1998)). The problem is solved by a systematic perturbation expansion of the tridiagonal eigenvalue problem for small ε. It turns out that – besides the usual damping term – two additional memory terms enter the Mathieu equation giving

$$0 = \ddot{\eta}_k + \nu k^2 \left[3 + \coth^2(kh)\right] \dot{\eta}_k + \left[\omega_0^2 + a\, k \tanh(kh) \cos(2\Omega t)\right] \eta_k +$$

$$\frac{2\sqrt{\nu k^2}}{\sqrt{\pi}\,\sinh(2kh)} \int_{-\infty}^{t} G(t-\tau) \left(\nu k^2 + \partial_\tau\right)^2 \eta_k(\tau)\, \mathrm{d}\tau + \tag{30}$$

$$\left(\nu k^2\right)^{3/2} \frac{\coth^3(kh) + \coth(kh) - 6\tanh(kh)}{\sqrt{\pi}} \int_{-\infty}^{t} G(t-\tau) \left(\nu k^2 + \partial_\tau\right) \eta_k(\tau)\, \mathrm{d}\tau\ ,$$

where $G(t) = \exp\left(-\nu k^2 t\right)/\sqrt{t}$ is the kernel of the memory term. The conventional damping $\propto \dot{\eta}_k$ is related to the dissipation in the bulk of the fluid while the two integrals represent the contributions from the bottom and top

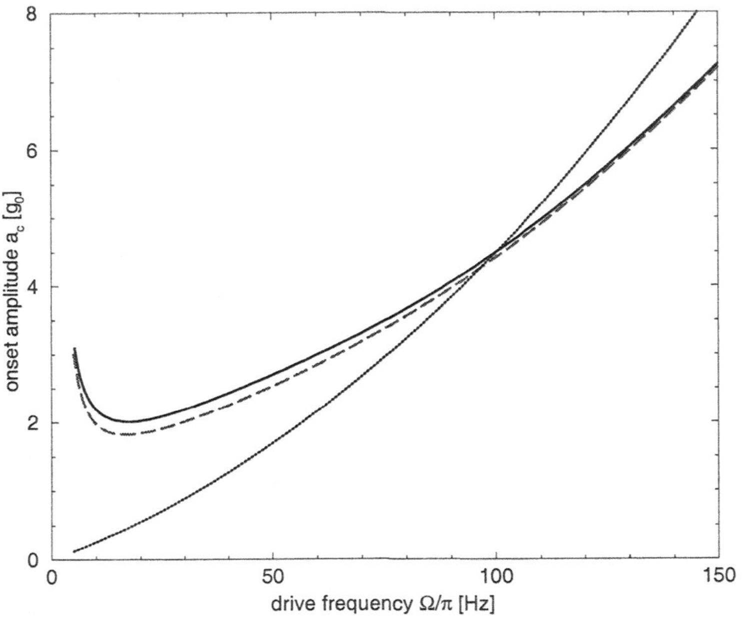

Fig. 3. The critical amplitude a_c for the onset of subharmonic Faraday waves. Solid line: exact numerical solution of the full hydrodynamic eigenvalue problem. Dotted line: stability boundary resulting from the heuristically motivated damped Mathieu equation. Dashed line: stability boundary according to the systematic perturbation expansion (31). Parameters are $\rho = 0.934\text{g/cm}^3$, $\nu = 10\text{mm}^2/\text{s}$, $\alpha = 20.1 \star 10^{-3}\text{N/m}$, $h = 1\text{mm}$.

viscous boundary layer, respectively. The first memory integral scales with $O(\nu^{1/2})$ and is the leading dissipative contribution, if the wavelength $2\pi/k$ compares with the depth of the layer (shallow water limit $kh = O(1)$). This contribution is associated with the damping in the bottom boundary layer, because it dies out for $h \to \infty$. The second integral remains in the infinite depth limit; it is related to the free surface boundary layer and provides an $O(\nu^{3/2})$-correction. This contribution has been derived earlier by Beyer and Friedrich (1995). Their analysis did not yield the bottom-contribution as they used free slip boundary conditions. The different scaling with ν of the two integral contributions arises because there is no-slip at the bottom but a free top boundary condition.

The critical acceleration amplitude a_{Sc} for the onset of subharmonic Faraday waves resulting from the modified Mathieu oscillator (30) is found to be

$$a_{\mathrm{Sc}} \simeq \frac{2\Omega^2}{k}\coth(kh)\left\{\frac{\varepsilon^{1/2}\sqrt{2}}{\sinh(2kh)} + \varepsilon\left[3+\coth^2(kh)\right]\right. \tag{31}$$

$$\left. + \frac{\varepsilon^{3/2}}{2\sqrt{2}}\left[-15\tanh(kh)+5\coth(kh)+2\coth^3(kh)\right]\right\}\Bigg|_{k=k_{\mathrm{Sc}}} + O(\varepsilon^2)\ .$$

The critical wavenumber k_{Sc} is the solution of the viscous dispersion relation

$$\omega_0^2(k)\Big|_{k=k_{\mathrm{Sc}}} \simeq \Omega^2 + \frac{\varepsilon^{1/2}\sqrt{2}}{\sinh(2kh)} \tag{32}$$

$$- \frac{\varepsilon^{3/2}}{2\sqrt{2}}\left[-15\tanh(kh)+5\coth(kh)+2\coth^3(kh)\right]\Bigg|_{k=k_{\mathrm{Sc}}}\ ,$$

where $\omega_0^2(k)$ is taken from (25). Figure 3 shows the analytic approximation according to (31) and (32) by a dashed line. Favourable agreement with the exact numerical solution (solid line) is achieved. The shortcomings of the traditional approach (dotted line) are eliminated. For high drive frequencies (31) and (32) are well approximated by the infinite depth limit $kh \to \infty$. In the opposite shallow water limit ($kh = O(1)$) we observe in Fig. 3 a sharp re-increase of the threshold amplitude towards smaller k indicating the increasing dissipation in the bottom boundary. As a result the governing subharmonic tongue $n = 1$ in the neutral chart (see Fig. 2) is raised up and the neighbouring harmonic tongue $n = 2$ eventually provides the absolute minimum. If this inverted situation occurs, the primary Faraday resonance turns out to be *harmonic*, i.e. *synchronous* with respect to the external drive. This situation, first predicted by Kumar (1996), is depicted in Fig. 4. The diagram shows that the threshold curves for the harmonic and the subharmonic Faraday instability intersect at intermediate drive frequency, resulting in bicriticality with two different rivaling wavelengths. Although the bi-critcality has been experimentally verified (Müller et al. (1997)), a systematic investigation of harmonically resonating Faraday patterns has not yet been undertaken. This is because the harmonic resonance occurs at rather low drive frequencies. The high modulation elevations necessary to generate Faraday waves in this frequency range exceed the performance of most commercial vibration exciters.

If the drive frequency is decreased even further or if the filling depth h is reduced to the depth of the viscous boundary layer $\delta = \sqrt{\nu/\Omega}$, viscous dissipation strongly increases. An analytic treatment of the strong dissipative limit $\varepsilon \gg 1$ has been presented by Cerda and Tirapegui (1997). They find that the instability mechanism is similar to the Rayleigh-Taylor instability (see also Lioubashevski et al. (1997)). Any of the infinite series of harmonic or subharmonic tongues in the neutral chart can define the threshold a_{c}. The critical wavelength $2\pi/k_{\mathrm{c}}$ appears to be almost independent of the drive frequency Ω/π, but scales with the filling depth h. This is in contrast to the weak dissipative limit where k_c obeys the dispersion relation. The strong dissipation assumption applies when the viscous boundary layer extends over the whole depth of the fluid. This case is called the lubrication limit. Clearly, the strong damping gives rise to enormous critical accelerations for the onset of surface waves. Experiments in this parameter region have not yet been performed.

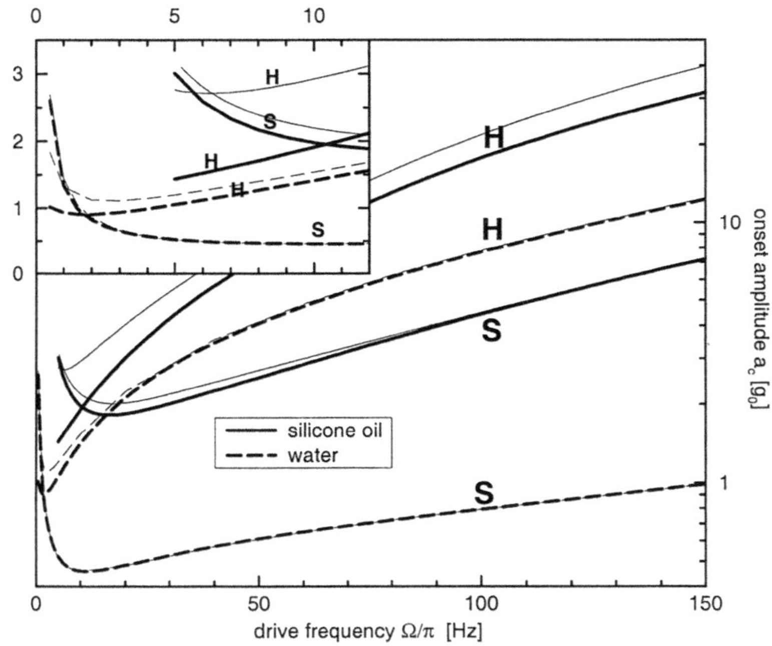

Fig. 4. The critical onset amplitude for the subharmonic (S) and harmonic (H) Faraday instability as a function of the forcing frequency Ω/π. Thick lines correspond to the analytic result (31), thin lines are obtained numerically as described in Sec. 3.3. Parameters for water (dashed): $\rho = 1.0\text{g/cm}^3$, $\alpha = 72.4 \star 10^{-3}\text{N/m}$, $\nu = 1\text{mm}^2/\text{s}$, $h = 1\text{mm}$; for silicone oil (solid): $\rho = 0.934\text{g/cm}^3$, $\alpha = 20.1\star 10^{-3}\text{N/m}$, $\nu = 10\text{mm}^2/\text{s}$, $h = 1\text{mm}$. The figure is taken from Müller (1997).

3.5 Comparison with experiments

Only a few experiments allow a quantitative comparison between empirical threshold data and the theoretical predictions. Kumar and Tuckerman

(1994) could bring their theory into good agreement with the experiments of Edwards and Fauve (1994) by using the surface tension and viscosity as fit parameters. They also compared with the experiments of Fauve et al. (1992) on a liquid-vapour interface. Far from the critical temperature T_c of the phase transition favourable agreement is achieved, however, closer to T_c the correspondence is less.

For low-viscosity fluids the comparison of the linear stability analysis with experiments is difficult, since parasitic dissipation beyond the idealized theoretical assumptions often plays a substantial role. Miles (1967) gives an account of the damping mechanisms which take place in closed basins. Important are the damping in the viscous boundary layers along the sidewalls, and dissipation due to the moving contact line of the meniscus and due to the surface viscosity of surfactant films. Bechhoefer et al. (1995) present a careful investigation of the threshold acceleration. By a proper choice of the geometry, fluid, and drive frequency they demonstrate that parasitic damping effects can be reduced to a negligible amount and that quantitative agreement with the stability theory can be achieved within a few percent. Christiansen et al. (1995) also performed a quantitative experimental investigation of the damping rate, the acceleration threshold, and dispersion relation over a wide frequency range. They identified damping from the moving contact line as a considerable contribution which reaches $10-20\%$ of the bulk dissipation. In order to estimate this effect they used a formula given by Milner (1991), which contains an open parameter, the microscale s over which the no-slip condition is violated. The value of this slip length depends on the length of the fluid molecules and the roughness of the wall. Christiansen et al. used this parameter to fit their experimental data. They found good agreement with the theory for values of s which are $10-100$ times higher than an estimate given by Milner.

Experimental measurements of the dispersion relation in large aspect ratio containers and in low-viscosity fluids usually agree very well with the theory. This is because the dispersion is – at leading order – independent of the viscosity and thus insensitive to dissipative mechanisms.

4 Nonlinear dynamics in finite geometries

Pattern formation in dissipative systems of finite extension has attracted much interest in the last two decades. Here we use *finite extent* to mean that the sizes of the structures which are generated are comparable to the systems size. These systems are well described in terms of a few spatial modes. The temporal evolution of the patterns is then related to an evolution of the amplitudes of the modes. Since modes varying on small spatial scales are highly damped (damping rate $\gamma = 2\nu k^2$), one expects that the dynamics is governed by a few degrees of freedom, which are commonly denoted as order parameters. Extensive investigations have been devoted to a determination of the order parameters and their dynamics on the basis of the hydrodynamic equations (see e.g. Cross and Hohenberg (1993)).

Order parameter equations can be derived for systems which are close to an instability (Haken (1983),(1987)). The order parameters are the amplitudes of the critical modes. Generally, there is one *real* order parameter in the case of a steady-state bifurcation or a *complex* one for the case of a Hopf-bifurcation. It is well known that low-dimensional chaotic temporal behaviour may also arise in such systems. One therefore expects that in these cases more than one dynamically active variable exists. A strategy to derive such order parameter equations from the basic equations in a rigorous way is to consider instabilities involving several critical modes. In order to obtain such instabilities one must have several control parameters, or, in mathematical terms, one has to deal with bifurcations of higher co-dimension. In principle, the order parameter equations can be derived in a rigorous way from the basic equations. In the case of the Faraday experiment, however, all of the order parameter equations stated in Sec. 4 are *assumed*, based on general dynamical systems and symmetry principles, and not at all derived from the underlying hydrodynamics.

The Faraday experiment has been performed in finite containers to explore such questions. In an experiment using a cylindrical container (radius $R = 6.35$cm, depth $h = 1$cm) Ciliberto and Gollub (1984) demonstrated that the surface deformations can exhibit chaotic temporal behaviour in certain regions of parameter space. Furthermore, the measurements show that the dynamics is essentially determined by the nonlinear interaction of a few spatial modes.

Digital imaging of the surface suggests that in accordance with the cylindrical symmetry the deformation $\eta(\mathbf{r}_\perp, t)$ can be decomposed into normal modes according to

$$\eta(\mathbf{r}_\perp, t) = \sin(\Omega t) \sum_{l,m} A_{l,m}(t)\; J_l(k_{l,m} r)\; \mathrm{e}^{\mathrm{i}l\Theta} + \text{c.c.} + \text{h.o.t.} \tag{33}$$

Here, polar coordinates $\mathbf{r}_\perp = (r, \Theta)$ are used. $J_l(k_{l,m} r)$ denotes the Bessel function of order l. The allowed wavenumbers $k_{l,m}$ are determined by the boundary conditions at $r = R$. The excited modes close to the instability are characterized by their azimuthal wavenumbers l. The amplitudes $A_{l,m}(t)$ are slowly varying in time, since the fast temporal oscillation is separated.

The first instability usually leads to a standing wave with subharmonic time dependence. The surface deformation then has the form

$$\eta(r, \Theta, t) = A(t)\; \sin(\Omega t)\; \mathrm{e}^{\mathrm{i}l\Theta}\; J_l(k_{l,m} r) + \text{c.c.} + \text{h.o.t.} \tag{34}$$

and the amplitude $A(t)$ obeys the equation

$$\dot{A}(t) = \mu\, A(t) - \beta\, |A(t)|^2 A \ , \tag{35}$$

where μ measures the strength of the supercritical drive. The spatial structure is characterized by the azimuthal wavenumber l. The linear growth rate μ as well as the Landau coefficient β could in principal be calculated from the basic equations. The above equation describes a relaxational dynamics towards a purely subharmonically oscillating pattern.

In order to consider mode interactions, which in the rigorous mathematical treatment have to be studied in the realm of bifurcations with higher codimensions, two control parameters are varied in the experiments. The first control parameter is, as usual, the driving amplitude a of the vertical oscillations. An increase of a leads to the instability of the plane surface. The second control parameter is the driving frequency Ω/π, which allows a selection of a specific mode characterized by the values (l, m). According to (33) l denotes the azimuthal wavenumber of the pattern whereas its radial dependence, i.e. the number of nodes in the radial direction, is indicated by m. Figure 5 exhibits the phase diagram evaluated in the experiments of Ciliberto and Gollub. The boundaries for single-mode instabilities of differ-

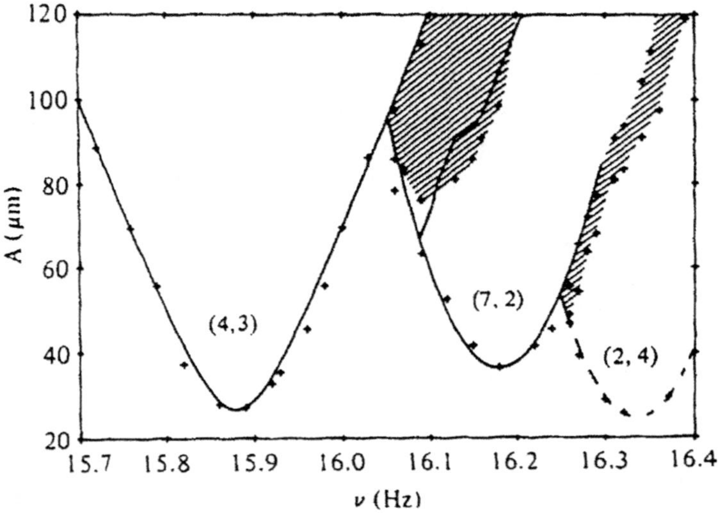

Fig. 5. Phase diagram of the experiment of Ciliberto and Gollub (1984). Here A labels the driving amplitude and ν the driving frequency, denoted by a and Ω/π in the text. In the regions labeled by (4,3), and (7,2) stable subharmonically oscillating patterns are observed.

ent values of the azimuthal wavenumber (l) intersect at certain values of the control parameters $a, 2\Omega$. In the experiments, the modes with ($l = 4, m = 3$) and ($l = 7, m = 2$) have been examined. Close to these codimension-two points, various dynamical regimes can be observed. In the shaded areas the amplitudes exhibit, in addition to the fast oscillation $e^{i\Omega t}$, a slow temporal evolution. This temporal evolution can be strictly periodic. However, a regime with chaotic temporal behaviour is also observed. In Ciliberto and Gollub (1985) a detailed investigation of the experimental data, using methods of dynamical systems theory (i.e. determination of fractal dimensions of the attractor, proof of the existence of one positive Lyapunov exponent etc.) has been reported.

Ciliberto and Gollub (1984) additionally present a model in order to explain the observed chaotic behaviour. They consider a surface deformation of the form

$$\eta(r,\Theta,t) = a_l(t)\, e^{il\Theta}\, J_l(k_{l,m}r) + a_{l'}(t)\, e^{il'\Theta}\, J_{l'}(k_{l',m'}r) + \text{c.c.} + \text{h.o.t.} \quad (36)$$

This ansatz is due to the experimental observation that the patterns are composed of modes with two different azimuthal wavenumbers l, l'. Starting point is the fact that the (fast varying) amplitude $a_l(t)$ of a mode with azimuthal wavenumber l in the linearized case follows a Mathieu equation (24). Close to a codimension-two point two modes with l and l' must be taken into account. In the following we shall suppress the radial wavenumbers in the notation and denote the two amplitudes by a_l and $a_{l'}$, respectively.

$$\begin{aligned} \ddot{a}_l + \gamma_l\, \dot{a}_l + [\omega_l^2 - a\,\psi_l \cos(2\Omega t)]\, a_l &= \zeta_l\, a_l^3 \ , \\ \ddot{a}_{l'} + \gamma_{l'}\, \dot{a}_{l'} + [\omega_{l'}^2 - a\,\psi_{l'} \cos(2\Omega t)]\, a_{l'} &= \zeta_{l'}\, a_{l'}^3 \ . \end{aligned} \quad (37)$$

The cubic terms are included in order to saturate the mode amplitudes above threshold. The nonlinear couplings between the two modes are contained in terms $\psi_l, \psi_{l'}$, which are assumed to be of the form

$$\begin{aligned} \psi_l &= \bar{\psi}_l + \beta_{l,l'}\, a_{l'}^2 \ , \\ \psi_{l'} &= \bar{\psi}_{l'} + \beta_{l',l}\, a_l^2 \ . \end{aligned} \quad (38)$$

The coefficients are chosen in such a way as to fit the experimentally determined instability boundaries for single-mode excitations. Numerical integration of the dynamical system (37) yields periodic as well as chaotic behaviour in the same regions as obtained in the experiments. However, it should be kept in mind that (37) are only model equations. There is no systematic derivation.

Meron and Procaccia (1986) extended the model of Ciliberto and Gollub (1984) by establishing the normal form equations for the codimension-two instability under consideration. They assume that the surface deformation is essentially determined by two modes according to

$$\eta(r,\Theta,t) = \sin(\Omega t)\, \left[A(t)\, e^{il\Theta}\, J_l(k_{l,m}r) + B(t) e^{il'\Theta} J_{l'}(k_{l',m'}r) + \text{c.c.}\right] + \text{h.o.t.} \quad (39)$$

The quantities $A(t)$, $B(t)$ are the slowly varying complex amplitudes of the two critical modes $(l, m), (l', m')$. They obey the coupled amplitude equations:

$$\dot{A} = [\mathrm{i}\,(\Omega_A - \Omega) - \lambda_A]\,A + \mathrm{i}\,\gamma_1 A^* + \mathrm{i}\,\gamma_2 |A|^2 A + \mathrm{i}\,\gamma_3 |B|^2 A + \mathrm{i}\,\gamma_4 A^* B^2 \ ,$$
$$\dot{B} = [\mathrm{i}\,(\Omega_B - \Omega) - \lambda_B]\,A + \mathrm{i}\,\delta_1 B^* + \mathrm{i}\,\delta_2 |B|^2 B + \mathrm{i}\,\delta_3 |A|^2 B + \mathrm{i}\,\delta_4 B^* A^2 \ . \qquad (40)$$

Here, Ω_i denote the eigenfrequencies of the modes, and λ_i are the corresponding damping rates. The mode coupling coefficients γ_i, δ_i were calculated from the inviscid equations (10), (11), (12), and (13). A numerical investigation of the coupled amplitude equations using these calculated coefficients, however, failed to reproduce the experimental observations for the case of mode competition between $(4, 3)$ and $(7, 2)$: No chaotic behaviour was obtained. Nevertheless, the normal form equations apply, although with different values of the mode coupling coefficients γ_i, δ_i. Taking suitable coefficients Meron and Procaccia were able to reproduce the major characteristics of the experimentally observed data. The most advanced study has been performed by Umeki and Kambe (1989). Their numerical predictions of the nonlinear stability boundaries near the critical point are in qualitative agreement with the experiments of Ciliberto and Gollub.

A systematic study of mode competition was performed by Simonelli and Gollub (1989) who experimentally investigated mode competition in square and rectangular containers. They investigated the dynamics of two modes, which are degenerate in the case of a square layer. In a slightly rectangular cell the modes are separated in frequency. Mode competition produces periodic as well as chaotic states. The experimental set up allowed Simonelli and Gollub (1989) to explore the phase space of the normal form equations belonging to this kind of mode competition.

Crawford et al. (1990) performed an analysis of the mode competition problem using group theoretic methods. To this end they considered a stroboscopic map of the state vector (39), i.e. they considered the amplitudes $A_n = A(t_n)$ and $B_n = B(t_n)$ at multiples of the driving period $T = \pi/\Omega$, $t_n = nT$. Close to the codimension-two point there exists a four dimensional map of the form

$$[A_{n+1}, B_{n+1}] = [F_A(A_n, B_n), F_B(A_n, B_n)] \qquad (41)$$

which, due to the fact that the experiment is performed in a cylindrical vessel, is equivariant with respect to $O(2)$-symmetry. A rotation of the system in the azimuthal direction by an arbitrary angle Θ_0 leads to a transformation of the amplitudes A_n, B_n, according to (39), of the form

$$[A_n, B_n] \to [\mathrm{e}^{\mathrm{i}l\Theta_0} A_n, \mathrm{e}^{\mathrm{i}l'\Theta_0} B_n] \ . \qquad (42)$$

Reflection $\Theta \to -\Theta$ leads to the transformation

$$[A_n, B_n] \to [A_n^*, B_n^*] \ . \qquad (43)$$

Let us consider the case $l + l'$ even. The following quantities are invariants under the transformations (42), (43):

$$\alpha = |A|^2 \quad , \quad \beta = |B|^2 \quad , \quad \sigma = A^{l'}(B^*)^l + (A^*)^{l'} B^l \ . \tag{44}$$

Furthermore, each invariant function can be represented as a function of these three expressions. The most general form of the map (41), which is invariant with respect to the transformations (42), (43) then takes the form

$$\begin{aligned} F_A &= p(\alpha, \beta, \sigma)\, A + q(\alpha, \beta, \sigma)\, (A^*)^{l'-1}\, B^l \ , \\ F_B &= r(\alpha, \beta, \sigma)\, B + s(\alpha, \beta, \sigma)\, A^{l'}\, (B^*)^{l-1} \ , \end{aligned} \tag{45}$$

where p, q, r, s are arbitrary functions of the basic invariants (44). This map has two simple stationary solutions, $A \neq 0$, $B = 0$, $A = 0$, $B \neq 0$ which correspond to the single-mode standing waves with azimuthal wavenumbers l and l', respectively. For suitable functions p, q, r, s each of these standing wave solutions undergoes a secondary instability leading to standing waves, which are superpositions of the two modes with wavenumbers l, l'. These standing waves also show subharmonic response. From this mixed mode, a tertiary bifurcation in the form of a Hopf bifurcation may occur (for suitably chosen functions F_A, F_B). This solution should be identified with the behaviour denoted as periodic in the phase diagram of Fig. 5. It leads to a slow periodic modulation of the subharmonic response. Numerical calculations indicate that by further increasing the driving amplitude this invariant circle undergoes a transition to chaotic behaviour. The advantage of the group-theoretic approach is the fact that it leads to a rather complete classification of the bifurcation structure contained in the normal forms. An extensive application of group theoretic methods to mode competition in square cells has been performed by Crawford et al. (1993), where theoretical predictions are compared with experimental findings.

In summarizing, experimental work together with the various theoretical approaches has shown convincing evidence that the Faraday experiment performed in small geometries is governed by a few temporal degrees of freedom which are related to amplitudes of certain spatial modes. Nonlinear interaction of these modes can lead to secondary and higher instabilities which eventually may result in chaotic temporal behaviour. However, the theoretical approaches again are based either on the approximation of ideal fluids with subsequent phenomenological introduction of viscous effects, as in the work of Meron and Procaccia, or on a classification of the bifurcation structure of the normal form (45) using group theoretic arguments. However, a quantitative connection between experiment and theory has not yet been achieved due to the fact that the coefficients occuring in the normal forms have not been determined from the fluid dynamic equations. The necessary calculations seem to be quite hard since, as the reader should realize, even the modes of the linear problem are difficult to determine due to viscous boundary layers at the boundaries. However, qualitative analysis of the normal form equations leading to classification of possible dynamical behaviours can give valuable information on the mode interaction problem close to multiple bifurcation points.

5 Nonlinear patterns in the large system limit

5.1 Theoretical background

The theoretical investigation of nonlinear pattern selection is considerably facilitated if the influence of sidewalls can be ignored and the analysis can be performed for a horizontally infinite system. This large system limit is relevant to the problems of nonlinear pattern selection, secondary instabilities and spatio-temporal chaos. Usually sidewall effects are assumed to be negligible in large aspect ratio systems, $Lk \gg 1$, where L is the lateral size of the container and k the wavenumber of the pattern. This condition, however, is not sufficient for experiments in low-viscosity fluids. In this case the width $\Delta k(\mu)$ ($\mu = (a - a_c)/a_c$) of the neutral resonance tongue just above onset (see Fig. 2) is very narrow. The resulting geometrical mode quantization is only negligible if L is much larger than the linear correlation length $\pi/\Delta k$. In the following we assume that both conditions are fulfilled. This is the so-called "large system limit".

Due to the rotational symmetry of the infinite horizontal plane, critical wavevectors $\mathbf{k}_i$ can appear in any lateral orientation. To describe the surface deformation we consider a superposition of N standing wave modes with wavevectors $\pm\mathbf{k}_i$, $|\mathbf{k}_i| = k_c$, and $i = 1, \ldots, N$ of the form

$$\eta(\mathbf{r}_\perp, t) = \eta_{k_c}(t) \sum_{i=1}^{N} A_i(t) \exp(\mathrm{i}\,\mathbf{k}_i \cdot \mathbf{r}_\perp) + \text{c.c.} \tag{46}$$

The case of the *subharmonic* Faraday resonance is being investigated here. $\eta_{k_c}(t)$ is the real subharmonic eigenfunction, which solves the linear stability problem. The complex amplitudes A_i are assumed to vary slowly in time with respect to the surface oscillation. Up to cubic order the amplitude equations read

$$\partial_t A_i = \mu A_i - \sum_{j=1}^{N} \beta(\theta_{ij}) |A_j|^2 A_i, \qquad i = 1, \ldots, N \ , \tag{47}$$

where $\mu = (a - a_c)/a_c$ is the relative distance from threshold. The form of this equation can be derived from simple symmetry arguments. The function $\beta(\theta)$ defines the cross-coupling coefficients, which depend on the separation angle θ_{ij} between $\mathbf{k}_i$ and $\mathbf{k}_j$. The evaluation of this function clearly requires a rigorous small amplitude expansion of the hydrodynamic equations as done by Chen and Viñals (1997) and Chen and Viñals (1998). The absence of quadratic terms in (47) results from the subharmonic symmetry of the time dependence $\eta_{k_c}(t+\pi/\Omega) = -\eta_{k_c}(t)$, which requires (47) to be sign invariant. Note: this symmetry argument does not apply to the *harmonic* Faraday resonance. Nor does it apply if the subharmonic symmetry is deliberately broken by e.g. perturbing the monochromatic excitation signal $\cos(2\Omega t)$ by a second frequency component, which is an odd multiple of Ω. This situation occurs in Faraday experiments with multi-frequency drive (see Sec. 5.3).

As sidewall effects can be ignored in the large system limit, pattern selection is solely determined by nonlinearity, i.e. by the coupling coefficients $\beta(\theta_{ij})$. Although being very cumbersome, the function $\beta(\theta)$ can be derived by a perturbation expansion of the hydrodynamic equations (see Chen and Viñals (1997), Chen and Viñals (1998)). As (47) is of gradient form $\partial_t A_i = -\partial \mathcal{F}/\partial A_i^*$ the stability of symmetric $2N$-fold patterns of arbitrary N can be checked by the following reasoning: That value of N which minimizes the average $(1/N)\sum_{j=0}^{N-1} \beta(j\pi/N)$ provides the smallest "free energy" $\mathcal{F}$ and thus determines the selected structure. Patterns with $N = 1, 2, 3$ correspond to stripes, squares, and hexagons (or triangles), respectively. Higher values of N describe structures with a $2N$-fold orientational order but without a translational symmetry. Those patterns are denoted as quasi-periodic.

5.2 Ordered patterns at single-frequency drive

Faraday and Lord Rayleigh figured out experimentally that for low-viscosity fluids ($\nu \simeq 30 - 50\text{mm}^2/\text{s}$ depending on the excitation frequency) the prevailing pattern close above the onset is a square structure. This observation was verified by many authors, e.g. Ezersky et al. (1986), Edwards and Fauve (1992), Edwards and Fauve (1993), Edwards and Fauve (1994), Daudet et al. (1995). For higher viscosity, lines have been found (Edwards and Fauve (1994)). A similar observation at a CO_2-interface near the liquid-vapour critical temperature T_c also suggests that high dissipation (closer to T_c) may be responsible for the selection of stripes. Depending on the initial conditions, stripes can also occur in form of spirals (Kiyashko et al. (1996)).

Christiansen et al. (1992) investigated capillary ripples in a low-viscosity fluid at drive frequencies beyond 300Hz. At 5% above the onset the primary instability, the convective pattern in the middle of the cell consists of $N = 4$ wavevectors separated by mutual angles of $\pi/4$. This is one of the first experiments in which a "two-dimensional quasi-crystal" has been observed in a macroscopic non-equilibrium dynamical system. By increasing the drive amplitude Christiansen et al. observed a transition to hexagonal ($N = 3$) and later to square patterns before spatio-temporal chaos occurred. Hexagonal structures in single-frequency Faraday experiments have also been found by Kumar and Bajaj (1995). Usually the occurence of hexagonal patterns is understood as a result of a triad wavevector interaction mediated by quadratic nonlinearities in the amplitude equation (47) (c.f. hexagons in non-Boussinesq Rayleigh-Bénard convection). However, as such quadratic terms are excluded by the subharmonic symmetry, the selection mechanism must be based on the cubic interaction. The necessary conditions for the coupling function $\beta(\theta)$ to enforce hexagons (or more complicated quasi-periodic structures with $N \geq 4$) have been worked out by Malomed et al. (1989) and Newell and Pomeau (1993). Based on this idea Müller (1993b) presents two-dimensional model equations with a relaxational dynamics towards quasi-periodic patterns (see Sec. 5.8). It is an immediate consequence of the cubic selection mechanism that patterns with $N = 3$ can either be hexagons or triangles (see Sec. 5.3 for a more detailed discussion). Kumar and Bajaj (1995) observed both forms. They also found slow temporal oscillations between triangles and hexagons. This phenomenon arises from the dynamics of an almost neutrally stable phason mode.

A comprehensive experimental classification of the $\nu - \Omega$ parameter space into regions of different appearing patterns has been published by Kudrolli and Gollub (1996). Their systematic investigation covers viscosities between $\nu = 10 - 100\text{mm}^2/\text{s}$ and excitation frequencies up to 200Hz. Besides the usual well-ordered structures (stripes, squares, hexagons) they also observe coexistence and competition. Secondary instabilities as well as order-disorder transitions towards spatio-temporal chaos are also considered. Particularly interesting is the "melting" of hexagonal patterns and a quantification of the decay of the orientational order. Transversal amplitude modulation instability plays a crucial role in the transition of stripe patterns towards chaotic states.

In order to come to a theoretical understanding of weakly nonlinear ordered surface patterns, the coupling function $\beta(\theta)$ must be computed. In the Faraday problem the angular dependence of $\beta(\theta)$ is strongly influenced by interactions of the critcal modes with neighbouring, almost critical resonance tongues. Milner (1991) derives amplitude equations from the inviscid evolution equations for capillary waves and introduces damping heuristically as discussed in Sec. 3.2. Since he neglects dissipation of the nonlinearly excited modes via the following mechanism, a divergence of β at $\theta_{12} \simeq 75°$ arises. This angle separates two critical subharmonic modes $\mathbf{k}_1$, $\mathbf{k}_2$ (with $|\mathbf{k}_1| = |\mathbf{k}_2| = k_c$) interacting with a harmonic mode $\mathbf{k}_1 + \mathbf{k}_2$ by the resonance condition $\omega_0(\mathbf{k}_1 + \mathbf{k}_2) = 2\omega_0(k_c)$. The given angle results if the capillary dispersion relation $\omega_0^2(k) = \alpha k^3/\rho$ is inserted. Since the divergence is towards $+\infty$ energy is pumped from the subharmonic modes into the harmonic one and the system tries to avoid this interaction angle. By evaluating the free energy for various N, Milner found, in agreement with experiments, that squares are the preferred structure in low-viscosity capillary waves. Zhang and Viñals (1996) and Zhang and Viñals (1997a) evaluated the coupling coefficient in the framework of their quasi-potential approximation, which applies in the low-damping limit $\varepsilon \ll 1$. Their results show that the avoided interaction angle continuously decreases from $\theta_{ij} \simeq 75°$ down to $0°$ when the drive frequency Ω/π is lowered into the region of gravity-capillary waves (note: for pure gravity waves resonance between subharmonic and harmonic modes does not work). In a narrow frequency window around 30Hz Zhang and Viñals predict by the free energy selection criterion quasi-periodic patterns up to order $N = 7$. However, in their numerical simulations they could only verify patterns up to $N = 4$. In a recent experiment on gravity-capillary waves in a large aspect ratio vessel Binks and van de Water (1997) demonstrate a series of periodic and quasi-perodic surface patterns up to tenfold ($N = 5$) rotational symmetry. The corresponding parameter range is in good agreement with the theoretical prediction of Zhang and Viñals. Recently, a systematic nonlinear perturbation expansion based on the full hydrodynamic equations has been undertaken by Chen and Viñals (1997) and Chen and Viñals (1998). The results are similar to those of Zhang and Viñals but they apply to fluids of arbitrary viscosity. This allows a prediction of the parameter region where stripes instead of squares are to be expected. All in all the work of Chen and Viñals based on the full hydrodynamic equations is more satisfactory than the work of Zhang and Viñals, which makes use of the uncontrolled quasi-potential approximation.

5.3 Ordered patterns at multi-frequency drive

One of the advantages of Faraday waves is that there is no experimental limitation of the choice of the forcing function $W(t)$. Since the vibration is always vertical the spatial symmetries of the system are preserved. It is therefore tempting to impose more complicated excitation signals in order to influence or control nonlinear pattern formation. In order to generate stationary patterns the drive signal must be periodic. As a first step in this direction, Edwards and Fauve (1992), (1993), (1994) performed an experimental study, in which they used a pair of rationally related frequencies. In the drive signal

$$W(t) = a\left[r\cos(m\Omega t) + (1-r)\cos(n\Omega t + \psi)\right] \tag{48}$$

r describes the relative amplitude, n, m are integers and relatively prime, and ψ denotes a possible phase shift. The point is that with two frequencies a bicritical point arises with two co-existing critical wavenumbers (see Beyer and Friedrich (1995), Besson et al. (1996) for a linear stability analysis). In the case of single-frequency drive, it has been pointed out that the subharmonic time dependence prohibits quadratic terms in the amplitude equations. With two-frequency forcing this argument needs no longer be true. Consider e.g., the case $m = 2$ and $n = 1$ with r close to unity: The second frequency contribution in (48) can thus be considered as a perturbation: As the excitation $\cos(2\Omega t)$ dominates the excitation signal, the spectrum of the surface response mainly consists of odd multiples of Ω (see the discussion in Sec. 3.3). However, due to the parametric coupling between surface elevation and drive signal, the spectrum also contains even frequencies with small amplitudes proportional to $(1 - r)$. In turn these components break the subharmonic symmetry by producing quadratic terms in the amplitude equation and enable triad-interactions among wavevectors which form equilateral triangles. This is the known mechanism which enforces subcritically

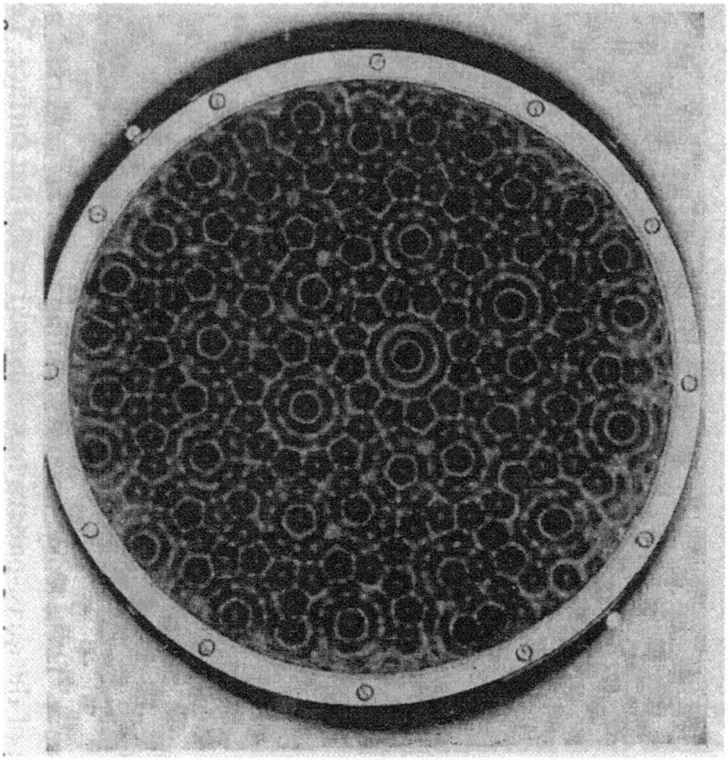

Fig. 6. Quasiperiodic pattern with 12-fold orientational order as observed by Edwards and Fauve in an experiment with two-frequency drive. The figure is taken from Edwards and Fauve (1994).

bifurcating hexagonal patterns (c.f. hexagons in non-Boussinesq convection). Similar arguments apply whenever an even frequency excitation signal is perturbed by an odd one. Edwards and Fauve investigated the case $m = 4$, $n = 5$ systematically. As the viscosity of the fluid is rather high ($\nu \simeq 100\text{mm}^2/\text{s}$) line patterns occur in this case. These line patterns survive when r is slowly increased since the subharmonic symmetry of the response is preserved (odd frequency perturbed by an even). This is contrary to the case where r is slightly smaller than unity. The broken subharmonic symmetry immediately leads to the generation of a subcritical hexagonal structure. By increasing the drive amplitude a a hexagon-to-line transition takes place over a finite hysteretic range of μ. This also coincides with the bifurcation behaviour resulting for the amplitude equation. Particularly interesting is the neighbourhood of the bi-criticality where the competition between the two instabilities gives rise to a twelvefold ($N = 6$) quasi-periodic pattern (see Fig. 6). By trying several container shapes Edwards and Fauve (1992), (1993), (1994) verified that the observed structure is extremely robust and independent of the geometry. This observation of a two-dimensional quasi-crystal was simultaneous with the experiments of Christiansen et al. (1992). Kudrolli et al. (1998) document additional two-frequency superlattice patterns that have provoked interesting theoretical work by Silber and Proctor (1977).

Another example of a two-frequency forcing experiment is the work of Müller (1993a) in which for the first time triangular patterns occured (see Fig. 7). On using an excitation signal with $m = 2$ and $n = 4$ subharmoni-

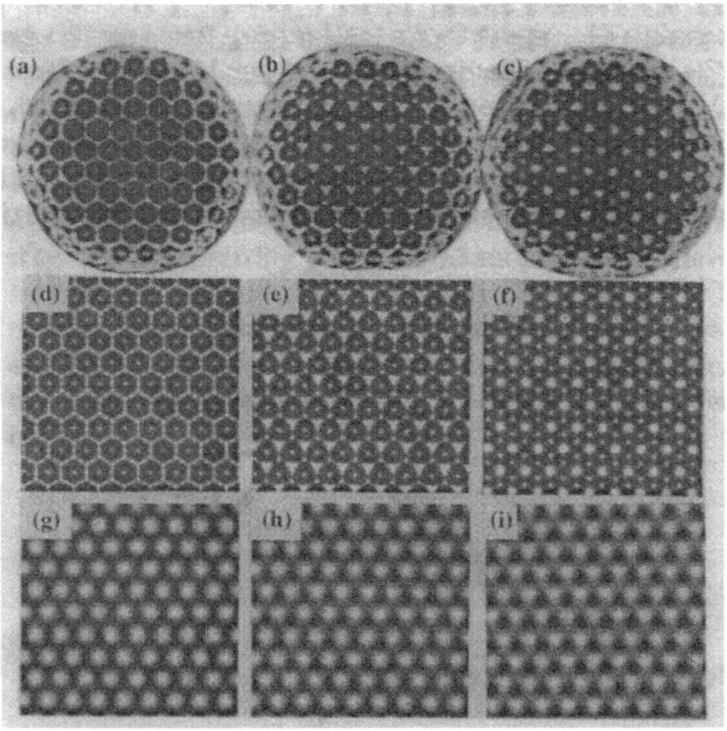

Fig. 7. (a)-(c): Photographs of experimental surface patterns. As comparison (g)-(i) show computer generated surface topographies (elevation translated into proportional grey levels) of hexagons with $\Phi = 0$ (g), triangles with $\Phi \simeq \pi/4$ (h) and regular triangles with $\Phi = \pi/2$ (i). By considering the experimental visualization technique the corresponding calculated video images are shown in (d)-(f). These pictures are to be compared with the experiment (a)-(c). The figure is taken from Müller (1993a).

cally responding hexagons and triangles have been observed (note: there are also no quadratic contributions in the amplitude equations). Although both hexagons and triangles are composed of $N = 3$ wavevectors separated by 60°, the distinction arises from different relative phases of the complex amplitudes A_1, A_2, A_3. By introducing $A_i = R\,\mathrm{e}^{\mathrm{i}\phi_i(t)}$ it is assumed that the dynamics of the fast modulus R has already settled and the focus is on the succeeding slow evolution of the phases $\phi_i(t)$. As the sum $\Phi = \phi_1 + \phi_2 + \phi_3$ remains invariant under spatial translations, it can be argued that $\Phi(t)$ describes a "phason mode" which changes the pattern. Indeed, by varying Φ the structure continuously changes from hexagons ($\Phi = 0, \pi$) to regular triangles ($\Phi = \pi/2, 3\pi/2$) as shown in Fig. 7 (g) and (i), respectively. Structures with intermediate values as depicted in (h) are called triangles. On the level of the cubic amplitude equations the phason dynamics turns out to be neutrally stable

$$\partial_t \Phi = 0 \ , \tag{49}$$

rendering hexagons, triangles, and regular triangles equally stable (note: if quadratic nonlinearities appeared in (47) one would obtain $\partial_t \Phi \propto \sin \Phi$, indicating that hexagons are the only stable pattern). In order to lift this continuous symmetry degeneration one has to proceed up to quintic order in the amplitude expansion. However, without an explicit calculation the form of the resulting phase equation can be written

$$\partial_t \Phi = \kappa \sin (2\Phi) \ . \tag{50}$$

The fixed points of (50) describe hexagons and regular triangles. The former (latter) are stable if $\kappa < 0$ ($\kappa > 0$). Zhang and Viñals (1997b) applied their quasi-potential approach to this problem and evaluated the nonlinear coupling function $\beta(\theta)$. For certain values of the phase shift ψ the two excitation frequency components resonate at an interaction angle close to 90°. Accordingly squares are suppressed and priority is given to structures with $N = 3$.

5.4 Secondary instabilities and natural patterns

Natural cellular patterns arising in nonequilibrium systems do not exhibit strict symmetries but may show spatially slow deformations of the cellular structures as well as defects and grain boundaries. Furthermore, there are long wavelength instabilities like the Eckhaus instability etc. Such phenomena are studied using evolution equations for amplitudes which are slowly varying in time as well as in space. These so-called envelope equations can be derived by the method of Newell and Whitehead (1969).

5.5 Amplitude equations

Amplitude equations have been used for investigating the Faraday experiment in large aspect ratio systems for low-viscosity fluids by Ezersky et al. (1986)

and Milner (1991). The starting point is the following superposition of plane waves for the surface deformation

$$\eta(\mathbf{r}_\perp, t) = \sum_j A_j(\mathbf{r}_\perp, t) \exp\left[\mathrm{i}(\mathbf{k}_j \cdot \mathbf{r}_\perp - \Omega t)\right] + \text{c.c.} + \text{h.o.t.} \tag{51}$$

Here, the case of a subharmonic response is considered. The assumption of a monochromatic time dependence is appropriate for low-viscosity fluids. An expansion similar to (51) applies to the velocity field. The amplitudes of the plane waves with wavevector $\mathbf{k}_j$ now depend on the horizontal coordinates. Denoting the amplitude of the plane wave with wavevector $-\mathbf{k}_j$ by A_{-j} the expansion procedure yields the following amplitude equations:

$$\begin{aligned}\dot{A}_j &= -\gamma^{(0)} A_j - a\frac{\mathrm{i}k}{4\Omega} A^*_{-j} - \left[v_g(\mathbf{e}_j \cdot \nabla) - \mathrm{i}\frac{v_g}{2k}\nabla^2 + \mathrm{i}\frac{v_g}{4k}(\mathbf{e}_j \cdot \nabla)^2\right] A_j \\ &\quad - \sum_k \left(\gamma^{(1)}_{jk} - \mathrm{i}T^{(1)}_{jk}\right) |A_k|^2 A_j - \sum_k \left(\gamma^{(2)}_{jk} - \mathrm{i}T^{(2)}_{jk}\right) A_k\, A_{-k}\, A^*_{-j} \ . \end{aligned} \tag{52}$$

Here, we restrict ourselves to the special case of capillary waves for which the group velocity takes the form $v_g = 3\Omega/2k$. For convenience we have introduced the unit vector ($\mathbf{e}_j = \mathbf{k}_j/k_j$). The nonlinear coupling coefficients $T^{(i)}_{jk}$ have been derived by Ezersky et al. and by Milner from the basic equations for inviscid fluids. The linear terms proportional to A_j reflect the linear dispersion relation for free surface waves. The term $-\gamma^{(0)}A_j$ accounts for viscous dissipation. The term $-a\frac{ik}{4\Omega}A^*_{-j}$ is due to the external excitation and thus proportional to the excitation amplitude a. It provides the coupling to the wave propagating in the opposite direction. Miles (for a review see Miles and Henderson (1990)) has considered these amplitude equations for the inviscid case. He has obtained the corresponding amplitude equations by a Hamiltonian formulation of the basic equations.

Milner (1991) has considered viscous damping on the linear and nonlinear level following a suggestion of Landau and Lifshitz (1987). Nonlinear dissipative effects are accounted for by the coefficients $\gamma^{(i)}_{jk}$. Their determination is briefly outlined as follows. In experiments with low viscosity fluids the viscous penetration or skin depth, $\delta = \sqrt{\nu/\Omega}$, is much smaller than the wavelength of the capillary waves: the flow is therefore ideal over almost all the volume. Accordingly, the contribution to the viscous dissipation from the skin depth is small compared to that in the bulk. The fluid motion can therefore be considered as a purely potential one, $\mathbf{v} = \nabla\phi$, and the viscous dissipation rate is given in terms of the velocity potential $\phi(\mathbf{r}_\perp, z, t)$ by

$$\dot{E} = -2\nu\rho \int \mathrm{d}V \left(\nabla_i \nabla_j \phi\right)^2 \ . \tag{53}$$

The velocity potential can be expanded in powers of the amplitudes A_j. Inserting this expansion into the expression for the viscous dissipation leads to:

$$\dot{E} = -D^{(0)} \sum_j |A_j|^2 - \sum_{jk} D^{(1)}_{jk} |A_j|^2 |A_k|^2 - \sum_{jk} D^{(2)}_{jk} A_j \; A_{-j} A^*_k \; A^*_{-k} \; , \quad (54)$$

where terms up to quartic order in the amplitudes have been retained. Therefore, one may assume that dissipation leads to the following contribution in the amplitude equation:

$$\dot{A}_j = -\gamma^{(0)} A_j + \sum_k \gamma^{(1)}_{jk} |A_k|^2 A_j - \sum_k \gamma^{(2)}_{jk} A_k \; A_{-k} \; A^*_{-j} \; . \quad (55)$$

The coefficients $\gamma^{(i)}_{jk}$ are then determined by the following reasoning: The amplitude equations without damping represent a Hamiltonian system. The Hamiltonian H is given by an expansion of the total mechanical energy of the system, i.e. kinetic energy plus surface tension, in powers of the amplitudes A_j:

$$H = \sum_j h^{(0)} |A_j|^2 + \sum_{jk} h^{(1)}_{jk} |A_j|^2 |A_k|^2 + \sum_{jk} h^{(2)}_{jk} A_j \; A_{-j} \; A^*_k \; A^*_{-k} \; . \quad (56)$$

The temporal derivative of the Hamiltonian must be equal to the dissipation rate $\dot{E}$. Since the coefficients h and D can be calculated and are known, the equality $\dot{H} = \dot{E}$ determines the damping coefficients $\gamma^{(i)}_{jk}$. All coefficients are specified in Milner (1991).

It is quite important that the amplitude equations (52) are not variational. This allows for complex spatio-temporal behaviour.

5.6 Quasi one-dimensional waves: Annulus experiments

Douady et al. (1989) performed a Faraday experiment using a fluid confined in an annulus (diamater 110mm, gap width 5mm, height 8mm). This geometry enforces nearly two dimensional wave patterns due to the small gap width. Thus, close to the instability the pattern consists of waves travelling along the annulus (x-direction). The surface deformation has the form

$$\eta(x,t) = A_+ \, \mathrm{e}^{[\mathrm{i}(\Omega t - kx)]} + A_- \, \mathrm{e}^{[\mathrm{i}(\Omega t + kx)]} + \text{c.c.} + \text{h.o.t.} \quad (57)$$

A_+, A_- are slowly varying amplitudes of right and left travelling waves, respectively. The corresponding amplitude equations take the form:

$$\begin{aligned} \partial_t A_+ + c\partial_x A_+ &= (-\gamma + \mathrm{i}\nu)A_+ + \alpha\partial_x^2 A_+ + (\beta|A_+|^2 + \delta|A_-|^2)A_+ + a\mu A^*_- \; , \\ & \qquad (58) \\ \partial_t A_- - c\partial_x A_- &= (-\gamma + \mathrm{i}\nu)A_- + \alpha\partial_x^2 A_- + (\beta|A_-|^2 + \delta|A_+|^2)A_- + a\mu A^*_+ \; . \end{aligned}$$

Due to the terms $a\mu A^*_-$ and $a\mu A^*_+$ the counter-propagating waves are coupled by the external forcing and no propagative solutions (e.g. $A_+ \neq 0, A_- =$

0) exist. However, there is a solution in the form of spatially homogeneous standing waves, described by an amplitude A:

$$[A_+, A_-] = A[1, \mathrm{e}^{\mathrm{i}\varphi}] \ , \tag{59}$$

where φ is the phase shift which depends on the damping. In their experiments Douady et al. observed these homogeneous standing waves close to the onset with a certain wavelength. When the driving frequency is increased (decreased) the system bifurcates abruptly into another standing wave pattern by nucleation (respectively, annihilation) of one wavelength. This is a signature of a side-band Eckhaus type instability. By increasing the driving amplitude the pattern undergoes an oscillatory instability to a compression mode of the cellular structure, i.e. a spatially phase modulated standing wave. An analysis including numerical integration of the amplitude equations (58) yields similar results as in the experiments. However, the coefficients have not been calculated from the basic equations, but have been set appropriately in order to meet the experimental results.

5.7 Instabilities of square patterns and the transition to spatio-temporal chaos

As indicated above a linear coupling exists between the amplitudes A_j and A_{-j} in the amplitude equations. This means that the counter-propagating waves are coupled to standing waves. From the amplitude equations Milner (1991) has derived equations for standing waves which are extensions of the equations of Sec. 5.5. The standing waves are solutions of (52) obeying

$$[A_j, A_{-j}] = \tilde{A}_j[1, \mathrm{e}^{\mathrm{i}\varphi}] \ , \tag{60}$$

$$\tau\partial_t \tilde{A}_i = \mu(\nabla_\perp)\,\tilde{A}_i - \sum_j^N \beta(\theta_{ij})|\tilde{A}_j|^2\tilde{A}_i \ . \tag{61}$$

The only difference is the slow spatial dependence of the amplitudes obtained from the linear growth rate $\mu(\mathrm{i}\mathbf{k})/\tau$ for plane waves with wavevector $\mathbf{k}$. Since the evolution equation is variational an optimal standing wave pattern is determined by the minimum of the corresponding potential. For the low-viscosity case this results in a square pattern. Furthermore, the wavelength of the optimal pattern can be calculated.

Milner has performed a stability analysis of the standing wave patterns. Solutions of the standing wave equation (61) exhibit the usual longitudinal Eckhaus and transverse phase instabilities. However, patterns of the full travelling wave equations (52) exhibit also a transverse amplitude modulation (TAM) instability, which has been described by Ezersky et al. (1986) as well. According to Milner (1991), when the driving amplitude is increased the optimal standing wave pattern undergoes such an instability. Milner has also computed the critical drive amplitude $a_{\mathrm{c,TAM}}$ as well as the wavenumber

$\mathbf{q}_{\mathrm{TAM}}$ of the critical mode. This instability seems to be a peculiar feature of parametrically excited pattern forming systems. In order to investigate the TAM instability further and especially to get access to the postcritical states Zhang and Viñals (1995) introduced a model equation for a complex field $\psi(x, y, t)$ (see Sec. 5.8). Experimental reports on secondary instabilities of square wave patterns have been given by Ezersky et al. (1986) as well as Tufillaro et al. (1989). The Faraday experiments performed by Tufillaro et al. (1989) using a low-viscosity fluid and a large aspect ratio system are devoted to an examination of higher instabilities: Close to the onset of instability the surface deformation exhibits subharmonic standing waves forming a well ordered square-symmetric pattern. These patterns are not influenced by the vertical boundaries. At a slightly higher excitation amplitude long wavelength modulations are observed. These modulations are believed to stem from the TAM instability discussed above. In fact, the predictions of Milner

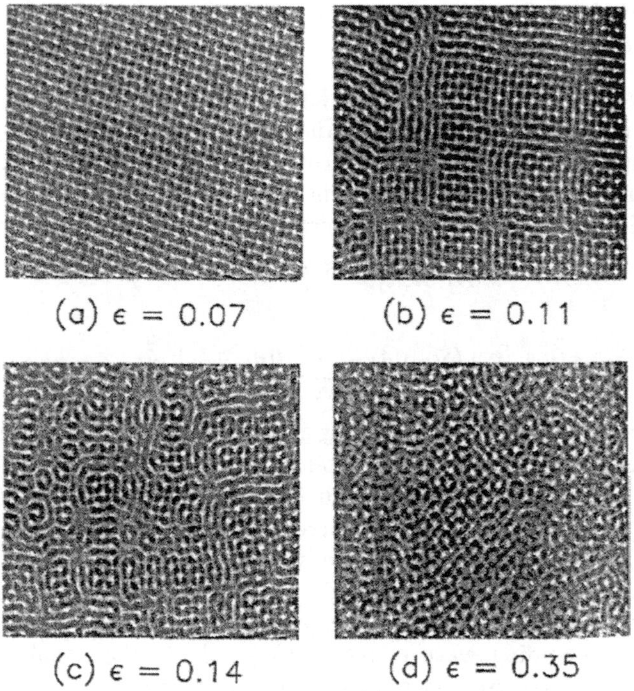

Fig. 8. Images of capillary wave patterns for various values of the drive amplitude $\epsilon = (a - a_c)/a_c$ (a_c denotes the critical drive amplitude). The sequence demonstrates the transition to disordered capillary ripples by increasing the drive amplitude. The figure is taken from Tufillaro et al. (1989).

agree qualitatively with these experiments (although there are quantitative discrepancies). A further increase of the driving amplitude leads to the appearance of spatially disordered patterns. Due to the TAM instability the amplitude of the waves can become small in certain regions allowing for the nucleation of defects. The nucleation of these defects subsequently leads to destruction of the long range orientational order. As a result, the patterns locally exhibit surface deformations in form of squares whose orientations, however, vary in space (c.f. Fig. 8).

Tufillaro et al. (1989) performed a quantitative characterization of the transition to these disordered patterns. To this end they digitized the refraction patterns to obtain an image $I(\mathbf{r}_\perp, t)$ of an instantaneous pattern. They computed the two-dimensional autocorrelation function $C(\mathbf{r}_\perp, \Delta\mathbf{r}_\perp)$ (the brackets denote a time average, i.e. an average over a long sequence of images):

$$C(\mathbf{r}_\perp, \Delta\mathbf{r}_\perp) = \langle\, \Delta I(\mathbf{r}_\perp, t)\, \Delta I(\mathbf{r}_\perp + \Delta\mathbf{r}_\perp, t)\, \rangle / \langle\, \Delta I(\mathbf{r}_\perp, t)^2\, \rangle \ , \qquad (62)$$

where $\Delta I(\mathbf{r}_\perp, t) = I(\mathbf{r}_\perp, t) - \langle\, I(\mathbf{r}_\perp, t)\, \rangle$ is the deviation from a mean time averaged pattern. The correlation as a function of $\Delta\mathbf{r}_\perp = (\Delta x, \Delta y)$ is examined along a certain direction, e.g. along the x-axis. In the regime of well-ordered patterns the correlation function is a non-decaying periodic function of Δx. However, in the disordered regime, one observes an exponential decay which allows to define a correlation length. This length as a function of the driving amplitude shows a sharp decrease at the order-disorder transition. In fact, the correlation length drops to the order of few wavelengths of the cellular patterns. A relationship to the melting of two dimensional crystals has been mentioned. Furthermore, a function measuring orientatial correlations was calculated. This demonstrated the presence of a long-range orientational order on a scale comparable to the system size.

Gluckman et al. (1993) and Gluckman et al. (1995) performed a statistical analysis of disordered Faraday waves. By extracting digitalized images from the experiment they determined various statistical quantities like time averaged patterns, spatial as well as temporal correlation functions. However, no comprehensive understanding of the complex spatio-temporal patterns has yet been achieved.

Bosch and van de Water (1993) and Bosch et al. (1994) tried to find analogies between fully developed turbulent fluid motion and Faraday ripples. To this end they evaluated the spectral fluctuations of the capillary wave fields and showed that they exhibit intermittency: The probability distribution of the empiric spatial intensity spectrum deviates from a Gaussian distribution. The authors attribute the existence of these deviations to the local creation and annihilation of coherent structures, which are regions with ordered surface deformations.

5.8 Model equations

The investigation of model equations of the type first introduced by Swift and Hohenberg (1977) plays an important role in the understanding of pattern formation in nonlinear systems. In these models a scalar order parameter field $\psi(\mathbf{r}_\perp, t)$ is introduced, which obeys a partial differential equation of the form

$$\dot{\psi}(\mathbf{r}_\perp, t) = \Lambda(\nabla_\perp^2)\, \psi(\mathbf{r}_\perp, t) - N[\psi(\mathbf{r}_\perp, t)] \ . \qquad (63)$$

The linear operator $\Lambda(\nabla_\perp^2)$ is constructed to reproduce the linear stability properties of the underlying physical system. The nonlinearity N is a functional of the field $\psi(\mathbf{r}_\perp, t)$. In general, it can be derived from the basic equations invoking adiabatic elimination of the stable modes. However, usually N turns out to be a complicated, sometimes nonlocal functional and one resorts therefore to approximations. This procedure is discussed by Bestehorn and Friedrich (1997).

The Rayleigh-Bénard system is a classical example for which the approach via model equations has proven to be very useful (Neufeld and Friedrich (1994)). A variety of different models for different phenomena has considerably improved our understanding of nonlinear pattern formation, secondary instabilities and the statistical properties of spatially disordered patterns (Neufeld and Friedrich (1995)).

The observation of spatially quasi-periodic patterns in the Faraday experiment has motivated the search for models which exhibit solutions of this type. The purpose is to identify the relevant selection mechanisms. Newell and Pomeau (1993) originally introduced this idea to derive model equations for spatially chaotic solutions, which they called turbulent crystals.

Müller (1993b) considered a Swift-Hohenberg equation with a nonlinear coupling term of the form

$$N[\psi] = a\,\psi^3 + b\,\psi\left(\nabla_\perp^2\right)^2 \psi^2 + \psi\left[c\left(\nabla_\perp^2\right)^2 - d\left(\nabla_\perp^2\right)^3\right]^2 \psi^2 \ . \qquad (64)$$

The form of the nonlinearity has been chosen in such a way that it allows for the existence of a stable solution corresponding to a quasi-periodic pattern of an eight-fold orientational symmetry. The dynamics of this equation can be derived from a Lyapunov functional, which guarantees a relaxational dynamics towards the state of minimum free energy. The nonlinearity has been constructed to generate a potential, whose absolute minimum corresponds to the desired quasi-periodic solution.

An alternative model (Müller (1993b)) with two coupled generalized SH-equations has been shown to allow a robust solution of a 12-fold quasi-periodic pattern. Here the two equations account for two different unstable wavelengths which occur in an experiment where the excitation signal is a superposition of two sinusoidal signals. Recently, Lifshitz and Petrich (1997) followed this idea and introduced a model equation containing a more complicated linear operator of the form

$$\dot{\psi}(\mathbf{r}_\perp,t) = \epsilon\,\psi(\mathbf{r}_\perp,t) - c\,\left(1+\nabla_\perp^2\right)^2\left(q^2+\nabla_\perp^2\right)^2\psi(\mathbf{r}_\perp,t) \\ +\alpha\,\psi(\mathbf{r}_\perp,t)^2 - \psi(\mathbf{r}_\perp,t)^3\ . \qquad (65)$$

The linear operator gives rise to bicriticality at two different unstable length scales. The authors report regions in the control parameter space where stable steady state solutions with 2-, 4-, 6- and 12-fold symmetry occur. They relate their model to the two-frequency driving experiment of Edwards and Fauve (1994). However, as the instability of (65) is stationary the cubic nonlinearity does not correctly reflect the interaction between Faraday modes of subharmonic and harmonic time dependence.

Zhang and Viñals (1995) introduced the following model equation for a complex order parameter field $\psi(\mathbf{r}_\perp,t)$:

$$\dot{\psi} = -\gamma\,\psi + \mathrm{i}a\,\psi^* + \frac{3\mathrm{i}}{4}\left(1+\nabla_\perp^2\right)\psi + (\mathrm{i}-\gamma_n)\,|\psi|^2\psi\ . \qquad (66)$$

Here, the term with the real coefficint γ_n is a nonlinear damping term. Unlike the previous functions ψ, this field is assumed to be complex, since it describes only the slowly varying envelope of the fast oscillating surface deformation. The point was to investigate the TAM instability further and to get access to the postcritical states. However, the disadvantage of the model is that, due to the simple nonlinearity the primary instability leads to line patterns instead of the experimentally observed squares. Nevertheless, the authors analytically demonstrate the existence of the TAM instability of the line pattern as an amplitude instability in addition to the usual phase instabilities denoted as Eckhaus and zigzag. The numerical evaluation of the model equation yields complex disordered patterns beyond the instability, which are dominated by defects nucleated due to the TAM instability.

The quasi-potential approximation of Zhang and Viñals (1997a) reviewed in Sec. 2 can also be understood as a model for parametrically driven surface waves. Extensive numerical simulations of these two coupled equations exhibit solutions in form of standing square wave patterns for capillary waves as well as hexagonal patterns and a sequence of quasi-priodic patterns for gravity-capillary waves.

In order to obtain a closer relationship between experiment and the model equations of the form (63) one might think of considering a class of model equations, where the nonlinear terms are parametrized by a number of free coefficients. Applying a method similar to the one descibed in Uhl and Friedrich (1995) definite values of these coefficients could be extracted directly from data of the experimentally observed patterns.

6 Summary, open problems, perspectives

The interest in Faraday waves during the last decade has concentrated on three principal issues: (i) Linear stability analysis on fluids of finite viscosity, (ii) the low-dimensional dynamics due to mode interactions in small aspect ratio systems, and (iii) ordered regular structures in large systems, including their transition to disordered structures and spatio-temporal chaos.

A full numerical treatment of the linear stability problem has been developed and applies to systems with arbitrary viscosities. In the case of weak and strong dissipation analytical approximations show that the problem can be reduced to a slightly damped or respectively overdamped Mathieu oscillator. It turns out that besides the subharmonic surface resonance which is commonly observed in experiments, response in synchronous resonance with the external drive is also possible. This phenomenon occurs if the vortical boundary layers dominate the dissipation.

A great deal of experimental and theoretical investigations is dedicated to the low-dimensional dynamics which takes place when the spatial structure of the pattern is confined by the geometry. In this situation, the wavelength of the pattern is of the same order as the size of the container. Several model equations have been proposed to account for the interplay between two competing container modes, when the setup is operated in the vicinity of a codimension-two bifurcation. Although several characteristic features, in particular the transition into the chaotic regime, are reproduced by the models, a quantitative agreement is still lacking.

Considerable progress has been made in developing an understanding of pattern formation in large aspect ratio systems. In order to guarantee that the vertical boundaries do not influence the system's behaviour, the container sizes should exceed a suitably defined correlation length. This correlation length rises steeply as the viscosity of the fluid decreases. If the large aspect ratio condition is met, pattern selection is then dominated by nonlinear interactions in the bulk of the fluid. In the weak dissipative limit the pattern selection mechanism close above onset can be derived from a potential. This allows a prediction of the most favoured symmetry of the pattern.

Faraday experiments with multi-frequency excitation have had great impact on research and opened the possibility for a richer dynamics. Especially, a multicritical situation can be achieved allowing for mode competition with different wavelengths. Historically, those experiments were the first in which quasi-periodic patterns or patterns with a triangular tesselation were observed.

Most theoretical investigations of Faraday waves use the so-called "almost Hamiltonian limit". In this approximation dissipation is introduced heuristically, neglecting the vortical boundary layer close to the container walls. This approach is appropriate for low-viscosity fluids like water or alcohols. However, it becomes uncontrollable if for instance viscous silicone oils or concentrated water-glycerin mixtures are used. The theoretical analysis has revealed the importance of a resonant three-wavevector interaction between the neutral subharmonic modes and the neighbouring almost critical waves with harmonic time dependence. In the region of gravity-capillary waves this mechanism leads to quasi-periodic patterns. This has also been confirmed experimentally.

At higher drive amplitudes the well ordered patterns become unstable due to secondary bifurcations. Here, the most important destabilization mechanism is a transversal amplitude modulation, which appears to be of non-variational origin. Therefore, it is not captured by the amplitude equations for standing waves. The dynamics of the standing wave mode of opposite temporal phase (which is stable at onset) must be taken into account. The secondary instabilities usually give rise to disordered structures. The first steps towards a characterization of the spatio-temporal chaotic states have been undertaken by averaging methods and the evaluation of temporal or spatial correlations.

In spite of experimental advantages (simple setup, short relaxation times) the mathematical treatment of parametrically driven surface waves has turned out to be very complicated. Difficulties arise due to the nonlinear boundary conditions, the presence of several intrinsic length scales (wavelength, viscous skin depth, correlation length) and due to the parametric drive, which renders the problem non-autonomous. Even the linear stability analysis for fluids of arbitrary viscosity must be performed numerically. In the case of large damping (large viscosity, small filling depth) the linear eigenfunction exhibits a rich frequency spectrum with considerable contributions of higher harmonics. None of the available nonlinear theories presently is appropriate to account for these situations. In the low-viscosity limit the spectral power is concentrated in the basic subharmonic frequencies $\pm\Omega$. The frequency contributions $\pm 3\Omega, \pm 5\Omega$ carry small amplitudes of order ν, ν^2 etc. The viscosity also gives rise to a small phase shift between the excitation signal and the surface response. It turns out that both the phase shift and the 3Ω frequency components contribute at leading order to the nonlinear saturation coefficient. The question of whether nonlinear dissipative contributions also must be taken into account is still beeing discussed controversially in the literature. In summary, it can be stated that the nonlinear saturation mechanism for Faraday waves is not yet fully understood.

Important steps have been made in comparing theoretical predictions with experimental findings. However, a comprehensive experimental survey is lacking because a *quantitative* evaluation of the surface patterns is much more difficult than a simple visualization of the planforms. Improved quantitative measurements are also desired for a better understanding of secondary instabilities and the transition to spatio-temporal chaos. Although this transition is one of the main topics of theoretical studies, appropriate tools for a reasonable characterization of the phenomena are still under development. The investigation of suitable model equations of the form discussed in Sec. 5.8

may have an improtant impact. Apparently, these models can reproduce the main characteristics of the transitions to disordered structures and can be used to gain considerable insight into the underlying physical mechanisms. Furthermore, the appearing free model parameters can be used to fit the experimental data. This procedure promises to provide realistic equations for an interpretation of experiments in the transition regimes towards spatio-temporal chaos.

References

J. Bechhoefer, V. Ego, S. Manneville, and B. Johnson, J. Fluid Mech. **288**, 325 (1995).

T. B. Benjamin and F. Ursell, Proc. R. Soc. Lond. A **225**, 505 (1954).

T. Besson, W. S. Edwards, and L.S.Tuckerman, Phys. Rev. E **54**, 507 (1996).

M. Bestehorn and R. Friedrich, in J. Parisi, S. C. Müller, W. Zimmermann (eds.), *A Perspective Look at Nonlinear Media in Physics, Chemistry and Biology*, Springer-Verlag, Berlin (1997).

J. Beyer and R. Friedrich, Phys. Rev. E **51**, 1162 (1995).

D. Binks and W. van de Water, Phys. Rev. Lett. **78**, 4043 (1997).

E. Bosch, H. Lambermont, and W. van de Water, Phys. Rev. E **49**, R3580 (1994).

E. Bosch and W. van de Water, Phys. Rev. Lett. **70**, 3420 (1993).

E. Cerda and E. Tirapegui, Phys. Rev. Lett. **78**, 859 (1997).

P. Chen and J. Viñals, Phys. Rev. Lett. **79**, 2670 (1997).

P. Chen and J. Viñals, J. Fluid Mech. (1998), to appear.

B. Christiansen, P. Alstrøm, and M. T. Levinsen, Phys. Rev. Lett. **68**, 2157 (1992).

B. Christiansen, P. Alstrøm, and M. T. Levinsen, J. Fluid Mech. **291**, 323 (1995).

S. Ciliberto and J. P. Gollub, Phys. Rev. Lett. **52**, 922 (1984).

S. Ciliberto and J. P. Gollub, J. Fluid Mech. **158**, 381 (1985).

J. D. Crawford, E. Knobloch, H. Riecke, Physica D 44, 340 (1990).

J. D. Crawford, J. P. Gollub, D. Lane, Nonlinearity **6**, 119 (1993).

M. C. Cross and P. C. Hohenberg, Rev. Mod. Phys. **65**, 851 (1993).

L. Daudet, V. Ego, S. Manneville, and J. Bechhoefer, Europhys. Lett. **32**, 313 (1995).

S. Douady, J. Fluid Mech. **221**, 383 (1990).

S. Douady, S. Fauve, and O. Thual, Europhys. Lett. **10**, 309 (1989).

W. S. Edwards and S. Fauve, C. R. Acad. Sci. Paris **315-II**, 417 (1992).

W. S. Edwards and S. Fauve, Phys. Rev. E **47**, R788 (1993).

W. S. Edwards and S. Fauve, J. Fluid Mech. **278**, 123 (1994).

A. B. Ezersky, M. I. Rabinovich, V. P. Reutov, and I. M. Starobinets, Sov. Phys. JETP **64**, 1228 (1986).

M. Faraday, Philos. Trans. R. Soc. London **52**, 319 (1831).

S. Fauve, K. Kumar, C. Laroche, D. Beyens, and Y. Garrabos, Phys. Rev. Lett. **68**, 3160 (1992).

B. J. Gluckman, C. B. Arnold, and J. P. Gollub, Phys. Rev. E **51**, 1128 (1995).

B. J. Gluckman, P. Marcq, J. Bridger, and J. P. Gollub, Phys. Rev. Lett. **71**, 2034 (1993).

H. Haken, *Synergetics. An Introduction*, 3rd ed., Springer-Verlag, Berlin (1983); H. Haken, *Advanced Synergetics*, 2nd ed., Springer-Verlag, Berlin (1987).

S. V. Kiyashko, L. N. Korzinov, M. I. Rabinovich, and L. S. Tsimring, Phys. Rev. E **54**, 5037 (1996).

A. Kudrolli and J. P. Gollub, Physica D **97**, 133 (1996).

A. Kudrolli, B. Pier, and J. P. Gollub, Physica D (1998), to appear.

K. Kumar, Proc. R. Soc. London A **452**, 1113 (1996).

K. Kumar and K. Bajaj, Phys. Rev. E **52**, R4606 (1995).

K. Kumar and L. Tuckerman, J. Fluid Mech. **279**, 49 (1994).

L. Landau and E. M. Lifshitz, *Fluid Mechanics*, 2nd edn. Pergamon Press (1987).

R. Lifshitz and D. M. Petrich, Phys. Rev. Lett. **79**, 1261 (1997).

O. Lioubashevski, J. Fineberg, and L. S. Tuckerman, Phys. Rev. E **55**, R3832 (1997).

T. Lundgren and N. Mansour, J. Fluid Mech. **194**, 479 (1988).

B. A. Malomed, A. A. Nepomnyashchii, and M. I. Tribelskii, Sov. Phys. LETP **69**, 388 (1989).

E. Meron and I. Procaccia, Phys. Rev. A **34**, 3221 (1986).

J. W. Miles, Proc. Roy. Soc. Lond. A **297**, 459 (1967).

J. Miles and D. Henderson, Ann. Rev. Fluid Mech. **22**, 143 (1990).

S. T. Milner, J. Fluid Mech. **225**, 81 (1991).

H. W. Müller, Phys. Rev. Lett. **71**, 3287 (1993a).

H. W. Müller, Phys. Rev. E **49**, 1273 (1993b).

H. W. Müller, H. Wittmer, C. Wagner, J. Albers, and K. Knorr, Phys. Rev. Lett. **78**, 2357 (1997).

H. W. Müller, in J. Parisi, S. C. Müller, W. Zimmermann (eds.), *A Perspective Look at Nonlinear Media - From Physics to Biology and Social Sciences*, Springer-Verlag, Berlin (1998).

A. H. Nayfeh and D. T. Mook, *Nonlinear oscillations*, Wiley (1979).

M. Neufeld and R. Friedrich, Ann. Phys. Fr. **19**, 721 (1994).

M. Neufeld and R. Friedrich, Phys. Rev. E **51**, 2033 (1995).

A. C. Newell and Y. Pomeau, J. Physique A **26**, L429 (1993).

A. C. Newell and J. A. Whitehead, J. Fluid Mech. **38**, 278 (1969).

F. Simonelli and J. P. Gollub, J. Fluid Mech. **199**, 471 (1989).

Lord Rayleigh, Phil. Mag. **16**, 50 (1883).

K. Ruvinsky, F. Feldstein, and G. Freidman, J. Fluid Mech. **230**, 339 (1991).

M. Silber and M. R. E. Proctor, preprint (1977).

J. B. Swift and P. C. Hohenberg, Phys. Rev. A **15**, 319 (1977).

N. B. Tufillaro, R. Ramshankar, and J. P. Gollub, Phys. Rev. Lett. **62**, 422 (1989).

C. Uhl, R. Friedrich, and H. Haken, Phys. Rev. **E 51**, 3890 (1995).

M. Umeki and T. Kambe, J. Phys. Soc. Jpn. **58**, 140 (1989).

W. Zhang and J. Viñals, Phys. Rev. Lett. **74**, 690 (1995).

W. Zhang and J. Viñals, Phys. Rev. E **53**, R4286 (1996).

W. Zhang and J. Viñals, J. Fluid Mech. **336**, 301 (1997a).

W. Zhang and J. Viñals, J. Fluid Mech. **341**, 225 (1997b).

Pattern Formation in an Inhomogeneous Environment

Walter Zimmermann[1,2], Ben Painter[3] and Robert Behringer[3]

[1] Institut für Festkörperforschung and FORUM Modellierung, Forschungszentrum Jülich, D-52425 Jülich
[2] Theoretische Physik, Universität des Saarlandes, D-66041 Saarbrücken
[3] Department of Physics and Center for Nonlinear Science and Complex Systems Duke University, Durham NC, 27708-0305, USA

Abstract. A major step in understanding pattern formation has been achieved by studying idealized systems. However, in nature inhomogeneous systems are much more abundant than their idealized homogeneous counterparts. Here we report about experimental results on pattern formation in two inhomogeneous systems, thermal convection in porous media and Taylor-vortex flow between a rough and a smooth cylinder. Several aspects of heterogeneity effects in pattern formation are theoretically investigated for model equations and analytical descriptions are given for a few phenomena.

1 Introduction

During recent decades, pattern formation has experienced enormous progress [1-5] and important insights were contributed by studies of idealized fluid systems such as Rayleigh-Bénard convection (RBC) [6], Taylor-Couette flow (TCF) [7] and electroconvection [8, 9]. In many cases these experiments were carefully prepared and essentially homogeneous. Close to the onset of flow, there is now a general classification scheme of bifurcations and simple normal forms, so–called Ginzburg–Landau equations, provide a good qualitative description of some region near the onset of flow [4]. In fluid systems, quantitative understanding exists for many phenomena in pattern formation [4]. They are characterized by *control parameters*, which apply everywhere within the system. With increasing control parameter, ϵ, additional instabilities occur and lead to more and more complex spatio-temporal dynamics in these ideal systems. Typically, these complex flows are at best only partly understood, and are among the current challenges for dynamicists.

A much less studied issue is pattern formation in systems which are intrinsically inhomogeneous. In nature, pattern forming systems are in most cases inhomogeneous and their homogeneous isotropic counterparts are often prepared in laboratory experiments. An intrinsically inhomogeneous system is convection in porous media [10]. In fluid systems, as mentioned above, inhomogeneities may also arise from irregularities at the container boundaries. Often these irregularities are small compared to the extension of the fluid containers and the length scale of the pattern and the related effects are below the experimental resolution. However, in recent experiments, the size of the container has, in some cases, been considerably reduced and/or the experimental resolution strongly enhanced. Both effects make it more likely that roughness effects arise. These effects are unlikely to be explained by models assuming ideal flat boundaries. Here we describe experiments where irregularities are introduced in a controlled manner, either at the container boundaries or in the bulk. Such investigations allow a more systematic analysis of typical effects of irregularities. For instance, they allow an estimate of the robustness of phenomena which are usually studied under ideal conditions. But, the reduced symmetry and greater spatial complexity of inhomogeneous systems make them more difficult to study.

Linear waves, such as acoustic or electromagnetic waves in disordered media are already a subject extensively studied in recent years [11]. Perhaps most famous is Anderson localization, where disorder in a linear media causes an exponential decay of the transmission coefficient (see e.g. Ref. [11, 12]). Nonlinearity may act counter to linear disorder effects, thus enhancing the transmission coefficient. Hence, solitons and spatially periodic patterns in model systems stay undistorted if nonlinearities are strong[13, 14] .

A subclass of inhomogeneous systems are spatially periodic patterns which are spatially periodic modulated. These have been investigated already to some extent in one and two spatial dimensions. The competition between the wave length of the forcing and the naturally occurring wave length of the pattern leads to a great variety of phenomena. In one spatial dimension it leads to commensurate/incommensurate transitions, phase solitons etc. [15, 16, 17, 18] and in two spatial dimensions to complex patterns [19, 20, 21]. Single periodically forced systems and statistically heterogeneous systems share a number of common phenomena. For instance, in both cases the up-down symmetry can be broken and the bifurcation might be imperfect. However, it is a major difference between both that in heterogeneous systems the spectrum of forcing is broad. As discussed here, the broad spectrum may lead to localization effects and nonconstant weight for the wave lengths in the spectrum may lead to rounding of the bifurcation which is not mainly due to the imperfect bifurcation. Other heterogeneous and extensively investigated examples are coupled oscillators [22, 23], where aspects of synchronization are in the focus of investigations and rough surfaces used for drag reduction (see e.g. [24]).

Here, we consider two spatially inhomogeneous systems experimentally. In Section 2 porous medium convection (PMC) and in Section 3 Taylor-Couette flow (TCF) where the inner cylinder is roughened by machining. In section 4 effects of spatially varying coefficients are analyzed theoretically for one of the simplest models describing spatially periodic patterns, the Swift-Hohenberg model and effects of spatially varying boundaries in Rayleigh-Bénard convection.

2 Porous media convection (PMC)

We illustrate PMC in Fig. 1, and discuss why PMC naturally falls into a class of intrinsically inhomogeneous nonlinear pattern forming systems. PMC is nominally similar to RBC, except that the region containing the fluid is shared with a solid medium. More specifically, the fluid resides in the open connected channels of a porous solid. If a thin horizontal layer of fluid-saturated porous medium is heated from below, leading to a temperature difference, ΔT, convection will occur if the porous Rayleigh number [10]

$$Ra = \alpha g \gamma d / \kappa \nu \tag{1}$$

exceeds a critical value. In the conventional picture of PMC as modeled by Darcy's law, $Ra_c = 4\pi^2$. Here, α, κ, ν and d are the appropriate thermal expansion coefficient, thermal diffusivity, kinematic viscosity and heights of the porous layer, and γ, defined below, is the permeability. At onset, the expected pattern is straight rolls of wavelength λ.

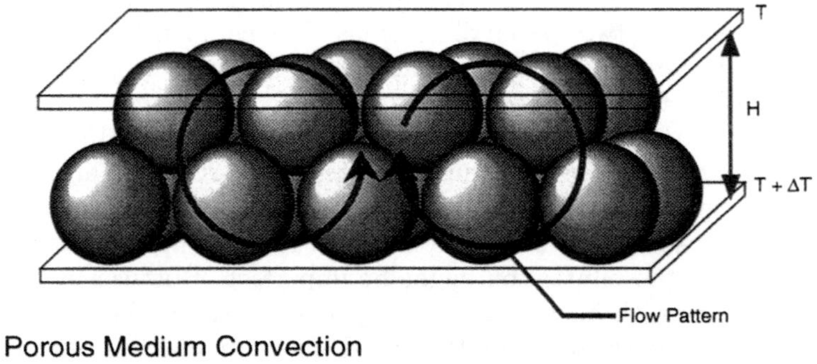

Figure 1: *Illustration of convection in a porous medium*

Although the conventional Darcian picture of PMC may give an approximate picture, it also assumes that the intrinsic structure of the medium is irrelevant. Recent experiments have shown that this structure can have a significant impact on pattern selection and heat transport [25, 26, 27]. It is this structure which can make a porous medium spatially inhomogeneous. In principle, one might construct an experiment for which the pore scale, δ, is arbitrarily small relative to the height of the porous layer, d. In this case the inhomogeneity resulting from the pores should be negligible. However, for conventional fluids like water, this is not really feasible. Here, we explore why that is the case by examining the conventional Darcy picture and its predictions. On some scale, a porous medium may be thought of as a continuum which, for slow flow speeds is characterized by Darcy's law:

$$\mathbf{v} = -\gamma/\eta \cdot \nabla P. \tag{2}$$

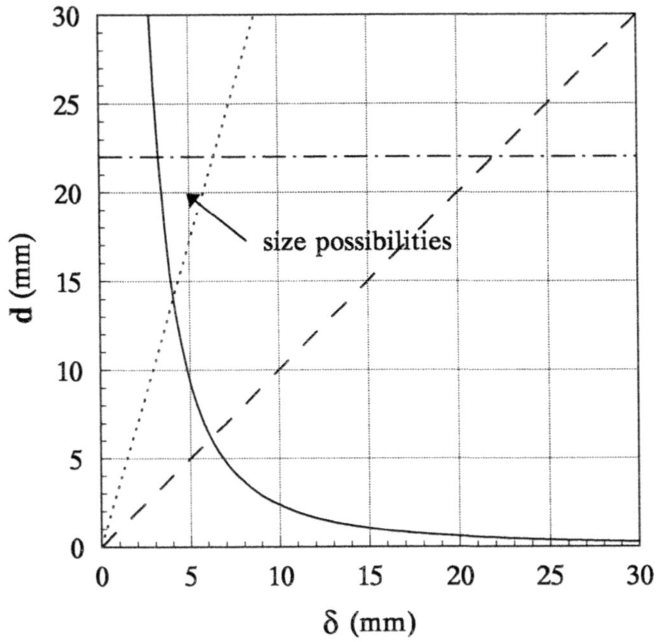

Figure 2: *Constraints for porous media convection (PMC), as represented by the pore scale, δ and the layer height, d. The hyperbolic-like curve applies for water, where $\Delta T = 10$ K at onset, assuming that this corresponds to Boussinesq convection. The region below this curve is non-Boussinesq. The horizontal line corresponds to a vertical thermal relaxation time of $\tau_v = 30$ mins — roughly a maximal acceptable time. In addition, to be even remotely a porous medium, δ must be no larger than d, dashed line. A spatially homogeneous medium should satisfy $d/\delta >> 1$ or alternatively $Da = \gamma/d^2 << 1$. If we choose $Da = 2 \times 10^{-5}$ then one must stay beyond the dotted line. When all these constraints are imposed, only the small triangular region remains as acceptable on all accounts. This region does not satisfy $\delta << d$.*

Here, γ is the permeability and η is the shear viscosity. More sophisticated continuum descriptions exist, but they all share the assumption that there is a length scale, L, intermediate between the pore size, δ, and the overall system size, S, such that an average of the flow over a representative elementary volume (REV), $\sim L^3$, is described to a good degree of accuracy by Darcy's law. In a typical lab-scale system, the pore size cannot be made to satisfy the condition that $\delta << \lambda \simeq 2d$. This limitation arises because of two competing constraints. First, the pore size cannot be made very small, since otherwise, the temperature difference at and above onset is too large to have Boussinesq flow. Second, the height of the system cannot be made very large without incurring impractical relaxation times. The issue is captured in Fig. 2. This figure shows the typical pore scale, δ vs. the layer height, d, with fluid parameters pertaining to water. Boussinesq convection can be obtained above the hyperbolic-like curve (assumes $\Delta T_c = 10\,\mathrm{K}$). Acceptable relaxation times ($\tau = d^2/\kappa$) for vertical heat diffusion of less than 0.5 hr apply below

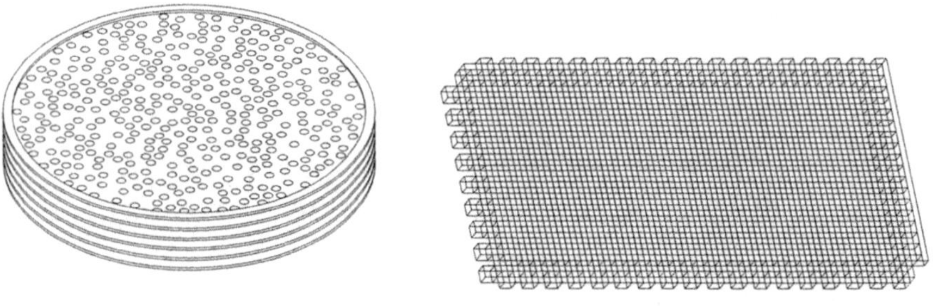

Figure 3: *Specially constructed porous media*

the horizontal line. And of course to be a sensible porous medium, we must have $\delta < d$.

The bottom line of this analysis is that although δ can be set less than d, the usual assumption, $\delta << d$ is hard to met. Therefore, there will be microscopic structure, which we can think of as local variation in R, with fluctuations on scales not too much smaller than the convective wavelength. Therefore, this system is a natural one for probing the effects of spatial fluctuations on **weakly nonlinear** pattern formation. Other systems have been studied which have slow spatial variations. Examples are TCF [29] and RBC with a spatial ramp [30] or RBC and EC with spatially single periodic boundary variations [31, 16, 32]. PMC is different in that inhomogeneities occur in the bulk and that the spectrum of spatial variations is broad and has significant strength at wave vectors larger (spatial scales smaller) than the characteristic system scale, $2\pi/d$. In our examples, these variations may be either random or periodic.

The PMC experiments which we describe here were carried out using two rather different techniques, discussed in more detail in Refs. [25, 26, 27]. Each technique is an experimental solution to the dilemma that in most cases, it is impossible to optically visualize the flow in porous media. This difficulty arises for several reasons. First, most porous media are not transparent. Second, even when the medium consists of transparent solid particles, differences in the indices of refraction of the fluid and solid plus the random curved or rough interfaces between the fluid and solid rapidly scatter the light. The first experiment described here uses specially constructed porous media for which there is a line of sight from above which is not affected by the problems above. Two types (out of an infinite collection of such materials) are sketched in Fig. 3. One is spatially random (by design) and consists of randomly perforated, stacked, plastic sheets as shown in Fig.3a). The other is actually spatially periodic and is constructed by regular stackings of plastic bars,cf. Fig. 3b) [25, 27].

The second experiment uses Magnetic Resonance Imaging (MRI), a noninvasive technique which can directly measure both the local fluid density and velocity [26]. Here, we present results for the vertical velocity. Each tech-

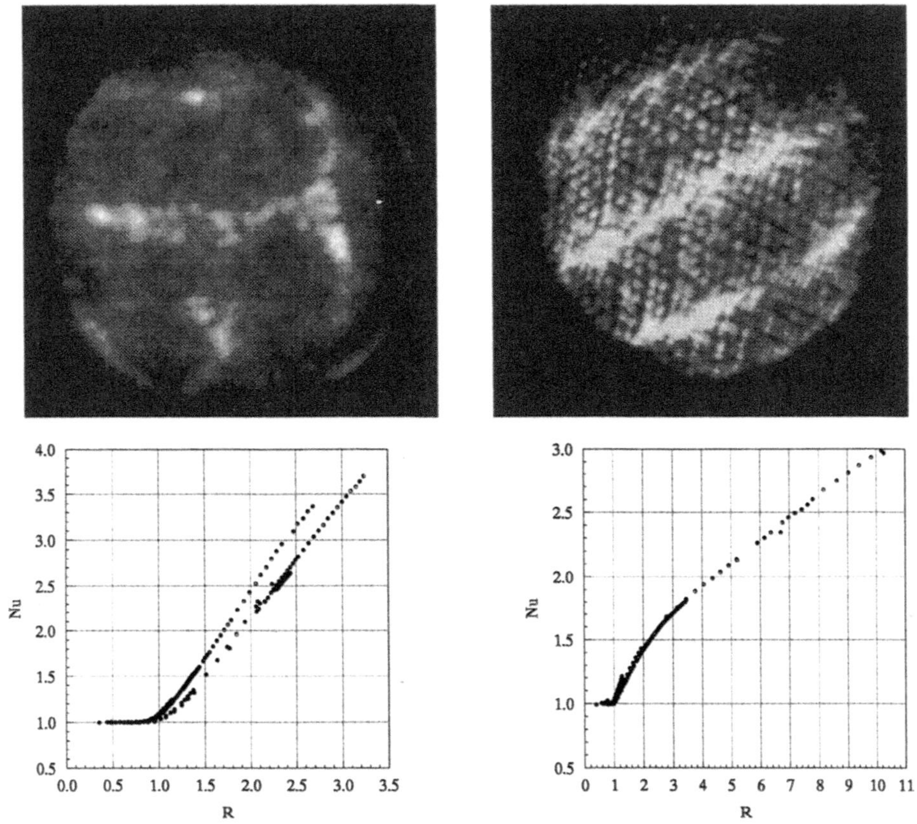

Figure 4: *In a) a typical pattern observed for random medium is shown, cf. Fig. 3a), and in b) a typical pattern observed for a rectangular grid porous medium, cf. Fig. 3b). In c) and d) the respective data for the heat transfer (Nusselt number Nu) are shown as function of the reduced Rayleigh-number $R = Ra/Ra_c$.*

nique has advantages and disadvantages. The first is restricted to a subset of possible porous media, but can also yield very high precision data for the heat transport – Nusselt number Nu. The second can be used on virtually every medium (providing it is not magnetic or electrically conducting) but it is more difficult to obtain data for the heat transport.

Data from the first type of experiment, including the two different media are given in Figs. 4. For the random medium shown in Fig. 3a) the fluid flow occurs via a cellular pattern which begins locally and with increasing R fills in the complete fluid layer, cf. Fig. 4a). This is also obvious from the Nusselt data which show a rounded transition, cf. Fig. 4c). By contrast, the experiment constructed with a periodic stacking of bars with a rather small pore scale, cf. Fig.3b), shows an instrumentally sharp transition to convection, cf. Fig 4b),d). Thus, for the rather small pore scales of both media, $\delta < d$, the rounding in Nu in Fig. 4c) is clearly associated with the random spatial variations in the porous structure.

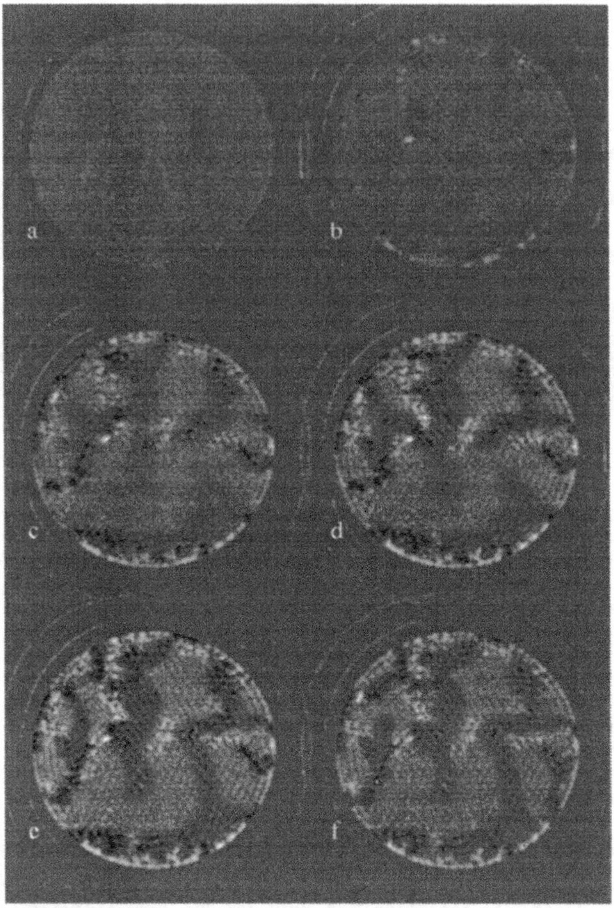

Figure 5: *Vertically averaged, vertical velocity for a circular cell with disordered packing of spheres. In a) for a vanishing temperature difference between the top and bottom boundary,* $\Delta T = 0.0$. *From part b) to f) the velocity pattern is shown for different values of the reduced temperature difference,* $\varepsilon = (\Delta T - \Delta T_c)/\Delta T_c$. *b)* $\varepsilon = -0.5$, *c)* $\varepsilon = 1.0$, *d)* $\varepsilon = 3.0$, *e)* $\varepsilon = 5.0$ *and f)* $\varepsilon = 3.0$, *but after first decreasing* ΔT *to* 0 *and then raising it back to* 3.0.

This is seen as well in the experiments using PMC. Here, we contrast in Figs. 5,6 data for a nominally random packing of spheres, a hexagonal packing of spheres in a rectangular container, and a hexagonal packing of spheres in a hexagonal container. In this sequence of different containers, we move from random to increasingly ordered structures. In the disordered packing, the flow starts in regions where there are defects in the packing (the source of the disorder), and with increasing ΔT convection fills the cell, similarly to the first case above. In the ordered packings, well defined convection rolls form instead, cf. Fig. 6. In the case of the rectangular container, there is another interesting feature associated with local variations in Rayleigh number.

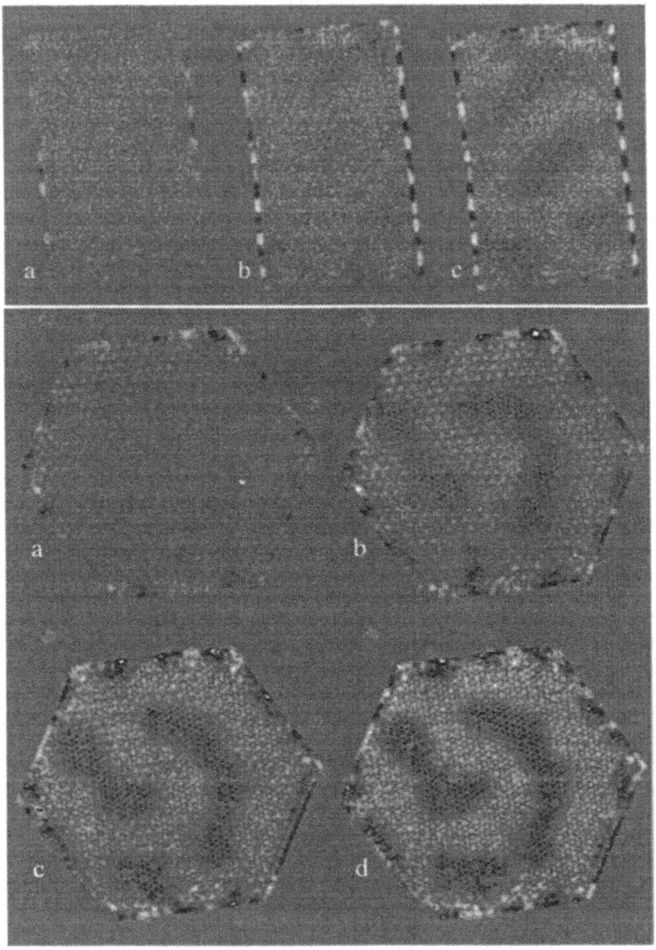

Figure 6: *Vertically averaged, vertical velocity for a packing of the spheres with hexagonal symmetry. Top: Rectangular container for different values of the reduced temperature a)* $\varepsilon = -0.5$*, b)* $\varepsilon = 0.2$ *and c)* $\varepsilon = 0.4$*. Bottom: For a hexagonal cell at different values of* ε *below and above critical.*

Note that packing spheres with a hexagonal pattern in a rectangular container means that along one pair of walls there will be a particularly large porosity/permeability and hence locally large Rayleigh number. In fact, convection does first start in this highest-γ region, although the small rolls which start there remain highly localized and do not seem to strongly affect the flow at higher values of the Rayleigh number.

3 Taylor-Couette Flow with Random Spatial Variations

A somewhat simpler example of a system with random spatial variations is a conventional Taylor–Couette (TCF) apparatus as shown in Fig. 7 [7], for which the gap size is variable as indicated in Fig. 8. In our experiment[28] the inner cylinder has been roughened by machining axisymmetric grooves along the length of the cylinder. The location of the grooves is random by design, and the spectrum of wavelengths, Fig. 8, is roughly flat up to the wavenumber corresponding to the axial width of the grooves. The inner surface of the outer cylinder is smooth, so that the grooves in the inner cylinder uniquely impart random axial variation in the Taylor number. Unlike earlier experiments

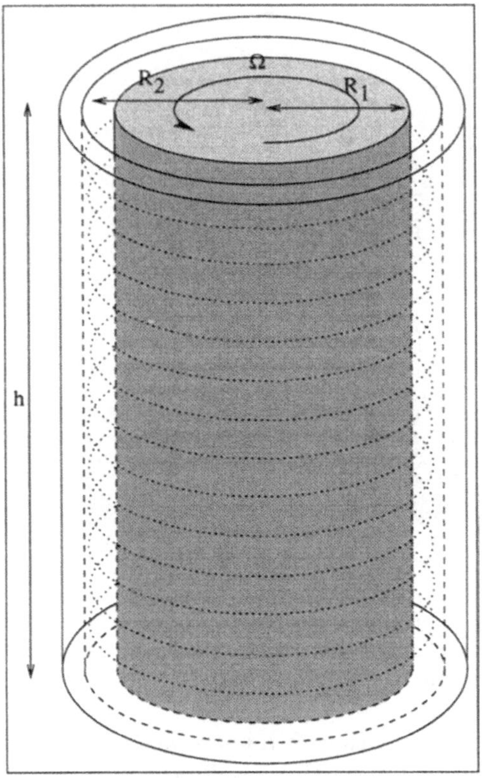

Figure 7: *The conventional Taylor-Couette apparatus is shown. The gap between the two cylinders at R_1 and R_2 is filled with water and the inner cylinder is rotated. Beyond a critical angular frequency Ω_1 the linear shear profile (Couette flow) becomes unstable and periodic Taylor vortices occur (see e.g. [7]).*

on fluid flow in containers with boundary variations, such as ramps [29, 30] or spatially periodic variations [16, 32], the experiment we consider here is similar in character to PMC in that the variations are random and have a broad spectrum with a significant power for wavelengths less than d.

In order to characterize the effect of spatial variability in Reynolds number Re,

$$Re = \frac{\Omega_1 R_1 (R_2 - R_1)}{\nu}, \tag{3}$$

it is important to measure the local velocity field throughout the experiment. We do so by particle imaging velocimetry (PIV). The fluid is seeded with a very low density of Kalariscope particles and then imaged with a short depth-of-field using a small CCD camera mounted on a stage which we transport vertically. The field of view is restricted to the outer 1/4 of the fluid, and from computer analysis of the particle tracks, we obtain the vertical velocity. For example, Fig. 9 shows a representative example of the tracks generated by particles, and Fig. 10 shows an example of the vertical velocity field v of Taylor vortices for a smooth inner cylinder.

The velocity measurements with the rough inner cylinder show clearly the effects of localization of the flow, Fig. 10. The initial flow clearly occurs in the regions of locally largest Re which is intuitively obvious. With increasing Ω_1, the flow spreads to other regions of the fluid and the localization becomes weaker, as indicated by the decreasing width of the power spectrum for the Taylor vortices.

For homogeneous systems, the bifurcation of the pattern, here the Taylor vortices, is characterized either by the Re-dependence of the local vertical velocity at a fixed position or by the Re–dependence of the spatially (axially) averaged velocity $\langle v^2 \rangle$. For inhomogeneous systems, such as for a rough inner cylinder, the functional dependence of the vertical velocity v on Re will change with the position of the measurement. Hence it is more appropriate to characterize the bifurcation by the spatially averaged velocity $< v^2 >$, as done in Fig. 11. For the smooth inner cylinder $< v^2 >$ is a linear function of Re but for the rough inner cylinder it behaves over a wide range of Re nearly exponentially.

Since no ab–initio calculation for the Taylor system with a rough inner cylinder is yet available, it is interesting to ask which generic model best fits the measurements. Inspection of the power spectrum of the roughness, cf. Fig. 8, tells us that the wave number at the first peak, q_{n1}, is close to the critical wave number q_c of the Taylor vortices. Hence, one expects a resonance $1 : 1 = q_c/q_{n1}$, which can be described by an appropriate Ginzburg–Landau-equation with an additive constant [31, 34]. In the power spectrum in Fig. 8 there are also strong contributions at larger wave numbers. Therefore, there could also be higher order resonance effects, such as $1:2$, $1:3$ and $1:4$ resonances. The modification of a supercritical bifurcation according to such resonances has been described by Coullet [17]. For a $1:2$ and a $1:4$ resonance the former supercritical bifurcation is still sharp. A $1:2$–resonance just shifts the threshold to smaller values and the $1:4$ reduces the prefactor β in the bifurcation relation $v^2 = \beta\epsilon$. Assuming that the rounded bifurcation in Fig. 11 is mainly due to these resonant contributions of the power spectrum of the roughness, the part of the spectrum contributing to the $1:1$ and the $1:3$

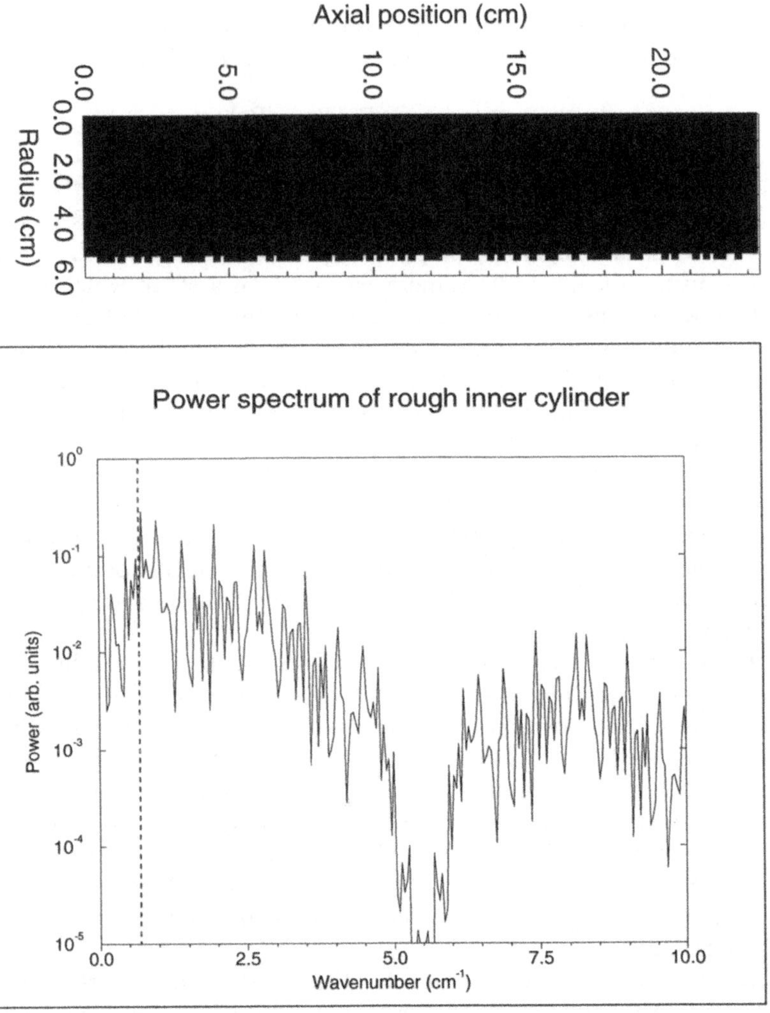

Figure 8: *The top part shows half of the rough inner cylinder of the TCF apparatus. The profile of the inner cylinder was constructed by cutting 2mm deep by 1.8mm wide azimuthally symmetric grooves. The profile is random, and the Radius of the inner cylinder fluctuates between the two values* $R_1 = 5.18cm$ *and* $R_1 = 5.38cm$ *with equal probability. The radius of the outer cylinder is* $R_2 = 5.38cm$. *The figure at bottom gives the power spectrum related to the roughness of the inner cylinder, whereby the dashed line corresponds to the wavenumber of the Taylor vortices without the roughness.*

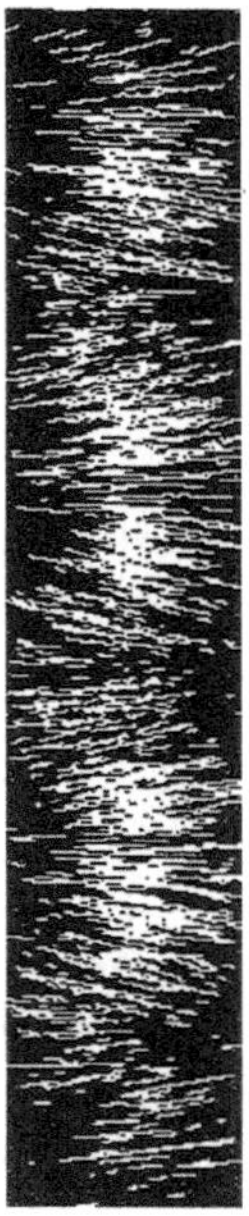

Figure 9: *Particle tracks in Taylor Couette Flow.*

resonance can primarily induce the rounding. Both resonances have been considered separately for single periodic forcing in Ref. [17].

The deviation between the local velocity v^2 and the averaged one $< v^2 >$ is correlated with the localization of the pattern, whereby the localization and the deviation between a local and global measurement is expected to decrease with increasing amplitude of the velocity of the vortices.

Accordingly, the model describing both resonances simultaneously reads as

$$\partial_t A = \epsilon A + bA^2 - cA^3 + h, \tag{4}$$

if any spatial dependence apart the resonance effects is neglected. The solid line in Fig. 11 corresponds to a least squares of stable steady solution of this model to the measurement. The quality of the fit is fair, but a nearly exponential law in an intermediate range is not covered by this simple imperfect bifurcation.

We contrast the data for the rough inner cylinder with data for a nearly ideal experiment, cf. Fig. 11. Within experimental resolution we see a sharp bifurcation (the Eckmann vortices visible in Fig. 10 represent a weak imperfection in this case). Moreover, for a smooth inner cylinder the data for $< v^2 >$ are well fitted by a linear function of Re.

The model of Eq.(4) completely neglects the influence of wave numbers in the power spectrum, aside from 1 : 1 and 1 : 3 resonances. However, the other contributions are crucial to explain localization of velocity field as shown in

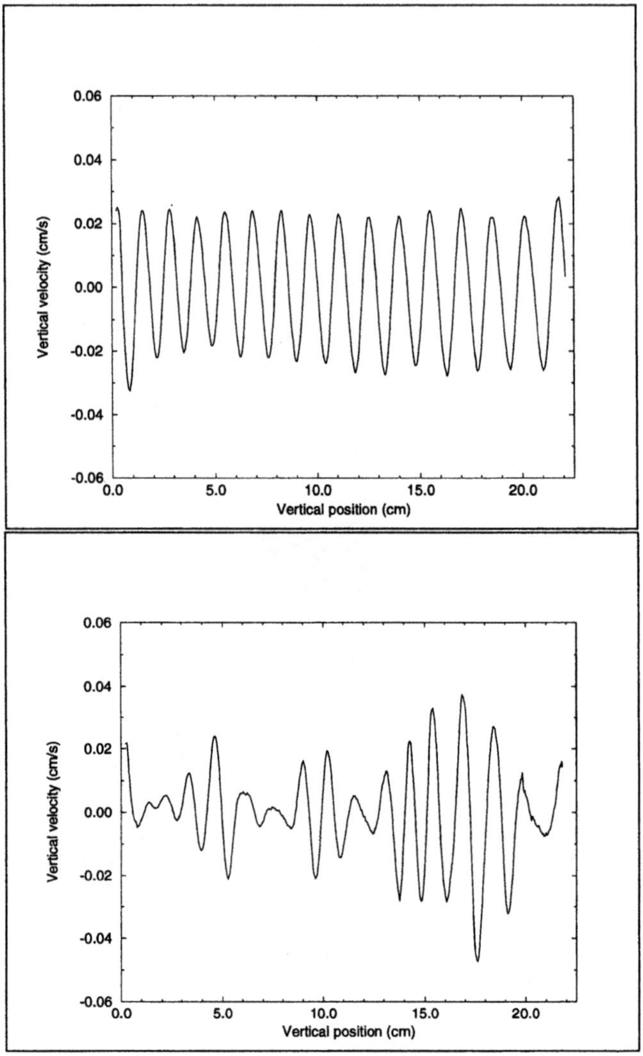

Figure 10: *Vertical velocity field of Taylor vortices with a smooth inner cylinder (top) and for the rough inner cylinder as shown in Fig.8 (bottom).*

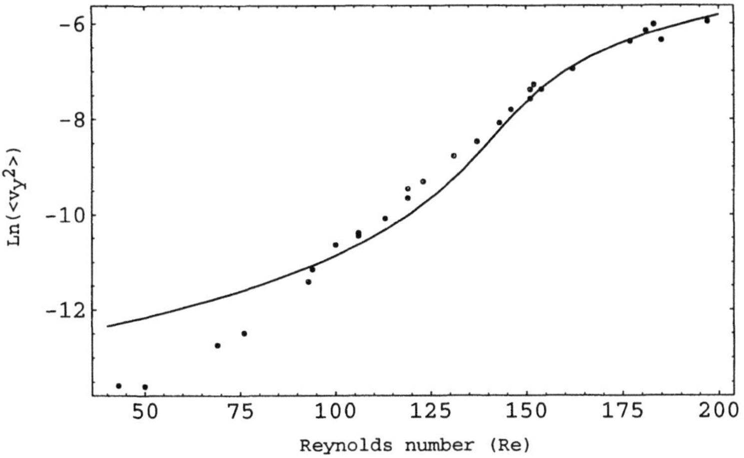

Figure 11: *The spatially averaged and squared axial velocity* $log(\langle v^2 \rangle)$ *is shown as function of the Reynolds number* $Re = \Omega_1 R_1 (R_2 - R_1)/\nu$, *dots, and the solid line is a fit with the model given in Eq. (4).*

cf. Fig. 10. Localization effects are only expected for a perturbation with a broad spectrum, never for single periodic resonant forcing as assumed in the model.

Usually, rounding of the bifurcation as shown in Fig.11 is interpreted as an imperfect bifurcation as it s meet for instance close to a 1 : 1 resonance. However, it has been shown by a numerical study of the Swift–Hohenberg equation with a random perturbation of the control parameter, that one can have both, such a rounded bifurcation and a sharp threshold for the onset of the pattern simultaneously [14]. This and other theoretical aspects of random perturbations are discussed in the next section 4.

4 Simple Models for Inhomogeneities in pattern formation

Thermal convection has been theoretically investigated for inhomogeneous conditions. For spatial variations at the container boundaries [31, 35, 34, 36, 30, 32] (including modulations in a rotating cylindrical annulus [33]), for a weakly heterogeneous porous medium [37] and for homogeneous porous media with boundary imperfections [38, 39, 40, 41]. For these examples, well established equations of motion are available and approximately solved, whereby contributions in the disorder spectrum are emphasized, which are in 1 : 1 resonance with the spatial periodicity of the pattern. This type of resonance induces an imperfect bifurcation and hence leads to a rounded bifurcation shape. However, the broad wave number spectrum of heterogeneities is crucial for a number of disorder effects. An example is localization of the pattern, which is absent for single periodic forcing. Localization of a pattern has been shown for the Swift–Hohenberg equation with a random contribution to the control parameter [14].

The Swift–Hohenberg equation is one of the simplest models describing a bifurcation from a homogeneous initial state to a stationary and spatially periodic pattern [42, 4]. For this model, the bifurcation is sharp but rounded. That is, the no-flow state exists and is stable up to a critical value of the control parameter, and above the bifurcation, the dependent field shows upward curvature. Here we show by a perturbational calculation that this rounding is partly due to nonresonant contributions in the spatially varying control parameter. In a further section we emphasize that spatially varying drifts induce a time dependence of spatially periodic patterns.

4.1 Swift–Hohenberg with spatially varying control parameter

In this section a generalized Swift–Hohenberg model is studied, which has a broken spatial translational symmetry. This system,

$$\partial_t u = \left[\varepsilon + M_1(x) - (q_0^2 + \partial_x^2)^2\right] u - u^3 + M_3(x) \quad , \tag{5}$$

was discussed first in Ref. [14]. ε is the control parameter that is similar to the reduced Taylor or Rayleigh number in real systems, and q_0 is the critical wave number at threshold of the unperturbed case, $M_i = 0$.

Various perturbations affect the pattern selection in this system in different ways. A spatial and singly periodic perturbation $M_1(x)$, with a modulation wave number k_1 of the order of the wave number of the pattern, q_0, leads to commensurate/incommensurate transitions in one spatial dimension [17, 43] and to a new pattern in two spatial dimensions [20, 21]. A random function $M_1(x)$ leads to localization of the pattern and to a rounding of the bifurcation, which is still sharp [14]. An additive function $M_3(x)$ breaks the $\pm u$-symmetry in equation (5). In two spatial dimensions this broken symmetry forces hexagonal patterns. Thermal fluctuations are a function of space and time and occur also additively, such as $M_3(x, t)$.

Numerical calculations of Eq.(5) have shown for $M_3(x) = 0$ that the bifurcation from the basic state, $u \equiv 0$, is still sharp, but shifted to lower values [14]. In addition the integral

$$N(\varepsilon) = \frac{1}{L} \int_0^L dx\, u(x)^2 \tag{6}$$

as function of the control parameter, ε, has a rounded shape similar to an imperfect bifurcation [31].

If we assume a small modulation , $M_1(x) = \eta \bar{M}_1(x) \propto O(1)$ ($\eta << 1$) then the threshold shift as well as the rounding of the bifurcation can be calculated by a perturbational calculation as indicated in the following two sections.

4.1.1 Threshold shift - $M_3(x) = 0$

We expand the control parameter as well as the solution of the linear part of Eq.(5) in powers of the modulation strength η:

$$\varepsilon_c = \varepsilon_c^{(0)} + \eta\varepsilon_c^{(1)} + \eta^2\varepsilon_c^{(2)} + \dots , \quad (7)$$

$$u(x) = u_0(x) + \eta u_1(x) + \eta^2 u_2(x) + \dots . \quad (8)$$

With the definition of the linear operator,

$$\mathcal{L}_0 = \varepsilon_c^{(0)} - (q_0^2 + \partial_x^2)^2 , \quad (9)$$

the following hierarchy of equations is obtained from Eq.(5) :

$$\eta^0 : \quad \mathcal{L}_0 u_0 = 0 , \quad (10)$$

$$\eta^1 : \quad \mathcal{L}_0 u_1 = -[\bar{M}_1(x) + \varepsilon_c^{(1)}]u_0 , \quad (11)$$

$$\eta^2 : \quad \mathcal{L}_0 u_2 = -[\bar{M}_1(x) + \varepsilon_c^{(1)}]u_1 - \varepsilon_c^{(2)}u_0 . \quad (12)$$

From these equations, threshold shifts can be calculated at the respective orders $\varepsilon_c^{(1)}, \varepsilon_c^{(2)}$. The final expressions are in general lengthy and depend on the choice of the stochastic properties of $\bar{M}_1(x)$. However, the essential features can be demonstrated with a periodic perturbation,

$$\bar{M}_1(x) = 2\bar{G}\cos(kx) . \quad (13)$$

If we start with an unperturbed solution,

$$u_0(x) = A_0 \cos(q_1 x) , \quad (14)$$

then we obtain $\varepsilon_c^{(0)} = (q_0^2 - q_1^2)^2$ from the marginal stability condition Eq.(10).

No resonance $k \neq 2q_1$: When the external modulation is not in resonance with the pattern, then the product $\bar{M}_1(x)u_0$ includes no contribution proportional to the unperturbed solution, u_0. In this case the first order equation (11) can be solved for $\varepsilon_c^{(1)} = 0$:

$$u_1(x) = -\left(\frac{\bar{G}\cos((q_1+k)x)}{\varepsilon_c^{(0)} - (q_0^2 - (q_1+k)^2)^2} + \frac{\bar{G}\cos((q_1-k)x)}{\varepsilon_c^{(0)} - (q_0^2 - (q_1-k)^2)^2} \right) A_0 . \quad (15)$$

In Eq.(12), the product $\bar{M}_1(x)u_1(x)$ includes a contribution proportional to $u_0(x)$ and $u_0(x)$ belongs to the kernel of the operator $\mathcal{L}_0$. At the right hand side in Eq.(12) the part $\propto u_0(x)$ vanishes only when

$$\varepsilon_c^{(2)} = \frac{\bar{G}^2}{\varepsilon_c^{(0)} - (q_0^2 - (q_1+k)^2)^2} + \frac{\bar{G}^2}{\varepsilon_c^{(0)} - (q_0^2 - (q_1-k)^2)^2} , \quad (16)$$

which gives the threshold shift in second order. This expression simplifies to

$$\varepsilon_c^{(2)} = -\frac{\bar{G}^2}{k^2}\left(\frac{1}{(2q_0+k)^2} + \frac{1}{(2q_0-k)^2}\right) \quad (17)$$

when the wavelength is adjusted at the minimum of the unperturbed threshold, $q_1 = q_0$. For small wave numbers $k << q_0$, this expression becomes even more compact:

$$\varepsilon_c^{(2)} = -\frac{2\bar{G}^2}{4\,q_0^2\,k^2}\,. \tag{18}$$

Hence a spatially periodic modulation (13) reduces the threshold for the bifurcation, whereby a divergence occurs in Eq.(17) both for $k \to 0$ and $k \to 2q_1$. The latter resonant case is now considered separately.

Resonance $k = 2q_1$: Here, the product $\bar{M}_1(x)u_0$ on the right hand side of Eq.(11) includes a contribution proportional to u_0 and Eq.(11) can be fulfilled only when $\varepsilon_c^{(1)} = -\bar{G}$. For this resonant case, the solution $u_1(x)$ includes only the first contribution of the expression in Eq.(15), and the whole expression for the threshold ε_c is given by

$$\varepsilon = \varepsilon_c^{(0)} - \eta\bar{G} - \eta^2 \frac{\bar{G}^2}{\varepsilon_c^{(0)} - (q_0^2 - 9q_1^2)^2} + \dots\ . \tag{19}$$

This short calculation shows that the threshold is shifted linearly by contributions to $\bar{M}_1(x)$ which are resonant to the pattern $k = 2q_1$ and that the shift is quadratic otherwise.

4.1.2 Rounding of the bifurcation – $M_3(x) = 0$

For the homogeneous case, $M_1 = 0$, N in Eq.(6) increases linearly with ε beyond threshold. Here we address the question of how $N(\varepsilon)$ is modified by a spatially periodic and nonresonant modulation as given in Eq. (13). The technical steps become simpler for a long wave length modulation $k << q_0$. In this limit, the Swift–Hohenberg equation reduces further to an amplitude equation (see Appendix B of Ref.[4]) with a spatially periodic coefficient,

$$\partial_t A = \left[\varepsilon + M_1(x) + 4q_0^2\partial_x^2 - 3\mid A\mid^2\right]A\,, \tag{20}$$

and with $u(x,t) = A(x,t)\exp(iq_0x) + A^*(x,t)\exp(-iq_0x)$. Repeating the steps of the perturbation expansion for the Swift–Hohenberg equation in Sec.4.1.1, one finds again the threshold shift for a long wavelength modulation as given in Eq.(18). Eq.(20) has for $M_1 = 0$ a constant solution $A^2 = \varepsilon/3$. We assume again small modulation amplitudes $M_1(x) \propto \eta$ and we calculate the corrections to this constant nonlinear solutions:

$$A = \frac{1}{\sqrt{3}}\left[A_0 + \eta A_1 + \eta^2 A_2 \dots\right]\ . \tag{21}$$

With this ansatz we expand Eq.(20) with respect to powers of the small parameter η and we obtain:

$$\eta^0:\quad (\varepsilon + 4q_1^2\partial_x^2)A_0 - \mid A_0\mid^2 A_0 = 0\ , \tag{22}$$

$$\eta^1: \quad (\varepsilon + 4q_1^2\partial_x^2)A_1 - A_0^2\,(2A_1 + A_1^*) = -2\bar{G}\cos(kx)A_0\ , \tag{23}$$

$$\eta^2: \quad (\varepsilon + 4q_1^2\partial_x^2)A_2 - A_0^2\,(2A_2 + A_2^*) = -2\bar{G}\cos(kx)A_1 + A_0\left(A_1^2 + 2\mid A_1\mid^2\right)\ . \tag{24}$$

The solution of Eq.(23) is

$$A_1 = \frac{2\bar{G}A_0}{4q_1^2k^2 + 2\varepsilon}\ \cos(kx) = B_1\cos(kx)\ , \tag{25}$$

which vanishes in the limit of large values of ε such as $1/\sqrt{\varepsilon}$. The spatially independent contribution to A_2 is

$$\tilde{A}_2 = \frac{A_0}{\varepsilon}\,\frac{\bar{G}^2(4q_1^2k^2 - \varepsilon)}{(4q_1^2k^2 + 2\varepsilon)^2}\ . \tag{26}$$

These solutions A_i allow an expansion of N up to the leading orders, and one has

$$\begin{aligned} N &= 2\mid A\mid^2 = \frac{2}{3}\Big[A_0^2 + \eta^2(B_1^2/2 + A_0\tilde{A}_2) + \ldots\Big] \\ &= \frac{2}{3}\Big[\varepsilon + 2\eta^2\frac{\bar{G}^2\,q_1^2k^2}{(2q_1^2k^2 + \varepsilon)^2}\cdots\Big]\ . \end{aligned} \tag{27}$$

N is already finite at the threshold for the unperturbed SH-equation, $\varepsilon = 0$ and at threshold of the perturbed problem, $\varepsilon \propto -\eta^2$, ε in the denominator can be neglected and N vanishes precisely at the shifted threshold given in Eq.(18). Far beyond threshold N approaches the linear ε–dependence of the unperturbed situation. The initial slope of N,

$$\frac{\partial N}{\partial \varepsilon}(\varepsilon = \varepsilon_c) = \frac{2}{3}\Big[1 - 8\eta^2\frac{\bar{G}^2}{16q_1^4k^4}\cdots\Big]\ , \tag{28}$$

is reduced by the external perturbation. This reduction of the initial slope and the vanishing correction for N at larger values of ε indicate the rounding of $N(\varepsilon)$, as can easily be seen by plotting the expression (27) as a function of ε. Calculations for a perturbation $M_1(x)$ with a broad power spectrum lead to qualitatively similar results. This ε–dependent slope of $N(\varepsilon)$ is an analytical explanation of the numerically obtained rounding of the sharp bifurcation in Ref.[14]. Hence, not only 1 : 1 resonance leads to rounding of the bifurcation. The effects on N discussed here are proportional to A^2. Hence the corrections are $\propto \eta$ close to 1 : 2–resonance and $\propto \eta^2$ out of resonance. Whether the resonant or nonresonant corrections dominate, depends on the wave number distribution of disorder and on the value of the control parameter ε. This issue will be discussed in more detail elsewhere.

4.2 Effects of spatially varying drifts

It has been shown for a Rayleigh–Bénard experiment in Ref.[32] that a simultaneous modulation of the top and bottom container boundary leads to drifting convection rolls when the modulation wavelength is the same at the top and bottom and when both modulations are phase shifted relative to each other. This situation has been generalized to modulations of the top and bottom boundary with different wave lengths [21, 44], and a so-called *Hopf-bifurcation by frustrated drifts* has been predicted for such an inhomogeneous situation.

The simplest pattern forming model describing the effects induced by frustrated drifts, is the Swift-Hohenberg equation with a spatially varying drift coefficient $M_2(x)$:

$$\partial_t u = \left[\varepsilon - (q_0^2 + \partial_x^2)^2 + M_2(x)\partial_x\right]u - u^3 . \qquad (GSH) \tag{29}$$

A constant drift, $M_2(x)\partial_x$, with $M_2(x) = const.$, can be removed by transforming into the comoving coordinate frame $x \to x + Mt$. For a spatially periodic modulation, $M_2(x)$, with vanishing spatial mean value, $\int dx\, M_2(x) = 0$, the local drifts are *frustrated*. Above some critical amplitude of $M_2(x)$, the bifurcation from the trivial state, $u \equiv 0$, is oscillatory and one finds the so-called *Hopf bifurcation by frustrated drifts* [45, 46]. If, for instance, $M_2(x)$ is the sum of two harmonic functions, then even a fourfold degenerated Hopf-bifurcation might occur [46].

Next, we assume that $M_2(x)$ is harmonic and we consider the following two explicit forms:

$$M_2(x) = 2G_2\cos(2kx + \varphi), \tag{30}$$
$$M_2(x) = 2G_2\cos(4\hat{k}x + \varphi) + 2G_3\cos(6\hat{k}x + \varphi_2). \tag{31}$$

The first expression is periodic with respect translations $x \to x + \pi/k$, and the second one, given in Eq. (31), is periodic with respect to $x \to x + \pi/2\hat{k}$. In order to keep the results comparable later on, we choose $k = 2\hat{k}$. Stochastic drifts as well as a stochastic contribution in the control parameter ε are discussed elsewhere in more detail [14].

Because of the periodicity of $M_2(x)$, the solutions of the linear part of Eq.(29) can be divided into two classes. One class of solutions is symmetric with respect to a translation by $\lambda = \frac{\pi}{k}$, $u_s(x) = u_s(x + \lambda)$, and the other one is antisymmetric, $u_a(x) = -u_a(x + \lambda)$. $u_s(x)$ corresponds to solutions which are harmonic and $u_a(x)$ which are sub-harmonic with respect to translations $x \to x + \lambda$.

Part of the main effects can be shown again in the limit of long-wavelength modulations in Eq.(30) ($k << q_0$). In this limit equation (29) can be reduced further to an amplitude equation with a spatially varying frequency,

$$\partial_t A = [\varepsilon + iq_0 M_2(x) + 4q_0^2\partial_x^2]A - 3 \mid A \mid^2 A , \tag{32}$$

(i is the imaginary unit). For this derivation we employed the multiple scale ansatz $u = A\exp(iq_0x) + cc$, where $A(x,t)$ and $M(x)(\approx O(\varepsilon))$ are slowly varying on the length scale $2\pi/q_0$ ($k \approx O(\varepsilon^{1/2}q_0)$). Solutions to Eq.(32) have the same behavior as the starting equation (29), apart from minor differences.

Similar generalized GL-equations are expected for systems with comparable broken symmetries. An example is discussed below for Rayleigh-Bénard convection in Sec. 4.3. According to the relation between Eq.(29) and the rather general equation (32) the described properties are generic and not specific to the GSH-equation (29) with frustrated drifts. For instance, equations with random frequencies occur extensively in investigations on coupled oscillator systems [23, 22].

The first question to be investigated is the bifurcation from the trivial state $u \equiv A \equiv 0$. We do this again with the assumption of small modulation amplitudes $M_2(x) \propto \eta$ and expand the field $A(x)$, the control parameter ε and the frequency ω with respect to powers of η:

$$\begin{aligned} \varepsilon_c &= \varepsilon_c^{(0)} + \eta\varepsilon_c^{(1)} + \eta^2\varepsilon_c^{(2)} + \dots\,, && (33) \\ \omega_c &= \omega_c^{(0)} + \eta\omega_c^{(1)} + \eta^2\omega_c^{(2)} + \dots\,, && (34) \\ A(x) &= A_0(x) + \eta A_1(x) + \eta^2 A_2(x) + \dots\,. && (35) \end{aligned}$$

With the definition,

$$\mathcal{L}_0 = \varepsilon_c^{(0)} + 4q_0^2\partial_x^2, \tag{36}$$

this leads to the hierarchy of equations :

$$\begin{aligned} \eta^0 :&\quad \mathcal{L}_0 A_0 = 0\,, && (37) \\ \eta^1 :&\quad \mathcal{L}_0 A_1 = -[i\bar{M}_2(x) + \varepsilon_c^{(1)} + i\omega_c^{(1)}]A_0\,, && (38) \\ \eta^2 :&\quad \mathcal{L}_0 A_2 = -[i\bar{M}_2(x) + \varepsilon_c^{(1)} + i\omega_c^{(1)}]A_1 - [\varepsilon_c^{(2)} + i\omega_c^{(2)}]A_0\,. && (39) \end{aligned}$$

In the following we only discuss explicitly the case of a single periodic modulation (30). There are harmonic (symmetric) and subharmonic (antisymmetric) solutions with respect to the modulation $M_2(x)$. Accordingly, we start in the perturbation hierarchy either with constant (harmonic) solution, $A_0 = const$, or with a subharmonic one, $A_0 \propto \cos(kx)$.

Harmonic solutions: In this case, we start the expansion with $A_0 =$ *constant* and therefore $\varepsilon_c^{(0)} = 0$ follows from the marginal stability condition. If the solution $B_1(x)$ is periodic, then the integral of (38) over a whole period $2\pi/k$ vanishes. From this solubility condition, we obtain $\varepsilon_c^{(1)} = \omega_c^{(1)} = 0$ and for $\bar{M}_2(x) = 2\bar{G}_2\cos(2kx)$ (with $G_2 = \eta\bar{G}_2$),

$$A_1 = \frac{iq_0}{16q_0^2k^2}\,\bar{M}_2(x)A_0\,. \tag{40}$$

Also, the integral of (39) over a whole period $2\pi/k$ vanishes which leads to the threshold condition

$$\varepsilon_{h,c}(G_2,k) = \frac{G_2^2}{8k^2} + O(G_2^4), \qquad \omega_c = 0\,. \tag{41}$$

$\omega_c = 0$ means that there is a stationary bifurcation from $A \equiv 0$ for the branch of harmonic solutions. The modulation $iM_2(x)$ increases the threshold, whereas modulations of the control parameter, $\varepsilon + \xi(x)$, reduce the threshold as shown in the previous section 4.1.1.

Subharmonic solutions: To describe the subharmonic solutions, the expansion (35) has to be started with

$$A_0 = F_1 \exp[ikx] + F_{-1} \exp[-ikx]. \qquad (42)$$

The marginal stability condition, $\mathcal{L}_0 A_0 = 0$, gives with $\varepsilon_c^{(0)} = 4q_0^2k^2$ a higher threshold than for the harmonic case above. One has now for the subharmonic solutions $\mathcal{L}_0 \exp[\pm ikx] = 0$, and therefore the coefficients of $\exp[\pm ikx]$ on the rhs of Eq.(38) have to vanish. This requires

$$\varepsilon_c^{(1)} = 0, \qquad \omega_c^{(1)} = \pm \bar{G}_2\, q_0. \qquad (43)$$

The solution at next order is

$$A_1(x) = \frac{i\bar{G}_2}{32q_0k^2}\left[F_1 e^{3ikx} + F_{-1}e^{-3ikx}\right]. \qquad (44)$$

On the rhs of Eq.(39) the coefficients of $\exp[\pm ikx]$ have to vanish too, which allows the determination of $\varepsilon_c^{(2)} = \bar{G}_2^2/32k^2$. The threshold and the Hopf-frequency up to leading order are

$$\varepsilon_{s,c}(G_2,k) = 4q_0^2k^2 + \frac{G_2^2}{32k^2} + O(G_2^4)\ , \qquad (45)$$

$$\omega_{s,c}(G_2,k) = q_0G_2 + O(G_2^3)\ . \qquad (46)$$

At this subharmonic branch, the imaginary part differs from zero for every finite modulation amplitude G_2. However, the Hopf-bifurcation of the subharmonic branch is only relevant when its threshold drops below the threshold of the harmonic branch, cf. Eq. (41).

Codimension-2 bifurcation Comparing the thresholds for the harmonic branch, Eq.(41), and the subharmonic branch, Eq. (45), it is easy to see that the $\varepsilon_{h,c}$ grows faster with G_2 than $\varepsilon_{s,c}$ and crosses $\varepsilon_{s,c}$ at some critical value $G_{2_{CTP}}$. Then the subharmonic Hopf-bifurcation has the lower threshold. At the critical value $G_{2_{CTP}}$,

$$G_{2_{CTP}} = q_0k^2\sqrt{\frac{128}{3}},$$

$$\omega_{CTP} = q_0^2k^2\sqrt{\frac{128}{3}}, \qquad (47)$$

where the threshold of the harmonic and the subharmonic branch coincide, i.e., one has a codimension-2 point (**CTP**). Both instabilities compete at this point and the nonlinear interaction between both modes determines which mode survives. For increasing values of G_2, the threshold of the subharmonic Hopf-bifurcation becomes lower than that of the stationary harmonic solutions, which is a surprising result described earlier in a short note [45].

To point out that a periodic modulation $M_2(x)$ has a different consequence for a homogeneous bifurcation than for a periodic bifurcation, we discuss for completeness the generalized real Ginzburg-Landau equation:

$$\partial_t u = \left[\varepsilon + \partial_x^2 + M_2(x)\partial_x\right]u - u^3 \ . \qquad (GGL) \qquad (48)$$

With $M_2(x) = 0$, this equation has spatially homogeneous solutions. For spatially harmonic modulation, $M_2(x)$, the drifts in this equation are frustrated as well. However, in equation (48) the frustration does not lead to a Hopf bifurcation. Hence the interplay of a spatially harmonic drift, $M_2(x)$, and a periodic solution, such as in Eq.(29), is crucial for the phenomenon *Hopf bifurcation by frustrated drifts.* This phenomenon is also relevant for coupled oscillator systems with randomly varying frequencies [23, 47, 4, 48], but it will only occur when the phase approximation, usually employed for coupled oscillator systems, is dropped.

If the wave number of $M_2(x)$ is of the order of q_0, the behavior of the GSH-equation (29) and the GGL-equation (48) is changed in a similar manner as for a modulation, $M_1(x)$, of the control parameter ε - only the threshold is shifted. If the modulation wave-number k is much smaller than q_0, say $k < q_0/5$, then a periodically modulated part of the control parameter $M_1(x)$ and a periodic drift $M_2(x)$ lead to qualitatively different behavior of the GSH-equation (29) and periodic patterns in general.

We next describe the typical spatiotemporal behavior beyond a Hopf bifurcation by frustrated drifts. In Fig.12a) spatially modulated standing waves are shown, which occur immediately beyond the bifurcation for a single periodic modulation $M_2(x) = 2G_2 \cos(2kx)$ with parameters $\varepsilon = 0.05$, $G_2 = 0.05$, and $k = q_0/16$. With increasing ε, the temporal behavior of the solution at each space point becomes more and more anharmonic as shown in Fig.12b). This anharmonicity in the time dependence takes place especially near the secondary bifurcation back to the stationary solution which is stable at larger values of ε.

A doubly periodic drift term $M_2(x) = 2G_2 \cos(2kx + \pi/2) + 2G_3 \cos(3kx + \pi/4)$ leads, instead of standing waves, to aperiodic and chaotic motion as indicated in Fig. 13 for the parameters: $G_2 = 0.045$, $G_3 = 0.046$, $k = q_0/16$ and $\varepsilon = 0.068$). More details are given in Ref.[46].

Essential for the occurrence of the Hopf bifurcation by frustrated drifts is the interplay between the periodic instability and the spatially periodic drift (spatially periodic frequency), whereas the wavelength of the periodic drift is much larger than that of the instability ($k << q_0$). The similarities of Eq.(32) with model equations for coupled oscillator families [47, 4], especially for a random function $M_2(x)$, is obvious and will be discussed elsewhere.

4.3 Modulated Rayleigh Bénard Convection

The phenomenon of Hopf bifurcation by frustrated drifts as discussed for models in sec.4.2 can be found in Rayleigh-Bénard convection too. According to detailed calculations described in Ref. [44] this can occur for instance

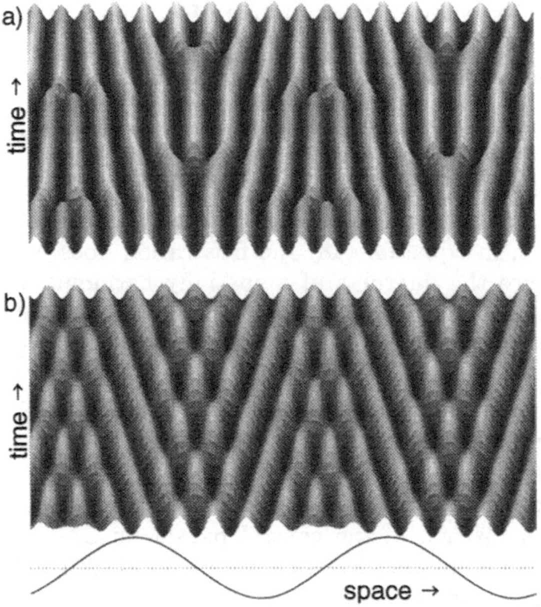

Figure 12: *The solution $u(x,t)$ of the GSH-equation (29) is shown above a Hopf bifurcation as function of space and time for a single periodic drift. In b) the solution is given just above threshold at $\varepsilon = 0.024$, whereas the threshold value for the Hopf bifurcation is at $\varepsilon = 0.0224\ldots$. In a) the solution is shown at $\varepsilon = 0.340$, just below the secondary bifurcation into the stationary solution. The modulation amplitude was fixed at $G_2 = 0.03$, and the modulation wave number at $k = q_0/16$. $M_2(x)$ is plotted at the bottom.*

for a wavy top boundary and a simultaneous temperature modulation at the bottom boundary as indicated in Fig.14. The undulated top boundary maybe located at $z = d(1 + H_0)$, with $H_0 = F_0 \cos(k_0 x)$ and the temperature maybe modulated at the bottom boundary ($z = 0$), as $T_b = \bar{T}_b + H_2(x)(\bar{T}_b - T_u)/R_c^f$, with the dimensionless modulation function $H_2(x) = F_2 \cos(k_2 x + \varphi_2)$ and $R_c^f = 27\pi^4/4$. These modulations at the top and bottom boundary then lead to an oscillatory bifurcation from a primary state. Such a Hopf–bifurcation by frustrated drifts takes place for various combinations of the modulation amplitudes F_i and the modulation wave numbers k_i [44, 49]. An example is $k_2/k_0 = 2$, $\varphi_2 = \pm\pi$ and the amplitudes F_i beyond a critical threshold [44].

For such an experimental design and for small modulation amplitudes it is also possible to derive an equation for the envelope $A_0(x,t)$ of thermal convection

$$\vec{u}(x,z,t) = \left[A_0 e^{iq_c x} + A_0^* e^{-iq_c x}\right] \vec{U}_0(z) \tag{49}$$

by using a well established perturbation scheme [50, 51, 4]. Herein the vector $\vec{u}$ contains the temperature field and the velocity potential as components and q_c is the critical wave number at onset of convection. For the expansion,

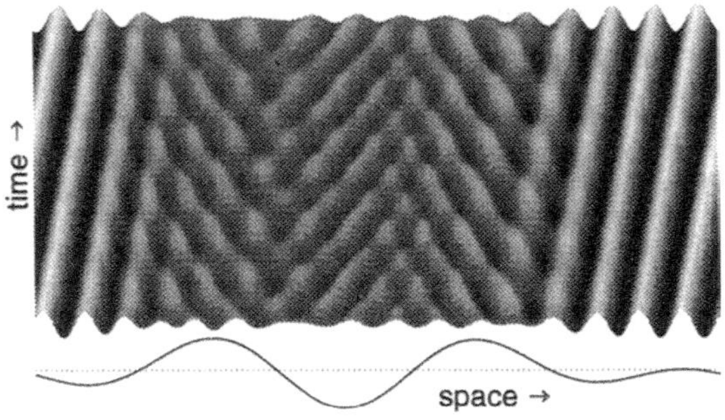

Figure 13: *The solution $u(x,t)$ of the GSH-equation (29) is shown above a Hopf bifurcation as function of space and time and for a doubly periodic drift* $M_2(x) = 2G_1 \cos[2kx + \pi/2] + 2G_2 \cos[3kx + \pi/4]$. *$M_2(x)$ is plotted at the bottom.*

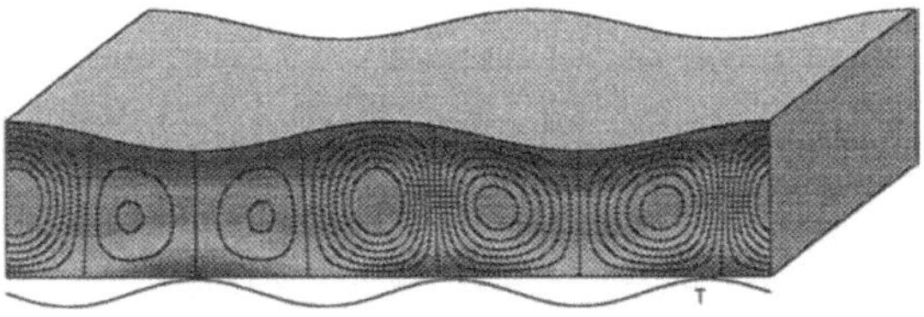

Figure 14: *An experimental design for modulated Rayleigh-Bénard Convection with a combination of an upper plate of corrugated form and a temperature modulation at the lower plate. The ratio of the wave numbers of the periodic undulation of the top plate (k_0) and the temperature modulation at the bottom plate (k_2) is* $k_0 : k_2 = 2 : 3$. *The gray scale in the fluid layer indicates the temperature field of the primary flow, and along the solid lines the velocity potential of the primary flow is constant [44]. The curve below the cell indicates the temperature variation at the bottom plate.*

one uses as a small parameter $\varepsilon = (\Delta T - \Delta T_c)/\Delta T_c$, which measures the distance from the critical temperature difference at convective onset in the unmodulated case. Assuming additionally small modulation wave numbers, $(k/q_c) \propto \varepsilon^{1/2}$ and $H_i(x) \propto \varepsilon^{1/2}$, one can derive for this modulated convection cell, as shown in Fig.14, a generalized amplitude equation of the form [52],

$$\begin{aligned} \tau_0 \partial_t A_0 &= \Big[\varepsilon - \frac{7}{3} H_0^2 + 3H_0 + \Big(1 + \frac{5}{3} H_0\Big)\frac{H_2}{R_c} \\ &+ i \frac{\sqrt{2}}{\pi}\Big(\frac{271}{576}\,\partial_x H_0 - \frac{5}{192 R_c}\,\partial_x H_2\Big)\Big] A_0 \\ &+ \Big[i\frac{2\sqrt{2}}{3\pi}\Big(H_0 - \frac{1}{R_c} H_2\Big)\partial_x + \xi_0^2 \partial_x^2 - \alpha \mid A_0 \mid^2\Big] A_0 \,, \end{aligned} \tag{50}$$

with $\xi_0^2 = \alpha = 8/3\pi^2$ and $\tau_0 = 9\pi^2/2R_c$. Here the coefficients in the equation are calculated for stress–free boundary conditions, but they have been determined for realistic boundary conditions too [46].

An analysis of the linear part of Eq.(50), as well as a full analysis of the Navier-Stokes equations [44] shows that by raising the modulation amplitudes in Rayleigh-Bénard convection, the onset of a spatially periodic convection pattern can be changed via Hopf bifurcation by frustrated drifts from a stationary to an temporarily oscillating one. There is good agreement between results obtained by starting from the full equations and the amplitude equation (50) [44, 49]. Similar as for the Swift-Hohenberg equation in (29), immediately above the Hopf bifurcation by frustrated drifts, standing waves instead of traveling waves are the preferred and stable solutions. With increasing values of the control parameter, the solutions of (50) become at first aperiodic and then spatiotemporal complex, before a regime of very slowly traveling waves is reached at large values of ε [49].

Traveling waves are known in binary fluid mixtures and in electroconvection. In both systems traveling waves bifurcate subcritically (discontinuous) and perturbation methods cannot be applied for the determination of the nonlinear solutions behavior. Since the nonlinear coefficient, α, in equation (50) is left unchanged by small modulations at boundaries, the Hopf bifurcation by frustrated drifts occuring in modulated thermal convection is supercritical as well. This justifies the application of perturbation methods for the characterization of the nonlinear solutions.

4.4 Concluding remarks

We have shown that heterogeneities reduce the symmetry and thereby induce a number of interesting phenomena in pattern forming systems, such as localization of wave pattern, cellular patterns, semi-log scaling and dynamical states via a Hopf-bifurcation by frustrated drifts.

So far examples have been studied where heterogeneities modify the bifurcation from a trivial basic state to a spatially periodic pattern. Another interesting question is how heterogeneities modify the bifurcation to traveling waves or to spatio-temporal chaotic patterns, such as Spiral–Defect–Chaos in thermal convection [53, 54]. In the latter system, the reduced symmetry may lead for instance to new frozen patterns and surprising domain–wall patterns.

The Hopf bifurcation by frustrated drifts shows that broken symmetries related to inhomogeneities also allow a controlled modification of the bifurcation type, for example, a change from a stationary bifurcation to an oscillatory bifurcation or even to spatio–temporal complex behavior.

An additional feature, which is also seen in porous media convection and in Taylor vortices flow: the presence of spatial randomness suppresses the higher order instabilities, i.e. wavy Taylor–Vortex–Flow or in the case of PMC, cross rolls or similar 3D structures. These and a number of similar questions not touched so far are addressed in future works.

5 Acknowledgment

Part of this work was done during fruitful collaborations with Lawrence Howle, Rainer Schmitz, Mark Shattuk and Mark Steen. The work of RPB and BP was supported by the US NSF under grant number CTS-9610248 and the work of WZ was supported by the DFG.

References

[1] *Propagation in Systems far from Equilibrium*, edited by J. E. Wesfreid *et al.* (Springer, New York, 1988).

[2] *Nonlinear Evolution of Spatio-Temporal Structures in Dissipative Continuous Systems*, edited by F. H. Busse and L. Kramer (Plenum Press, New York, 1990).

[3] *Pattern Formation in Complex Dissipative Systems*, edited by S. Kai (World Scientific, Singapore, 1992).

[4] M. C. Cross and P. C. Hohenberg, Rev. Mod. Phys. **65**, 851 (1993).

[5] *Spatio-Temporal Patterns in Nonequilibrium Complex Systems*, Vol. XXI of *Santa Fe Institute Studies in the Sciences of Complexity*, edited by P. Cladis and P. Palffy-Muhoray (Addision-Wesley, New York, 1995).

[6] F. H. Busse, in *Hydrodynamic Instabilities and the Transition to Turbulence*, Vol. 45 of *Topic in Applied Physics*, edited by H. L. Swinney and J. P. Gollub (Springer, New York, 1986).

[7] R. C. DiPrima and H. L. Swinney, in *Hydrodynamic Instabilities and the Transition to Turbulence*, *Topic in Applied Physics*, edited by H. L. Swinney and J. P. Gollub (Springer, New York, 1981), p. 139.

[8] W. Zimmermann, Mat. Res. Bulletin **16**, 46 (1991).

[9] L. Kramer and W. Pesch, Annu. Rev. Fluid Mech. **27**, 515 (1995).

[10] D. A. Nield and A. Bejan, *Convection in Porous Media* (Springer, Berlin, 1991).

[11] I. M. Lifshitz, S. A. Gredeskul, and L. A. Pastur, *Introduction to the Theory of Disordered Systems* (Wiley, New York, 1988).

[12] P. W. Anderson, Phys. Rev. **109**, 1492 (1958).

[13] Y. S. Kivshar, S. A. Gredeskul, A. Sanchez, and L. Vázquez, Phys. Rev. Lett. **64**, 1693 (1990).

[14] W. Zimmermann, M. Sesselberg, and F. Petruccione, Phys. Rev. E **48**, 2699 (1993).

[15] M. Lowe, J. P. Gollub, and T. Lubensky, Phys. Rev. Lett. **51**, 786 (1983).

[16] M. Lowe, B. S. Albert, and J. P. Gollub, J. Fluid Mech. **173**, 253 (1986).

[17] P. Coullet, Phys. Rev. Lett. **56**, 724 (1986).

[18] P. Coullet, L. Gil, and D. Repaux, in *Instabilities and Nonequilibrium Structures II*, edited by E. Tirapequi and D. Villaroel (Kluwer, Dordrecht, 1987), pp. 189–205.

[19] P. Coullet and D. Walgraef, Europhys. Lett. **10**, 525 (1989).

[20] W. Zimmermann *et al.*, Europhys. Lett. **24**, 217 (1993).

[21] A. Ogawa, W. Zimmermann, K. Kawasaki, and T. Kawakatsu, J. Phys. II (Paris) **6**, 305 (1996).

[22] Y. Kuramoto, Prog. Theor. Phys. **94**, 321 (1995).

[23] G. B. Ermentrout, Physica (Nonlinear Phenomena) D **41**, 219 (1990).

[24] P. Carpenter, Nature **388**, 713 (1997), L. Sirovich and S. Karlsson, Nature **388**, 753 (1997).

[25] L. Howle, R. Behringer, and J. Georgiades, Nature **362**, 230 (1993).

[26] M. Shattuck, R. Behringer, G. Johnson, and J. Georgiadis, J. Fluid Mech. **332**, 215 (1997).

[27] L. Howle, R. Behringer, and J. Georgiadis, J. Fluid Mech. **332**, 247 (1997).

[28] B. Painter and R. Behringer *Effects of spatial disorder on the transition to Taylor-Vortex Flow* (to be published).

[29] D. S. Cannell, M. A. Dominguez-Lerma, and G. Ahlers, Phys. Rev. Lett. **50**, 1365 (1983).

[30] I. Rehberg, E. Bodenschatz, B.Winkler, and F. H. Busse, Phys. Rev. Lett. **59**, 282 (1987).

[31] R. E. Kelly and D. Pal, J. Fluid Mech. **86**, 433 (1978).

[32] G. Hartung, F. H Busse, and I. Rehberg, Phys. Rev. Lett. **66**, 2741 (1991).

[33] J. Herrmann and F. H. Busse, J. Fluid Mech. **350**, 209 (1997).

[34] P. Coullet and P. Huerre, Physica (Nonlinear Phenomena) D **23**, 27 (1986).

[35] J. Tavantzis, E. L. Reiss, and B. J. Matkowsky, SIAM J. Appl. Math. **34**, 322 (1978).

[36] L. Kramer and H. Riecke, Z. Physik B **59**, 245 (1985).

[37] C. Braester and P. Vadasz, J. Fluid Mech. **254**, 345 (1993).

[38] N. J. O'Sullivan and R. McKibbin, Transport Porous Media **1**, 293 (1986).

[39] D. A. S. Rees and D. S. Riley, J. Fluid Mech. **199**, 133 (1989).

[40] D. A. S. Rees and D. S. Riley, Proc. R. Soc. London **A 421**, 303 (1989).

[41] P. Vadasz and C. Braester, Acta Mechanica **91**, 215 (1992).

[42] J. B. Swift and P. C. Hohenberg, Phys. Rev. A **15**, 315 (1977).

[43] P. Coullet and D. Repaux, Europhys. Lett. **3**, 573 (1987).

[44] R. Schmitz and W. Zimmermann, Phys. Rev. E **53**, 5993 (1996).

[45] W. Zimmermann and R. Schmitz, Phys. Rev. E **53**, R1321 (1996).

[46] W. Zimmermann, submitted to Physica D, 1998.

[47] P. C. Matthews, R. E. Mirollo, and S. H. Strogatz, Physica (Nonlinear Phenomena) D **52**, 293 (1991).

[48] N. Nakagawa and Y. Kuramoto, Physica (Nonlinear Phenomena) D **80**, 307 (1995).

[49] R. Schmitz and W. Zimmermann, Complex pattern in modulated Rayleigh-Bénard convection, submitted to J. Fluid Mech., 1998.

[50] A. C. Newell and J. A. Whitehead, J. Fluid Mech. **38**, 279 (1969).

[51] A. Newell, T. Passot, and J. Lega, Annu. Rev. Fluid Mech. **25**, 399 (1992).

[52] W. Zimmermann, Effects of Inhomogeneities in Pattern Formation, Habilitationsschrift, RWTH Aachen (unpublished), 1994.

[53] S. W. Morris, E. Bodenschatz, D. S. Cannell, and G. Ahlers, Phys. Rev. Lett. **71**, 2036 (1993).

[54] M. Assenheimer and V. Steinberg, Phys. Rev. Lett. **70**, 3888 (1993).

Electrically Driven Instabilities in Smectic Liquid Crystal Films

Harald Pleiner[1], Ralf Stannarius[2] and Walter Zimmermann[3,4]

[1] Max-Planck-Institute for Polymer Research, D-55021 Mainz, Germany
[2] Universität Leipzig, Fakultät der Physik and Geowissenschaften, D-04103 Leipzig, Germany
[3] Institut für Festkörperforschung and FORUM Modellierung, Forschungszentrum Jülich, D-52425 Jülich, Germany
[4] Max-Planck-Institute for Physics of Complex Systems D-01187 Dresden, Germany

Abstract: Various electric field driven instabilities in smectic liquid crystal films are discussed. Some of them resemble instabilities well-known from nematic liquid crystals. These are, for instance, the orientational Frederiks transition and electroconvection. However, the restriction of these two instabilities to two spatial dimensions can only be studied in free-standing smectic films, because free standing nematic liquid crystals are unstable. In addition several instabilities, being distinctive features of smectic liquid crystals are described. These are the undulational instability of the smectic layer structure and film buckling. All these different instability types are coupled in smectic C liquid crystals. For smectic C* liquid crystals films, the polarization Frederiks transition and the subharmonic regime of electroconvection are discussed. Electroconvection experiments with smectic C and C* films are reported that show the surface charge dominated 'vortex mode' and convection of the Carr-Helfrich type, respectively.

1 Introduction

Investigations of instabilities in liquid crystals, especially in nematic liquid crystals have become more and more a common subject of nonlinear science [1, 2, 3, 4, 5]. For a few nematic substances not only qualitative understanding, but also a quantitative agreement between theory and experiment of several instabilities could be achieved during the recent decade. While in nematics the hydrodynamic equations of motion are well established for a long time and the required phenomenological parameters are known to some extent, the situation for smectic liquid crystals is rather different.

Liquid crystals in general are an interesting class of systems to study various aspects of pattern formation. They offer the opportunity to investigate for instance electrically (and magnetically) driven bifurcations, as is well-known for nematic liquid crystals, also in various smectic phases under different symmetry conditions. Additionally in smectic phases new types of instabilities occur as discussed below.

In the nematic, smectic A (SmA), smectic C (SmC), and smectic C* (SmC*) liquid crystal phases there is an increasing molecular and macroscopic order (fig. 1), which gives rise to additional internal degrees of freedom. In the isotropic phase there is no preferred order of the ellipsoidal organic molecules. In the nematic phase one has an orientational order of the molecules described by a director $\boldsymbol{n}$ and its reorientation can be influenced by external fields. In the smectic A phase (SmA) there is an additional positional order in one spatial direction as indicated in fig. 1a). Layer deformations can be achieved by external fields.

In the smectic C phase the molecules are in addition tilted with respect to the smectic layers, cf. fig. 1a) and the projection of the director $\boldsymbol{n}$ onto the smectic layer plane, the $\boldsymbol{c}$ director, again shows a nematic degree of freedom in the plane of the smectic layers. Hence, the properties in the layer are isomorphic to a two-dimensional nematic, if the layers are assumed to be rigid. The azimuthal position of the tilt direction is not fixed.

In the chiral smectic C* phase there is an in-plane polarization $\boldsymbol{P}$ perpendicular to the (conic helical) director field $\boldsymbol{n}$. How the orientation of $\boldsymbol{n}$ and $\boldsymbol{P}$ change gradually from layer to layer is indicated in fig. 1b) [1]. Accordingly the polarization averages out in the bulk. When the formation of the helix is suppressed by geometrical constraints or can be neglected (i.e. in very thin films), it has similar transport properties as the SmC phase, e.g. an anisotropy of the electric conductivity.

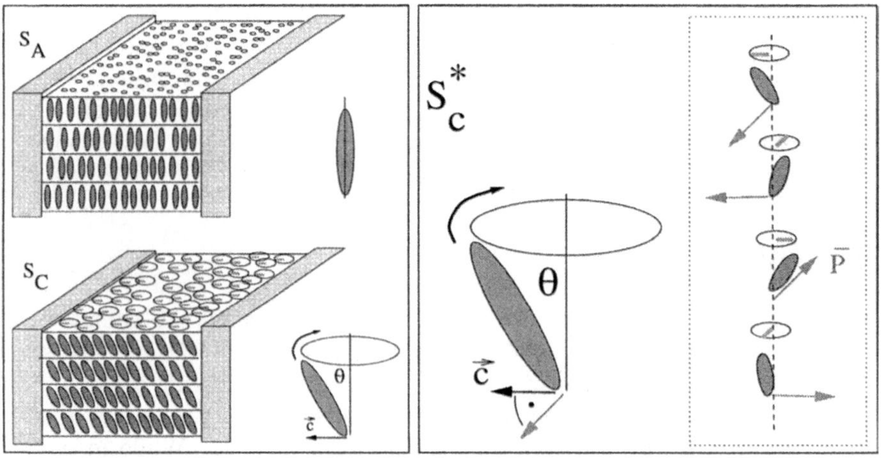

Figure 1: *Sketch of the molecular order in three different smectic liquid crystal phases. The layered structure of the smectic* A *phase* (SmA) *and smectic* C *phase* (SmC) *with the tilted order of molecules on the left and on the right the smectic* C* *phase, where the tilt direction rotates gradually from layer to layer, and where an in-plane polarization is rigidly coupled perpendicularly to the director* $\boldsymbol{n}$.

One consequence of the orientational order in nematic liquid crystals are anisotropic material properties, such as anisotropic susceptibilities, orientation-elastic moduli and viscosities. According to the anisotropic susceptibilities the director can be switched in external fields. Most famous is the switching of the orientational order in the common liquid crystal displays (see e.g. ref. [6]). It has been observed first in experiments by Frederiks and Tsvetkov already in 1934 [7]. In smectic liquid crystals the orientational order couples to the layer structure which gives rise to additional instabilities such as the Helfrich-Hurault instability (fig. 2), studied up to now in smectic A liquid crystals [1, 8, 9] or general layer deformations [10, 11].

Instead of studying smectic materials in the bulk, freely suspended films are a suitable alternative, which have been discovered about 20 years ago [12, 13, 14]. Like soap films, such smectic films with only a few nanometers or micrometers in thickness can have lateral extensions of several centimeters. The smectic layers are perfectly stacked in the film plane. Figure 1a) visualizes the geometry of freely suspended smectic A and C films. Such stable free standing films allow the design of experiments which are essentially two–dimensional. In contrast, thin nematic or cholesteric films are usually unstable.

Putting free standing smectic films in external fields the preferred orientation is switched, and according to the coupling of the mean orientation of molecules to the orientation of the smectic layers, the smectic layers are also deformed and they ultimately couple also to the shape of the film. Therefore field induced buckling may arise in SmC as discussed in sec. 2.5. In thin films the polarization in the SmC* gives rise to an additional coupling and indeed a novel transition, the so-called polarization Frederiks transition, has been predicted recently [15]. This is described in more detail in sec. 2.6. These different field induced deformations in smectic liquid crystals are usually not independent from each other, but are coupled and give rise to hybrid instabilities as discussed in sec. 2.4.

Electroconvection (EC) has an equally long history in nematic liquid crystals. 30 years after the discovery of electric field effects in thin cells of nematic liquid crystals [7] spatially periodic patterns have been observed by Williams [16] and, at higher applied voltages, the so-called dynamic scattering mode by Kapustin and Larionova [17]. Theoretical understanding of the basic mechanism of electroconvection has been achieved by Carr [18] and Helfrich [19]. The Helfrich theory was later extended by Dubois-Violette [20] to finite frequencies of the AC voltage and to two spatial dimensions in ref. [21]. A number of three-dimensional phenomena have then been observed in experiments as reviewed in refs. [2, 22]. The so-called oblique rolls observed as first instability by Joets and Ribotta [23, 24] triggered an extension of previous theoretical calculations to three dimensions [25, 26], at first linear and then more and more into the nonlinear regime of electroconvection in nematic liquid crystals. The further development in this field during the recent decade has been reviewed in several recent articles [27, 28, 29, 30, 31] and in this volume [32].

Research on electroconvection in smectic liquid crystals is less advanced. The formation of a striped texture induced by electroconvection in smectic

A phase ([33, 34, 35] and references therein) and in the smectic C phase [36, 37, 38] has been observed. In contrast to nematics the possibility of convection depends crucially on the orientation of the layers with respect to the external field. Electric fields perpendicular to the smectic layers hardly induce electroconvection because the layered structure strongly impedes flows perpendicular to the layers. On the other hand electric fields parallel to the smectic layers may induce convective motion essentially restricted to the smectic layers, which enables studies on convection in ideally two-dimensional systems. This has been shown recently for thin SmA films [39, 40], where the convective motion in the thin film can be easily visualized by dust particles, cf. sec. 4.2.

Thin films of smectic C liquid crystals are isomorphic to two-dimensional nematics (if the layers are fixed) and the convective motion can be analyzed more precisely than in three dimensional nematic layers. Thin films of smectic C* resemble two-dimensional nematics with an additional polarization perpendicular to the director. For both systems there is a preferred direction in the plane of the film and one may expect that anisotropic transport properties, typical for liquid crystals systems, play a crucial role for electroconvection. Experimental and theoretical results on both systems are described in secs. 3 and 4.3, 4.4.

2 Static Instabilities

In liquid crystals external electric and magnetic fields can induce various static instabilities. In phases with nematic degrees of freedom (e.g. uniaxial nematics) there is an orientational transition of the director, the so-called Frederiks transition [1, 7]. For layered systems (e.g. smectic A) the Helfrich–Hurault effect is possible, where the smectic layers are undulated relatively to each other. In (smectic) films the film can buckle as a whole (buckling instability). In smectic C films all these possibilities are present simultaneously and the instabilities never occur independently, but are always coupled. Nevertheless we discuss them at first independently, and then look step by step how they are coupled.

In this section we discuss *static* instabilities, which would mean, strictly speaking, that no flow (or any other dissipative process) is induced and the stability is governed by a free energy functional. Such purely static instabilities are possible for very clean systems and geometries only, where charges do not have to be considered. Thus, the instabilities discussed below are *static* as an approximation only for dirty systems, insofar the flow and its effect on other degrees of freedom are small and flow is not a relevant part of the instability mechanism itself. The opposite case, where flow is a necessary part of the mechanism, will be discussed in sec. 3.

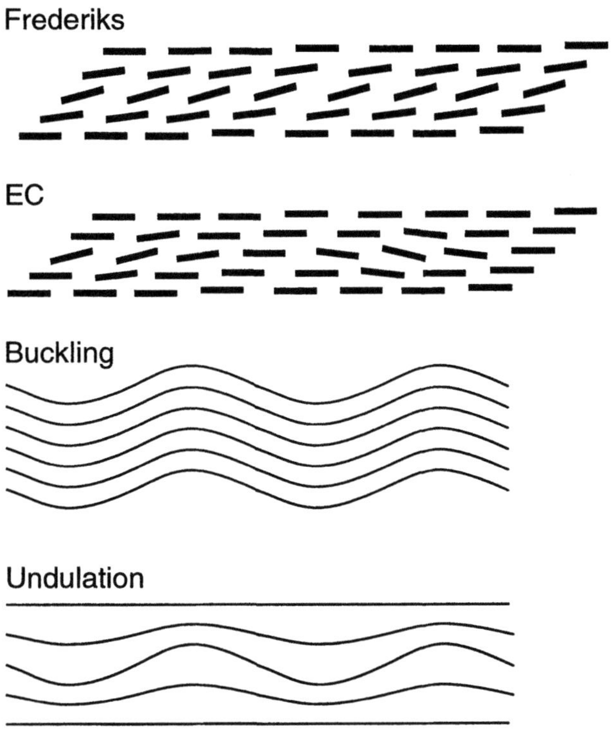

Figure 2: *Sketch of the four instabilities discussed in this article: The orientational transition (Frederiks transition), the buckling instability of smectic layers, the undulation instability (Helfrich–Hurault) in section 2 and the electrohydrodynamic instability (*EC*) in section 3.*

2.1 Theoretical Formulation

In liquid crystals a few continuous symmetries are spontaneously broken and the related hydrodynamic fields are introduced in the following. In the nematic phase the rotational symmetry is spontaneously broken due to the ordering of the molecules. The preferred mean direction is described by the director field $\boldsymbol{n}(\boldsymbol{r}, t)$, with $\boldsymbol{n}^2 = 1$ and the constraint that $\boldsymbol{n}$ is equivalent to $-\boldsymbol{n}$ in all equations.

The smectic layering is due to a small periodic modulation of the density along one direction (the layer normal). This denotes a spontaneously broken translational symmetry in one direction. Within the layers a smectic A phase is isotropic and liquid. The layering is described by a scalar 'phase' variable $\phi(\mathbf{r}, t)$, which can be thought of numbering all layers (it is integer in the middle of each layer). The layer normal is then defined by [41]

$$\boldsymbol{k} = \frac{\boldsymbol{\nabla}\phi}{|\boldsymbol{\nabla}\phi|} \qquad \leftrightarrow \qquad \boldsymbol{k} \cdot \delta\boldsymbol{k} = 0\,. \tag{1}$$

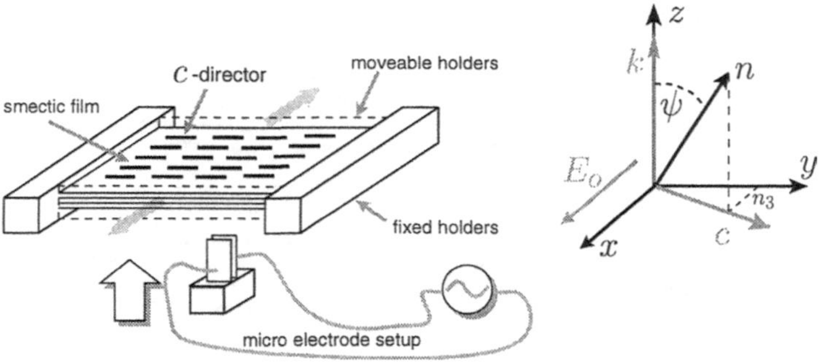

Figure 3: *Sketch of an experimental arrangement of a* SmC *film. The notation is as introduced in the text.*

Generally the area $\phi(\boldsymbol{r},t) = const.$ can have any shape, although in the thermodynamically stable case these are flat layers. Without restriction of generality one can choose

$$\phi_{\,\mathbf{flat}}(\boldsymbol{r}) \;=\; z \;, \tag{2}$$

while deformations of this flat ground state are described by a displacement vector $\boldsymbol{u}(\boldsymbol{r},t)$, which is the appropriate variable according to the broken translational symmetry,

$$\phi \;=\; z - \boldsymbol{u}\cdot\boldsymbol{k} \;. \tag{3}$$

However, only displacements perpendicular to the layers are relevant, since the layers themselves are fluid. Thus

$$\boldsymbol{k}\times\boldsymbol{u} = 0 \quad \leftrightarrow \quad \delta\boldsymbol{k}\times\boldsymbol{u} + \boldsymbol{k}\times\delta\boldsymbol{u} = 0 \quad . \tag{4}$$

If the deformations of the flat layers are small (and we will assume that case in the following), all expressions can be expanded into gradients of the relevant variable $u = \boldsymbol{u}\cdot\hat{\boldsymbol{e}}_z$ and we have

$$\boldsymbol{\nabla}\phi = \hat{\boldsymbol{e}}_z - \boldsymbol{\nabla}u \;, \tag{5}$$

$$|\boldsymbol{\nabla}\phi| - 1 = \sqrt{1 - 2\nabla_z u + (\boldsymbol{\nabla}u)^2} - 1 \approx -\nabla_z u + \frac{1}{2}(\boldsymbol{\nabla}_\perp u)^2 \;, \tag{6}$$

where $\boldsymbol{\nabla}_\perp = (\nabla_x, \nabla_y, 0)$ denotes gradients perpendicular to the (undistorted) layer normal.

In the smectic C phase (SmC) the director is tilted with respect to the layer normal

$$\cos\theta = \boldsymbol{k}\cdot\boldsymbol{n} = \text{const.} \quad \leftrightarrow \quad \boldsymbol{k}\cdot\delta\boldsymbol{n} - \boldsymbol{n}\cdot\delta\boldsymbol{k} = 0. \tag{7}$$

The (normalized) projection of $\boldsymbol{n}$ onto the layers is called the $\boldsymbol{c}$ director. Thereby rotational symmetry within the layers is spontaneously broken and SmC is biaxial. Thus SmC combines both, a smectic degree of freedom (u) and one that is nematic-like [42],

$$n_3 = \frac{\boldsymbol{n}\times\boldsymbol{k}}{|\boldsymbol{n}\times\boldsymbol{k}|}\cdot\delta\boldsymbol{n} \;, \tag{8}$$

and which describes rotations of the director about the layer normal (thus conserving the tilt angle θ). In fig. 3 the geometry is illustrated. The external electric field is applied perpendicular to both, the undistorted layer normal and the director.

Static distortions of the SmC structure subject to an external field are given by the free energy density [1, 43]

$$\begin{aligned} 2\tilde{\varepsilon} &= -\epsilon_1 \, (\boldsymbol{E}\cdot\boldsymbol{n})^2 - \epsilon_2 \, (\boldsymbol{E}\cdot\boldsymbol{k})^2 - \epsilon_3 \, \boldsymbol{E}^2 - 2\,\epsilon_4 \, (\boldsymbol{E}\cdot\boldsymbol{n})(\boldsymbol{E}\cdot\boldsymbol{k}) \\ &\quad + B\left(\nabla_z u - \frac{1}{2}(\boldsymbol{\nabla}_\perp u)^2\right)^2 + \mathcal{T}_{ijkl}\,(\nabla_i\nabla_j u)(\nabla_l\nabla_k u) \\ &\quad + \mathcal{F}_{ij}\,(\nabla_i n_3)(\nabla_j n_3) + 2\,\mathcal{C}_{ijk}(\nabla_i n_3)(\nabla_j\nabla_k u)\,, \end{aligned} \tag{9}$$

where the first line describes the dielectric coupling of the field to both, the layer normal and the director. The second line gives the elastic distortion energy due to layer distortions, while the third one contains the orientational-elastic energy due to director distortions and the static couplings between director and layer distortions, respectively. The materials tensors $\mathcal{F}_{ij}$, $\mathcal{T}_{ijkl}$, $\mathcal{C}_{ijk}$ contain 4 (F_{11}, F_{22}, F_{33}, F_{13}), 6 (C_1,...,C_6), and 9 (T_1,...,T_9) coefficients, respectively, according to the monoclinic symmetry of SmC [42].

The dynamics of such distortions is described by balance equations for the director as well as for the layers. They are of the scalar form

$$\frac{d}{dt}u + X \;=\; \frac{\partial}{\partial t}u + \boldsymbol{v}\cdot\boldsymbol{\nabla}u + X \;=\; 0\,, \tag{10}$$

$$\frac{d}{dt}n_3 + Y \;=\; \frac{\partial}{\partial t}n_3 + \boldsymbol{v}\cdot\boldsymbol{\nabla}n_3 + Y \;=\; 0\,, \tag{11}$$

with $\boldsymbol{v}$ the velocity field. The quasi-currents X and Y contain the dynamic couplings to the additional hydrodynamic degrees of freedom, like mass density, momentum density and energy density [42, 44, 45]. For static distortions the explicit form of X and Y is not needed and we will discuss them in sec. 3 only.

2.2 Frederiks Transition

The most famous transition in liquid crystals is the Frederiks transition [1, 7], which is an orientational transition of the director driven by an external field via the anisotropies in the dielectric (and diamagnetic) susceptibilities. This orientational transition is crucial for the functioning of the modern liquid crystal displays [6].

In smectic C liquid crystals director rotations are described by n_3. Neglecting all (static and dynamic) cross-couplings to other variables the free energy (9) has an extremum if

$$\left(F_{11}\,\frac{\partial^2}{\partial y^2} + F_{22}\,\frac{\partial^2}{\partial x^2} + F_{33}\,\frac{\partial^2}{\partial z^2} + 2\,F_{13}\,\frac{\partial^2}{\partial z\,\partial y} + \epsilon_1 E_o(t)^2\right) n_3 = 0\,. \tag{12}$$

The F_{ij}'s are generalized Frank constants [43] and ϵ_1 is one of the susceptibility anisotropies. The symmetry of eq. (12), $x \to -x$ and $(y, z) \to (-y, -z)$, reflects the reduced symmetry of the SmC phase. Stability of the ground state ($n_3 = 0$) is probed with respect to periodic distortions

$$n_3 = \left[n_{3,c} \cos(k_y \pi y) \cos\left(\frac{z\pi}{d_z}\right) + n_{3,s} \sin(k_y \pi y) \sin\left(\frac{z\pi}{d_z}\right) \right] \cos(k_x \pi x) \ . \quad (13)$$

This simple ansatz does not fix n_3 at the top and bottom of the film ($z = \pm d_z/2$); a z-dependence is assumed in order to describe the 3-dimensionality of the system (layer undulations are taken into account more realistically below); the y-direction is assumed to be infinite, so k_y is continuous and has to be determined by minimizing the threshold; since the electrodes are assumed to be rigid boundaries for the director, the lowest possible value for k_x is $k_x = 1$ in dimensionless units. The threshold condition follows from inserting the ansatz (13) into eq. (12) and reads [46]

$$E^2_{fred\ 1,2} = \left(F_{11} k_y^2 + F_{22} k_x^2 + F_{33} \frac{1}{d_z^2} \pm 2 F_{13} k_y \frac{1}{d_z} \right) \frac{\pi^2}{\epsilon_1} \ . \quad (14)$$

The critical wavelength is given by

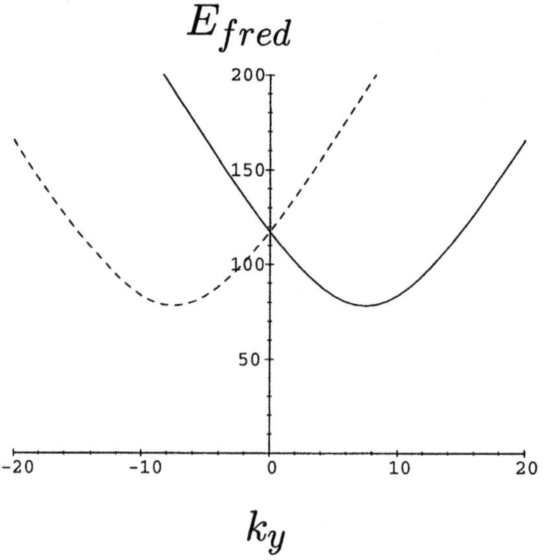

Figure 4: *Frederiks threshold in thin films with* ($F_{ii} = 1$, $F_{13} = 0.75$, $k_x = 1$, $d_z = 0.1$)*, cf. eq. (14)* [46].

$$| k_{yc} | = \frac{| F_{13} |}{F_{11}} \frac{1}{d_z} \quad (15)$$

In fig. 4 the threshold field strength is plotted against the wavelength k_y. The minimum at non-zero wavelength indicates that the variation of the director field depends on all three coordinates in contrast to nematics, where F_{13} is zero and the director field varies only along the x-direction. Although the simplified analytical treatment of this section cannot lead to quantitatively correct results, the qualitative result of 3-dimensional patterns survives a more complete treatment taking into account layer undulations, additionally.

2.3 Undulational Instability

We will now come to the opposite special case, where the director is kept fixed but the layers can move. In this subsection we assume that the smectic film stays flat on average, i.e. the layers undulate such that either the upper and lower layer or the middle one is fixed. An overall buckling of the film is considered in sec. 2.5.

An external field tends to reorient the layer normal according to the dielectric anisotropy. Since the layers cannot rotate homogeneously, they undulate, thus creating regions, where the layer normal is rotated. This happens if the electric field is strong enough to overcome the elastic and bending-elastic energy involved. Neglecting dynamic effects the state is determined by minimizing the free energy (9), while keeping only contributions involving the layer distortion u

$$\bar{\epsilon} E_o^2(t) \frac{\partial^2 u}{\partial x^2} - B \frac{\partial^2 u}{\partial z^2} + 6\, T_4 \frac{\partial^4 u}{\partial x^2 \partial z^2} + T_3 \frac{\partial^4 u}{\partial x^4} + T_1 \frac{\partial^4 u}{\partial z^4} + T_2 \frac{\partial^4 u}{\partial y^4} + 2\, T_5 \frac{\partial^4 u}{\partial y^2\, \partial x^2}$$
$$+12\, T_7 \frac{\partial^4 u}{\partial z\, \partial y\, \partial x^2} + 4\, T_9 \frac{\partial^4 u}{\partial z\, \partial y^3} + 2\, T_6 \frac{\partial^4 u}{\partial z^2\, \partial y^2} + 4\, T_8 \frac{\partial^4 u}{\partial z^3\, \partial y} = 0. \qquad (16)$$

with $\bar{\epsilon} = \epsilon_2 + 2\, \epsilon_4 \cos\psi + \epsilon_1 \cos^2 \psi$. For a simplified analytic discussion we choose an ansatz for u

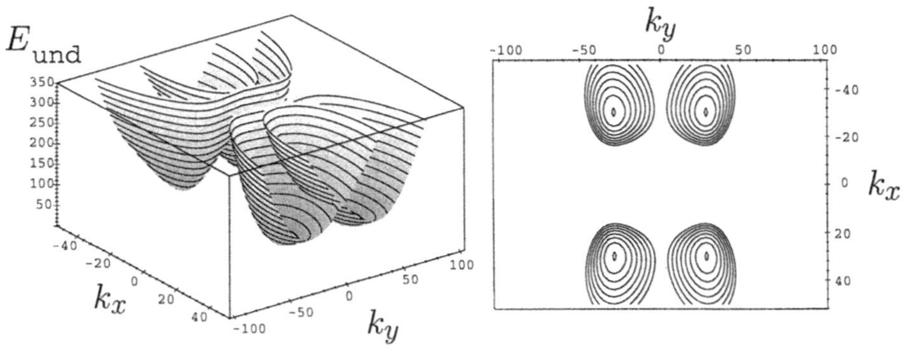

Figure 5: *Undulational threshold (18) for thin films $d_z = 0.1$ with $T_i = 1$. The undulation is periodic also in y-direction* [46].

$$u = \left[u_s \sin(k_y \pi y) \sin\!\left(z \frac{\pi}{d_z} \right) + u_c \cos(k_y \pi y) \cos\!\left(z \frac{\pi}{d_z} \right) \right] \cos(k_x \pi x), \qquad (17)$$

that allows 3-dimensional patterns, is compatible with the smectic C symmetry, and denotes no-slip conditions for the layers at the electrodes, but cannot accommodate fixed boundary conditions for the first and last layers ($z = \pm d_z/2$). For more realistic boundary conditions numerical methods are necessary. This immediately leads to the threshold conditions

$$\begin{aligned}\bar{\epsilon}\, d_z^2\, k_x^2 \pi^{-2} E^2_{und\,1,2} &= B\pi^{-2} + T_1\, d_z^{-2} + T_2\, k_y^4\, d_z^2 + 6\, T_4\, k_x^2 + T_3\, k_x^4\, d_z^2 \\ &+ 2\, T_6\, k_y^2 + 2\, T_5\, k_y^2\, k_x^2\, d_z^2 \pm 4 k_y d_z (3\, T_7\, k_x^2 + T_8 d_z^{-2} + T_9\, k_y^2)\end{aligned} \tag{18}$$

for the two eigenvalues $\{u_s, u_c\} = \{1, \pm 1\}$. The true threshold is the minimum of E^2_{und} with respect to k_x and k_y (strictly speaking, $k_x = n + 1/2$ with $n \geq 1$ is discrete). Analytically this leads to rather involved equations so we present the graphical solution in fig. 5, where one can clearly see that the lowest threshold value is obtained at a finite critical wave vector k_y indicating truly 3-dimensional patterns in contrast to the Helfrich-Hurault instability in smectics A.

For thicker films, however, the critical k_y decreases with the film thickness and finally an almost 2-dimensional pattern is found (fig. 6). In that case the

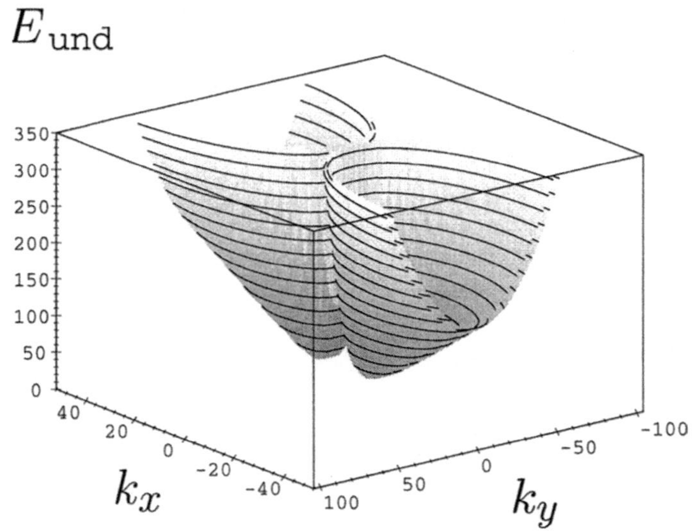

Figure 6: *Undulational threshold (18) for thick films $d_z = 1$. The wave vector k_y tends to zero, while k_x remains finite, but small. ($T_i = 1$)* [46].

threshold condition (18) can be simplified into

$$\bar{\epsilon}\, d_z^2\, k_x^2 \pi^{-2} E^2_{und\,1,2} = B\pi^{-2} + T_1\, d_z^{-2} + 6\, T_4\, k_x^2 + T_3\, k_x^4\, d_z^2 \tag{19}$$

Minimization of this expression with respect to k_x leads to the critical value

$$k_{xc}^2 = \frac{1}{\pi d_z^2} \sqrt{\frac{T_1 \pi^2 + B d_z^2}{T_3}}\,. \tag{20}$$

The film thickness dependence of the threshold field (fig. 7) is slightly different from the $E^2 \propto 1/d_z$ behavior in smectics A.

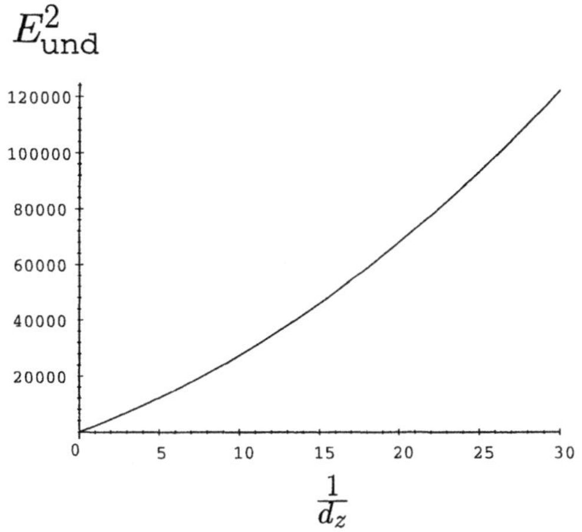

Figure 7: *Dependence of the undulational threshold on the (reciprocal) film thickness for thick films.* $(T_i = 1;\ B = 10^5,\ \epsilon_1 = -0.072,\ \epsilon_2 = 1,\ \epsilon_4 = 0)$ [46].

2.4 Interaction between Instabilities

In the last two subsections we have discussed director and layer instabilities separately by deliberately neglecting coupling terms. Here we will give an example how the coupling terms can affect the instabilities. We could combine the two types of ansatz (13) and (17) and look for combined layer and director distortions. However, this is still too complicated for the analytical treatment, since it leads to 4x4 matrix problem. Thus, we choose the even more simplified mode ansatz

$$\begin{aligned} n_3 &= n_o \cos(k_x \pi x) \cos(k_y \pi y) \cos(z \frac{\pi}{d_z}) \,, \\ u &= u_o \sin(k_x \pi x) \cos(k_y \pi y) \cos(z \frac{\pi}{d_z}) \,, \end{aligned} \tag{21}$$

which, however, is only sufficient, if we assume in addition

$$C_1 = C_4 = T_7 = T_8 = T_9 = F_{13} = 0 \,. \tag{22}$$

The conditions (22) mean that we have eliminated all contributions responsible for the 3-dimensionality of the resulting instability pattern. Indeed $k_y = 0$ is found for the critical wave vector in y-direction. Of course, a more sophisticated treatment would regain 3-dimensional structures.

The threshold condition is obtained by minimizing the free energy using eq. (21) and can be written in the form $\det(L_{ij}) = 0$, where the elements of the 2x2 matrix are,

$$
\begin{aligned}
L_{11} &= \pi^2 \left[k_y^2 F_{11} + k_x^2 F_{22} + d_z^{-2} F_{33}\right] - \epsilon_1 E_o(t)^2 \qquad (23)\\
L_{12} = L_{21} &= \pi k_x [\epsilon_1 \cos\psi \, E_o(t)^2 \; + \; \epsilon_4 \, E_o(t)^2 \\
&\quad + \pi^2 k_y^2 C_2 + \pi^2 d_z^{-2} C_3 + 2\,\pi^2 d_z^{-2} C_5 + k_x^2 \pi^2 C_6] \\
L_{22} &= \pi^4 \left[k_y^4 T_2 + 6\, k_x^2 d_z^{-2}\, T_4 + 2\, k_y^2 d_z^{-2}\, T_6 + k_x^4\, T_3 + d_z^{-4}\, T_1 \right.\\
&\quad \left. + 2\, k_x^2 k_y^2\, T_5\right] + \pi^2 \left[d_z^{-2} B + \bar{\epsilon}\, k_x^2 E_o(t)^2\right]
\end{aligned}
$$

The coupling of the layer and director instabilities is provided by the (remaining) coupling coefficients C_i as well as by the dielectric anisotropy $\epsilon_{1,4}$. We will not write down the analytical threshold condition, but give the graphical representation of the neutral curve for different film thicknesses.

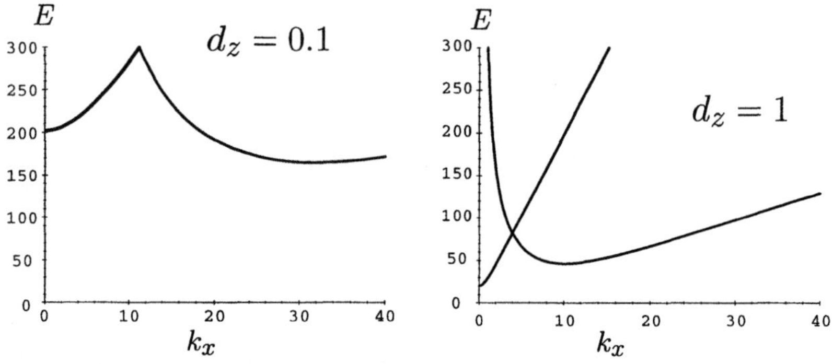

Figure 8: *Neutral curves of the combined layer and director instabilities for different film thicknesses ($C_i = 0.9$, $F_{ii} = 3$, $T_i = 1$)* [46].

In both cases two distinct instabilities are visible, the Frederiks-like one at zero wave vector and the undulation-like one at finite k_x. However, the unstable mode always contains both, layer and director distortions. Nevertheless the qualitative nature (the basic mechanism) of the instabilities is still preserved. The reason for this "weak" coupling is the large difference of critical wave vectors of the underlying decoupled instabilities. Note that the minimum of the threshold switches from one instability to the other by changing the film thickness.

In order to get a more realistic treatment one has to take into account, firstly, all variables, e.g. the electrical potential ϕ^{el}, defined by $\boldsymbol{E} = \boldsymbol{E}_0 - \boldsymbol{\nabla}\phi^{el}$. Secondly, the full dynamics has to be considered and finally more spatial modes (which obey realistic boundary conditions) have to be included in order to get a better characterization of the patterns. In the case of external AC fields (with frequency ω) also temporal modes are needed. Then only a numerical treatment is possible. The basic procedure is described in the EC section below.

For the 3-dimensional smectic C case and for a special set of parameters, the following neutral stability curves are obtained (fig. 9). For large Frank-like

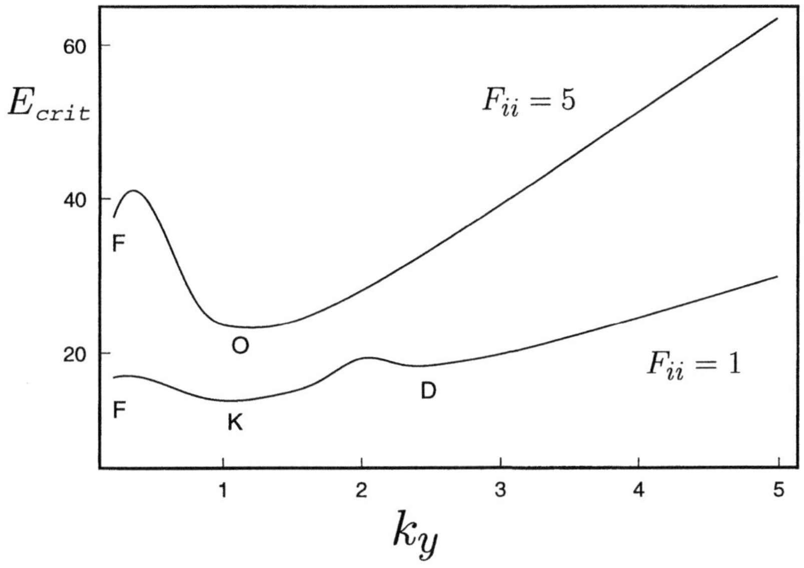

Figure 9: *Numerical neutral curves for different values of the Frank-like constants (F_{ii} = 1,5) and positive dielectric anisotropy ($\epsilon_1 = +0.0724$, $d_z = 1$, $\omega = 100$, free boundary conditions)* [46].

constants there are two minima (instabilities), which again can be characterized as Frederiks- (F) and undulation-like (O). The corresponding eigenvectors are visualized in figs. 10 and 11, respectively. There, the contributions of the variables (including the velocity components vx= v_x, vz= v_z and the electric potential Phi= ϕ^{el}, which have been neglected before) to the eigenvector are shown in a grey scale (white = zero), either in the x/y plane or the temporal evolution along a line $y = const.$ (t/x plane). Across the film suitable mean values have been taken, i.e. $u(x,y) = (\int \mid u(x,y,z) \mid^2 dz)^{1/2}$.

For small Frank-like constants ($F_{ii} = 1$) the undulation-type instability is gone and two new ones (K and D) occur. They are related to electroconvection.

2.5 Buckling Instability

In the previous subsections we have assumed that the film as a whole stays flat when subjected to an external field and only the layers inside the film are undulated and compressed or dilated. Now we will consider the opposite case that the film as a whole buckles and all layers inside follow homogeneously (cf. fig. 2). Of course, one could think of a mixed case, where buckling and undulations occur simultaneously, but it is hard to see, how the system could gain from that.

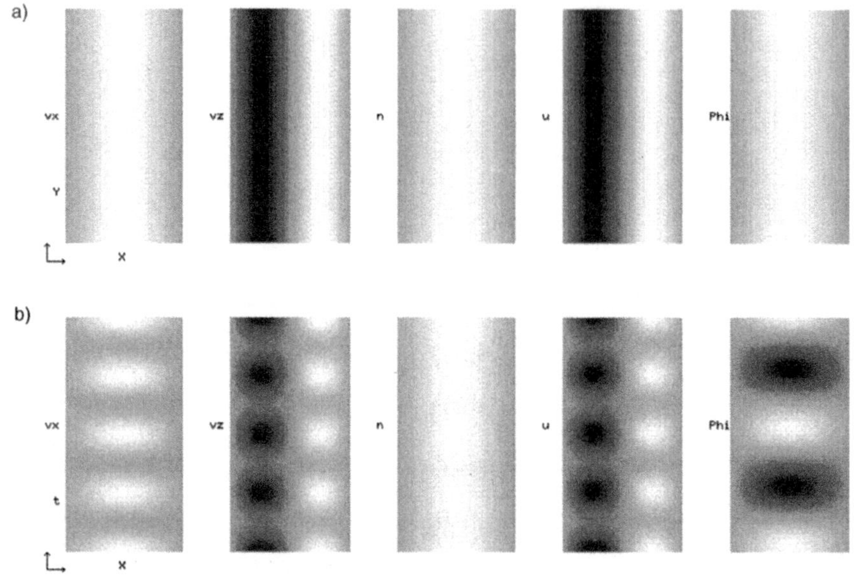

Figure 10: *Frederiks-like instability with layer undulations (and flow), $k_y = 0$, corresponding to minimum* F *in fig. 9 ($F_{ii} = 5$)* [46].

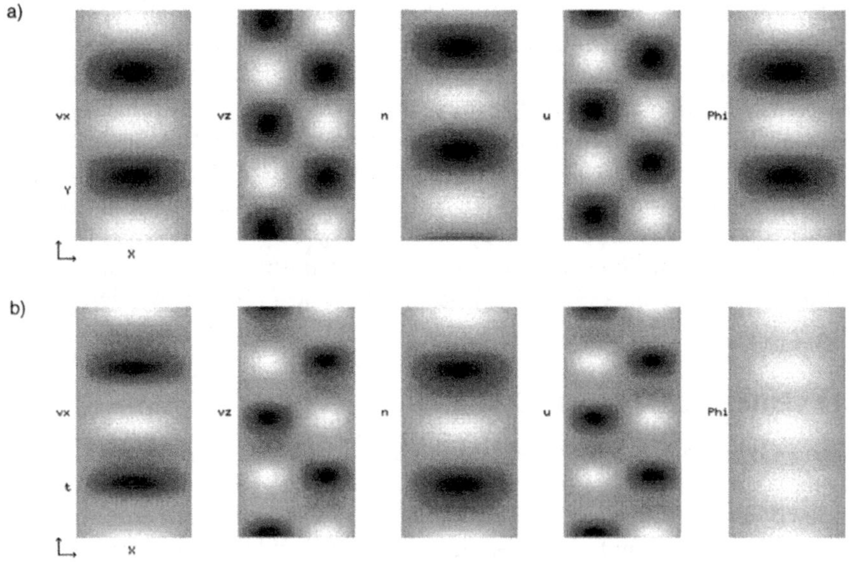

Figure 11: *Undulation-like instability with director reorientation (and flow), $k_y \neq 0$, corresponding to minimum* O *in fig. 9 ($F_{ii} = 5$)* [46].

For the homogeneous buckling there is no z-dependence and the system can be treated two-dimensionally. Because of that the cost in elastic energy seem to be smaller for buckling than for undulations and one could infer that the threshold field for layer buckling should always be lower than that for undulations. However, there is another important point to be considered with buckling. If the film is and stays flat, the electrostatics is trivial. The electric field is homogeneous (like in a plate capacitor) inside and outside the film (we always assume infinitely large electrodes). However, when the film buckles this is no longer true and the electric potential will be distorted by the curved boundary between the (dielectric) film and the air above and below. Thus there is additional dielectric energy involved due to the buckling of the film, which can drive the buckling threshold beyond that for undulations.

The proper treatment for the buckling instability would be to solve the hydrodynamic equations together with the electric boundary problem. Since the boundaries are moving according to the solution of the whole problem, this is a very intricate problem. In order to get some insight analytically, we will simplify it by assuming that the influence of the curved boundaries can be treated electrostatically and in a coarse grained fashion. First we notice that the electrical potential ϕ^{ind}, induced by the film buckling, is proportional to $\nabla_x u$. Since ϕ^{ind} has to vanish at the electrodes and since it is the solution of Laplace's equation outside the film (and a somewhat more complicated equation inside, because of the dielectric anisotropy) the appropriate ansatz for film buckling is

$$u = u_o \cos^2(k_x \pi x)\, \cos(k_y \pi y)\,. \tag{24}$$

Neglecting all cross-couplings to other variables and taking into account only the elastic and dielectric anisotropy energy (9), but neglecting for the moment the buckling induced electric energy, the neutral curve for buckling would be [46]

$$4\,\bar{\epsilon}\,k_x^2 \pi^{-2}\, E_{buck}^2(noinduced) = 3\,k_y^4\,T_2 + 16\,k_x^4\,T_3 + 8\,k_x^2 k_y^2\,T_5 \tag{25}$$

It is easy to notice that the critical wave vector k_y is zero, while k_x cannot be smaller than 1. This describes a 1-dimensional pattern above the threshold (the neglected cross-couplings would recover a finite k_y, i.e. a 2-dimensional pattern)

$$E_{buck}^2(k_x = 1; k_y = 0; noinduced) \;=\; 4\,\frac{T_3\,\pi^2}{\epsilon_z - \epsilon_x}\,, \tag{26}$$

with the abbreviations

$$\epsilon_x = \epsilon_3\,, \qquad \epsilon_z = \bar{\epsilon} + \epsilon_3 = \epsilon_2 + \epsilon_3 + 2\,\epsilon_4 \cos\psi + \epsilon_1 \cos^2\psi\,. \tag{27}$$

However, we still have missed the extra dielectric energy due to electric potential induced by the curved boundaries. Taking a buckled film described by a fixed displacement (24) with $k_y = 0$ the electrostatic boundary problem can be solved by standard means. In the limit of thin films and averaging over space the electric energy density within the film at a given buckling amplitude

u_0 is found to be

$$\tilde{\varepsilon}'_{buck-el} = -\frac{1}{2}\,\epsilon_x E_o^2 + \frac{\pi^2}{4}\,E_o^2\,u_o^2\,(\epsilon_x - \epsilon'_z)\;, \tag{28}$$

with

$$\epsilon'_z = \epsilon_z + \frac{1}{\pi d_z \sqrt{\epsilon_x \epsilon_z}}\Big[\frac{1}{2}\,\sinh\left(2\pi\sqrt{\frac{\epsilon_x}{\epsilon_z}}\,d_z\right)(\epsilon_z - 1)^2 + 2\sinh\left(\pi\sqrt{\frac{\epsilon_x}{\epsilon_z}}\,d_z\right)(\epsilon_z - 1)(1 - \frac{\epsilon_z}{\epsilon_x})\epsilon_x\Big]\,. \tag{29}$$

This electric energy density (28) has exactly the same form as the dielec-

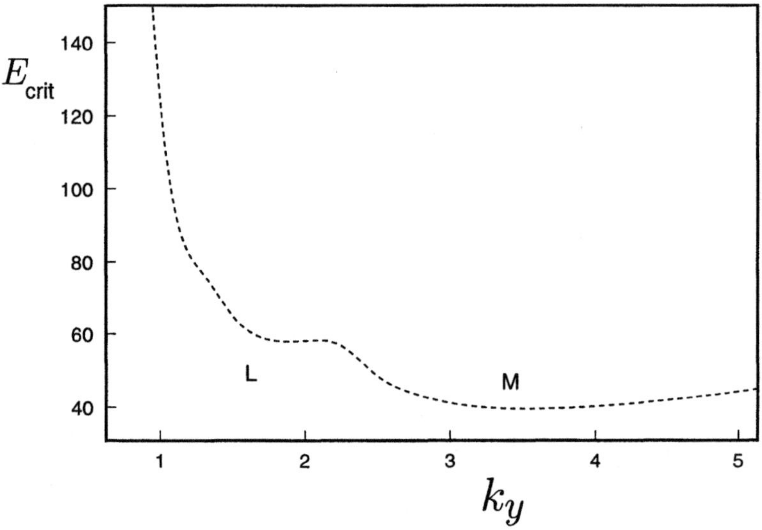

Figure 12: *Neutral instability curve involving film buckling, director reorientation and EC (flow) for negative dielectric anisotropy. The patterns are 2-dimensional and minimum* M *corresponds to an EC-like instability, while minimum* L *is qualitatively new* [46].

tric anisotropy energy density (i.e. the electric energy density without the buckling-induced part) except that there ϵ_z occurs instead of ϵ'_z. Vice versa this means that in this approximation the buckling-induced electric energy can be accounted for by replacing ϵ_z by ϵ'_z in the free energy expression (9). Then the buckling threshold is obtained from (26) as

$$E^2_{buck}(k_x = 1; k_y = 0) = 4\,\frac{T_3\,\pi^2}{\epsilon'_z - \epsilon_x}\;. \tag{30}$$

In the numerical stability analysis one can take into account all cross-couplings, the dynamics, more spatial (and temporal) modes for a given parameter set, but still treats the buckling-induced electric energy as described above. It is found that buckling couples "strongly" to the other types of instability in the sense that not only all basic instability types occur in mixed form, but that new instabilities arise that are not present in any of the limiting cases. E.g. the minimum L in fig. 12 (which can be the absolute minimum for a different parameter set) is neither present in the pure buckling, nor in the Frederiks, nor in the pure EC case and is, thus, qualitatively different from those. A more detailed discussion of the numerical stability analysis of smectic layers under external electric field will be given elsewhere.

2.6 Polarization Frederiks Transition

In the previous sections we have discussed exclusively achiral smectic C films. We now switch to the chiral smectic C* phase, where not only is the director conic helical but also there is an in-plane polarization always parallel to the director. Due to the helical nature of this polarization the electrostatic boundary problem at the electrodes is non-trivial even in the ground state. We therefore assume the helical pitch much larger than the film thickness, which allows us to neglect the helix. The problem is then homogeneous across the film. Since we want to concentrate on the influence of the polarization on the Frederiks transition, we also assume the layers to be fixed. The problem is then strictly 2-dimensional. The advantage is that within this approximation not only the linear stability problem is feasible, but also the complete nonlinear stability problem can be obtained.

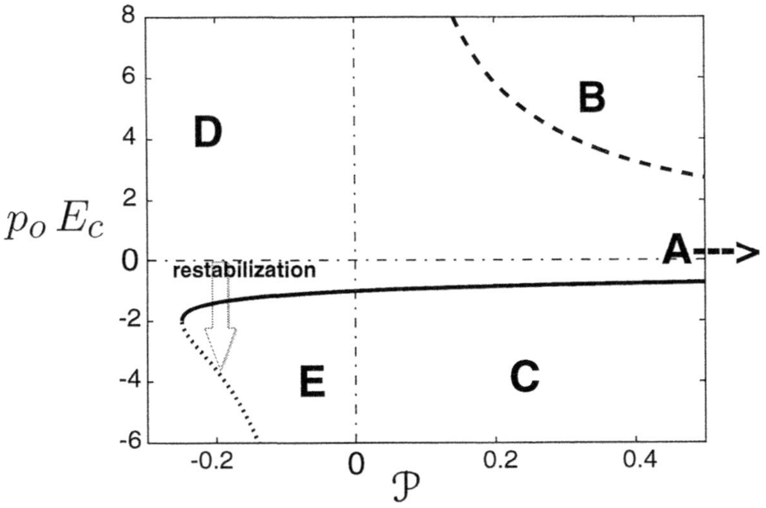

Figure 13: *The threshold field E_c of the generalized Frederiks transition as function of the material parameter combination $\mathcal{P} = \epsilon_1/p_o^2$. The region* E *allows for a possible restabilization* [15].

In the achiral case the sign of the dielectric anisotropy ϵ_1 governs the stability. In the geometry chosen ($\boldsymbol{n} \perp \boldsymbol{E}$) positive (negative) ϵ_1 characterizes a dielectric unstable (stable) situation. In the chiral case the polarization stabilizes (destabilizes) the structure according to the ferroelectric coupling, if it is parallel (antiparallel) to the external field. Since the in-plane polarization is rigidly connected to the director both (in)stability mechanisms work together, either constructively or destructively. Most interesting is the case, where the situation is dielectrically stable but ferroelectrically unstable. Since the ferroelectric coupling is linear in the external field, it will win for small fields, while the dielectric stabilization gains at higher field strengths. Thus one can expect – with increasing field – first an instability (the combined structure of director and polarization rotates) to occur, while then either a restabilization to the ground state occurs, or the structure rotates further so that the director and the polarization have turned by π (except at the boundaries), which is a genuinely stable situation. The detailed theory [15, 47] shows on what parameters the different scenarios depend.

First we have to add the ferroelectric energy to eq. 9,

$$\tilde{\varepsilon}_p = \tilde{\varepsilon} - P_i E_i \ . \tag{31}$$

Since we can neglect flow, we are left with two variables, the induced electric potential ϕ^{el}, defined by $\boldsymbol{E} = \boldsymbol{E_0} - \boldsymbol{\nabla}\phi^{el}$ and the rotation of the in-plane director $\boldsymbol{c}$. With the ansatz

$$\boldsymbol{c} = (\sin\theta,\ \cos\theta,\ 0)\,, \qquad\qquad \boldsymbol{P} = p_o\,(\cos\theta,\ -\sin\theta,\ 0)\,, \tag{32}$$

one gets two partial differential equations for θ and ϕ^{el} at the threshold postulating the electric current as well as the torque on the director (and on the polarization) to vanish [48]

$$\begin{aligned}&\left(F_{11}\sin^2\theta + F_{22}\cos^2\theta\right)\,\nabla_x^2\theta - p_0\,\sin\theta\,(E_o - \nabla_x\phi^{el})\\ &+\left(\,(F_{11}-F_{22})(\nabla_x\theta)^2 + \epsilon_1(E_o - \nabla_x\phi^{el})^2\,\right)\sin\theta\,\cos\theta = 0\,,\end{aligned} \tag{33}$$

$$\begin{aligned}&\sigma_{xx}\left(2\cos\theta\,\sin\theta\,(\nabla_x\theta)(E_o - \nabla_x\phi^{el}) - \cos^2\theta\,(\nabla_x^2\phi^{el})\right)\\ &+\sigma_{yy}\left(2\cos\theta\,\sin\theta\,(\nabla_x\theta)(E_o - \nabla_x\phi^{el}) - \sin^2\theta\,(\nabla_x^2\phi^{el})\right) = 0\,.\end{aligned} \tag{34}$$

where σ_{xx} and σ_{yy} are the in-plane electric conductivities. The external field is chosen to be static, in order to prevent the ferroelectric effect to be averaged out. Planar boundary conditions are chosen such that the ground state is given by $\theta = 0 = \phi^{el}$.

In order to get the linear stability threshold it is enough to consider θ with the ansatz $\theta(x) = A_o\,\cos\pi x$. Equation (33) immediately leads to [15]

$$p_o E_{c1,2} = \frac{1}{2\mathcal{P}}\left(1 \pm \sqrt{1 + 4\mathcal{P}F_{22}}\,\right)\,, \qquad \text{with} \qquad \mathcal{P} \equiv \frac{\epsilon_1}{p_o^2}\,, \tag{35}$$

which is visualized in fig. 13. Case A corresponds to the well-known achiral case ($\mathcal{P} \to \infty$, or $\epsilon_1 \to \infty$, $p_0 \to 0$), with the two thresholds $E_c = \pm\sqrt{F_{22}/\epsilon_1}$.

In region B (C) the ferroelectric torque stabilizes (destabilizes) and the Frederiks transition threshold is increased (decreased) compared to the achiral case. In region D both effects are stabilizing and no instability results. Region E (ferroelectrically unstable, dielectrically stable) is the most interesting one showing a possible restabilization. An instability is possible for $p_0 > \sqrt{-4\epsilon_1 F_{22}}$ only. For that region we have performed a weakly nonlinear analysis resulting in an amplitude equation of the form

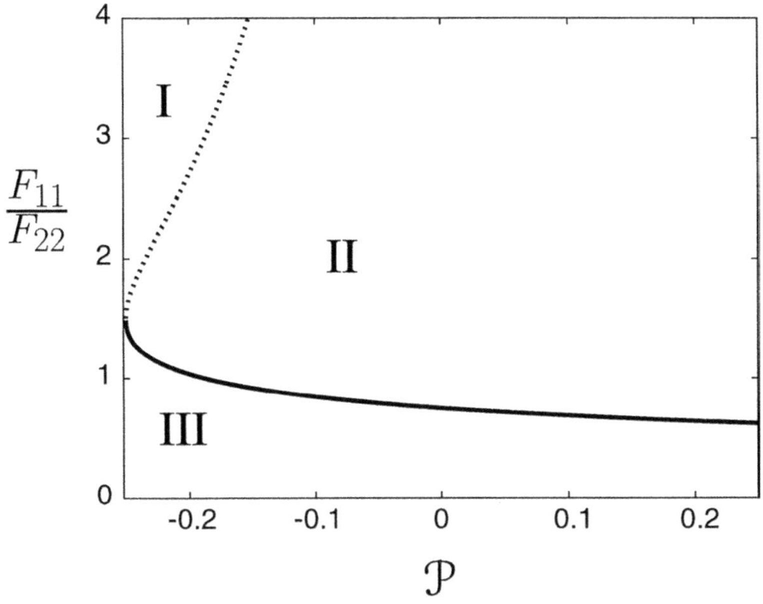

Figure 14: *The zeros of $g_{1,2}$ denoting the boundaries, where the instabilities change their nature (see text)* [15].

$$\tau_o \partial_t A = \left(\eta_{1,2} - g_{1,2} A^2\right) A\,, \tag{36}$$

with

$$g_{1,2} = \frac{3}{8}\,\frac{1 + \frac{8}{3} F_{11} \mathcal{P} \pm \sqrt{1 + 4\mathcal{P} F_{22}}}{1 + 4\mathcal{P} F_{22} \pm \sqrt{1 + 4\mathcal{P} F_{22}}}\,, \qquad \tau_o = \frac{2\mathcal{P}}{1 + 4\mathcal{P} F_{22} \pm \sqrt{1 + 4\mathcal{P} F_{22}}}\,. \tag{37}$$

where $\eta_{12} = (E_o - E_{c1,2})/E_{c1,2}$ are the reduced relative threshold fields for the Frederiks instability and the restabilization, respectively. The upper (lower) sign refers to the stabilization (restabilization) curve. The nature of the bifurcation changes (continuous / discontinuous, forward / backward), where the cubic coefficient $g_{1,2}$ changes sign. The zeros of $g_{1,2}$ are shown in fig. 14 as function of the relevant parameters. In region I both instabilities are continuous, in region II the restabilization has become discontinuous, while in region III both instabilities are discontinuous.

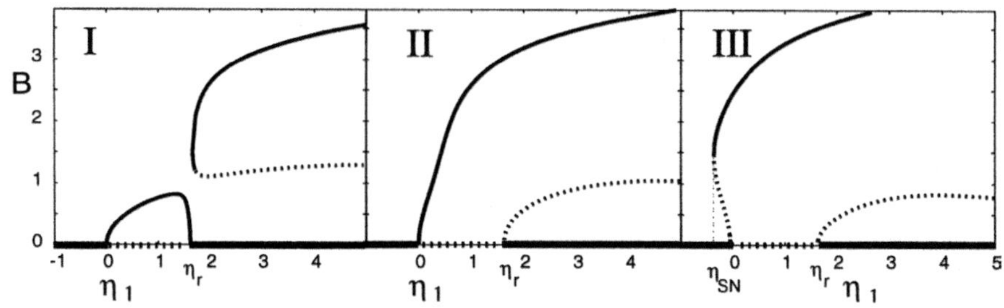

Figure 15: *The amplitude B as function of the reduced external field $\eta_1 = (E_0 - E_{c1})/E_{c1}$ for different values of F_{11}/F_{22} = 2.9, 1.5, 0.5. Dotted lines show unstable states* [15]. *The ground state is stable again for $\eta_1 > \eta_r = (E_{c2} - E_{c1})/E_{c1}$.*

The weakly nonlinear analysis is valid only near the threshold, where the amplitudes are still small. For large amplitudes, i.e. in order to investigate the restabilization behavior, we have performed a numerical integration of eqs. 33 and 34. With fixed boundary condition $\phi^{el}(x = \pm 1/2) = \theta\,(x = \pm 1/2) = 0$ a standard shooting procedure is used. The states are characterized by an amplitude B

$$B = \frac{1}{2}\int_{-1/2}^{1/2} \theta(x)\,dx\;, \tag{38}$$

which reduces to the amplitude A_0 introduced in the linear case. In the ground state and the restabilized state $B = 0$. For special values of F_{11}/F_{22} the three regimes I, II, and III are regained (fig. 15). As one can see from this figure, the large amplitude behavior (restabilization or not) is very closely related to the small amplitude behavior (forward or backward bifurcation). In region I, where both bifurcations (at $\eta_1 = 0$ and $\eta_1 = \eta_r$) are continuous, the restabilized ground state above η_r is actually reached. In regimes II and III,

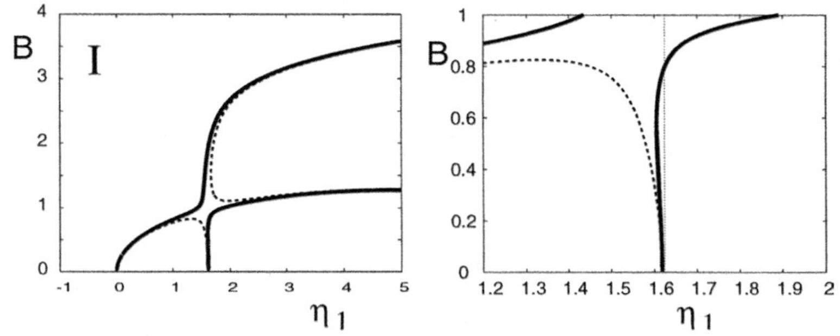

Figure 16: *Example of regime* I *(restabilization bifurcation is continuous), where the restabilized ground state is not reached (F_{11}/F_{22} = 2.75). The dotted lines show the generic case (F_{11}/F_{22} = 2.8) for that regime.* [46]

where at least one bifurcation is discontinuous, the restabilized ground state is not reached and instead a state with $\theta = \pi$ (except near the electrodes) is reached. In region III the Frederiks transition is first order like. Only in a very small parameter range the small amplitude and the large amplitude behavior are asynchronous. There the restabilization instability is still continuous, but the stabilized ground state is not reached (fig. 16), showing that the amplitude equation is qualitatively correct for most parameter values, but not really everywhere.

Since the fully nonlinear solutions are obtained only numerically, there would be some advantage for having an analytical model that reproduces the numerical results. The amplitude eqs. (36) are really 2 different equations (for each instability a separate one) so they cannot describe the full bifurcation scenarios. A suitable model is the following phenomenological amplitude equation

$$\partial_t A = \eta\,(1-\eta)\,A - g(\eta)\,A^3 - A^5\,, \tag{39}$$

which can be derived from a potential P via $\partial_t A = -\frac{\delta P}{\delta A}$.

$$P = \frac{1}{2}\left[\frac{1}{3}A^6 + \frac{1}{2}g(\eta)A^4 - \eta\,(1-\eta)A^2\right] \tag{40}$$

With the choice $g(\eta) = S\,(C-\eta)$ the three regimes I, II, and III are reobtained by $C > 1$, $0 < C < 1$, and $C < 0$, respectively. The fine-tuning can be done by choosing S. An example for the potential P in regime I is shown in fig. 17.

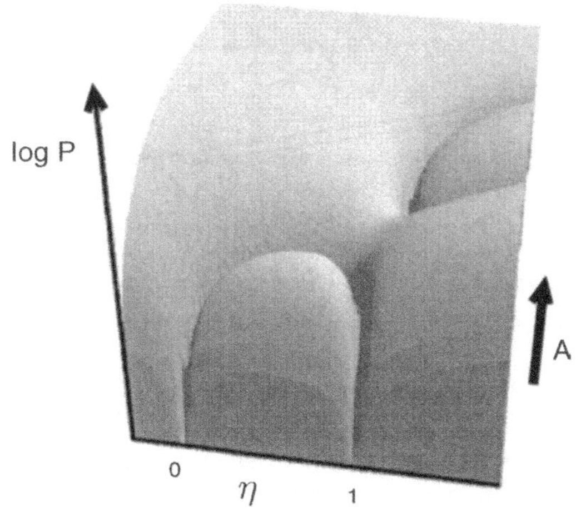

Figure 17: *Model potential P suitable for regime* I

3 Electroconvection Theory in Smectic C and Smectic C* Films

Most of the experimental and theoretical work on electroconvection (EC) in liquid crystals is devoted to nematic liquid crystals. This part has been reviewed in several works [2, 29, 30, 31, 32, 49]. In comparison rather little is known about electroconvection in smectic phases. Electroconvection observed in experiments on smectics so far is mainly driven by mechanisms which are rather independent of the Carr-Helfrich mechanism and the liquid crystalline symmetries. Some of those mechanisms usually work also in the isotropic phase of the material as described in more detail in sec. 4.1.

It is an interesting question which distinctive features of electroconvection might be related to the symmetries of smectic liquid crystals. Application of electric fields to a smectic C sandwich cell with the layers stacked perpendicular to the cell plane (as shown in fig. 18) can in principle induce convection structures driven by a mechanism similar to the Carr-Helfrich mechanism in the nematic phase. Figure 18 sketches a side view of such a cell with the smectic layers in bookshelf geometry. The $\boldsymbol{c}$ director, denoted by the arrows in the fig. 18, plays the same role here as the director $\boldsymbol{n}$ in the nematic phase. The $\boldsymbol{c}$ director is fixed at the glass plates but can reorient in the smectic layer plane.

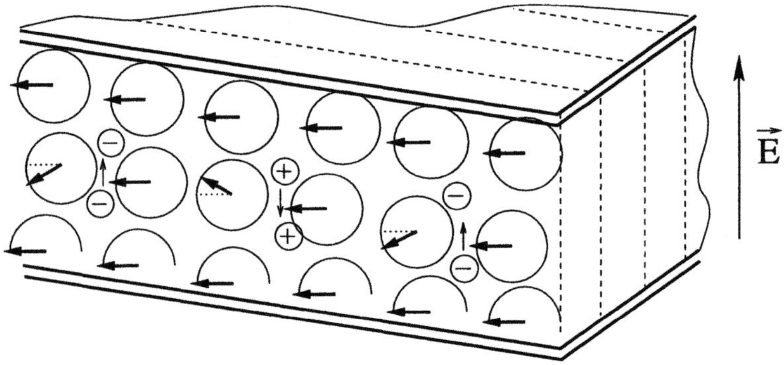

Figure 18: *Geometry of smectic* C *sandwich cell with electroconvection domains, layers in bookshelf geometry. The arrows denote the* **c** *director, the projection of the tilted director* **n** *onto the layer plane.*

Here we describe mechanisms for electroconvection which are specific for liquid crystals, in particular smectic C and smectic C* films. We do not discuss surface charge driven instabilities, because they are not specific to liquid crystals. Such mechanisms can be avoided using large parallel capacitor like electrodes. We restrict the theoretical formulation to fixed smectic layers (i.e. rigid film geometry $\mathcal{T}_{ijkl} \to \infty$, $u = 0, v_z = 0$). It is also assumed that neighboring smectic layers behave equally, hence all gradients ∂_z vanish.

Under these conditions electroconvection in free standing smectic C films is isomorphic to a two dimensional nematic liquid crystal. Investigations on this system are rather promising to gain further insight into the mechanism of electroconvection, because the motion can be visualized in more detail in a film.

As discussed in sec. 2.6 the SmC* is biaxial for film thicknesses small compared to the pitch. The number of coefficients contained in the material tensors such as the tensors for electric conductivity, for the dielectric tensor and elasticity is much larger than for nematic liquid crystals. However, this is neglected in the following consideration and a SmC* film is described being isomorphic to a two-dimensional nematic but with an additional spontaneous electric polarization **P**, which is coupled rigidly to the in-plane director . The same geometry as in sec. 2.6 and in fig. 3 is considered.

3.1 Linear Equations

The stability of the convection–free, planar basic state $\mathbf{c} = (0,1)$, $\mathbf{v} = 0$ and $\phi^{el} = 0$ is investigated in this section. To test the stability of this state, the equations of motion are linearized with respect to small deviations from it. Due to the 2D description there are no gradients in z–direction and all vectors have vanishing z–components (i.e. $v_z = 0$). The component v_y is expressed by v_x via the incompressibility condition.

After elementary algebraic operations, described in more detail in refs. [46, 47] one ends up with three coupled two–dimensional equations

$$\begin{aligned} 0 = &\left((\epsilon_1 \sin^2\psi + \epsilon_3)k_y^2 - \epsilon_3 \nabla_x^2\right) \partial_t \phi^{el} + (-\epsilon_1 \sin\psi\, E_o(t) + p_0)\, k_y\, \partial_t n_3 \\ &+ \left(\sigma_{yy} k_y^2 - \sigma_{xx} \nabla_x^2\right) \phi^{el} - \Big(\epsilon_1 \sin\psi\, (\,\partial_t E_o(t)\,) + \frac{\sigma_{yy} - \sigma_{xx}}{\sin\psi}\, E_o(t)\Big) k_y\, n_3\,, \end{aligned} \tag{41}$$

$$\begin{aligned} 0 = &-k_y \partial_t n_3 + (p_0 - \epsilon_1 \sin\psi\, E_o(t))\, k_y^2\, \phi^{el} - \left(\alpha_2 k_y^2 + \alpha_3 \nabla_x^2\right)\, v_x \\ &+ \left(\epsilon_1 \sin\psi\, E_o^2(t) - p_0\, E_o(t) - F_{11}\, k_y^2 + F_{22} \nabla_x^2\right)\, k_y n_3\,, \end{aligned} \tag{42}$$

$$\begin{aligned} 0 \;=\; &\rho_m (k_y^2 - \nabla_x^2)\, \partial_t v_x + \left(\eta_c\, k_y^4 - 2\,\hat{\eta}\, k_y^2\, \nabla_x^2 + \eta_b\, \nabla_x^4\right)\, v_x \\ &+ k_y \left(\alpha_2 k_y^2 + \alpha_3 \nabla_x^2\right) \partial_t n_3 + \left(\epsilon_1 \sin\psi\, E_o^2(t) - p_0 E_o(t)\right)\, k_y^3\, n_3 \\ &+ \left(\epsilon_3 \nabla_x^2 - (\epsilon_3 + \epsilon_1 \sin^2\psi)\, k_y^2\,\right)\, k_y^2\, E_o(t)\, \phi^{el}\,. \end{aligned} \tag{43}$$

Herein the following abbreviations for the viscosities have been used:

$$\begin{aligned} 2\,\alpha_2 &= -1 - \lambda^{(3)}\,, & 2\,\alpha_3 &= 1 - \lambda^{(3)}\,, \\ \eta_b &= \nu_3 + \frac{1}{4}\,(1 - \lambda^{(3)})^2\,, & \eta_c &= \nu_3 + \frac{1}{4}\,(1 + \lambda^{(3)})^2\,, \\ \hat{\eta} &= \frac{1}{\gamma_1}(\nu_1 + \nu_2 - \nu_3) + \frac{1}{4}(1 - {\lambda^{(3)}}^2), \end{aligned} \tag{44}$$

where the ν's are ordinary viscosities [42], while $\lambda^{(3)}$ is a reactive transport parameter [50]. These equations can be rescaled and reformulated in dimensionless form as discussed in ref. [47]. They are similar to the linear equations for planarly aligned nematic liquid crystals [26, 47, 51], but contain in addition the effects of the polarization p_0. Accordingly, the eqs. (41-43) have a different symmetry, which is described in more detail in sec. 3.3. They can be written in a more compact matrix form,

$$\mathsf{B}(t, \nabla_x) \cdot \partial_t \boldsymbol{U}(x,t) = \mathsf{L}(t, \nabla_x) \cdot \boldsymbol{U}(x,t)\,, \tag{45}$$

with the (formal) vector field,

$$\boldsymbol{U} = (\phi^{el}, \theta, v_x). \tag{46}$$

Either DC or periodic AC driving voltages are assumed in the following. For periodic voltages $V(t) = V(t+T)$, with $T = 2\pi/\omega$, the matrices B and L are periodic in time and eq. (45) shares some similarities with the well known Mathieu equations. In the following we assume additionally

$$V(t + T/2) = -V(t). \tag{47}$$

3.2 Boundary Conditions

A specification of the boundary conditions is needed when the linear equations (41-43) are analyzed. Usually films are considered which are infinitely extended in y-direction (or take periodic boundary conditions). In x-direction the film is confined between electrodes located at $x = \pm\frac{1}{2}$, with the consequence that the velocity v_x, perpendicular to the electrodes, has to vanish at this surface

$$v_x(\pm\frac{1}{2}) = 0\,. \tag{48}$$

In addition either stress-free or rigid boundary conditions

$$\partial_x^2 v_x(\pm\frac{1}{2}) = 0\,, \qquad \partial_x v_x(\pm\frac{1}{2}) = 0\,, \tag{49}$$

are assumed, where the latter case is closer to real experimental conditions.

The induced potential has to be zero at the surface and for the director $\boldsymbol{c}$ we consider planar boundary conditions ($\boldsymbol{c} = (0,1)$)

$$\theta(\pm\frac{1}{2}) = 0\,, \qquad \phi^{el}(\pm\frac{1}{2}) = 0\,. \tag{50}$$

3.3 Symmetries and Floquet Analysis

The solutions of the linearized eqs. (45) can be classified with respect to their symmetry properties under spatial reflections and translations in time. Choosing the line $x = 0$ to be the middle of the film, i.e. $-1/2 < x < 1/2$, where 1 is the rescaled film width in fig. 3, eq. (45) is invariant under reflection with respect to $x \to -x$.

This symmetry of the equations, together with the boundary conditions chosen, allows to characterize all solutions to be either symmetric, $\boldsymbol{U}(x) = \boldsymbol{U}(-x)$, or antisymmetric, $\boldsymbol{U}(x) = -\boldsymbol{U}(-x)$, with respect to x. For periodic voltages that fulfill eq. (47), the linear homogeneous eq. (45) is invariant under time translations

$$t \to t + nT \qquad (n \text{ integer, positive}) . \tag{51}$$

According to the spectral method of Floquet [52] this leads to a general solution of the form,

$$\boldsymbol{U}(z,t) = \hat{\boldsymbol{U}}(z,t)e^{\sigma t}, \tag{52}$$

with the Floquet exponent σ and the periodic function $\hat{\boldsymbol{U}}$. Due to the symmetry (51) the solutions $\hat{\boldsymbol{U}}$ are grouped according to the integers n into harmonic ones ($n = 1$) and subharmonic ones ($n = 2$). The harmonic ones are invariant under a single T-translation $t \to t+T$, i.e. $\hat{\boldsymbol{U}}(t) = \hat{\boldsymbol{U}}(t+T)$, while the subharmonic ones change sign $\hat{\boldsymbol{U}}(t) = -\hat{\boldsymbol{U}}(t+T)$ under that translation. The full solutions $\boldsymbol{U}(t)$ do not have these symmetries because of the Floquet multiplier $\exp(\sigma t)$. However, the prefactor produced by a time translation, i.e. $\exp(\sigma T)$ for a T-translation, is irrelevant, since the amplitude of a solution $\boldsymbol{U}(t)$ is not determined by the linear, homogeneous equation (45).

The Floquet exponent σ is a function of the external (V_0, ω) and internal (material) parameters of the system and depends on the transverse wave vector k_y. It governs the linear stability of the basic, non-convective state against inhomogeneous perturbations, which is stable, if $Re(\sigma) < 0$, and unstable for positive growth rates $Re(\sigma) > 0$. Thus, from the condition,

$$Re\left[\sigma(V_0, k_y, ...)\right] = 0\,, \tag{53}$$

those parameter values are determined, which separate the linear stable regime from the unstable one. Equation (53) allows for instance the determination of V_0 as a function of k_y, the so called neutral curve, $V_0(k_y)$, at which the real part of the Floquet exponent changes its sign. The absolute minimum of $V_0(k_y)$ at a critical k_{yc}, $V_c = V_0(k_{yc})$, gives the threshold for the onset of convection in linearized stability analysis.

For vanishing p_0 (SmC phase) and if additionally eq. (47) holds, e.g. the driving voltage is sinusoidal, or of square or triangular wave form, there is an additional symmetry with respect to time translations, since eq. (45) is invariant under the replacement,

$$t \to t + \frac{T}{2}, \quad \text{if} \quad \begin{pmatrix} \phi^{el} \\ \theta \\ v_x \end{pmatrix} \to \pm\gamma \begin{pmatrix} -\phi^{el} \\ \theta \\ v_x \end{pmatrix}, \tag{54}$$

where γ is an irrelevant constant. This symmetry requires the solutions $\hat{\boldsymbol{U}}$ to be harmonic, since by applying (54) twice (i.e. after a T-translation) $\hat{\boldsymbol{U}}$ is mapped to itself. Thus, if eq. (54) is valid, subharmonic solutions are ruled out. The upper and lower sign in (54) belong to two different solutions, corresponding to the so-called conductive and dielectric regime, respectively, known from EC in nematic liquid crystals [1]. In SmC* however, this symmetry (54) is lifted, since $p_0 \neq 0$. In that case subharmonic solutions are possible

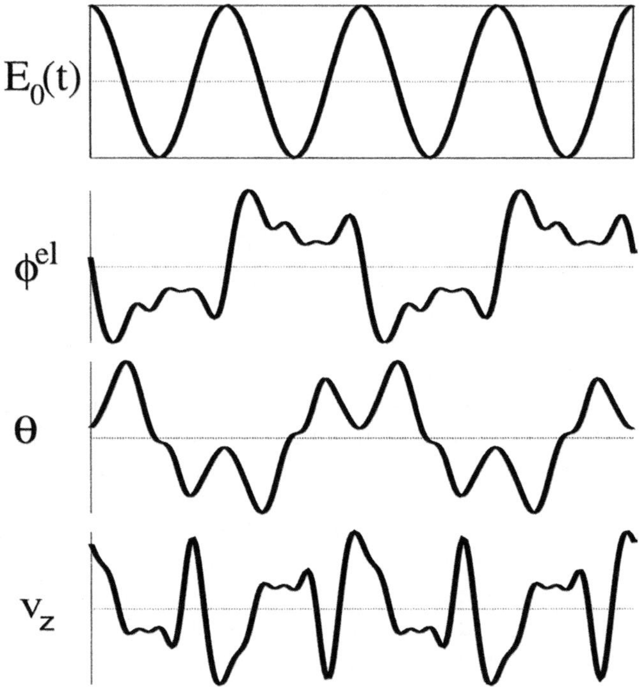

Figure 19: *The time dependence of the field variables is shown at the onset of EC in the SmC* phase (with macroscopic polarization). The fields at the cell center ($x = 0$) are plotted in arbitrary units for the frequency $\omega = 390$ of the applied voltage. The dynamics of the system is $2T$-periodic only, although the driving force $E_0(t)$ is T-periodic* [47].

(fig. 19) and the two harmonic regimes – although still discernible – are no longer purely "conductive" nor purely "dielectric". This is discussed in more detail in sec. 3.5.

3.4 Onset of Convection for DC Voltage

As for the polarization Frederiks transition described in sec. 2.6, the existence of a finite polarization in the SmC* phase has also a profound influence on the DC electroconvective instability. Without a polarization p_0, as in the smectic C phase, the sign of the DC voltage is arbitrary and the threshold for the DC electroconvection depends on V_0^2 only. This behavior is changed by the presence of the polarization **P** in SmC*. Now **E** parallel to **P** ($V_0 > 0$) is energetically preferred to the case **E** antiparallel to **P** ($V_0 < 0$) (cf. fig. 3) indicating that the basic state is more stable in the former situation than in the latter one. Thus, one can expect the DC electroconvection threshold to increase (decrease) with p_0 for $V_0 > 0$ (< 0), which indeed has also been found numerically for rigid boundaries [47].

This can be understood analytically as well. Assuming stationary instabilities (a Hopf bifurcation has not been found in our studies), we get a quadratic equation for $V_0(q)$, i.e. $b_1 V_0^2 + b_2 p_0 V_0 + b_3 = 0$, which is written down explicitly for the special case $\epsilon_a = 0$ and stress-free boundary conditions in ref. [47]. The interesting point is that the linear term of the polynomial is proportional to p_0. Thus, in contrast to the smectic C case, in the smectic C* phase the neutral curve for DC EC is no longer symmetric w.r.t. $V_0 \to -V_0$.

This simple threshold condition also explains, why for finite values of p_0 the range of existence of the DC electroconvection is larger than for zero p_0. Real values for V_c are only possible, if $p_0^2 b_2^2 > -4 b_1 b_3$. Thus, for zero p_0 the product $b_1 b_3$ must be negative, while for large enough p_0 also positive values of $b_1 b_3$ lead to an instability threshold. In particular, for a large enough and destabilizing polarization (**E** antiparallel to **P**, $V_0 < 0$), an electroconvective instability is possible, even if $\epsilon_1 (= \epsilon_a) \leq 0$ and $\sigma_1 (= \sigma_a) = \sigma_{yy} - \sigma_{xx} < 0$. In this situation the planar basic state is stable within the simple Carr-Helfrich mechanism ($p_0 = 0$), but the destabilization due to p_0 overcomes the stabilization due to negative dielectric and conductive anisotropies. An interesting competition between the static Frederiks transition and the EC may be found, as discussed in more detail in ref. [47].

3.5 Onset of Convection for AC Driving Voltage in SmC* Films

For an AC driving voltage, in the smectic C phase ($p_0 = 0$) the harmonic solutions (cf. sec. 3.3) are each decomposed into two independent classes (called A and B in [49, 51]) representing the conductive and dielectric regime, respectively. These two regimes are well known from EC in nematic liquid crystals [1, 30, 32]. Hence SmC liquid crystals are isomorphic to two-dimensional nematics in the present approximation.

The presence of the finite polarization in SmC* allows, in addition, for a subharmonic ($2T$-periodic) solution (figs. 19, 20, 21), thus breaking spontaneously the discrete time translational symmetry of the (T-periodic) driving voltage. The subharmonic regime is strictly prohibited, if the symmetry (54) holds.

For finite polarization, besides the neutral curves for the (former) conduction and the (former) dielectric regime, an additional neutral curve for the subharmonic regime arises at intermediate frequencies, as is shown in fig. 20. The former conductive and dielectric regime are shifted by the polarization. Changing the frequency ω the three minima of the neutral curve in fig. 20 are shifted relative to each other, such that any of them, can be the absolute minimum, i.e. the threshold V_c for a certain frequency range. The critical threshold voltages, $V_c(\omega)$ and the associated critical wave vectors $k_{yc}(\omega)$ are plotted for each regime as a function of frequency ω in fig. 21 for two different

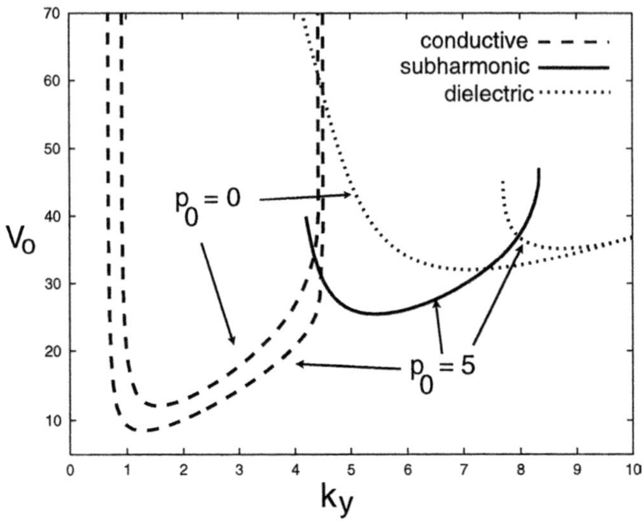

Figure 20: *Neutral curves $V_0(k_y)$ are shown for two different values of the polarization ($p_0 = 0$ and 5 according to the arrows) at a fixed frequency ($\omega = 180$) of the applied voltage. The conduction regime (dashed lines) and the dielectric regime (dotted lines) are present for any p_0. Only for finite polarizations the additional subharmonic regime occurs at intermediate k_y values (solid line).*

values of the polarization p_0.

For sufficiently high values of p_0 the neutral curve belonging to the subharmonic regime has the lowest minimum for a certain frequency range (solid line in fig. 21). It always appears at intermediate frequencies, between the conductive and dielectric regime. Figure 21b shows the critical wavenumber k_{yc} as a smooth function of the external frequency within a given regime, but with discontinuous jumps when the marginal stability switches from one regime to another. Again only two regimes are present for $p_0 = 0$, but three for large enough p_0. Although there is no simple mechanism that could explain, why a subharmonic regime exists as the marginally stable solution, the following remarks may help the intuition. With $p_0 = 0$ the sign of V_0 is undefined and it is rather V_0^2 that governs the instability thresholds. A finite p_0, however, introduces contributions linear in V_0 as has been shown explicitly in the DC case (sec. 3.4). Thus for sinusoidal voltages $V(t)$ the fields may oscillate either with half the frequency of $V(t)^2$ or $V(t)$, depending on the influence of the different destabilizing forces. For very large threshold voltages, such as in the dielectric regime, contributions in the director relaxation time being quadratic in the voltage, V^2, win over those effects linear in V. At small frequencies in the conduction regime the threshold is relatively small and the impurity charges can follow the action of the external field immediately while leaving the director orientations roughly unchanged. In both cases the fields oscillate

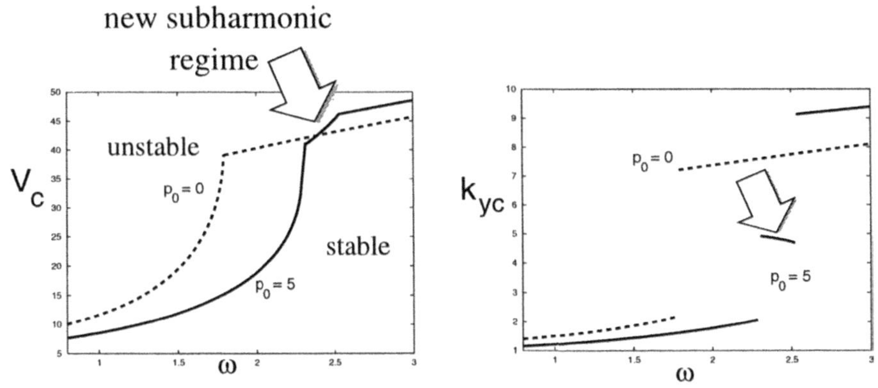

Figure 21: *Onset of instability: The critical voltages $V_c = V(k_{yc})$ (left) the critical wavelengths k_{yc} (right) of the three regimes as a function of the applied AC frequency. For low or vanishing polarization p_0 there are only two regimes, while for $p_0 > 4.5$ (for the parameters chosen) there is in addition the subharmonic regime at intermediate frequencies. At those frequencies, where the instability switches from one regime to the other, the critical wave vector shows a jump* [47].

in a subharmonic fashion with respect to V^2.

The frequency range of the conduction regime is always bounded from above by the inverse charge relaxation time. Beyond that frequency the charges cannot follow the external field. If immediately beyond that frequency the contribution $p_0 V$ to the director relaxation time is more important than those proportional to V^2, then the subharmonic regime may have a lower threshold than the dielectric one. With further increasing frequency also the threshold increases, and therefore at large frequencies V^2 dominates the V effect and the dielectric regime is favored. Using these qualitative arguments the subharmonic regime can only occur as an additional regime between the conduction and the dielectric regime.

3.6 Codimension-3 Bifurcation

The value of p_0 can be seen as a third control parameter of the system in addition to the amplitude, V_0, and frequency, ω, of the applied voltage. For large polarizations the subharmonic threshold curve $V_c(\omega)$ intersects both, the conductive and the dielectric threshold curve, at two codimension-2 points (indicated in fig. 21). If p_0 is lowered, these two codimension-2 points approach each other and finally merge (cf. fig. 22). We have found numerically that this coalescence happens just at the point, where also the conductive and dielectric thresholds intersect. Here all three regimes coexist and a competition of three solutions having different wavelengths should be seen in experiments. For $p_0 < p_3$, the subharmonic regime disappears. (The actual value of p_3 depends obviously on the material parameters.) This scenario seems to be generic, since near this codimension-3 point the threshold curves $V_c(\omega)$ are nearly straight lines, where the conductive and dielectric curve have the largest and smallest slope, respectively. Increasing however the polarization p_0 well

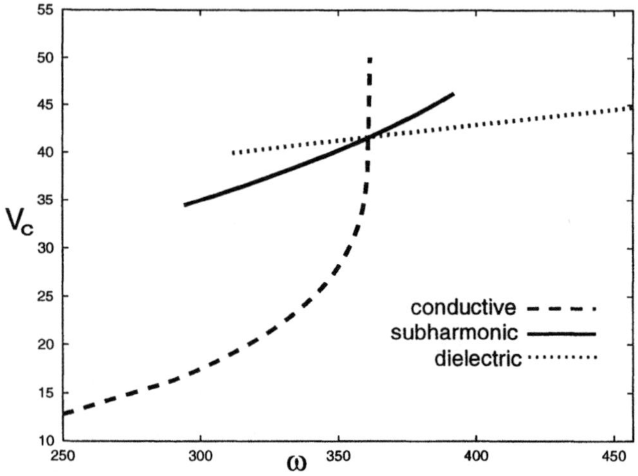

Figure 22: *Codimension-3 point: At the minimal value of the polarization* ($p_0 = 4.5$ *for the parameters chosen), for which the subharmonic regime exists, the three different regimes have equal critical voltages at a certain external AC frequency* $\omega_3 \approx 361$. *This is a codimension-3 point* [47].

beyond p_3 the cutoff frequency of the conductive regime is shifted to higher values and the threshold and critical wavelength at a fixed frequency is lowered further. The subharmonic regime appears always after the cutoff of the conductive regime for high values of p_0 and is extended to much higher frequencies at the cost of the dielectric regime. For large polarizations p_0 in the conductive regime we find typically rather small threshold values $k_{yc} \approx 0.15$ and $V_c \approx 0.5$ at $\omega = 200$. The observation of such very small wavenumbers should not be confused with the Frederiks transition ($k_y \equiv 0$, which does not occur for $\epsilon_1 < 0$).

3.7 Parameter Dependencies

The frequency range, where the subharmonic regime has the lowest threshold, depends not only on the polarization but also on the values of the other material parameters and on the film width. To design an experiment for investigating the subharmonic regime it is helpful to know, for which geometric and material parameters the subharmonic regime can be observed most likely. All results described in the previous sections of this work have been calculated for the values given in Appendix A of ref. [47]. The electroconvective instability is especially sensitive to changes in the electrical anisotropies $\epsilon_a = \epsilon_1$ and $\sigma_a = \sigma_{yy} - \sigma_{xx}$ [25, 26, 49, 51]. Therefore we consider the influence of variations of these quantities on the existence range of the subharmonic regime.

As an example the dependence on σ_a has been investigated by keeping σ_{xx} constant and changing the value of σ_{yy}. Increasing values of σ_a enforce the ability of space charges to follow the applied alternating field up to higher frequencies according to the Carr-Helfrich mechanism [1, 19]. Thus the cut-off frequency of the conductive regime increases approximately proportional to σ_a while the dielectric regime is only slightly affected. This is shown for $p_0 = 0$ by the dashed line in fig. 23. Similar behavior is found for a finite polarization (solid lines) although for $\sigma_a/\sigma_{xx} < 0.55$ (for the parameters chosen) the subharmonic regime squeezes in between the two other regimes. The threshold voltages of the conductive and the dielectric regimes diverge by approaching $\sigma_a \to 0$ according to the Carr-Helfrich mechanism causing smaller values of the anisotropy σ_a to favor the subharmonic regime. Furthermore the subharmonic regime exists even for (slightly) negative values of the anisotropy σ_a. The same has been found for an applied DC-voltage (sec. 3.4). Increasing p_0 will again lead to a larger frequency range of the subharmonic instability and the subharmonic regime will exist to even larger values of σ_a than indicated in fig. 23.

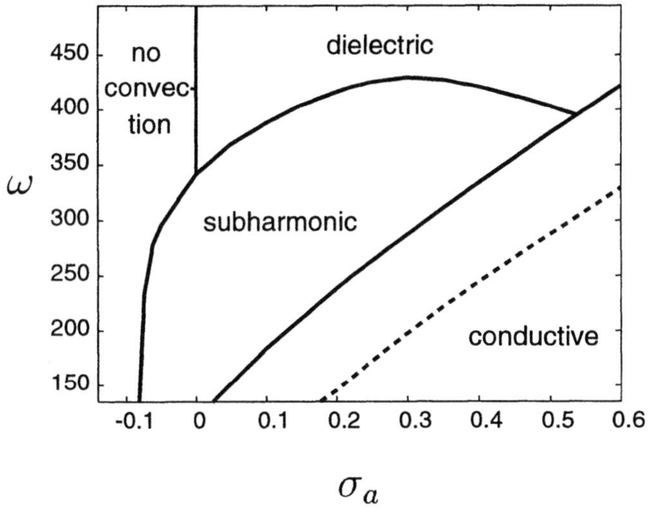

Figure 23: *The solid lines indicate the existence ranges of the different instability regimes at various AC frequencies as a function of the conduction anisotropy σ_a for $p_0 = 5$ and $\sigma_{xx} = 133$ and $\epsilon_x = 5.25$; $\epsilon_a = -0.38$. The dashed line describes the codimension-2 line between the conductive and the dielectric regimes for $p_0 = 0$.*

Another interesting dependence, the dependence of variations on the film width has also been discussed in ref. [47].

4 Electroconvection Experiments on Smectic Films

4.1 General

When pattern formation by electroconvection in freely suspended smectic films is discussed, at least three different mechanisms have to be considered. The first type of instability bases on the Carr-Helfrich mechanism known from nematic liquid crystals. Here an in-plane anisotropy of the conductivity is required, which is present in the tilted smectic C and C* phases, but not in the A phase. A modification of this mechanism by an additional polarization **P** has been discussed theoretically in the previous section.

A second mechanism works via charge injection at the electrodes and has been reported for isotropic liquids already in 1936 [53]. Since then this instability has been studied in a variety of systems (e.g. [54, 55, 56, 57, 58]). Another mechanism is due to surface charge layers ("diffusion layers") [59, 60, 61]. These different mechanisms in general do not necessarily exclude each other, a competition or interplay of them is very likely in smectic films. In some cases, the special choice of substances or experimental conditions may allow a separate study of the individual effects.

In the first experiments reported on electroconvection in freely suspended films nematic liquid crystals have been used [59, 60]. Faetti and his coworkers used the nematic liquid crystal 4-n-butyl-N-(4-methoxybenzylidene)-aniline (MBBA). The nematic director in these films is anchored at the smectic-air interface at some small angle with respect to the film normal. An electric field applied parallel to such films leads to characteristic convection vortices which can be visualized in the optical transmission texture.

Faetti et al. discovered two different convection modes in their nematic films. In thin layers, the threshold for onset of electroconvection depends linearly on the film thickness. For an explanation of this so-called *vortex mode* (VM) the authors developed a model which bases on the interactions of surface charges on the film with the electric field. This vortex mode occurs also in the isotropic phase of a material and therefore the Carr-Helfrich mechanism could be excluded as the driving effect. Moreover, the experiments were also performed with insulated and conducting electrodes to exclude convection driven by charge injection.

With increasing film thickness, the threshold for the vortex mode increases and consequently the authors observed the crossover to a different convection type [59] in thick films (several micrometers). The threshold voltage of this structure, the so-called *domain mode* (DM), is independent of the film thickness. This regime is driven by bulk charges and two frequency regimes of convection have been observed for AC electric fields, similar as in electroconvection in nematics between two glass plates. Accordingly the domain mode has been also explained by the Carr-Helfrich mechanism.

The latter is characterized by an oscillating charge field at low frequencies and another at high frequencies with oscillating director field. The viewing direction in the geometry of nematic films is into the rolls, in contrast to conventional nematic sandwich cells. In the isotropic liquid phase, only the vortex mode remains, the domain mode requires the anisotropic electric properties of the nematic phase.

4.2 Experiments on Freely Suspended Smectic A Films

In isotropic liquids and nematic phases a molecular layer structure as in smectic liquid crystals is absent. Therefore such nematic films are unstable, their thickness is inhomogeneous and subject to considerable fluctuations and their lifetime is restricted. Moreover, the velocity field is very complex and not necessarily restricted to a two-dimensional flow.

In contrast, the layer structure of smectics guarantees well-defined homogeneous film thickness and the flow takes place essentially in planes of the smectic layers. Morris and coworkers [39, 40] investigated free-standing smectic A films under lateral electric fields and discovered a vortex mode analogous to the nematic phase. Faetti's theoretical explanation which is independent of the anisotropy in the smectic layers, was adapted to smectic films and developed further. Analytical expressions for the threshold voltage of vortex electroconvection has been derived by Daya et al. [61]

$$U_c = 2\frac{d_z}{\epsilon_0}\sqrt{\sigma\eta R_c}\ . \tag{55}$$

σ is the in-plane electric conductivity, η is the viscosity of the smectic material, d_z is the film thickness and R_c denotes a constant, which depends on the electrode shape. Plate electrodes with $R_c \approx 92$ and wire electrodes with $R_c \approx 77$ lead to slightly different threshold fields. Characteristic for the vortex electroconvection is the linear dependence of the onset voltage on film thickness, which is confirmed very well in experiments. Smectic A films are isotropic within the layers. Hence, the Carr-Helfrich mechanism does not work in these materials. For this isotropic mechanism in smectic A the theory has been also extended recently by a perturbation calculation into the weakly nonlinear regime [62].

4.3 Experiments on Freely Suspended Smectic C Films

Electroconvection in smectic C films has been studied quite recently in experiments by Becker et al. [63] and Langer et al. [64, 65]. The principal difference to the smectic A phase is the in-plane anisotropy. With polarizing microscopy, a flow induced optical texture (SmC *schlieren*) can be observed. Vortex flow in the smectic C phase can be immediately traced from the texture changes in the optical reflection or transmission image. Two effects are superimposed, the first is a texture transport by the mass flow in the film plane. The second effect is the reorientation of the $\boldsymbol{c}$ director by flow coupling, in an inhomogeneous flow field, a hydrodynamic torque acts on $\boldsymbol{c}$. Figure 24 shows typical convection textures in a smectic C film driven at low frequency AC (polarized light, no analyzer). Between adjacent dark and bright stripes, the $\boldsymbol{c}$-director is rotated by 90°.

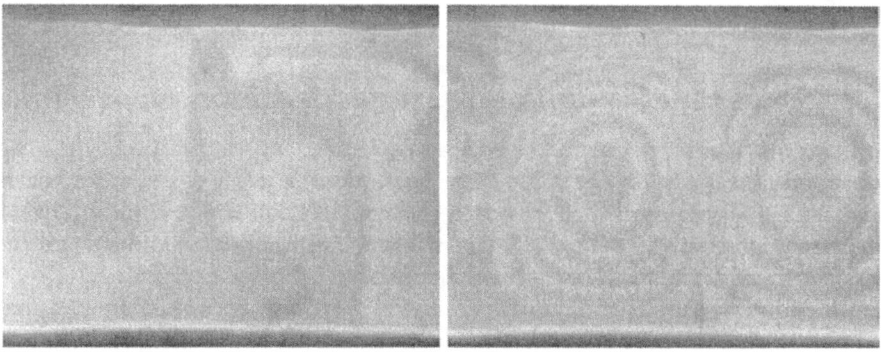

Figure 24: *Reflection texture of* EC *vortices in a smectic* C *film. The electrodes are at the upper and lower edges, with 0.55mm gap. Vortices appear at 22V and 0.5Hz (left), while ring structures are seen at the field reversal (30V) and 0.3Hz (right).*

At DC excitation the flow is unidirectional, the director walls are carried with the flow field and wind up to spirals until the counter-action of increasing elastic deformations establishes a stationary asymptotic state of the director fields.

At AC excitation, the director walls wind up during one half cycle (fig. 24 left) but the flow is reversed in the next half cycle. The right image in fig. 24 shows the moment of field reversal, where the walls have formed a concentric ring system. When threshold fields for convection are measured in dependence upon the film thickness, a qualitative and quantitative correspondence with the vortex mode in SmA is found [65]. Also for a smectic C mixture, the threshold increases rather linearly with the film thickness as shown in fig. 25. This indicates that the observed pattern in SmC is of the same type as the vortex convection found in smectic A films. It is driven by surface charges and decorated by the texture flow. The deformation of the $\boldsymbol{c}$-director field by the flow field is huge, while in the opposite direction the influence is negligibly small. In the smectic C material, in principle a Carr-Helfrich type of convection could occur as discussed in sec. 3, but the surface driven vortex flow effects have been suppressed. In these experiments on SmC the vortex mode has obviously a lower threshold field. Carr-Helfrich driven convection is conceivable only at high frequencies or in very thick films ($d_z \gg 1\mu m$). A cross-over between both types as detected in nematic films [60] has not yet been described for smectics.

4.4 Electroconvection in smectic C* films

Electroconvection in ferroelectric smectic C films have been performed so far only in materials with high spontaneous polarization $\boldsymbol{P}$ [65]. In this material,

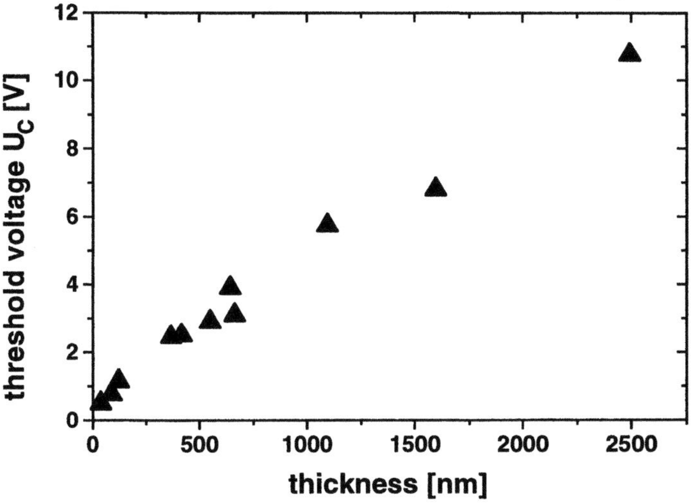

Figure 25: *Onset voltage for electroconvection in a smectic* C *mixture in dependence of the film thickness d_z (50% 5-n-Octyl-2-(4-n-hexyloxyphenyl)-pyrimidin and 50% 5-n-Octyl-2-(4-n-octyloxyphenyl) -pyrimidin with phase sequence* $< -10°C$ S_C $50.5°C$ S_A $53°C$ N $68°C$ I, *electrode gap 0.5mm , T=25° C,* $\sigma = 8.92 \cdot 10^{-12} A/(Vm)$) [65]

the interaction of $\boldsymbol{P}$ with the external electric field $\boldsymbol{E}$ is dominant. The $\boldsymbol{c}$ director aligns perpendicular to $\boldsymbol{E}$. A convective vortex flow leads to periodic deflections of the $\boldsymbol{c}$ director from its fixed alignment but is too weak to create disclination walls. The vortices are visualized with crossed polarizers slightly oblique to the electrode geometry which decorates oppositely rotating vortices with different reflection intensities. Figure 26 shows the reflection texture of a 2.1 μm thick film in the SmC* phase observed with crossed polarizers. A small angle $< 20°$ was adjusted between the electrode and the polarizer. The oppositely rotating vortices are visible as rectangular areas of nearly uniform brightness.

In the SmC* films one observes DC electroconvection already at very low voltages. Figure 27 gives the flow velocity as a function of the applied voltage. Note that the onset threshold in SmC was 0.5V even in the thinnest films. For SmC* a non-zero critical voltage cannot be derived from these experiments, it is below the experimental resolution. Moreover, it is characteristic that the convective flow velocity is independent of the film thickness d_z. This indicates that the convection in these films is not primarily driven by the film surface.

The interpretation of these observations is not unambiguous. On the one hand, many arguments characterize the observed effect as a Carr-Helfrich type convection as predicted in ref. [47], cf. sec. 3. The well defined ground state with the $\boldsymbol{c}$ director parallel to the electrode edges exists, the periodic director deflection is small ($\ll 90°$) and the observed flow velocity is film thickness independent over a large thickness range. The flow field itself is very similar

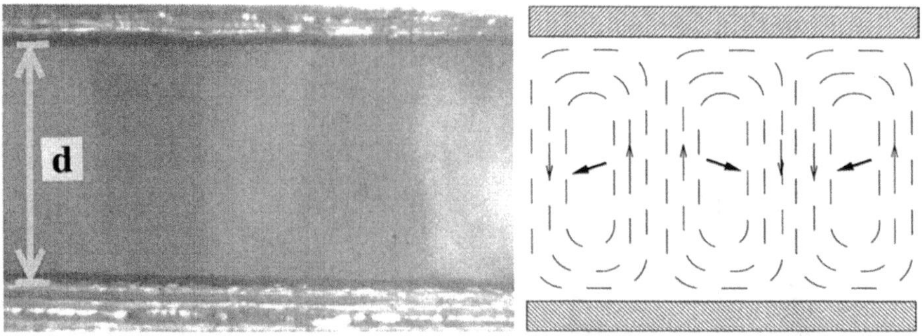

Figure 26: *Reflection texture of* EC *in a smectic* C* *film* ($d_z = 2.1\mu m$), *with electrode gap 0.5mm, DC voltage 60V and crossed polarizers slightly rotated out of the electrode direction.* [66]

to that in the surface driven structures in SmA and SmC. A direct proof for or against one of both mechanisms has not yet been achieved (cf. the nematic films where a clear crossover between both mechanisms can be observed).

Experiments with AC electric fields can elucidate more of the driving mechanisms. However, in the ferroelectric smectic films the electroconvection is disturbed by the switching of the $\boldsymbol{c}$ director at field reversal, it reorients by 180° as the spontaneous polarization aligns with the external field $\boldsymbol{E}$. The disturbance of the EC vortices by ferroelectric switching is small only at very low frequencies (a few Hz). In these experiments one observes that the vortices change their sense of rotation with the driving field. Speaking in terms of the Carr-Helfrich mechanism, this means that the charges remain at their positions while the director reverses sign with the external field, this is characteristic for the so-called dielectric regime in EC of nematics (cf. also sec 3.5).

Finally we summarize the progress made in the study of electroconvection in thin quasi two-dimensional films and sketch the perspectives of future work. The investigation of freely suspended films of tilted smectic phases has revealed new electroconvection scenarios and to a certain extend it has improved understanding of the electrohydrodynamic effects, but there are still many problems to be solved before a comprehensive description will be obtained. Theory and experiment are still a few steps apart, it should be a goal of future experimental investigations to prepare very low $\boldsymbol{P}$ ferroelectric samples, and an important theoretical task will be the combination of works in refs. [61, 47], which should include both surface effects and an in-plane electric anisotropy of the smectic material. In view of the many potential applications of ferroelectric smectics the understanding of electroconvection effects in these materials is not only of pure academic interest.

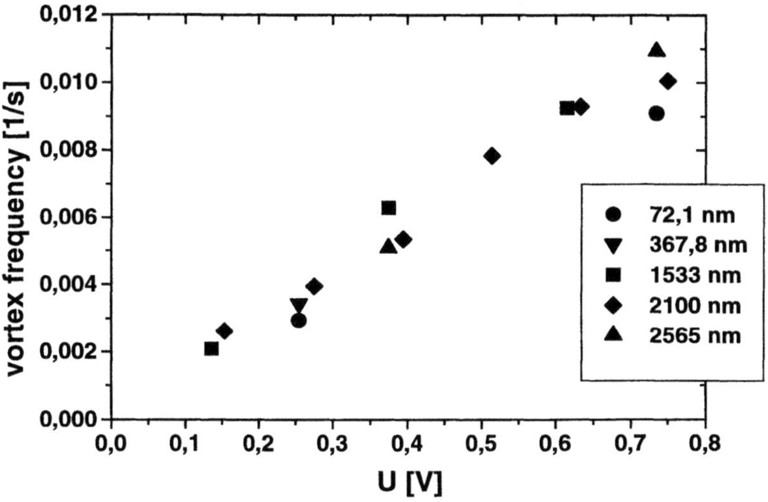

Figure 27: *Vortex velocity of convective flow in an* SmC* *film (ZLI 4237-100,* Merck*); films of different thickness d_z are compared (gap width 0.5mm, T=25° C).* [64]

Acknowledgement

H.P. and W.Z. thank S. Ried, on whose PhD thesis ([46]) sec. 2 is based, and H.R. Brand for valuable discussions. R.S. thanks Christian Langer for preparation of figures, W. Weissflog for synthesis of the SmC material and Stephen Morris for valuable discussions. This work has been supported by the Deutsche Forschungsgemeinschaft through Schwerpunkt "Strukturbildung in dissipativen kontinuierlichen Systemen – Experiment und Theorie im quantitativen Vergleich".

References

[1] P. G. de Gennes and J. Prost, *The Physics of Liquid Crystals* (Clarendon, Oxford, 1993).

[2] L. M. Blinov, *Electro-optical and magneto-optical properties of liquid crystals* (John Wiley, New York, 1983).

[3] M. C. Cross and P. C. Hohenberg, Rev. Mod. Phys. **65**, 851 (1993).

[4] *Spatio-Temporal Patterns in Nonequilibrium Complex Systems*, Vol. XXI of *Santa Fe Institute Studies in the Sciences of Complexity*, edited by P. Cladis and P. Palffy-Muhoray (Addision-Wesley, New York, 1995), .

[5] *Pattern Formation in Liquid Crystals*, edited by A. Buka and L. Kramer (Springer, Berlin, 1996).

[6] B. Bahadur, *LIQUID CRYSTALS, Applications and Uses* (World Scientific, Singapore, 1993).

[7] V. Frederiks and V. Tsvetkov, Sov. Phys. **6**, 490 (1934).

[8] W. Helfrich, Appl. Phys. Lett. **17**, 531 (1970).

[9] J. P. Hurault, J. Chem. Phys. **59**, 2068 (1973).

[10] A. Rapini, J. Phys. (Paris) **33**, 237 (1972).

[11] M. Hareng, S. L. Berre, and J. J. Metzger, Appl. Phys. Lett. **27**, 575 (1975).

[12] C. Y. Young, R. Pindak, N. A. Clark, and R. B. Meyer, Phys. Rev. Lett. **40**, 773 (1978).

[13] R. A. Pelcovits and B. I. Halperin, Phys. Rev. B **19**, 4614 (1979).

[14] C. Rosenblatt, R. B. Meyer, R. Pindak, and N. A. Clark, Phys. Rev. A **21**, 140 (1980).

[15] W. Zimmermann, S. Ried, H. Pleiner, and H. R. Brand, Europhys. Lett. **33**, 521 (1996).

[16] R. Williams, J. Chem. Phys. **39**, 384 (1963).

[17] A. P. Kapustin and L. S. Larionova, Kristallografiya **14**, 741 (1964).

[18] E. F. Carr, E. A. Hoar, and W. T. McDonald, J. Chem. Phys. **9**, 297 (1968).

[19] W. Helfrich, J. Chem. Phys. **51**, 4092 (1969).

[20] E. Dubois-Violette, P. G. deGennes, and O. Parodi, J. Phys. (Paris) **32**, 305 (1971).

[21] P. A. Penz and G. W. Ford, Phys. Rev. A **6**, 414 (1972).

[22] S. Kai and K. Hirakawa, Prog. Theor. Phys. Suppl. **64**, 212 (1978).

[23] A. Joets and R. Ribotta, in *Cellular Structure in Instabilities*, edited by J. E. Wesfreid and S. Zaleski (Springer, Heidelberg, 1984).

[24] R. Ribotta, A. Joets, and L. Lei, Phys. Rev. Lett. **56**, 1595 (1986).

[25] W. Zimmermann and L. Kramer, Phys. Rev. Lett. **55**, 402 (1985).

[26] E. Bodenschatz, W. Zimmermann, and L. Kramer, J. Phys. (Paris) **49**, 1875 (1988).

[27] I. Rehberg, S. Rasenat, and V. Steinberg, Phys. Rev. Lett. **62**, 756 (1989).

[28] S. Kai and W. Zimmermann, Prog. Theor. Phys. Suppl. **99**, 458 (1989).

[29] W. Zimmermann, Mat. Res. Bulletin **16**, 46 (1991).

[30] L. Kramer and W. Pesch, Annu. Rev. Fluid Mech. **27**, 515 (1995).

[31] L. Kramer and W. Pesch, in *Pattern Formation in Liquid Crystals*, edited by A. Buka and L. Kramer (Springer, Berlin, 1996).

[32] U. Behn and W. Pesch, Electroconvection in nematic liquid crystals, this volume, 1998.

[33] F. Aliev and K. F. Abbasov, Sov. Phys. Kristallogr. **30**, 442 (1985).

[34] J. P. Petrov, A. G. Petrov, and G. Pelzl, Liquid Crystals **11**, 865 (1992).

[35] A. B. Davey and W. A. Crossland, Mol. Cryst. & Liq. Cryst. **263**, 2267 (1995).

[36] M. Goscianski and L. Léger, J. Phys. (Paris) Colloq. C1 **8**, 231 (1975).

[37] L. K. Vistin and A. P. Kapustin, Kristallografiya **13**, 349 (1968).

[38] A. L. Bermain, E. Gelerinter, and A. de Vries, Mol. Cryst. & Liq. Cryst. **33**, 55 (1976).

[39] S. W. Morris, J. R. de Bruyn, and A. May, Phys. Rev. Lett. **65**, 2378 (1990).

[40] S. W. Morris, J. R. de Bruyn, and A. May, Phys. Rev. A **44**, 8146 (1991).

[41] M. Kléman, Rep. Progr. Phys. **52**, 1455 (1989).

[42] P. C. Martin, O. Parodi, and P. Pershan, Phys. Rev. A **6**, 2401 (1972).

[43] A. Saupe, Mol. Cryst. Liq. Cryst. **7**, 59 (1969).

[44] H. Pleiner and H.R. Brand, in *Pattern Formation in Liquid Crystals*, edited by A. Buka and L. Kramer (Springer, Berlin, 1996) p. 15.

[45] H. Pleiner and H. R. Brand, to be published

[46] S. Ried, Elektrisch getriebene Konvektion in smektischen Filmen, PhD Thesis, University of Essen, (unpublished), 1997.

[47] S. Ried, H. Pleiner, W. Zimmermann, and H. R. Brand, Phys. Rev. E **53**, 6101 (1996).

[48] H. R. Brand and H. Pleiner, Phys. Rev. A **35**, 3122 (1987).

[49] L. Kramer *et al.*, Liquid Crystals **5**, 699 (1989).

[50] H.R. Brand and H. Pleiner, J. Phys. (Paris) **43**, 853 (1982).

[51] W. Zimmermann, in *Defects, Singularities and Patterns in Nematic Liquid Crystals, NATO Advanced Study Institute Series*, edited by J. Coron, F. Helen, and J. Ghidaglia (Kluwer, Dordrecht, NL, 1991).

[52] G. Iooss and D. D. Joseph, *Elementary Stability and Bifurcation Theory* (Springer, Berlin, 1980).

[53] D. Avsec and M. Luntz, Compt. Rend. Acad. Sci Paris **203**, 1140 (1936).

[54] A. J. Rich, J. L. Sproston, and G. Walker, in *Eighth International Conference on Conduction and Breakdown in Dielectric Liquids*, edited by G. Molinari and A. Vivani (IEEE, New York 1984) p. 366.

[55] M. Suzuki, Phys. Rev. A **31**, 2548 (1985).

[56] P. Atten and L. Eloudie, J. Electrostatics **34**, 279 (1995).

[57] F. Argoul and A. Kuhn, Physica A **213**, 209 (1995).

[58] R. Chicon, A. Castellanos, and E. Martin, J. Fluid Mech. **344**, 43 (1997).

[59] S. Faetti, L. Fronzoni, and P. Rolla, J. Chem. Phys. **79**, 1427 (1983).

[60] S. Faetti, L. Fronzoni, and P. Rolla, J. Chem. Phys. **79**, 5054 (1983).

[61] Z. A. Daya, S. W. Morris, and J. R. de Bruyn, Phys. Rev. E **55**, 2682 (1997).

[62] V. B. Deyirmenjian, Z. A. Daya, and S. W. Morris, Phys. Rev. E **56**, 1706 (1997).

[63] A. Becker, S. Ried, R. Stannarius, and H. Stegemeyer, Europhys. Lett. **39**, 257 (1997).

[64] C. Langer, R. Stannarius, A. Becker, and H. Stegemeyer, Proc. ECLC '97 Zakopane, 1998.

[65] C. Langer and R. Stannarius, Phys. Rev. E, 1998.

[66] Figure by courtesy of Christian Langer.

Electrohydrodynamic Convection in Nematics

W. Pesch[1], U. Behn[2]

[1] Universität Bayreuth, D-95447 Bayreuth, Germany
[2] Universität Leipzig, D-04109 Leipzig, Germany

Abstract. The purpose of this review is to present a status report on the electrohydrodynamic convection in nematic liquid crystals. The considerable progress achieved in the past two years is emphasized.

1 Introduction

Pattern formation in fluid systems driven away from equilibrium is a common phenomenon in nature [1, 2]. A famous canonical example is the Rayleigh-Bénard convection (RBC) in simple isotropic fluids which continues to be the subject of numerous experimental and theoretical studies (see e.g. [3]). More recently the rich variety of dynamical structures found in liquid crystals (LCs) has attracted considerable and still growing attention.

LCs are materials made up of highly anisotropic organic molecules in a state that reflects the anisotropy [4, 5]. Thus LCs have become a prime model for the study of pattern formation in anisotropic systems. There are several classes among which nematic LCs (nematics) play a dominant role in this article. Nematics are fully liquid without long-range translational, but with long-range uniaxial orientational ordering of the molecules. As a result of the coupling of the molecular alignment axis (described by the director $\hat{\mathbf{n}}$) with mass flow, electric and thermal currents, the hydrodynamic equations involve numerous nonlinearities (see Sect.2), which easily lead to instabilities when a state of nonequilibrium is maintained (see e.g. [6]). Convective flows can be driven either electrically through space charges that naturally arise in an anisotropic conductor in the presence of spatial variations (electrohydrodynamic convection, EHC) or thermally through buoyancy forces (Rayleigh-Bénard convection, RBC).

EHC has attracted most of the attention and will be exclusively discussed in this review. In the typical thin-layer geometry shown in Fig.1a the nematic is sandwiched between glass plates (separation $d \sim 10 - 100\mu$) with transparent electrodes. The surfaces are treated to provide uniform anchoring of the director, in most cases along the x direction (*"planar"* or "homogeneous" alignment), but sometimes also in the z direction (*"homeotropic"* alignment).

Above an applied (critical) voltage $V_c \sim 10V$ (typically low-frequency ac) convection rolls appear with associated director distortions, which are easily detected optically. The spacing of the rolls is of order d except in the higher-frequency "dielectric range". Fig.1b shows a typical pattern with *normal* rolls, i.e. normal to the undistorted director in the x direction.

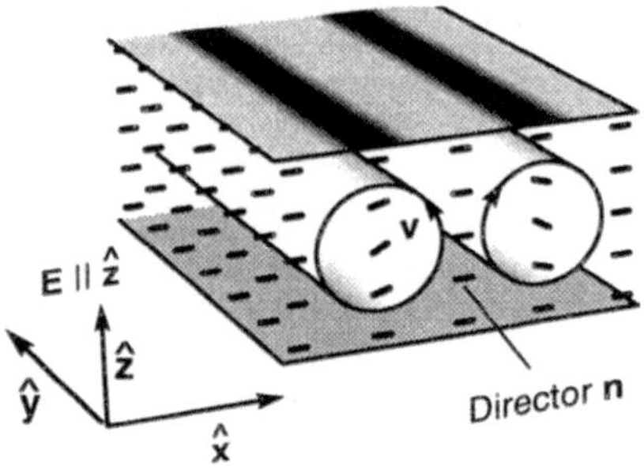

Fig.1a. Cell geometry with section of a roll pattern for EHC (planar configuration). **E** = electric field, **v**= velocitey.

Fig.1b. Normal roll pattern for EHC with a dislocation.

From the experimental point of view EHC is attractive because of the easily accessible control parameters like electric and magnetic fields. The characteristic times are typically short (μ sec) and the extension of the patterns large (up to 1000 perfectly aligned convection rolls). For the theory it is important that spatio-temporal complexity appears quite often already in the vicinity of the convection onset. Thus common perturbational schemes, like the order parameter approach can be put on a sound basis and their reliability can easily be evaluated.

Liquid crystals are complicated and an intuitive understanding of all the mechanisms hidden in the nematohydrodynamic equations is not easy. In fact, the remarkable progress in the last two decades is not imaginable without the particularly close collaboration between experimental and theoretical groups in this field. There are numerous examples (only some of which we will mention explicitly) where experimental findings have motivated the theoretical efforts and vice versa. For a classical review of convective instabilities in LCs, see [7] and for EHC one may consult the books of Blinov [8] and Pikin [9] and the review articles [10, 11, 12, 13, 14]. In recent overviews [15, 16] in particular the progress achieved in the years from 1984 - 1996 was emphasized.

The investigation of convection scenarios in LCs is also a problem of materials. In experiments the standard reference materials MBBA (4-methoxybenzylidene-4'-n-butyl-aniline) or a mixture, Merck Phase 5, have typically been used (sometimes doped with an ionic substance). These are the only room-temperature nematics with dielectric anisotropy, $\epsilon_a < 0$, where virtually all the material parameters have been measured (for tabulated values: see e.g. [17] for MBBA, [18] for Phase 5 and references therein). Since these nematics are similar in their properties, the investigation of other classes is highly desirable and promising. A recent successful example is the new very stable material I52 (4-ethyl-2-fluoro-4'-[2-(trans-4-n-pentylcyclohexyl)-ethyl]-biphenyl) doped with iodine [19, 20]. Even at onset qualitatively new localized structures ("worms") [21] have been detected, which provide a new challenge for the theory. This applies also to materials where one can switch continuously from the nematic to the smectic phase by decreasing the temperature [22].

In the present article we will concentrate on the most recent results in various realizations of EHC. In addition we will focus on an interesting topic in detail which could be addressed only briefly in the previous reviews [15, 16] because of space limitations. The restoring forces on the director in LCs are not very strong and its dynamics is susceptible to very weak external bias. In hindsight it is therefore not too surprising that for the first time the influence of thermal noise on a continuous nonequilibrium phase transition was successfully analyzed in nematics [23, 24]. In this paper it will be demonstrated that a *controlled* application of noise is very attractive both from the experimental and theoretical point of view.

After the introduction and explanation of the basic equations (Sect.2) the theoretical concepts (the linear and weakly nonlinear analysis) will be sketched in Section 3. Section 4 deals with various new aspects of EHC in the planar configuration and in Section 5 the homeotropic configuration is discussed. Section 6 is devoted to the noise-driven EHC near threshold. Finally, in the General Conclusions (Sect. 7), we shall mention some perspectives for future work. Furthermore a brief connection to other hydrodynamic instabilities in nematics is made.

2 Basic equations and instability mechanisms

The dynamics of liquid crystals is described by a well accepted set of macroscopic equations (see e.g. [4, 5, 25, 26]). Here we will sketch only the most simplest version pertaining to nematic LCs.

In the nematic state the isotropy is spontaneously broken and the averaged molecular orientation is described by the director **n**, i.e. a unit vector where $\pm\mathbf{n}$ are equivalent. The dynamical **balance of torques** on the director **n** is determined by:

$$\gamma_1 \mathbf{n} \times \dot{\mathbf{n}} = \mathbf{n} \times (\boldsymbol{h}_{el} + \mathbf{h}_{vis}) \tag{1}$$

where the dot stands for the material derivative $\frac{d}{dt} + \mathbf{v} \cdot \nabla$. In Eq. (1), the elastic molecular field, $\mathbf{h}_{el}$, derives from the elastic free energy density f_{el}:

$$f_{el} = \frac{1}{2}[\,k_{11}(\nabla \cdot \mathbf{n})^2 + k_{22}(\mathbf{n} \cdot \nabla \times \mathbf{n})^2 + k_{33}(\mathbf{n} \times \nabla \times \mathbf{n})^2\,] \qquad (2)$$
$$-\frac{1}{2}\mu_0 \chi_a (\mathbf{n} \cdot \mathbf{H})^2 - \frac{1}{2}\epsilon_0 \epsilon_a (\mathbf{n} \cdot \mathbf{E})^2$$

according to the relationship

$$(\mathbf{h}_{el})_i = -\frac{\partial f_d}{\partial n_i} + \partial_j \frac{\partial f_d}{\partial(\partial_j n_i)} \;. \qquad (3)$$

The k_{11}, k_{22}, k_{33} terms are associated with splay, twist and bend distortions of the director field. The importance of electric ($\mathbf{E}$) and magnetic ($\mathbf{H}$) contributions is determined by the anisotropy of the magnetic and electric susceptibilities $\chi_a = \chi_{\|} - \chi_{\perp}$ and $\epsilon_a = \epsilon_{\|} - \epsilon_{\perp}$, respectively.

The viscous part of the molecular field, $\mathbf{h}_v$, can be written as:

$$\mathbf{h}_v = -\alpha_2 \mathcal{D} \cdot \mathbf{n} - \alpha_3 \mathbf{n} \cdot \mathcal{D} \qquad (4)$$

where the tensor $\mathcal{D}$ characterizes the velocity shear ($\mathcal{D}_{ij} = \partial v_i / \partial x_j$). Since $\alpha_2 < 0$, and $\gamma_1 = \alpha_2 - \alpha_3 > 0$ (see Eq.(1) with $|\alpha_2| \gg \alpha_3$) the torque tends to rotate the director in order to avoid the director-transverse velocity gradients $\mathbf{n} \times \mathcal{D}\mathbf{n}$.

The **momentum balance** results in the (generalized) Navier-Stokes equation for the velocity field

$$\rho_m \frac{d\mathbf{v}}{dt} = \mathbf{f}_B + \nabla \cdot \mathbf{T} \qquad (5)$$

with the bulk force $\mathbf{f}_B$ to be discussed below and the stress tensor

$$T_{ij} = -p\delta_{ij} - \frac{\partial f_{el}}{\partial n_{k,i}} n_{k,j} + t_{ij} \qquad (6)$$

where ρ_m denotes the mass density and p the pressure. The viscous contribution t_{ij} contains the six Leslie shear viscosity coefficients α_i [27]

$$t_{ij} = \alpha_1 n_k n_m A_{km} n_i n_j + \alpha_2 n_i N_j + \alpha_3 n_j N_i + \alpha_4 A_{ij} + \alpha_5 n_i n_k A_{kj} + \alpha_6 n_j n_k A_{ki}. \qquad (7)$$

with the positional strain tensor $A_{i,j} = 1/2(\mathcal{D}_{i,j} + \mathcal{D}_{j,i})$. There exists also an orientational strain ($\sim \alpha_2, \alpha_3$, cf Eq.(7)) which is associated with the director dynamics relative to the local solid body rotation (with the rate $\frac{1}{2}$curl$\mathbf{v}$):

$$\mathbf{N} = \frac{d\mathbf{n}}{dt} - \frac{1}{2}(\mathrm{curl}\mathbf{v} \times \mathbf{n}). \qquad (8)$$

With the use of an Onsager relation $\alpha_6 - \alpha_5 = \alpha_2 + \alpha_3$ the number of independent parameters can be reduced [28]. Incompressible flow is assumed

($\nabla \cdot \mathbf{v} = 0$), which is guaranteed by the introduction of the toroidal and poloidal velocity potentials (see [29]).

Crucial for the occurrence of EHC is the bulk force $\mathbf{f}_B$ in the Navier-Stokes equation (5):

$$\mathbf{f}_B = \rho_{el}\mathbf{E} + (\mathbf{P} \cdot \nabla)\mathbf{E}, \quad \mathbf{P} = (\epsilon - 1)\mathbf{E}. \tag{9}$$

The first term in $\mathbf{f}_B$ is the classical Coulomb force on the charge density ρ_{el}. Much less important is ponderomotive force in the second term which contains the macroscopic polarization $\mathbf{P}$ (for the dielectric tensor $\boldsymbol{\epsilon}$ see Eq.(11)). In principle there exist also flexoelectric contributions to $\mathbf{P}$ [30, 31]. They seem to be negligible except for very thin cells and for not too large ac frequencies [32, 16, 26]. The equation determining ρ_{el} is obtained from charge conservation and Poisson's law

$$\frac{d\rho_{el}}{dt} + \nabla \cdot (\boldsymbol{\sigma} \cdot \mathbf{E}) = 0, \quad \rho_{el} = \nabla \cdot (\boldsymbol{\epsilon} \cdot \mathbf{E}). \tag{10}$$

Here the dielectric and conductivity tensors $\boldsymbol{\epsilon}$ and $\boldsymbol{\sigma}$ (typical for uniaxial anisotropy) are given by:

$$\epsilon_{i,j} = \epsilon_\perp + \epsilon_a n_i n_j; \quad \sigma_{i,j} = \sigma_\perp + \sigma_a n_i n_j. \tag{11}$$

The assumption of an anisotropic, but fixed ohmic conductivity is characteristic for the standard model. A more general ansatz (the "weak-electrolyte" model, WEM [33]) is discussed in Section 4.2.

According to the standard model (SM) described above one can distinguish between three relaxation-time scales for the director (τ_d), the velocity (τ_{visc}) and the charge (τ_q). The following expressions are easily derived:

$$\tau_d = \gamma_1 d^2/(\pi^2 k_{11}), \quad \tau_{visc} = 2\rho_m d^2/(\pi^2 \alpha_4), \quad \tau_q = \varepsilon_0 \varepsilon_\perp/\sigma_\perp. \tag{12}$$

τ_d is typically the longest time ($1 - 10\,\mathrm{sec}$) followed by τ_q ($\sim 10^{-3}s$). Even shorter is τ_{visc} ($10^{-5}s$). Thus the velocity field can usually be treated adiabatically (neglect of "inertial terms").

Now we are in a position to discuss the basic driving mechanism for EHC. The important point is that in almost all nematics σ_a is substantially positive (typically $\sigma_a/\sigma_\perp \sim 0.3 - 1$). Choosing materials with negative or only slightly positive dielectric anisotropy ϵ_a (here the materials show great diversity) one easily sees that in the presence of an applied field $\mathbf{E}$ and with a (small) spatial fluctuation of the director $\mathbf{n}$ a space charge ρ_{el} results (no solution of Eqs. (10) with $\rho_{el} = 0$). Roughly speaking the charges are focused at locations where the director bends. The bulk force in the Navier-Stokes equation (5) may then overcome viscous stresses and drive a velocity field $\mathbf{v}$. Via the viscous coupling (see Eq.(1)) this may enhance the spatial variation of the director and thus generate an instability. For low frequencies and for materials with not too large dielectric anisotropy the threshold voltage is of the order $V_c \approx \sqrt{\pi^2 k_{11}/(\sigma_\perp \tau_q)}$ and the introduction of the reduced control parameter $R = V^2 \frac{\sigma_\perp \tau_q}{\pi^2 k_{11}}$ is often useful.

3 Elements of the Theoretical Analysis

The theoretical methods for analyzing pattern forming instabilities are quite extensively discussed in the literature (see e.g. [2, 34, 35] and [36] in particular for nematics) The set of macroscopic equations as presented in the previous section can be written in the following symbolic form:

$$\mathcal{L}\mathbf{V} + \mathbf{N}_2(\mathbf{V}|\mathbf{V}) + \mathbf{N}_3(\mathbf{V}|\mathbf{V}|\mathbf{V}) + \cdots = (\mathcal{B}_0 + \mathcal{B}_1(\mathbf{V}) + \mathcal{B}_2(\mathbf{V}|\mathbf{V}))\frac{\partial \mathbf{V}}{\partial t}. \quad (13)$$

$\mathbf{V} = (\phi, \mathbf{n}, \mathbf{v}, ...)$ stands for the collection of field variables involved in EHC; they are choosen in such a way that $\mathbf{V} = 0$ corresponds to the nonconvecting ("basic", "primary") state. The components of the vector operators $\mathbf{N}_2$, $\mathbf{N}_3 \ldots$ are quadratic, cubic... in $\mathbf{V}$ and its spatial derivatives, whereas the quantities $\mathcal{L}$ and $\mathcal{B}_i$ represent matrix differential operators of the order in $\mathbf{V}$ indicated. A reformulation of Eq.(13) in Fourier space is often appropriate with respect to the horizontal directions $\mathbf{x} = (x, y)$ of "infinite" extent. In the transverse direction (z) the boundary conditions are satisfied by expanding $\mathbf{V}$ with respect to a suitable complete set of test functions (Galerkin method).

The onset of the instability is obtained from a standard linear stability analysis of the basic (primary) state. For a certain Fourier mode $\mathbf{U}(\mathbf{q})$ with the wave vector $\mathbf{q} = (q, p)$ one arrives from Eq. (13) at an eigenvalue problem:

$$\lambda \mathcal{B}_0(i\mathbf{q}, \partial_z, R)\mathbf{U}_{\mathbf{q}}(z) = \mathcal{L}(i\mathbf{q}, \partial_z, R)\mathbf{U}_{\mathbf{q}}(z), \quad (14)$$

where R denotes the main control parameter (e.g. the squared voltage in the case of EHC).

The eigenvalue $\lambda(\mathbf{q}, R) = \sigma(\mathbf{q}, R) \pm i\Omega(\mathbf{q}, R)$ with the largest real part, determines the growthrate σ and the frequency Ω of planforms near onset. The condition $\sigma(\mathbf{q}, R) = 0$ defines the neutral surface $R = R_0(\mathbf{q})$. Minimizing $R_0(\mathbf{q})$ with respect to $\mathbf{q}$ gives the threshold $R_c = R_0(\mathbf{q}_c)$ with the critical wavevector $\mathbf{q}_c = (q_c, p_c)$ and the critical frequency $\Omega_c = \Omega(\mathbf{q}_c)$, which vanishes for a stationary bifurcation (the more common case) but differs from zero for a Hopf (oscillatory) bifurcation.

In an axially anisotropic system like planar EHC, one speaks of "normal" rolls (see Fig.1a) if $\mathbf{q}_c$ is parallel to the preferred direction ($p_c = 0$). If $\mathbf{q}_c$ is at an oblique angle, one speaks of "oblique" rolls (see Fig.2a below). Clearly one then has the two symmetry-degenerate directions ("zig" and "zag") which may superpose to give rectangles. In the case of a Hopf bifurcation one has a degeneracy between waves traveling in opposite directions, which may also superpose to give standing waves. In the oblique-roll case even four degenerate modes are involved.

In EHC, for the usual case of driving with a pure ac field of angular frequency $\omega = 2\pi f$, the eigenvector $\mathbf{U}_{\mathbf{q}}$ of Eq.(14) inherits the additional periodic time dependence and the eigenvalue λ becomes a Floquet coefficient. Then there is an additional discrete symmetry $(z, t) \to (-z, t + 1/(2f))$ and each component of $\mathbf{U_q}$ has a definite parity: Generally the *"conductive"* mode (nonvanishing time average of all fields except the induced electric potential ϕ) destabilizes first at low frequencies f. Above a "crossover frequency" f_d the *"dielectric"* mode with the opposite parity determines the threshold. The existence of these two regimes planar EHC was first pointed out by the Orsay group [37, 38] (for further details see [12, 13, 16]).

A good understanding of the regime slightly above threshold is conventionally achieved by the weakly nonlinear analysis [39, 40, 1, 34]. The basic idea in its rather general form (for a recent more detailed presentation see [41, 42, 36]), is to reduce the phase-space dimension of the system by expanding $\mathbf{V}$ in an appropriate basis of states, characterized as the "dynamically active" ones [34]. At first the linear modes in Eq. (14) of positive or slightly negative growthrate are included in this set. Their expansion coefficients $A(\mathbf{q})$ correspond to order parameters, which vanishes at threshold. By a systematic expansion up to cubic order in A one arrives at the order parameter equations for the $A(\mathbf{q})$ [36, 43, 41], which permit the calculation of roll solutions and their stability at threshold. In particular the universal features become more transparent when the order parameter equations are reformulated in real space in terms of amplitude (envelope) or Ginzburg-Landau equations (GLE), for which the fast spatial variations ($\sim q_c^{-1}$) are separated out. This real-space formulation is essential when it comes to the description of more complex spatio-temporal patterns with disorder and defects, which have been studied extensively in EHC slightly above threshold (see Fig.1b). One ends up with the famous (slightly generalized) Ginzburg Landau equation [44]

$$\partial_t A = \lambda(\mathbf{q}_c - i\nabla, \epsilon)A - \gamma|A|^2 A \tag{15}$$

where λ is the linear growth rate of Eq.(14). Clearly λ is zero at threshold $\epsilon = (R - R_c)/R_c = 0$, $\nabla = 0$ and should be expanded in both arguments. At threshold it is sufficient to keep the following terms

$$\lambda(\mathbf{q}_c - i\nabla, \epsilon) \approx \epsilon + \xi_1^2\partial_x^2 + 2a\xi_1\xi_2\partial_x\partial_y + \xi_2^2 W\partial_y^2 - iZ\xi_1\xi_2^2\partial_x\partial_y^2 - \xi_2^4\partial_y^4. \tag{16}$$

The various constants in Eq.(16) determine the curvature of the neutral curve at $\epsilon = 0$.

On this level the expansion can be cast into an overall expansion scheme in terms of $\epsilon^{1/2}$, or equivalently A. In the anisotropic case, where there is no continuous degeneracy of the critical mode(s), one may in general assume $\epsilon \sim A^2 \sim \partial_t \sim \partial_x^2 \sim \partial_y^2$, so that the higher order terms $\sim \partial_x\partial_y^2$ and ∂_y^4 drop out ($W = O(1)$). The terms in Eq. (15) thus become uniformly of order $\epsilon^{3/2}$.

Going back to Eqs.(15, 16) with $a = 0$ it is easy to see that a decrease of W from positive to negative value owing to the change of some secondary control parameter, like the frequency in EHC describes a transition from normal to oblique rolls. Details of this transition, which is the analog of a Lifshitz point in the theory of equilibrium phase transitions, have been discussed elsewhere [45, 17, 46, 47]. The corresponding uniform scaling $\sim \epsilon^{3/2}$ as in Eq.(15) is recovered with $W \sim \epsilon^{1/2} \sim \partial_x$ and now $\partial_y \sim \epsilon^{1/4}$. This corresponds to the scaling adopted in isotropic media. In fact the well-known Newell-Whitehead-Segel amplitude equation for isotropic systems [48, 49] can now be obtained as the special case $W = 0$ and $Z = \pm 2$ in Eq.(16). In the case of a Hopf bifurcation (as observed in EHC) all the coefficients of Eq.(15,16) become complex. The resulting celebrated "complex Ginzburg-Landau equation" (CGLE) exhibits transitions (e.g. at the Benjamin-Feir instability) to various forms of spatio-temporal chaos and is presently studied intensely (for general reviews see e.g. [2, 34]).

The general structure of GLE's can be deduced a priori from the symmetries in a system as for instance the invariance properties with respect to space- and time translations or suitable reflections. The symmetries also manifest themselves in the linear growth-rate function (16). In simple cases very few coefficients determine the GLE, which can be extracted by comparison with the experiments. But in general for a quantitative comparison with experiments the often tedious calculation of the coefficients is inevitable [17, 12, 41, 50, 33, 16].

The GLE(15) determines only the stability of rolls with respect to disturbances of the generalized Eckhaus type, i.e. slow modulations of the roll spacing and undulations along the roll axis [45]. For the description of secondary bifurcations and modulated roll patterns away from threshold the weakly nonlinear analysis must be extended through the consideration of additional modes. For instance nonanalytic "mean flow " contributions of the order $\sim A^2$ have been included [43, 51].

In the simplest version the equation for the mean-flow amplitude B assumes the form:

$$\left(c_1\partial_x^2 + c_2\partial_y^2\right) B = q_1\partial_x\partial_y|A|^2 + q_4\partial_y\left(iA^*\partial_y^2 A + c.c.\right) + ... \qquad (17)$$

Eq.(15) is supplemented by a coupling term term $\sim A\partial_y B$. The mean flow is excited by long-wavelenght modulations of the pattern $A(x, y, t)$. Since the field B satisfies an anisotropic Poisson equation, its long-range character is evident. In nematics the mean flow turns out to enhance transverse modulations ($q_4 > 0$) in distinct contrast to isotropic fluids in most cases ($q_4 < 0$) [43].

Very recently is has been emphasized that in EHC other slowly damped modes have to be included in the set of active modes besides mean flow. Their inclusion leads to additional equations, which have a structure similar to Eq.(17). Most important are the twist modes of the director $\mathbf{n}$, which correspond to a rotation of $\mathbf{n}$ in the plane of the layer. They are weakly damped and can easily be excited near threshold. In the case of homeotropic alignement with rotational invariance this is obvious [52]. Any configuration with a finite in-plane component of the director spontaneously breaks a continuous symmetry, namely the isotropy. The growthrate of the associated Goldstone modes, which tend to restore the isotropy by a rotation of the in-plane director approaches zero in the homogenous case and is consequently small for slow modulations. But in the planar case inspite of the elastic torques due to anchoring of the director at the boundaries the damping of the twistmode is also weak [56].

The coupling of a short-wavelength patterning mode to a weakly damped long-wavelength one (like a Goldstone mode) readily leads to complex-spatio temporal patterns at threshold. One should mention a 1d model for seismic waves, where a spontaneous symmetry breaking is also important[53] or the coupling to a concentration mode in binary fluids [54]. The occurence of localized "worm" structures [21] and their modelling by a slow "charge-mode" [55] will be discussed in Sect.4.2.

An order-parameter approach for the regime further above threshold is certainly less systematic than the derivation of Eq.(15) since for instance the ϵ-scaling is no longer benificial. Therefore fully nonlinear calculations are indispensable. One follows the approved method in isotropic RBC: Galerkin expansions are applied and the resulting coupled highly nonlinear equations for the expansion coefficients are solved by Newton methods. Afterwards the solutions are tested for stability.

On the basis of numerical results and of comparison with experiments it has been demonstrated that EHC in nematics is one of the very few cases where secondary (and even higher bifurcations) can be captured semi-quantitavely by an order-parameter approach (see e.g. [56]). A further very useful simplified description makes use of a suitable phase-diffusion equation ([57]), which allows for a transparent description of the quite intricate bifurcation scenarios in EHC [58]. Convection instabilities in nematics are therefore an important paradigm for the validation of a reduced dynamical description of pattern forming systems in general.

4 EHC in the planar configuration

4.1 General background

The mechanism for the instability in EHC is based mainly on space charges generated by preferential conduction along $\hat{\mathbf{n}}$ ("charge focusing", see Sect.2). The basic idea has been suggested by Carr [59] and incorporated into a first one-dimensional model by Helfrich [60]. A first generalization to include the common case of *ac* driving followed almost immediately [37, 38]. It became clear that one has to distinguish between the low-frequency "conductive" and the "dielectric" regime above a certain transition frequency f_d. In the following sections we will first concentrate on the conductive regime.

For example, the need to go beyond one dimensional models became evident on the basis of the observation of specific three-dimensional structures, called **oblique** rolls [61, 62, 63, 64]. At low frequencies they can nucleate already at threshold, but otherwise appear at a secondary bifurcation. A representative pattern is shown in Fig.2a. One observes domains with two symmetry-degenerate directions ("zig" and "zag") separated by walls. Note that such a kind of spontaneously broken chiral symmetry is unique for anisotropic systems. In fact, the fully three dimensional linear stability analysis of the standard nematohydrodynamic equations (see Sect.2 and ,e.g., [17, 16] for references) has yielded the properties at onset of the instability: the threshold voltage and the wavevector of the twodimensional patterns. The agreement between experimental and theoretical results is typically quite satisfactory (see Fig.2b).

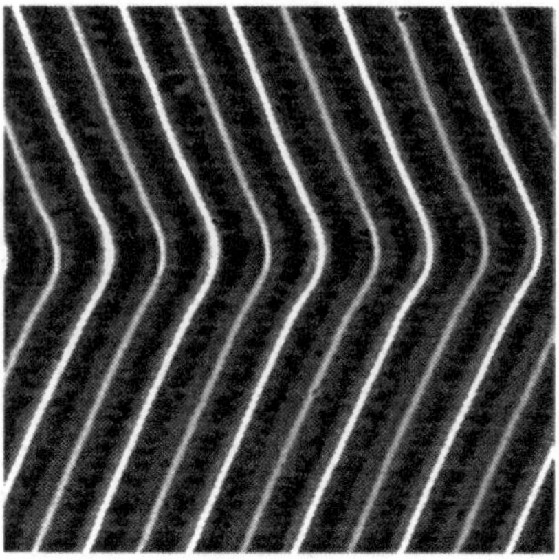

Fig.2a. Zig-zag pattern after increasing the voltage in Fig.1b.

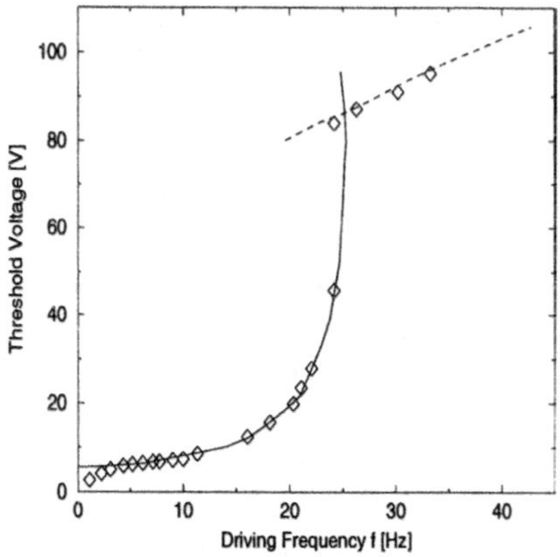

Fig.2b. Threshold curve for EHC as function of frequency. Experimental points from [65].

However the "standard model",SM, as described in Sect.2 ,where LCs are treated as ohmic conductors is incapable of explaining the Hopf bifurcation leading to travelling waves, which are often observed in sufficiently thin layers (below about 50μ) and clean materials (low conductivity) [63, 66, 67, 68]. An extension (the "weak-electrolyte model", WEM), where electric transport in the nematic is described in terms of two mobile ion species (of opposite charge) which are coupled via a slow dissociation-recombination reaction[69, 33] has proven to be successful (see Sect.4.2).

In the further progress of the theory the weakly nonlinear analysis in terms of Ginzburg-Landau amplitude equations [17, 12] have provided information on the Eckhaus stability boundaries of periodic roll patterns, in good agreement with experiments [70, 71, 72]. An even more sensitive test is the analysis of the dynamics and interaction of defects (dislocations) in ideal patterns, which compares well with experiments [73, 74, 75, 76] too. It was realized that mean-flow effects are very important in the transition from ordered periodic to weakly turbulent patterns [47, 77, 78, 43] (for a phenomenological treatment, see [79, 80, 81, 82]).

Fully nonlinear theoretical studies have followed [83], which put the concept of **abnormal** rolls, at first observed in homeotropic EHC, [84] on a sound basis. At low frequency their bifurcation is usually preceded by a long-wavelength zig-zag instability; at higher frequency they appear after a secondary bifurcation of normal rolls. The wave vector of abnormal rolls is parallel to the preferred x -direction, but the underlying director field has experienced a spontaneous rotation in the $x-y$ plane. Previous experimental findings in EHC in different geometries are now understood for the first time and new experimental activities have been motivated. The abnormal rolls will be discussed in detail in Sect.4.3. Very recently the dielectric regime has attracted considerable attention. For instance a first convincing explanation for the common chevron pattern has been given (see Sect.4.4).

4.2 The weak electrolyte model (WEM)

The WEM model [33, 69] has provided the basis for the understanding of the Hopf bifurcation observed quite frequently at threshold [85, 86]. For a Hopf bifurcation two processes that compete on a comparable time scale are necessary. One might think of the director and charge relaxations, but they do not compete. Instead they support each other usually; in addition the time scales (12) are very different; the director relaxation is much slower than all other processes, and thus determines the dynamics.

According to the WEM model it is assumed that a slow process is connected with the relaxation of the mobile ion densities n^+ and n^- on the time scale τ_{rec}, which may results from a dissociation-recombination reaction. One thus obtains for singly charged ions

$$\rho_{el} = e(n^+ - n^-), \qquad \boldsymbol{\sigma} = \sigma\boldsymbol{\sigma}', \tag{18}$$

where

$$\sigma = e(\mu_\perp^+ n^+ + \mu_\perp^- n^-), \quad \sigma'_{ij} = \delta_{ij} + \frac{\sigma_a}{\sigma_\perp} n_i n_j. \tag{19}$$

Here $\mu_\perp^\pm$, $\mu_\|^\pm$ are the ionic mobilities perpendicular and parallel to the director, respectively. For simplicity the anisotropies have been assumed to be the same for both types of ions so that $\sigma_a/\sigma_\perp = \mu_\|^\pm/\mu_\perp^\pm - 1$. Thus σ is an additional variable now. From the balance equations for n^+ and n^- one easily recovers Eq.(10) and in addition one obtains

$$\begin{aligned} &\tfrac{d\sigma}{dt} + \nabla \cdot \left[\left((\mu_\perp^+ + \mu_\perp^-)\sigma + \mu_\perp^+\mu_\perp^-\rho_{el}\right)\right] \qquad (20)\\ &= \tfrac{\sigma_{eq}}{2\tau_{rec}}\left[1 - \tfrac{(\sigma+\mu_\perp^-\rho_{el})(\sigma-\mu_\perp^+\rho_{el})}{\sigma_{eq}^2}\right] \approx \tfrac{1}{\tau_{rec}}\left[(\sigma-\sigma_{eq}) + \tfrac{\mu_\perp^+-\mu_\perp^-}{2}\rho_{el}\right], \end{aligned}$$

where $\sigma_{eq} = e(\mu_\perp^+ + \mu_\perp^-)n_{eq}$ contains the equilibrium ion density n_{eq}. The last expression is obtained by linearization in the quantities $n^+ - n_{eq}$ and

$n^- - n_{eq}$. In this model the effect of ion accumulation on the conductivity is thus included, whereas ionic diffusion is neglected as in the SM. The charge accumulation counteracts the standard (Helfrich) mechanism of generation of space charges. If the time scale τ_{rec} is sufficiently slow one can expect to find an oscillatory behavior of the system at threshold, i.e. a Hopf bifurcation.

The detailed linear stability calculations for the WEM model have been performed [69, 33] using the same approximations that led to the analytic threshold formulas within the SM [87, 17, 12, 16]. It is found that there is an upward shift in the stationary threshold, which may be quite small, and more importantly, that indeed a Hopf bifurcation occurs [85, 18] with critical frequency

$$\omega_H \tau_d = 2\pi f_H \tau_d = \frac{R_c \tilde{\alpha} C}{1+\omega'^2} \sqrt{1 - \left(\frac{(1+\omega'^2)\lambda_\sigma \tau_d}{R_c \tilde{\alpha} C} \right)^2}, \tag{21}$$

when the expression under the square root is positive. Here $R = V^2 \frac{\sigma_\perp \tau_q}{\pi^2 k_{11}}$ is the reduced control parameter (see Sect.2), $\tilde{\alpha}^2 = \mu_\perp^+ \mu_\perp^- \gamma_1 \pi^2 / (\sigma_\perp d^2)$ is proportional to the geometric means of the mobilities and $\omega' = \omega \tau_q \beta$ with $\beta = \frac{(\sigma_\| / \sigma_\perp) q'^2 + p'^2 + 1}{(\epsilon_\| / \epsilon_\perp) q'^2 + p'^2 + 1}$ is a reduced frequency. Moreover $\lambda_\sigma = -[\tau_{rec}^{-1} + \tau_d^{-1} R \tilde{\alpha}^2 \beta / (1 + \omega'^2)]$ (< 0) is the damping rate of the (new) WEM mode. Its dominant contribution is usually just determined by the ion recombination rate $1/\tau_{rec}$. The factor C [33] contains only SM quantities and is about one (see Ref. (28) in [86] for a misprint in [33]).

For the Hopf bifurcation to occur, i.e. for a positive argument of the square root in Eq.(21), the quantity $\tau_d / (\tilde{\alpha} \tau_{rec})$ must be sufficiently small. This requires that the recombination of ions is sufficiently slow and that the layer is sufficiently thin and clean. Note, however, that for materials with negative dielectric conductivity, where R_c diverges at the cutoff frequency f_d (in the approximation used), the Hopf condition is always satisfied near the cutoff, and the Hopf frequency, which is then just given by the prefactor of the square root in Eq.(21), becomes large there. This appears to be consistent with the experiments [63, 66, 67]. Moreover, the prediction of the theory, that for materials with vanishing dielectric anisotropy the Hopf frequency becomes essentially independent of the external frequency, has been verified experimentally using the material I52 [85]. I52 has the property that ϵ_a changes from negative to positive values when the temperature is increased through $T \approx 60°$C.

A quantitative test of Eq.(21) with experiments has been performed recently for the nematic Phase 5, where (almost) all material parameters have been measured. In Fig.3a the Hopf frequency (21) is shown as function of the ac frequency (solid line). The units are chosen according to the predicted theoretical scaling behavior: $\omega_h \sim d^{-3} \sigma_\perp^{-1/2}$ is easily obtained from Eq. (21), if the $d-$dependencies of the director relaxation time τ_d, the threshold R_c

and of $\tilde{\alpha}$ are combined; the proportionality $\sim \sigma_{\perp}^{-1/2}$ is due to $\tilde{\alpha}$. The ac-frequency scale in ω' is set by τ_q. The agreement with the WEM theory is very convincing.

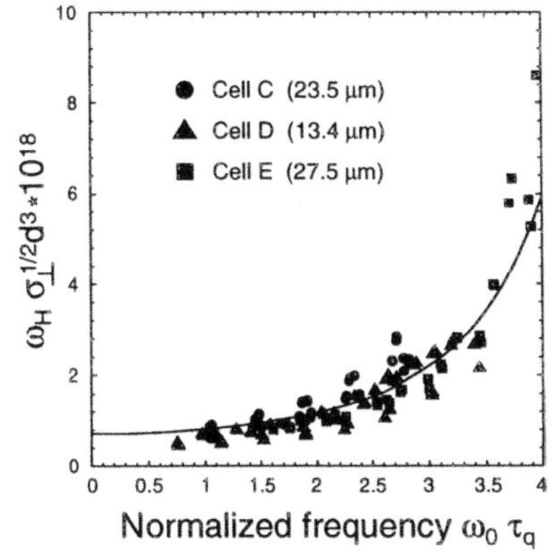

Fig.3a. The Hopf frequency of travelling rolls multiplied by $(\sigma_{\perp}^{eq})^{1/2} d^3$ (see text) .

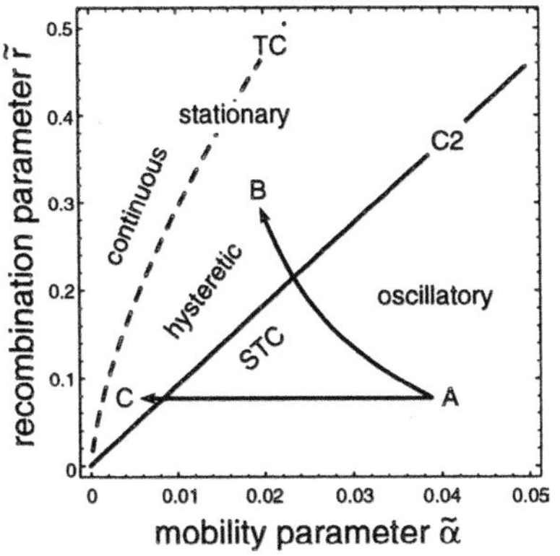

Fig.3b. Phase diagram in the $\alpha - r$ plane (see text)

Subsequently a weakly nonlinear analyis of the WEM model has been performed [86] in order to determine the bifurcation type at threshold. Some results are shown Fig. 3b. In dependence on the mobility parameter $\tilde{\alpha}$ and the recombination parameter $\tilde{r} = \tau_d/\tau_{rec}$ one finds below the curve labelled $C2$ a supercritical Hopf bifurcation. In the regime bounded by C2 and the curve labelled TC the bifurcation is stationary and hysteretic, while above TC one finds a continuous stationary bifurcation. In the limit $\tilde{r} \longrightarrow \infty$ the standard model is recovered. The point A corresponds to the experiments discussed in [21] for the material I52 at 30^0C. The arrow to point C indicates a change of the bifurcation type as the temperature is increased to 60^0. In contrast to the SM the bifurcation scenarios depend also on the cell thickness d: The arrow to B corresponds to a doubling of d.

As has already been mentioned, the WEM model has first been compared with experiments for the new nematic I52 [85]. Recently two-dimensional localized states, called 'worms' (see the snapshots in Fig.4), have been described in this material [88, 21]. They are localized and have a unique, small width in the y direction but are of much a greater varying length along $\hat{\mathbf{x}}$, i.e. the anisotropy axis. Rapidly traveling waves in one direction $\parallel \hat{\mathbf{x}}$ pass under a slowly moving envelope moving in the opposite direction. Their number increases with increasing voltage and they tend to grow by coalescing until at higher voltage the cell is filled with convection.

The theoretical analysis of the worms is challenging. One-dimensional pulses have been experimentally discovered in binary fluids [89, 90] and have been obtained as solutions of the one-dimensional CGL [91] (for further references, see [55]). Two-dimensional localization in such hydrodynamic pattern-forming systems is not well documented in experiments. From theoretical studies (see e.g. [92, 93]) one expects that 2d-localized structures can be obtained in the case of a subcritical Hopf bifurcations. Thus an ad hoc model [94] has been proposed in the form of an anisotropic Swift-Hohenberg equation with adjustable complex coefficients and a quintic nonlinearity to capture the subcritical bifurcation. It is not too surprising that indeed localized structures similar to the experiments could be produced. However, the basic assumption of the subcritical Hopf bifurcation in planar EHC is in conflict with theory and experiments [95].

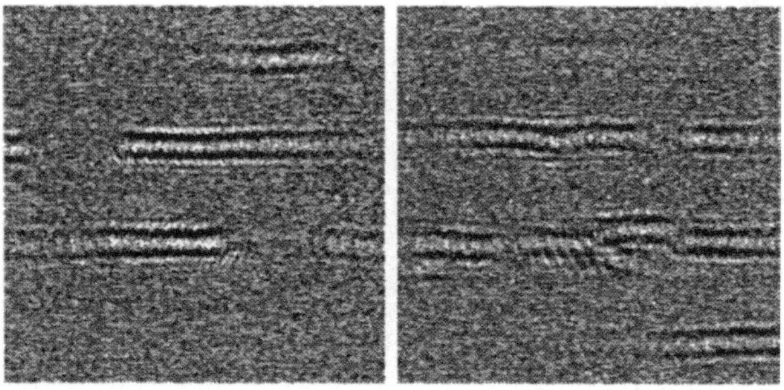

Fig.4. Localized worm structures for $\epsilon = 0.012$ left and for $\epsilon = 0.057$ (right) (Courtesy G. Ahlers)

Therefore a more realistic model was recently proposed [55] where two amplitude equations for oblique travelling rolls were coupled to a slow mode, which might be identified with the charge-accumulation mode. Even details of the experiments could be reproduced satisfactorily. From a theoretical point of view the model is also very interesting, since it involves a new mechanism for the localized states. One has a forward Hopf bifurcation for extended cell-filling patterns in agreement with the "microscopic" theory [86]. However, the transition to localized states is predicted to be hysteretic in line with the recent experimental results [95]. The final step, i.e. the derivation of the new equations from the WEM model, has not yet been accomplished.

While the localized worms are certainly spectacular in EHC for I52, the cell filling-states developing from the supercritical Hopf bifurcation are also very interesting. The system is below a Lifshitz point, obviously with a bifurcation of four degenerate oblique-roll solutions (zig-zag, left-right traveling).

Their interaction can easily lead to various types of spatio-temporal chaos (STC) near threshold [96], which can be controlled through a variation of the conductivity of the nematic. The experimental have stimulated efforts of theoretical modeling [97].

4.3 Abnormal rolls

Conventional normal rolls (NR) can only exhibit point defects (dislocations). Oblique rolls (OR) break the reflection symmetry $y \to -y$ and thus can show in addition line defects (grain boundaries) separating domains with wavevectors (q, p) ("zig") and $(q, -p)$ ("zag") [98, 99]. In Fig.5a,b typical grain boundaries along $\hat{\mathbf{x}}$ and $\hat{\mathbf{y}}$ are shown for the nematic Phase 5.

Surprisingly, in an apparently NR structure, line defects have been observed as well. With increasing voltage (i.e. ϵ) the angle between the roll axis and $\hat{\mathbf{y}}$ is found to systematically decrease and may even reach zero. In this process the grain boundary of Fig.5b is transformed into the wall of Fig.5c. That the two domains on either side of the wall are not equivalent becomes more evident as ϵ is further increased. Indeed, one then observes the branching of an additional wavevector $(k_x, +k_y)$ on the right side, and $(k_x, -k_y)$ on the left side (see Fig.5d).

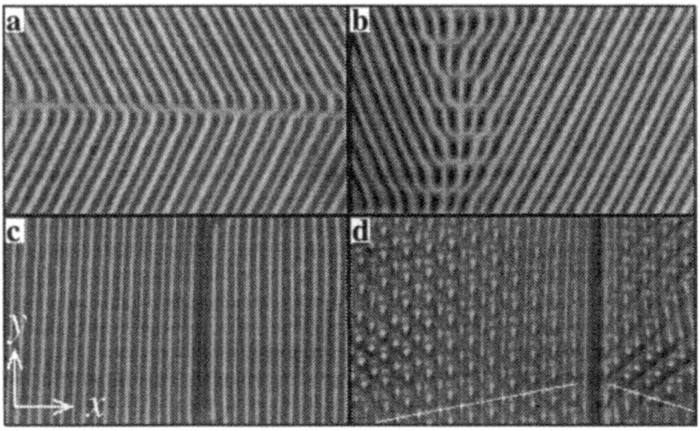

Fig.5. Representative snapshots of the experimental evolution of electroconvection in a cell of thickness 5μm, at $\omega\tau_0 = 0.3$. a, b: zig-zag structures near threshold ($\epsilon \sim 0.02$), with "horizontal" and "vertical" grain boundaries. c: wall at $\epsilon \sim 0.40$ originating from the grain boundary of b. d: varicose structures at $\epsilon \sim 0.70$, originating from the structure c. The two modulation directions are indicated by the white lines.

The existence of a two-variant state of normal rolls indicates a new symmetry breaking. Those rolls have been named *"abnormal rolls"* (AR), a term introduced in the context of homeotropically aligned cells [84] for rolls with an optically detected symmetry breaking, see Sect.5 below.

AR cannot be understood by the standard weakly nonlinear approach where the roll structure is characterized unambiguously by its wavevector and the linear eigenvector at threshold (see Sect.3).

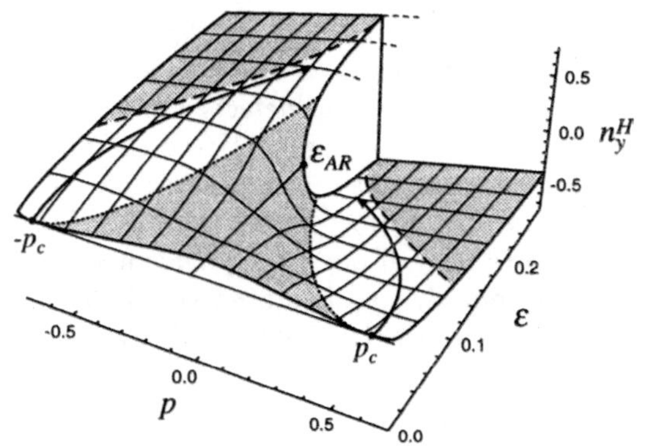

Fig.6 Amplitude n_y^H of the twist mode in roll patterns of wavevector $\mathbf{q} = (q_c, p)$, as a function of p and ϵ at $\omega\tau_0 = 0.3$ (parameters of Phase 5). Unstable regions in gray (Eckhaus boundary dotted, Bimodal Varicose boundary dashed). Arrows: sketch of the experimental evolution of $p(\epsilon)$. In the experiments also q evolves with ϵ, which leads to quantitative changes of the diagram.

The key result of a fully nonlinear Galerkin analysis is presented in Fig.6. It shows the amplitude n_y^H of a twist mode corresponding to a *homogeneous rotation* (i.e. independent of x, y) of the director in the x, y plane as function of ϵ and p for roll solutions with $\mathbf{q} = (q_c, p)$. This rotation is symmetric with respect to the mid plane of the layer, and is largest there. At $p = 0$, n_y^H vanishes if ϵ is sufficiently small, in agreement with the predictions of the weakly nonlinear theory. The novel and striking feature is the pitchfork bifurcation at a secondary transition point $\epsilon = \epsilon_{AR}$ to a state with $n_y^H \neq 0$ (at $p = 0$!). In these AR the $y \mapsto -y$ symmetry is spontaneously broken without tilting of the rolls. In addition the amplitude of the (periodic) n_z component, which increases like $n_z \sim \sqrt{\epsilon}$ for small ϵ, as expected from the weakly nonlinear theory, is found to remain nearly constant for $\epsilon \geq \epsilon_{AR}$. For $p \neq 0$ the director "prefers" to rotate towards the axis of the rolls: e.g. $n_y^H < 0$ for $p > 0$ (see the lower sheet in Fig.6).

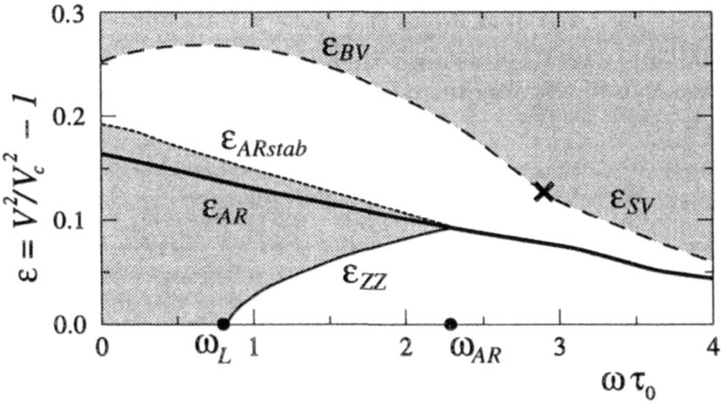

Fig.7. Stability diagram for rolls in the normal direction ($p = 0$) in the ω, ϵ plane (unstable regimes in gray). ϵ_{AR} (thick line): NR $\rightarrow$ AR bifurcation. ω_L = Lifshitz frequency, ϵ_{ZZ} (thin line): zig-zag instability. ϵ_{ARstab} (dotted line): restabilization of the abnormal rolls. The upper dashed line is at low frequencies a short-wavelength varicose instability (ϵ_{BV}) that transforms into a long-wavelength skewed-varicose instability (ϵ_{SV}) at $\omega\tau_0 \simeq 2.8$ ($\times$).

In Fig.7 the stability regime (in white) for rolls with $\mathbf{q} = (q_c, 0)$ is shown in more detail in the ϵ, ω plane. For $\omega < \omega_L$, where the primary bifurcation is to OR, NR are unstable near threshold ($\epsilon \simeq 0$) against long-wavelength undulations along the roll axis ("zig-zag" instability). The mechanism becomes less efficient for AR, where the $y \mapsto -y$ symmetry is broken. Thus AR become stable when $|n_y^H|$ is large enough, for $\epsilon > \epsilon_{ARstab}(\omega)$ (dotted line). When ϵ is increased further the AR are destabilized at $\epsilon_{BV}(\omega)$ (dashed line), now by a short-wavelength instability with a new wavevector $|\mathbf{k}| \simeq q_c$ roughly parallel to the homogeneous part of the in-plane director (n_x^H, n_y^H) in the AR. The bimodal patterns in Fig.5d then obviously result from the destabilisation of AR at $\epsilon = \epsilon_{BV}$ (see Fig.7).

In an intermediate frequency range $\omega_L < \omega < \omega_{AR}$, NR are stable for $\epsilon \leq \epsilon_{ZZ}(\omega)$ (solid line). At ϵ_{ZZ} a zig-zag instability (ZZ) develops [100], which derives continuously from the oblique rolls at onset for low ω: $\epsilon_{ZZ}(\omega) \rightarrow 0$ for $\omega \rightarrow \omega_L$. Above ϵ_{ZZ} the situation is analogous to the low-frequency regime, i.e. one gets stable AR for $\epsilon > \epsilon_{ARstab}$. When ϵ is increased further the AR experience the varicose destabilisation at ϵ_{BV} as before.

The stability limits $\epsilon_{AR}, \epsilon_{ARstab}$ and ϵ_{BV} decrease with increasing ω as seen in Fig.7. Above the frequency ω_{AR}, where ϵ_{ZZ}, ϵ_{ARstab} meet the line ϵ_{AR}, the bifurcation NR $\rightarrow$ AR occurs *in the stable range*. Along the varicose line $\epsilon_{BV}(\omega)$, the modulation wavevector $\mathbf{s} = \mathbf{k} - \mathbf{q}$ approaches zero with increasing ω, whereas the ratio s_y/s_x stays finite ($s_y/s_x \simeq \pm 2.7$). Above

$\omega\tau_0 \simeq 2.8$ (see the cross in Fig.7), the varicose instability thus becomes a long-wavelength modulational instability of the skewed-varicose type [101]. The substance MBBA has no Lifshitz point. Therefore the part of the stability diagram for $\omega > \omega_L$ in Fig.7 is shifted to lower ω.

The detailed bifurcation sequences can be discussed with the help of Fig.8 at a representative frequency $\omega\tau_0 = 0.5$ for patterns with wave vectors $\mathbf{q} = (q_c, p)$. In contrast to Fig.6 the amplitude n_y^H is not shown. For $\epsilon \leq \epsilon_{zz} \sim 0.05$ normal rolls ($p = 0$) are stable. Above ϵ_{zz} a bifurcation to oblique rolls with finite p takes place, where a whole p band is available except for the unstable " bubble" marked by the dashed line. At ϵ_{ARstab} (in the ZZ-unstable regime within the bubble) abnormal rolls bifurcate. They become ZZ-stable at ϵ_{ARstab}.

It seems that in experiments when slowly increasing ϵ beyond ϵ_{ZZ} the patterns follow at first more or less the boundary curve of the unstable bubble. Thus the angle $\alpha = \arctan(q/p)$ of the wavevector with respect to the x-axis does not become too large. The pattern appearing beyond the ZZ-instability can be described at first as rolls with a smooth undulation along their axis. If ϵ is increased further (presumably above ϵ_{AR}) zig-zag patterns with sharp boundaries as shown in Fig.2a develop. The sequence NR-undulated rolls-OR has been first described in Ref. [62]. For higher $\epsilon \leq \epsilon_{BV}$ it is a pattern selection problem whether the system actually uses the opportunity to return to AR patterns with $p = 0$ or not. In most experiments reported up to now p decreases but $p = 0$ is not reached (except in the case of a wall as in Fig.5c). However, there is a new very promising approach to map the stability diagram in Figs. 7,8. directly. A strip-like cell is used where straight rolls along the long side seem to be strongly preferred such that above ϵ_{AR} perfect AR are recovered. The whole scenario is quite complicated and involves modulated roll patterns, which can exist in the unstable bubble when decreasing ϵ starting from above ϵ_{ARstab}. There exist already first results based on a phase diffusion equation which seem to explain the rather intricate phenomena [58, 57].

In Ref. [83] further examples are given which prove that the concept of abnormal rolls is very important for EHC in nematics. There exists semi-quantitative agreement with experiments, which now allow for a new interpretation. This applies for instance to various kinds of (dynamic) defect structures [98, 99, 102], which are presumably triggered by the SV-instabilites in Fig.7. The bifurcation phenomena in particular associated with abnormal rolls appear to be generic for planar nematic convection, since they are also found in Rayleigh-Bénard convection of nematics [56].

Finally one might ask why abnormal rolls have escaped the attention of experimentalists until quite recently. The direct observation of a homogeneous n_y distortion in the planar configuration is difficult. The optical axis of the uniaxial nematics is the director. Maximal optical contrast is achieved in extraordinary light, when the polarization of the light is parallel to the

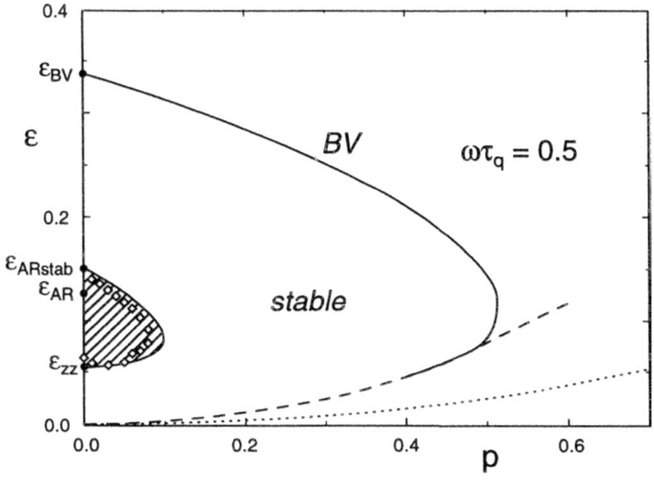

Fig.8. Stability diagram (MBBA) for rolls escaping in the oblique direction at fixed ($q = q_c$) The unstable bubble was tested with Galerkin expansions truncated at four (shaded) and six terms (diamond)

director at the plates, $\hat{\mathbf{x}}$. The polarization then follows adiabatically any n_y excursion in the bulk and exits parallel to $\hat{\mathbf{x}}$ again, as long as n_y varies slowly over a wavelength of the light (Mauguin's principle). Therefore only n_z is monitored, whereas the n_y distortion has almost no effect on the propagation of the rays. An exception is the grain boundary in Fig.5c with opposite n_y^H orientations on either side. The resulting strong n_y gradients inside the grain, coupled to a localized peak of n_z, explain the observed optical contrast of the wall. However, a small modification of the optical setup permits to visualize the small differences between clockwise and counterclockwise twisted domains in general. The trick is to lift the (optical) equivalence according to Maugin's principle through the introduction of a quarter-wave plate at a 45^0 orientation to $\hat{\mathbf{x}}$ [103]. It is also very illuminating to observe the slow relaxation time of the twist mode. At first a strong in-plane rotation is seeded in an abnormal roll pattern for $\epsilon \geq \epsilon_{AR}$. Then ϵ is suddenly quenched to zero. While the pattern itself vanishes almost immediately, it takes much longer time for the director to turn back in the x direction.

4.4 Dielectric regime

In the dielectric regime (see Sect.3) the director and the velocities oscillate with the external frequency about the basic state while the polarity of the space charge remains time independent in leading order. The theoretical analysis of the patterns follows in principle the techniques already approved in the conduction regime.

With respect to the linear regime it is not difficult to determine numerically the critical wavevector and the onset voltage. The salient features are however already contained in an approximate analytic threshold formula, which can be derived with the help of appropriate lowest-order test functions [16, 104]. In comparison with the conductive regime convection sets in at a higher voltage, which increases continuously with the ac-frequency ω. In order to overcome the stabilizing dielectric torque larger in-plane gradients are required for the convection destabilisation and thus the wavelength increases a function of ω. Unless the critical wavelenght becomes comparable to the cell thickness the dielectric instability is a "bulk instability" which is insensitive to the boundary conditions. The wavelenght is determined by the diffusion length of the bend deformation $k_{33}/(\gamma_1\omega)$, i.e. by material parameters and not by the vertical distance d of the cell boundaries. The convection rolls are very thin and the pattern looks like a planar lamellar structure. Consequently the appropriate order parameter in the dielectric regime is the d-independent electric field instead of the voltage as in the conductive regime.

In addition one finds from the analytical formulation that the threshold stays finite for $\epsilon_a \rightarrow 0$, and in fact remains so when ϵ_a becomes positive. This appears to be in conflict with early treatments [38], but improved formulas have been developed later [105, 7]. Very recent and particularly careful measurements of the linear properties in the nematic Phase 5 agree very well with the theory [106]. In MBBA the measured threshold exceeds typically the theoretical prediction. The reason might be, that MBBA is a quite unstable nematic and thus its conductivity increases with time due to molecular dissociation processes. In addition dopants are often added to increase the conductive frequency range. It has never been tested systematically whether the commonly used material parameter may not have to be modified accordingly. Also the flexoelectric effects could be more important than in the conduction regime [104]. On the other hand it is reassuring that quite subtle properties like the phase shift of the director oscillations with respect to the applied voltage compare favourably with experiment [107].

Passing to the weakly-nonlinear dielectric regime the lamellar-type structure is only weakly anchored at the boundaries and it should be easily destabilized by (slow) transverse modulations. Thus it is not surprisising that again the twist mode is found to be responsible for a secondary bifurcation to abnormal rolls, typically at a very small $\epsilon \sim 10^{-2}$ [104]. The results are consistent with recent experiments in Phase 5, where the dielectric regimes takes over at a small frequency ($f_d = 26Hz$) [108]. The secondary instability is found at a rather small $\epsilon_{sb} \leq 0.08$. As already stated abnormal rolls at a very small ϵ were also found directly in an particularly designed experimental setup [103].

With increasing ϵ the situation becomes simpler at least from the experimental point of view. We again refer to the recent experiment [108]; above ϵ_{sb} the rolls become slightly undulated and a persistent dynamics is introduced by the appearance of well separated dislocations. The number of these dislocations increases continuously and above $\epsilon = 0.24$ one observes periodically ordered chains of dislocations oriented along the normal-roll direction. The polarity of the defects is the same along each chain and alternates from chain to chain. The defects move in alternating directions (for a snapshot see Fig.9). The name "chevron structure" has been coined for this pattern, which is known for many years (see e.g [109]).

Fig.9. Chevron pattern at $\epsilon = 0.26$ in Phase 5 [108]

Recently a first theoretical explanation has been proposed for the chevrons [110]. The starting point are Eqs.(24, 25) (see Section 5) originally derived for homeotropic EHC. They describe the coupling of the patterning mode A to the order parameter of the abnormal rolls (i.e. the in-plane director $\mathbf{c}$ with an angle ϕ relative to the x axis). In the dielectric as well as in the homeotropic case the convection starts in a situation where the electric field is perpendicular to the in-plane director $\mathbf{c}$, which is subject to a very weak external torque. The boundary effects in the dielectric case can be identified with the action of a small magnetic field $\mathbf{H}$ parallel to $\mathbf{x}$ in the homeotropic case. In some parameter regimes defect-chaotic solutions (see Figs.12, 13) are indeed observed in simulations of Eqs.(24, 25), when $h^2 = H^2/(-2\Gamma\epsilon)$

drops below a critical value $O(1)$ (corresponding to small anchoring). With increasing ϵ the defects become then ordered in chevron-like stripes as in Fig. 9 [111].

The quite regular chevrons at larger ϵ can then be captured theoretically by a coarse-grained description in terms of reaction-diffusion equations [111, 110]. The key to the formation of chevrons is the "anomalous" torque on $\mathbf{c}$, which already leads to the abnormal-roll bifurcation at small ϵ and to defect chaos with increasing ϵ. When the density of defects is sufficiently high a topological charge field can be defined as the difference between the densities of defects with positive and negative polarity. The chaotic motion of the defects leads to a finite diffusivity of the topological charge density. In the theoretical description the angle ϕ plays the role of an activator, which enhances the (local) topological charge imbalance. The reason is that the rolls tend to follow the rotation of $\mathbf{c}$. The resulting reorientation of the roll pattern entails the motion of defects with oppositely charges in opposite directions, which then play the roll of an inhibitor. Elastic effects are responsible for diffusion in ϕ. Once having arrived at the level of a reaction-diffusion systems the possible scenarios are well known: One can have either oscillatory instabilities or a steady spatially periodic Turing pattern, which, in the present system, is identified with the chevron structure.

5 EHC in the homeotropic configuration

In this chapter we will discuss in some detail the electrohydrodynamic instabilities in the homeotropic configuration. The general setup is very similar to Fig.1a except that the director is initially oriented parallel to the electric ac field $\mathbf{E} \parallel \mathbf{z}$ by an appropriate surface treatment of the confining plates. The system is now isotropic in the plane of the nematic layer. We will address the case of negative dielectric anisotropy $\epsilon_a < 0$, where, for energetic reasons, the director has the tendency to orient perpendicular to $\mathbf{E}$ in order to minimize the electric torque. For the case of positive ϵ_a, see e.g. [112, 113, 16]. If $\mathbf{E}$ (or the applied voltage $V = |\mathbf{E}|d$) is strong enough to overcome the opposing elastic torque at first the homogeneous Freédericksz transition [4, 5] shows up, which is a equilibrium transition. The isotropy is then spontaneously broken and a definite orientation of the in-plane director component is singled out, which we call the $x-$ axis. In some cases an additional magnetic field $\mathbf{H} \parallel \mathbf{x}$ is applied perpendicular to $\mathbf{E}$ to lift the rotational invariance and to single out a definite orientation. Under the combined action of H and V the Freédericksz transition takes place if

$$\left(\frac{V}{V_F}\right)^2 + \left(\frac{H}{H_F}\right)^2 \geq 1. \tag{22}$$

is fullfilled [4]. V_F and H_F are defined as:

$$V_F = \pi\sqrt{\frac{k_{33}}{|\epsilon_a|\epsilon_0}}, \quad H_F = \frac{\pi}{d}\sqrt{\frac{k_{33}}{|\chi_a|\mu_0^{-1}}} \tag{23}$$

where ϵ_a and χ_a are the dielectric and diamagnetic anisotropies, respectively, and k_{33} is the elastic bend constant (see Sect.2). By testing experimentally the transition lines according to the inequality (22) one can check for imperfections in the anchoring of the director at the cell boundaries [106].

Above the Freédericksz transition a planar layer develops in the central region of the cell. In view of the obvious analogy to the planar configuration it is not too surprising that it can become unstable against a convection instability above a second threshold. The transitions are sketched in Fig.10.

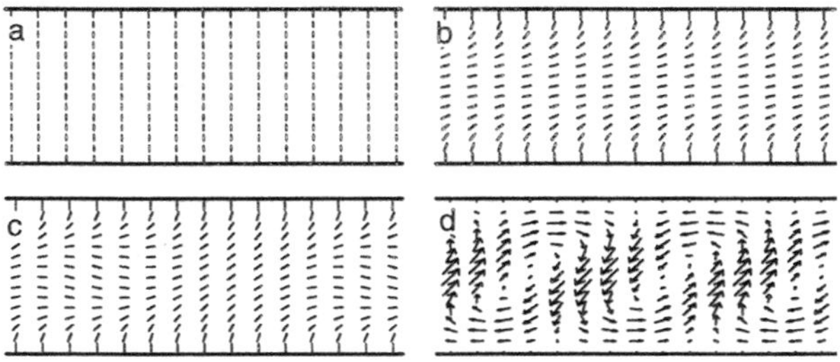

Fig.10 a) Homeotropic groundstate, b) Director configuration after the Freédericksz transition c) Modulated director configuration in the EHC state. d) Velocity field in the EHC state.

The theoretical analysis of the bifurcations is more complicated than in the planar configuration due to the inhomogeeous ground state. It has been achieved with the use of Galerkin-expansions [112] which are numerically demanding. The linear analysis reproduces the scenarios already known from planar EHC. At low frequencies oblique rolls, at higher frequencies, above a Lifshitz point ω_L, normal rolls are found. By further increasing the frequency one switches at $\omega = \omega_f$ from the conductive to the dielectric regime. The agreement with experiments is very satisfactory [106].

In Fig.11a the secondary destabilisation of normal rolls at $\omega\tau_0 = 1$ is shown as function of an imposed planar magnetic field H on the basis of a full Galerkin calculation (GAL). For a sufficiently strong magnetic field a long wavelength zig-zag instability is obtained. The details become clearer from Fig.11b, where the secondary destabilisation ϵ_{sb} is shown as function of $\omega\tau_0$ for a fixed magnetic field ($H = 1$). Starting from the Lifshitz point (near $\omega\tau_0 = 0.6$) the zig-zag line increases continuously. Near $\omega\tau_0 = 1.5$ there is an abrupt change to the homogeneous abnormal-roll bifurcation. The bifurcation scenario is in perfect analogy to the planar case (see Figs.7,8). The bifurcation lines do not join smoothly at the Lifshitz point. This is explained by the fact that the stability of patterns with the critical wavevector $\mathbf{q_c} = (q_c, p_c)$ is shown. Approaching ω_L from above ithe bifurcation point ϵ_{ZZ} (the lowest point of the bubble in Fig.8), moves down until it vanishes at ω_L. Below ω_L the bubble is deformed and increasingly less important for increasing p; the patterns remain stable up to much higher ϵ.

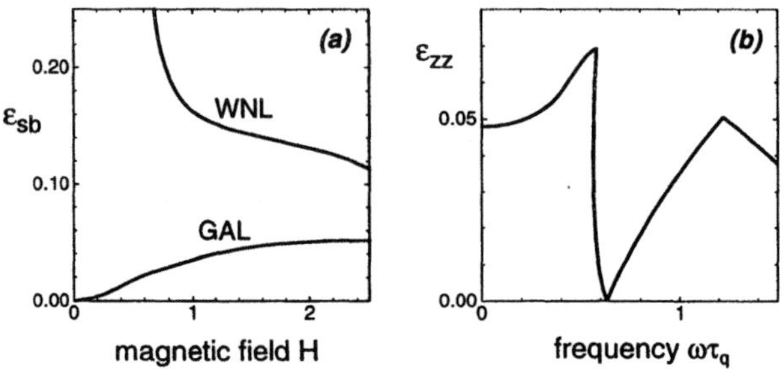

Fig.11 a) Secondary destabilisation ϵ_{sb} of normal rolls at bandcenter for $\omega\tau_0 = 1.0$ (GAL: Full Galerkin analysis; WNL: weakly nonlinear analysis. b)Secondary destabilisation of normal rolls at bandcenter as function of the frequency $\omega\tau_q$ for the reduced magnetic field $h_x = 1$, from [112].

However, in the limit of zero magnetic field, when the preferred axis, i.e the in-plane director $\hat{\mathbf{c}}$, is not fixed externally, the analogy to the planar case breaks down. This is already evident from Fig.11a where normal rolls become unstable immediately at threshold. The reason is that transverse modulations lead to a torque on $\hat{\mathbf{c}}$ which cannot be compensated. Oblique roll solutions do not even exist because of this torque [114]. Even with a finite magnetic field the standard weakly nonlinear analysis breaks down (see the curve WNL in Fig.11a) because the rotation of $\hat{\mathbf{c}}$ becomes too large.

The important difference between the homeotropic and the planar case has been explained in Sect.3 in terms of the Goldstone mode associated with the spontaneously broken isotropy due to the Freédericksz transition. The action of the Goldstone mode corresponds to a rotation of the director in the plane, i.e. to a finite n_y component, with a nonvanishing z-average. There is iin fact a close analogy to the AR bifurcation and the abnormal rolls have been identified at first in the homeotropic case [84].

The absence of stable roll patterns has led to the conjecture of a new type of spatio-temporal chaos at onset [114]. A convincing experimental confirmation is shown in Fig.12. On the right side the oblique-roll pattern for a frequency below the Lifshitz point (upper panel) and normal rolls (lower panel) for the frequency above the Lifshitz point are stabilized by a magnetic field. Switching off the magnetic field leads within seconds to the disordered patterns shown on the left hand side.

Fig.12 Experimental pictures from [115]. Typical convection patterns with (left panel) and without magnetic field (right panel) are shown at two different frequencies,. 60 Hz (above) and 300 Hz (below)

A clear understanding of the disordered patterns near threshold has been achieved from a novel weakly nonlinear description that incorporates the critical convection mode together with the Goldstone mode [52].

The general form of the equations is governed by symmetry considerations (we write $\hat{\mathbf{c}}_\perp = \hat{\mathbf{z}} \times \hat{\mathbf{c}}$):

$$\tau\partial_t\tilde{A} = \Big[\varepsilon + \xi_{xx}^2(\hat{\mathbf{c}}\cdot\partial_{\mathbf{r}} - iq_c)^2 + \xi_{yy}^2(\hat{\mathbf{c}}_\perp\cdot\partial_{\mathbf{r}})^2 - g|\tilde{A}|^2 + i\beta_y\hat{\mathbf{c}}_\perp\cdot\nabla\varphi\Big]\tilde{A}, \tag{24}$$

$$\gamma_1\partial_t\varphi = K_3\hat{\mathbf{c}}_\perp\cdot\nabla^2\hat{\mathbf{c}} + (K_1 - K_3)\hat{\mathbf{c}}_\perp\cdot\nabla(\nabla\cdot\hat{\mathbf{c}}) + \chi_a(\hat{\mathbf{c}}\cdot\mathbf{H})(\hat{\mathbf{c}}_\perp\cdot\mathbf{H}) + \frac{\Gamma}{4}(-iq_c\tilde{A}^*(\hat{\mathbf{c}}_\perp\cdot\partial_{\mathbf{r}})\tilde{A} + \text{c.c.}). \tag{25}$$

The various coefficients in Eqs.(24,25) can be obtained from the Galerkin code used for Fig.11.

At first one recognizes a generalized Swift-Hohenberg equation for the pattern amplitude A in Eq.(24) where the local coordinate system is determined by the (local) anisotropy axis $\hat{\mathbf{c}} = (\mathbf{cos}(\phi), \mathbf{sin}(\phi))$. The torque due to spatial modulations exerted on the in-plane director is included in Eq. (25). where in addition the magnetic torque is taken into account.

These equations have been used for numerical simulations. In Fig.13 a representative snapshots of $\psi = \tilde{A}+c.c.$ (left side) is compared with experimental observations (right side) for $\varepsilon = 0.02$, $H = 0$ [116]). Material parameters as in [114] have been used used to calculate the coefficients of (24,25) (for details see [52]). Evidently the patterns in experiments and simulations look very similar. Though in the normal roll regime some defects appear, the rolls are locally aligned along a main direction. The oblique roll regime is dominated by a superposition of zig and zag. Again the preferred axis changes only over large distances. A persistent time dependence is observed in the simulations in agreement with the experiments in the oblique roll case. However, the experiments in the normal roll regime at higher frequency reveal two different regimes near threshold [106, 117, 118]. First a frozen-in disordered pattern is observed which looks like the one shown in Fig.13 (lower left panel). When the control parameter is increased a crossover to a time dependent disordered patterns is found. It is not yet clear whether an important term is missing in Eqs. (24,25), or whether small inhomogeneities in the experimental cell suppress a persistent dynamics at threshold.

Equations (24,25) represent normal-form equations for quasi-2D pattern-forming systems with a novel kind of symmetry, and thus should be of general interest. Other realizations might be found in convection instabilities in smectic-C liquid crystals, where a $\hat{\mathbf{c}}$-director exists ab initio , and in Rayleigh-Bénard convection of homeotropically aligned NLCs with an additional electric field. In this case the fields that drive the Freédericksz transition and the convection instability, respectively, can be varied independently, which should permit access to a large parameter range of Eqs. (24,25).

For $H = 0$ and $\Gamma < 0$ Eqs. (24,25) yield STC at onset. Other examples are the Küppers-Lortz instability in rotating Rayleigh-Bénard convection [119, 120] and systems undergoing a Hopf bifurcation. An example for the latter is the Benjamin-Feir destabilisation mechanism (see e.g. [2]) or the so-called dispersive chaos [19], where a description by a simple Ginzburg-Landau equation should be possible [121]. The origin of chaotic behavior is of course very different in the various systems and their detailed comparison will be fruitful.

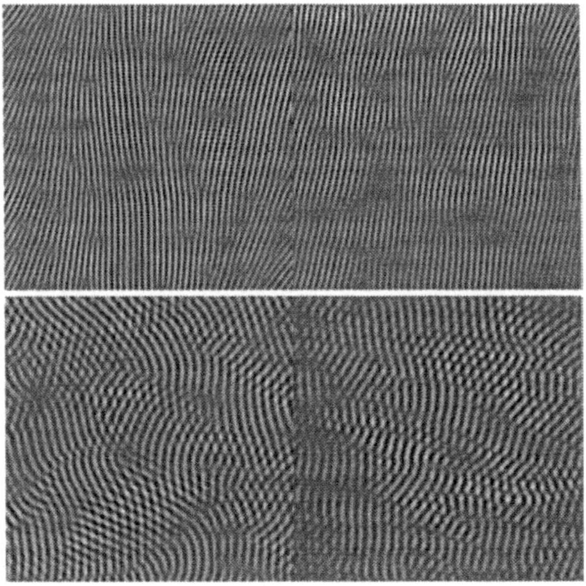

Fig.13 Simulations of Eqs. (24,25) (left) vs. experiment [116] (right) for normal (top) and oblique (bottom) rolls (see text).

6 EHC driven by multiplicative noise

6.1 Experimental situation and motivation

The influence of stochastic temporal modulation of external parameters in spatially extended systems is an interesting topic of current research [123]. Among the possible effects of the external noise are: shifts of thresholds, the appearance of new bifurcation types, modifications of the bifurcation sequences (e.g. a direct transition towards chaos), or a change from a continuous to hysteretic onset of the pattern-forming instabilities.

The understanding of the various phenomena has been considerably promoted by the investigation of the electrohydrodynamic instabilities in nematic liquid crystals (EHC). The starting point is the common planar configuration (see Fig. 1a), however the applied electric field across the nematic layer is now a superposition of a deterministic component E_{det} and a stochastic one, E_{stoch}. In the experiments a low-frequency periodic E_{det} was used, i.e. in a regime where the deterministic threshold curve as a function of frequency is almost flat (see Fig. 2b). This justifies to use in the theoretical description almost exclusively a time-independent E_{det} [124, 125, 126, 127, 128, 129, 130, 131, 132, 133, 134, 135, 136, 137, 138, 139, 140], cf. however

[141]. The stochastic field E_{stoch} is mainly characterized by its strength and the correlation time τ_{stoch} of its auto-correlation function.

In the first experiments a 'fast' stochastic component was superimposed [124, 125, 126], then subsequent work was also devoted to 'slow' stochastic driving [127, 128, 129]. Slow and fast refer here to the relation of τ_{stoch} to the characteristic times of the liquid crystal, which govern the relaxation of space charge and director in absence of external electric fields.

In these early experiments [124, 125, 126, 127, 128, 129] a highly doped nematic MBBA was used, where the electric conductivity is quite large. It has been demonstrated that adding a fast stochastic field E_{stoch} with its strength below a certain critical value E_c leads to a stabilisation of the homogeneous state, i.e. the threshold increases in comparison to a pure deterministic case. However beyond E_c the stochastic component is destabilizing and a *discontinuous* behaviour of the threshold curve as a function of stochastic-field strength is found (direct transition). With increasing stochastic driving the planforms at the convection threshold changed as well from normal rolls to a grid pattern and finally to a chaotic state (the dynamical scattering mode). The direct transition towards chaos at very high strength of the external noise was described to occur via intermittent bursts of spatially incoherent structures embedded in a homogeneous background [126]. The frequency of the bursts, as well as their duration and the area filled with spatially disordered structures increase with increasing deterministic voltage until the whole cell is filled with a strongly fluctuating spatially disordered structure. The superposition with a slow stochastic field on the other hand would always decrease the deterministic threshold. There were also some indications that the transition might be slightly hysteretic [129, 130]. For a comprehensive review see [129].

In later experiments undoped MBBA (low conductivity) [136, 137] and a different, chemically more stable nematics (Mischung 5) [138] was used. It was at first confirmed that slow external noise was destabilizing, in contrast to a not too strong, fast external noise, which was stabilizing. With increasing strength destabilizing effects take over, i.e. the threshold was eventually found to decrease *continuously* with the strength of the stochastic field.

In these experiments, for pure stochastic driving ($E_{det} = 0$), a new scenario was revealed: With decreasing correlation time a transition from a "conductive" to a "dielectric" regime was observed at a critical τ_{stoch}. In the latter regime the threshold was no longer sharp, since stripe pattern would appear and disappears randomly in time [137, 138, 140]. The pattern exhibits a characteristic blinking which corresponds to the phenomenon of *on-off*-intermittency, recently discussed for low-dimensional systems randomly driven just below their threshold of stability, cf. e.g. [142]. The probability distribution for the duration of the laminar phases, i.e., off-periods is governed by a power-law over several orders of magnitude [140].

An important issue for the following is the appropriate description of the stochastic process. A common approximation is Gaussian white noise, which applies to situations, where the correlation time of the driving stochastic field, τ_{stoch}, is kept much shorter than all characteristic time scales in the system. One might argue, that due to the critical-slowing-down the time scale at the onset of EHC is always long, but in fact it turns out that the dynamics of other important processes is not well separated in time from τ_{stoch}. Since the system couples to the square of the field strength there is another -more formal- reason which forbids to use a Gaussian white noise: The square of such a process is not a well defined mathematical quantity.

For these reasons the modelling of the stochastic process by a white noise is excluded: One has to use a stochastic model process with finite correlation time, such as the Ornstein-Uhlenbeck process or the dichotomous Markov process. The latter one allows a rigorous treatment of linear problems and can easily be generated in the experiment.

An interesting mathematical problem is posed by the stability of a stochastic trajectory. Different criteria have been discussed in the literature such as the bifurcation of the maximum of the probability density [143], the stability of the first or higher moments [144], or the asymptotic stability of the stochastic trajectory with a probability one (sample stability) [145, 146]. It is plausible that as long as the noise is fast compared to the time scales of the system these stability criteria should give similar results. However, as will be demonstrated for EHC, the correlation time τ_{stoch} is not in all cases short. Thus the different criteria give indeed different results except in some limits. Typically, for a fixed noise strength, the tendency towards a destabilisation increases with increasing order of the moments. This feature can be understood, since large but rare fluctuations of a stochastic process are progressively enhanced in the high-order moments. In any case, the sample-stability criterion is consistent with numerical simulations and appears as the natural proper choice [139].

The standard tool to describe the convection instability of the quiescent state is the linear stability analysis. For a deterministic driving voltage the procedure is not difficult to carry through and is well understood (see Sect. 3). However, for stochastic driving already the linear regime presents a strong challenge. Due to the inherent difficulties of the problem further approximations have to be introduced, which amount to a simplified treatment of all convection fields with respect to their vertical (z-) dependence, which is captured by one single mode. This treatment leads to the so called two-dimensional model [148, 134, 135] and its simplified one-dimensional version [38, 105]. In the case of deterministic driving it has been shown in detail that the exact critical properties can thus be satisfactorily reproduced within a few percent [17, 16].

In the sequel at first the model for stochastically driven EHC is presented (Sect. 6.2), before the criteria for stochastic stability are discussed in Sect. 6.3. The detailed results of the linear analysis in comparison with experiments are presented in Sect. 6.4. It will turn out, that the theory can explain well a variety of different experimental findings. A discussion of related topics and a outlook to future work can be found in some Concluding remarks.

6.2 Modelling of stochastically driven EHC

In the following at first the basic equations and its formal solutions are presented. Furthermore several criteria of stochastic stability are discussed.

The linearized equations of the two-dimensional model. The stability investigation of the quiescent state against the formation of normal rolls follows the standard method (see Sect. 3). The anisotropy axis of the system is parallel to $\mathbf{x}$, and all field variables vary only in x and z. The starting point are the basic equations as discussed in Sect. 2. In the horizontal directions with infinite extent a Fourier transformation (wavenumber k_x) is applied; with respect to the transverse direction one Galerkin mode (with wavenumber k_z) for each field (velocity, director etc) is used. Their choice is dictated by the boundary conditions at the confining horizontal plates. We have used stress-free boundary conditions, which simplify the calculations and are known to give satisfactory results [17, 149]. The director distortions, the induced space charge, and the quantities v_z, $\partial_z v_x(z)$ have to vanish at the boundaries, where v_x, v_z denote the velocity components parallel and perpendicular to the confining plates, respectively. The velocity field can then be adiabatically eliminated because of the small viscous time scale.

One ends up with a system of two coupled ODE's, which describe the dynamics of the space charge q and the spatial variation of the angle θ between the director and electrode plates, $\psi = \partial_x \theta$ [148], given as:

$$\dot{\mathbf{z}} = \mathbf{C}(t)\mathbf{z}, \tag{26}$$

where $\mathbf{z} = (q, \psi)^T$, and

$$\mathbf{C}(t) = -\begin{pmatrix} 1/T_q & \sigma_H E_t \\ aE_t & \Lambda_1 - \Lambda_2 E_t^2 \end{pmatrix} . \tag{27}$$

The total driving electric field $E_t = E_{det} + E_{stoch}$ is the superposition of a constant component E_{det} and a dichotomous Markovian component $E_{stoch} = E_t^{DMP}$ which takes randomly the values $\pm E$ and has the autocorrelation $< E_t^{DMP} E_{t'}^{DMP} >= E^2 \exp[-2\nu(t-t')]$. ν determines the inverse correlation time (i.e. $\tau_{stoch} = 1/2\nu$) and describes the mean number of jumps in unit time. One sees immediately that the *square* of the stochastic field enters, which excludes the use of Gaussian white noise as already mentionned in the introduction.

The parameters T_q, σ_H, a, $\Lambda_{1/2}$ are explicitely given by:

$$T_q = \varepsilon_0(\varepsilon_{\parallel} k_x^2 + \varepsilon_{\perp} k_z^2)/(\sigma_{\parallel} k_x^2 + \sigma_{\perp} k_z^2), \tag{28}$$

$$\sigma_H = (\sigma_\parallel \varepsilon_\perp - \varepsilon_\parallel \sigma_\perp)(k_x^2 + k_z^2)/(\varepsilon_\parallel k_x^2 + \varepsilon_\perp k_z^2) , \tag{29}$$

$$a = \frac{1}{f}\left[\frac{1}{2}\cdot\frac{(\gamma_1-\gamma_2)k_x^4 + (\gamma_1+\gamma_2)k_x^2k_z^2}{\alpha_1 k_x^2 k_z^2 + (k_x^2+k_z^2)(\eta_1 k_x^2 + \eta_2 k_z^2)} - \frac{\varepsilon_a k_x^2}{\varepsilon_\parallel k_x^2 + \varepsilon_\perp k_z^2}\right] , \tag{30}$$

$$\Lambda_1 = \left[K_{33}k_x^2 + K_{11}k_z^2\right]/f , \tag{31}$$

$$\Lambda_2 = \varepsilon_0 \varepsilon_a \varepsilon_\perp (k_x^2 + k_z^2)/\left[f(\varepsilon_\parallel k_x^2 + \varepsilon_\perp k_z^2)\right] . \tag{32}$$

where $\eta_{1/2} = [\alpha_4 \mp \gamma_2 + \alpha_5 + \alpha_3]/2$ are the well-known Miesowicz coefficients constructed from the viscous constants $\alpha_i, i = 1,5$ ($\gamma_1 = \alpha_3 - \alpha_2$; $\gamma_2 = \alpha_3 + \alpha_2$, see e.g. [17]). For the sake of brevity we have introduced

$$f = \gamma_1 - \frac{1}{4}\cdot\frac{\left[(\gamma_1-\gamma_2)k_x^2 + (\gamma_1+\gamma_2)k_z^2\right]^2}{\alpha_1 k_x^2 k_z^2 + (k_x^2+k_z^2)(\eta_1 k_x^2 + \eta_2 k_z^2)} . \tag{33}$$

The coefficients depend on the wavenumber k_x, which is to be determined by minimizing the threshold voltage. From the vertical boundary condition $k_z = \pi/d$ holds (odd solutions with $k_z = 2\pi/d$ play no role like in the deterministic case). Note that both $1/T_q$ and σ_H are proportional to $\sigma_\parallel$ if $\sigma_\parallel/\sigma_\perp = const$. The one-dimensional model is readily obtained setting $k_z = 0$, i.e. by neglecting the influence of the vertical boundaries. As the characteristic length λ_x of the pattern in the low frequency conduction regime is set by the width of the nematic layer, the 1d theory fails to predict λ_x; one has to insert $\lambda_x \approx d$ as a fit parameter to obtain reasonable values for thresholds. However note, that in the high-frequency dielectric regime the critical wavelength becomes an intrinsic bulk property [147, 16] and is already determined within the one-dimensional approach.

For a deterministic driving the undistorted state remains stable against the formation of normal rolls if the solution of (26) converges to zero. For stochastic driving there exist several stability criteria to be discussed below.

Formal solution. To obtain a formal solution of Eq. (26) for a given realization of E_t^{DMP} with jumps at times t_ν, $\nu = 0, 1, \ldots, n$, where $t_n > t_{n-1} > \cdots > t_1 > t_0$ we first consider a time interval between two jumps where $\mathbf{C}(t) = \mathbf{C}^\sigma = const, \sigma = \pm$ and diagonalize $\mathbf{C}^\sigma$ by a unitary transformation

$$\mathbf{U}^\sigma \mathbf{C}^\sigma (\mathbf{U}^\sigma)^{-1} = \begin{pmatrix} \lambda_1^\sigma & 0 \\ 0 & \lambda_2^\sigma \end{pmatrix} \equiv \mathrm{diag}\,(\lambda_i^\sigma) . \tag{34}$$

Introducing $\mathbf{W}^\sigma = \mathbf{U}^\sigma \mathbf{z}$ one finds $\mathbf{W}^\sigma = \mathrm{diag}\,(\lambda_i^\sigma)\mathbf{W}^\sigma$ which is solved by $\mathbf{W}^\sigma(t) = \mathrm{diag}\,\{\exp[\lambda_i^\sigma(t-t')]\}\,\mathbf{W}^\sigma(t')$. The inverse transformation leads to

$$\mathbf{z}_\sigma(t) = \mathbf{T}^\sigma(t-t')\mathbf{z}_\sigma(t'), \tag{35}$$

where $\mathbf{T}^\sigma$ is the time evolution matrix

$$\mathbf{T}^\sigma(\tau) = (\mathbf{U}^\sigma)^{-1} \operatorname{diag}[\exp(\lambda_i^\sigma \tau)]\mathbf{U}^\sigma . \tag{36}$$

For a given realization of the driving process with jumps at the random times t_ν iteration of (35) gives the formal solution [139]

$$\mathbf{z}_{\sigma_n}(t) = \mathbf{T}^{\sigma_n}(t-t_n) \cdots \mathbf{T}^{\sigma_1}(t_2-t_1)\mathbf{T}^{\sigma_0}(t_1-t_0)\mathbf{z}_{\sigma_0}(t_0). \tag{37}$$

The stability of the stochastic trajectory $\mathbf{z}_{\sigma_n}(t)$ is determined by the largest Lyapunov exponent λ_1 of the product of random matrices in (37) in the limit $n \to \infty$. If this exponent has a positive real part the trajectory diverges, otherwise it converges to zero. The evaluation of infinite products of random matrices is a notorious difficult problem, which appears also in a number of different fields in statistical physics [150] (for further references see e.g. [139]). Eq. (37) is used for the numerical simulations in order to determine stability thresholds (to be compared with analytical results) [139] and also the probability density of the duration of quiescent periods just below the threshold [140].

6.3 Stochastic stability criteria

In the following we will present and compare several stability criteria.

Mean field decoupling. We first describe a simple approximation [125] which rests on the assumption that the characteristic time of the driving stochastic process is fast compared to all other characteristic times. Then the system will "feel" only the average value of the stochastic field

$$E_t \to < E_t >= E_{det}, \; E_t^2 \to < E_t^2 >= E_{det}^2 + E^2. \tag{38}$$

This physical picture corresponds to a simple mean-field type decoupling of the averages

$$< E_t \mathbf{z} > \to < E_t >< \mathbf{z} > . \tag{39}$$

With the above replacements in (27) one obtains from det $\mathbf{C} = 0$ the threshold

$$E_{det,th}^2 = \frac{\Lambda_1 - \Lambda_2 E^2}{\sigma_H a T_q + \Lambda_2}, \tag{40}$$

which increases ($\Lambda_2 < 0$) in a linear way with the strength of the stochastic field E^2. In the mean field approximation it is therefore impossible to explain neither the experimentally found discontinuous behaviour of the threshold [126, 127, 128] nor the dependence on the correlation time of the noise [128].

Stability of moments. The equations of motions for the first moments form a closed system for $<\mathbf{z}>$ and $<E_t^{DMP}\mathbf{z}>$. One uses a theorem of Shapiro and Loginov [151]

$$(\partial/\partial t + 2\alpha) < E_t^{DMP}\mathbf{z} > = < E_t^{DMP}\dot{\mathbf{z}} >, \tag{41}$$

and exploits in addition $(E_t^{DMP})^2 = E^2 = const.$ The quantity $\dot{\mathbf{z}}$ on the RHS of Eq. (41) is replaced by Eq. (26). (Similar equations hold for the higher moments.) The *exact* threshold condition for the stability of moments is nonlinear in the field strength thus opening the possibility for a qualitative (though not quantitative) understanding of the behaviour of the threshold. The standard model in its one-, two- and three-dimensional versions and the 1d model including flexoeffect have been extensively studied within this approach [132, 133, 134, 135].

Sample stability. For the following it is crucial that one is able to reduce the complicated dynamics of EHC to a system of two coupled stochastic equations (Eqs. (26)). In that case there exists a standard method to analyze the asymptotic stability of the trajectory (Eq. (37)) for almost all realizations of the driving process (sample stability) [145]. The first step is a transformation from (q, ψ) to polar coordinates (r, φ), which leads from (26) to the pair of equations

$$\dot{r} = g(E_t, \varphi)r, \tag{42}$$

$$\dot{\varphi} = h(E_t, \varphi). \tag{43}$$

Equation (43) depends only on φ (the system is skew symmetric) and it is possible to find the stationary solution $P_\sigma(\varphi)$ of the associated Kolmogorov forward equation for the joint process (E_t, φ)

$$\dot{P}_\sigma = \frac{d}{d\varphi}(h_\sigma P_\sigma) - \alpha(P_{-\sigma} - P_\sigma), \sigma = \pm, \tag{44}$$

where h_σ is a shorthand notation for $h(E_\sigma, \varphi)$. Equation (42) is linear in r and can be solved for a given trajectory of the driving process. This leads to an expression for the leading Lyapunov exponent [145]

$$\lambda_1 = \lim_{t\to\infty} \frac{1}{t} \int_0^t d\tau\, g(E_\tau, \varphi) = \int_{\text{supp}} d\varphi \sum_{\sigma=\pm} P_\sigma(\varphi) g(E_\sigma, \varphi). \tag{45}$$

The second equality holds due to the multiplicative ergodic theorem of Oseledec [152] with *Prob* 1, i.e., for almost all trajectories.

For the nondegenerate (the matrices $\mathbf{C}^+$ and $\mathbf{C}^-$ do not have an eigenvector in common) and nonrotational (the eigenvalues of $\mathbf{C}^\sigma$ are real and distinct) case which is of interest here, λ_1 can be found up to quadratures [145]

$$4\lambda_1 = -\sum_{\sigma=\pm} \mathrm{Sp}\, \mathbf{C}^\sigma - \frac{\int_{\mathrm{supp}} d\varphi\, F \cdot \frac{d}{d\varphi} \ln |h_+/h_-|}{\int_{\mathrm{supp}} d\varphi\, F/h_+}, \tag{46}$$

where

$$F(\varphi, \varphi_0) = \exp\left[-\alpha \int_{\varphi_0}^{\varphi} d\varphi \left(\frac{1}{h_+} + \frac{1}{h_-}\right)\right], \ \varphi_0 \in \mathrm{supp}. \tag{47}$$

The two Lyapunov exponents of the system $\lambda_1 > \lambda_2$ are related by

$$\lambda_1 + \lambda_2 = 1/2 \sum_{\sigma=\pm} \mathrm{Sp}\, \mathbf{C}^\sigma. \tag{48}$$

For pure stochastic excitation ($E_{det} \equiv 0$), the quadratures in (46) and (47) have been evaluated explicitly in terms of generalized hypergeometric functions both for the 1d [134] and 2d model [139]. Also for a constant, finite field E_{det} analytical results exist for the 1d model, whereas for the 2d model the integrals in (46) and (47) were evaluated numerically [139].

Numerical simulations. The numerical simulation folloes trajectories $\mathbf{z}_{\sigma_n}(t)$ starting from a nonzero but small initial value $\mathbf{z}(0)$ for a given realization of the driving stochastic process, cf. Eq. (37). A trajectory with N jumps (corresponding to an average time N/α with constant E_t intervalls) is considered as diverging if $\psi = \partial_x \theta \sim k\theta = (\pi\theta)/d > \pi^2/(4d)$, i.e. for $\theta > \theta_c \approx \pi/4$. This critical value for θ is choosen, because for $\theta \gtrsim \pi/4$ the linearization of $\sin\theta \approx \theta$ on which Eq. 27 is based becomes invalid. The thus determined thresholds are virtually independent of N and ψ_c for $N \gtrsim 10^4$, i.e. they vary only within a limit of less than one percent when N and ψ_c are varied both over a range of several orders of magnitude [139].

In the following the theoretical results are exemplified in detail and also compared with experimental results.

6.4 Results

Comparison of different stability criteria. The thresholds for the appearance of normal rolls calculated in the one-dimensional model according to the different criteria discussed above are compared in Fig. 14. The total voltage is the superposition of a constant deterministic part ($U_1 = E_{det}$) and a stochastic one ($U = E_{stoch}$) with fast and slow stochastic driving frequency ν, respectively. In this part the material parameters are taken in general from [153] (referred to as MBBA I in [139])), but we allow for modifications of the electrical conductivities. The cell thicknes is fixed to 100μ in all calculations.

For small values of the stochastic voltage U the thresholds from all criteria coincide as must be expected irrespective of ν. The insert in Fig.14 addresses the case of slow driving calculated for $\sigma_{\|} = 1.5 \cdot 10^{-8} \Omega^{-1}\,\mathrm{m}^{-1}$ and $\sigma_{\|}/\sigma_{\perp} =$

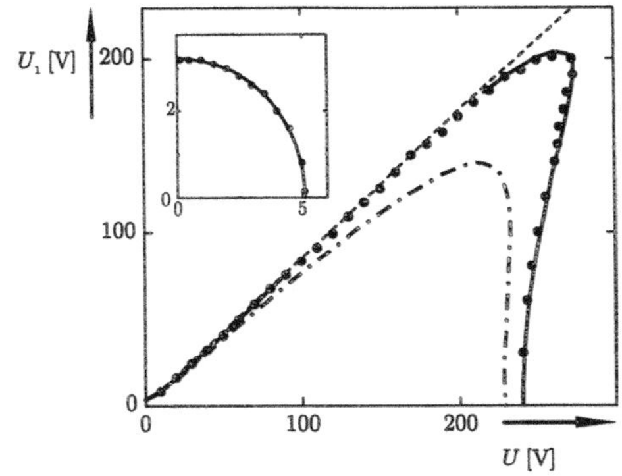

Fig. 14. Thresholds for the appearance of normal rolls calculated in the one-dimensional model for a superposition of constant voltage U_1 and a fast stochastic dichotomous voltage U ($\nu \simeq 1000\,\mathrm{s}^{-1}$). The thresholds are obtained from: mean-field decoupling (dashed line), stability of first moments (dash-dotted line), sample stability (solid line), and numerical simulation (•). The insert shows the case of a 'slow' stochastic voltage ($\nu \simeq 100\,\mathrm{s}^{-1}$) for which all criteria (besides the mean field decoupling) give the same result (from [139]; for the material parameters: see text).

1.3. This corresponds to a cut-off frequency $f_d \approx 62$Hz in the case of an applied deterministic ac voltage, which is comparable with $\nu = 100$Hz. The results for fast driving are shown (for $\sigma_{\|} = 6 \cdot 10^{-9} \Omega^{-1}\,\mathrm{m}^{-1}$ and $\sigma_{\|}/\sigma_{\perp} = 1.3$) are shown in more detail. The stochastic frequency is now much larger than the corresponding cut-off frequency $f_d = 26$Hz. For a wide range of U the threshold from the mean field decoupling is very close to that of the sample stability, but for U beyond the threshold ($\approx 240V$) for pure stochastic excitation ($U_1 = 0$) they are drastically different.

The results can be understood through an comparison of the characteristic times of the system and of the noise, cf. Fig. 15. The characteristic times of the stochastically driven system are given by the modulus of the inverse of the Lyapunov exponents λ_1 and λ_2. At the threshold we have $\lambda_1 = 0$, the corresponding time τ_1 diverges and is thus well separated from τ_{stoch}. The second characteristic time for the electric fields at the threshold can be determined exactly [139]

$$\tau_2 = |1/\lambda_2| = \left(1/T_q + \Lambda_1 - \Lambda_2(E_{det}^2 + E^2)\right)^{-1}, \tag{49}$$

it decreases with increasing values of the threshold fields and may reach the order of τ_{stoch}. In this case the mean-field decoupling is not justified and the threshold obtained from the stability of the moments differs quantitatively from the sample-stability threshold. If instead τ_{stoch} is clearly separated from

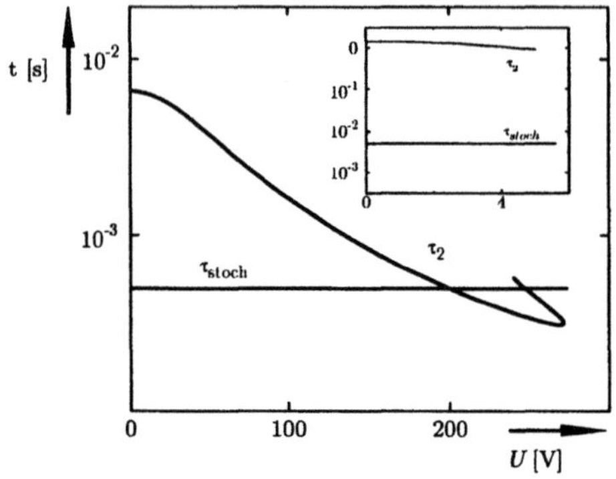

Fig. 15. The characteristic time $\tau_2 = |1/\lambda_2|$ at the sample stability threshold shown in 14 decreases with increasing voltage [139]. The initially clearly separated time scales of τ_2 and $\tau_{stoch} = 1/2\nu$ become of the same order at sufficiently high stochastic voltage. The insert shows the case where the time scales are well separated.

at least one of the characteristic times of the system, both moment's and sample stability criteria give similar results, see inserts in Figs. 14 and 15. The numerical simulations confirm the results obtained from the criterion of sample stability, which was therefore used in [139, 140].

The region describing the stability of the moments is always smaller than that for sample stability since the divergence of the trajectories is sufficient for the divergence of the moments: For large times one finds

$$r_t = \mathrm{e}^{<\lambda[E_r]>t} \leq <\mathrm{e}^{\lambda[E_r]t}>=<r_t>, \tag{50}$$

where the inequality holds due to the convexity of the exponential function. Finally one should mention that the thresholds for stochastic excitation are always below that for deterministic excitation (for otherwise identical parameters and wave numbers). This is plausible since generally the thresholds increase with the increasing frequency and a stochastic trajectory with a given mean number of jumps always contains lower Fourier components.

Pure stochastic excitation. The stability diagram for pure stochastic excitation shows a topological difference between 'slow' and 'fast' driving, cf. Fig. 16. In the former case we have an unstable island in the left lower corner of Fig. 16a. The corresponding mode (low threshold and small wavenumber) is called *conductive*. With increasing mean frequency ν of the driving field this island shrinks continuously. There is a sharp transition if it disappears: The instability is now towards a mode with a higher threshold and a much larger wave number as it is typical for the *dielectric* regime, cf. Fig. 16b.

We remark that also for deterministic driving the stability diagram has analogous properties. Note, that the different temporal symmetries of dielectric and conductive mode in the deterministic case (see Sect. 3) are not crucial here.

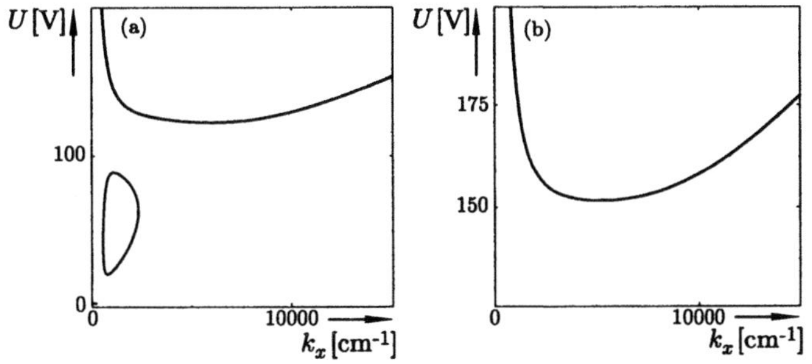

Fig. 16. Mode selection for pure stochastic excitation [139]. Shown are the neutral curves for the two-dimensional model in (a) the conductive regime ($\nu = 100\ \mathrm{s}^{-1}$) and in (b) the dielectric regime ($\nu = 400\ \mathrm{s}^{-1}$). The unstable region in the left lower corner of (a) corresponds to the conductive mode and shrinks with increasing mean frequency of the driving process and is then absent in the dielectric regime (b).

The "critical" wave number k_x increases as function of the mean frequency ν with a slope much smaller than for deterministic driving. This is also found in experiments [138], cf. Fig. 17, and can be qualitatively understood by an argument similar to that given above according to which stochastic driving contains always contributions of lower Fourier modes.

For stochastic driving one observes in the region indicated in Fig. 17 by open rectangles phenomena which resemble *on-off*-intermittency: Normal roll patches appear and disappear in an irregular way at voltages already below the sample stability threshold [137, 138, 140]. The frequency of the bursts increases with the strength of the stochastic field, their duration depends on the characteristic time τ_{stoch} of the noise. Since the undistorted state is well described by the linear theory this phenomenon can be captured by simulations based on Eq. (37). We have found when approaching the threshold ($\lambda_1 = 0$) from below that the probability distribution for the duration τ_{off} of the undistorted state (i.e the laminar or off-periods) is governed by a power-law $p(\tau_{off}) \sim \tau_{off}^{-3/2}$ over several orders of magnitude, in good agreement with the experiment [140].

Superposition of deterministic and stochastic field. The superposition of a deterministic voltage U_1 with a 'fast' stochastic voltage U with increas-

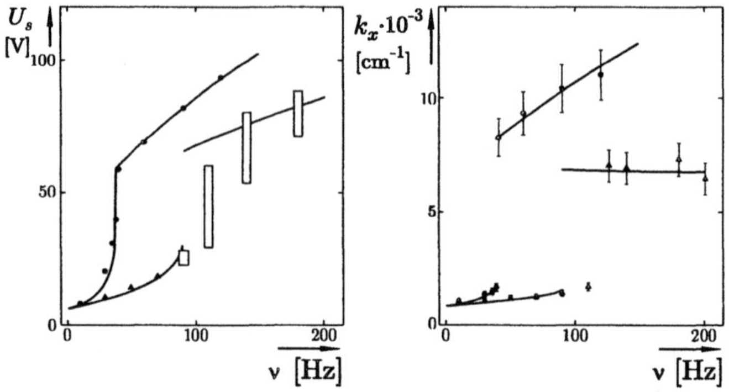

Fig. 17. Theory versus experiment: Thresholds and selected wave numbers of normal rolls depending on the (mean) frequency of the driving field [138]. As explained in the text, the curves for stochastic driving are always below those for deterministic driving. Experimentally determined thresholds (• deterministic driving, triangles: stochastic driving) are compared with those obtained from the two-dimensional theory (sample stability criterion for the stochastic case). The experiment is performed with the nematic "Mischung 5". The open rectangles indicate the range where for stochastic driving phenomena were observed which resemble *on-off*-intermittency (see text).

ing strength stabilizes at first the undistorted state. This is shown in Fig. 18, where the threshold increases initially with increasing U. This remains true up to a certain "critical" value of U beyond which the threshold curve bends down (U becomes destabilizing). The threshold curves are determined experimentally following a suitable protocol involving changes in U, U_1 inside the stable regime until a point on the linear stability curve is hit.

The region where the undistorted state is stable *may* extend beyond the threshold values of deterministic or stochastic driving alone, thus forming a stable tongue in the U-U_1 plane (i.e. the upper curve in Fig. 18 bends back at large U). This explains why in experiments [126, 127, 128, 129] a discontinuous behaviour of the threshold was observed, when a stochastic voltage U of a given strength was applied first and then the deterministic voltage U_1 was increased up to the instability of the homogeneous phase. In other words, below a certain value U_c ($\approx$ 230V in Fig. 18) the threshold is found at a finite deterministic voltage U_{1th}, whereas for $U > U_c$ convection sets in immediately at $U_{1th} = 0$.

The appearance of the stable tongue depends on the material parameters. Roughly speaking, the tendency towards its formation decreases with increasing Helfrich parameter ζ^2 (independent of the stability criterion, i.e. for moment's stability [132] as well as for sample stability [139]. The Hel-

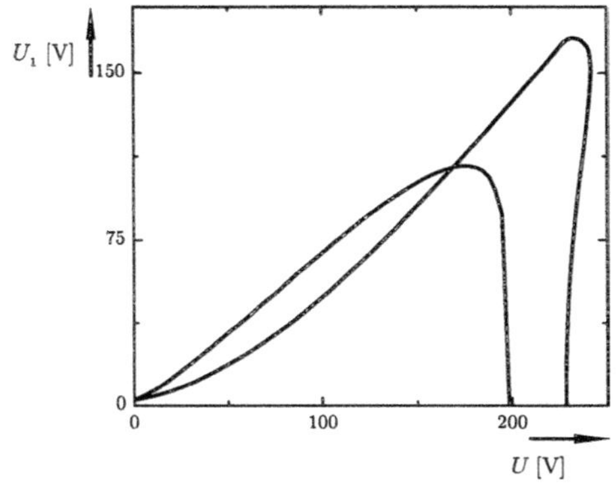

Fig. 18. Comparison of sample stability thresholds for different values of the conductivity at fixed $\nu = 1000\text{s}^{-1}$ in the one-dimensional model [139]. The tendency towards formation of a stable tongue increases with increasing conductivity. Shown are the cases $\sigma_{\|} = 1.11 \cdot 10^{-10} \Omega^{-1}\,\text{cm}^{-1}$ and $\sigma_{\|} = 2.22 \cdot 10^{-10} \Omega^{-1}\,\text{cm}^{-1}$, $\sigma_{\|}/\sigma_{\perp} = 1.5$. In the latter case there is a stable tongue so that, following the measuring procedure described in the text, the threshold curve appears discontinuous. The mean number of jumps is $\nu = 1000\text{s}^{-1}$ (Material parameters MBBA II from [148, 139]).

frich parameter (which alone determines f_d in units of the charge relaxation time τ_q (see Eq. (12)) depends only on the ratio $\sigma_{\|}/\sigma_{\perp}$. A further important parameter is in fact the absolute value of the conductivity $\sigma_{\|}$ which may differ from sample to sample considerably and which can be changed easily by doping. The tendency towards formation of a stable tongue increases with increasing $\sigma_{\|}$ while $\sigma_{\|}/\sigma_{\perp}$ is kept constant (cf. Fig. 18). Note that $f_d \sim \sigma_{\|}$ in physical units.

The early experiments [126, 127, 128, 129] were performed presumably with highly doped MBBA ($f_d = 360\text{Hz}$ in [129]). In later experiments with undoped MBBA [136, 137] ($f_d \approx 170\text{Hz}$) and a different, chemically more stable nematics (*Mischung 5*) [138] no stable tongue and correspondingly no discontinuous behaviour of the threshold was found.

As already mentioned the superposition of a deterministic field with a 'slow' stochastic one leads always to a monotonous decrease of the threshold with increasing stochastic field (see insert in Fig. 14) in agreement with experiments.

The difference of the thresholds for the one-dimensional and two-dimensional versions of the standard model are in general small. The selected wave numbers, however, differ significantly, cf. Fig. 19. This is evident, since k_x is always chosen to be π/d in the one-dimensional version, whereas it increases in fact with increasing strength of the stochastic field. This is consistent with

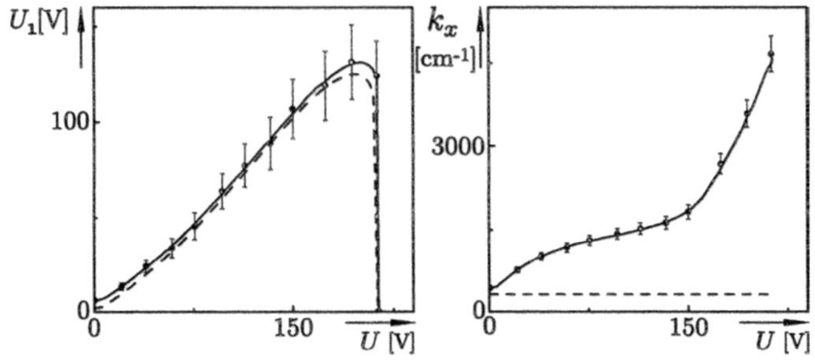

Fig. 19. Thresholds and selected wave numbers for the superposition of a constant voltage U_1 with a stochastic dichotomous voltage U obtained for the one-dimensional (dashed line) and the two-dimensional model (solid line) from the sample stability criterion and from numerical simulation (•) [139]. Material parameters from [148], $\nu = 1000\ \mathrm{s}^{-1}$.

experimental findings [136, 137, 138]. A detailed quantitative comparison of two-dimensional theory and experiment is in preparation.

6.5 Concluding remarks and outlook

The theoretical treatment presented in this review has obviously the capability for a convincing explanation of a variety of experimental findings. This has been demonstrated for the following phenomena: (i) The discontinuous behaviour of the threshold at a critical strength of the noise, (ii) the change from discontinuous to continuous behaviour of the threshold with increasing correlation time of the noise, (iii) the change from continuous to discontinuous behaviour of the threshold with increasing conductivity, (iv) the change from stabilizing to destabilizing effect of the noise if its correlation time becomes comparable to the correlation time of the system, and (v) the on-off intermittency observed in very recent experiments. In the following we will mention some issues that certainly need further investigation.

In the numerical simulations of the two-dimensional model it was observed that the fluctuations in the distribution of Lyapunov exponents for trajectories of finite length (which mean fluctuations of the threshold) increase if the characteristic times of noise and system become comparable, cf. Fig. 19. Fluctuations of thresholds were also observed in experiment for sufficiently large noise [136, 137, 138, 140]. The theoretical treatment of these fluctuations leads to the problem of generalized Lyapunov exponents [150, 154].

The conductivity enters the standard model considered here as a material parameter. Recently, Treiber and Kramer [33] have developed a more sophisticated model (WEM, see Sect. 4.2), where ionic migration, diffusion and dissociation-recombination processes are included. Thus the electric conductivity becomes a variable which introduces new time and length scales. In the deterministic case the primary bifurcation phenomena become much richer (Hopf, subcritical). and the analysis of stochastic driving within this model will be certainly rewarding.

In the present work the stability of the undistorted state against *one* mode describing roll patterns of a fixed wave number was considered. If the characteristic time of the noise is of the order of the inverse growth rate of a typical mode, the process of mode selection will not be completed until the next jump of the noise. Thus a band of wave numbers might be involved. This could lead to a sort of dynamical pattern as observed in experiments. At least in a conceptually different system (of Hamiltonian or dissipative gradient type) it has been recently shown, that a continuous band of wave numbers might remain relevant for the long-time behaviour [162].

The inherent difficulties of stochastically driven EHC substantially increase if it comes to the description of the nonlinear regime. A numerical treatment of the full set of nonlinear electrohydrodynamic equations seems at present not to be feasible even in the deterministic case. However, a reliable description is possible in the weakly nonlinear regime in terms of order parameter equations (see Sect. 3). The general problem of systematically deriving amplitude equations in the stochastic case has been addressed in few investigations, see e.g. [157, 158, 159, 160, 161]. However, it is an open question, which approach applies to the case of multiplicative noise in EHC with not always well separated time scales and finite correlation time of the noise. One might speculate that for sufficiently fast driving a kind of order parameter equations with "averaged" coefficients could result, while in the opposite case the coefficients might be stochastic quantities. One should also be aware of the fact, that a deterministic supercritical bifurcation can change to a subcritical one under the influence of noise [155, 156].

At the moment nonlinear partial differential equations, e.g. of Ginzburg-Landau or Swift-Hohenberg type, with additive or multiplicative Gaussian white noise (*introduced ad hoc*) are a subject of intense studies [163, 164, 165, 166, 167, 168, 169, 170, 171, 172, 173, 174]. In this context, Becker and Kramer [175, 176] found a controllable approximation to determine the threshold of sample stability without knowledge of the stationary distribution. A zero-dimensional version, the Stratonovich model, was solved rigorously for Gaussian white noise [177, 178] and dichotomous noise [179]. More recently, a 1d model with multiplicative noise that is a product of a Gaussian white noise in time and a spatial periodic function was considered [180].

Here we have concentrated on the case of stochastically driven electrohydrodynamic convection, i.e. a multiplicative, parametric, external noise. One should note that the very interesting and intriguing problem of thermal fluctuations, i.e. the case of additive, internal noise has also motivated many experimental and theoretical activities [24].

7 General Conclusions

In this review we have described to some extent the present status of research in EHC. In near future it is to be expected, that the combined application of electric and magnetic fields will be explored in more detail. Any additional control parameter is obviously very useful to enfold bifurcation scenarios and to accentuate specific mechanism (for a recent example see [181]). But also very unexpected and not yet understood phenomena can show up like the dendritic growth of EHC patterns (see [182]). Many other problems ask for further investigation. Among them are the properties of the abnormal rolls, in particular the defect and wall structures in this state, which certainly determine the patterns not too far from threshold. EHC is a nice paradigm for the formation of chevrons, i.e. an super-structure of particle-like defects, which develops out of a gas of "free" defects. The formation of defect chains is also observed in other cases for instance in simulations of the complex Ginzburg-Landau equation [183].

Another issue is the characterization of spatial temporal chaos (STC), which is the subject of current research in general. STC can easily be produced in EHC experiments by increasing the driving voltage. The system is apparently well suited to study such statistical issues, because of the short characteristic timesi and large aspect ratios. For example it has been stressed recently that time-averages of the patterns unlike a single snapshot might contain useful information about characteristic times and length of a system [184] as function of the control parameter. This has been confirmed in recent experimental studies [185], where the transition and the possible origin of the STC has been characterized.

Furthermore one should keep in mind, that typical bifurcation scenarios in EHC can be found as well in RBC of nematics [186]. It is for instance easy to find a Hopf bifurcation at threshold in the homeotropic case (for recent experiments see e.g. [187]). The close analogy between EHC and RBC is also reflected in the theoretical analysis [56].

Though not directly related, one should mention finally that interesting instabilities in nematics can also appear under the influence of shear flow [188]. Then a viscous torque is exerted on the director, which was part of the destabilisation loop in EHC, where the director is situated in the shear plane. The flow instabilities in this geometry are currently under investigation, under the application of oscillatory flows. The scenarios are unexpectedly rich. One finds spatially homogeneous oscillations of the director in Couette flow [189] as well as the bifurcation to convection-roll patterns under Poisseuille flow [190].

Acknowledgements We have benefitted a lot from discussions with G. Ahlers, H. Brand, A. Buka, S. Kai, I. Rehberg and W. Zimmermann. Exceptional is the role of L. Kramer, who has been always strongly involved in the many aspects of EHC, which have been discussed in this review. Crucial for any scientific progress is the effort of motivated coworkers. We are in particular indebted to H. Amm, W. Decker, A. Hertrich, Th. John, A. Lange, R. Müller, E. Plaut, A. Rossberg and M. Treiber. They and G. Ahlers, A. Buka, I. Rehberg have also kindly provided graphs of their results. We are grateful to F. Schmoegner for his help preparing the manuscript.

References

1 P. Manneville, *Dissipative Structures and Weak Turbulence*, Academic Press, (New York 1990).
2 M. C. Cross and P. C. Hohenberg, Rev. Mod. Phys. **65**, 851 (1993).
3 F. H. Busse, Fundamentals of thermal convection, in W. R. Peltier, editor, *Mantle convection, Plate Tectonics and Global Dynamics*, page 23, Gordon and Breach, (1989).
4 P. de Gennes and J. Prost, *The Physics of Liquid Crystals*, Clarendon Press, (Oxford 1993).
5 S. Chandrasekhar, *Liquid Crystals*, University Press, (Cambridge 1992).
6 A. Buka and e. L. Kramer, *Pattern formation in liquid crystals*, Springer-Verlag, (New York 1995).
7 E. Dubois-Violette, G. Durand, E. Guyon, P. Manneville, and P. Pieranski, Instabilities in nematic liquid crystals, in L. Liebert, editor, *Solid state physics, Suppl. 14*, page 147, Academic Press, New York, (1978).
8 L. M. Blinov, *Electrooptical and Magnetooptical Properties of Liquid Crystals*, John Wiley, (New York 1983).
9 S. A. Pikin, *Structural Transformations in Liquid Crystals*, Gordon and Breach Science Publishers, (New York 1991).
10 I. Rehberg, B. L. Winkler, M. de la Torre, S. Rasenat, and W. Schöpf, Adv. Solid State Physics **29**, 35 (1989).
11 S. Kai and W. Zimmermann, Prog. Theor. Phys. Suppl. **99**, 458 (1989).
12 L. Kramer, E. Bodenschatz, W. Pesch, W. Thom, and W. Zimmermann, Liquid Crystals **5**(2), 699 (1989).
13 W. Zimmermann, MRS Bulletin **XVI**, 46 (1991).
14 S. Kai, Forma **7**, 189 (1992).
15 L. Kramer and W. Pesch, Annu. Rev. Fluid Mech. **27**, 515 (1995).
16 L. Kramer and W. Pesch, Electrohydrodynamic instabilities in nematic liquid crystals, in A. Buka and L. Kramer, editors, *Pattern formation in liquid crystals*, Springer, (New York 1995).
17 E. Bodenschatz, W. Zimmermann, and L. Kramer, J. Phys.(Paris) **49**, 1875 (1988).
18 M. Treiber, N. Eber, Á. Buka, and L. Kramer, J. Phys. France II **7**, 649–661 (1997).
19 M. Dennin, G. Ahlers, and D. Cannell, Measurement of material parameters of the nematic crystal i52, in P. Cladis and P. Palffy-Muhoray, editors, *Spatio-temporal patterns in nonequilibrium complex systems*, Santa Fe Institute Studies in the Sciences of Complexity **XXI**, Addison-Wesley, New York, (1994).
20 M. Dennin, D. Cannell, and G. Ahlers, Mol. Cryst. Liq. Cryst. **261**, 337 (1995).

21 M. Dennin, G. Ahlers, and D. S. Cannell, Phys. Rev. Lett. **77**, 2475 (1996).
22 H. R. Brand, C. Fradin, P. L. Finn, W. Pesch, and P. E. Cladis, Physics Lett. A **235**, 508 (1997).
23 I. Rehberg, S. Rasenat, M. de la Torre Juarez, W. Schöpf, F. Hörner, G. Ahlers, and H. Brand, Phys.Rev.Lett **67**, 596 (1991).
24 M. Treiber, Thermal fluctuations in pattern forming instabilities, in A. Buka and L. Kramer, editors, *Pattern formation in liquid crystals*, Springer, (New York 1995).
25 M. J. Stephen and J. P. Straley, Rev. Mod. Phys. **46**, 617 (1974).
26 H. Pleiner and H. Brandt, Hydrodynamics and electrohydrodynamics of liquid crystals, in A. Buka and L. Kramer, editors, *Pattern formation in liquid crystals*, Springer, (New York 1995).
27 F. M. Leslie, Quart. J. Mech. Appl. Math. **19**, 357 (1966).
28 O. Parodi, J. Phys. (Paris) **31**, 581 (1970).
29 F. H. Busse and E. W. Bolton, J. Fluid Mech. **146**, 115 (1984).
30 R. B. Meyer, Phys. Rev. Lett. **22**, 918 (1969).
31 N. Madhusudana and G. Durand, J. Phys.Paris **46**, L195 (1985).
32 W. Decker, PhD thesis, Universität Bayreuth, (1995).
33 M. Treiber and L. Kramer, Mol. Cryst. Liq. Cryst. **261**, 311 (1995).
34 A. C. Newell, T. Passot, and J. Lega, Annu. Rev. Fluid Mech. **25**, 399 (1993).
35 G. Dangelmayer and L. Kramer, in F. Busse and A. Müller, editors, *this volume*, Springer, (New York 1998).
36 W. Pesch and L. Kramer, General mathematical description of pattern forming instabilities, in A. Buka and L. Kramer, editors, *Pattern formation in liquid crystals*, Springer, (New York 1995).
37 Orsay Liquid Crystal Group, Phys. Rev. Lett. **26**, 1642 (1970).
38 E. Dubois-Violette, P. G. de Gennes, and O. J. Parodi, J. Phys.(Paris) **32**, 305 (1971).
39 H. Haken, *Synergetics*, Springer-Verlag, (1978).
40 M. C. Cross, Phys. Fluids **23**, 1727 (1980).
41 Q. Feng, W. Pesch, and L. Kramer, Phys. Rev. A **45**, 7242 (1992).
42 Q. Feng, W. Decker, W. Pesch, and L. Kramer, J. Phys. France **II** , 1303 (1992).
43 M. Kaiser and W. Pesch, Phys. Rev. E **48**, 4510 (1993).
44 V. Ginzburg and L. Landau, Zh.Eksp.Teor.Fiz. **20**, 1064 (1950).
45 W. Pesch and L.Kramer, Z.Phys. B **63**, 121 (1986).
46 L. Kramer, E. Bodenschatz, W. Pesch, and W. Zimmermann, in W. Güttinger and G. Dangelmayr, editors, *The Physics of Structure Formation*, Springer, (Berlin 1987).
47 E. Bodenschatz, M. Kaiser, L. Kramer, W. Pesch, A. Weber, and W. Zimmermann, Patterns and defects in liquid crystals, in P. Coullet and P. Huerre, editors, *New Trends in Nonlinear Dynamics and Pattern Forming Phenomena: The Geometry of Nonequilibrium*, Plenum Press, (1990). NATO ASI Series.
48 A. C. Newell and J. A. Whitehead, J. Fluid Mech. **38**, 279 (1969).
49 L. A. Segel, J. Fluid Mech. **38**, 203 (1969).
50 A. Buka and L. Kramer, Theory of transient patterns in the splay freedericksz transition of nematics, in S. Kai, editor, *Pattern formation in complex dissipative systems*, World Scientific, (Kitakyushu, Japan 1991).

51 W. Decker and W. Pesch, J.Phys.II France **4**, 493 (1994).
52 A. G. Rossberg, A. Hertrich, L. Kramer, and W. Pesch, Phys. Rev. Lett. **76**, 4729–4732 (1996).
53 M. I. Tribelsky and K. Tsuboi, Phys. Rev. Lett **76**, 1631 (1996).
54 H. Riecke, Phys. Rev. Lett. **68**, 301 (1992).
55 H. Riecke and G. D. Granzow, to appear in Phys. Rev. Lett., (1998).
56 E. Plaut and W. Pesch, to appear in Phys. Rev. E (1998).
57 H. Zhao and L. Kramer. in preparation, (1998).
58 H. Zhao, L. Kramer, I. Rehberg, and A. Rudroff. submitted to Phys. Rev. Lett, (1998).
59 E. F. Carr, Mol. Cryst. Liq. Cryst. **7**, 253 (1969).
60 W. Helfrich, J. Chem. Phys. **51**, 4092 (1969).
61 R. Ribotta and A. Joets, Defects and interactions with the structure in ehd convection in nematic liquid crystals, in J. E. Wesfreid and S. Zaleski, editors, *Cellular Structure in Instabilities*, page 249, Springer, (Berlin 1984).
62 A. Joets and R. Ribotta, J. Phys.(Paris) **47**, 595 (1986).
63 S. Kai and K. Hirakawa, Prog. Theor. Phys. Suppl. **64**, 212 (1978).
64 S. Kai, N. Chizumi, and M. Kohno, Phys. Rev. A **40**, 6554 (1989).
65 S. Kai. unpublished, (1988).
66 A. Joets and R. Ribotta, Phys. Rev. Letters **60**, 2164 (1988).
67 I. Rehberg, S. Rasenat, and V. Steinberg, Phys. Rev. Lett. **62**, 756 (1989).
68 I. Rehberg, S. Rasenat, M. de la Torre-Juarez, and V. Steinberg, Phys. Rev. Lett. **61**, 2448 (1988).
69 M. Treiber, *On the Theory of the Electrohydrodynamic Instability in Nematic Liquid Crystals near Onset*, PhD thesis, Universität Bayreuth, (1996).
70 M. Lowe and J. Gollub, Phys.Rev.Lett **55**, 2575 (1985).
71 S. Rasenat, E. Braun, and V. Steinberg, Phys. Rev. A **42**, 5728 (1991).
72 S. Nasuno and S. Kai, Europhys. Lett. **14**(8), 779 (1991).
73 S. Nasuno, S. Takeuchi, and Y. Sawada, Phys. Rev. **A40**, 3457 (1989).
74 G. Goren, I. Procaccia, S. Rasenat, and V. Steinberg, Phys. Rev. Lett. **63**, 1237 (1989). It appears that there is no merit to the theory presented in this paper, see L. Kramer, E. Bodenschatz and W. Pesch, Phys. Rev. Lett. **64**, 2588 (1990)
75 S. Rasenat, V. Steinberg, and I. Rehberg, Phys. Rev. A **42**, 5998 (1990).
76 E. Braun and V. Steinberg, Europhys. Lett. **15**(2), 167 (1991).
77 W. Pesch, W. Decker, Q. Feng, M. Kaiser, L. Kramer, and A. Weber, Weakly nonlinear analysis of pattern formation in nematic liquid crystals, in J. M. Coron, F. Helein, and J. M. Ghidaglia, editors, *Nematics: Mathematical and Physical Aspects*, page 291, Kluwer Academic Publishers, (Dordrecht 1991). NATO ASI Series.
78 M. Kaiser, W. Pesch, and E. Bodenschatz, Physica D **59**, 320 (1992).
79 S. Sasa, Prog. Theor. Phys. **83**, 824 (1990).
80 S. Sasa, Prog. Theor. Phys. **84**, 1009 (1990).
81 E. Braun, S. Rasenat, and V. Steinberg, Europhys. Lett. **15**, 597 (1991).
82 S. Nasuno, O. Sasaki, S. Kai, and W. Zimmermann, Stability diagram in electrohydrodynamic convection of nematics electrohydrodynamic convection in nematics: the homeotropic case, in S. Kai, editor, *Pattern formation in complex dissipative systems and Global Dynamics*, page 275, World scientific, (1992).

83 E. Plaut, W. Decker, A. Rossberg, L. Kramer, and W. Pesch, Phys. Rev. Lett. **79**, 2376, (1997).
84 H.Richter, A.Buka, and I.Rehberg, in P. Cladis and P. Palffy-Muhoray, editors, *Spatio-temporal patterns in nonequilibrium complex systems*, Santa Fe Institute Studies in the Sciences of Complexity **XXI**, Addison-Wesley, New York, (1994).
85 M. Dennin, M. Treiber, L. Kramer, G. Ahlers, and D. Cannell, Phys. Rev. Lett. **76**, 319 (1996).
86 M. Treiber and L. Kramer, to appear in Phys. Rev. E (1998).
87 W. Zimmermann and L. Kramer, Phys. Rev. Lett. **55**, 402 (1985).
88 M. Dennin, G. Ahlers, and D. S. Cannell, Science **272**, 388 (1997).
89 E. Moses, J. Fineberg, and V. Steinberg, Phys. Rev. A **35**, 2757 (1987).
90 R. Heinrichs, G. Ahlers, and D. S. Cannell, Phys. Rev. A **35**, 2761 (1987).
91 S. Fauve, S. Douady, and O. Thual, J. Phys.(Paris) **II 1**, 311 (1991).
92 R. Deissler and H. Brand, Phys. Rev. A **44**, R3411 (1991).
93 R. Deissler and H. Brand, Physica D **117** ,95 (1998).
94 Y. Tu, Phys. Rev. E **56**, R6554 (1997).
95 U. Bisang and G. Ahlers, Phys.Rev.Lett **80**, 3061 (1998).
96 M. Dennin, D. S. Cannell, and G. Ahlers, Phys. Rev. E **57**, 638 (1997).
97 H. Riecke and L. Kramer, submitted to Physica D (1998).
98 A. Joets and R. Ribotta, J. Stat. Phys. **64**, 981 (1991).
99 X. Yang, A. Joets, and R. Ribotta, in J. E. Wesfried, H. R. Brand, P. Manneville, G. Albinet, and N. Boccara, editors, *Propagation in Systems far from Equilibrium*, Springer, (Berlin 1988).
100 R. Ribotta, A. Joets, and L. Lin, Phys. Rev. Lett. **56**, 1595 (1986).
101 F. H. Busse, Rep. Prog. Phys. **41**, 1929 (1978).
102 S. Nasuno, O. Sasaki, S. Kai, and W. Zimmermann, Phys. Rev. A **46**, 4954 (1992).
103 H. Amm, R. Stannarius, and A. G. Rossberg, submitted to Physica D (1998).
104 A. Lindner, A. G. Rossberg, and L. Kramer. in preparation.
105 I. Smith, Y. Galerne, S. Lagerwall, E. Dubois-Violette, and G. Durand, J. Phys. Colloq. France **36**, C1–237 (1975).
106 P. Tóth, Á. Buka, J. Peinke, and L. Kramer, Phys. Rev. E (1997, submitted).
107 U. Schneider, M. de la Torre-Juarez, W. Zimmermann, and I. Rehberg, Phys. Rev. A **46**, 1009 (1992).
108 M. Scheuring, L. Kramer, and J. Peinke, Phys. Rev. E (1998, submitted).
109 S. Kai, Y. Adachi, and S. Nasuno, Stability diagram, defect turbulence, and new patterns in electroconvection in nematics, in P. E. Cladis and P. Palffy-Muhoray, editors, *Spatio-Temporal Patterns in Nonequilibrium Complex Systems*, Addison-Wesley Publishing Company, (1994). SFI Studies in the Sciences of Complexity.
110 A. G. Rossberg and L. Kramer, Physica D **submitted** (1997).
111 A. Rossberg, PhD thesis, Universität Bayreuth, (1997).
112 A. Hertrich, PhD thesis, Universität Bayreuth, (1995).
113 L. Kramer, A. Hertrich, and W. Pesch, Electrohydrodynamic convection in nematics: the homeotropic case , in S. Kai, editor, *Pattern formation in complex dissipitative systems and Global Dynamics*, page 238, World scientific, (1992).
114 A. Hertrich, W. Decker, W. Pesch, and L. Kramer, J. Phys. France **II 2**, 1915 (1992).

115 H. Richter, N. Klöpper, A. Hertrich, and A. Buka, Europhys. Lett. **30**, 37 (1995).
116 H.Richter, A.Buka, and I.Rehberg. unpublished.
117 S. Kai, K. Hayashi, and Y. Hidaka, J. Phys. Chem. **100**, 19007 (1996). A
118 S. Kai, K. Hayashi, and Y. Hidaka, preprint (1997).
119 G. Küppers and K. Lortz, J. Fluid Mech. **35**, 609 (1996).
120 Y. Hu, R. E. Ecke, and G. Ahlers, Phys. Rev. Lett. **74**, 5040 (1995).
121 P. Kolodner, S. Slimani, N. Aubry, and R. Lima, Physica D **85**, 165 (1995).
122 L. Kramer, E. Bodenschatz, and W. Pesch, Phys. Rev. Lett. **64**, 2588 (1990).
123 *Noise in nonlinear dynamical systems*, Vols. 1-3, edited by F. Moss and P. V. E. McClintock (Cambridge University Press, Cambridge, 1989).
124 S. Kai, T. Kai, M. Takata, and K. Hirakawa, J. Phys. Soc. Jpn. **47**, 1379 (1979).
125 T. Kawakubo, A. Yanagita, and S. Kabashima, J. Phys. Soc. Jpn. **50**, 1451 (1981).
126 H. R. Brand, S. Kai, and S. Wakabayashi, Phys. Rev. Lett. **54**, 555 (1985).
127 S. Kai, T. Tamura, S. Wakabayashi, M. Imasaki, and H. R. Brand, IEEE-IAS Conf. Records **85CH**, 1555 (1985).
128 S. Kai, H. Fukunaga, H. R. Brand, J. Phys. Soc. Jpn. **56**, 3759 (1987).
129 S. Kai, H. Fukunaga, H. R. Brand, J. Stat. Phys **54**, 1133 (1989).
130 S. Kai, in: Ref. [123], 22-76.
131 H. R. Brand, in: Ref. [123], 77-89.
132 U. Behn and R. Müller, Phys. Lett. **113A**, 85 (1985).
133 R. Müller and U. Behn, Z. Phys. B **69**, 185 (1987).
134 R. Müller and U. Behn, Z. Phys. B **78**, 229 (1990).
135 A. Lange, R. Müller, and U. Behn, Z. Phys. B **100**, 477 (1996).
136 H. Amm, U. Behn, N. Klöpper, and R. Stannarius, 25. Freiburger Arbeitstagung Flüssigkristalle (1996).
137 H. Amm, U. Behn, Th. John, and R. Stannarius, Mol. Cryst. Liq. Cryst. **304**, 525 (1997).
138 Th. John, Diploma thesis (Universität Leipzig, 1997).
139 U. Behn, A. Lange, and Th. John, Phys. Rev. E **58**, 2047 (1998).
140 U. Behn, Th. John, and R. Stannarius, On-off intermittency in electrohydrodynamic convection in nematics driven by multiplicative noise, in preparation.
141 M. Fricke, A. Kühnel, and U. Behn, preprint NTZ 47/1997.
142 N. Platt, E. A. Spiegel, and C. Tresser, Phys. Rev. Lett. **70**, 279 (1993); J. F. Heagy, N. Platt, and S. M. Hammel, Phys. Rev. **49**, 1140 (1994).
143 W. Horsthemke and R. Lefever, Noise induced transitions (Springer-Verlag, Berlin, 1983).
144 R. C. Bourret, U. Frisch, and A. Pouquet, Physica **65**, 303 (1973).
145 L. Arnold and P. Kloeden, SIAM J. Appl. Math. **49**, 1242 (1989).
146 L. Arnold and R. Khasminskii, in: *Stochastic Analysis*, Proceedings of Symposia in Pure Mathematics, Volume 57, pp 543, edited by M.C. Cranston and M.A. Pinsky (American Mathematical Society, Providence, Rhode Island, 1995).
147 E. Dubois-Violette, J. Phys. (Paris) **33**, 95 (1972).
148 W. J. A. Goossens, in: *Advances in liquid crystals*, Vol. 3, pp 1, edited by G. H. Brown (Academic Press , New York, 1978).

149 The two-dimensional model with free boundary conditions can be adapted to rigid boundary conditions in the one-mode approximation. Since the coefficients look then much less transparent all explicit results were calculated for free boundaries.
150 A. Crisanti, G. Paladin, and A. Vulpiani, *Products of Random Matrices in Statistical Physics*, Vol. 104 of *Springer Series in Solid-State Sciences*, edited by M. Cardona, P. Fulde, K. von Klitzing, and H.-J. Queisser (Springer-Verlag, Berlin, 1993).
151 V. E. Shapiro and V. M. Loginov, Physica A **91**, 563 (1978).
152 V. I. Oseledec, Trans. Moscow Math. Soc. **19**, 197 (1968).
153 N. V. Madhusudana, V. A. Raghunathan, and K. R. Sumathy, Pramāna, J. Phys. **28**, L311 (1987).
154 J. M. Deutsch, Phys. Rev. E **48**, R4179 (1993).
155 M. V. Feigel'man and I. E. Staroselsky, Z. Phys. B **62**, 261 (1986).
156 R. Müller, K. Lippert, A. Kühnel, and U. Behn, Phys. Rev. E **56**, 2658 (1997).
157 R. Graham, Phys. Rev. A **10**, 1762 (1974).
158 J. Swift and P. C. Hohenberg, Phys. Rev. A **15**, 319 (1977).
159 P. C. Hohenberg and J. Swift, Phys. Rev. A **46**, 4773 (1992).
160 G. De Nigris, G. Nicolis, and H. Frisch, Phys. Rev. A **34**, 4211 (1986).
161 M. Treiber and L. Kramer, Phys. Rev. E **49**, 3184 (1994).
162 D.R. Kurtze, Phys. Rev. Lett. **77**, 63 (1996).
163 J. Viñals, E. Hernández-García, M. San Miguel, and R. Toral, Phys. Rev. A **44**, 1123 (1991).
164 T. C. Elston and R. F. Fox, Phys. Rev. A, **44**, 8403 (1991).
165 E. Hernández-García, M. San Miguel, R. Toral, and J. Viñals, Physica D **61**, 159 (1992).
166 E. Hernández-García, J. Viñals, R. Toral, and M. San Miguel, Phys. Rev. Lett. **70**, 3576 (1993).
167 J. García-Ojalvo, A. Hernández-Machado, and J. M. Sancho, Phys. Rev. Lett. **71**, 1542 (1993).
168 L. Ramirez-Piscina, A. Hernández-Machado and J. M. Sancho, Phys. Rev. B **48**, 119 (1993).
169 L. Ramírez-Piscina, A. Hernández-Machado and J. M. Sancho, Phys. Rev. B **48**, 125 (1993).
170 J. M. Sancho, A. Hernández-Machado, L. Ramírez-Piscina, and A. M. Lacasta, Acta Physica Polonica B **24**, 733 (1993).
171 P.-M. Lam and D. Bagayoko, Phys. Rev. E **48**, 3267 (1993).
172 T. Yamada, Phys. Lett. A **161**, 489 (1992).
173 T. Yamada, K. Fukushima, and H. Fujisaka, Physica A **204**, 755 (1994).
174 C. Van den Broeck, J. M. R. Parrondo, J. Armero, and A. Hernández-Machado, Phys. Rev. E **49**, 2639 (1994).
175 A. Becker and L. Kramer, Phys. Rev. Lett. **73**, 955 (1994).
176 A. Becker and L. Kramer, Physica D **90**, 408 (1996).
177 A. Schenzle and H. Brand, Phys. Rev. A **20**, 1628 (1979).
178 R. Graham and A. Schenzle, Phys. Rev. A **25**, 1731 (1982).
179 A. Teubel, U. Behn, and A. Kühnel, Z. Phys. B **71**, 393 (1988).
180 J. Röder, H. Röder, and L. Kramer, Phys. Rev. E **55**, 7068 (1997).
181 A. Hertrich, W. Pesch, and J. T. Gleeson, Europhys. Lett. **44**, 417 (1996).

182 J. T. Gleeson, Nature, **356**, 511,(1996)
183 R. Faller and L. Kramer, Phys. Rev. E, **57**, R6249 (1998).
184 B. J. Gluckmann, C.B. Arnold and J.P. Gollub Phys. Rev,. E, **51**, 1128, (1995), I. Ning, Y. Hu, R.E. Ecke and G. Ahlers Phys. Rev. Lett. **71**, 2216, 1993.
185 S. Rudroff and I. Rehberg, Phys. Rev. E, **55**, 2742 (1998).
186 G. Ahlers, Experiments on thermally driven convection, in A. Buka and L. Kramer, editors, *Pattern formation in liquid crystals*, Springer, (New York 1995).
187 L. Thomas, W. Pesch and G. Ahlers, to appear in Phys.Rev.E, 1998.
188 E. Dubois-Violette and P. Manneville, Flow instabilities in Nematics, in A. Buka and L. Kramer, editors, *Pattern formation in liquid crystals*, Springer, (New York 1995).
189 A. P. Krekhov and L. Kramer, Phys. Rev. E, **53**, 4925 (1996).
190 T. Börzsönyi,A. Buka, A. P. Krekhov and L. Kramer, Phys. Rev. E, submitted (1998) (1996).

Dynamics of Patterns of Chemical Reactions on Surfaces

R. Imbihl[1], H. Engel[2], and M. Eiswirth[3]

[1] Institut für Physikalische Chemie und Elektrochemie, Universität Hannover, Callinstr. 3 - 3a, D-30167 Hannover, Germany
[2] Institut für Theoretische Physik, Technische Universität Berlin, Hardenbergstr. 36, D-10623 Berlin, Germany
[3] Fritz-Haber-Institut der Max-Planck-Gesellschaft, Faradayweg 4 - 6 , D-14195 Berlin, Germany

Correspondence to: eiswirth@fhi-berlin.mpg.de; phone: (030) 8413-5145; fax: (030) 8413-5106.

1 Introduction

Nonlinear dynamics in heterogeneous catalysis has been studied extensively over the past decades (see [1-14] for review articles). A large number of reactions, different types of catalysts and reactor designs have been used. In general, an understanding of pattern formation on catalytic surfaces requires not only detailed knowledge of the reaction and diffusion processes on the surface, but also needs to assess the influence of temperature changes, the flow of the relevant species (usually in the gas, sometimes in the liquid phase) and mixing effects in the fluid phase. The 'chemical part' of the problem can be isolated by working at very low pressure (typically below 10^{-4} mbar) on single crystal surfaces (in order to minimize the effects of catalyst inhomogeneities). Naturally, the reaction rate is much smaller at low than at high pressure so that experiments can be conducted under strictly isothermal conditions and the heat balance need not be taken into account. Moreover, at 10^{-4} mbar there are typically 10^5 site changes of an adsorbed particle for 1 adsorption event, i.e., diffusion is fast enough to guarantee a locally well-mixed adsorption layer. As a consequence, the pattern formation can be successfully modeled by reaction-diffusion equations. In this article we restrict ourselves to typical aspects of surface reactions which distinguish them from other reaction-diffusion systems (such as reactions in the liquid phase or in gels). These aspects include threshold kinetics (i.e., a reaction does not occur until a certain concentration exceeds a finite value), anisotropy (in particular dynamical changes of anisotropy) and global coupling (which occurs due to good mixing of the gas phase). In each case, it is described how these effects occur naturally in surface reactions, and their impact on pattern formation is discussed.

2 Threshold Kinetics

Usually, chemical reactions proceed at a rate proportional to some (small) power of the reactant concentration, i.e., the rate remains finite for finite concentrations. We speak of threshold kinetics if a concentration has to exceed a finite critical value before a certain process sets in. Such behavior arises naturally, e.g., in the kinetics of phase transitions.

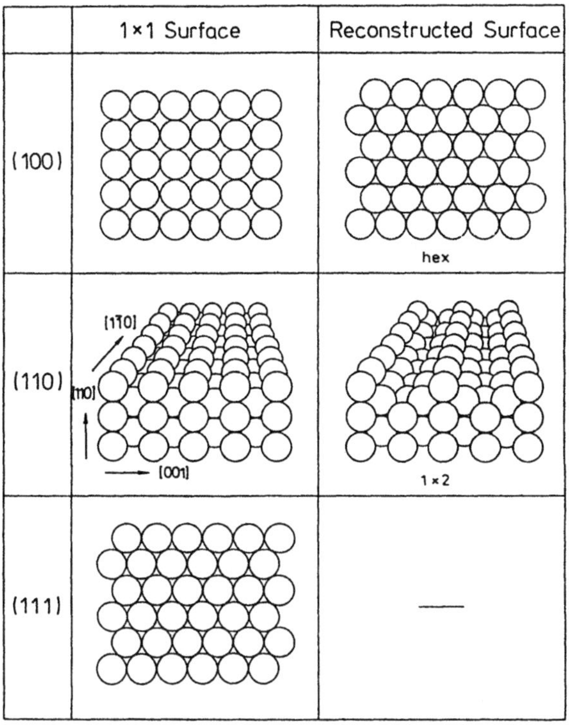

Fig. 1. Structures of the low-index planes of *Pt*. The densely packed *Pt*(111) surface does not reconstruct.

2.1 Structural transformations on *Pt* surfaces

Platinum exhibits a fcc structure, consequently its low-index planes should exhibit a structure as shown on the left-hand side of Fig. 1. However, the clean (100) and (110) surfaces reconstruct, forming a 5x20 ("hex") resp. 1x2 missing row structure (Fig. 1). The driving force for these reconstructions is the lowering of the surface energy achieved when the topmost *Pt* atoms regroup to form a denser layer. However, upon chemisorption of certain species like CO, these Pt atoms have another binding partner so that the reconstruction is lifted, i.e., the bulk-like 1x1 structure forms again. The beginning of this process is shown in Fig. 2 for $Pt(110)$ (on which we shall concentrate in the following). Studies by (low-energy electron diffraction (LEED) and by scanning tunneling microscopy (STM) [15] revealed that the structural change does not take place until CO reaches a coverage of 0.2, while the reconstruction is completely lifted for coverages above 0.5. In between, islands of both structures coexist.

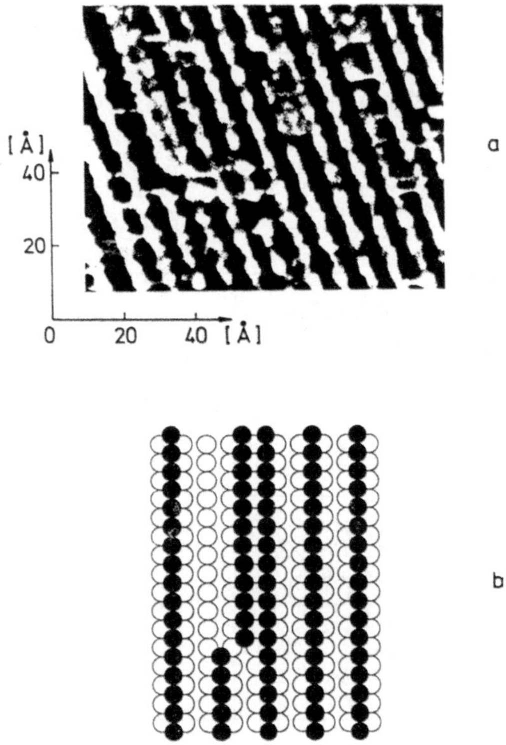

Fig. 2. Initial stage of the lifting of the 1 × 2 reconstruction of $Pt(110)$ upon CO adsorption. The scanning tunneling microscopic picture (a) indicates that rows of Pt atoms are moved, as depicted in (b).

2.2 Reconstruction model of CO oxidation on $Pt(110)$

Since during the course of chemical oscillations or during the passage of a chemical wave on a surface the CO coverage changes significantly, the coverage-induced structural changes play an important role for the dynamical behavior and have to be included for a successful model. Thus, for CO

oxidation not only the coverages of CO and O, u and v, have to be taken into account, but also the degree of 1x1 coverage w is included. The terms describe the adsorption of CO and O, the desorption of CO, the reaction between CO and O, and the diffusion of CO (desorption and diffusion of O can be neglected). The structural variable w depends on the CO coverage and is fitted using a sigmoid- shaped function $f(u)$ which is zero below 0.2, unity above 1 and differentiable everywhere.

$$\partial_t u = k_1 p_{CO}(1-u^3) - k_2 u - k_3 uv + D\nabla^2 u \tag{1a}$$

$$\partial_t v = k_4 p_{O_2}(1-u-v)^2 - k_3 uv \tag{1b}$$

$$\partial_t w_t = k_5(f(u) - w) \tag{1c}$$

with

$$k_4 = w s_1 + (1-w)s_r. \tag{1d}$$

See [8,16] for details and experimental determinations of the parameters. Eq. (1d) gives rise to a negative feedback loop since $s_1 < s_r$. Thus, a high CO coverage causes a structural change to the 1x1 phase w which exhibits a higher sticking coefficient for oxygen. The resulting faster oxygen adsorption in turn lowers the CO coverage by reactive removal.

The fact that the slow variable w does not change until the CO coverage u reaches a threshold can lead to situations in which the corresponding nullcline intersects more than once with the middle part of the fast nullcline (see Fig. 3). As pointed out by Koper [17], this is typical of dynamical systems which do not exhibit a cross-shaped phase diagram, but the oscillatory region is oriented asymmetrically with respect to the bistable one (Fig. 3b). An important consequence of this asymmetry is that homoclinic bifurcations of stable limit cycles can occur, which was indeed observed experimentally in the CO oxidation on $Pt(110)$ [16].

2.3 Backfiring, spiral breakup and turbulence

In the excitable regime the coexistence of two unstable fixed points with the stable rest state also has consequences for the spatiotemporal pattern formation [18,19]. While standard systems just cease to be excitable when the time scales of activator and inhibitor become comparable, in the present case excitability persists even as the refractory zone shrinks and finally vanishes (Fig. 4a). Consequently, a traveling pulse can backfire, i.e., emit another pulse traveling in the opposite direction, eventually giving rise to chemical turbulence in one spatial dimension (Fig. 4b).

In two spatial dimensions over a wide range of parameters spiral waves exist [20,21], but as the time scale of the inhibitor increases (corresponding to increasing catalyst temperature), the spiral arm breaks near the center [18,19], leaving two open ends which tend to spiralize and in turn break, leading to defect-mediated turbulence (Fig. 5). An experimental observation of such a turbulent state is reproduced in Fig. 6, obtained using photoemission electron microscopy (PEEM).

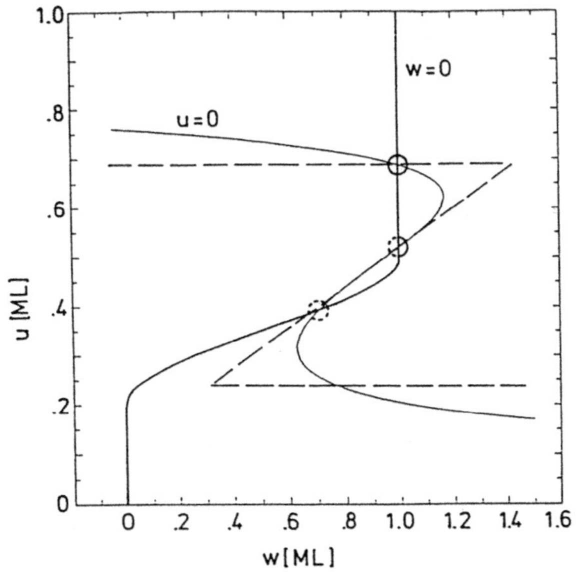

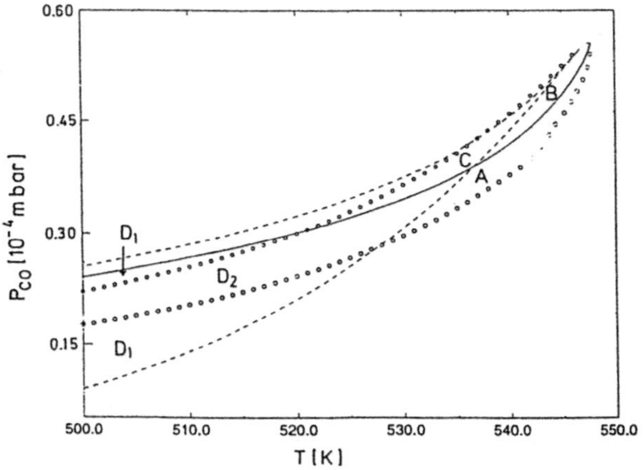

Fig. 3. (a) Nullclines of activator u and inhibitor w of the reaction part of Eqs. (1a-1d) at $p_{O_2} = 1.33 \times 10^{-4}$ mbar, $p_{CO} = 4.5 \times 10^{-5}$ mbar and temperature T = 540 K. (b) Bifurcation diagram of Eqs. (1a-1d) at $p_{O_2} = 1.33 \times 10^{-4}$ mbar. Regions A and C are excitable, B is oscillatory and D is bistable. The latter is further subdivided depending on whether one (D_1) or two fronts (D_2) exist (full line: hopf bifurcation, dashed line: saddle-node bifurcation; points limit the existence region of pulses [19]).

3 Anisotropic wave patterns

3.1 Pattern formation in the liquid phase and on surfaces

In contrast to reaction systems in the liquid phase where diffusion is isotropic, diffusion on surfaces is in general anisotropic. On surfaces the diffusion an-

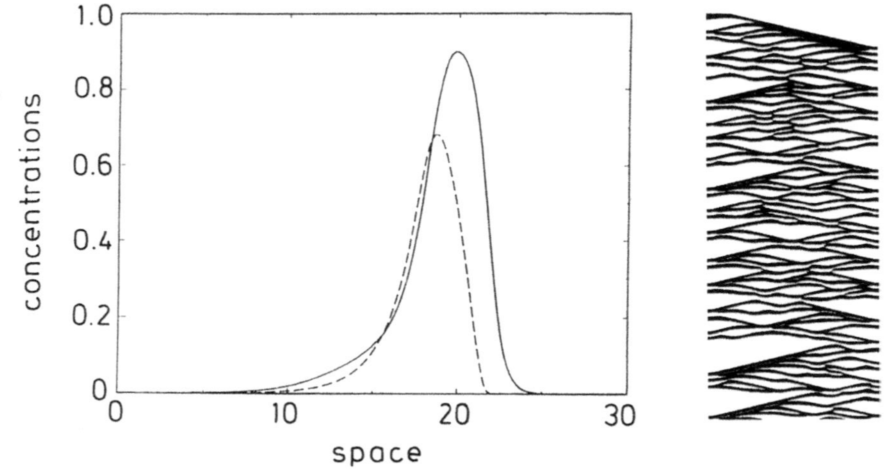

Fig. 4. (a) Profile of a pulse with vanishing refractory zone (full line: activator, dashed: inhibitor). (b) Space-time plot of turbulence developing from a single pulse (computations carried out with Eqs. (1a- 1d), dimensionless units).

isotropy is mainly determined by the corrugation of the substrate (neglecting the corrugation of the electronic surface). Surfaces like the (110) planes of a fcc metal with their trough-like surface geometry will accordingly exhibit a strongly anisotropic surface diffusion. Even on ideally isotropic surfaces like fcc(100) and fcc(111) diffusion can be anisotropic if, due to a deviation from the ideal cut angle, atomic steps with a preferential orientation are present.

From simple anisotropy, i.e., an anisotropy which can be removed by rescaling the coordinates, one expects an elliptical deformation of chemical wave patterns. Such elliptically deformed target patterns and spiral waves have, in fact, been observed in the systems $Pt(110)/CO$+O_2 and $Rh(110)/NO$-+H_2 [22-25]. A rescaling is, however, no longer possible if the anisotropy is not constant but varies depending on the adsorbate coverages. Such a state-dependent anisotropy can be generated when the substrate geometry changes with coverage, a situation given when adsorbate-induced surface reconstructions exist and/or when attractive or repulsive interactions between the adparticles lead to a coverage-dependent anisotropy of surface diffusion. Both effects can be realized in surface reaction systems, and in the examples given below it will be demonstrated that rectangular patterns and wave fragments may develop when such a state-dependent anisotropy of surface diffusion exists.

The symmetry of the substrate also becomes important when the atomic adsorbates start to penetrate into the bulk forming subsurface species. In such a case, we observed that triangular-shaped reaction fronts developed on a surface which is isotropic with respect to surface diffusion [26]. The

Fig. 5. Development of spiral turbulence in a simulation of Eqs. (1a-1d), parameters correspond to region C of Fig. 3.

unusual shape of the front is tentatively attributed to elastic stress which is generated between surface phases with different periodicities and which effectively makes the reaction anisotropic. Anisotropic diffusion, elastic stress and attractive or repulsive interactions between the diffusing particles are effects which do not exist in the liquid phase and which lead to new types of pattern formation on surfaces. Most of the observed new phenomena could be reproduced with general mathematical models which had been modified to include the a. m. effects [25,27]. The development of realistic models remains a challenging task because the mechanistic details of the various reaction systems are quite complicated. Nevertheless, based on a number of adsorption studies and on *in-situ* experiments, realistic models could be formulated and simulations with these models are in progress.

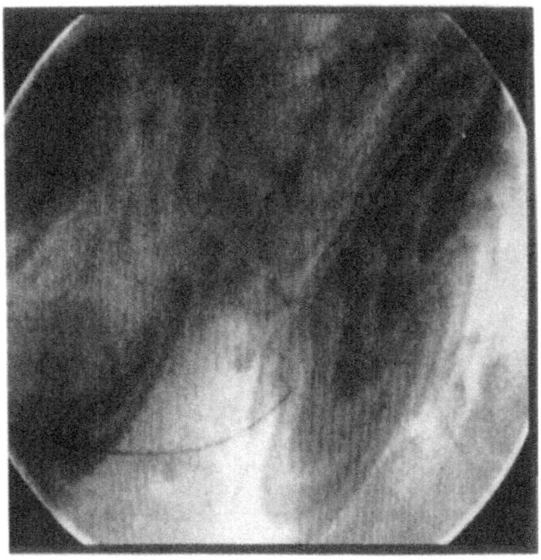

Fig. 6. PEEM snapshot of a turbulent state (diameter 180 μ m) in the CO oxidation on $Pt(110)$ at T = 540 K, $p_{O_2} = 3 \times 10^{-4} mbar, p_{CO} = 1 \times 10^{-4}$ mbar.

3.2 Rectangular patterns and wave fragments in the $NO + H_2$ reaction on $Rh(110)$

The reaction studied here is the catalytic reduction of NO with hydrogen on a $Rh(110)$ single crystal surface. Rhodium is an essential constituent of the automotive catalytic converter due to its ability of effectively dissociating adsorbed NO [28]. Both, H_2 and NO, thus absorb dissiociately. The atomic oxygen reacts with hydrogen to form H_2O, while atomic N can either combine to form N_2 or react with H to NH_3. All three products, H_2O, NH_3 and N_2, desorb immediately. At higher temperature (at $T > 500K$) only atomic oxygen and nitrogen are left on the surface, and both of these adsorbates cause a restructuring of the metal surface [29]. Depending on the adsorbate and on the coverage, different reconstruction geometries are obtained. For the system $Rh(110)/O, N$ a whole zoo of different reconstruction geometries exists, but all of these structures can be built up with only a few construction principles. These are illustrated in Fig. 7 showing structural models of the reconstructions generated by atomic oxygen and nitrogen, i.e., the c(2x6)-O and c(2x8)-O of adsorbed oxygen, the (2x1)-N and the (3x1)-N of adsorbed nitrogen, as well as the mixed c(2x4)-2O, N structure. Oxygen induces reconstructions of the "missing row" type which exhibit trough-like structures along the [110] direction, while nitrogen generates reconstructions with $-Rh - N - Rh$ chains in the [001] direction.

Given that the anisotropy of surface diffusion follows the geometric corrugation, a structural transition from the c(2x6)-O of oxygen to the (2x1)-N of nitrogen causes the anisotropy to rotate by 90° . Hence, the anisotropy varies with the adsorbate coverages, and this leads to the following consequences with respect to pattern formation. In the excitable $Rh(110)/NO + H_2$ sys-

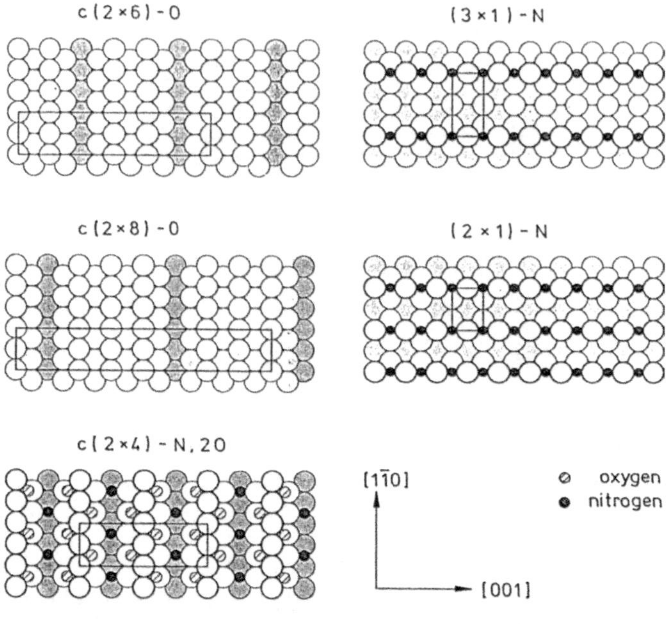

Fig. 7. Structural models of the adsorbate-induced reconstructions which presumably are responsible for the coverage-dependent variation of the diffusion anisotropy during the $NO + H_2$ reaction on $Rh(110)$: the c(2x6)-O of oxygen and the (2x1)-N of nitrogen. The oxygen adatoms are not shown in the model of the c(2x6)-O. Indicated by thin lines are the unit cells of the structures.

tem one observes at $T = 540K$ elliptical target patterns [24]. With increasing temperature the patterns become more and more deformed until finally they have the rectangular shape depicted in Fig. 8a [24,25,30-32]. The dark area in these images can be associated with an oxygen-covered surface, while the bright area can be attributed to a nitrogen-covered surface. Since the N- and O-covered surfaces exhibit different reconstructions, it is evident that the anisotropy of the surface changes along the profile of a chemical wave. The unusual rectangular shape of the patterns is apparently a consequence of the anisotropy being state-dependent.

In the experiment another unusual effect was seen which is displayed in Fig. 8b. If after generation of a rectangular pattern pH_2 was decreased, those parts of the wave pattern oriented in [001] dissolved, leaving isolated stripes in the [110] direction. The chemical waves thus exhibit a propagation failure in one direction, and one obtains wave fragments traveling with constant velocity along a fixed crystallographic direction [33]. A PEEM image showing a large number of these traveling wave fragments is reproduced in Fig. 8b.

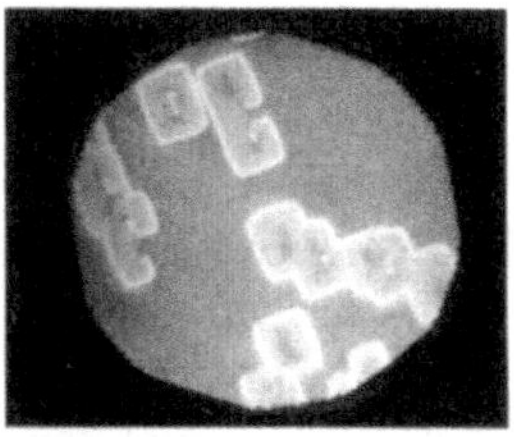

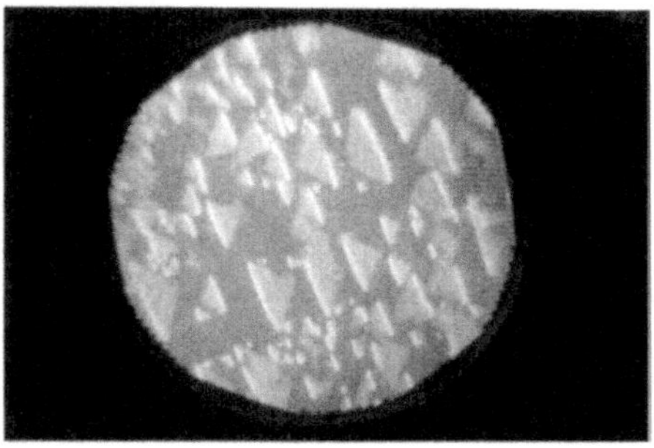

Fig. 8. PEEM images of the $NO + H_2$ reaction on $Rh(110)$ showing new types of chemical wave patterns in a medium with state-dependent anisotropy [25]. (a) Rectangularly shaped patterns. Experimental conditions: T = 620 K, $p_{NO} = 1.8 \times 10^{-6}$ mbar, $p_{H_2} = 5 \times 10^{-6}$ mbar. (b) Traveling wave fragments. Experimental conditions: T = 560 K, $p_{NO} = 1.8 \times 10^{-6}$ mbar, $p_{H_2} = 5.3 \times 10^{-6}$ mbar. The diameter of the imaged area is roughly 400 μ m in (a) and (b).

Depending on pH_2, the fragments either expand or contract in the [110] direction while they travel, but they do not spiral up as one would expect in an isotropic excitable medium.

It was first shown by Mikhailov, based on rather general considerations for the evolution of a front geometry in a medium in which the front velocity varies in different directions, that patterns with sharp corners can develop [34]. The condition for the formation of cusps is that the angular dependence of the front velocity, $c(\alpha)$, obeys the following relation

$$c(\alpha) + \frac{\partial^2 c(\alpha)}{\partial \alpha^2} < 0 \tag{2}$$

where α represents the angle with respect to a fixed direction on the surface. Subsequently, it was demonstrated by Gottschalk et al. who used a standard activator-inhibitor model describing pattern formation in excitable media that practically all of the observed patterns can be reproduced if state-dependent diffusion anisotropy is introduced into the equations of the FitzHugh-Nagumo model [25,37]. The equations then read

$$\begin{aligned} \partial_t u &= (u - u^3/3 - v)\varepsilon + \nabla[D(v)\nabla u], \\ \partial_t v &= u + b - av \end{aligned} \tag{3}$$

$$D_x = D_0 \quad D_y(v) = \begin{cases} D_1, & v \leq v_{crit} \\ D_2, & v > v_{crit}. \end{cases} \tag{4}$$

In this model the diffusion anisotropy of the activator u depends on the inhibitor concentration v. As the inhibitor concentration v crosses a threshold value, v_{crit}, the diffusion constant of the activator in the y-direction, D_y, is switched to a different value while the diffusion constant in the x-direction, D_x, remains constant. Simulations with the model reproduced the rectangular target patterns and spiral waves seen in the experiment, the traveling wave fragments and the coexistence of two different front geometries under identical conditions. The latter phenomenon, which is displayed in Fig. 9a, is observed in the so-called range of dynamic bistability [25,36]. An elliptical oxygen island (black) forms, followed by nucleation of a second front (white) with almost rectangular shape in the interior of the island. The experiment is accurately reproduced by the simulation displayed in Fig. 9b. The cross section through the simulated pattern in Fig. 9c explains the origin of the unusual behavior. In the transition white-black the switching of the anisotropy only occurs after passage of the front, and therefore this front exhibits the usual elliptical geometry. In contrast, in the transition black-white in the second front the switching happens in the immediate vicinity of the steepest gradient of the activator. In this case, the state-dependent anisotropy will affect the shape of the front leading to the rectangular front visible in Fig. 9b.

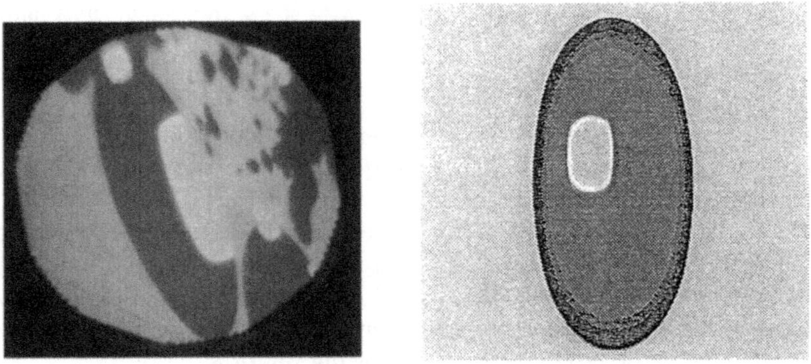

Fig. 9. (a) Simultaneous presence of different front geometries in chemical wave patterns [25]. (a) Experiment: PEEM image recorded during the $NO + H_2$ reaction on $Rh(110)$ at T = 620 K, $p_{NO} = 1.8 \times 10^{-6}$ mbar, $p_{H_2} = 5.2 \times 10^{-6}$ mbar. The diameter of the imaged area is roughly 400 μ m. (b) Simulation of the experiment in (a) with the model Eqs. (3, 4) described in the text.

Simulations with the modified FitzHugh-Nagumo model demonstrated that also the origin of the traveling wave fragments lies in the state-dependent anisotropy of the medium [27]. The formation of these fragments can also be explained within the context of the kinematic theory [33]. It should be added

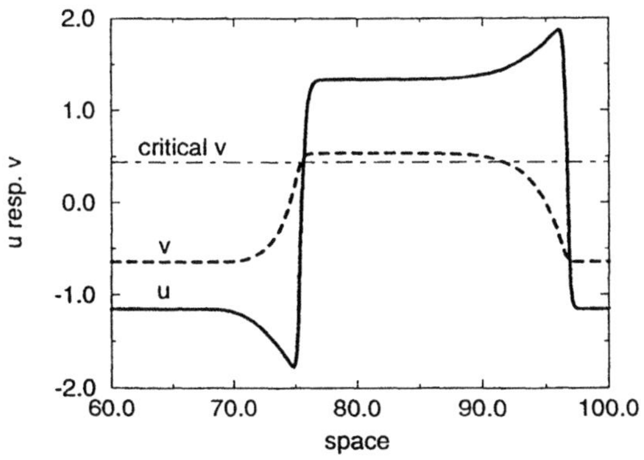

Fig. 9. (c) Concentration profile through part of the simulated pattern in (b) demonstrating the origin of the different front geometries. The right part shows the transition from white to black, the left part the transition from black to white. The dashed line indicates the critical value of the inhibitor concentration at which the diffusional anisotropy of the activator switches.

that the "soliton-like" oxygen pulses seen earlier in catalytic CO oxidation on $Pt(110)$ in principle represent the same phenomenon as the traveling wave fragments on $Rh(110)$ [35]. A state-dependent anisotropy in that case is given by a different diffusion anisotropy of CO on the 1x1 and 1x2 phases of $Pt(110)$ [23].

In order to investigate anisotropic front propagation and the excitation mechanism in more detail, two systems related to $Rh(110)/NO + H_2$ were studied: the bistable $Rh(110)/O_2{+}H_2$ system and the excitable $Rh(111)/NO{+}H_2$ reaction. Remarkably, in the $O_2 + H_2$ reaction on $Rh(110)$ the anisotropy of the reaction fronts varied with the experimental parameters [36]. The origin of this unusual behavior could not be clarified so far but several factors indicate that the energetic interactions of the adparticles play a decisive role. In the $NO + H_2$ reaction on $Rh(111)$ one finds traveling pulses similar to $Rh(110)$ but with isotropic propagation [37].

3.3 Triangular wave patterns

Although in the $NO{+}H_2$ reaction on $Rh(110)$ and $Rh(111)$ part of the atomic adsorbate will dissolve in the bulk, the essential pattern forming processes are still restricted to the surface layer. The situation becomes different, however, when large amounts of a subsurface species are formed. In that case, pattern formation is influenced by the formation of a bulk phase, as demonstrated by the following experiment. When we expose the $Rh(111)$ surface to a large dose of oxygen (ca. 10^6 L), about 5-10 monolayers of oxygen will dissolve in the subsurface region. With LEED we see a (8×8)-O superstructure (surface oxide). During the titration of the oxygen phase with hydrogen we observe the formation of triangular reaction fronts, as demonstrated by the PEEM image in Fig. 10a [5]. The sides of the triangle which mark the interface between bare $Rh(111)$ surface and the surface oxide are oriented along close-packed rows of the (111) surface, as indicated in Fig. 10b. The front geometry depends on the experimental parameters, and for high temperatures or low hydrogen pressures ($p < 5 \times 10^{-7}$ mbar) the usual circular reaction fronts are found.

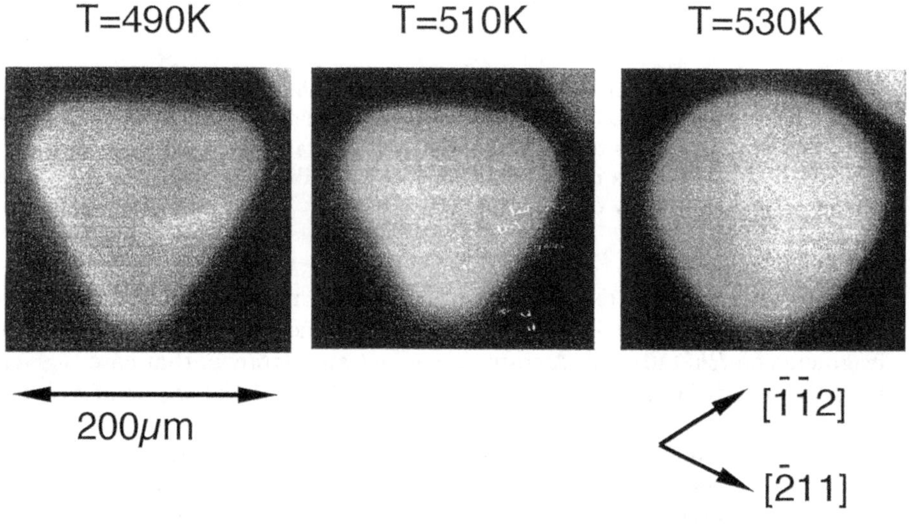

Fig. 10. (a) PEEM images showing triangular and circular front geometries during the titration of the (8× 8)-O structure on $Rh(111)$ with hydrogen ($p_{H_2} = 5.6 \times 10^{-7}$ mbar) at varying temperatures. The (8 × 8)-O structure was prepared by exposing the $Rh(111)$ surface at T = 770 K to $p_{O_2} = 2 \times 10^{-4}$ mbar for 1h [26].

The $Rh(111)$ surface is isotropic with respect to surface diffusion. Hence, one expects circular reaction fronts, and these fronts are seen in experiments in which no large amounts of subsurface oxygen are present. Therefore, the origin of the triangular fronts has to be closely related to the formation of a surface oxide phase. Since the surface oxide phase and the bare $Rh(111)$ surface exhibit different periodicities, strong elastic stress will exist at the interface between the two phases, i.e., in the reaction front itself. Elastic stress is known to modify the transport properties of materials, and stress can, of course, also modify the activation barriers for a reaction [38]. Although a detailed microscopic picture still needs to be developed, these arguments strongly suggest that elastic stress plays a key role in the formation of the

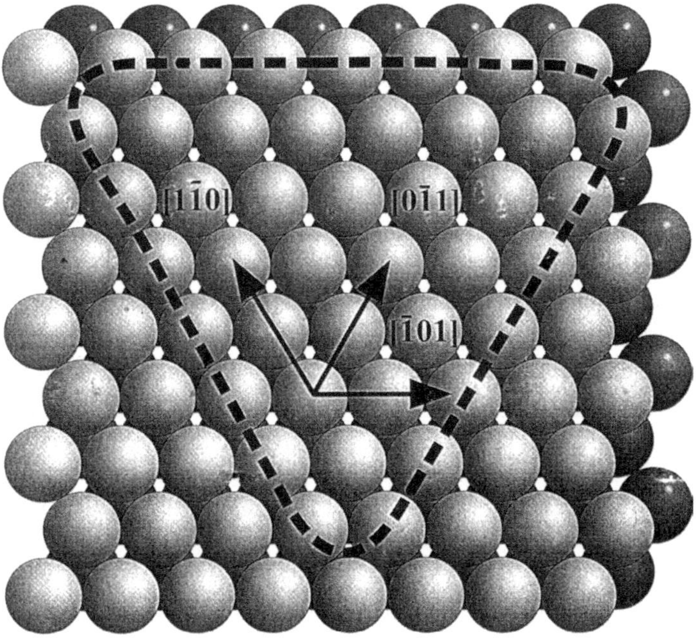

Fig. 10. (b) Structural model of the $Rh(111)$ surface with the directions indicated along which the sides of the triangular fronts in (a) are oriented.

triangular fronts. Since diffusion on $Rh(111)$ is isotropic, the triangular fronts can apparently only be explained if the elastic stress makes the reaction effectively anisotropic. The anisotropy in the reaction is an effective anisotropy because several mechanistic steps are involved in the propagation of the reaction front. It remains to be shown whether this tentative explanation can be substantiated by experiments showing the microscopic processes that participate in front propagation.

3.4 Mechanistic aspects and formulation of realistic models

The formulation of mechanistic models for the pattern forming $NO + H_2$ reaction on $Rh(110)$ and $Rh(111)$ is not a simple task because of two basic problems: Several different reconstruction phases are involved in the excitation cycle on $Rh(110)$, and in addition, adsorbed atomic oxygen and nitrogen can penetrate into the bulk forming subsurface species [37,39]. From our current knowledge it looks as if neither the reconstruction mechanism nor the subsurface species are essential for the excitation mechanism but that it is the destabilizing effect of coadsorbed oxygen on nitrogen which plays the key role [40,41]. An alternative model in which an NH_x species is essential has been proposed for the system $Rh(111)/NO + H_2$ [42,43].

In the $NO+H_2$ reaction on $Rh(110)$ a number of different reconstruction phases have been identified with LEED, and their stability and reactivity in a reacting environment have been investigated in adsorption experiments [30]. Based on these experiments and on the results of detailed adsorption and coadsorption studies conducted earlier, the following mechanistic scheme has been proposed [40]:

(1) Starting with an oxygen-covered surface which inhibits hydrogen adsorption, a few vacant sites are created by some defect. Hydrogen adsorption causes an autocatalytic increase in the number of vacant sites, and this leads to a propagating reaction front which reactively removes the oxygen adlayer.

(2) On the bare surface NO can adsorb and dissociate freely but, since the oxygen reacts off with hydrogen, atomic nitrogen accumulates.

(3) On the N adlayer which is formed, NO can still adsorb and dissociate leading to a mixed c(4×2)- $2O, N$ phase. Nitrogen is destabilized in this phase due to repulsive interactions between oxygen and nitrogen. Nitrogen therefore desorbs, and the initial situation of an oxygen-covered surface is established again.

Experiments with H/D isotopes showed that the velocity of the chemical waves and the existence ranges of the wave patterns exhibit very strong isotope effects [44]. This demonstrates that the diffusing species which is rate-limiting in the propagation of the chemical waves is, in fact, atomic hydrogen. Based on the above mechanistic scheme, a realistic mathematical model for the $NO+H_2$ reaction on $Rh(110)$ was formulated [45]. The same model with some small modifications should also apply for $Rh(111)/NO + H_2$ because the pulses there involve a similar change between oxygen and nitrogen as on $Rh(110)$, and similar repulsive interactions between the two adsorbates are present on $Rh(111)$ [37].

Although not believed to be an essential part of the excitation mechanism, subsurface oxygen formation on $Rh(111)$ is presumably responsible for some unusual additional features we find in the chemical wave patterns on $Rh(111)$ [37]. When two pulses in the $NO+H_2$ reaction on $Rh(111)$ collide, we observe in PEEM that the area in between them starts to become increasingly bright with decreasing separation of the pulses. This situation is depicted in Fig. 11. The brightness increase (i.e., lowering of the work function) is attributed to the formation of a subsurface oxygen species. Due to its location below the surface plane, subsurface oxygen exhibits a dipole moment opposite to that of chemisorbed oxygen. A similar process has also been observed in catalytic CO oxidation on $Pt(100)$ and $Pt(110)$ and was also attributed to subsurface oxygen [46]. A remarkable feature of this subsurface oxygen species is that it shows up in PEEM only under pattern forming reaction conditions but not in adsorption experiments. The subsurface oxygen species is observed in the $O_2 + H_2$ and in the $NO + H_2$ reaction on $Rh(111)$, and it is also present on a $Rh(110)$ surface when prior to exposure to reaction conditions structural defects are generated by sputtering [47].

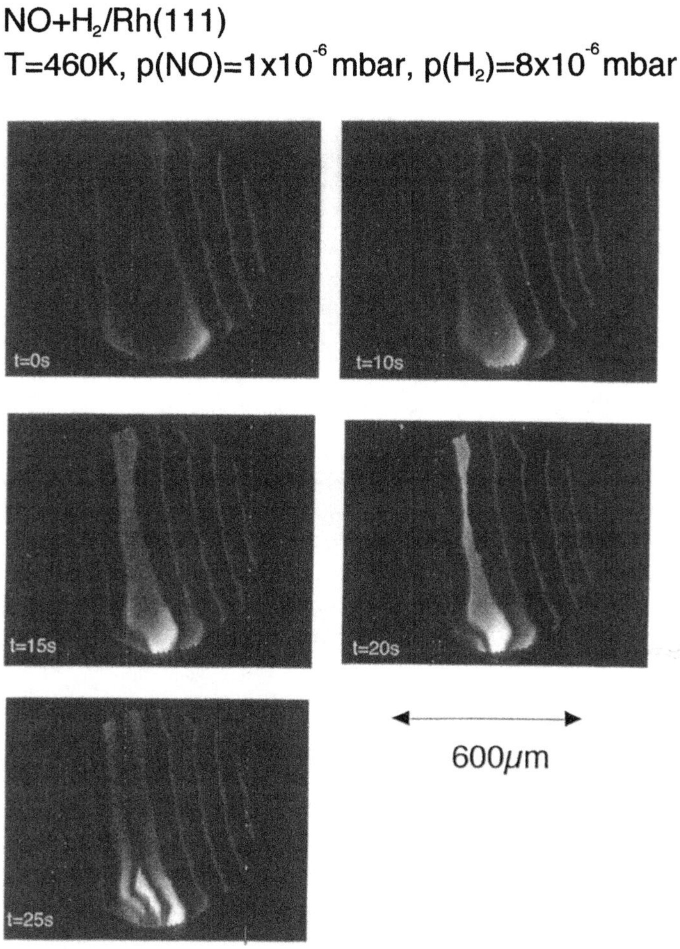

Fig. 11. PEEM images showing the formation of a new species in the $NO + H_2$ reaction on $Rh(111)$ as two wave trains undergo collision. The increase in brightness between the two pulses is attributed to subsurface oxygen [37].

4 Global coupling

Besides local coupling provided by surface diffusion, in surface chemical reactions under low pressure conditions global coupling through the gas phase may arise. In the oscillatory regime of CO oxidation on $Pt(110)$, for example, periodic variations of the partial pressure of CO have been observed, which are due to the changes in the overall adsorption and desorption rates. Global coupling is expected to have a synchronizing effect stabilizing the homogeneously oscillating state. Depending on the experimental conditions, however, global coupling may also induce instabilities and generate new spatiotemporal patterns. As examples in the following we discuss the breakdown of uniform oscillations into several synchronously oscillating domains and the formation of standing waves.

4.1 Complex Ginzburg-Landau equation (CGLE) with global coupling

Independently from the considered reaction-diffusion system, near the threshold of oscillations the dynamics can be reduced to the CGLE which represents a normal form close to a supercritical Hopf bifurcation. In the presence of global coupling this equation for the complex oscillation amplitude η reads [48]

$$\frac{\partial \eta}{\partial t} = \eta - (1 + i\beta) \mid \eta \mid^2 \eta + (1 + i\varepsilon)\nabla^2\eta + \mu e^{i\chi}\langle\eta\rangle \tag{5}$$

Here β determines the nonlinear frequency shift of local oscillators, and ε characterizes the dispersion of traveling waves. These parameters are determined by the particular surface reaction. In the additional integral term, χ describes the phase shift between the global oscillation amplitude and the driving force, while μ denotes the coupling strength. The brackets denote spatial averaging over the surface according to

$$\langle\eta\rangle = \frac{1}{A}\int\limits_A dz\eta(z,t). \tag{6}$$

An important class of solutions are uniform oscillations

$$\eta(z,t) = Ae^{i\Omega t} \tag{7}$$

with amplitude A and a frequency Ω given by, respectively,

$$A = (1 - \mu\cos\chi)^{\frac{1}{2}} \quad , \quad \Omega = -\beta - \mu(\sin\chi - \beta\cos\chi). \tag{8}$$

From the stability analysis with respect to small perturbations $\rho\exp[i(kx + \Omega t) + \lambda t]$, the following expression has been obtained for the growth rate $\lambda(k)$ [49]

$$\lambda(k) = -k^2 + 2\mu\cos\chi - 1 \pm [-(\beta - \mu(\sin\chi + \beta\cos\chi) \\ +\varepsilon k^2)^2 + (1+\beta^2)(1-\mu\cos\chi)^2]^{\frac{1}{2}}. \tag{9}$$

Without global coupling, $\mu = 0$, from Eq. (9) we recover the Benjamin-Feir criterion $1 + \varepsilon\beta = 0$ for the onset of phase turbulence in oscillatory media. In the presence of global couplings it follows from Eq. (9) that uniform oscillations become unstable if

$$2\mu\cos\chi - 1 > 0 \tag{10}$$

or, when condition (10) does not hold, if

$$\mu[2 + \cos 2\chi + \beta \sin 2\chi] - 2(\cos\chi + \beta \sin\chi) < 0. \qquad (11)$$

Note that due to the global term in the CGLE the growth rate of uniform perturbations, $\lambda(k = 0)$, is different from the long wavelength limit of the growth rate of nonuniform perturbations: $\lambda(k = 0) \ldots \lambda(k = 60)$.

In the parameter range where Eqs. (10, 11) are satisfied, uniform oscillations are stable with respect to uniform perturbations, $Re\lambda(k = 0) < 0$, but unstable with respect to long wavelength perturbations, $Re\lambda(k = 60) > 0$. Numerical solutions of the globally coupled CGLE show that in this situation the uniform oscillatory state decays into clusters. Each cluster consists of oscillators that have the same amplitude and phase, which differ, however, from one cluster to another. The period of oscillation is the same in all clusters, but between the oscillations in different clusters a constant phase shift exists. Cluster formation also occurs in networks of coupled oscillators when there is no local coupling between them. In this case, Eqs. (10, 11) have been shown to be equivalent to a criterion for the onset of clustering derived previously by Hakim et al. [49,50]. Therefore, we conclude that clustering expresses the dominance of global over local coupling. Under the influence of local coupling homogeneously oscillating spatial domains develop, which oscillate with a constant phase difference from each other. These spatial cluster patterns have no intrinsic characteristic length scale (Fig. 12a).

Other spatiotemporal structures generated by global coupling are modulated standing waves. These structures exhibit an intrinsic wave number. They have been observed in a parameter range of the globally coupled CGLE where the growth rate $\lambda(k)$ exhibits a positive maximum, $Re\lambda(k_{max}) > 0$, at a certain finite wave number k_{max} given by

$$k_{max}^2 = -\frac{1}{\varepsilon}\left[\beta - \mu(\sin\chi + \beta\cos\chi) + \left(\frac{1+\beta^2}{1+\varepsilon^2}(1 - \mu\cos\chi)^2\right)^{\frac{1}{2}}\right]. \qquad (12)$$

Additionally, we have $\pi/2 < \chi < 3\pi/2$, i.e., the global coupling is destabilizing. For solutions of the CGLE representing a superposition of a uniform oscillation and a standing wave according to

$$\eta(z,t) = \eta_0(t) + \eta_k(t)(e^{ikz} + e^{-ikz}), \ \ \eta_0 = E_0 e^{-i\Omega t}, \ \ \eta_k = E_k e^{-i(\Omega t - \gamma)} \qquad (13)$$

in Eq. (6) analytical expressions have been derived for the amplitude E_0 and the frequency Ω of the oscillatory mode, for the amplitude of the standing wave E_k, and for the phase shift γ between both modes. An example of a modulated standing wave solution of the globally coupled CGLE is shown in Fig. 12b. Standing waves exist only beyond critical values for ε and μ. The threshold for μ can be found from the condition $Re\lambda(k = 60) = 0$, for $\mu > \mu_{cr}$ we have $Re\lambda(k = 60) < 0$. The threshold for ε follows from the condition $Re\lambda(k_{max}) = 0$ with $Re\lambda(k_{max}) > 0$ for $\varepsilon > \varepsilon_{cr}$.

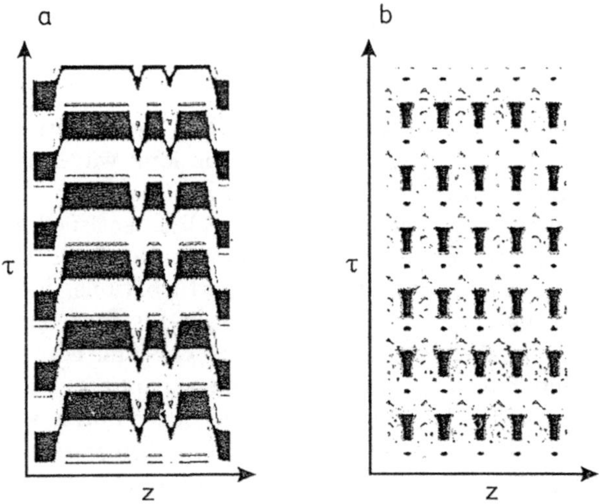

Fig. 12. Space time plot of spatiotemporal patterns resulting from numerical simulations of the globally coupled CGLE. Here and in the following figures τ and z denote the dimensionless time and space coordinate, respectively. Bright areas correspond to high values of the real part of the oscillation amplitude, dark areas to low values of Re η. (a) Cluster pattern obtained at parameters $\beta = -1.8, \varepsilon = 0, \chi = 4.21, \mu = 9.0$. (b) Standing wave solution ($\beta = -1.5, \varepsilon = -5, \chi = 4.587, \mu = 4.0$).

Summarizing, the analysis based on the globally coupled CGLE shows that in the Benjamin-Feir stable situation global coupling may induce instabilities of the uniform oscillatory state accompanied by cluster formation or, e.g., the emergence of modulated standing waves. In systems with anisotropic diffusion (different values for ε in different directions, as typical for catalytic reactions involving a adsorbate-driven surface reconstruction), the standing wave instability can generate stripe patterns on an oscillatory background. This occurs when the diffusion coefficients are chosen in a way that allows the instability in one direction, suppressing it in the other. The stripes are oriented perpendicular to the direction with faster diffusion [49]. Thus, stripe patterns may be understood as resulting from the interplay of the standing wave instability due to global coupling and anisotropic diffusion.

Another possible regime described by the CGLE are turbulent patterns. The phase diagram for turbulence in the CGLE has been constructed recently in [51]. In great detail the turbulent regime in the globally coupled CGLE has been discussed in [52]. It has been shown, for example, that in the Benjamin-Feir unstable parameter region a pattern showing phase turbulence under sufficiently high global coupling is stabilized into a standing wave pattern [53].

4.2 Globally coupled reconstruction model for CO oxidation on Pt (110)

After discussing the impact of global coupling in the framework of the CGLE, we now consider the same phenomena using a modified reconstruction model whose applicability is not restricted to the vicinity of the oscillation threshold. We will find that both models lead to qualitatively equivalent results.

Global coupling through the gas phase arises because adsorption and desorption of CO produce local differences of the partial pressure of carbon monoxide that balance immediately (within 10^{-3} s compared to the time scale of the oscillations of about 5 s). Therefore, the reconstruction model Eqs. (1a-1d) has to be supplemented by a balance equation for the pressure of carbon monoxide in the gas phase [54]

$$\frac{dp_{CO}}{dt} = \frac{J_{i0}}{V}\left\{p_{COE} - p_{CO}\left\{1 + \frac{V_{ML}}{J_{i0}}\int_A \frac{dr^2}{A}\left\{k_1 p_{CO}(1-u^3) - k_2 u\right\}\right\}\right\} \tag{14}$$

(the partial pressure of oxygen changes on a much smaller scale and can be considered as fixed in a first approximation). Eq. (15) describes the gas supply, exhaust and the adsorption and desorption. The constant J_{i0} denotes the in- and outflow of gas in the reactor, V is the volume of the vacuum chamber, and p_{COE} is the CO pressure in the gas inlet. The integral term is taken over the whole catalytic surface of area A. The parameter V_{ML} corresponds to the volume one monolayer of CO would have under the applied conditions (computed from the ideal gas law). The amplitude of the p_{CO} variations in the gas phase is a measure for the strength of global coupling: The larger the amplitude the larger their influence is on the local dynamics. At least for small values of V_{ML} the p_{CO} oscillation amplitude is proportional to V_{ML}. Therefore, we consider V_{ML} as the bifurcation parameter associated with the strength of global coupling. With increasing coupling strength five qualitatively different types of spatiotemporal patterns have been observed in the globally coupled reconstruction model: uniform oscillations, cluster formation, standing waves, phase flips and turbulent patterns. The corresponding phase diagram in the $p_{O_2} - V_{ML}$ parameter plane is shown in Fig. 13 [55, 56].

The stability of the uniform oscillatory state has been studied numerically in [55]. In the parameter region with positive Floquet exponent cluster patterns were found. Each cluster has a certain size, that is the number of oscillators belonging to the cluster. Oscillators in the same cluster have the same amplitude and phase. Beyond the cluster instability the catalytic surface becomes separated in homogeneous domains, which oscillate at the same frequency but with a constant phase difference from each other. The main characteristic features of a two-cluster state are the size ratio, the amplitude ratio and the phase shift. These quantities are plotted against the strength of gas phase coupling in Fig. 14. For comparison, the results calculated both

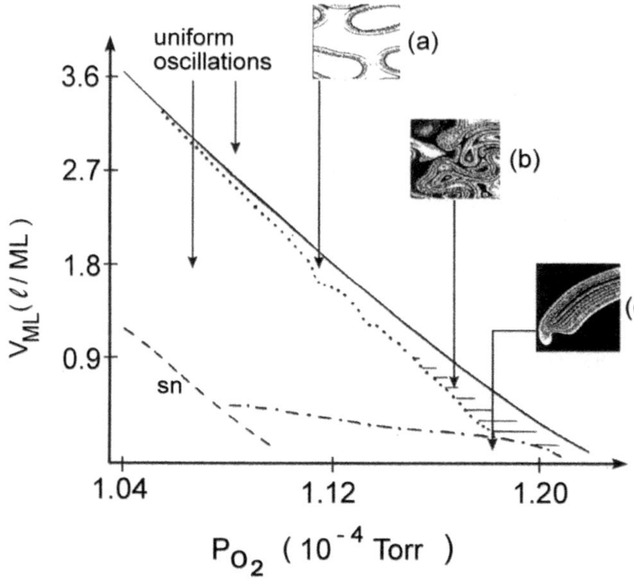

Fig. 13. Phase diagram of spatio-temporal structures obtained for the globally coupled version of the reconstruction model in the $p_{O_2} - V_{ML}$ parameter plane at $p_{COE} = 5.3 \times 10^{-5}$ mbar, T = 545 K, J_{io} = 360 l s^{-1}, and V = 55 l. The inserts show typical pattern obtained (a) during cluster formation, (b) in the turbulent regime (hatched area), and (c) in the region of phase flips. Outside the bounded regions we have uniform oscillations.

with and without diffusion of CO are presented. The differences between the two cases are small. The threshold of the cluster instability was the same regardless of whether surface diffusion was considered or not. Thus, again we find that cluster formation is predominantly a result of global coupling over the gas phase.

Numerical simulations of the globally coupled reconstruction model in a small transition region between cluster states and turbulent states result in standing wave solutions with wavelengths ($\sim 50\mu m$) comparable to the experimental values. Fig. 15 gives an example. This pattern coexists with uniform oscillations. For a detailed discussion compare [54].

Now let us include the anisotropy of surface diffusion. Due to the adsorbate-induced surface reconstruction, the diffusion coefficients in the [110] and [001] direction, D_1 and D_2, respectively, are different, and D_2 becomes w-dependent. We propose a cubic dependence of D_2 on w according to [57] such that the diffusion operator in eq. (1a) is substituted by

$$D_1 \frac{\partial^2 u}{\partial x^2} + \frac{\partial u}{\partial y}\left[D_2(w)\frac{\partial u}{\partial y}\right] \quad \text{with } D_2(w) = D_2\left[1 + \left(\frac{D_1}{D_2} - 1\right) w^3\right]. \qquad (15)$$

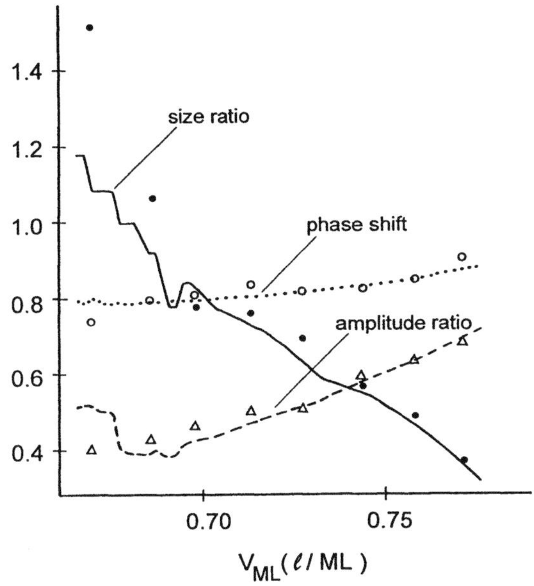

Fig. 14. Characteristic features of a two-cluster state vs the strength of global coupling measured by V_{ML}. The lines show the results for pure global coupling (no diffusion), the corresponding symbols are obtained with both local and global coupling.

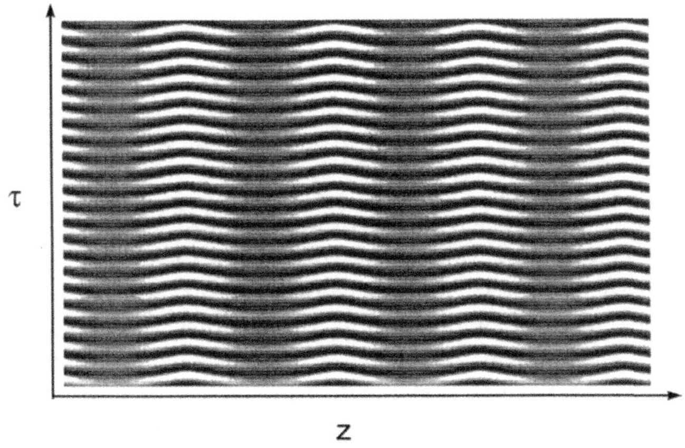

Fig. 15. Standing wave solution of the globally coupled reconstruction model at $p_{O_2} = 1.5 \times 10^{-4}$ mbar, $p_{COE} = 5.3 \times 10^{-5}$ mbar, $V_{ML} = 0.75l/ML$, T = 545 K, $J_{io} = 360$ l s^{-1}, and V = 55 l. Bright (dark) gray level represents an area with high (low) CO coverage. The length of the system is 0.677 mm, the oscillation period is 4.21 s.

In this situation global coupling causes five different regimes of pulse propagation on the surface which are illustrated in Fig. 16. For sufficiently large values of the coupling strength ($V_{ML} > V_{ML,4}$ in Fig. 16) pulse propagation is impossible. Uniform oscillations are the only stable regime. For $V_{ML,4} < V_{ML} < V_{ML,3}$ pulses can travel only in the [001] direction, and if they have an open end, they shrink. Below $V_{ML,3}$ pulses can travel in both crystallographic directions. For $V_{ML,2} < V_{ML} < V_{ML,1}$ pulses traveling in the [001] direction can develop spiral waves at their open ends. Before spiral formation sets in, the pulse aligns in the [110] direction, i.e., vertical to the direction with the smallest propagation velocity. Below $V_{ML,1}$ all pulses with open ends form spiral waves.

Recently, the chemical wave patterns which accompany the rate oscillations in $Pt(110)/CO + O_2$ have been investigated with two new microscopic techniques: mirror electron microscopy (MEM) and low- energy electron microscopy (LEEM). The two techniques are based upon a different contrast mechanism than PEEM and, compared to the former PEEM experiments, the pressure range in these studies was about one order of magnitude lower, i.e., the pressure was in the 10^{-5} mbar range. Both MEM and LEEM use electrons which are reflected off the surface for imaging and are therefore sensitive to topographical features but, in addition, diffraction contrast in LEEM makes this technique also sensitive to structural changes of the surface.

With MEM it was demonstrated that the oscillating surface may synchronize to form a cellular structure (Fig. 17). The network structure undergoes a periodic cycle parallel to the rate oscillations. CO islands nucleate simultaneously all over the surface. They grow coalescing finally into a network structure with a mesh size of ca. $1-5\mu m$ which then dissolves again. The network structure strongly deviates from the ideal hexagonal structure probably due to surface defects.

In a different experiment conducted under almost identical conditions the formation of standing wave patterns was observed with LEEM [58]. LEEM which, in contrast to PEEM, is sensitive to structural changes of the surface showed that the surface roughens at the nodes of the standing waves. The roughening is a consequence of the mass transport of Pt atoms that is associated with the $1 \times 1 - 1 \times 2$ phase transition; earlier experiments conducted in the 10^{-4} mbar range showed that this roughening may even turn into real (micro)faceting of the surface. The roughening which may be regarded as the initial stage of faceting is reversible. A small decrease in p_{CO} causes the parallel stripes of roughened surface to transform into a channel structure inside which oxygen pulses propagate. Global coupling persists, and thus, standing waves still periodically expand and contract in space in between the channels. The LEEM experiments [59] demonstrate that very complex patterns can develop when reversible roughening becomes part of the pattern forming process effectively introducing a kind of memory effect into the system.

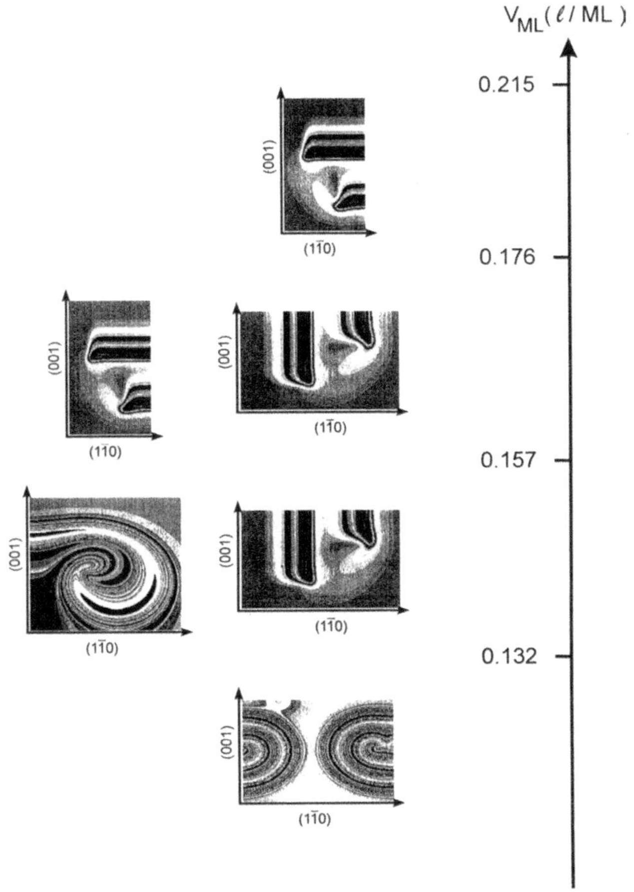

Fig. 16. Different regimes of pulse propagation obtained within the globally coupled reconstruction model with state dependent surface diffusion according to Eqs. (1a-1d,15,16). At parameter values $p_{O_2} = 1.5 \times 10^{-4}$ mbar, $p_{COE} = 5.3 \times 10^{-5}$ mbar, T = 545 K, J_{io} = 360 l s^{-1}, and V = 55 l we have $V_{ML,4}$ = 0.215 l/ML, $V_{ML,3}$ = 0.176 l/ML, $V_{ML,2}$ = 0.157 l/ML, and $V_{ML,1}$ = 0.132 l/ML.

5 Summary

Surface reactions under isothermal conditions give rise to the formation of a variety of patterns in the micrometer down to the submicrometer range, which can be studied experimentally using different types of electron microscopy techniques. The pattern formation can be rationalized in terms of reaction-diffusion models, into which special features such as anisotropy and global coupling are introduced.

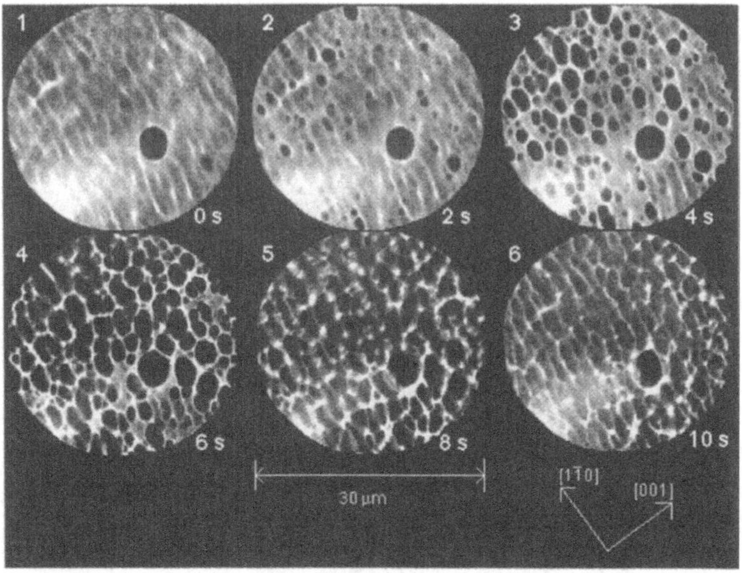

Fig. 17. Cellular structure on $Pt(110)/CO + O_2$ observed with mirror electron microscopy (MEM). T = 430 K, $p_{CO} = 5.4 \times 10^{-6}$ mbar, $p_{O_2} = 1.3 \times 10^{-5}$ mbar; diameter of image is 30 μm.

References

[1] M. Sheintuch and R.A. Schmitz, Catal. Rev. Sci. Eng. 15 (1977) 107.

[2] M.G. Slinko and M.M. Slinko, Catal. Rev. Sci. Eng. 17 (1978) 119.

[3] R.A. Schmitz, G.A. DÆNetto, L.F. Razon and J.R. Brown, in: Chemical Instabilitities, Eds. G. Nicolis and F. Baras. NATO ASI Series C. Mathematical and Physical Sciences, Vol. 120 (Reidel, Dordrecht, 1984).

[4] L.F. Razon and R.A. Schmitz, Catal. Rev. Sci. Eng. 28 (1986) 89.

[5] R. Imbihl, in: Optimal Structures in Heterogeneous Reaction Systems, Ed. P. Plath, Springer Series in Synergetics, Vol. 44 (Springer, Berlin, 1989).

[6] G. Ertl, in: Spatial Inhomogeneities and Transient Behavior in Chemical Kinetics, Eds. P. Gray, G. Nicolis, F. Baras and S.K. Scott (Manchester University Press, Manchester, 1990).

[7] G. Ertl, Phys. Bl. 46 (1990) 339.

[8] M. Eiswirth, K. Krischer and G. Ertl, Appl. Phys. A 51 (1990) 79.

[9] G. Ertl, Adv. Catal. 37 (1990) 79.

[10] M. Eiswirth, in: Chaos in Chemistry and Biochemistry, Eds. R.J. Field and L. Györgyi (World Scientific, Singapore, 1993).

[11] F. Schüth, B.E. Henrz and L.D. Schmidt, Adv. Catal. 39 (1993) 51.

[12] R. Imbihl, Prog. Surf. Sci. 44 (1993) 183.

[13] M. Eiswirth and G. Ertl, in: Chemical Waves and Patterns, Eds. R. Kapral and K. Showalter (Kluwer, Dordrecht, 1994).

[14] M. Eiswirth, M. Bär and H.H. Rotermund, Physica D 84 (1995) 40.

[15] T. Gritsch, D. Coulman, R.J. Behm and G. Ertl, Appl. Phys. A 49 (1989) 403.
[16] K. Krischer, M. Eiswirth and G. Ertl, J. Chem. Phys. 96 (1992) 9161.
[17] M.T.M. Koper, J. Chem. Phys. 102 (1995) 5278.
[18] M. Bär and M. Eiswirth, Phys. Rev. E 48 (1993) R1635.
[19] M. Bär, N. Gottschalk, M. Hildebrand and M. Eiswirth, Physica A 213 (1995) 173.
[20] M. Bär, N. Gottschalk, M. Eiswirth and G. Ertl, J. Chem. Phys. 100 (1994) 1202.
[21] S. Nettesheim, A. von Oertzen, H.H. Rotermund and G. Ertl, J. Chem. Phys. 98 (1993) 9977.
[22] S. Jakubith, H.H. Rotermund, W. Engel, A. von Oertzen and G. Ertl, Phys. Rev. Lett. 65 (1990) 3013.
[23] A. v. Oertzen, H.H. Rotermund and S. Nettesheim, Surf. Sci. 311 (1994) 322.
[24] F. Mertens and R. Imbihl, Nature 370 (1994) 124.
[25] N. Gottschalk, F. Mertens, M. Bär, M. Eiswirth and R. Imbihl, Phys. Rev. Lett. 73 (1994) 3483.
[26] A. Schaak and R. Imbihl, Chem. Phys. Lett. 283 (1998) 386.
[27] N. Gottschalk, diploma thesis, TU Berlin 1994.
[28] K.C. Taylor, in: Automotive Catalytic Converters (Springer, Berlin 1984).
[29] M. Kiskinova, Chem. Rev. 96 (1996) 1431.
[30] F. Mertens and R. Imbihl, Surf. Sci. 347 (1996) 355.
[31] F. Mertens and R. Imbihl, J. Chem. Phys. 105 (1996) 4317.
[32] A. Schaak and R. Imbihl, J. Chem. Phys. 107 (1997) 4741.
[33] F. Mertens, N. Gottschalk, M. Bär, M. Eiswirth, A. Mikhailov and R. Imbihl, Phys. Rev. E 51 (1995), R5193.
[34] A.S. Mikhailov, Phys. Rev. E 49 (1994) 5875.
[35] H.H. Rotermund, S. Jakubith, A. von Oertzen, and G. Ertl, Phys. Rev. Lett. 66 (1990) 3083.
[36] F. Mertens and R. Imbihl, Chem. Phys. Lett. 242 (1995) 221.
[37] N.M.H. Janssen, A. Schaak, B.E. Nieuwenhuys and R. Imbihl, Surf. Sci. 364 (1996) L555.
[38] H. Schmalzried, Chemical Kinetics of Solids (VCH Weinheim, 1995).
[39] D.F. Padowitz and S.J. Sibener, Surf. Sci. 254 (1991) 125.
[40] F. Mertens and R. Imbihl, J. Chem. Phys. 106 (1997) 4319.
[41] M. Kiskinova, S. Lizzit, G. Comelli, G. Paolucci and R. Rosei, Appl. Surf. Sci. 64 (1993) 185.
[42] A.G. Makeev, M.M. Slinko, N.M.H. Janssen, P.D. Cobden and B.E. Nieuwenhuys, J. Chem. Phys. 105 (1996) 7210.
[43] A.G. Makeev, N.M.H. Janssen, P.D. Cobden, M.M. Slinko, and B.E. Nieuwenhuys, J. Chem. Phys. 107 (1997) 965.
[44] A. Schaak, S. K. Shaikhutdinov and R. Imbihl, in preparation.
[45] B. Rubinstein, R. Imbihl and L. Pismen, in preparation.
[46] J. Lauterbach, K. Asakura and H.H. Rotermund, Surf. Sci. 313 (1994) 52.
[47] S.K. Shaikhutdinov, A. Schaak and R. Imbihl, Surf. Sci. 391 (1997) L1178.
[48] F. Mertens, R. Imbihl, and A. Mikhailov, J. Chem. Phys. 99 (1993) 8668.
[49] M. Falcke, H. Engel, M. Neufeld, Phys. Rev. E 52 (1995) 763; Phys. Rev. E 53 (19956) 6589E.
[50] V. Hakim, V.J. Rappel, Phys. Rev. A 46 (1992) R7437.

[51] H. Chaté, P. Manneville, Physica A 224 (1996) 348.
[52] D. Battogtokh, A. Mikhailov, Physica D 90 (1996), 84; D. Battogtogh, A. Preusser, A. Mikhailov, Physica D 106 (1997) 327.
[53] F. Mertens, R. Imbihl, and A. Mikhailov, J. Chem. Phys. 101 (1994) 9903.
[54] M. Falcke, Strukturbildung in Reaktions-Diffusions-Systemen mit globaler Kopplung, Dissertation (Wissenschaft & Technik Verlag, Berlin, 1995).
[55] M. Falcke, H. Engel, J. Chem. Phys. 101 (1994) 6255.
[56] M. Falcke, H. Engel, Phys. Rev. E 50 (1994) 1353.
[57] M. Falcke, H. Engel, Phys. Rev. E 56 (1997) 635.
[58] K.C. Rose, D. Battogtokh, A. Mikhailov, W. Engel, and A.M. Bradshaw, Phys. Rev. Lett. 76 (1996) 3582.
[59] K.C. Rose, B. Berton, R. Imbihl, W. Engel, and A.M. Bradshaw, Phys. Rev. Lett. 79 (1997) 3427.

Convection and Pattern Formation Induced by Autocatalytic Chemical Reactions

H.R. Brand[1] and S.C. Müller[2]

[1] Theoretische Physik III, Universität Bayreuth, D-95440 Bayreuth, F.R. Germany
[2] Otto-von-Guericke-Universität Magdeburg, Institut für Experimentelle Physik, D-39106 Magdeburg, F.R. Germany

September 16, 1998

Abstract. A brief overview is given on chemical patterns that form due to the coupling between nonlinear reactions with transport processes. Emphasis lies on experimental reaction-diffusion systems such as the methyleneblue-sulfide-oxygen (MBO)reaction, in which stationary Turing patterns are observed, and the Belousov-Zhabotinsky (BZ) reaction, which is a frequently used example for an excitable medium showing propagating oxidation waves. When traveling in a liquid solution without a gel, such chemical waves are known to induce a convective flow in the reaction medium. We summarize the different types of flow behavior in a thin layer of BZ solution. Large chemical wavelengths lead to individual flow patterns, each associated with a single wave front. Short chemical wavelengths induce a cooperative phenomenon in the form of large global flow waves traveling through the fluid layer. The chemical reaction can induce this convection by producing local density changes as well as by generating gradients of surface tension due to local changes in the chemical composition of the medium. These phenomena are modeled by coupling reaction-diffusion equations to the underlying hydrodynamic equations. Close to the instability onset envelope equations can be derived provided the spatial variations are sufficiently smooth. Using this technique we demonstrate that chemically driven convection can stabilize Turing patterns. By direct numerical simulation of the basic equations we show that a large scale collective hydrodynamic flow can be generated by the concentration dependence of the surface tension. In this case rapid spatial variations of the concentrations preclude a description in terms of envelope equations.

1 Introduction

The research on nonlinear system dynamics comprises several types of pattern forming systems with spatial degrees of freedom. Among these, simple fluids play a major role in which hydrodynamic instabilities occur, due to which spatio-temporal patterns arise in the form of specific flow fields. One of the most established phenomena of this type is the Rayleigh-Bénard convection [1, 2]. In a solution layer between two horizontal plates, where the lower one has a higher temperature, characteristic convection rolls appear if the temperature difference exceeds a critical threshold. Convective systems are amply discussed in the Chapter by Busse et al. of this book. In a system with a free surface hexagonal convection cells form due to the Marangoni instability [3]. In this case local changes in the surface tension destabilize the former roll pattern. In our natural environment Rayleigh-Bénard type convection occurs, for example, in the atmosphere, forming large convection rolls or in the Earth's core, largely influencing weather phenomena and the continental drift, respectively.

In reaction-diffusion systems, on the other hand, pattern formation is addressed from the more chemical side. In this case, traveling excitation waves can be observed [4-6] as well as stationary periodic patterns of the Turing type [7-9] and, as shown very recently, self replicating spots [10]. Representative systems are the Belousov-Zhabotinsky (BZ) reaction [4], the chloride-iodide-malonic acid (CIMA) reaction [11] and the methyleneblue-sulfide-oxygen (MBO) reaction [12]. If these reactions are prepared in a continuously stirred tank reactor (CSTR), the spatially homogeneous case is realized, where characteristic temporal oscillations between the active (usually the oxidized) and the resting (reduced) state occur. In a spatially extended layer the CIMA and MBO reactions typically give rise to stationary stripe or spot patterns, whereas the BZ reaction is frequently prepared as an excitable medium that supports excitation waves traveling by diffusion of reagents [6,13]. In a two-dimensional system circular as well as spiral geometries of the waves are observed (Fig. 1a). The BZ reaction has become a model system for excitable media because it can be prepared and observed easily. Media of this type play an important role, especially in biology. One example is the propagation of action potentials on neurons [14] or, in systems with higher dimensionality, excitation waves on neural tissue like in spreading depression [15] or on the heart muscle [16]. On the other hand, investigation of the CIMA and MBO reactions as chemical examples for stationary Turing patterns may be helpful to understand the potential role of this type of organization in the living nature, since it is possibly involved in pattern formation in morphogenesis, e.g. in mollusks or on the skin of animals [17, 18].

A system that combines effects arising both from hydrodynamic instability and reaction-diffusion coupling is a BZ reaction proceeding under certain conditions in a thin layer of liquid solution. In the case of a reaction dish without cover, one can observe propagation of excitation waves as well as convection of the Marangoni type due to the evaporative cooling of the surface. The convection cells lead to a visible deformation of the chemical wave pattern [18, 19] (see Fig. 1b). If evaporation is suppressed by covering the reaction dish, leaving just a small air gap, convection induced by another driving force can occur: now the chemical waves themselves force the reaction medium to move. This flow is referred to as chemically driven convection [20,21]. From the hydrodynamical point of view this type of convection represents a fascinating type of intrinsic forcing of the hydrodynamic system, leading to interesting new types of flow patterns. For the reaction-diffusion

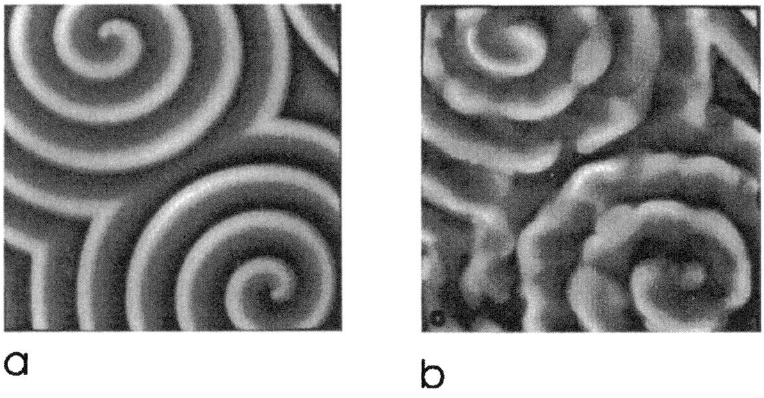

Fig. 1. (a) An undisturbed pair of spiral waves in a liquid layer of BZ solution (left picture), observed at 490 nm with a 2D spectrophotometer (image side length: 1 cm). (b) A pair of spiral waves in a liquid BZ layer deformed by natural Marangoni convection (right picture). A description of the measurement is given in ref. [20].

system the convection acts as a second transport mechanism in addition to the diffusion, accelerating the front velocity, deforming the shape of the wave fronts, and inducing additional interesting and complex dynamics.

In this Chapter, we will summarize some of our recent knowledge about stationary and traveling patterns in (mostly liquid, homogeneous) reaction-diffusion systems, including a short description of the complex world of three-dimensional topologies and a cross-link to the vast field of similar chemical patterns in heterogeneous situations (cf. also the Chapter by Imbihl et al. of this volume). A large portion of our attention will then be focussed on the intriguing structural and dynamic effects that occur when convective transport is allowed to act in these reaction-diffusion systems. Experimental results are presented as well as theoretical approaches to elucidate the mechanisms at work under these rather complex circumstances.

2 Patterns in Reaction-Diffusion Systems

2.1 Stationary Turing patterns

There has been an increasing interest in the experimental study of stationary spatial pattern formation during the past five years. Already in 1952 A. Turing proposed that the coupling between nonlinear chemical reactions and diffusion processes may break spatial symmetry to lead to stable, stationary structures [22]. Activating and inhibiting chemical species must have different diffusion coefficients for the emergence of Turing patterns; in particular, the inhibitory species must diffuse faster than the activating species. The first experimental observation of such structures (in the chlorite-iodide-malonic aicd (CIMA) reaction) dates from 1990 [5, 6]. In the CIMA reaction different diffusivities are realized by binding the activator I_2 as triiodide (I_3^-) to immobilized starch in a gel matrix and thus slowing down the diffusion of I_2. In this reaction Turing patterns have been observed in an open system [5, 6] as well as in a gradient-free closed system [23]. Recently, Watzl and Münster have reported that another chemical reaction-diffusion system involving methyleneblue, sulfide, sulfite and molecular oxygen in a polyacrylamide gel matrix (the PA-MBO-system) readily displays Turing-like patterns [9, 24].

This nonlinear chemical oscillator was discovered in 1984 by Burger and Field [12]. The underlying basic chemical reaction is the methyleneblue-catalyzed oxidation of HS^- by O_2 in an unbuffered water solution. The organic dye molecule methleneblue exists in an oxidized blue form MB^+ (absorption maximum at 668 nm) and a reduced colorless form MBH. A mechanistic proposal for the MBO has been given by Resch et al. [25] which involves 14 reactions with several ionic and radical species. This mechanism has been simplified by Zhang and Field [26]. For the experiments, as described in Ref. [24], a petri dish with a diameter of 90 mm is placed on a thermostated ($22 \pm 0.5°$ C) plexiglass vessel and illuminated from the bottom. The chosen setup allows the exchange of oxygen between the reaction mixture and the air. Commonly, in order to avoid convective instabilities, one uses a polyacrylamide gel in these experiments. However, it has been shown that the polyacrylamide gel, which consists of crosslinked polymer chains and has specific mechanical properties, additionally plays an active role in the pattern formation process. In a typical experiment the gel components were mixed together with the solutions of the MBO reaction (methyleneblue, sodium sulfide and sodium sulfite) according to Ref. [9] resulting in a gel layer of about 2 mm thickness. The emerging patterns were monitored in transmitted light with a CCD camera (black-and-white).

Beyond certain values of the methyleneblue- and sulfide-concentrations transient spatial patterns emerge spontaneously from the initial homogeneous state. The patterns are transient because the system is open only to oxygen but closed with respect to the other reactants. Typically it takes about 30 min to build up the first visible structures and within another 60 min the pattern has fully developed. A "blue" hexagonal pattern is shown in Fig. 2a. The black color in this picture corresponds to a high concentration of MB^+ in the gel matrix (blue in the real experiment), the light color corresponds to the reduced form MBH. During an experiment the conditions slowly change due to the semi batch experimental setup and the spots eventually turn into short stripes and zig-zag patterns (Fig. 2b). However, both types of patterns can coexist in one experimental run.

For other initial concentrations, e.g increased Na_2S concentration, one obtains transient striped patterns with small defects instead of hexagons. A striped pattern is shown in Fig. 2c.

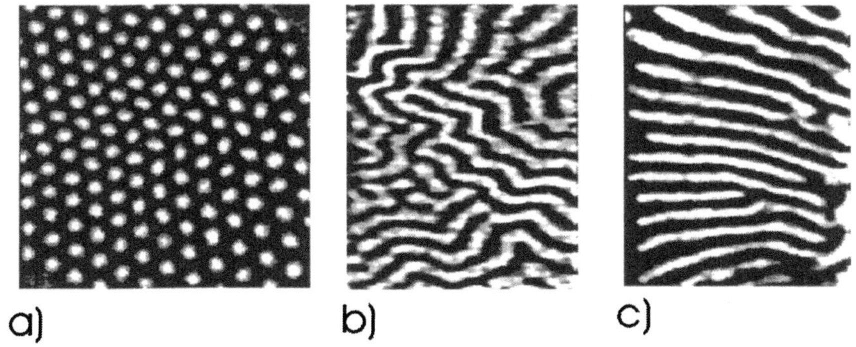

Fig. 2. Stationary Turing patterns in the PA-MBO system. The black and white color corresponds to the oxidized or reduced state of methyleneblue, respectively. (a) Hexagonal spots, (b) zig-zag pattern, (c) striped pattern. Typical pattern wavelength: 2mm (from [24]).

2.2 Excitation waves

Traveling waves of excitation are abundant in biology. In addition to the examples already given in the Introduction (spreading depression in chicken retina [13], electrophysiological activity of the heart muscle [14]) we mention here the observation of signal transmitter waves during the aggregation of slime mould amoebae [27] and calcium waves in frog oocytes [28]. These waves are also plentiful in physical and physico-chemical systems and recent overviews are available [13]. These waves exhibit, due to their nonlinear nature, an amazing richness in dynamic shapes and patterns.

The best known laboratory system for the study of excitation waves is the Belousov-Zhabotinsky (BZ) reaction which is the bromination of malonic acid in acid medium. For the experiments the reaction solution is prepared as a mixture of the components malonic acid, bromate, bromide, sulfuric acid, and a catalyst and redox indicator, in our case the metal-ion complex ferroin. The malonic acid acts as the substrate. During the reaction it is consumed and degraded to other organic products, CO_2 and water. This reaction can switch between different reaction paths, which leads to oscillations in concentrations of the intermediates. Due to the catalyst these oscillations are indicated by a clearly visible color change between red (the reduced state ferroin) and blue (the excited, oxidized state ferriin). Some of the details of the mechanism have been clarified and are frequently taken into account in terms of the simplified Oregonator scheme [29] which is a three- or two-variable kinetic model of the reaction.

For the observation of chemical oscillations the reaction solution is filled into a stirred vessel to obtain spatial homogeneity. The solution exhibits periodic or, in special cases, aperiodic oscillations. In order to solve the problem of the aging of the system one can perform such experiments in a continuously fed stirred tank reactor (CSTR), where the system receives permanently fresh reactants. As a result, period doubling events and transitions to chaotic oscillations are obtained as a function of the flow rates, for instance [30].

Unstirred, spatially inhomogeneous systems may exhibit chemical oscillations, as well, with perhaps some spatial phase differences in the oscillatory state, forming kinematic or phase waves. Of more interest here are the "trigger waves" which are formed by reaction-diffusion coupling, whereby the main chemical species involved do not necessarily have different diffusivities as in the Turing case. Much experimental and numerical work has been devoted to study these wave phenomena in thin (0.5 - 1.0 mm) layers of excitable reaction mixtures. In these systems a local excitation (by immersing a silver wire, for instance) influences the solution in the surroundings via diffusion of chemical components, turning it also to the excited state. As a result, this effect leads to the propagation of a chemical wave front. Behind this front the solution returns to the reduced state, remaining unexcitable for some time (refractory period) which is the reason for the annihilation of colliding wave fronts. Exciting the system in a single point leads to a circular wave which propagates away from this point with the same velocity in each direction.

Such systems also support rotating spiral waves that develop from broken wave ends. Figure 1a shows a pair of counter-rotating spirals in the ferroin-catalyzed BZ reaction. To observe color changes, the petri dish containing the patterned layer is illuminated homogeneously with parallel blue light (490 nm) from below. This wavelength is chosen to obtain the largest possible contrast between ferroin and ferriin. Thus, the wave crests, that is the excited areas of highest catalyst oxidation, appear bright in transmitted light, because at the chosen observation wavelength only the reduced form of the catalyst absorbs light substantially, while the oxidized form does not.

The transmitted light is observed with a video camera and the images are stored on video tape. With a calibrated setup it is possible to obtain a direct relation between light absorption and concentration of the oxidized catalyst ferriin. After digitization of image sequences, it is possible to estimate various system parameters using image processing techniques [31] , for example, wave profiles, front curvature, or the position and traces of the tip of spiral waves [32,33].

In the images of Fig. 3 we show two examples of excitation waves found in biological systems that obviously have structural features quite analogous to the chemical waves in Fig. 1a. Image (a) is a dark-field photograph of aggregation waves in colonies of slime mould amoebae (Dictyostelium discoideum). The pattern is due to differences in light scattering behavior of moving and resting amoebae cells. With each turn of the spirals a chemotactic response to the spiral-shaped distribution of the signal transmitter molecule cAMP results in a gradual displacement of the cell population towards the spiral centers. With persisting rotation of this concentration distribution a signaling mechanism is set up that gathers many single cells in the central spiral

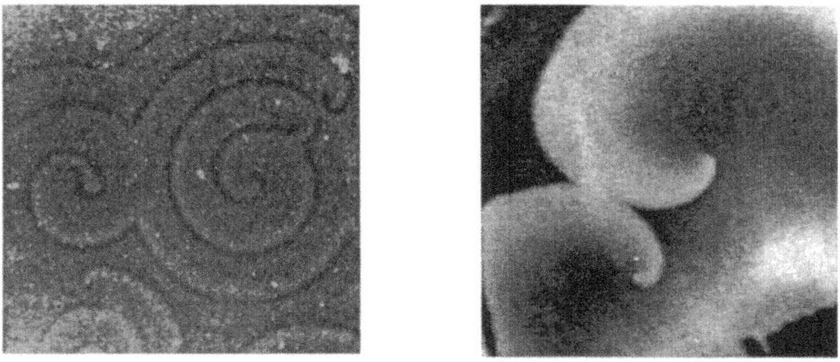

Fig. 3. (a) Darkfield photograph of an early stage in the aggregation of slime mould amoebae. Dark and bright lines indicate the spiral-shaped, chemotactic motion field directed towards the spiral center. Image area: 1.9×1.9 cm^2 (from [34]). (b) Colliding pair of Spreading Depression (SD) spiral waves in the retina of chicken. The waves are visualized by scattered light due to propagating fronts of increased turbidity that indicate strong structural changes of neuronal tissue.

region where they grow together to form a multicellular slug. This is an important step in the life cycle of this population. In the second image of this Figure (b) a colliding spiral pair in the Spreading Depression (SD) phenomenon is shown. This phenomenon is detected on the chicken retina as a zone of increased turbidity visualized by white scattered light. Spreading depression may occur in many types of neural tissue. A depression of electric activity of the neurons is associated with a dramatic failure of ion homeostasis, which may be linked to the aura of classical migraine [35].

For all these systems, a particular interest lies in the study of the tip dynamics of the spirals, because the trajectory of the tip can follow quite different traces depending on the preparation of the system in terms of its excitability and threshold of excitation [36]. The most commonly known regime is that of rigid rotation of the tip, as observed in certain standard preparations of the BZ reaction (Fig. 4a). Also the spirals in the slime mould shown in Fig. 3a follow such a rigid regime. It turns out, however, that the more frequent case is a hypocycloidal (or "meandering") tip motion in which at least 2 frequencies - a small and a much larger one - determine the resulting compound trajectory (Fig. 4b). These floral trajectories may have a quite different number of lobes under different initial conditions. In extreme cases the cycloidal motion turns into a 'loopy line' along which the spiral drifts towards the boundary of the dish (Fig. 4c). When the excitability of the system is decreased, the regularity of the pattern dynamics may be replaced by a rather irregular motion (see the example in Fig. 4d). These changing aspects of the intrinsic spiral dynamics have been systematically studied in the BZ reaction and in

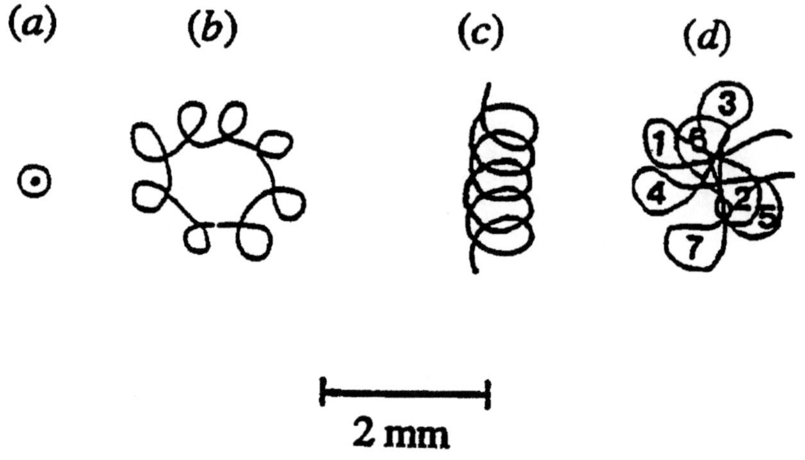

Fig. 4. Typical trajectories of spiral tip motion measured in the BZ-system. (a) rigid rotation, (b) hypocycloidal trajectory (meander), (c) loopy line, (d) irregular motion (from [33]).

numerical simulations on the basis of appropriate mathematical models (see corresponding chapters in ref. [13]).

Nowadays there is a large interest in subjecting the chemical waves, in particular spiral-shaped ones, to external forces in order to find ways how to influence and control their intrinsic dynamics. Two methods provide efficient tools for such an external influence: application of an electric field acting on the motion of charged species in the BZ reaction, and illumination of a photo-sensitive variant of this reaction that reduces its excitability with the applied light intensity. This opens a vast field of applications which, in the context of this Chapter, cannot be further discussed.

2.3 Three-dimensional topologies

Spiral waves, as commonly observed in thin layers of the excitable BZ reaction, may be regarded as suitably oriented cross-sections of so-called scroll waves in a three-dimensional medium (see review by Tyson and Strogatz [37]). As a particularly interesting example we show in Fig. 5 (a-d) the spectrophotometric projection of a "scroll ring" at four different stages during one cycle of spiral rotation. The reactive BZ medium is contained in a rectangular cell of both 10 mm width and thickness. The circular edge of the scroll ring rotates periodically around a curve called filament, which in this case forms a closed ring (see closed ellipse in the projections of Fig. 5b and c). The propagation of the wave edge is slightly faster during phase (b) as compared with (d) due to the different sign of curvature along the filament. Thus the filament gradually contracts and finally, once a critical filament diameter is reached, opposed parts of the wave edges collide already during the phase corresponding to Fig. 5b, before proceeding through the opening of the filament as in (c).

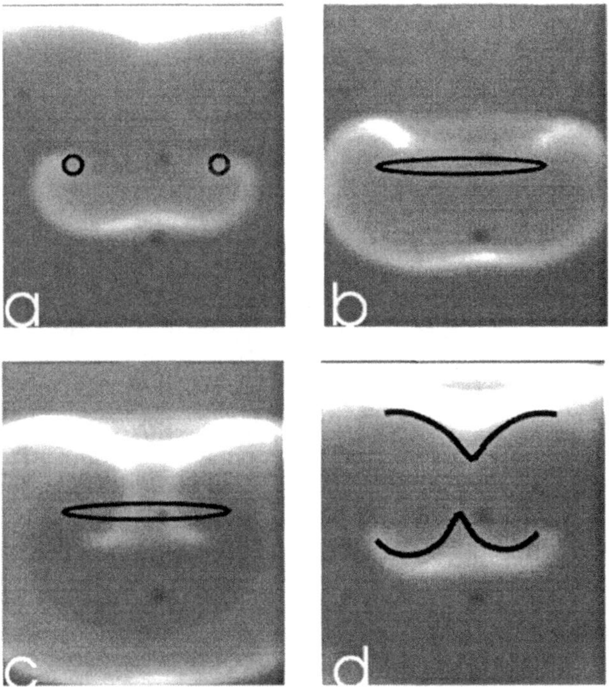

Fig. 5. Scroll ring in the ferroin-catalyzed BZ reaction. Sample thickness and width 10mm, respectively. As seen in projection, the wave resembles a pair of spiral waves undergoing rigid rotation (circles in a). At any phase the actual 3-D wave structure has rotational symmetry. The leading wave edge is scrolling around a circular filament (slim ellipse in b and c). After the wave edge has passed through the filament, certain segments of the wave front approach the symmetry axis (passing through the ellipse center). Due to collision the leading part of the wave segregates (in d) to start a new cycle (period about 3 minutes).

A quantitative study of scroll waves with less symmetry than the scroll ring is not possible by only using projections under a fixed angle of view. Recently, optical tomographic techniques have been established (as sketched in Fig. 6) that provide 3D reconstructions of chemical waves of arbitrary shape in the BZ reaction [38,39]. These techniques are suitable to present the full spatial distribution of quantitative data on the catalyst concentration.

Thus, further experiments with more complicated filament dynamics will be facilitated. But the preparation of suitable initial conditions for more complex 3D waves is still an unsolved experimental problem. These structures are, for example, characterized by a knotted filament or two circular rings linked to one another and were already investigated in numerical simulations (see [40] and references therein). In contrast to experimental observations of scroll rings, these do not finally undergo self-annihilation, but stabilize at a certain size and shape. The balancing repulsion against curvature-induced filament contraction is putatively attributed to an interaction with a twist (i. e. a phase shift) of the wave edge along the filament, but this effect remains to be investigated experimentally.

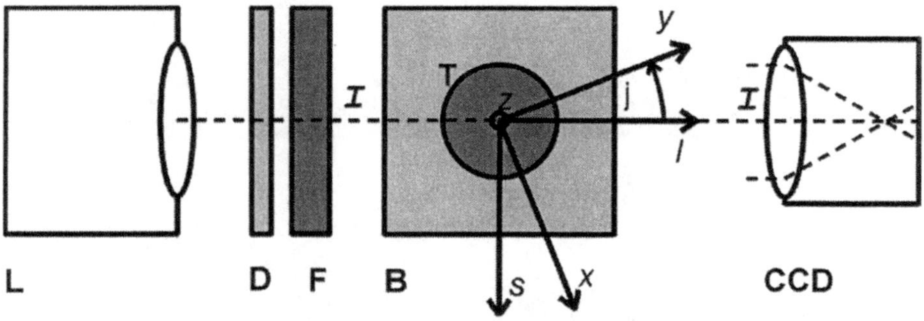

Fig. 6. Sketch of the experimental setup for optical tomography. The BZ reaction gel is contained in a glass tube T (diameter 18mm) that is immersed in a waterbath B and rotated around its symmetry axis z. Incident light I_0 from a tungsten lamp L has been diffused (D) and filtered (F, $\lambda \approx 490$ nm). Parallel projections $I(x', z)$ of the BZ gel are recorded with a CCD camera (cf. [38]).

2.4 Waves and patterns in heterogeneous systems

Spatiotemporal chemical patterns occur also under heterogeneous conditions as amply documented, for instance, in the contribution by Imbihl et al. to this Volume on surface structures in heterogeneous catalysis. Here we briefly mention some of the increasing number of experimental [41] and theoretical [42] investigations on electrochemical systems. Otterstedt, Jaeger, Plath and Hudson [43] report experimental results on propagating electrochemical waves on a ring electrode during the oscillatory electrodissolution of cobalt in buffered 1.0 M phosphoric acid. The ring is the area between two non-concentric circles and its width thus varies continuously with angle. This is reflected by variations in the velocity of the rotating wave. The experiments were controlled potentiostatically with the tip of the capillary of the reference

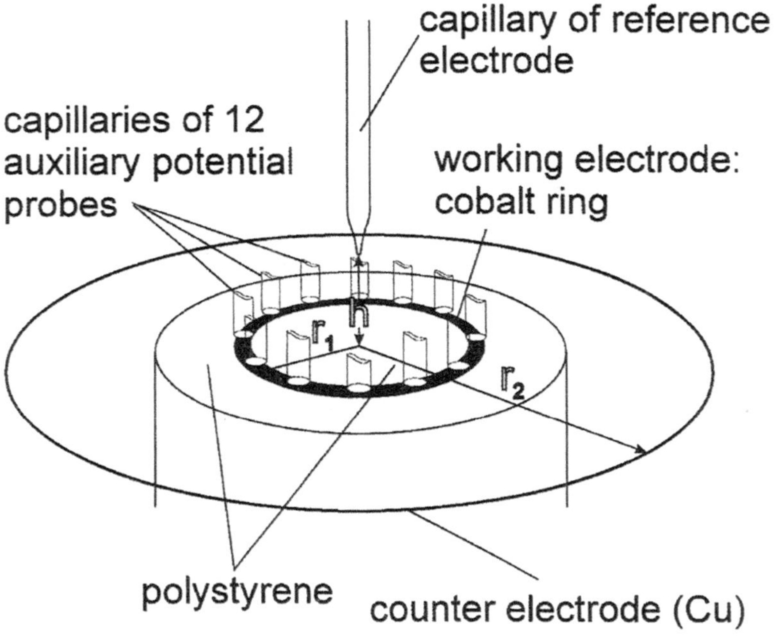

Fig. 7. Experimental setup for electrochemical pattern detection. The electrode geometry is located in a cylindrical vessel which is not shown in the Figure. The 12 capillaries of the auxiliary reference electrodes are placed about 0.5 mm above the surface of the cobalt ring working electrode which is embedded in a polysterene resin. The distance h between the capillary and the plane of the ring is about 1 mm. Outer radius r_1 is 6.25 mm, thickness varies smoothly between 1.0 and 1.5 mm, radius of counter electrode r_2 is 35 mm (from [43]).

electrode placed in the center of the ring 1mm above the surface (Fig. 7). Under these conditions rotating waves, with and without modulation, occur as shown in Fig. 8. The standard three-electrode setup for kinetic electrochemical measurements was adapted in order to study and control non-local coupling through the electric field which strongly influences the spatiotemporal dynamics of electrochemical systems. Positive nonlocal coupling drives the passive-active transition. Even though the applied potential is set to a value cathodic to the Flade potential, the whole electrode surface is not activated (The Flade potential is the potential at which the transition from the active to the passive state occurs). The occurrence of rotating waves is due to the action of negative non-local coupling which provides for an electrode potential farther in the passive region than the imposed potential opposite to the area where the activation reaction occurs. These waves produce an integral current that varies with time in a roughly periodic fashion.

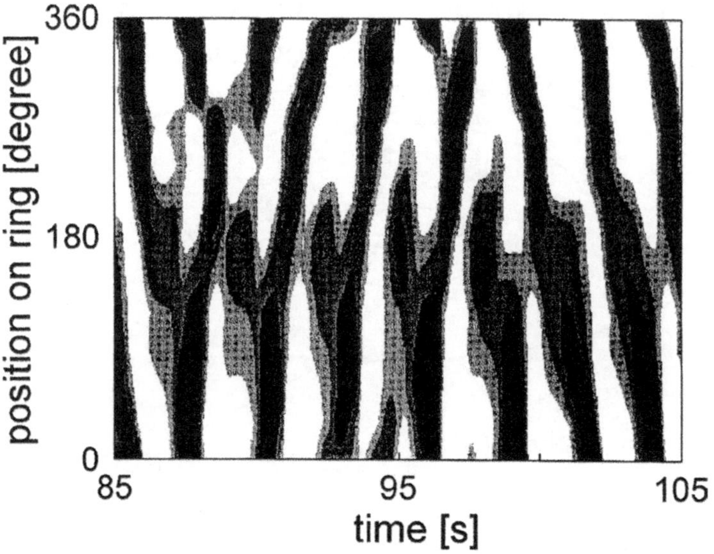

Fig. 8. Grey scale representation (4 shades) of the potential difference between the working electrode and the 12 auxiliary reference electrodes as a function of space and time. The darker the grey shade, the smaller the potential difference, i.e. the more active the electrode surface. Rotating, slightly breathing waves occur with occasional change of the direction of propagation. Applied potential: 900 mV/Ag/AgCl. (Courtesy of R.D. Otterstedt, N.I. Jaeger, P.J. Plath).

3 Chemically Driven Convection

One of the major topics of this Chapter is the complex interplay between reaction-diffusion waves and convection induced by the autocatalysis in the wave front. Many experiments in this respect have been performed in the BZ reaction. We now present the applied methods and the observations made in this system.

3.1 Chemical preparation and observation method

Nowadays the spatially distributed BZ reaction is frequently prepared in a gel matrix to avoid mechanical disturbances of the system and the influence of hydrodynamic flow [44]. Especially in continuously fed systems a gel matrix is necessary to avoid disturbances due to the inflow of the reactants [7,8]. However, for the observation of chemically driven convection the system must be prepared without a gel. Therefore, the reactants are mixed in a vessel and filled in a specially cleaned petri dish, where the solution forms a thin layer ($< 1 mm$). Using a suitable recipe and a well prepared dish, the solution remains in the reduced state for at least 20 min, indicated by the red color of the ferroin. A local disturbance, for example with a silver wire, induces the system at that point to undertake one oscillation to the oxidized state, indicated by the blue color of the oxidized form of the catalyst ferriin. This is the starting point for a circular wave. Spiral-shaped waves and their properties, as well as the optical setup to observe their propagation in transmitted light, have been described in Section 2.2.

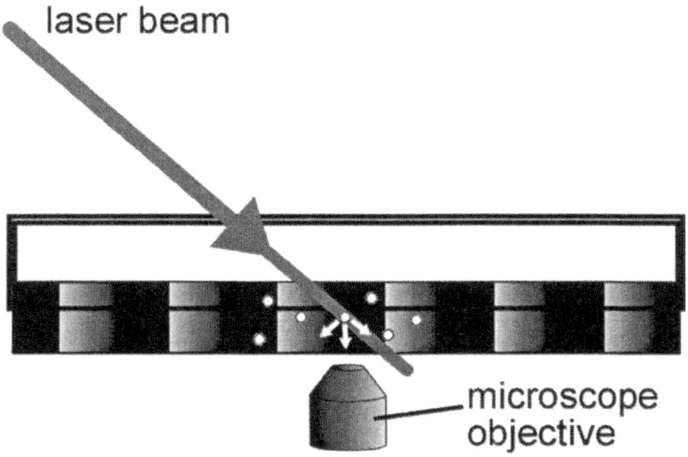

Fig. 9. Principle of the flow measurements. The Petri dish is filled with the BZ-solution with suspended latex particles. These particles are illuminated by a He-Ne-laser and the scattered light is observed with an inverse microscope.

We now turn to the description of effects related to the occurrence of flow fields in aqueous BZ layers without a gel. The local hydrodynamic state of the system can be detected by observing suspended small tracer particles (standard dow latex, 0.5 μm) with a microscope (Fig. 9). Due to the large magnification the depth of focus is short enough to obtain a resolution up to 30 μm in vertical direction. On the other hand, the high magnification restricts the horizontal region for velocity measurements to nearly a single point as the microscope image has a side length of only 0.4 mm. With special software for 2D velocimetry it is possible to calculate the local flow velocity from the recorded motion of the particles automatically [45]. Thus, this set-up allows to measure the flow velocity at one point of the system as a function of time.

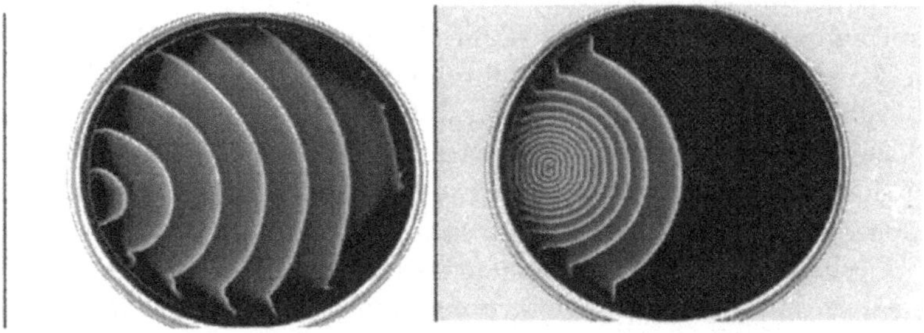

Fig. 10. (Left) A target pattern in a liquid Belousov-Zhabotinsky system, consisting of 4 ml BZ solution, filled in a Petri dish with a diameter of about 7 cm. The figure shows the pattern about 20 min after the preparation of a system with a pacemaker at the left boundary. (Right) A spiral wave in a liquid BZ system with parameters like those for the target pattern. To get the pair of spirals, initially a circular wave was induced. After a while its wave front was broken with a gentle blast of an air stream. The figure shows the pattern about 5 min after the induction of the spirals.

3.2 The observed flow behavior

The observed flow behavior in chemically driven convection depends highly on the underlying chemical wave pattern, in particular on the wavelength. Thus, the flow in a system with circular waves is quite different from that in a system with spiral waves. In the case of circular front geometries the wavelength is usually comparatively long and therefore each single wave front creates its individual flow field [21]. The preparation of a system with circular wave propagation for the flow measurements contains only one center of wave generation. Usually this pacemaker is induced externally near the boundary of the dish by use of a silver wire (Fig. 10, left picture) and flow measurements are carried out in the center of the dish. At the surface of the fluid layer a strong flow towards the approaching front is measured that exceeds in some measurements a velocity of 100 μm/s. This is close to the velocity of chemical wave propagation. As the front passes by there is a rapid change in flow direction. Near the bottom the behavior is in principle the same, but the flow velocity is smaller and its direction is opposite. After the wave has passed the measuring point, there is a smooth decrease in the flow velocity (Fig. 11). Including velocity data in the vertical direction these measurements reveal a strong descending flow in the chemical wave front and a moderate ascending flow in the outer regions. Hence each chemical wave front generates a pair of convection rolls and carries them through the solution. The strong descending flow in the front indicates that this is the region where the main coupling between reaction and convection takes place.

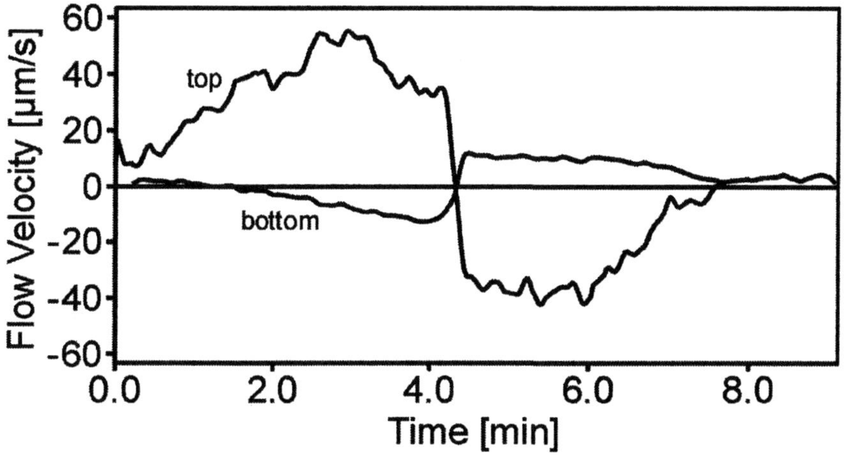

Fig. 11. Observed flow velocity in the case of a single circular wave front. The wave is generated in the middle of the dish and the flow is measured at the top and near the bottom of the fluid layer half the way to the boundary to get rotational symmetric conditions. Passage of the wave front at the measuring point t=4.1 min.

In a system with spiral wave propagation the flow behavior is different. For the flow measurements the preparation of the system contains just one pair of spirals with a distance of about 0.5 cm between their centers. Usually the spiral centers are induced about 1 cm from the boundary of the petri dish by breaking a wave front of a circular wave with a gentle blast of air (Fig. 10, right picture). Flow measurements are carried out at the surface of the fluid layer during the evolution of the spiral waves after the initial wave break. The first few wave trains generated by the spirals show a quite long wavelength ($> 3mm$), but, as a consequence of the dispersion relation of chemical waves, the wave trains move closer and closer together until they reach a stationary distance that is slightly less than 1 mm. After about 15 min the whole petri dish is filled densely with chemical waves.

During this evolution process the flow behavior changes as well and passes through several regimes [21] (Fig. 12). In the early stage of the experiments, when the wavelength is long, the flow is similar to that found for circular wave propagation. But when the wavelength gets shorter, this kind of flow stops and a second phase with almost no surface flow starts. In a third phase, after 15-20 min, when the whole dish is filled with wave trains, a completely new flow behavior occurs spontaneously. Now the flow direction at the surface changes periodically, about once with every passing wave front. This so-called

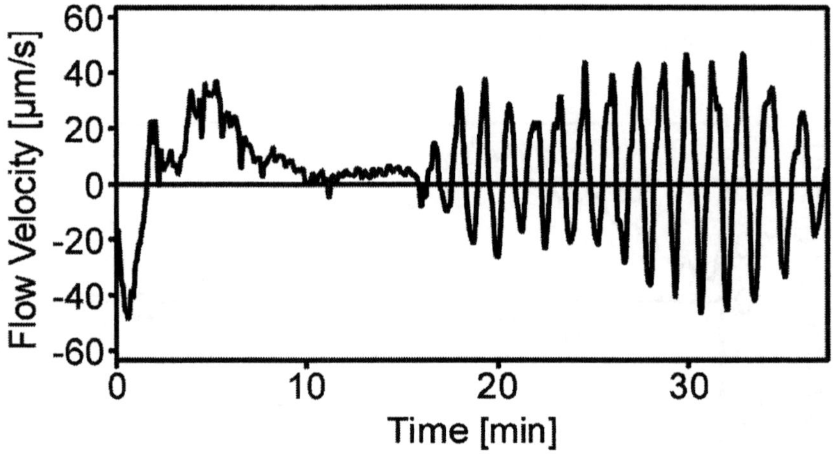

Fig. 12. Surface flow velocity in the middle of the Petri dish during the evolution of a spiral wave. After a discrete phase as observed in circular wave propagation and a rest phase the flow reaches an oscillatory behavior, when the whole dish is filled with very close wave trains.

oscillatory flow is the typical flow behavior in a fully developed spiral wave. After about 40 min the system dies due to the aging of the solution and the oscillatory flow stops.

In the case of circular waves one period of the chemical oscillation corresponds exactly to one period of the convection rolls. Hence the ratio of about 2:1 in the case of spiral waves is remarkable. Here two chemical oscillations are necessary to induce one complete hydrodynamic period. However, it is important to note that this correlation is not perfectly fulfilled. The period ratio is only close to 2:1 which indicates that there is no more a precise correlation between single wave fronts and the local convection. Consequently, the observed flow appears to be a cooperative phenomenon of a larger group of wave fronts in the dish.

For this complex flow behavior it is impossible to reconstruct the three-dimensional flow field just from the measurements at a single point. Unfortunately the presented method of direct flow measurement does not allow to obtain simultaneously the velocity distribution in the whole dish, but we found that it is possible to acquire more information about the global features of the flow, if one takes into account the influence of the hydrodynamics on the shape of the chemical wave fronts. In fact, it has been shown that the convection deforms the wave fronts in a characteristic manner, depending on the flow direction relative to the direction of wave front propagation [46]. The deformation of the wave fronts changes the local light transmission in the system, as can be visualized by image processing procedures. Regions where the surface flow is directed against the direction of wave propagation appear about 5% darker compared to the case without convection. The flow in the opposite direction results in a higher light transmission [47]. This way one can obtain information about the global flow behavior just by observing the transmitted light over the whole dish. The oscillatory flow turns out to feature large global flow waves extending over about ten chemical wavelengths and traveling through the dish towards the center of wave generation (Fig. 13). This very complex connection between a large group of dense chemical waves and global flow in the whole petri dish is a unique phenomenon and theoretically only partially understood (compare section 7).

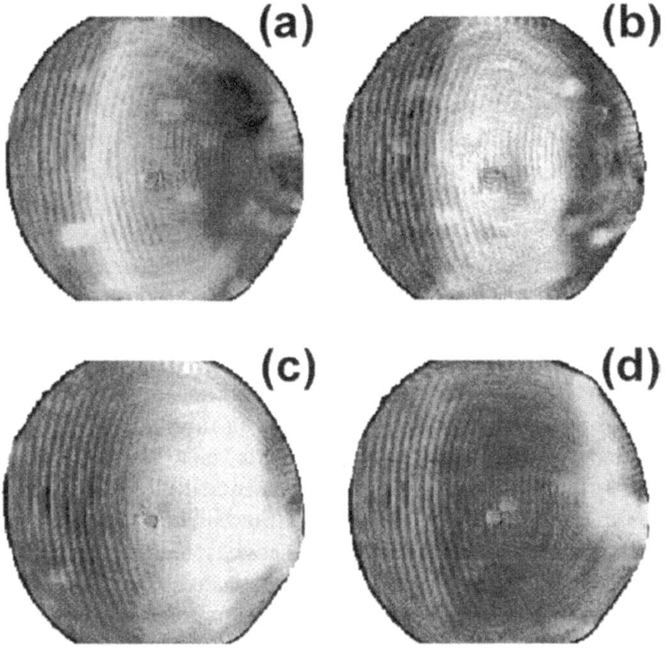

Fig. 13. Global flow waves in the BZ reaction shown in four snapshots of the petri dish during the oscillatory flow. Time interval between the images: 16.8 s. In the dish a spiral wave is induced near the right boundary. The images are processed by background subtraction, moving averaging to eliminate the structure of the chemical waves, and contrast enhancement. The bright and dark stripes represent a surface flow in and against the direction of wave propagation respectively (from [47]).

4 The coupling mechanism between reaction and convection

The question now arises about the coupling mechanism between chemical waves and convective motion arises. How can the waves induce the hydrodynamic flow in the reactive medium? There are two main regions, where such a coupling can take place: the bulk, where the chemical reaction can change the local density $\rho(x)$, and the surface, where the reaction may influence the local surface tension $\sigma(x)$. Density gradients as well as surface tension gradients may be caused by changes in temperature or directly by the alterations of the local chemical composition of the medium. From calorimetric measurements it is well known, that the BZ reaction is exothermic and most of the heat is produced in the oxidized state [48]. Hence temperature gradients can be expected in the medium. In addition, the existence of different types of ions, especially the ferroin and ferriin ions, should lead to surface tension gradients and, via different hydration, to gradients in density [49].

To clarify the role of these different types of coupling mechanisms , it is necessary to measure the magnitude of the effects. The temperature changes between the oxidized and the reduced state in the BZ reaction are estimated by Körös et al. in a calorimeter, i.e. a spatially homogeneous system. From these measurements, a maximum temperature difference of 0.1 K can be assessed [48]. Preliminary measurements in a spatially inhomogeneous system were done recently using a thermopile-IR-detector [50]. As a consequence of heat diffusion the measured values are smaller and show that the wave fronts are accompanied by a local temperature maximum with a temperature difference of about 50 mK. This maximum results in a local density minimum as well as in a minimum of the surface tension near the chemical wave front. Both effects would force the solution to an ascending flow near the wave front, which is in contradiction to the observed flow behavior. Consequently these results show that the temperature produced by the reaction does not play a constructive role in the chemically driven convection. The observed flow must be generated by surface tension or density gradients caused directly by the chemical composition of the medium.

Yoshikawa et al. measured the surface tension changes in the BZ-reaction [51]. The authors also used a spatially homogeneous system by preparing the BZ reaction in a stirred vessel and used for their measurements the Wilhelmy plate method. As a result, surface tension changes in the order of 2 mN/m between the oxidized and the reduced state of the reaction were obtained. In a spatially inhomogeneous system the chemical wave fronts are expected to be accompanied by a local maximum in surface tension. This will lead to a descending flow near the wave front, which is consistent with our observations. Thus, the good agreement of the measured surface tension changes with the flow behavior supports the view that the chemically driven convection is of the Marangoni type.

For the density changes there exist no reliable measurements. Pojman et al. only calculated values for the different hydration of ferroin and ferriin [49]. Consequently, there is no definite decision yet, whether the observed convection phenomena are density or surface tension driven or a superposition of both effects. Further experiments related to density changes are on the way to answer this question.

5 Chemically driven convection - a simple model

In the next two sections we describe several approaches to model chemically driven convection. We follow to a large extent refs. [52-54]. Section 6 is dedicated to the derivation of envelope equations for the stationary instability associated with Turing patterns. In section 7 we study experimentally and theoretically the influence of the concentration dependence of surface tension.

In ref. [52] we discussed in the framework of a two species reaction - diffusion model coupled to the hydrodynamic equations how a chemical reaction drives convective motion in a fluid layer for an oscillatory instability with vanishing wavenumber at onset. We derived from the model equations an envelope equation of Ginzburg-Landau type with complex coefficients. To cubic order in the envelope the generated convective velocity is slaved to the concentration variables and has no independent dynamics. A feedback from the flow on the chemical reaction arises at higher order in the amplitude. We stress, however, that although the flow does not enter the coefficients of the envelope equation to cubic order, it is generated to quadratic order in the envelope. Thus there is a *qualitative difference* between the purely chemical case and chemically driven convection already *to second order in the envelope.*

The envelope equation derived in ref. [52] contains to cubic order in the amplitude only terms that come from the reaction-diffusion dynamics, that is there is no back reaction of the fluid motion on the reaction-diffusion-dynamics. The fluid motion is just slaved to the chemical concentrations and follows them instantaneously. The result of detailed calculations up to fifth order in ϵ for generalized free-free boundary conditions shows that even to this order there are no contributions of the fluid motion in the corresponding corrections to the envelope equation. There are two reasons for the passive role of the fluid velocity. First of all it has to be generated by the density gradients created by the internal dynamics of the chemical concentrations since there is no external stress applied to the fluid layer. And since the primary instability is the instability of the chemical reaction-diffusion-system, the response of the fluid to the instability only arises in $O(\epsilon^2)$ which together with the scaling of the convective terms means that there are no convective contributions at all in the $O(\epsilon^3)$ equations. The second reason is that due to the relatively high order of the convective terms the lowest order nonlinearity is the quadratic part of the chemical reaction terms and not like in a classical convection problem a convective term. A detailed analysis of the higher order contributions coming from the convective terms and the autocatalytic nonlinearities shows - making use of the symmetry properties of the appropriate terms - that only to seventh order convective terms contribute to the envelope equation. Thus for generalized free-free, symmetric boundary conditions for the stream function feedback between chemical wave fronts and the flow only arises in seventh order in the amplitude.

However if one considers different boundary conditions for the stream function (for example free-rigid) at the top and bottom surface, which are more realistic considering the experiments of Miike et al. [21,55], the situation changes. The lowest order solution for the stream function has no definite symmetry in this case and therefore there will be contributions to the envelope equation from the convective terms as soon as they enter the equations, that is in $O(\epsilon^5)$. To this order also nonlinear gradient terms contribute to the resulting envelope equation.

6 Chemically driven convection can stabilize Turing Patterns

In this section we study the effects of chemically driven convection on Turing structures. We consider a model where convective motion is generated by density gradients due to concentration variations of the Turing structures in a thin horizontally infinite layer of a solute reaction-diffusion system. We study this system with envelope equations and find that the coupling to chemically driven convection modifies the nonlinear coefficients in these equations. For a roll structure we show that this coupling can stabilize pure Turing patterns. An experimental set-up to investigate these effects is proposed.

In 1952 Turing [22] demonstrated that the competition of chemical reactions and diffusion can give rise to a rich variety of pattern forming instabilities. In particular Turing showed that the wavelength of the resulting stationary patterns is independent of the characteristic lengths in the system, a feature that is qualitatively different from the case of many hydrodynamic instabilities for which the wavelength of the patterns is often given by the geometry of the set-up. This is for example the case for Rayleigh-Bénard convection and for Taylor vortex flow. Apart from chemical systems this mechanism is also relevant to pattern forming biological systems [18] which were Turings main motivation. Only quite recently experimental evidence for the existence of Turing structures [7,8] in comparatively simple chemical systems emerged. These structures were observed in reaction-diffusion systems embedded in a gel strip where one species of the reaction diffusion system is effectively trapped in the gel to provide a big difference in the diffusion coefficients of activator and inhibitor which tends to increase the volume of parameter space where stationary instabilities can occur. Another reason for using gel as a medium for reaction-diffusion pattern forming systems is to avoid the development of convective instabilities which are believed to destroy the emerging Turing patterns. There seems to be no evidence for the existence of Turing patterns in an aqueous solution where convection is possible. But it has been shown that the gel does not seem to be crucial in providing a difference in the diffusion coefficients of the activator and inhibitor species [56]. This difference — for example in the CIMA reaction — rather seems to depend on the concentration of starch molecules in which one of the reacting species can be trapped and diffuses therefore slower since starch is a fairly big molecule with a diffusion coefficient smaller than that of usual species in water by one order of magnitude. Without the gel Turing like patterns could still be observed [56]. In this case convective motion is possible which leads to the question how these Turing patterns will interact with the convective flow they induce via concentration or temperature gradients produced by the chemical reactions.

While for the case of oscillatory unstable or excitable reaction diffusion systems there are numerous studies on the interactions of the oscillatory chemical patterns and convection in aqueous solutions [49,57-60], triggered by experimental evidence [21,55,61,62], there has not been much work on the interaction of Turing structures and the convective flows they induce in aqueous solutions. The works published so far are mainly numerical studies of the problem in a restricted geometry [63]. There has however been some work on the interaction of Turing patterns and an imposed hydrodynamic flow [64,65], which was not driven by the Turing patterns. Here we investigate the emergence of Turing patterns under chemically driven convection close to the onset of the stationary instability using amplitude expansions.

To be definite consider an isothermal solute reaction-diffusion system in a thin horizontally infinite layer of thickness L with free-free boundary conditions for the fluid motion and Dirichlet boundary conditions for the concentrations in z-direction. The system is described by the following set of equations

$$\partial_i v_i = 0 \tag{1}$$

$$\rho_0 \left(\partial_t v_i + v_j \partial_j v_i\right) = -\partial_i p - \rho g \delta_{i3} + \nu \nabla^2 v_i \tag{2}$$

$$\partial_t c_i + v_j \partial_j c_i - D_i \nabla^2 c_i = f_i\left(c_i\right), \tag{3}$$

where we assume incompressibility. ρ is a function of the concentrations of the two chemical species of the reaction-diffusion-system, whereas the density ρ_0, the viscosity ν and the diffusivities D_i of the two chemical species c_i are assumed to be constant (generalized Boussinesq approximation). The functions $f_i\left(c_i\right)$ describe the kinetics of the chemical reaction of the two species which are assumed to be in general polynomials in c_1 and c_2. As we will demonstrate in the following density gradients that are created and maintained by the chemical reaction will create fluid motion because of the density dependent gravity term in the momentum equation for the fluid. This is somewhat similar to the case of a container heated from the sidewalls in thermal convection [66], where however the stress exerted on the fluid layer now is of an intrinsic nature.

We consider the stability of the stationary constant concentration solutions $(c_{10}, c_{20}, \mathbf{v} = 0)$ to infinitesimal perturbations $(c_1, c_2, \mathbf{v})$ where we assume that the density ρ depends on the concentrations and their small perturbations as follows

$$\rho(c_i) = \rho_0(c_{10}, c_{20}) + \alpha c_1 + \beta c_2. \tag{4}$$

The linear stability analysis shows that the system will undergo a stationary, e.g. a Turing instability, if the condition

$$\frac{(K_{11}D_2 - K_{22}D_1)^2}{4D_1D_2} = -K_{12}K_{21} \tag{5}$$

is fulfilled. Another requirement is that the system has not developed an oscillatory stability before the threshold for the stationary instability is reached. Here k_c is the wavenumber of the most unstable wavelength given by

$$k_c^2 = \frac{K_{11}D_2 + K_{22}D_1}{2D_1D_2} - \pi^2 \tag{6}$$

and the K_{ij} are the components of the Jacobian of the kinetic functions, $\partial f_i/\partial c_j$, for the chosen equilibrium concentrations (c_{10}, c_{20}). These can be thought of as being controlled via temperature or the concentrations of basic ingredients in an experiment. There is a critical wavenumber if $k_c^2 > 0$ or otherwise the wavenumber that first becomes unstable is zero corresponding to an infinite wavelength. We shall be interested in finite wavelengths here. It is important to keep in mind that the critical wavenumber is of an intrinsic nature, determined by the details of the reaction-diffusion system and not by the size of the experimental setup as is typically the case for hydrodynamic systems like for example in Rayleigh-Bénard convection. In an experiment undertaken to check the results of this analysis, one must therefore match the system size, e.g. the thickness of the layer, with that intrinsic wavelength. Otherwise effective two dimensionality is no longer preserved and the system we are looking at becomes more complicated. An apparatus similar to the one investigated in this context has already been proposed [67].

To study the effect of convection on Turing structures we proceed to look at the example of roll like structures. We look at roll patterns in the weakly nonlinear regime close to the onset of the instability using standard perturbation expansion methods [68,69,70] to derive an envelope equation that describes the long time and large scale behavior of these roll-like solutions. In the first order of the perturbation expansion the solution for the concentration perturbation is

$$\begin{pmatrix} c_1 \\ c_2 \end{pmatrix} = W(X, Y, T)\, e^{ik_c x} \cos(\pi z) \begin{pmatrix} 1 \\ \eta \end{pmatrix}, \tag{7}$$

$$\eta = \frac{D_1\left(k_c^2 + \pi^2\right) - K_{11}}{K_{12}} \tag{8}$$

where we have introduced the amplitude of the roll-like solution which only depends on the slow time coordinate T and the rescaled horizontal coordinates $\{X, Y\}$ whose dependence on these coordinates has to be determined by solving the equations for the higher orders in the perturbation expansion. The corresponding solution for the flow velocities are

$$v_{x1} = ik_c \chi W(X,Y,T) e^{ik_c x} \cos(\pi z) + c.c. \tag{9}$$

$$v_{y1} = 0 \tag{10}$$

$$v_{z1} = -\frac{k_c^2}{\pi} \chi W(X,Y,T) e^{ik_c x} \sin(\pi z) + c.c. \tag{11}$$

the first nontrivial contribution to v_y,

$$v_{y2} = \chi \partial_Y W(X,Y,T) e^{ik_c x} \sin(\pi z), \tag{12}$$

being of the next order in the perturbation expansion and where

$$\chi = \frac{g\pi(\alpha + \eta\beta)}{\nu(k_c^2 + \pi^2)^2}. \tag{13}$$

As one sees the first contributions to the velocity in the expansion are of the same order as the first contributions to the chemical concentration variations that drive convection. This is qualitatively different from the case of the oscillatory instability discussed in ref. [52], where the first nontrivial contributions to the velocity were only of second order.

The resulting envelope equation for rolls is then given by

$$\partial_T W + \frac{D_1 D_2}{\delta} \left(\partial_Y^2 + 2ik_c \partial_X\right)^2 W \tag{14}$$

$$+ (\gamma_{11} K_{22} + K_{11}\gamma_{22} - K_{12}\gamma_{21} - \gamma_{12} K_{21}) W/\delta = b_3 |W|^2 W$$

where

$$\delta = D_2(k_c^2 + \pi^2) - K_{22} + D_1(k_c^2 + \pi^2) - K_{11} \tag{15}$$

The γ_{ij} are the variations of the K_{ij} when crossing into the part of the two dimensional parameter space where stationary instabilities are possible [69]. The cubic coefficient in the amplitude equation contains contributions from the convective term in the reaction-diffusion-system whereas the advection terms of the velocity equations do not contribute to the cubic coefficient as is the case for Rayleigh-Bénard convection [68,70]. It should be noted that this contribution to the cubic coefficient will always lower the value of the cubic coefficient and hence increase the stabilizing contributions to the cubic term in the envelope equation. This is due to the explicit mathematical form of the terms that describe the coupling of convection to the chemical concentrations and vice versa. They have the same form as in the case of the coupling of the temperature field to convection and vice versa in simple Rayleigh-Bénard convection where they result in the cubic coefficient which gets only one contribution from the advective term in the temperature equation always saturating the envelope equation for the roll solution [68,70]. Therefore convection enhances the stability of the roll solution. Namely, if convection results in a negative cubic coefficient where, without convection, there would be a positive cubic coefficient, a roll structure that would otherwise be unstable, is stabilized, if convection is present, and therefore some other pattern that exists in the absence of convection might be replaced by the roll solution under convection.

To demonstrate this possibility we have calculated the coefficient in question for a particular model reaction diffusion system — the Schnakenberg model [71,72]. The equations (1)-(3), when specialized to the Schnakenberg model and after suitable reduction of parameters, read

$$\partial_i v_i = 0 \tag{16}$$

$$\partial_t v_i + v_j \partial_j v_i - \nabla^2 v_i + \partial_i p = -Sc\,(R_1\, c_1 + R_2\, c_2)\,\delta_{i3} \tag{17}$$

$$\partial_t c_1 + v_j \partial_j c_1 - \frac{1}{Sc}\nabla^2 c_1 = \frac{\gamma}{Sc}\left(a - c_1 + {c_1}^2 c_2\right) \tag{18}$$

$$\partial_t c_2 + v_j \partial_j c_2 - \frac{1}{dSc}\nabla^2 c_2 = \frac{\gamma}{Sc}\left(b - c_1 + {c_1}^2 c_2\right). \tag{19}$$

where

$$R_1 = \frac{g\alpha L^3}{D_1 \nu}\left(\frac{k_2}{k_3}\right)^{1/2} \qquad R_2 = \frac{g\beta L^3}{D_1 \nu}\left(\frac{k_2}{k_3}\right)^{1/2} \tag{20}$$

$$Sc = \frac{\nu}{D_1} \qquad d = \frac{D_1}{D_2} \qquad \gamma = \frac{L^2 k_2}{D_1} \tag{21}$$

ν is the viscosity of the fluid in question, the parameter d is the ratio of the diffusion coefficients D_1 and D_2. The R_i are somewhat analogous to the Rayleigh number and Sc is the Schmidt number associated with species 1. a and b are the suitably rescaled pool species concentrations, which are controlled and held constant in an experiment and determine the type of instability the system can undergo. We introduced the thickness of the layer as relevant length scale. The k_i are the coefficients of the Schnakenberg model's chemical reaction terms.

The contributions of the convection to the cubic term in the amplitude equation can lead to the stabilization of a roll-type structure as can be seen in Fig. 14, where convection results in a shift of the cubic coefficient for the relevant parameter value $b = 0.35$ towards a negative value which assures saturation of the roll amplitude to cubic order. Stabilization has been observed for different sets of parameters. The parameters here are evaluated using $\nu = 10^{-2} cm^2/s$, $D_1 = 5 \cdot 10^{-6} cm^2/s$ and assuming $L \propto 1mm$. This leads to $Sc = 2000$ and we subsequently chose $\gamma = 80$. The parameters d, R_1, R_2 were varied in the calculations.

Fig. 15 demonstrates a situation where the cubic coefficient is already negative, but becomes more negative when convection is taken into account.

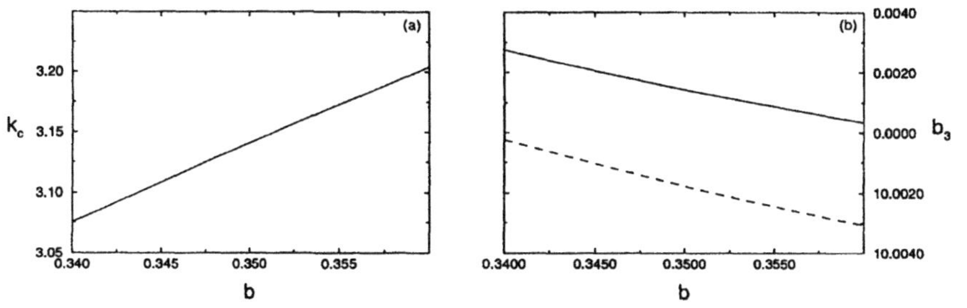

Fig. 14. Fig. 14a is a plot of the critical wavenumber k_c as a function of b where $d = 0.3$, $\gamma = 80$, $Sc = 2000$, $R_1 = 1 \cdot 10^{-4}$ and $R_2 = 0$. The critical wavenumber $k_c = \pi$ that matches the layer depth is reached at approximately $b = 0.35$. Fig. 14b is a plot of the cubic coefficient b_3 where the solid line denotes the case without convection and the dashed line the case with convection.

Furthermore one can see that a contribution of the second species to the density variation ($R_2 \neq 0$) can reduce the effect, because the concentration c_2 ist out of phase with the concentration c_1 ($\eta < 0$). But the effect is larger again, when R_2 is bigger than a certain value that leads to a vanishing contribution to the cubic coefficient. All this is due to the fact that the coupling parameters R_i appear in the relevant contribution to the cubic coefficient in the form $(R_1 + \eta R_2)^2$.

We have thus demonstrated that Turing patterns and the convection induced by them are of the same order in the perturbation expansion in the Schnakenberg model. Therefore convection is important already close to onset of the instability. The interaction of Turing patterns and convection results in the stabilization of roll patterns as compared to the situation when convection is absent, e.g. in Turing structures in gel strips. This result indicates that coexistence of chemically driven convection and Turing patterns driving them is possible at least in the weakly nonlinear regime.

We have chosen the geometry of the system we were looking at already in the spirit of what might be a promising experimental set-up to study these effects. In matching the thickness of the fluid layer with the intrinsic wavelength of the chemical instability we arrive at a quasi two-dimensional pattern forming system that is very similar to the situation encountered in Rayleigh-Benard convection or in electroconvection in liquid crystals. A layer much thicker than the critical intrinsic wavelength would result in three-dimensional Turing structures [73] whose interactions with convection are likely to be much more complicated and difficult to observe than the system described here. So matching the intrinsic wavelength of the system and the thickness of the apparatus we are using to study chemically driven convection in Turing patterns is essential to control the system and study it. We think that it is also more likely to observe ordered structures in the set-up propo-

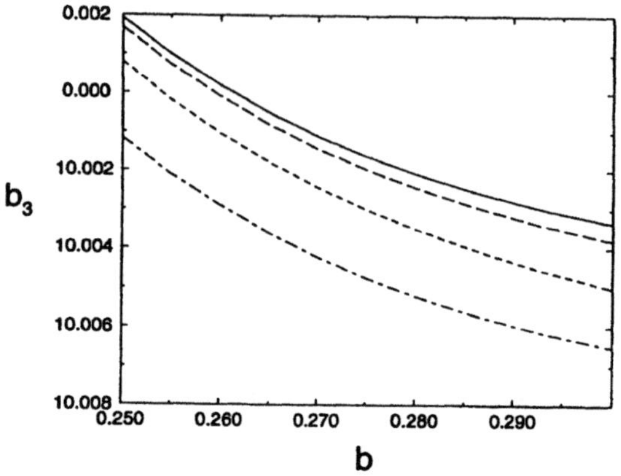

Fig. 15. This is a plot of the cubic coefficient b_3 as a function of b where $d = 0.5$, $\gamma = 80$, $Sc = 2000$, $R_1 = 1 \cdot 10^{-4}$. The critical wavenumber $k_c = \pi$ that matches the layer depth is reached at approximately $b = 0.27$. The solid line denotes the case without convection. The short dashed line denotes the case with convection and $R_2 = 0$, whereas the long dashed line denotes the case with convection, but with $R_2 = 1.5 \cdot 10^{-5}$ and the dot dashed line denotes the case with convection with $R_2 = 1 \cdot 10^{-4}$. The partial cancellation of the effects of the two different concentrations is due to η being negative.

sed here. To reduce experimental difficulties it might be interesting to start studying the interaction of Turing patterns and chemically driven convection in sol-like media where the strength of convective flows could be reduced to a level that is easier to control.

7 Oscillatory hydrodynamic flow due to concentration dependence of surface tension

In this section, using a two species reaction - diffusion model in the excitable regime coupled to the hydrodynamic equations, we discuss how a group of chemical waves drives an oscillatory convective motion in a fluid layer. The experiments show an oscillatory behavior of the surface flow with the maximum velocity opposite to the direction of wave propagation. The mechanism considered is a variation of the surface-tension along a free surface due to an inhomogeneous concentration distribution. Our results lead to an understanding of the observed oscillatory flow as well as of the large scale features observed.

The study of autocatalytic chemical reactions is an active field of research, which has attracted increasing attention by the recent observation of Turing instabilities in gels [7,8]. Much less work has been devoted to investigate to what extent a chemical reaction drives convection and whether this convective motion can feed back and alter the chemical reaction. Experimental observations and a discussion of some of the qualitative features of the phenomena observed have been reported in refs. [58,74]. In [75] the convective flow induced by a reaction front in a vertical cylinder has been analyzed. Wilke performed numerical simulations of single reaction-fronts and their interaction with convective motion [76,77].

Quantitative measurements of the flow in systems with dense wave trains have been reported by Miike et al. [21,55] for a thin layer of BZ-reagent with an open surface. They observed the appearance of an oscillatory surface flow with a period that is about twice the period of the chemical waves. Matthiessen et al. [47] found that this oscillatory flow is forming global flow waves with an extension over about ten chemical wave fronts traveling towards the center of the spiral. These results have never been modeled. Here we present the results of our numerical study of a two species autocatalytic chemical reaction coupled to the appropriate nonlinear hydrodynamic equations [78], namely the balance equations for the two concentrations and the Navier Stokes equation. We will assume that the convective flow is driven by surface tension variations due to an inhomogeneous distribution of one of the chemical species, a property that was shown to have a marked effect on the surface tension in a stirred tank experiment [51]. For convection driven by density variations a Ginzburg-Landau equation with complex coefficients [69] was derived, for the case of an oscillatory instability with vanishing wave number at onset [52]. However, here we work in the excitable regime and our numerical study captures the steep fronts that are present in these systems, which a Ginzburg-Landau equation is not well suited to describe.

An excitable BZ-solution was prepared as described in refs. [21,55]. A volume of 4 ml was filled in a Petri dish, where it formed a layer with a depth of about 0.8 mm in the center. The dish was covered by a glass plate to suppress evaporation, leaving an air gap of about 1 cm over the solution. Wave trains with a short wavelength were generated by initiating a counterrotating pair of spiral waves near the boundary according to a procedure given in ref. [31]. The flow velocity was detected via the observation of suspended small latex particles with a diameter of 0.5 μm. The experimental set-up and the automatic velocity analysis are described in refs. [21,47,55].

Fig. 16 is a plot of the flow velocity perpendicular to the chemical wave fronts as measured at the surface in the middle of the petri dish. The observed flow shows typical oscillations with a period of about 85 s. For comparison the figure shows also the local light transmission in the point of flow measurement, representing the local chemical state of the system. The hydrodynamic oscillation has a period that is about twice the period of the chemical oscillation.

The flow at the surface was measured in additional experiments following talcum particles sprinkled over the solution. In Fig. 17 we show the trace of one of these particles for 140s. Other particles in the observed area were

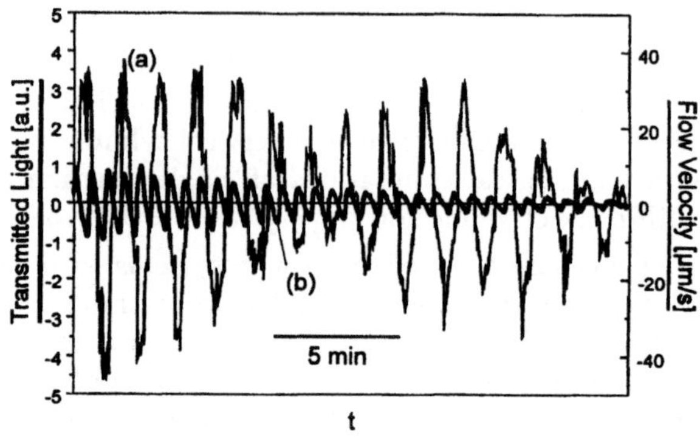

Fig. 16. Oscillatory surface flow (a) and local light transmission through the point of measurement (b) for a system with a dense chemical wave pattern.

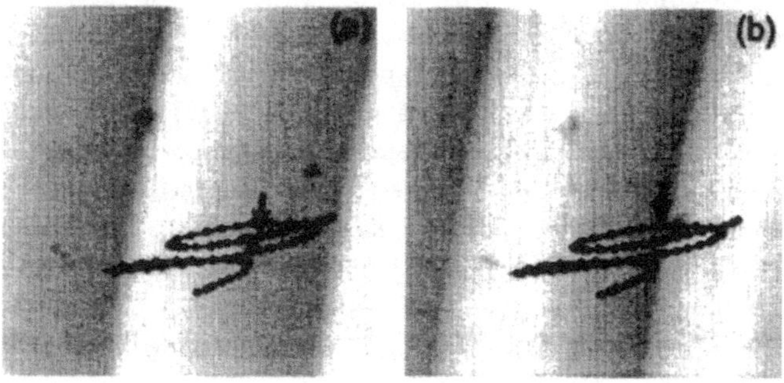

Fig. 17. Trace of a talkum particle on the surface of the fluid layer during the oscillation of the hydrodynamic flow. The images represent in a frame of $2.6 \times 2.6\ mm^2$ the appearance of the wave fronts traveling to the left at the instants of time marked on the trajectory by arrows.

moving synchronously. The two snapshots of the chemical waves correspond to the time instants marked on the particle's trace and cover about one full oscillation of the hydrodynamic flow. The particle's velocity is indicated by the distance of the dots on the trajectory representing equal time intervals. It reaches a maximum, when the particle moves opposite to the direction of wave propagation (b). The shape of the wave fronts changes during the hydrodynamic oscillation as already shown in ref. [47]. When the flow is directed opposite to the wave propagation (b) the front appears broader and loses sharpness as compared to the case of flow in the direction of wave propagation (a). This behavior can be interpreted as a deformation of the front due to the surface flow.

Measurements of the height dependence of the hydrodynamic motion in a single wave front presented in ref. [77] show that any significant hydrodynamic motion is limited to almost only the upper half of the fluid layer. In the case of the dense wave trains induced by a spiral this feature is even more distinct. This underscores the importance of surface tension effects for the oscillatory flows in BZ-spirals.

We model the behavior of the coupled reaction-diffusion-convection system by using the Oregonator [79] coupled to the hydrodynamic equations. It turns out to be sufficient to consider the dynamics in a quasi two-dimensional set-up discarding velocity components parallel to one of the horizontal coordinates and the dependence of the concentrations and the velocity fields on the corresponding horizontal coordinate. Thus we can introduce a stream function Ψ and a vorticity function ω.

The set of equations we consider reads

$$(\partial_t u + \mathbf{v} \cdot \nabla u) = \epsilon \nabla^2 u + \frac{1}{\epsilon}\left(u - u^2 - fv\frac{u-q}{u+q}\right) \tag{22}$$

$$\partial_t v + \mathbf{v} \cdot \nabla v = \epsilon \nabla^2 v + u - v \tag{23}$$

$$\partial_t \omega + \mathbf{v} \cdot \nabla \omega = \nu' \Delta \omega \tag{24}$$

which has already been rescaled as in [79] and for which rate constants have been estimated [80]. ω and Ψ are defined by $\omega = \epsilon_{ijk}\partial_j v_k$ and $\Delta\Psi = -\omega$ and the flow velocities are given by $v_x = -\partial_z \Psi$ and $v_z = \partial_x \Psi$. ϵ, f and q are the dimensionless parameters of the Oregonator model which were taken to be $\epsilon = 0.005$, $f = 2$ and $q = 0.002$. We have assumed isothermal conditions and u and v correspond to the concentrations of the two chemical species $[HBrO_2]$ and $[ferriin]$. The rescaled viscosity $\nu' = \frac{\nu k_5 B}{k_3 A}$ is equal to 6.7 for the particular model used here, where the k_i, A and B were taken from [79] and ν is the kinematic viscosity of the solution. We consider a system with vertical sidewalls and finite thickness in the vertical z-direction. For the chemical concentrations we choose no-flux boundary conditions which correspond to the experimental conditions. For the hydrodynamic flow we consider rigid boundaries for the bottom of the box and free boundaries for the sidewalls since it was found that in this way the perturbations from the sidewalls can be kept at a minimum. The upper boundary of the box corresponds to the free surface in a typical BZ-experiment and therefore we chose Marangoni boundary conditions [78]:

$$\partial_z v_x = \frac{1}{\eta}\frac{\partial \alpha}{\partial v}\partial_x v = S\partial_x v, \quad S = \frac{1}{\eta}\frac{\partial \alpha}{\partial v}, \qquad v_z = 0 \tag{25}$$

where α is the surface tension depending on the concentration of the inhibitor concentration v and η is the dynamic viscosity. For the simulations we have set $S = -100$, because the ferriin concentration was found to increase the surface tension [51]. We will assume that v is the sole driving force of the hydrodynamic motion, because of the strong effect the ferriin concentration has on the surface tension [51], although there seems to be a contribution of the density variations due to concentration variations as well [52]. The influence of temperature effects in the system was found to be rather insignificant [50]. The presented set of equations was solved and its time-evolution studied using standard implicit-explicit finite difference methods for the concentration and vorticity equations and a relaxation method to compute the stream function [81].

To study the behavior of a wave train consisting of tightly spaced reaction-fronts in the excitable regime, we prepared the system using initial concentration distributions that resembled the wave trains familiar from experiments. We let them evolve according to the Oregonator model until they had reached a certain characteristic shape that would not change any more. Now the coupling to the hydrodynamic system was added and the time evolution was studied in the rest frame of the chemical wavefronts.

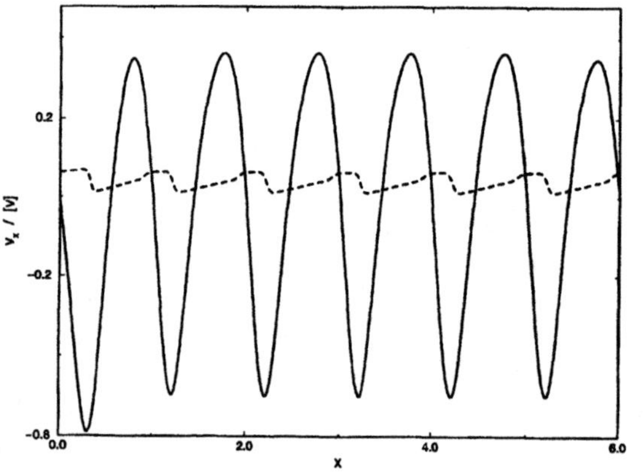

Fig. 18. Spatial variation of the inhibitor concentration $[v]$ (dotted line) and the horizontal velocity v_x (solid line) along the surface from the simulation at one instant of time.

We observe numerically a strong flow in the direction opposite to the propagation direction of the reaction-fronts that can be traced back to the sharp fronts in the ferriin (v) concentration which develop only when the reaction-fronts are tightly spaced (see Figure 18). This back-flow is rather strong in agreement with the experiments (see Fig. 16). Fig. 19 gives the overall topology of the flow structure observed numerically. An important observation from the numerical simulations is the strong distortion of the reaction-fronts

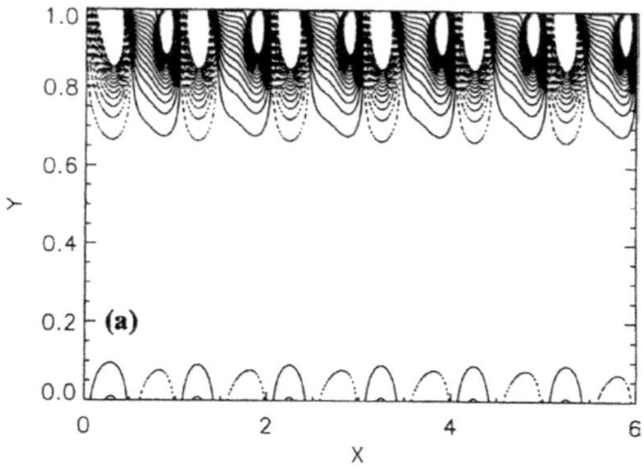

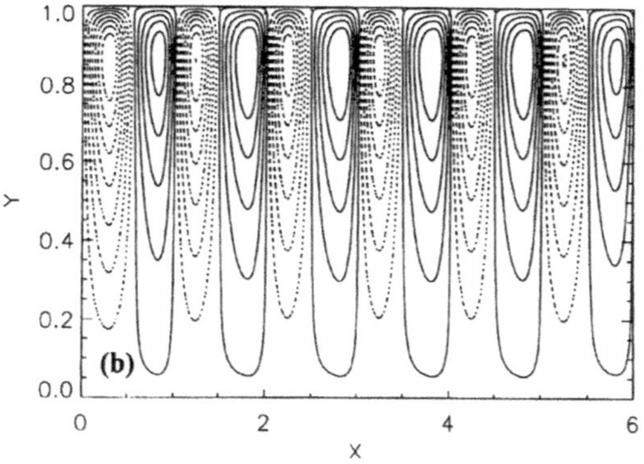

Fig. 19. a) Contour plot of the vorticity ω. Dark regions correspond to velocities in propagation direction and the broader light regions are associated with flow in the counterpropagating direction. b) Contour plot of the stream function Ψ corresponding to a). One sees very clearly that the gradients become much steeper towards the surface as the flow becomes stronger.

near the surface: the reaction-front gets disconnected from the free surface and the lateral gradients in the ferriin concentrations are smoothed out. The hydrodynamic flows have thus lost their driving power and therefore the layer can relax to a motionless state. We were able to demonstrate that the disrupted reaction-front will reconnect to the free surface and reestablish the strong gradients in the inhibitor concentration along the open surface, once the strong flow has relaxed to a motionless state. Clearly, this starts over the whole dynamics.

Comparing this to the experiment we find striking correspondence: In the experiments the strong flow driven by the concentration gradients exists only until it has smoothed the reaction fronts out along the surface. A smaller mean flow appears following the direction of wave propagation. After a period of time the reaction-front has reestablished its original shape and is connected to the free surface again starting another strong back flow that leads to a rapid motion of the tracer particles. On the basis of this mechanism an oscillatory flow appears whose period is independent of the period of the chemical waves. In the experiments the period ratio between convective flow and chemical reaction is about 2:1 [21,55].

Now we want to give an interpretation of some of the most pronounced effects in the BZ-convection experiment in terms of the simulations presented here. First of all, the fact that surface tension driven convection leads to a disruption and a smoothing out of the sharp reaction-fronts along the surface can account for the large scale feature that was observed to travel towards the spiral's core [47] and corresponds to a region of several wavelengths where there is a broad concentration distribution instead of the rather sharp individual reaction-fronts. This can easily be explained by the mechanism observed in the numerical simulations where the surface tension driven flows lead to a disruption and smoothing out of the reaction fronts on and near the free surface. We emphasize that the convective flows couple the individual reaction-fronts. So the mechanism described above is not an individual feature of the separate wavefronts, but is communicated to the nearest neighbors of the reaction-fronts as well. This coupling leads to a collective behavior of the wavefronts which all undergo a periodic wave front disruption due to convective flow, but not necessarily in phase. The wave observed in the experiment is the natural consequence of this collective phenomenon.

This novel mechanism can account for the rather unusual correlation of reaction-fronts and changes of flow-direction in the experiments where the flow velocities near but below the free surface were studied: what one is looking at in the experiment is a superposition of convective motions with different time scales. The wavelike excitation of the surface flow contributes a strong, but temporary velocity component. In addition there are convection rolls in between the reaction fronts that are driven by density and surface tension variations and persist even after the surface flow has stopped. These rolls are dragged along by the wavefronts and thereby their spatial flow structure is mapped to a temporal variation in the measurements.

To sum up, we have numerically studied a model coupling autocatalytic chemical reactions to hydrodynamics driven by surface tension. We found a strong surface flow that is due to sharp gradients in the ferriin concentrations. These are clearly present in the experiment where the surface flow and the behavior of the reaction-fronts near the surface were studied. This strong flow leads to a disruption of the wavefronts and inhibits itself by smoothing out the concentration gradients. The disrupted reaction-fronts will reconnect to the free surface of the layer as soon as the hydrodynamic flow has stopped which will start the strong surface flow again. This behavior is a natural explanation for the global wave observed by Matthiessen and Müller [47] and also explains the unusual ratio of wave train-passages and flow reversal. Thus this kind of strong surface tension driven convection is due to the tight spacing of the wavefronts that forces the concentration profile of the inhibitor to develop sharp fronts.

Acknowledgements
This work was partially supported by the Deutsche Forschungsgemeinschaft and the European Space Agency, Paris. The authors thank A.F. Münster and M. Watzl for three of their images and N.J. Jaeger, R.D. Otterstedt and P.J. Plath for two of their plots shown in section 3 of this article. They are grateful to K. Matthiessen, M. Diewald, M. Böckmann, H. Miike, T. Sakurai and S. Kai for having shared the enthusiasm of the authors on the topic of chemically driven convection.

References

[1] H. Bénard, Rev. Gen. Sci. Pure Appl. **11**, 1261 and 1309 (1900)
[2] M.C. Cross and P.C. Hohenberg, Rev. Mod. Phys. **65**, 851 (1993)
[3] M. J. Block, Nature **178**, 650 (1956)
[4] A. N. Zaikin and A. M. Zhabotinsky, Nature **225**, 535 (1970)
[5] A.T. Winfree, Science **175**, 634 (1972)
[6] R. J. Field and M. Burger (Eds.) Oscillations and Traveling Waves in Chemical Systems (Wiley, New York, 1971)
[7] V. Castets, E. Dulos, J. Boissonade and P. DeKepper, Phys. Rev. Lett. **64**, 2953 (1990)
[8] Q. Ouyang and H. L. Swinney, Nature **352**, 610 (1991)
[9] M. Watzl and A. F. Münster, Chem. Phys. Lett. **242**, 273 (1995)
[10] K.-J. Lee, W. D. McCormick, J. E. Pearson and H. L. Swinney, Nature **369**, 215 (1994)
[11] Q. Ouyang, J. Boissonade, J. C. Roux and P. DeKepper, Phys. Lett. A **124**, 282 (1989)
[12] M. Burger and R. J. Field, Nature **307**, 720 (1984)
[13] R. Kapral and K. Showalter (Eds.), Chemical Waves and Patterns (Kluwer, Dordrecht (1995)
[14] R. Fitzhugh, Biophys. J. **1**, 445 (1961)
[15] N. A. Gorelova and J. Bures, J. Neurobiol. **14**, 353 (1983)
[16] J. M. Davidenko, A. V. Pertsov, R. Salomonsz, W. Baxter and J. Jalife, Nature **355**, 349 (1992)
[17] H. Meinhard and M. Klinger, J. Theor. Biol. **126**, 63 (1987)
[18] J. D. Murray, "Mathematical Biology" (Springer-Verlag, Berlin, Heidelberg, New York, (1989) p. 435
[19] K. I. Agladze and V. I. Krinsky, A. M. Pertsov, Nature **308**, 834 (1984)
[20] S. C. Müller, Th. Plesser and B. Hess, "Chemical Waves and Natural Convection" in:"Physical Hydrodynamics: Interfacial Phenomena" , ed. by M.G. Velarde (Plenum Press, New York, 1988) p. 423
[21] H. Miike, S. C. Müller and B. Hess, Phys. Lett. A **141**, 25 (1989)
[22] A. M. Turing, Trans. R. Soc. Lond. B **237**, 37 (1952)
[23] I. Lengyel, S. Kadar and I. R. Epstein, Science **259**, 493 (1993)

[24] A. F. Münster, M. Watzl and F. W. Schneider, Physica Scripta **T67**, 58 (1996)
[25] P. Resch A. F. Münster and F. W. Schneider, J. Phys. Chem. **95**, 6270 (1991)
[26] Y. X. Zhang and R. J. Field, J. Phys. Chem. **95**, 723 (1991)
[27] F. Siegert and C. Weijer, Proc. Natl. Acad. Sci. USA **89**, 6433 (1992)
[28] J. Lechleiter, S. Girard, E. Peralta and D. Clapham, Science **252**, 123 (1991)
[29] R.J. Field and R.M. Noyes, J. Chem. Phys. **60**, 1877 (1974)
[30] V. Petrov, V. Gaspar, J. Masere and K. Showalter, Nature **361**, 240 (1993)
[31] S. C. Müller, Th. Plesser and B. Hess, Physica D **24**, 71 (1987)
[32] G. S. Skinner and H. L. Swinney, Physica D **48**, 1 (1991)
[33] Zs. Nagy-Ungvarai, J. Ungvarai and S. C. Müller, Chaos **3**, 15 (1993)
[34] P. Foerster, S. C. Müller and B. Hess, Development **109**, 11 (1990)
[35] M. A. Dahlem and S.C. Müller, Exp. Brain Res. **115**, 319 (1997)
[36] W. Jahnke and A.T. Winfree, Int. J. Bifurcation and Chaos **1**, 445 (1991)
[37] J. J. Tyson and S. H. Strogatz, Int. J. Bifurcation and Chaos **1**, 723 (1991)
[38] D. Stock and S. C. Müller, Physica D **96**, 396 (1996)
[39] A.T. Winfree, S. Candle, G. Chen, P. McGuire and Z. Szilagyi, Chaos **6**, 617 (1996)
[40] A. T. Winfree, Physica D **84**, 126 (1995)
[41] R.D. Otterstedt, P.J. Plath, N.I. Jaeger, J.C. Sayer and J.L. Hudson, Phys. Rev. E **54**,3744 (1996)
[42] G. Flätgen and K. Krischer, J. Chem. Phys. **103**, 5428 (1995)
[43] R.D. Otterstedt, N.I. Jaeger, P.J. Plath and J.L. Hudson, in: A Perspective Look at Nonlinear Media , ed. by J. Parisi, S.C. Müller and W. Zimmermann (Springer, Berlin Heidelberg, 1997), p. 283
[44] T. Yamaguchi, L. Kuhnert, Zs. Nagy-Ungvarai and S.C. Müller, J. Phys. Chem. **95**, 5831 (1991)
[45] Y. Aizu and T. Asakura, Appl. Phys. B **43**, 209 (1987)
[46] H. Miike and S.C. Müller, Chaos **3**, 21 (1993)
[47] K. Matthiessen and S.C. Müller, Phys. Rev. E **52**, 492 (1995)
[48] E. Körös, M. Orbàn and Zs. Nagy, Acta, Chim. Acad. Sci. Hung. **100**, 29 (1979)
[49] J.A. Pojman and I.R. Epstein, J. Phys. Chem. **94**, 4966 (1990)
[50] T. Böckmann, B. Hess and S.C. Müller, Phys. Rev. E **53**, 5498 (1996)
[51] K. Yoshikawa, T. Kusumi, M. Ukitsu and S. Nakata, Chem. Phys. Lett. **211**, 211 (1993)
[52] M. Diewald and H.R. Brand, Chem. Phys. Lett. **216**, 566 (1993)
[53] M. Diewald and H.R. Brand, Phys. Rev. E **51**, 5200 (1995)
[54] M. Diewald, K. Matthiessen, S.C. Müller and H.R. Brand, Phys. Rev. Lett. **77**, 4466 (1996)
[55] H. Miike, S.C. Müller and B. Hess, Chem.Phys.Lett. **144**, 515 (1988); Phys.Rev.Lett. **61**, 2109 (1988)
[56] K. Agladze, E. Dulos and P. De Kepper, J.Phys.Chem.**96**, 2400 (1992)
[57] I. Nagypál, G. Bazsa and I.R. Epstein, J. Am. Chem. Soc. **108**, 3635 (1986); J.A. Pojman, I.R. Epstein, T.J. McManus and K. Showalter, J.Phys.Chem. **95** 1299 (1991)
[58] J. Rodriguez and C. Vidal, J.Phys.Chem. **93** (1989) 2737.
[59] Th. Plesser, H. Wilke and K.H. Winters, Chem.Phys.Lett. **200**, 158 (1992)

[60] J. W. Wilder, B. F. Edwards and D. A. Vasquez, Phys. Rev. A**45**, 2320 (1992); D. A. Vasquez, J. W. Wilder and B. F. Edwards, J. Chem. Phys. **98**, 2138 (1993) and references therein

[61] H. Miike, H. Yamamoto, S. Kai and S.C. Müller, Phys.Rev. E**48**, 1627 (1993); and private communication

[62] H. Miike, H. Yamamoto, S. Kai and S.C. Müller, Phys. Rev. E **48**, 1627 (1993); S. Kai and H. Miike, Physica A **204**, 346 (1994)

[63] D. A. Vasquez, J. W. Wilder and B. F. Edwards, Phys. Rev. Lett. **71**, 1538 (1993)

[64] C. R. Doering and W. Horsthemke, Phys. Lett. A **182**, 227 (1993)

[65] S. Ponce Dawson, A. Lawniczak and R. Kapral, J. Chem. Phys. **100**, 5211 (1994)

[66] J. E. Hart, J. Atmos. Sci. **29**, 687 (1972)

[67] Z. Nosticzius, H. Farkas, A. Schubert, J. Swift, W. D. McCormick and H. L. Swinney, in "Spatio-Temporal Organization in Nonequilibrium Systems", S.C. Müller and Th. Plesser, Eds., Projekt Verlag, Dortmund, 1992, p. 190

[68] A.C. Newell, in: Nonlinear Wave Motion, Lectures in Applied Mathematics, Vol.15, ed. A.C. Newell (American Mathematical Society, Providence, Rh.I., 1974) p.157

[69] Y. Kuramoto and T. Tsuzuki, Prog.Theor.Phys. **54**, 687 (1975); Prog.Theor.Phys. **55**, 356 (1976)

[70] R. Graham, Phys. Rev. A **10**, 1762 (1974)

[71] J. Schnakenberg, J. Theor. Biol. **81**, 389 (1979)

[72] V. Dufiet and J. Boissonade, J. Chem. Phys. **96**, 664 (1992)

[73] A. De Wit, G. Dewel, P. Borckmans and D. Walgraef, Physica D **61**, 289 (1992)

[74] I. Nagypál, G. Bazsa and I.R. Epstein, J.Am.Chem.Soc. **108**, 3635 (1986); J.A. Pojman, I.R. Epstein, T.J. McManus and K. Showalter, J.Phys.Chem. **95**, 1299 (1991)

[75] B.F. Edwards, J.W. Wilder and K. Showalter, Phys.Rev. A**43**, 749 (1991); D.A. Vasquez, B.F. Edwards and J.W. Wilder, Phys.Rev. A**43**, 6694 (1991); D.A. Vasquez, B.F. Edwards and J.W. Wilder, Phys.Fluids A**4**, 2410 (1992); J.W. Wilder, B.F. Edwards and D.A. Vasquez, Phys.Rev. A**45**, 2320 (1992); D.A. Vasquez, J.W. Wilder and B.F. Edwards, J.Chem.Phys. **98**, 2138 (1993)

[76] H. Wilke, Physica **D 86**, 508 (1995)

[77] K. Matthiessen, H. Wilke and S. C. Müller, Phys. Rev **E 53**, 6056 (1996); Phys.Lett. **A 141**, 25 (1989)

[78] L.D. Landau and E.M. Lifshitz, Hydrodynamics (Pergamon, New York, 1959)

[79] J. P. Keener and J. J. Tyson, Physica **21 D**, 307 (1986)

[80] J. J. Tyson, in: Oscillations and Traveling Waves in Chemical Systems, ed. R. Field and M. Burger, Wiley N.Y. (1986), p. 307

[81] C.A.J. Fletcher, Computational Techniques for Fluid Dynamics, Vol. 2, Berlin 1988

Formation of Spatio-Temporal Structures in Semiconductors

E. Schöll[1], F.J. Niedernostheide[2], J. Parisi[3], W. Prettl[4] and H.G. Purwins[2]

[1] Technische Universität Berlin, Institut für Theoretische Physik, Hardenbergstr. 36, 10623 Berlin, Germany
[2] Universität Münster, Institut für Angewandte Physik, Corrensstraße 2/4, 48149 Münster, Germany
[3] Universität Oldenburg, Fakultät für Physik, 26111 Oldenburg, Germany
[4] Universität Regensburg, Institut für Experimentelle und Angewandte Physik, 93040 Regensburg, Germany

Abstract. We review recent theoretical and experimental advances in the formation of spatio-temporal structures in semiconductors. Charge transport in semiconductors has been demonstrated to exhibit many features of complex nonlinear spatio-temporal dynamics and self-organized pattern formation which occur in other dissipative systems as well. Recent progress has been obtained, for instance, in the theoretical description of such phenomena in terms of microscopic transport models or reaction-diffusion systems and activator-inhibitor kinetics as well as in the discovery of various patterns, e.g., in Si, GaAs, and ZnS based materials and advanced quantitative experiments. Detailed computer simulations in one and two spatial dimensions have enabled a quantitative comparison between theory and experiment. Among the particularly instructive examples are investigations of current filaments and their higher bifurcations such as travelling, rocking, breathing, or spiking filaments in layered semiconductor structures and in doped semiconductors in the regime of low temperature impurity breakdown for different contact geometries, as well as recent studies of localized two-dimensional structures in ZnS films.

Keywords: semiconductors, pattern formation, current filaments, high-field transport, impact ionization, nonlinear spatio-temporal dynamics

1 Introduction

Semiconductors are dissipative dynamical systems due to the dissipation of electrical energy in charge transport. They can be considered as reaction systems due to the generation and recombination of charge carriers [1], thus providing an analogy with chemical reaction kinetics which has been noted long ago [2, 3]. For diluted systems the reaction rates follow the mass action law and are sums of products of the various concentrations; therefore semiconductors are nonlinear systems in general. Finally they must be considered as continuous media because of spatial coupling due to diffusion. From these properties various spatial, temporal and spatio-temporal patterns are expected in semiconductors [1, 4, 5, 6, 7, 8]. These patterns and their description are expected to be similar to other dissipative systems treated in this book, e.g., surface reactions, or autocatalytic chemical reaction systems. In the case of charged particles, the coupling with electric fields provides an important nonlinear feedback mechanism. Systems of this type are often adequately described in terms of activator-inhibitor kinetics. This concept was applied to the selforganisation of living organisms by various authors, e.g., [9, 10, 11, 12]. An activator-inhibitor system is most simply modelled as a system of nonlinear reaction-diffusion equations

$$\tau_a \frac{\partial a(x,t)}{\partial t} = f(a,u) + l^2 \frac{\partial^2 a}{\partial x^2} \tag{1}$$

$$\tau_u \frac{\partial u(x,t)}{\partial t} = g(a,u) + L^2 \frac{\partial^2 u}{\partial x^2} \tag{2}$$

where a and u denote the local activator and the inhibitor, respectively, τ_a,τ_u are time scales, l, L are characteristic diffusion lengths, and f and g are appropriate functions of a and u. The production of the inhibitor u increases with activator concentration a and saturates with increasing u. The activator a is increased by autocatalytic reproduction, and is inhibited by u. In the generic case, $f(a,u)$ is a non-monotonic function of a, increasing in a certain interval, and is linearly decreasing with u, while g may be of the form $g = a - u$. Depending upon the intersection between the null-clines $f = 0$ and $g = 0$, e.g., either bistable, excitable, or oscillatory (Turing-Hopf) behaviour is found. While a description of transport in terms of the simple reaction-diffusion model (1), (2) is possible for various semiconductor structures like pnpn-diodes, pin-diodes, or other layered structures [13, 14, 15, 16, 17, 18, 19], some semiconductor systems may require refined modelling in terms of more sophisticated transport models , e.g., drift-diffusion equations combined with Poisson's equation for the electric potential, hydrodynamic balance equations for the carrier densities, mean carrier momenta and mean carrier energies, or even a kinetic approach in terms of the semiclassical Boltzmann equation or a full quantum transport theory [20]. Often a combination of these approaches is used.

Semiconductor transport instabilities are characterized by a nonlinear dependence of the current density j upon the electric field $\mathcal{E}$. This relation is often non-monotonic and displays negative differential conductivity (*NDC*). Two important cases are described by an *N*-shaped or an *S*-shaped $j(\mathcal{E})$ characteristic, and denoted by *NNDC* and *SNDC*, respectively. However, more complicated forms like *Z*-shaped, loop-shaped, or disconnected characteristics are also possible [21]. In case of *NNDC*, the *NDC* branch is often but not always - depending upon exernal circuit and boundary conditions - unstable against the formation of field domains, while in the *SNDC* case current filamentation generally occurs [22]. These primary selforganized spatial patterns may themselves become unstable in secondary bifurcation leading to periodically or chaotically breathing, rocking, moving, or spiking filaments or domains, or even solid-state turbulence and spatio-temporal chaos.

In applying the activator-inhibitor principle to the description of semiconductor transport, one first has to identify the activator and inhibitor variables. The activator is generally an internal dynamical variable which triggers the instability via an autocatalytic process. Examples are provided by the carrier (electron or hole) density in doped semiconductors. Impact ionisation of electrons or holes either across the bandgap or from localised levels is an autocatalytic process which leads to a multiplication of the number of carriers. In doped semiconductors, impurity impact ionization is essential in explaining the *SNDC* observed in the regime of low-temperature impurity breakdown. Another activator variable may be provided by the interface charge density ρ_s in semiconductor heterostructures and low-dimensional structures, where abrupt junctions between different materials on an atomic length scale cause conduction band discontinuities resulting in potential barriers and quantum wells. The local charge accumulations in these wells, together with nonlinear transport processes across the barriers may induce the positive nonlinear feedback between the carrier distribution and the electric potential distribution characteristic for an activator.

The inhibitor is often associated with the electric field $\mathcal{E}$. Dielectric relaxation of the field acts as an inhibiting process which reduces the internal space charge field according to the current continuity equation $\epsilon\dot{\mathcal{E}} = J - en\mu\mathcal{E}$ where J is the total current density composed of the displacement current density $\epsilon\dot{\mathcal{E}}$ and the drift current density j with mobility μ, and ϵ is the permittivity. The dot denotes the temporal derivative. A more general form of this inhibitor equation [21] which also includes circuit components like a series load resistor R and a parallel external capacitor C_{ext}, and an applied bias voltage U_0 is given by $C\dot{U} = \frac{U_0 - U}{R} - I(a, U)$ where $C = C_{int} + C_{ext}$ is the sum of the internal (geometric) and the external capacitance, U is the sample voltage, and $I(a, U)$ is the sample current which is given in the absence of charge accumulations by $I(a, U) = \int_A j(x, y, t)dxdy$, integrated over the current cross section A, representing a global coupling which is also typical for long range inhibitors.

Within the last decade the introduction of concepts and methods from nonlinear dynamics and pattern formation has stimulated a large amount of experimental and theoretical work in a variety of semiconducting materials, e.g. p-Ge or n-GaAs at liquid Helium temperatures, or layered semiconductor structures like pin, pnpn, or heterostructure hot electron diodes. Recent progress has been obtained in elaborate experimental techniques to detect spatially and temporally resolved structures like scanning electron microscopy [23, 24], scanning laser microscopy [25, 26, 27], potential probe measurements [15, 28], or quenched photoluminsescence [29]. Detailed computer simulations in one and two spatial dimensions have enabled a quantitative comparison between theory and experiment. While in earlier work irregular and chaotic *temporal* behavior was in the center of interest [30, 31], the focus has now shifted towards more complex *spatio-temporal* dynamics including the dynamics of solitary filaments and multifilamentary states, spatio-temporal chaos, higher bifurcations of the elementary dissipative structures, interaction of defects and structural imperfections with pattern formation, and complex two-dimensional sample and contact geometries.

In this overview these concepts will be illustrated by some exemplary model systems which are presently of particular interest. Firstly, we consider the dynamics of current filaments in the regime of low-temperature impurity breakdown in n-GaAs. Recent computer simulations are reviewed in Sect. 2. In section 3 we report on experimental investigations of current density filaments in high-purity n-GaAs epitaxial layers using state-of-the-art visualization techniques. In section 4 diverse types of filament oscillations and bifurcations observed experimentally in Si multilayered devices are reviewed and compared with numerical results based on a two-layer model. A survey of patterns arising in thin ZnS films is given in section 5. A common feature of these exemplary systems is that the activator-inhibitor concept can be fruitfully applied, and that they exhibit generic spatio-temporal structures irrespectively of their particular microscopic nature. They all represent bistable SNDC systems where close comparison between theory and experiment has recently been achieved.

Other semiconductor systems where nonlinear transport phenomena and pattern formation have been found include parallel transport in modulation–doped heterostructures associated with real-space transfer [32, 33], i.e., the real-space analog of the Gunn effect, and vertical transport in a superlattice [34, 35, 36], as well as field-enhanced trapping in semi-insulating GaAs [37]. In all those systems the current-field characteristic is N-shaped and leads to the formation of electric field domains and oscillatory instabilities. Nonlinear spatio-temporal carrier dynamics in low-dimensional semiconductor structures like quantum wires and quantum dots has also become an active field recently [38, 20].

2 Simulation of current filaments in n-GaAs

Impact ionization of charge carriers is a key process in inducing instabilites and spatio-temporal structures in semiconductors. It is a process in which a charge carrier with high kinetic energy collides with a second charge carrier, transferring its kinetic energy to the latter which is hereby lifted to a higher energy level. The result of the process is carrier multiplication which may induce electrical instabilities at sufficiently high external electric fields [4]. Impact ionization from shallow donors or acceptors is responsible for impurity breakdown at low temperatures where almost all carriers are bound to shallow impurities [39]. Being a nonlinear autocatalytic process, it induces a nonequilibrium phase transition [1] between a low conductivity and a high conductivity state and may lead to a menagery of spatio-temporal instabilities including self–oscillations and chaos as first observed by Aoki [40] and

Teitsworth et al [41]. If a critical threshold voltage is applied to the semiconductor sample, the free carrier concentration grows in an avalanche-like manner, and the current increases by orders of magnitude at an almost constant voltage due to current filamentation and lateral growth of filaments.

Current filamentation is usually attributed to the bistability of an S-shaped dependence of the current density j on the electric field $\mathcal{E}$, i.e. SNDC. The spatial coexistence of the high-conductivity and the low-conductivity state at a fixed voltage leads to the formation of current filaments. In generation-recombination models [1] the bistability is caused by impact ionization of multilevel impurities. The generation-recombination cycle starts with impact ionization from the ground state and is afterwards sustained by impact ionization of much less strongly bound excited states. S-type $j(\mathcal{E})$ relations may also result, taking into account the impurity ground state only, due to hot electron runaway and increased screening of electron scattering caused by ionized impurities [42]. Recent computer simulations of n-GaAs have shown, however, that the electron temperature in a nascent current filament is only slightly elevated compared to the low conducting outside regions [43].

In the past, most theoretical work has been restricted to one-dimensional treatments. Breathing filaments [44, 45] and the intermittent and chaotic behavior of laterally travelling filaments [46, 47] have been studied in one-dimensional simulations where only the transverse spatial coordinate perpendicular to the current flow was taken into account [47]. The spatio-temporal modes and the stochastics of the breakdown process have been investigated in a one-dimensional longitudinal model for p-Ge [48]. Longitudinally travelling charge density waves have been found in simulations for n-GaAs [49]. The dipolar electric field between two point contacts was included in a phenomenological, effectively one-dimensional model for current filaments in n-GaAs [50]. Much progress has been made in the microscopic analysis of low-temperature impurity breakdown in terms of single-particle [51] and many-particle Monte Carlo (MC) simulations [52] for p-Ge and for n-GaAs [53, 54]. In order to study the nascence of current filaments the transverse and the longitudinal degrees of freedom have been combined for a realistic two-dimensional sample geometry with appropriately modelled point contacts, and detailed microscopic information on the generation-recombination kinetics obtained from MC simulations has been included [55, 56]. The widening of the filaments in a weak transverse magnetic field in the direction of the Lorentz force has also been simulated [57] and the effects of external circuits have been studied [58].

2.1 Monte Carlo simulation of impact ionization breakdown

We consider a doped semiconductor at temperatures of liquid helium. In the following we shall assume n-type material; p-type semiconductors can be treated analogously. The carrier density in the conduction band, and hence the current density, is determined by the generation-recombination (GR) processes of carriers between the conduction band and the donor levels. The experimentally observed S-shaped current density-field relation in the regime of impurity breakdown can be explained in terms of standard GR kinetics only if impact ionization from at least two impurity levels is taken into account [59]. Therefore we model the infinite hydrogenlike energy spectrum of the shallow donors by the ground state and an "effective" excited state close to the band edge. In this case the state of the system can be characterized by the spatial distribution of the carrier densities in the conduction band $n(\mathbf{r},t)$ as well as in the impurity ground state and excited state $n_1(\mathbf{r},t)$, $n_2(\mathbf{r},t)$, respectively, where $\mathbf{r}$ is the position vector and t denotes time.

In the following we shall adopt an approach which combines Monte Carlo (MC) simulations of the microscopic scattering processes and GR kinetics with rate equations for the macroscopic spatio-temporal dynamics of the carrier densities which occurs on a much slower time-scale than the microscopic dynamics. First, we will list the macroscopic constitutive equations.

The temporal evolution of n is governed by the continuity equation

$$\dot{n} = \frac{1}{e}\nabla\cdot\mathbf{j} + \phi(n, n_1, n_2, |\mathcal{E}|), \tag{3}$$

where the dot denotes the partial derivative with respect to time, e is the electron charge and $\mathcal{E}$ is the local electric field within the sample. Within the drift-diffusion approximation the current density $\mathbf{j}$ can be expressed as

$$\mathbf{j} = e(n\mu\mathcal{E} + D\nabla n) \tag{4}$$

with the field-dependent mobility μ and diffusion coefficient D. We assume the validity of the Einstein relation $D = \mu k_B T_L/e$, where k_B is Boltzmann's constant and T_L is the lattice temperature.

The rate ϕ of GR processes depends on the local values of the carrier densities in the conduction band and at the impurities, given by n, n_1, and n_2, respectively, and the strength of the electric field $\mathcal{E} = |\mathcal{E}|$. Analogously, rates ϕ_1, ϕ_2 determining the temporal evolution of n_1, n_2 can be defined as

$$\dot{n}_i = \phi_i(n, n_1, n_2, \mathcal{E}) \tag{5}$$

with $i = 1, 2$.

In an explicit two-level model [1], e.g., the GR rates are given by

$$\phi = X_1^S n_2 - T_1^S n p_t + X_1 n n_1 + X_1^* n n_2, \tag{6}$$

$$\phi_1 = T^* n_2 - X^* n_1 - X_1 n n_1, \tag{7}$$

$$\phi_2 = -\phi - \phi_1, \tag{8}$$

where $p_t = N_D - n_1 - n_2$ is the density of ionized donors, N_D is the total density of donors, X_1^S is the thermal ionization coefficient of the excited level, T_1^S is its capture coefficient, X_1, X_1^* are the impact ionization coefficients from the ground and excited level, respectively, X^*, T^* denote the transition coefficients from the ground level to the excited level and vice versa, respectively.

The electric field is coupled to the carrier densities via Gauss' law

$$\epsilon \nabla \cdot \mathcal{E} = e(N_D^* - n_1 - n_2 - n), \tag{9}$$

where ϵ is the dielectric constant and $N_D^* \equiv N_D - N_A$ holds with the compensating acceptor concentration N_A. From equations (3),(5),(8),(9) the charge conservation equation can be derived, given by $\nabla \cdot \mathbf{J} = 0$ with $\mathbf{J} = \epsilon \dot{\mathcal{E}} + \mathbf{j}$, where $\mathbf{J}$ is the total current density composed of displacement current and conduction current densities.

The essential nonlinearities of the constitutive model equations (3)-(9) in the regime of low-temperature impurity breakdown are contained in the dependence of the GR coefficients upon n, n_1, n_2, and $\mathcal{E}$. In order to derive these from a microscopic theory we have performed single particle MC simulations for a spatially homogeneous steady state [53].

The microscopic rates of all band-impurity processes depend upon the carrier densities in the band and impurity states, which in turn depend upon the nonequilibrium carrier distribution function. To obtain these carrier densities, the MC method has to be combined self-consistently with the rate equations (3)-(5) in the homogeneous steady state, where the GR coefficients X_1, X_1^*, T_1^S are calculated by averaging the microscopic transition probabilities (P_{ii}^1, P_{ii}^2, P_{rec} for impact ionization from the ground state, the excited state, and capture, respectively) over the nonequilibrium distribution function $f(\mathbf{k})$, which is extracted from the MC simulation at each step:

$$\begin{aligned} X_1(n, n_1, n_2, \mathcal{E}) &= \frac{1}{nn_1} \int d^3k\, f(\mathbf{k}; n, n_1, n_2, \mathcal{E}) P_{ii}^1(\mathbf{k}, n_1), \\ X_1^*(n, n_1, n_2, \mathcal{E}) &= \frac{1}{nn_2} \int d^3k\, f(\mathbf{k}; n, n_1, n_2, \mathcal{E}) P_{ii}^2(\mathbf{k}, n_2), \\ T_1^S(n, n_1, n_2, \mathcal{E}) &= \frac{1}{np_t} \int d^3k\, f(\mathbf{k}; n, n_1, n_2, \mathcal{E}) P_{rec}(\mathbf{k}, p_t). \end{aligned} \tag{10}$$

Note that f, and hence X_1, X_1^* and T_1^S, in turn depend parametrically on n, n_1, n_2 and $\mathcal{E}$. An iteration procedure, where n_1 and n_2 are expressed by their steady-state dependence on n and $\mathcal{E}$, is used to solve the above problem self-consistently [53].

It is possible to express the dependence of the GR coefficients upon $\mathcal{E}$ and n through the electron temperature $T_e(\mathcal{E}, n)$ [55]. Here we use the notion of electron temperature T_e in the usual sense of the mean electron energy $(3/2)k_B T_e = \langle E \rangle = (\hbar^2/2m)\langle k^2 \rangle$, i.e. the electron temperature is extracted from the MC data essentially as the second moment of the nonequilibrium distribution function $f(\mathbf{k})$. This should not be confused with the concept of a heated Maxwellian distribution function, where the electron temperature appears as a parameter. This latter concept is much more restrictive since it assumes a quasi-thermal equilibrium of the conduction band subsystem. The MC data yield a strong increase of T_e with rising electric field on the high-conductivity branch of the current-voltage characteristics as opposed to only a slight increase on the low-conductivity branch. Physically the strong increase is associated with a population inversion between the donor ground and the excited state on the upper branch. For $\mathcal{E} > 9\text{V/cm}$ impact ionization does no longer contribute to energy relaxation in a significant way since the donor ground states are almost completely ionized and impact ionization from the excited state dominates, which is now more strongly populated than the ground state. The GR cycle runs now between the donor excited state and the conduction band. Because of the much smaller energies involved, this process cools less efficiently.

In the following sections our strategy will be to insert fitted analytical representations of the MC data into the macroscopic equations (3)-(9). We use this approach in order to take into account as much detailed information as possible about the microscopic scattering processes, while still retaining manageable expressions. As a check of consistency, the spatially uniform stationary $n(\mathcal{E})$ characteristic which we obtain with the fit functions from (6)-(8) in the steady state using the condition of charge neutrality $N_D^* = n(\mathcal{E}) + n_1(\mathcal{E}, n) + n_2(\mathcal{E}, n)$, can be compared with the characteristic obtained by direct MC simulation. The S-shaped negative differential conductivity (SNDC) is successfully reproduced.

2.2 Current filament formation in point contact samples

In this section we shall present exemplary two-dimensional simulations for thin samples with point contacts, considering the longitudinal x-direction as well as the transverse z-direction. In this way it is possible to model the dynamics of current filaments. We consider a square sample with side lengths $L_x = L_z = 0.02\text{cm}$ representing a thin n-GaAs film (thickness $1.4 \times 10^{-3}\text{cm}$). We model point contacts by applying Dirichlet boundary conditions to two opposite regions of length $L_c = 8 \times 10^{-4}\text{cm}$ at the centers of the sample edges parallel to the z-axis. At the contacts n is fixed to a value $n_D = 5 \times 10^{15}\text{cm}^{-3}$ to model Ohmic contacts. All other boundaries are treated as insulating where the components of the current density $\mathbf{j}$ and the electric field $\mathcal{E}$ perpendicular to the boundaries vanish. Our numerical algorithm employs an implicit finite element scheme [60].

We study the nascence of current filaments when the applied voltage is switched rapidly to a value above breakdown threshold so that the semiconductor is forced from the nearly insulating state to a highly conducting state. Within 1ps the sample voltage is linearly increased from $U = 0\text{V}$ to $U = 0.48\text{V}$ corresponding to an average field $\mathcal{E}_{av} = 24\text{V/cm}$, i.e., above the threshold field $\mathcal{E}_{th} = 17\text{V/cm}$ of the homogeneous $n(\mathcal{E})$ characteristic. Initially the current density is very low (Fig. 1a). Subsequently impact ionization multiplies the electron concentration at the cathode and establishes a front

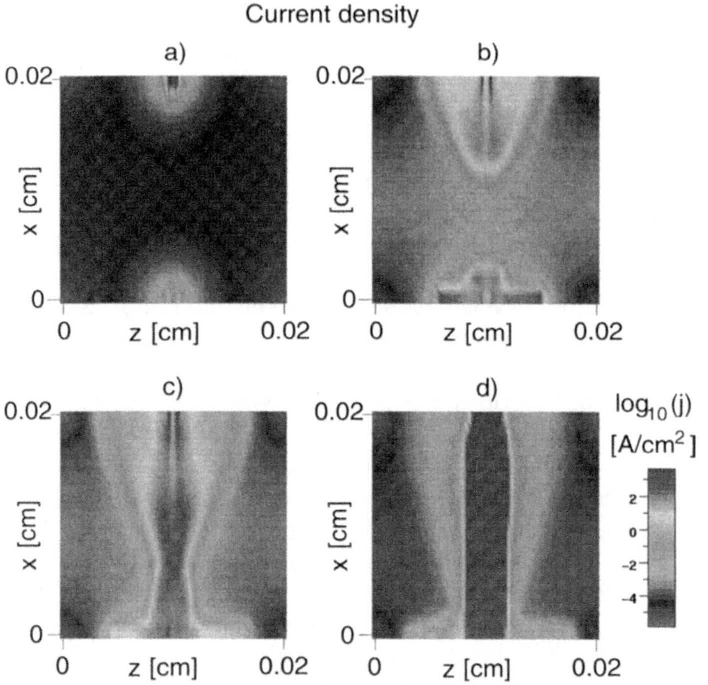

Fig. 1. Simulation of the formation of a current filament in n-GaAs. The temporal evolution of the current density $j(x, z)$ after application of a voltage $U = 0.48$V is shown as a function of the spatial coordinates x and z for a square sample with two point contacts. a) $t = 1$ps, b) $t = 0.5$ns, c) $t = 1.0$ns, d) $t = 2.5$ns. Note that the injecting cathode is at the top (calculated for $T_L = 4.2$ K with the parameters $N_D = 7.0\times10^{15}$ cm^{-3}, $N_A = 2.0\times10^{15}$ cm^{-3}, $\mu = 10^5$ cm^2/Vs, $\epsilon_r = 10.9$, and generation-recombination coefficients taken from a Monte Carlo simulation). After [56].

that moves towards the anode (Fig. 1b). The propagation of the front is accompanied by a high field domain associated with a slightly increased electron temperature T_e. Although the electric field behind the front is smaller than ahead of it, for reasons of current continuity the increased electron density in regions passed by the front is almost conserved because recombination is a much slower process than generation. Hence impact ionization downstream is encouraged, whereas further generation upstream is inhibited. When the front meets the region of increased carrier density around the anode, a rudimentary filament is formed (Fig. 1c), albeit with a carrier density several orders of magnitude lower than that corresponding to completely ionized donor states of density $N_D^* = 5 \times 10^{15}$cm^{-3}. Now donor impact ionization is becoming enhanced in the rudimentary filament because the excited level

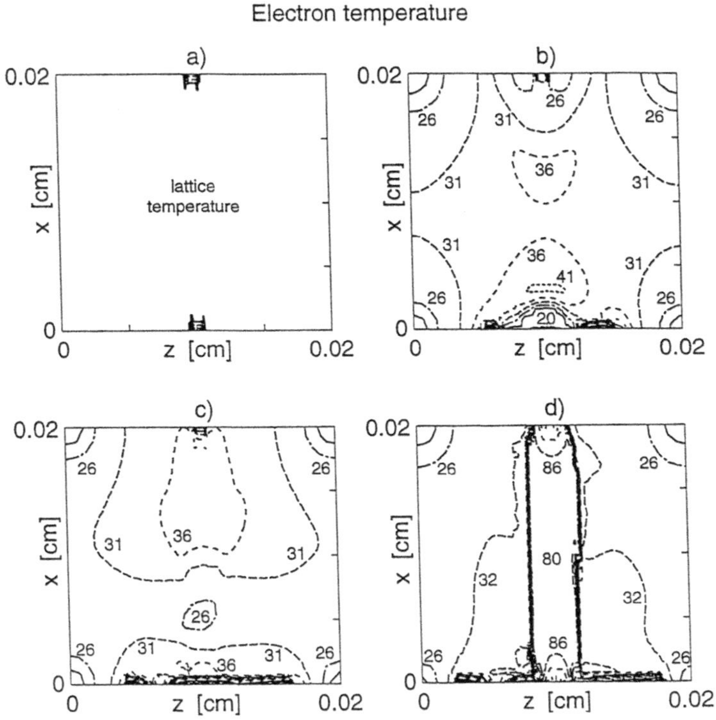

Fig. 2. Simulated temporal evolution of the electron temperature $T_e(x, z)$. Timesteps and voltage as in Fig. 1. Contour plots are shown, where the numbers denote T_e in K.

n_2 is increasingly populated . The current density and the electron temperature are significantly growing in the rudimentary filament. Finally impact ionization leads to a uniform increase of electron density until the filament reaches its mature state (Fig. 1d) where almost all donors are ionized, and the carrier density corresponds to the upper branch of the homogeneous steady state characteristic $n(\mathcal{E})$. Within the filament the excited donor level is much more highly populated than outside. Nevertheless still only about 2 percent of the band carriers are trapped in the excited level, while the ground level is *completely* depleted inside the filament. Thus the population ratio between ground and excited level is inverted in the filament. Also the current density j and the electron temperature T_e are much larger inside the filament than outside. The corresponding plots of the electron temperature and potential distributions are shown in Figs. 2 and 3.

Hence from our simulation we find three stages of impact ionization breakdown: a stage of front creation and propagation from the injecting contact, i.e. the cathode (stage I) followed by a stage of stagnation after the front

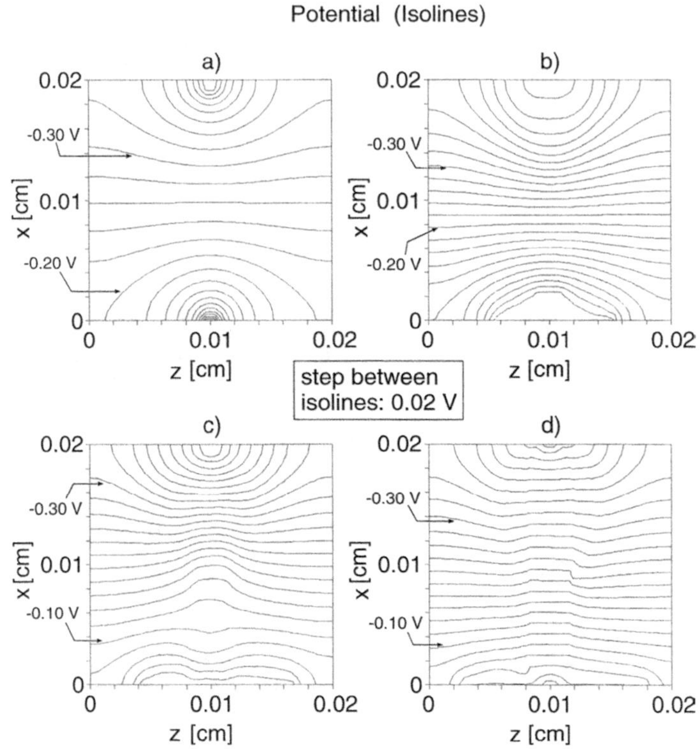

Fig. 3. Simulated temporal evolution of the electric potential $\Phi(x, z)$. Timesteps and voltage as in Fig. 1. Potential isolines are shown.

has reached the anode (stage II) and a final stage during which the rudimentary filament grows to a mature filament (stage III) [56]. Fig. 4 shows the fully developed stationary current filament for four different values of a series load resistance R. With increasing R the operating point on the characteristic is shifted towards lower sample voltage and current, and the width of the filament becomes markedly smaller, in good agreement with experiments [29, 43]. Close to the holding field of the breakdown (d) the filament becomes very thin and unstable.

It has been found experimentally that an external magnetic field applied perpendicular to the electric field can sensitively affect the spatio-temporal instabilities in the regime of impurity impact ionization [61, 62, 63]. The dynamic Hall effect has been proposed as a mechanism to induce chaotic temporal oscillations in the breakdown regime [64]. The transverse periodic or chaotic motion of a current filament under crossed electric and magnetic field has been numerically analyzed for p-type Ge by one-dimensional simula-

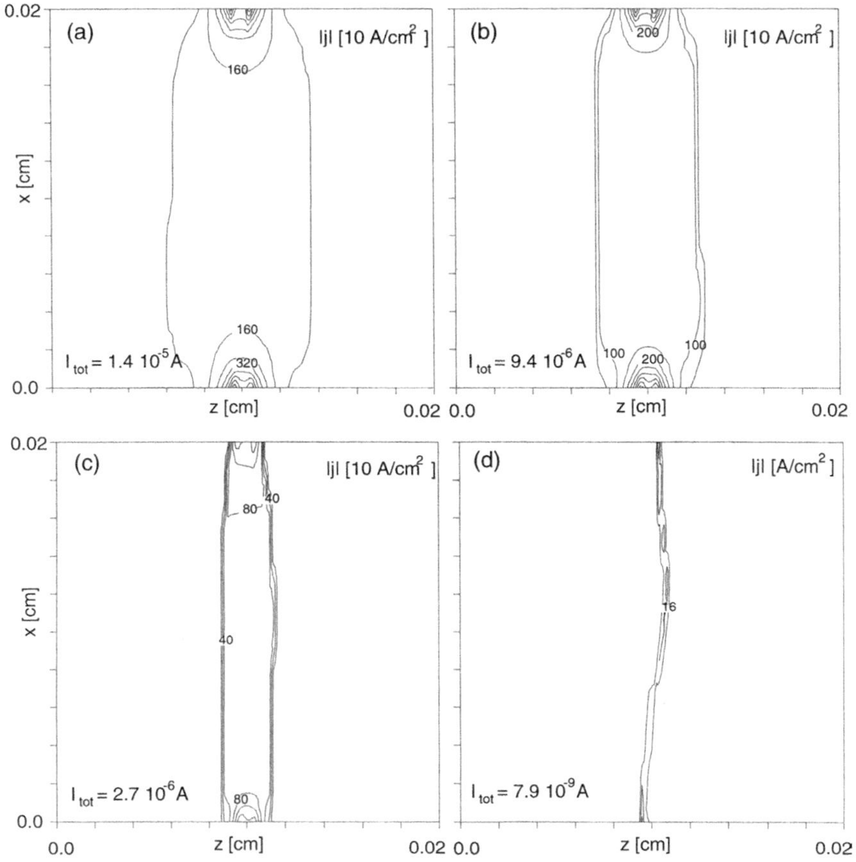

Fig. 4. Simulated current filaments in n-GaAs in a square sample with two point contacts for an applied voltage $U_0 = 0.48\text{V}$ and different load resistances R. The stationary current density $|\mathbf{j}(x, z)|$ is shown as a function of x and z as a contour plot. a) $R = 1\ \text{k}\Omega$, b) $R = 10\ \text{k}\Omega$, c) $R = 100\ \text{k}\Omega$. d) $R = 50\ \text{M}\Omega$ (calculated with the same parameters as Fig. 1.)

tions based on the semiclassical semiconductor-transport equations neglecting the longitudinal spatial dependence [47]. The spatial deformation of current filaments in the presence of a transverse magnetic field has been explained by an effectively one-dimensional drift-diffusion model assuming a dipole-like electric field between two point-contacts [50]. However, in order to fully understand the effects of a transverse magnetic field on the filament properties, two-dimensional analyses including the spatial dependence on both the longitudinal x- and the transverse z-direction are indispensable [57].

The current density $\mathbf{j}(\mathbf{B})$ in the presence of a magnetic field $\mathbf{B}$ normal to the current flow is given by

$$\mathbf{j}(\mathbf{B}) = \frac{\mathbf{j}(0) + \mu\mathbf{B} \times \mathbf{j}(0)}{1 + |\mu\mathbf{B}|^2}, \tag{11}$$

where $\mathbf{j}(0)$ is the current density for $B = 0$ as given by Eq. (4).

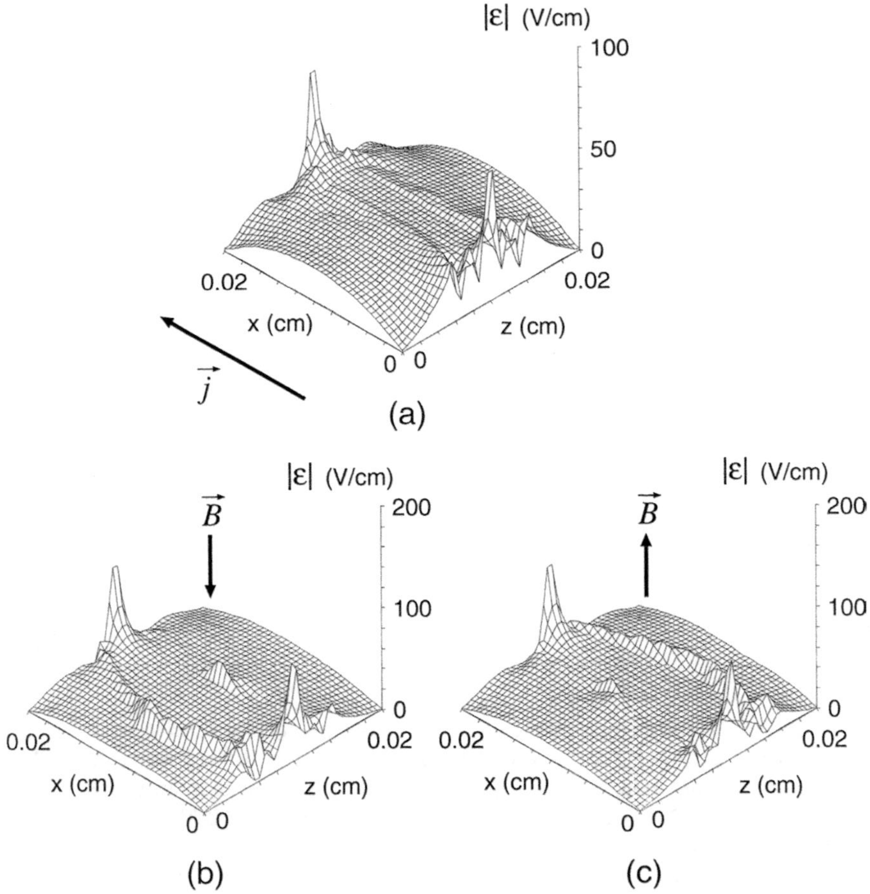

Fig. 5. Simulated spatial distribution of electric field strength $|\mathcal{E}(x, z)|$ for (a) $B = 0$, (b) $B = 5$ mT, and (c) $B = -5$ mT after $t \approx 2.5$ ns (calculated with the same parameters as Fig. 1.). A magnetic field pointing downwards is defined as positive and upwards as negative. After [57].

Upon application of a voltage ramp and a magnetic field of a few mT, first a thin, straight filament is formed as in the case without magnetic field. In the next stage, the straight filament is gradually widened and asymmetrically expanded in the direction of the Lorentz force [57]. The simulation shows that the electric fields at the filament border pointing in the direction of the Lorentz force are much stronger than at the opposite border (Fig. 5). These asymmetric fields are in good agreement with asymmetric ridge-like structures observed by scanning laser microscopy under a transverse magnetic field [25, 26, 50]. Also, the simulations and the experiments show that the width and curvature of the filaments become larger with increasing magnetic-field strength.

The physical processes involved in filament formation under a transverse magnetic field can be understood as follows. For $B=0$, the electric field profile is completely symmetric with respect to the center of the sample ($z=0.01$ cm) during all stages of filament formation. At first the electric field exhibits a gentle peak at the middle of the sample due to the dipolar electric field induced between the point contacts. The field profile then develops two maxima which arise from the space charge accumulated at the filament borders. A lateral electric field induced by this local space-charge dipole prevents electrons, generated by impact ionization, from diffusing outwards, thus stabilizing the filament.

For $B \neq 0$, the electric field also retains an almost symmetric profile until the stage of a rudimentary filament ($t \approx 1.0$ ns). However, once impurity breakdown sets in and a straight filament is growing, the electric field exhibits strong asymmetry between the opposite filament boundaries. This arises from the asymmetric lateral electric field, which may be conceived as a superposition of the dipole field due to the space charge already present in the filament walls at $B=0$, and the Hall field. On the filament border pointing in the direction of the Lorentz force, the diffusion of electrons out of the filament is enhanced. As a result more negative and positive space charges are generated by mobile electrons and by immobile ionized donors, respectively, on this border than for $B=0$, giving rise to a strong local lateral field. The high-field border is shifted by the Lorentz force inducing impact ionization in the region behind it. The lateral electric field serves as an attractive force compensating the outward Lorentz and diffusion forces, and if a balance holds between them, the filament border is stabilized under a transverse magnetic field.

On the other filament border, electrons diffusing outwards are impeded by the Lorentz force pointing into the opposite direction. Therefore less space charge and a weaker lateral field are created on this border than for $B=0$. This complementary action of lateral fields at the opposite boundaries can explain the asymmetric signals observed by laser scanning microscopy, where the signal of one filament border grows whereas the signal of the other border decays with increasing magnetic field [25]. We suggest that this asymmetric lateral field is also the source of the finite Hall voltage across the sample found in experiments [62] and it may play a crucial role in inducing chaotic oscillations [65].

This physical picture is consistent with the reasoning developed previously for a one-dimensional model [47] leading to transversally travelling filaments in the absence of a point-contact induced dipolar electric field. Fig. 6 shows simulations for triangular contacts on opposite sides of a rectangular sample [52]. For small magnetic fields the filament remains near the center of the sample, and exhibits asymmetrically breathing filament borders (a). For higher magnetic fields the breathing is transformed into regular transverse motion in the direction of the Lorentz force, where the filament travels some distance towards the sample boundary thereby becoming elongated, until it vanishes and a new filament is nucleated in the center (b). Small regular current and voltage oscillations arise due to the periodic nascence and destruction of traveling filaments. This motion becomes irregular and chaotic by a further increase of the magnetic field (c). Finally, at still higher magnetic fields, the filament gets pinned at the boundary after some transient traveling sequences (d). Such lateral excursions have been observed in p-Ge[66] and n-GaAs (see below).

2.3 Current filaments in Corbino disks

Next, we present simulations of a sample with circular contact symmetry (Corbino disk). The sample is confined by an outer ring electrode of radius 1050 μm (anode) and has a concentric inner point contact of 80 μm diameter (cathode) in the center of the disk. Such geometries offer the possibility to study pattern formation in samples without lateral boundaries, and are therefore of considerable interest. Various self-organized patterns of single and multiple current filaments have been observed experimentally in such samples [67].

In our simulation a constant voltage $U_0 = 2.4$ V is applied to the Corbino disk of thickness 3 μm via a load resistor $R = 10$ kΩ. Initially the sample is in the insulating state. Subsequently impact ionization multiplies the electron concentration at the injecting inner point contact and establishes a radially symmetric front that moves towards the circular anode (Fig. 7 (a)). At a certain radius the circular front breaks up into several modulated branches (b), each of which, upon reaching the anode, forms a rudimentary radial current filament (c). The symmetry of this pre-filamentary state, i.e. the number of rudimentary filaments, depends on sample parameters as well as on the applied voltage. Our simulations have shown that the system exhibits bistability: for equal initial conditions small numerical differences can lead to different symmetries of the pre-filamentary state. For the current and load conditions shown here, typically five or six filaments arise. In the final stage, all filaments except one are damped out, but due to the long recombination time scales they vanish only slowly (e). The surviving filament grows to a stationary, fully developed filament (f). This winner-takes-all dynamics is characteristic of a global coupling as imposed by the load resistance, and is familiar from general nonlinear self-organizing processes under global constraints [68]. For higher current densities, stable states with two, three,

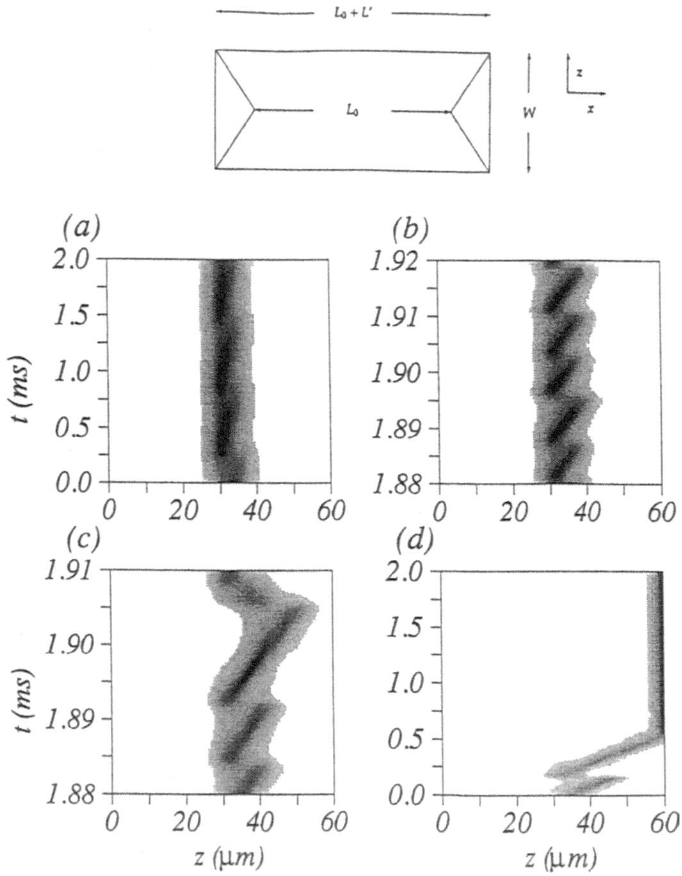

Fig. 6. Simulated lateral excursions of a filament in a transverse magnetic field for a sample with two triangular contacts (see inset). The carrier density is shown as a function of the transverse coordinate z and time t for (a) B=11 mT, (b) B=15 mT, (c) B=20 mT, (d) B=25 mT. (calculated with material parameters for p-Ge [52].)

or more, filaments are obtained, in good agreement with recent experiments, see Section 3 below (Figs.10,11,12).

The total current increases with time as shown in Fig. 8. The different stages of filament formation can be seen in the current–time characteristic in a way very similar to the case of a sample with point contacts. Also the time scales are similar.

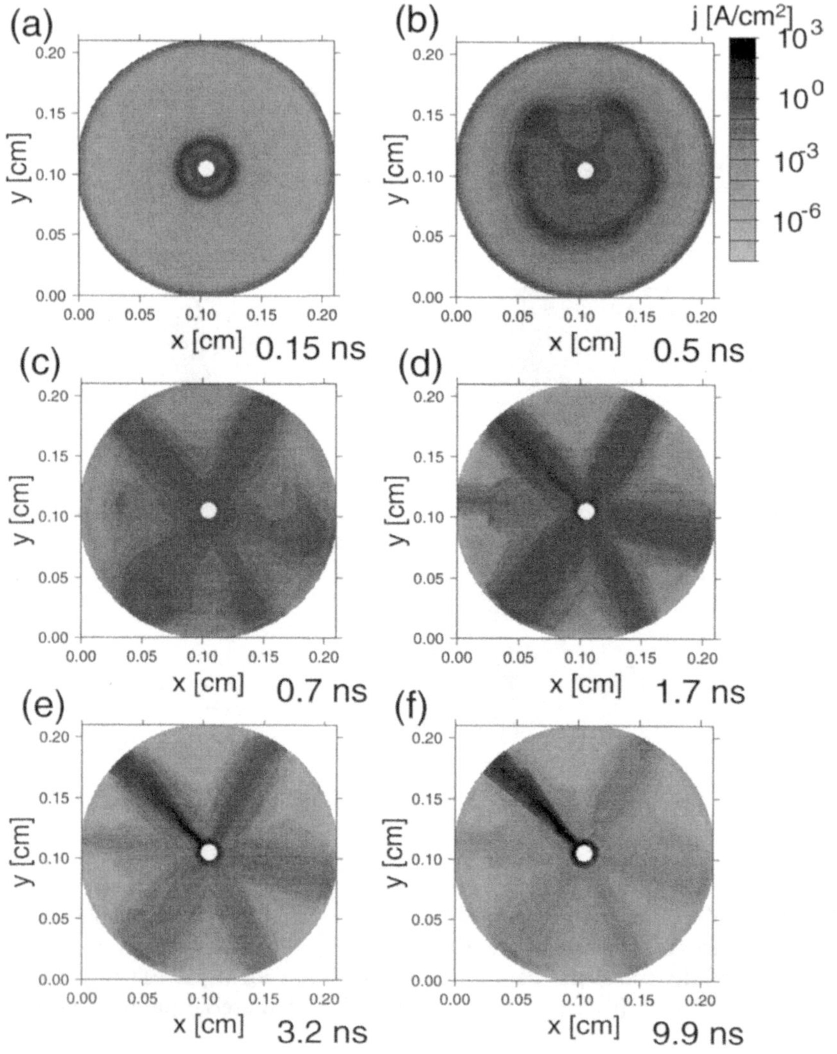

Fig. 7. Simulation of the formation of a current filament in a Corbino disk. The temporal evolution of the current density $|\mathbf{j}(x, y)|$ is shown as a function of the spatial coordinates x and y for a circular sample of diameter 1050 μm and an inner contact diameter of 80 μm. a) $t = 0.15$ ns, b) $t = 0.5$ ns, c) $t = 0.7$ ns, d) $t = 1.7$ ns e) $t = 3.2$ ns f) $t = 9.9$ ns. The applied voltage $U = 2.4$ V and the load resistance $R = 10$ kΩ correspond to an operating point with an average field $\mathcal{E}_{av} = 4.7$ V/cm, i.e., between the holding field $\mathcal{E}_h = 4.0$ V/cm and the threshold field $\mathcal{E}_{th} = 17$ V/cm of the homogeneous $n(\mathcal{E})$ characteristic (calculated with the same parameters as Fig. 1). After [69].

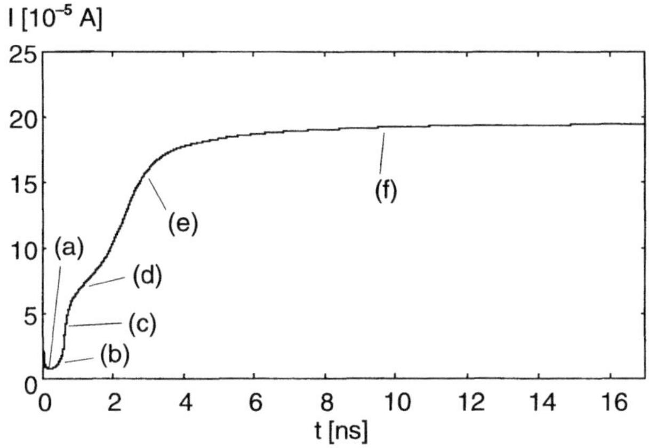

Fig. 8. Calculated total current $I(t)$ during the formation of a filament in a Corbino disk. The letters (a)–(f) correspond to the stages of filament formation shown in Fig. 7(a)–(f).

2.4 Spatio-temporal spiking

Finally, we mention another type of spatio-temporal pattern, which is associated with the repeated nascence and subsequent disappearance of current filaments, and which has been observed in computer simulations of various model systems: *Spatio-temporal spiking* of the current density distribution has been found in the heterostructure hot electron diode (HHED) [70] which exhibits an S-shaped current- voltage characteristic for vertical transport across a GaAs/AlGaAs heterojunction [71]. Similar spiking behaviour, including transient spatio-temporal chaos, occurs in a simple reaction-diffusion model of the type (1), (2) [72, 73] which holds more generally for a variety of layered semiconductor structures [19, 74, 75, 76]. Such behaviour was shown to arise generically near a codimension-two Turing-Hopf bifurcation [77, 78]. The effect of global coupling due to external circuits has been studied both analytically and numerically in these reaction-diffusion systems on arbitrary two-dimensional spatial domains with Neumann boundary conditions [79]. It was proved generally for the voltage-driven regime that in a convex domain any filament has at least one unstable linear eigenmode. Introducing a global coupling may either eliminate the unstable mode with the largest increment or induce oscillatory instabilities. Filaments with negative differential conductance can be stabilized by sufficiently strong global coupling, i.e. a sufficiently large load resistance. It has been proved under rather general conditions that Neumann boundaries attract filaments, in agreement with numerical results obtained for other reaction–diffusion systems [80], and competition between pinning of current filaments by seed inhomogeneities and attracting boundaries has been demonstrated. Such reaction–diffusion systems will be studied in more detail in Section 4.

3 Visualization of current filaments in n-GaAs

In the past several methods have been developed to reconstruct the spatial patterns of current distribution in semiconductors under non-equilibrium conditions at low temperatures. The first experimental proof of current filamentation in an extrinsic semiconductor was performed by Milnes et al. [81] applying a junction field effect transistor of cylindrical symmetry. A ring shaped pn-junction was formed on a cylindrical bar of n-Ge. Drain-source current-voltage characteristics for various biasing conditions allowed the authors to conclude the formation of a current filament.

In planar samples high-resolution images of impurity breakdown induced current filaments were obtained by using a scanning electron microscope [82, 23, 83]. The effect of an external magnetic field on current filaments has been investigated by a scanning laser microscope [25, 26]. In both cases the surface of a thin sample is scanned line by line by a focused electron or laser beam, respectively, and the current response is synchronously recorded as a function of the focal position. Irradiation destabilizes the current flow at the filament borders yielding structures in the spatially resolved response which indicate the extent of the current flow in the sample.

Scanning techniques are basically sequential and introduce a spatially inhomogeneous perturbation in the sample by the focused electron or laser beam which itself may change the spatial distribution of the current flow. A less invasive method to visualize impurity breakdown induced current filaments makes use of the quenching of photoluminescence. With spatially and temporally homogeneous luminescence excitation the symmetry of the system is not externally perturbed. Until now this method has been applied to n-GaAs because of the experimentally convenient wavelengths of photoluminescence in the near infrared which virtually allows to photograph filamentary structures [29]. For low excitation intensity photoluminescence consists of two narrow bands at $\lambda = 830$ nm and 820 nm caused by donor-acceptor pair recombination and radiative exciton extinction, respectively. The relative strength of these bands depends on the purity of the material. In very pure samples only luminescence due to excitons is observed whereas in higher doped material both bands are present. Inside a current-density filament donor states are depleted and excitons dissociate due to impact ionization. This leads to a reduced rate of radiative transitions [84, 85, 86] and, consequently, filaments appear as black stripes in an image of the photoluminescence distribution.

Luminescence at these wavelengths may easily be detected by Si-photodiodes or high sensitivity infrared video cameras. Thus, this technique allows rapid imaging of filamentary current structures. Additionally, using single photodiodes one can measure the local development of the electron density as a function of time which allows to determine filament movements in space.

3.1 Experimental techniques

Here we review experimental investigations of current density filaments formed by the generation-recombination instability discussed in section 2 in thin n-GaAs epitaxial layers. Nucleation of filaments and filament division scenarios [67, 87] will be shown in semiconductor films with contact geometries of different symmetries like pairs of point contacts, Corbino disks, concentric ring contacts and parallel stripe contacts. The effect of a magnetic field on filamentary structures will be demonstrated.

Several n-GaAs epitaxial layers of low doping level with various contact geometries have been investigated. The semiconductor films of about $3\,\mu$m thickness were grown by molecular beam epitaxy and doped either homogeneously or by a sequence of 60 Si δ-layers 50 nm apart. This distance between the layers is approximately the average dopant spacing within a δ-layer. These samples are thus practically homogeneously doped. The donor concentration of all samples reported here is in the range of $N_D \approx 5 \cdot 10^{15}$ cm^{-3} at 20% compensation, yielding a mobility at 77 K of about $4 \cdot 10^4$ Vs/cm^2. The samples were structured by photolithography and ohmic contacts were prepared by alloying AuGe. Point contacts and the center contacts of Corbino disks have a diameter of $80\,\mu$m.

Luminescence has been excited by the diffuse radiation of ten red light emitting diodes (emission peak at $\lambda \approx 630$ nm) fixed close to the sample inside the cryostat. The luminescent surface of the epitaxial layers has been imaged onto an image intensifier that includes an S25-photocathode and operates with video frequency. The exposure time of a single picture has been 20 ms. The spatial resolution is $10\,\mu$m $\times$ $10\,\mu$m which is substantially better than the previous technique using a one-dimensional diode-array scanning the two-dimensional luminescence-distribution in times of minutes [29]. An interference filter of wavelength $\lambda = 830$ nm ($\Delta\lambda = 10$ nm) is used to select the dominant donor- acceptor transitions and to block visible excitation light. The samples have been fixed in an optical cryostat in the center of a superconducting magnet and have been cooled to 1.8 K. A magnetic field, applied normally to the plane of the epitaxial layer by a superconducting solenoid, has been used to investigate lateral excursions of filaments. The samples were biased in series with load resistors by a high-precision adjustable constant-voltage source. All samples exhibit the well known nonlinear and hysteretic I-V characteristics and show self-sustained oscillations at the threshold of impact ionization breakdown [54]. Without an external magnetic field, these oscillations are periodic and may be attributed to repetitive ignition and extinction of a current filament([61, 88]. With increasing average current the oscillations vanish due to the formation of a stable current filament. All measurement have been carried out in this regime of the current-voltage characteristics.

Application of a magnetic field normal to the plane of the semiconductor layer destabilizes the filamentary current flow and leads to current and voltage oscillations which may be periodic but also aperiodic and even chaotic [62, 65].) In the case of periodic oscillations the trace of the filament motion can be obtained by sequential imaging of spots of the sample surface on a small photodiode. Then the signal of the photodiode is measured as a function of time and space with respect to the phase of the periodic sample voltage oscillations. Stroboscopic excitation of photoluminescence has been applied to samples with concentric circular contacts revealing filaments rotating in an external magnetic field in the direction of the Lorentz force.

3.2 Current filaments for various contact geometries

In Fig. 9 images of a rectangular sample with two point contacts at opposite edges are displayed for increasing current from top left to bottom right. At a

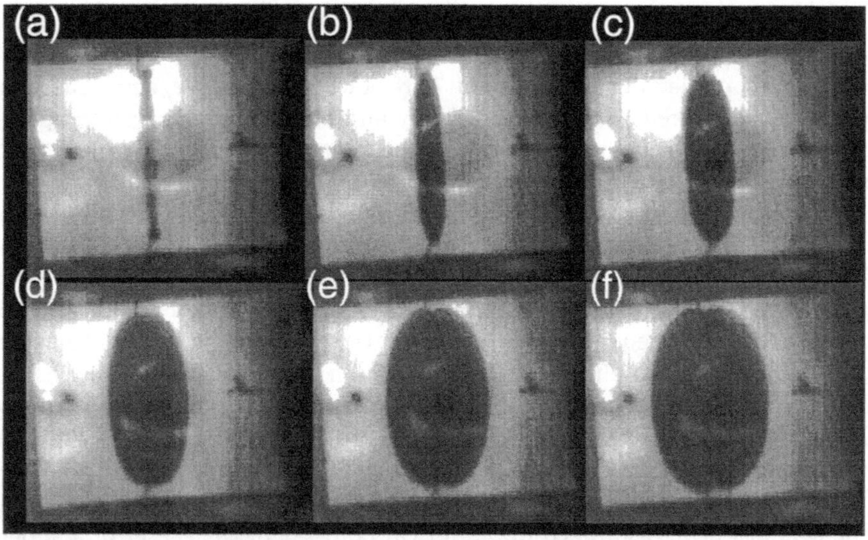

Fig. 9. Spatially resolved photoluminescence of a sample with two point contacts for various currents. The distance between the contacts is 4 mm. Dark areas in the images correspond to regions of high electron density due to quenching photoluminescence. Current rises from 0.4 mA (a) to 11.1 mA (b).

critical value of the dc-bias a thin filament of about 100 μm width is formed which increases in width with rising current. At formation and for small currents the borders of the filament are parallel in spite of the fact that the electric field between two point contacts is inhomogeneous. This seems to be a general feature of self-organized current filaments. Such a shape has been reproduced recently by Monte-Carlo simulations of filament formation carried out for the first time in two dimensions [43], cf. Figs.1 and 4. At higher currents (lower row) the filament borders tend to follow the dipole-like electric field distribution. Thus a transition takes place from a self-organized filament to a structure whose shape is controlled by an externally imposed inhomogeneous distribution of the electric field [50, 89]. This scenario is in good agreement with two-dimensional calculations based on a simplified model which takes into account a two-valued conductivity without considering its microscopic origin [90].

In Fig. 10 the nucleation of current filaments in a Corbino disk is displayed for various currents. Below breakdown a homogeneous luminescence distribution is seen indicating that the whole sample is in a homogeneous state of low carrier concentration. With rising current the uniform state bifurcates into a localized current-density filament appearing at a random position (Fig. 10 (a)). Figs. 10 (b) and (c) show recordings of the luminescence distribution for larger values of the current. Again the filaments have parallel borders. No broadening of the filament close to the outer ring electrode is observed, which would be expected if the electric field decreases with increasing radial distance. This indicates that self-organized filament formation results in large redistributions of space charges to enforce a filament with practically constant width along the radial direction. Again this is in good agreement with recent simulations as shown in Fig. 7.

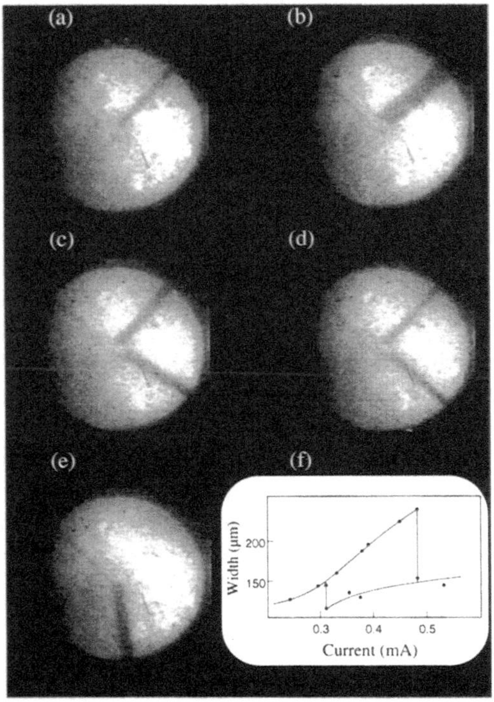

Fig. 10. (a)-(e) Corbino disk of 8 mm diameter of the outer ring contact. Recordings of photoluminescence distributions demonstrating filament splitting. Sample currents $I = 0.19, 0.48, 0.53, 0.36, 0.28$ mA for (a) to (e), respectively. (f) Width of a single filament as a function of current demonstrating a subcritical bifurcation.

At a further increase of the dc-bias the filament approximately doubles its width (Fig. 10 (b)) and a bifurcation from one filament to two filaments (Fig. 10 (c)) occurs. Typically a new filament is separated from the original one at an angular distance in the range of 25° to 90°. Larger angles than 90° have not been observed. This indicates the existence of an interaction between filaments, which force nucleation of additional filaments in preferred angular distances. As a result the filament division process depending on the current is not continuous. Several rearrangements of the angular distance occur in jumps much faster than the temporal resolution of the present measurement. Between these jumps the distance between the filaments is stable. Thus there is no continuous relation between the distance of filaments and the current. Rearrangement of the angular distance does not change the width of the filaments.

Figs. 10 (d) and (e) show recordings of the photoluminescence distribution for decreasing current. The two filaments are sustained at much lower currents than the critical current $I = 0.53$ mA at which they were formed. Both filaments reduce their widths simultaneously down to a minimum width of about 120 μm at a sample current $I = 0.28$ mA, at which one of them vanishes and the other one is broadened. The entire bifurcation cycle just described, which includes the broadening of a filament and its splitting into two filaments with increasing dc-bias and the simultaneous narrowing of filaments finally resulting in the extinction of one of the filaments with decreasing dc-bias, is subcritical and yields the hysteresis shown in Fig. 10 (f).

The subsequent generation of additional filaments at increasing dc- bias occurs in the same way. With increasing current all filaments grow in width and in most cases suddenly one of the filaments splits resulting in two filaments of smaller width. It is not predictable which filament is divided. After spontaneous rearrangements of the pattern the filaments have approximately the same widths. Fig. 11 shows some exemplary recordings of a bifurcation sequence. The first two images (Figs. 11 (a,b)) have been taken before and after the splitting of two filaments. A spontaneous rearrangement and further splitting of a filament lead to the distribution shown in Fig. 11 (c). The angular distribution of the filaments gets more and more symmetric for six and more filaments (see Figs. 11 (d-f)). For these cases a large region around the central contact with strongly reduced luminescence appears. This indicates an enhanced density of free electrons in the vicinity of the central contact. This region acts as a large central contact for further filament generations. Decreasing the dc-bias when many filaments are present reveals a very wide regime of hysteresis. Most of the filaments survive up to relatively low values of the sample current, usually much lower than the corresponding current at which they have been generated on increasing the dc-bias.

The creation of a new filament or a rearrangement of filaments is usually connected with a slight jump in the sample voltage (< 30 mV). As the details of filament nucleation and rearrangements are not predictable, the current-

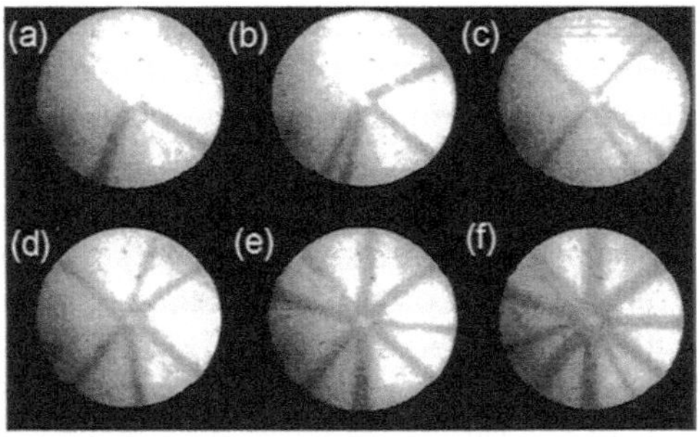

Fig. 11. Photoluminescence images showing successive generation of filaments by rising current from 0.32 mA (a) to 2.93 mA (f).

voltage characteristic may also differ when the experiment is repeated and thus is not predictable either.

The obtained images provide evidence for different dynamics controlling the formation of filamentary structures in the two contact geometries studied above. In point contact samples the filamentary current flow may be satisfactorily explained by assuming an S-shaped local conductivity-electric field characteristic of the material [1, 50]. Above the threshold field value a region of high conductivity is formed, which retains the external symmetry of the contact geometry. The sharp contrast of the luminescence image at the filament borders corresponds to a separation between the two stable phases. Numerical simulations of the dynamics of filament formation show that two nascent filaments are formed at both contacts which continuously grow until they merge in the center of the sample [56], cf. Fig. 1.

On the other hand, current filamentation in Corbino disks is not self-evident due to the circular symmetry of the electric field below breakdown and the decay of the field strength $\sim 1/r$ with the radius r. The electric field reaches its threshold value of instability simultaneously at all angular positions on the outer ring contact. According to the previous scenario this would yield a transition of the total sample into the high-conducting phase which conserves the circular symmetry of the electron distribution. This is obviously not the case, as symmetry is broken by the formation of a filament. The formation of additional filaments repeats this symmetry breaking. After the formation of six filaments the arrangement becomes more regular but is still of reduced symmetry. Thus, the observed filament formation in the Corbino geometry is a convincing proof of self-organization. Recent dynamical simulations of the nucleation of filaments in a Corbino disk [69] have shown that around the center electrode a circular area of elevated free carrier density is formed which grows in diameter. Out of this area at first several nascent filaments are created from fluctuations of the circumference. Then these rudimentary filaments proceed towards the ring contact, until finally one of them takes over the total current and survives, as shown in Fig. 7.

Photoluminescence patterns of the Corbino disk in comparison with a sample with concentric ring electrodes are shown in Fig. 12 (a-d) and (e-h), respectively. In the ring electrode structure basically the same bifurcation and filament splitting scenarios are observed as in a Corbino disk with a comparably very small central contact. The only difference is that between ring electrodes the width of filaments may be larger before splitting. At high currents the pattern of the Corbino disk shows a large region around the central contact with reduced luminescence (Fig. 12 (d)) and a large number of filaments. This indicates the existence of a circular area with enhanced electron density near the central contact. The diameter of the central high electron density region grows linearly with rising current. This means that the current density j at the edge of the circular domain is independent of the current being in the present sample $j \approx 65\mathrm{A/cm^2}$. The filament patterns are similar to those between ring contacts (Fig. 12 (h)) and single filaments are wider than in the case of lower current.

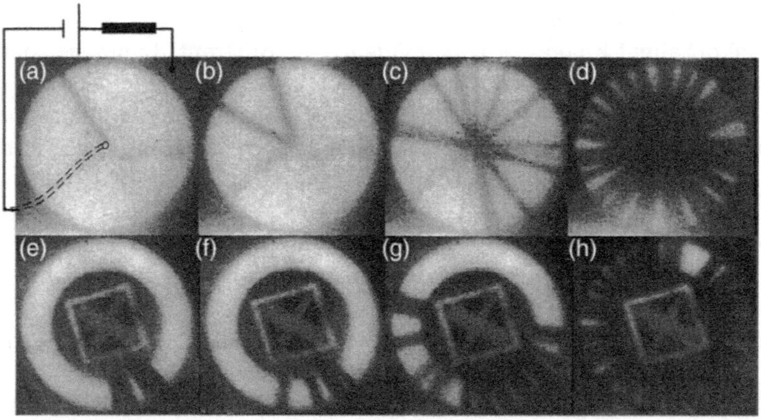

Fig. 12. Current filaments with rising current in a Corbino disk (a–d) and a sample with concentric ring electrodes (e–h). Diameter of the inner and outer electrodes: Corbino disk 80 μm and 2.1 mm, ring electrode sample 2 mm and 3 mm, respectively. Current rises from 0.17 mA (a) to 7.1 mA (d) and 1.5 mA (e) to 13.0 mA (h).

Finally, in Fig. 13 a sequence of images of filaments between two parallel stripe contacts are shown. Again the current increases from top left to bottom right. In this contact geometry filament division seems not to be the generic mechanism of carrying higher currents. Current filaments of any width have been observed. Filament splitting occurs, too, but may be attributed to imperfection of the contacts.

3.3 Lateral excursions of filaments in a magnetic field

In this section we present results on lateral motion of filaments in magnetic fields which lead to oscillations of the sample voltage and of the current

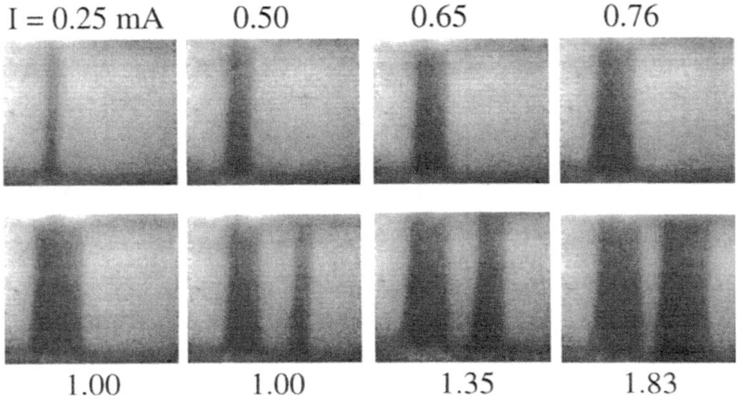

Fig. 13. Sequence of current filaments nucleating between parallel stripe contacts

through the sample. The discussion will be limited to thin filaments as they appear just after nucleation in samples with point contacts or are present in Corbino disks.

In samples with two point contacts a previously stable filamentary current flow is destabilized by the magnetic field leading to characteristic oscillations which are only observed with this contact geometry. The amplitude of these oscillation is about the same as that of relaxation oscillations but the frequency decreases with increasing current in contrast to increasing oscillation frequencies in the relaxation regime.

Luminescence images at constant magnetic field strength reveal that at lower currents the filament is laterally bent and is more and more stretched with increasing sample current(see Fig. 14 (b) and (c)), which is in agreement with previous investigations [26]. At sample currents in the oscillating regime no stable filament reconstruction is possible. Fig. 14 (a) shows the luminescence distribution averaged over many ($\approx 10^3$) electrical oscillations. A closer inspection of this image makes visible a half-moon like structure on the top half of the sample with slightly reduced photoluminescence intensity. This result indicates that a region of high electron concentration is oscillating over one half of the sample, in direction of the Lorentz force acting on free electrons. To get more detailed insight in the filament behavior in this unstable biasing regime the temporal evolution of the luminescence intensity at different positions on the sample surface was recorded with a single avalanche diode with respect to the phase of the magnetic field induced periodic voltage oscillation. The results are summarized in Fig. 15. In the top plate of this figure the oscillation of the sample voltage is displayed for five periods. Below the plate showing voltage oscillations a waterfall plot of the temporal evolution of the photoluminescence is displayed for different positions of the detector. The detector has been shifted along the perpendicular bisector to the connecting line of the two contacts. The range of detector positions

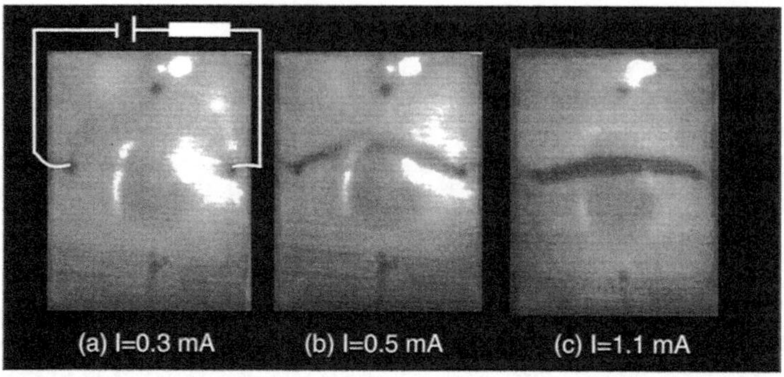

Fig. 14. Static photo-luminescence distribution at an applied magnetic field of 150 mT for three different average currents.

is indicated by the double arrow on an averaged luminescence image of the sample in the right plate of Fig. 15. The waterfall plot and the sample image are arranged in such a way that the mean of each luminescence trace corresponds to the location of the detector on the double arrow. From trace to trace the center of the detector is moved by 0.2 mm with respect to the sample geometry, detecting the luminescence of a spot of diameter 0.6 mm. The two black spots at top and bottom of the luminescence image are two unused contacts. The temporal evolution of the luminescence is strongly correlated to the oscillation of the sample voltage. Looking at the luminescence signal at the center of the sample, a drop of the luminescence is observed shortly after the drop of the sample voltage. When the detector is shifted in direction of the Lorentz force acting on free electrons (upward in the luminescence image), the drop in the luminescence signals occurs with increasing delay in time. Near the border of the sample no reduction of the luminescence can be detected.

The observed luminescence behavior clearly reveals the spatio-temporal dynamical behavior of a filament in crossed electric and magnetic fields. With the drop of the sample-voltage a filament is nucleated along a line between the contacts. Then the filament starts to move in direction of the Lorentz force. The reduction in the luminescence signal occurs with a delay increasing with distance from the connecting line. From the minima of the luminescence signal a lateral filament velocity in the order of $30\,\mathrm{ms}^{-1}$ has been estimated. During the lateral movement of the filament the sample voltage increases due to the lengthening of the filament and the corresponding rise of the resistance. Reaching the threshold voltage, the bent filament vanishes and a new one nucleates along the connecting line between the contacts. Similar lateral excursion and nucleation behavior have been obtained in computer simulations within a simplified model for samples with triangular contacts [47], see Fig. 6.

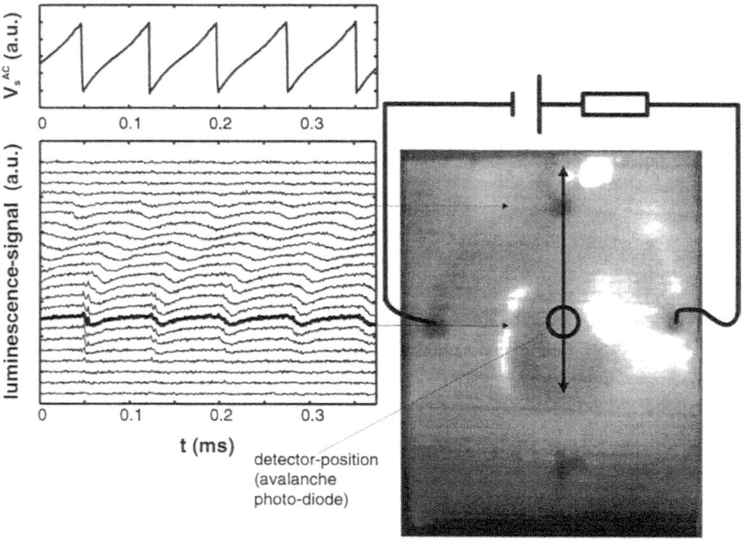

Fig. 15. Upper left plate: Sample voltage as a function of time. Lower left plate: Waterfall plot of the luminescence-signal as a function of time at different detector positions for the same time scale as the sample voltage. Right plate: Averaged luminescence picture. The different detector positions are indicated by the double arrow. The waterfall plot and the image of the sample are arranged so that the average of each luminescence trace corresponds to the location of the detector on the sample image.

In samples with circular symmetric contact arrangements rotating filament patterns and lateral movements of single filaments over finite angular distances are observed. Rotating filaments cause a strictly periodic oscillation of the sample voltage but of complex shape [91], see Fig. 17 (a)). The amplitude of these oscillations is very small being of the order of 10 mV compared to about 1 V dc bias. In contrast to the relaxation oscillations, these oscillations are intrinsic and do not depend on parasitic capacitances of the load circuit.

In Fig. 16 a set of images for various magnetic field strengths is shown, obtained from the Corbino disk (top row) and the ring electrode sample (bottom row). With rising magnetic field strength the filaments are bent in direction of the Lorentz force (Fig. 16 (a,b and e,f)). In higher magnetic fields the filament images are blurred and finally disappear leaving a homogeneous darkening of the whole or part of the sample luminescence (Fig. 16 (c,d and g,h)). In the darkened regions filament patterns rotate or move along a finite angular sector. This has been shown by tracing the photoluminescence with a single photodiode like in the above case of lateral filament vibrations in point contact samples. Additionally the rotation of filaments could be demon-

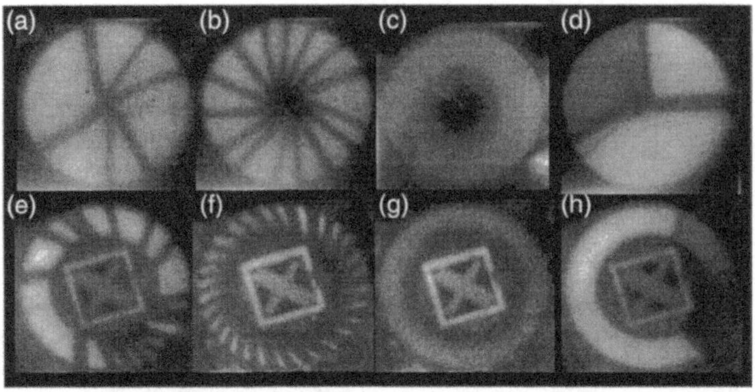

Fig. 16. Effect of magnetic field on filament patterns in circular symmetric contact geometries.

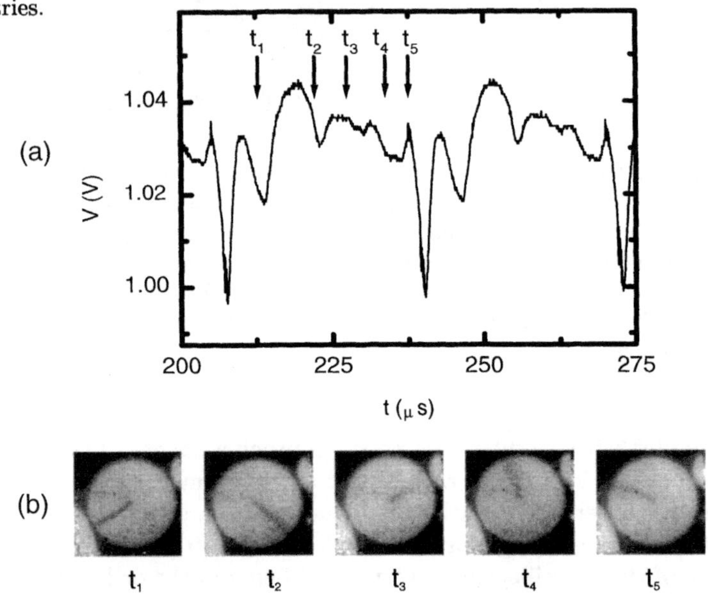

Fig. 17. Stroboscopic reconstruction of a rotating filament. (a) Periodic voltage oscillation due to a rotating filament. (b) Images taken at times t_i indicated in (a). $B = 147\,mT, I = 222\,\mu A$

strated by stroboscopic illumination of the sample triggered by the voltage oscillations. This is demonstrated for a single filament in Fig. 17.

The upper plate shows the periodic voltage oscillations of complex shape. Inverting the direction of the magnetic field or the polarity of the dc-bias reverses the shape of the oscillation and the rotations of the filament with respect to the time axis. The frequency of rotation increases with both the average sample current and the magnetic field strength, being in the range of a few kHz to hundred kHz. To induce moving filament patterns a minimum magnetic field strength is needed indicating that a pinning force acts on filaments.

In a perfectly homogeneous and symmetric sample no oscillation of the voltage drop across the sample is expected from rotating filaments. Thus, the observed oscillations of the sample voltage indicate imperfections like variations in the samples conductance at the momentary position of a current filament. This might be due to imperfections of the material and inhomogeneities of the contacts.

In case a sector between two previously stable filaments is darkened (see Fig. 16 (d) and (h)) it can be shown that one filament splits and the newly formed filament moves in direction of the Lorentz force until it approaches the second filament and merges with it. This process is periodically repeated leading to the blurred images which represent averages over many periods.

4 Pattern formation in silicon p^+-n^+-p-n^- devices

Silicon p^+-n^+-p-n^- devices represent an exemplary system of activator-inhibitor type [18]. They can be considered as a two-layer system with the particular feature that in one of the layers an autocatalytic charge carrier multiplication may occur, whereas the second layer provides an inhibiting mechanism. Note that two-layer arrangements have also been proposed for *p-i-n* diodes [14, 93], heterostructure hot electron diodes [70] and thyristors [94, 95]. In silicon p^+-n^+-p-n^- devices several generic oscillation modes of filaments can arise due to the competition between the activating and the inhibiting mechanism. Besides periodically oscillating and traveling filaments [28, 18, 96, 24] - both representing fundamental generic oscillation modes of filaments - irregular [97, 76] and quasiperiodic [8, 98] filament motions have been observed in experiments and also found in numerical solutions of the activator-inhibitor equations derived from the two-layer model [8, 99]. In what follows, we shortly describe the experimental set-up and the physical mechanisms governing the self-organized pattern formation and the observed filament dynamics in the p^+-n^+-p-n^- device. In the main part, some generic patterns and bifurcations observed in experiment are compared with numerically calculated solutions of the mentioned reaction-diffusion system.

4.1 Experimental details

When the p^+-n^+-p-n^- device is connected to a dc-bias via a load resistor (Fig. 18(a)) such that the p^+ layer is positively biased with respect to the n^- layer, the n^+-p junction is reverse-biased leading to a low-conductivity branch in the current-voltage characteristic $I(V)$ (Fig. 18(b)) at low sample currents. For a sufficiently large voltage drop across the sample, avalanche multiplication of charge carriers in the space charge region of that junction takes place by interband impact ionization. The generated electron-hole pairs are separated in the high field zone of the junction so that the electrons and holes move towards the p^+-n^+ junction and the p-n^- junction, respectively. Those carriers reaching the outer junctions induce an additional carrier injection there, leading to a further increase of carriers and causing the appearance of a branch with negative differential resistance in the $I(V)$-characteristic.

In order to characterize the spatio-temporal behavior of current-density filaments different measurement techniques have been applied. Here we refer to measurements of the electroluminescence and potential distributions. Additionally, probe measurements using an electron [24] and a laser beam [27] have been performed yielding information about filamentary states being not discussed in this contribution.

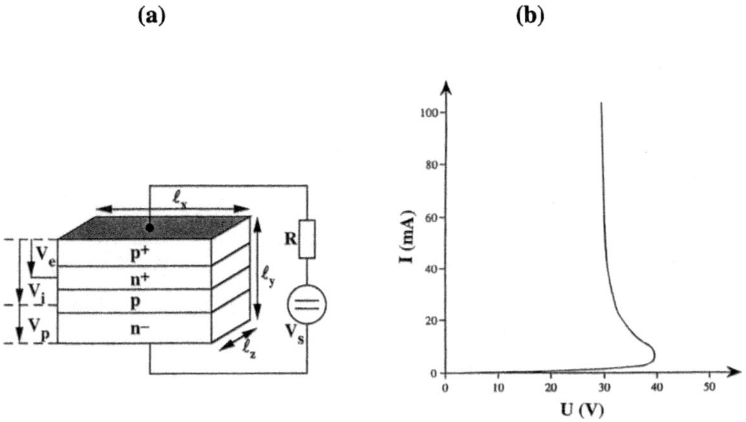

Fig. 18. (a) Experimental set-up. Typical dimensions of the sample are $l_x = 10$ mm, $l_y = 1$ mm, $l_z = 1$ mm. (b) Typical $I(V)$-characteristic.

The electroluminescence radiation emitted from one of the polished l_x-l_y surfaces of the sample [Fig. 18(a)] has been recorded by using a streak camera (Hamamatsu C1587) with an infrared sensitive S1 photocathode. Since the radiation is most effective near the p-n junctions a small stripe parallel to the anode comprising the p^+-n^+-p structure of the sample was focussed on the photocathode of the streak camera. Note that the electroluminescence distribution can be considered to reflect the current-density distribution.

The potential distribution has been measured on the l_x-l_y surface (Fig. 18(a)) between the two aluminium electrodes with a potential probe (Wentworth, Picoprobe 18B). The low input capacity of the probe of about 0.02 pF together with the contact resistance leads to a time resolution of about 1 μs. The probe is controlled by a micro-positioning system so that the potential can be measured with a spatial resolution of about 1 μm. All potential distributions presented here have been obtained by scanning the potential on a line parallel to the contacts in the n^- layer close to the p-n^- junction (cf. Fig. 18(a)).

4.2 Relations to activator-inhibitor systems

For a simple description of the device one may consider the positive feedback loop caused mainly by the injection of holes from the p^+-n^+ junction. Then, it is useful to imagine the device being composed by two parts, a p^+-n^+-p transistor and a p-n^- diode [18]. In order to get insight into the relevant physical processes let us consider the two device parts in more detail. Suppose that the voltage drop across the n^+-p junction is sufficiently large so that charge carrier multiplication takes place. Then, a small fluctuation of the voltage V_e of the p^+-n^+ emitter junction has the following consequences: Due to the essentially exponential dependence of the emitter current density on the emitter voltage, a fluctuation of V_e causes an additional injection of holes into the n^+ layer. Those holes which reach the high-field zone of the n^+-p junction generate electron-hole pairs which are separated due to the high electric field. The electrons move towards the p^+-n^+ junction and induce an additional injection of holes and, consequently, an increase of V_e. Because of this activating property of the transistor part, the emitter voltage V_e may be called an activator.

The coupling of the transistor part to the p-n^- diode leads to a counter-action to the autocatalytic increase of V_e: The current fluctuation caused by the fluctuation of V_e leads to an increased hole injection into the n^- layer and therefore to an augmentation of both the mean hole concentration p in the n^- layer and the voltage drop $V_p \propto \ln p$ across the p-n^- diode. Keeping the device voltage V constant means that the voltage drop $V_i = V - V_p$ across the transistor decreases if V_p increases. This inhibiting process in the p-n^- diode limits the autocatalytic process and, consequently, p can be viewed as inhibiting variable.

4.3 Results

The dc-bias is the most important parameter, because the sample is driven more and more away from the thermodynamical equilibrium state when the bias is increased. In what follows we compare the bifurcation sequences observed experimentally by varying the bias with numerical results obtained by solving the two-component set of reaction-diffusion equations that can be derived by treating the p^+-n^+-p transistor as avalanche transistor and the p-n^- diode as Shockley-diode. For details concerning the model and the derived equations see also Refs. [18, 99].

For low sample currents, the electroluminescence and, consequently, the current-density distributions as well as the potential distributions are uniform. This is illustrated in Fig. 19 for the measurements and the numerical

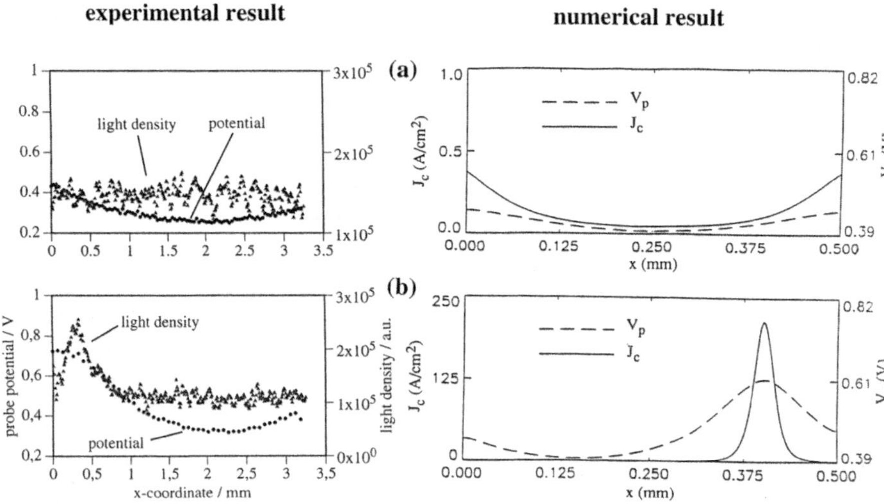

Fig. 19. Potential and electroluminescence/current-density distributions revealing the uniform low-current state (a) and a localized filament (b); experimental results are in the left, numerical results in the right column (from Ref. [96]).

solutions, respectively. In the numerical calculations Neumann boundary conditions have been used for the inhibitor, while Dirichlet boundary conditions have been applied to the activator; the latter causes the nonuniformities near the boundary in the numerically computed distributions. When the dc-bias is increased and avalanche multiplication occurs, a localized current-density filament evolves via a saddle-node bifurcation (Fig. 19). The bifurcation from the uniform state into the stationary localized filament can be evaluated quantitatively by plotting the widths of the measured and calculated distributions versus the sample current (Fig. 20). Obviously, there is a satisfactory correspondence between experimental and numerical results. The measured as well as the computed results reveal that the potential distributions are broader in comparison to the electroluminescence and the current-density distribution. This is due to the large carrier lifetime of about 100 μs in the n^- layer leading to a large diffusion length of the carriers there, which in turn causes the broad potential distribution.

The static filament is stable up to a certain threshold current. By exceeding this current the static filament becomes unstable in favor of a filament performing pendulum-like oscillations (Fig. 21(a)) via a supercritical Hopf bifurcation. These rocking filament oscillations are typically accompanied by oscillations of the sample current. They are caused by deformations the filament suffers near the turning points. In the experiment both the spatial filament oscillation and the current oscillation undergo a period-doubling cascade and finally result in a chaotic oscillation [97] with increasing sample

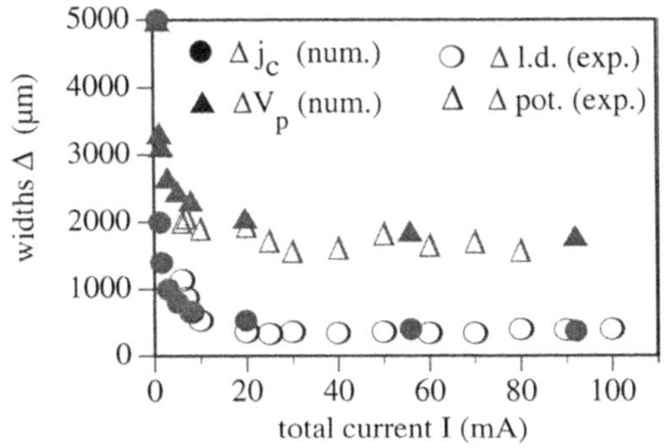

Fig. 20. Comparison of the experimentally measured and numerically calculated widths of the potential and the electroluminescence/current-density distributions (from Ref. [96]).

current. For even larger sample currents the rocking filament oscillation is transformed into a traveling filament motion (Fig. 21(b)). The numerically calculated scenario differs from the experimentally observed one in that the periodic rocking filament motion is transformed directly into a traveling filament (Fig. 21(b)). However, when Neumann boundary conditions are applied to both variables, the oscillating filament also evolves into a traveling filament via a period-doubling cascade and chaotic oscillations with increasing sample current [96]. We note that the experimentally observed velocities of the traveling filaments are in the range between 1 and 10 m/s while the calculated velocities of 30 m/s are slightly larger.

In Fig. 22(a) and (b) measured and computed potential lines are compared with linescans of the electroluminescence distribution and the current-density distribution, respectively, for a traveling filament. Both the experimental and the numerical results reveal that the potential distribution is broader than the current-density distribution. There is a characteristic precursor of the potential distribution running in front of the filament and an extended potential tail behind the filament. The latter is caused by the relatively large carrier life-time of about 100 μs in the n^- layer. The precursor together with the boundary conditions governs the reflection process of the filament near the sample border [18, 96]. Note that this traveling motion is different from that shown in Fig. 6, because there an additional magnetic field is necessary to induce a traveling filament motion. However, traveling filaments have also been obtained in simulations of impurity breakdown without magnetic field [46], indicating that this motion represents a generic filament mode.

Since the self-organized pendulum-like filament motion as well as the traveling filament motion are intrinsic oscillation modes of the filament and their

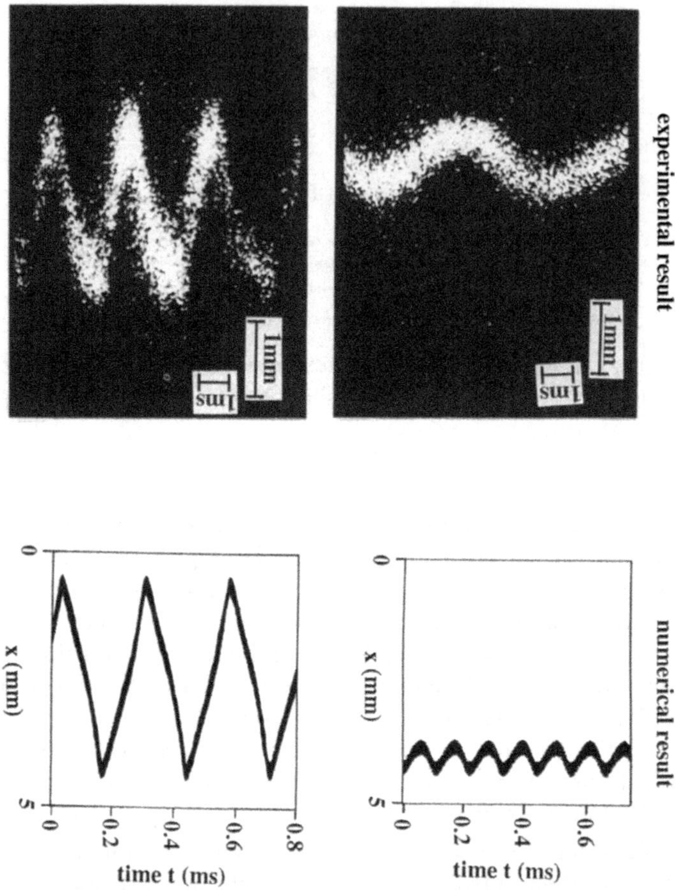

Fig. 21. Streak camera records (left) indicating a pendulum-like filament oscillation (upper image) and a traveling filament (lower image) and corresponding calculated filament motions (from Ref. [96]).

spatio-temporal behavior is experimentally accessible through measurement of the emitted electroluminescence radiation and the surface potential, they offer an excellent opportunity to study the influence of an external periodic drive on these basic modes of filament oscillation. In Ref. [98] it has been shown in detail for pendulum-like filament oscillations that depending on the frequency and the amplitude of the external drive frequency-locked and quasiperiodic filament oscillations or even bifurcations into chaotic filament motions can be observed when a sinusoidal ac voltage is superimposed on the dc-bias. Figure 23 shows the time series of the filament center constructed from a streak camera record and corresponding numerical calculations for a frequency-locked filament motion. The time series of the driving ac voltage is also shown for comparison. In both cases it can be clearly seen that the fila-

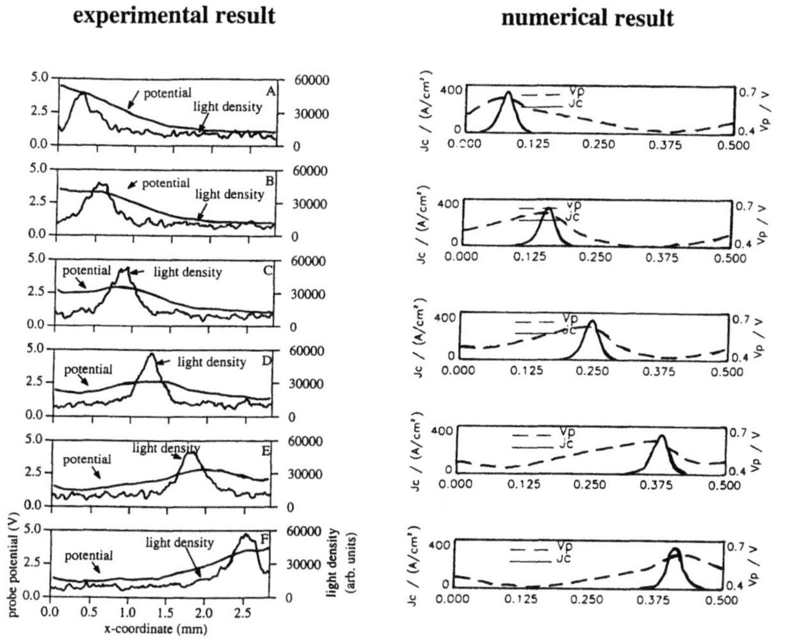

Fig. 22. Potential and electroluminescence/current-density profiles of a traveling filament at different times; experimental results (from Ref. [99]) and numerical results are shown in the left and right column, respectively.

ment performs two fundamental oscillations during three periods of the drive giving evidence of a 2:3 locking ratio of the filament oscillation with respect to the drive. It is remarkable that the filament adapts its spatial amplitude as well as its local velocity in order to achieve the locking state.

In Fig. 24 the results of a systematic variation of the drive frequency and amplitude are summarized in that experimentally measured locking ratios are marked. Obviously, equal locking ratios order in the shape of tongues as expected from the standard circle map. In between the tongues the filament performs pendulum-like quasiperiodic oscillations as has been shown by spatially and temporally resolved measurements of the potential [98]. For sufficiently large driving amplitudes the tongues overlap, i. e., there exist regions of bistability, in which the final filament motion depends sensitively on the way in which parameters are changed. Inside the tongues period-doubling cascades and transitions to chaotic filament motions have been observed in experiments. Finally, we note that quasiperiodic and chaotic filament motions can also be reproduced by the simplified model.

Further fundamental oscillation modes can be excited when a small capacitance is connected parallel to the sample. In particular spatially uniform oscillations of the current-density distribution and both periodic and non-

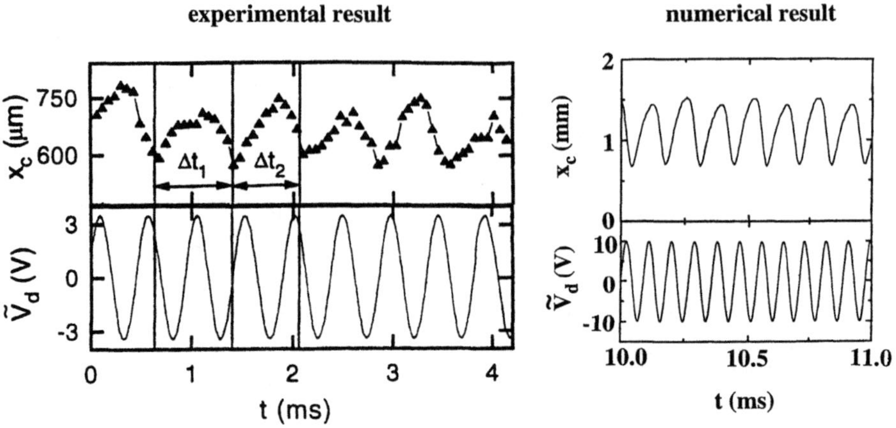

Fig. 23. Time series of the filament center x_c and the ac driving voltage V_d indicating a 2:3 locking of the filament motion with respect to the drive. Experimental (Ref. [98]) and numerical results are on the left and right column, respectively.

periodic spiking oscillations have been found in agreement with theoretical predictions [76].

Recent numerical studies of the mentioned model on two-dimensional domains reveal that new types of filament oscillation can appear due to the existence of the second spatial dimension [80]. In particular, the bifurcation of a static filament with a radial symmetry into an elliptically oscillating filament that finally splits into two traveling filaments with an increasing external control parameter has been found and even irregular and very complex motions of a filament may arise.

Further patterns, as e. g. static multi-filament distributions [28, 100] or static or moving fronts, have been observed in devices with suitably chosen doping profiles and it has been suggested to use the self-organized dynamics of fronts and filaments evolving in *p-n-p-n* devices to realize self-organizing learning algorithms [101].

Summarizing the results of this section, we point out that the investigated multi-layered device shows a variety of fundamental oscillation modes of localized structures that may appear also in other dissipative continuous media. By applying different measuring techniques we were able to characterize the spatiotemporal behavior of the localized structures and diverse bifurcation sequences in detail. The simple reaction-diffusion model derived has been successfully applied to reproduce many of the experimentally observed dynamic phenomena and bifurcations.

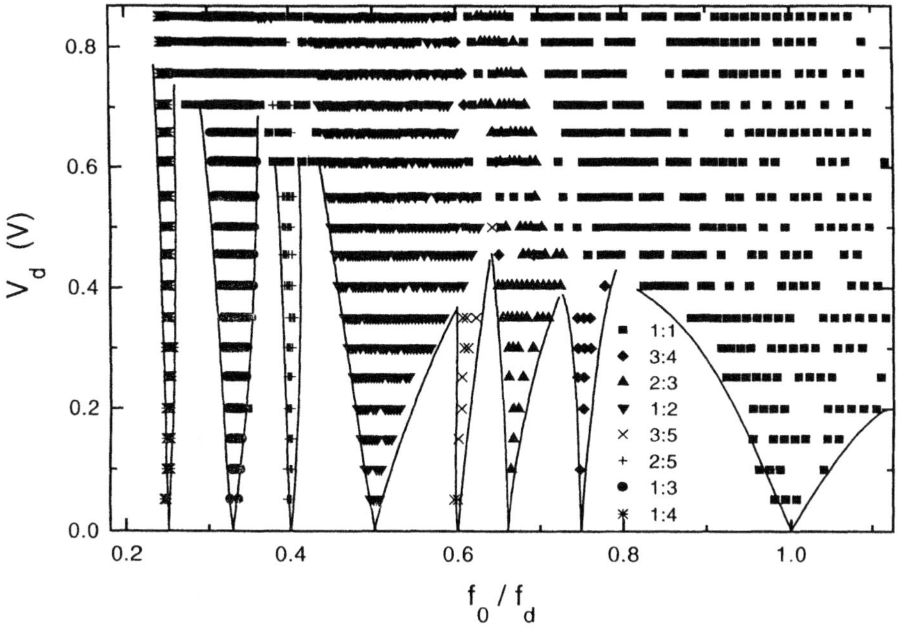

Fig. 24. Arnold tongues, in which frequency-locked pendulum-like filament oscillations have been observed experimentally (from Ref. [98])

5 Self-organization in thin ZnS:Mn films

Thin-film electroluminescence devices show typical features of excitable and bistable media. Self-organized pattern formation in these devices has first been studied in detail by Marello et al. [102], Rühle et al. [103, 104] and Rüfer et al. [105] for ac driven systems and by Beale et al. [106, 107] for dc driven films. Recently, first attempts have been made to describe the observed patterns in terms of reaction-diffusion systems of activator-inhibitor type [108, 109, 110, 111, 112]. Similar to the semiconductor systems studied in the previous sections the autocatalytic process is associated with charge carrier multiplication by impact ionization. In this section some typical patterns revealing the variety of structures that may evolve in these films are presented. We shall see that depending on the parameters basic structure elements as fronts, stationary and moving filaments may appear in the films with contact areas much larger than the typical dimensions of the structures. Beyond these structures encountered already in GaAs, heterodiodes and p^+-n^+-p-n^- diodes, further fundamental pattern elements such as spirals may be realized.

5.1 Experimental set-up

The experimental set-up including a schematic sketch of the samples is shown in Fig. 25. The thin-film consists of Mn-doped ZnS as semiconducting active layer that is sandwiched between two dielectric layers. More details concerning the preparation procedure can be found in Ref. [111].

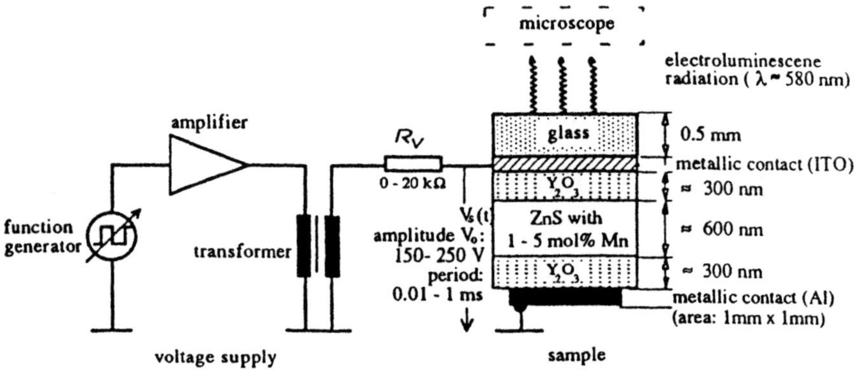

Fig. 25. Experimental set-up (from Ref. [111])

The thin-film structure is driven by an ac voltage source, which provides bipolar pulse trains or a sinusoidal signal. Current transport in the ZnS layer is essentially governed by the following mechanism: Electrons may tunnel from interface states, localized at the interface between the semiconductor and the dielectric layer, into the ZnS conduction band when the voltage across the ZnS layer is large enough and, consequently, the potential barrier width for the electrons is sufficiently small. In the ZnS layer the electrons are accelerated and may generate additional electron-hole pairs by impact ionization. According to a model introduced by Howard et al. [113] the holes are captured almost instantaneously by hole traps. Thus, a positive space charge is built up in the ZnS layer causing an additional electric field that in turn supports tunneling of further electrons from the interface states. When the electrons reach the counterelectrode, they are trapped again in interface states. Therefore with increasing charge transfer a field in opposite direction to the applied field builds up limiting the total transferred charge during one half-period. In the next half cycle, the polarity of the applied voltage changes and the described charge transport takes place in opposite direction.

5.2 Results and Discussion

Figures 26-30 show a selection of typical patterns that can be found in ZnS:Mn films for different amplitudes V_0 and repetition frequencies f of the driving voltage. Quite simple patterns in form of propagating circular domains can be observed for relatively low frequencies (Fig. 26). Typically, the system switches to the on-state at positions with slight inhomogeneities. From these positions circular luminescent domains evolve and propagate with a velocity of about 10 μm/s, and the velocity increases monotonically with increasing amplitude of the driving voltage [111].

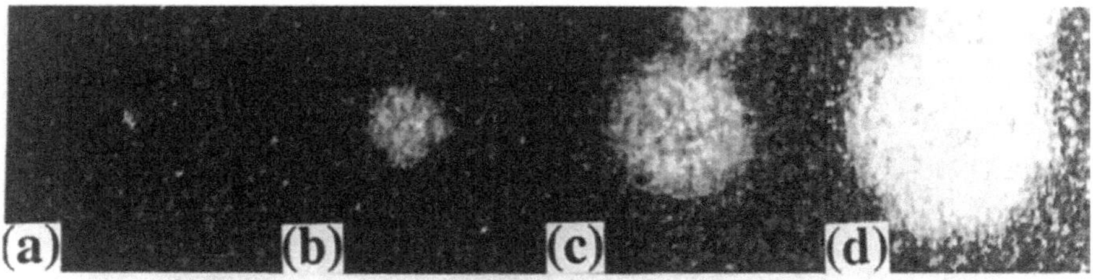

Fig. 26. Propagation of a domain. Bright regions correspond to regions of high luminescence intensity. Time between consecutive pictures is 400 ms. Parameters are: Sinusoidal excitation, $V_0 = 117.6$ V, $f = 1$ kHz, imaged device area = 30 × 30 μm^2 (from Ref. [111]).

The physical process determining the propagation of the on-state is governed by the charge in the semiconductor-insulator interface [108, 109]: The spreading of this charge along the interface causes a field-enhancement in the neighborhood of the front which in turn controls the switch on. The samples described here always showed a propagation of the on-state over the entire device domain; this is typical for a bistable system in which a diffusive autocatalytic variable dominates the system behavior. In contrast, Beletskii and Vlasenko report on the expansion of a domain in a similar film structure which becomes static at a certain size. This behavior is attributed to the influence of the resistance from the transparent electrodes which impedes a free domain expansion because of current limitation [109].

When the repetition frequency of the driving pulse trains is increased, moving wave structures appear. In this regime inhibiting processes counteracting the autocatalytic increase of the transferred charge are of importance. For example, a temperature increase caused by an increase of energy dissipation shifts the breakdown voltage to larger values and leads to a smaller hysteresis loop of the local brightness-voltage characteristic. The observed waves often have the form of target patterns. Their width is of the order of 5

μm. When the amplitude of the driving voltage is increased, the width, the velocity and the amplitude of the waves increase and, for sufficiently large voltages, the device motion becomes irregular. A typical snapshot of an irregular motion is shown in Fig. 27. We suppose that the irregular dynamic behavior results from the interaction of many waves moving in different directions, colliding and annihilating each other by their refractory zones. In addition, the interaction of waves with unavoidable inhomogeneities may cause the observed wave splitting.

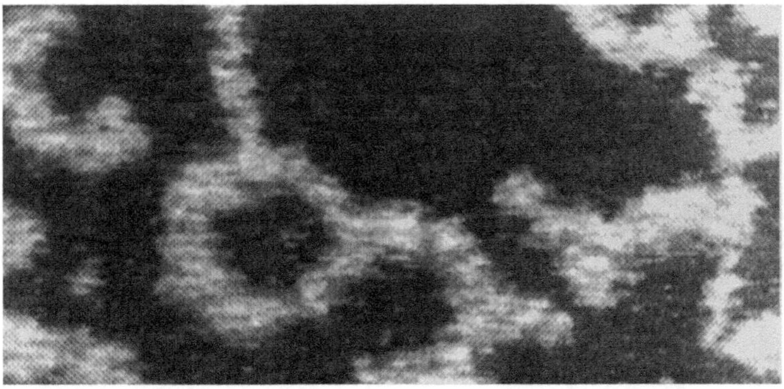

Fig. 27. Snapshot of irregularly moving waves. Parameters are: Pulsed excitation, pulse width = 13.5 μs, V_0 = 262 V, f = 1 kHz, imaged device area = 30 × 60 μm^2 (from Ref. [112]).

With increasing operating time of the device the maximum wave width decreases and reaches a saturation value of about 1-3 μm after an operation time of 1/2 h. Then a new type of pattern can be observed: mobile luminenscent strings coexisting with moving microfilaments (Fig. 29(b)). Our experiments have shown that the generation and disappearance of moving filaments and strings can take place in various ways [111]. Starting from a luminescence distribution with coexisting filaments and strings a decrease of the voltage amplitude causes the disappearance of the strings while the mobile filaments survive [Fig. 29(a)]. In contrast, an increase of the voltage amplitude favours the evolution of strings and a compact filament-free network of fluctuating strings is formed. When the voltage is increased further, the mean width of the mesh formed by the strings decreases, finally resulting in a uniform distribution. So far, this complex behavior is not well understood.

For high repetition frequencies (typically above 50 kHz) a luminescent spot with a diameter of several hundred μm can be stabilized. The spot size varies monotonically with the amplitude of the driving voltage. However, when a certain minimum spot size is reached by decreasing the amplitude, the

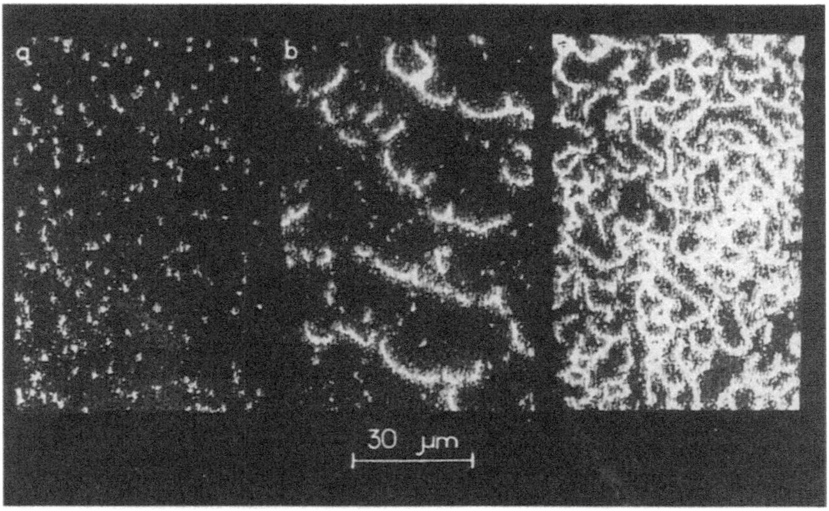

Fig. 28. Snapshots of a counterclockwise rotating spot. Parameters are: Sinusoidal excitation, $V_0 = 140$ V, $f = 71$ kHz, time between consecutive pictures is 1.04 s (from Ref. [111]).

spot starts moving on the domain. When the amplitude is further decreased, the spot motion becomes increasingly complex [111].

Figure 28 shows a rather simple example of a spot rotating counterclockwise. Note that the entire active area is imaged in the figure. Analyzing the spot on a microscopic scale one can discern several substructures as shown in Fig. 30. When the spot approaches the area viewed with the microscope, randomly moving filaments and strings evolve (Fig. 30(a-g)) transforming into a bright and uniform distribution (Fig. 30(h)). After the spot has passed, the bright domain dissolves in a fluctuating network of strings (Fig. 30(i,j)), which in turn transforms to moving filaments. Their density decreases in the course of time until the cycle repeats again. Information concerning more complex spot motions can be found in Ref. [111].

To summarize this section we point out that ZnS:Mn films reveal a rich variety of patterns comprising mobile microfilaments, strings and waves with spatial and temporal dimensions of a few μm and tenths of a second, respectively, as well as luminescent spots with a diameter of several hundred μm performing oscillations on the domain with a period of several seconds. Furthermore, there is a remarkable similarity with patterns appearing in chemical media and other reaction diffusion systems.

6 Conclusions

In this article we have presented concepts of spatio-temporal self-organization in semiconductors and illustrated these by some exemplary model systems

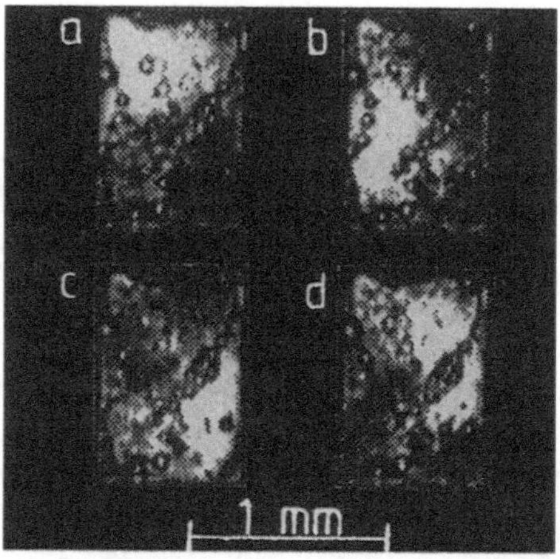

Fig. 29. Snapshots of luminescence patterns for different values of the driving amplitude V_0 revealing mobile filaments, $V_0 = 169.7$ V (a), the coexistence of mobile filaments and strings, $V_0 = 170$ V (b), and a fluctuating string network, $V_0 = 170.3$ V (c). Other parameters are: Pulsed excitation, pulse width $= 13.5$ μs, $V_0 = 262$ V, $f = 27.8$ kHz (from Ref. [111]).

which have been in the center of recent theoretical and experimental research activities. Investigations in diverse semiconductor materials have led to an advanced understanding of the complex dynamics of current filaments in one and two spatial dimensions.

A remarkable result is that inspite of the different microscopic nature of those systems some general principles of pattern formation in semiconductors can be worked out. Considering the semiconductor as an active medium which is governed by activator-inhibitor kinetics provides such a useful concept. Impact ionization of hot charge carriers plays the role of an autocatalytic process not only in the regime of low-temperature impurity breakdown (Sections 2, 3), but is also important in the multilayered semiconductor structures treated in Sect. 4 and in the ZnS films discussed in Sect. 5. Generic modes of filament dynamics such as spiking, breathing, rocking, or traveling filaments occur in a variety of systems.

In Sect. 2 a modeling approach is described which combines different levels of transport theory, and allows for a detailed understanding of both the microscopic scattering processes in the regime of impurity impact ionization breakdown and the macroscopic dynamics of the filaments for different two-dimensional (rectangular or circular) geometries. It is illustrated by computer simulations of thin n-GaAs epitaxial layers as studied experimentally in Sect. 3. There a very efficient method to visualize current filaments is presented utilizing photoluminescence quenching in spatial regions of high electron den-

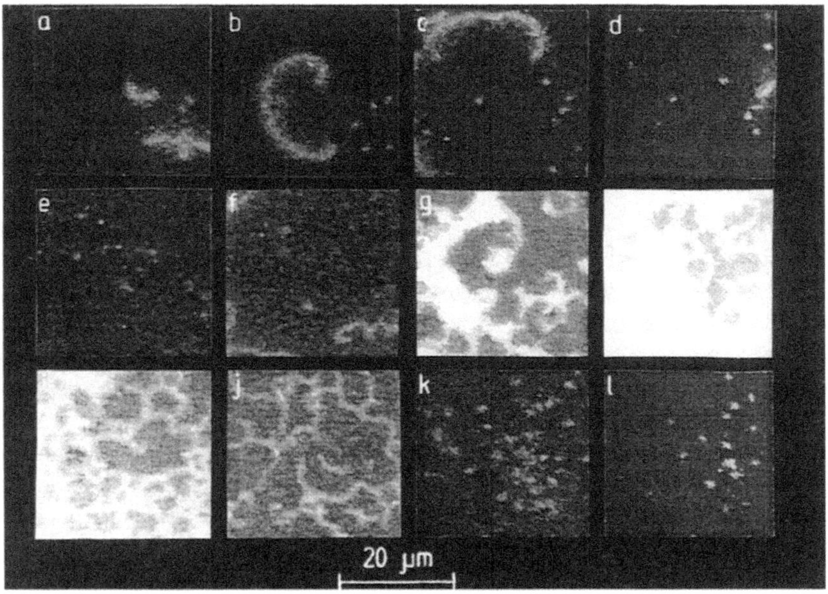

Fig. 30. Snapshots of luminescence distributions revealing substructures that are associated with a rotating spot as shown in Fig. 28. Parameters are sinusoidal excitation, $V_0 = 170$ V, $f = 52.6$ kHz, time between consecutive pictures is 180 ms (from Ref. [111]).

sity. The low intensity and the spatial homogeneity of the exciting irradiation minimizes parasitic effects which may affect current filamentation. Spatially resolved recordings of the luminescence yield images of the inhomogeneous electron distribution of high contrast, in good agreement with the computer simulations of Sect. 2. In samples with two point contacts filaments nucleate with sharp and parallel borders indicating self-organization. Additional strong evidence for the self-organized nature of current filamentation is obtained from measurements on samples with concentric circular electrodes. With these samples multifilamentary structures have been found which break the circular symmetry of the previously homogeneous and stationary state of the sample.

When the Lorentz force in a magnetic field exceeds a pinning force, lateral motion and rotation of filaments has been observed which cause characteristic magnetic field induced oscillations of the sample voltage. The modes of filament motion have been reconstructed by tracing the luminescence in space and time using a single photodiode and by taking stroboscopic images triggered by the characteristic oscillations.

For a variety of different semiconductor devices a successful theoretical description in terms of reaction-diffusion models with activator-inhibitor kinetics has been achieved. This has been demonstrated in particular for multilayered Si-diodes which have been investigated experimentally in detail through temporally resolved measurements of electroluminescence and potential distributions. The results reveal the activator and inhibitor kinetics, respectively, and permit comprehensive comparisons between experiments and model predictions.

The exemplary results presented in this chapter show that as far as pattern formation in one spatial dimension is concerned experiments and theoretical modelling lead to a far-reaching understanding of the basic mechanisms responsible for different generic modes of filament dynamics and bifurcations. Some of these results can be easily extended to systems with two spatial dimensions. However, the second spatial dimension provides additional degreees of freedom of filament dynamics as is obvious from numerical and analytical studies of multi-layered Si diodes and experiments in thin electroluminescent ZnS:Mn films where filaments, spots, and even spirals have been found. To achieve a complete understanding of these phenomena further experimental and theoretical work is necessary.

ACKNOWLEDGEMENTS

Financial support bei Deutsche Forschungsgemeinschaft (Schwerpunktprogramm *Strukturbildung*) is gratefully acknowledged. The authors thank S. Beizai, S. Bose, M. Gaa, R. Kunz, C. Lehmann, A. Reimann, P. Rodin, G. Schwarz, A. Wacker, H. Gajewski and R. Nürnberg, H. Kostial, V. Novák, J. Hirschinger, H. Klimenta, V.V. Belkov, W. Eberle, U. Margull, Ch. Wimmer, A.V. Gorbatyuk, C. Goßen, M. Kleinkes, B. Kukuk, M. Or-Guil, H.-J. Schulze, and S. Zuccaro for stimulating discussions and collaboration.

References

[1] E. Schöll, *Nonequilibrium Phase Transitions in Semiconductors* (Springer, Berlin, 1987).

[2] P. T. Landsberg and A. Pimpale, J. Phys. **C**, 1243 (1976).

[3] E. Schöll and P. T. Landsberg, Proc. Roy. Soc. A **365**, 495 (1979).

[4] M. P. Shaw, V. V. Mitin, E. Schöll, and H. L. Grubin, *The Physics of Instabilities in Solid State Electron Devices* (Plenum Press, New York, 1992).

[5] *Nonlinear Dynamics in Solids*, edited by H. Thomas (Springer, Berlin, 1992).

[6] J. Peinke, J. Parisi, O. Rössler, and R. Stoop, *Encounter with Chaos* (Springer, Berlin, Heidelberg, 1992).

[7] B. S. Kerner and V. V. Osipov, *Autosolitons* (Kluwer Academic Publishers, Dordrecht, 1994).

[8] *Nonlinear Dynamics and Pattern Formation in Semiconductors and Devices*, edited by F.-J. Niedernostheide (Springer, Berlin, 1995).

[9] A. Gierer and H. Meinhardt, Kybernetik **12**, 30 (1972).

[10] P. Fife, *Current Topics in Reaction-Diffusion Systems, Nonequil. Coop. Phenom. in Phys. and Related Fields*, Vol. 116 of *NATO ASI* (Plenum Press, New York, 1983).
[11] J. Smoller, *Shock Waves and Reaction-Diffusion Equations* (Springer, New York, Heidelberg, Berlin, 1983).
[12] J. D. Murray, *Mathematical Biology*, Vol. 19 of *Biomathematics Texts*, 1st ed. (Springer, Berlin Heidelberg, 1993).
[13] D. Jäger, H. Baumann, and R. Symanczyk, Phys. Lett. A **117**, 141 (1986).
[14] C. Radehaus, K. Kardell, H. Baumann, D. Jäger, and H.-G. Purwins, Z. Phys. B **65**, 515 (1987).
[15] H. Baumann, R. Symanczyk, C. Radehaus, H. Purwins, and D. Jäger, Phys. Lett. A **123**, 421 (1987).
[16] H.-G. Purwins, C. Radehaus, and J. Berkemeier, Z. f. Naturforsch. **43a**, 17 (1988).
[17] V. Vashchenko, B. Kerner, V. Osipov, and V. Sinkevich, Sov. Phys. Semicond. **24**, 1045 (1990).
[18] F.-J. Niedernostheide, B. S. Kerner, and H.-G. Purwins, Phys. Rev. B **46**, 7559 (1992).
[19] A. Wacker and E. Schöll, Z. Phys. B **93**, 431 (1994).
[20] *Theory of Transport Properties of Semiconductor Nanostructures*, edited by E. Schöll (Chapman and Hall, London, 1998).
[21] A. Wacker and E. Schöll, J. Appl. Phys. **78**, 7352 (1995).
[22] B. K. Ridley, Proc. Phys. Soc. **82**, 954 (1963).
[23] K. Mayer, J. Parisi, and R. Huebener, Z. Phys. B **71**, 171 (1988).
[24] A. Wierschem, F.-J. Niedernostheide, A. Gorbatyuk, and H.-G. Purwins, Scanning **17**, 106 (1995).
[25] A. Brandl, M. Völcker, and W. Prettl, Appl. Phys. Lett. **55**, 238 (1989).
[26] J. Spangler, B. Finger, C. Wimmer, W. Eberle, and W. Prettl, Semicond. Sci. Technol. **9**, 373 (1994).
[27] B. Kukuk, F. Reil, F.-J. Niedernostheide, and H.-G. Purwins, in *Selforganization in Activator-Inhibitor-Systems*, edited by H. Engel, F.-J. Niedernostheide, H. G. Purwins, and E. Schöll (Wissenschaft & Technik Verlag, Berlin, 1996), pp. 62–66.
[28] F.-J. Niedernostheide, M. Arps, R. Dohmen, H. Willebrand, and H.-G. Purwins, phys. status solidi (b) **172**, 249 (1992).
[29] W. Eberle, J. Hirschinger, U. Margull, W. Prettl, V. Novák, and H. Kostial, Appl. Phys. Lett. **68**, 3329 (1996).
[30] in *Special issue on Nonlinear and Chaotic Transport Phenomena in Semiconductors*, edited by Y. Abe (Applied Phys. **A 48**, 93-191, (1989)).
[31] E. Schöll, in *Handbook on Semiconductors*, 2nd ed., edited by P. T. Landsberg (North Holland, Amsterdam, 1992), Vol. 1.
[32] *Negative Differential Resistance and Instabilities in two-dimensional Semiconductors*, edited by N. Balkan, B. K. Ridley, and A. J. Vickers (Plenum Press, New York, 1993).
[33] Z. S. Gribnikov, K. Hess, and G. Kosinovsky, J. Appl. Phys. **77**, 1337 (1995).
[34] L. L. Bonilla, J. Galán, J. A. Cuesta, F. C. Martínez, and J. M. Molera, Phys. Rev. B **50**, 8644 (1994).

[35] F. Prengel, A. Wacker, and E. Schöll, Phys. Rev. B **50**, 1705 (1994), ibid **52**, 11518 (1995).

[36] H. T. Grahn, in *Semiconductor Superlattices, Growth and Electronic Properties*, edited by H. T. Grahn (World Scientific, Singapore, 1995), Chap. 4.

[37] F. Piazza, P. C. M. Christianen, and J. C. Maan, Phys. Rev. B **55**, 15591 (1997).

[38] *Quantum Transport in Ultrasmall Devices*, edited by D. K. Ferry, L. Grubin, C. Jacoboni, and A. P. Jauho (Plenum Press, New York, 1995).

[39] G. E. Stillman, C. M. Wolfe, and J. O. Dimmock, *Semiconductors and Semimetals* (Academic, New York, 1977), Vol. 12, p. 169.

[40] K. Aoki, T. Kobayashi, and K. Yamamoto, J. Physique Colloque C **7**, 51 (1981).

[41] S. W. Teitsworth, R. M. Westervelt, and E. E. Haller, Phys. Rev. Lett. **51**, 825 (1983).

[42] M. Kozhevnikov, B. M. Ashkinadze, E. Cohen, and R. Arza, Phys. Rev. B **52**, 4855 (1995).

[43] M. Gaa, R. E. Kunz, E. Schöll, W. Eberle, J. Hirschinger, and W. Prettl, Semicond. Sci. Technol. **11**, 1646 (1996).

[44] E. Schöll and D. Drasdo, Z. Phys. B **81**, 183 (1990).

[45] R. E. Kunz and E. Schöll, Z. Phys. B **89**, 289 (1992).

[46] G. Hüpper, K. Pyragas, and E. Schöll, Phys. Rev. B **47**, 15515 (1993).

[47] G. Hüpper, K. Pyragas, and E. Schöll, Phys. Rev. B **48**, 17633 (1993).

[48] R. E. Kunz and E. Schöll, Z. Phys. B **99**, 185 (1996).

[49] M. Gaa and E. Schöll, Phys. Rev. B **54**, 16733 (1996).

[50] V. Novák, C. Wimmer, and W. Prettl, Phys. Rev. B **52**, 9023 (1995).

[51] T. Kuhn, G. Hüpper, W. Quade, A. Rein, E. Schöll, L. Varani, and L. Reggiani, Phys. Rev. B **48**, 1478 (1993).

[52] W. Quade, G. Hüpper, E. Schöll, and T. Kuhn, Phys. Rev. B **49**, 13408 (1994).

[53] B. Kehrer, W. Quade, and E. Schöll, Phys. Rev. B **51**, 7725 (1995).

[54] H. Kostial, M. Asche, R. Hey, K. Ploog, B. Kehrer, W. Quade, and E. Schöll, Semicond. Sci. Technol. **10**, 775 (1995).

[55] M. Gaa, R. E. Kunz, and E. Schöll, in *Proc. 9th Int. Conf. on Hot Carriers in Semiconductors*, edited by K. Hess, J. P. Leburton, and U. Ravaioli (Plenum, New York, 1996), pp. 347–351.

[56] M. Gaa, R. E. Kunz, and E. Schöll, Phys. Rev. B **53**, 15971 (1996).

[57] K. Kunihiro, M. Gaa, and E. Schöll, Phys. Rev. B **55**, (1997).

[58] K. Kunihiro, M. Gaa, and E. Schöll, Electronics Letters **33**, 1261 (1997).

[59] E. Schöll, Z. Phys. B **46**, 23 (1982).

[60] R. E. Kunz, E. Schöll, H. Gajewski, and R. Nürnberg, Sol. State El. **39**, 1155 (1996).

[61] A. Brandl, T. Geisel, and W. Prettl, Europhys. Lett. **3**, 401 (1987).

[62] A. Brandl, W. Kröninger, W. Prettl, and G. Obermair, Phys. Rev. Lett. **64**, 212 (1990).

[63] U. Rau, W. Clauss, A. Kittel, M. Lehr, M. Bayerbach, J. Parisi, J. Peinke, and R. P. Huebener, Phys. Rev.B **43**, 2255 (1991).

[64] G. Hüpper and E. Schöll, Phys. Rev. Lett. **66**, 2372 (1991).

[65] A. Brandl and W. Prettl, Phys. Rev. Lett. **66**, 3044 (1991).

[66] M. Hirsch, A. Kittel, G. Flätgen, R. P. Huebener, and J. Parisi, Phys. Lett. A **186**, 157 (1994).

[67] J. Hirschinger, W. Eberle, W. Prettl, F.-J. Niedernostheide, and H. Kostial, Phys. Lett. A (1997) .

[68] L. Schimansky-Geier, C. Zülicke, and E. Schöll, Z. Phys. B **84**, 433 (1991).

[69] G. Schwarz and E. Schöll, Acta Techn. CSAV **42**, 669 (1997).

[70] A. Wacker and E. Schöll, Semicond. Sci. Technol. **9**, 592 (1994).

[71] K. Hess, T. K. Higman, M. A. Emanuel, and J. J. Coleman, J. Appl. Phys. **60**, 3775 (1986).

[72] A. S. Mikhailov, *Foundations of Synergetics 1*, 2 ed. (Springer, Berlin, 1994).

[73] *Self-Organization in Activator-Inhibitor-Systems: Semiconductors, Gas Discharge, and Chemical Active Media*, edited by H. Engel, F.-J. Niedernostheide, H. G. Purwins, and E. Schöll (Wissenschaft und Technik Verlag, Berlin, 1996).

[74] S. Bose, A. Wacker, and E. Schöll, Phys. Lett. A **195**, 144 (1994).

[75] A. Wacker, S. Bose, and E. Schöll, Europhys. Lett. **31**, 257 (1995).

[76] F.-J. Niedernostheide, H. Schulze, S. Bose, A. Wacker, and E. Schöll, Phys. Rev. E **54**, 1253 (1996).

[77] M. Meixner, S. Bose, and E. Schöll, Physica D **109**, 128 (1997).

[78] M. Meixner, A. D. Wit, S. Bose, and E. Schöll, Phys. Rev. E **55**, 6690 (1997).

[79] A. Alekseev, S. Bose, P. Rodin, and E. Schöll, Phys. Rev. E **57**, 2640 (1998).

[80] F.-J. Niedernostheide, M. Or-Guil, M. Kleinkes, and H.-G. Purwins, Phys. Rev. E **55**, 4107 (1997).

[81] I. Melngailis and A. G. Milnes, J. Appl. Phys. **33**, 995 (1962).

[82] K. Mayer, R. Gross, J. Parisi, J. Peinke, and R. Huebener, Solid State Comm. **63**, 55 (1987).

[83] K. Mayer, R. Huebener, and U. Rau, J. Appl. Phys. **67**, 1412 (1990).

[84] W. Bludau and E. Wagner, Phys. Rev. B **13**, 5410 (1976).

[85] O. Ryabushkin and V. Bader, *Proceedings of the International Conference on Optical Science and Engineering (Optics for Computer)* (Hague, the Netherlands, 1991), p. 1505.

[86] F. Karel, J. Oswald, J. Pastrňák, and O. Petřč ek, Semicond. Sci. Technol. **7**, 203 (1992).

[87] J. Hirschinger, F. J. Niedernostheide, W. Prettl, V. Novák, and H. Kostial, phys. stat. sol. **204**, 477 (1997).

[88] J. Spangler, U. Margull, and W. Prettl, Phys. Rev. B **45**, 12137 (1992).

[89] V. Novák and W. Prettl, in *Nonlinear Dynamics and Pattern Formation in Semiconductors*, edited by F.-J. Niedernostheide (Springer, Berlin, 1995).

[90] V. Novák, J. Hirschinger, W. Eberle, C. Wimmer, and W. Prettl, Acta Techn. CSAV **41**, 553 (1996).

[91] U. Margull, in *Selforganization in Activator-Inhibitor-Systems*, edited by H. Engel, F.-J. Niedernostheide, H. G. Purwins, and E. Schöll (Wissenschaft & Technik Verlag, Berlin, 1996), p. 50.

[92] B. Kerner and V. Sinkevich, Sov. Phys. - J. exper. theor. Phys. Letters **36**, 436 (1982).

[93] B.S. Kerner, D.P. Litvin, V.I. Sankin, Sov. Tech. Phys. Lett. **13**, 342 (1987).

[94] A.V. Gorbatyuk and P. Rodin, Proc. of the 6th Int. Conf. on Symp. Power Semiconductor Devices (Davos 1994), pp. 237–241, Z. Phys. B **104**, 45 (1997).

[95] M. Meixner, P. Rodin, and E. Schöll, phys. stat. sol. (b) **204**, 493 (1997).

[96] F.-J. Niedernostheide, M. Ardes, M. Or-Guil, and H.-G. Purwins, Phys. Rev. B **49**, 7370 (1994).
[97] F.-J. Niedernostheide, M. Kreimer, H.-J. Schulze, and H.-G. Purwins, Phys. Lett. A **180**, 113 (1993).
[98] F.-J. Niedernostheide, C. Brillert, B. Kukuk, H.-G. Purwins, and H.-J. Schulze, Phys. Rev. B **54**, 14012 (1996).
[99] F.-J. Niedernostheide, M. Kreimer, B. Kukuk, H.-J. Schulze, and H.-G. Purwins, Phys. Lett. A **191**, 285 (1994).
[100] A. V. Gorbatyuk and F.-J. Niedernostheide, Physica D **99**, 339 (1996).
[101] D. Ruwisch, M. Bode, H.-J. Schulze, and F.-J. Niedernostheide, in *Nonlinear Physics of Complex Systems, Lecture Notes in Physics, Vol, 476*, edited by J. Parisi, S. Müller, and W. Zimmermann (Springer, Berlin, Heidelberg, 1996), p. 194.
[102] V. Marello, W. Rühle, and A. Onton, Appl. Phys. Lett. **31**, 452 (1977).
[103] W. W. Rühle, V. Marello, and A. Onton, J. Elec. Mat. **8**, 839 (1979).
[104] W. W. Rühle, V. Marello, and A. Onton, J. Luminescence **18/19**, 729 (1979).
[105] H. Rüfer, V. Marello, and A. Onton, J. Appl. Phys. **51**, 1163 (1980).
[106] M. Beale, J. Kirton, and M. Slater, in *Electroluminescence, Springer Proceedings in Physics, Vol. 38*, edited by S. Shionoya and H. Kobayashi (Springer, Berlin, Heidelberg, 1989), p. 296.
[107] M. Beale, Philosophical Magazine B **68**, 573 (1993).
[108] A. Beletskii and N. Vlasenko, Tech. Phys. Lett. **19**, 13 (1993).
[109] A. Beletskii and N. Vlasenko, Tech. Phys. Lett. **19**, 798 (1993).
[110] F.-J. Niedernostheide, C. Goßen, and H.-G. Purwins, in *Advances in Synergetics*, edited by V. Kuvshinov and C. Krylov (Belarusian State University Press, Minsk, 1995), p. 33.
[111] C. Goßen, F.-J. Niedernostheide, and H.-G. Purwins, in *Nonlinear Dynamics and Pattern Formation in Semiconductors and Devices*, edited by F.-J. Niedernostheide (Springer, Berlin, 1995), pp. 113–132.
[112] M. Or-Guil, E. Ammelt, F.-J. Niedernostheide, and H.-G. Purwins, in *Nonlinear Dynamics and Pattern Formation in the natural environment*, edited by A. Doelman and A. van Harten (Longman, 1995), p. 223.
[113] W. Howard, O. Sahni, and P. Alt, J. Appl. Phys. **53**, 639 (1982).

Formation of Patterns in Granular Materials

André Betat[1], Christian M. Dury[2], Ingo Rehberg[1], Gerald H. Ristow[2], Michael A. Scherer[1], Matthias Schröter[1], and Gunther Straßburger[1]

[1] Otto-von-Guericke-Universität Magdeburg, Postfach 4120, D-39016 Magdeburg, Germany
[2] Philipps-Universität Marburg, Fachbereich Physik, Renthof 6, D-35032 Marburg, Germany

Abstract. In order to study the behaviour of fluidized granular material experimentally and numerically, we propose to investigate pattern-forming processes, because they can be easily observed and quantitatively characterized. We report measurements on five such systems, namely pattern formation of granular material i) by segregation, under the influence of ii) linear horizontal and iii) circular horizontal vibration, iv) water shear flow, and in v) a sedimenting suspension.

1 Introduction

Sand, when considered as a continuum, is an unusal one: Under some circumstances it behaves like a fluid, as is best demonstrated in a hour-glass. A stable pile of sand, on the other hand, demonstrates the limitations of this point of view: It does not flow apart. At present, a theoretical description of these two aspects or related peculiar behaviour of granular matter cannot be condensed into a coherent theoretical model (for a recent review see Jaeger et al. (1996b)). In order to describe granular flow, several complementary methods are currently being used. For sufficiently low densities and a sufficiently fast motion of the granulate, a hydrodynamic description is possible. Throughout the last few years, numerical methods have been successful in modelling the collective behaviour on a limited number of particles. In this chapter we will focus on those methods, and a number of experiments suitable for testing the limitations of theoretical models.

All the theoretical models rely on simplifications of the interaction of the grains. Thus, experimental tests are crucial. For some of the experiments presented in this chapter, such a test is indeed the primary motivation. The experimental difficulties of granular matter are due to the uncontrolled surface of the granulate, the influence of humidity, and the uncontrolled charging of the particles. Considering these difficulties it seems already a challenge in its own to find paradigmatic experimental systems that are suitable for a quantitative comparison with theoretical approaches describing granular material. The present chapter describes two systems which have this property: The problem of humidity and charging can either be overcome by using comparatively large particles, or by using a suspension of sand in water. Many of the experiments described below were chosen because we feel that they have a strong potential for clarifying fundamental questions concerning granular matter. It is not the intention of this article to present an overview of solved problems, we rather hope that our presentation helps to stimulate further research.

The paper is organized in the following way: In section 2, we mention briefly some of the most striking observations concerning the statics of granular matter. In section 3, some aspects of the dynamics are described. Special emphasis is given to fluidization, the necessary precursor of dynamical behaviour. For fluidized granulate we concentrate on two aspects of the dynamics, namely segregation and the formation of surface patterns. This allows an elegant approach to measure the quantitites of interest, namely image processing. A short overview on the numerical methods used to describe granular matter is given in the appendix.

2 Statics of Granulate

The force distribution in a static arrangement of granular particles is of technical interest, for example in silos. Here, unlike in a fluid, the pressure at the bottom of the container is not proportional to the height of the material resting on top of it, instead it reaches a nearly constant value (Janssen 1895), because the walls can support the extra weight. This effect also explains to some extend why the flow rate in a hour–glass is almost constant. The force distribution inside the granulate has an interesting structure, in particular, it shows spatial fluctuations. The fact that a worm survives in sand even some meters below the surface, where the pressure would be much too high for a fluid-like behaviour, is a manifestation of those fluctuations. Another interesting example is the pressure distribution at the bottom of a sand heap: The maximum pressure does not seem to be in the middle of the heap, where the sand is highest. There is no simple way to model this behaviour theoretically: It is not possible to create a heap from monodisperse spheres, the existence of the sand heap, and in consequence the complicated force distribution, is due to the more complicated shape of the sand particles.

3 Dynamics of Granular Flow

The important difference between granular material and the more classical many-body systems considered in statistical physics is dissipation - the collisions between two sand particles are inelastic. Therefore granular matter is normally at rest, energy must be supplied to observe dynamical behaviour. Both in nature (avalanches) and in technical applications (silos) this is often just gravitational energy. In such cases the flow is only of transient nature and thus not very convenient for laboratory experiments. In spite of this nuisance, the sedimentation experiment described below falls into this class.

Rotation - as in the segregation experiment described below - or vibration are the most often used methods to supply energy continuously. In particular, vertical vibration seems to be a popular method (Bizon et al. 1998), while we concentrate on horizontal vibrations in the experiments described below. In the third class of experiments we provide the energy to the granulate by means of rotational shear flow. In all experiments, the external driving has to overcome a threshold for a transition from a solid–like to a fluid– or gas–like behaviour of the granulate to occur. We have examined this transition in detail only for the case of linear horizontal vibration, which will be described in the following subsection.

3.1 Fluidization of Granular Matter

One of the most fascinating properties of granular matter is their transition from a solid-like behaviour to a fluid-like behaviour (the so–called fluidization, see e.g. Esipov and Pöschel (1997)), which appears when increasing the strength of external shear or the acceleration of vibration (Clément and Rajchenbach 1991, Gallas et al. 1992). In most experiments, this transition seems easy to pinpoint by visual inspection of the granulate. However, a precise determination is not possible: individual grains do move both below and above the fluidization point. An objective order parameter suitable for measuring this fascinating transition from solid-like to fluid-like behaviour is necessary.

We studied this transition experimentally and by computer simulations for horizontal vibration of a granulate and proposed two such order parameters, one based on the so-called granular temperature, the other extracted from the pair distribution function (Ristow et al. 1997). By using these order parameters, we have been able to determine the scaling of the fluidization point, and find that it scales with the *acceleration* of the vibration. This behaviour was also observed for the threshold of the free surface instability (Evesque and Rajchenbach 1989) and the spontaneous heaping instability (Clément et al. 1992) in sand piles undergoing vertical vibrations. Interestingly enough, according to our numerical simulations, the granular temperature scales with the *velocity* of the vibration. We cannot explain this scaling, but note that it is reminiscent of the scaling for the height of the center of mass of a granulate under vertical vibration (Warr et al. 1995, Luding et al. 1994), or the scaling of the granular temperature (Warr et al. 1995, Lee 1995).

The experimental apparatus used for the horizontal vibration of granulate is described in Ristow et al. (1997). The available amplitude was in the range of 0 mm – 50 mm, and the frequency was from 0.9 Hz – 4.3 Hz. We used a channel of 100 mm length in the direction of the vibration, 0.6 mm width and 20 mm height. The container was filled with granular material consisting of commercially available glass beads. The particles were close to spherical. We have developed a method to get rid of the grains which deviate considerably from this shape, by making use of the different rolling behaviour of spheres and non-spherical particles down an inclined plane. The diameter of the spheres was between 0.52 mm and 0.6 mm. The lower was determined by sieving, and the upper one by both a sieving procedure and the channel width. The channel was filled with a monolayer of spheres, and the height of the granulate was 3-5 particles, i.e. we use about 700 particles. No attempt was made to count this number exactly, because the solid-fluid transition seems to be fairly independent of the exact number. We used two methods to determine the particle motion.

By the first method the particle motion was measured in the co–moving frame by a CCD–camera connected to a frame grabber. The camera recorded the central 40 mm of the channel from the side, i.e. perpendicular to the direction of the oscillation. 32 snapshots were digitized with a sampling rate of 25 Hz. We obtained a binary image by giving pixels associated with the background a value of zero, and pixels associated with a particle the value of 1. In order to characterize the fluidization we subtracted consecutive binary images pixelwise, and took the mean square of the resulting difference. The resulting number is called *pixel change*. In order to obtain a temporal mean value over about one period of the driving, we took the mean value of about 7 (depending on the driving frequency) pixel changes. The resulting number is plotted in Fig. 1 as a function of the vibration frequency of the apparatus, for a fixed amplitude of 20 mm.

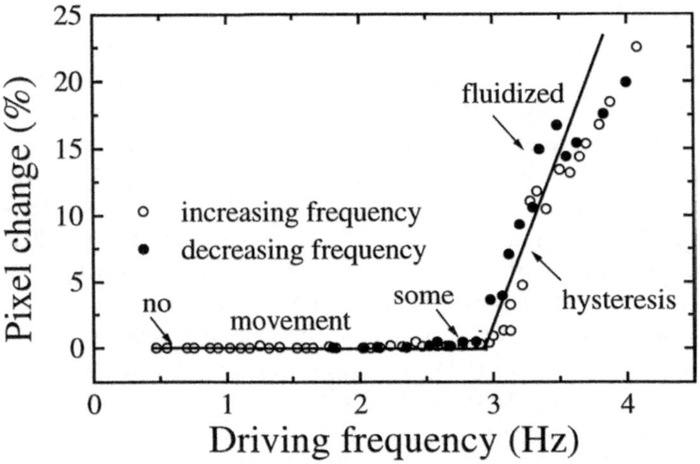

Fig. 1. Experimental system: relative pixel change as a function of the external frequency for an amplitude of 2 cm; increasing frequency: ∘, decreasing frequency: •; the solid line is a linear fit to determine the critical point f_c.

For low frequencies, the pixel change is zero, indicating that the particles do not move. Above 2.5 Hz the value increases slightly, which we attribute to the motion of a small fraction of particles. The dramatic increase of the pixel change around 3 Hz defines the transition to the fluidized phase. Open circles are drawn when the frequency is increased between the measurements, and solid circles represent data taken after a decrease of the driving frequency. One can easily distinguish a hysteresis for this transition. We determine the critical frequency f_c for the onset of fluidization by fitting a straight line to the supercritical values. Its intersection with the frequency axis yields $f_c = 2.9$ Hz. The onset of the motion of individual particles, denoted f_0, is clearly below this frequency (see Fig. 7 below).

In the second method a laser doppler velocimeter (Polytec LDV-380) was used to measure the particle velocity and the velocity distribution parallel to the direction of the vibration. These measurements served to verify the results obtained by image processing. The measurements were performed near the moment of maximal acceleration of the channel. We compare then the results with those determined by measuring the velocity of a layer of granulate glued in the channel (i. e.w̃ith the velocity distribution caused by finite measuring time and errors in the system). In Fig. 2 the variance of the velocity distribution is shown as a function of the driving frequency of the apparatus. The increase of the variance around 3 Hz is caused by the fluidization of the granulate. This frequency is in good agreement with the one determined by image processing (Fig. 1). The velocity distribution function itself is asymmetric. For frequencies above the transition point there are particles with velocities directed opposite to the driving.

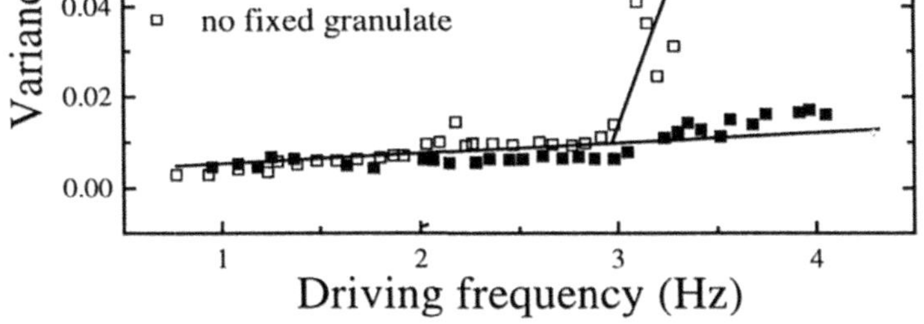

Fig. 2. Experimental system: variance of the velocity distribution function as a function of the frequency of vibration for an amplitude of 2 cm.

In order to study granular materials by computer simulations in two dimensions, we represent each particle i by a sphere with diameter d_i. Only contact forces during collisions are considered and particles are allowed to rotate. We use granular dynamics with a linear springdashpot model and only dynamic friction. The details of the numerical schemes are given in the appendix. In these simulations we have used a value of $\gamma_s = 1$ N s/m, which is sufficiently high so that the shear forces are dependent on μ only, and k_n was set to 10^6 N/m. The rest of the parameters were chosen from the detailed collision experiments using glass beads by Foerster et al. (1994) as $\gamma_n = 5.75$ N s/m which corresponds to a restitution coefficient of 0.75 in the normal direction and values of $\mu = 0.1$ for particle–particle and $\mu_w = 0.13$ for particle–wall contacts. In our model, we dissipate energy in the shear direction as well, characterized by a restitution coefficient of 0.75. The box length is 10.08 cm with periodic boundary conditions in the direction of shaking (x-direction) and contained 343 particles in two layers. We use a polydisperse diameter distribution ranging uniformly from 0.52 to 0.60 mm which corresponds to the beads used in our experimental system.

It is easier to determine the fluidization point in the numerical simulations, since the position and velocity information of all particles are given at any time. Noting that the pixel changes in Fig. 1 correspond to particles that have moved in the laboratory frame, we took as a quantitative measure for the phase transition in the numerical system the granular temperature, which is proportional to the velocity fluctuations, $T_g \equiv \frac{1}{N}\sum_{i=1}^{N}\frac{m_i}{2}(\mathbf{v}_i - <\mathbf{v}>)^2$ where $<\mathbf{v}> \equiv \frac{1}{N}\sum_{i=1}^{N}\mathbf{v}_i$. It is averaged over an integer number of full cycles of the external excitation frequency and shown in Fig.3 as function of frequency for the same amplitude as used in Fig.1.

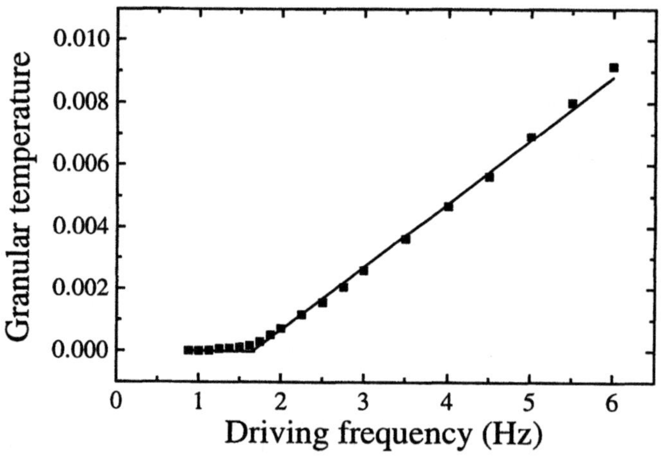

Fig. 3. Granular temperature for the numerical system as function of the external frequency for an amplitude of 2 cm.

The error bars are less than the symbol size. T_g is approximately zero for low frequencies and increases monotonically after a transition point which we estimate in a similar procedure as described above as $f^* = 1.7 \pm 0.3$ Hz for $A_0 = 2$ cm. The experimental and numerical system both show the same qualitative behaviour with a well defined transition point. Changing the restitution coefficient in normal direction in the numerical simulations did not change this transition point. On the other hand, changing the Coulomb friction coefficient μ has a dramatic effect. Setting it to zero lowers the transition point by 0.5 Hz and increasing it to one roughly increases the transition point by 1 Hz as well. We feel that the surface properties of our experimental system does lead to a higher value of μ than measured by Foerster et al. (1994) which will explain the slight shift. Varying the length of the box size did not change the transition point.

An alternative measure for the transition point can be obtained by looking at the pair distribution function (PDF) defined as (Allen and Tildesley 1990)

$$G(r) \equiv \frac{V}{N^2} < \sum_i \sum_{j \neq i} \delta(|\mathbf{r} - \mathbf{r}_{ij}|) > . \tag{1}$$

It measures the probability density to find a particle at a distance r from

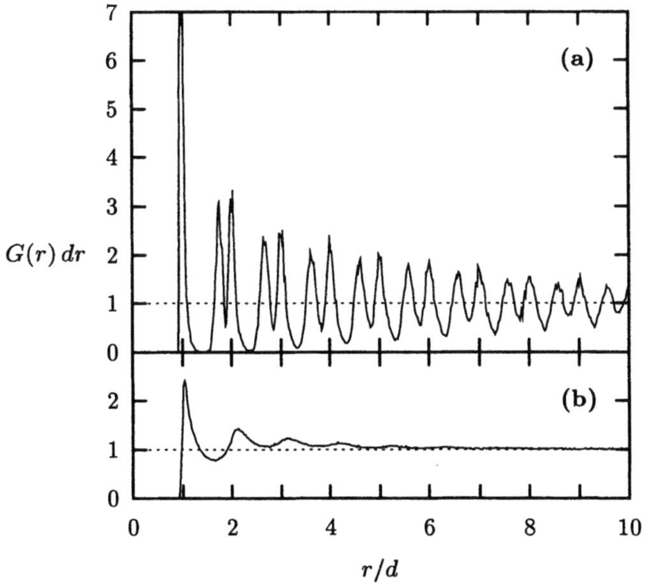

Fig. 4. Pair distribution function for the numerical system at a shaking amplitude of 2 cm: (a) at low frequency of 1.25 Hz and (b) at high frequency of 4.25 Hz.

another particle and is normalized to one for a completely random configuration. Clear double peaks are seen around integer values of r/d ($d = 0.56$ mm denotes the average particle diameter) when the system is at rest or only slightly excited, as shown in Fig. 4(a) for $A_0 = 2$ cm and $f = 1.25$ Hz, which is due to the two-dimensional hexagonal packing. Since only two layers of particles are present in our numerical system, we do not distinguish between surface and bulk particles but include all particle pairs in the calculation of the PDF. For the histogram we used a box width of dr. The height of the peaks decreases with increasing r because we have a finite system and polydisperse particles. If the external excitation is increased, either by increasing the amplitude or the frequency, the height and the number of clearly visible peaks decreases drastically, as shown in Fig. 4(b) for $f = 4.25$ Hz. This technique was e.g. successfully used to study the melting transition in plasma crystals (Melzer et al. 1996).

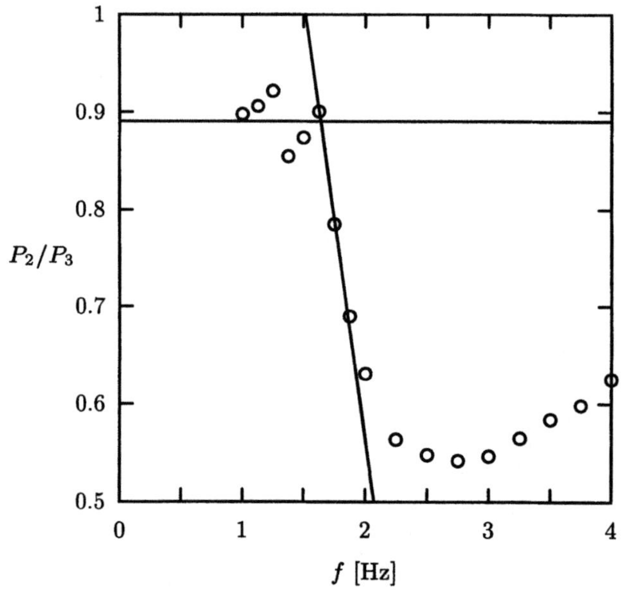

Fig. 5. Ratio of the second and third peak of the pair distribution function for the numerical system as function of the external frequency for an amplitude of 2 cm.

In order to quantify the decay of spatial order, we look at the height of the PDF at $r = \sqrt{3}d$ and $r = 2d$, which corresponds to two peaks in the hexagonal packing. The latter corresponds to order within the same particle layer whereas the first measures the order with respect to the neighbouring layer. In an infinite solid phase, both peaks have the same height, whereas the peak around $r \approx \sqrt{3}d$ decreases more rapidly when the excitation is increased and fluidization sets in, see e.g. Fig. 4(a). We propose as second order parameter the ratio of the heights, measured with respect to the constant background value of one, of these peaks, denoted by P_2/P_3. This is illustrated in Fig. 5 for a shaking amplitude of 2 cm. A qualitative change of the curve is seen around $f = 1.6$ Hz when the nearly constant ratio starts to decrease monotonically. This change corresponds to the transition point and its value, given by the intersection of the two straight lines in Fig. 5, is in perfect agreement with the one obtained using the granular temperature.

The increase with frequency of the kinetic energy or of the granular temperature varies much with the excitation amplitude. All curves collapse onto one universal curve when they are plotted as function of the dimensionless parameter $\Gamma \equiv A_0(2\pi f)^2/g$, where g denotes the gravitational constant, and the granular temperature is rescaled with the maximum excitation velocity to the power of 1.66. This might be related to the 5/3 law discussed by Taguchi (1993) and Ichiki and Hayakawa (1995). Here we use dimensionless quantities denoted by the superscript s using the average mass and diameter of a particle and also g, which is shown in Fig. 6 for three different shaking amplitudes. It is reminiscent of the scaling of the center of mass of a granulate under vertical vibration in two dimensional simulations where an exponent of 1.5 has been found (Luding et al. 1994). The error in our exponent is 0.05.

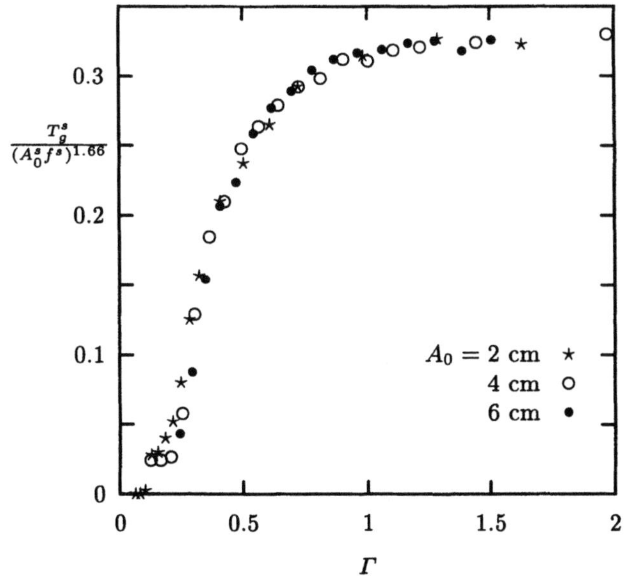

Fig. 6. Numerical simulations: rescaling of the granular temperature as function of $\Gamma \equiv A_0(2\pi f)^2/g$.

In order to find the correct scaling parameter for the horizontal shaking experiment, we look at a two-dimensional cross section of the system similar to the one used in the numerical simulation. Since the particles vary in diameter by only ± 7 %, a nearly hexagonal packing is found when the vibrations are switched off, i.e. one cylinder in the upper layer rests on two cylinders from the next layer below. For an estimate of the solid–fluid transition, one can then calculate the critical acceleration where a single cylinder would begin to move within the valley formed by these two cylinders to be $a_c = g \tan 30°$. This yields the following scaling of the critical frequency with the amplitude: $f_c \sim \sqrt{g/A_0}$.

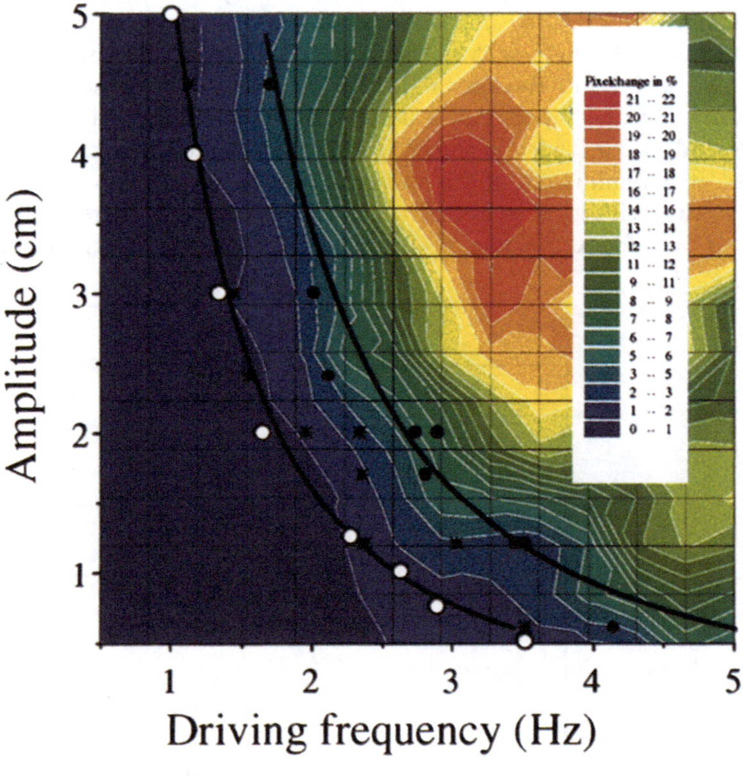

Fig. 7. Phase diagram for the solid–fluid transition: experimental points ⋆ (f_0) and • (f_c) with color scale image of the pixel change (see legend); data from numerical simulations ∘ with thick solid line as best fit. The thin solid line shows the theoretical curve.

Fig. 7 summarizes the results of the experimental, theoretical and numerical estimates for the solid–fluid transition. In order to avoid the ambiguity caused by the rounded transition from a solid to a fluid like state, we present the amount of the pixel using a color scale. The transition from no movement to a 1% change corresponds to the transition denoted f_0 above. It seems to agree fairly well with the analytical argument and with the results of the molecular dynamics simulation. The experimentally determined solid–fluid transition, denoted f_c, is obtained at slightly larger values of the driving acceleration. The difference between theory and experiment might be due to the additional friction caused by the side walls in the experimental setup.

3.2 Segregation of Granular Materials

One of the most puzzling phenomena encountered in granular matter is segregation of a polydisperse mixture of particles. In spite of much work, relatively little is known about the basic physical processes involved in the dynamics of the segregation of granular media and many puzzles remain to be solved in this field. Apart from posing numerous fundamental and difficult questions from a theoretical point of view, knowledge of segregation is needed for many industrial applications. The segregation of particles with different properties is a ubiquitous process of major importance in areas as agriculture, geophysics, material science, and almost all areas of engineering, i.e. involving preparation of food, drugs, detergents, cosmetics, and ceramics. Segregation also appears during industrial processes such as drying and coating of granular material in rotating kilns.

A common feature of all these processes is the dynamical interplay of polydisperse granular particles. Segregation can be brought about by many processes including pouring, shaking, vibration, shear and fluidization. For such systems the random mixed state is not stable and the different particle types tend to separate. In most cases the particle size is by far the most important property controlling segregation and size segregation is observed even in processes designed for particle mixing (Donald and Roseman 1962, Williams 1976, Bridgewater 1976). However, all these articles describe mechanisms for the phenomena rather than an explanation of the effects; i.e. they tell us how the particles move and not why. Size segregation seems to contradict equilibrium statistical mechanics since the density of the overall packing decreases with the amount of segregation, i.e. entropy is reduced by segregation. Segregation can occur whenever a mixture of particles of different sizes is disturbed in such a way that a rearrangement of the particles occurs; i.e. the mixture gets fluidized or expanded. There gaps between particles will occur, allowing a small particle to traverse through whereas for large particles the gaps are too narrow.

Segregation Driven by Vertical Vibration. A common way to achieve segregation is shaking in vertical direction, where one can observe the so called "Muesli effect". When a box of muesli is shaken, the large nuts and raisins are moving to the top of the muesli and do not get mixed. In general, a large ball placed at the bottom of a vibrated bed will rise to the surface. This effect occurs even if the large particle is denser than the fine particles. Two main driving mechanisms have been proposed.

One, for low accelerations is that the granular material becomes dilated with free volume introduced during each shaking cycle. The smaller particles fall relatively free; while the larger particle are prevented from doing so by statistical unlikelihood of collective motion forming such voids. Therefore the large particle rises and segregates. In this regime, segregation is interpreted as the result of competition between independent particle dynamics and collective reorganizations (Williams 1976, Dippel and Luding 1995).

The other mechanism, preferably for high accelerations, is the formation of bulk convective flow of the bed particles which, at the same time, carry the intruder (Knight et al. 1993, Cooke et al., 1996). When the large particle reaches the top, it is not able to penetrate the top layer to move downwards with the flow and therefore stays at the top. Boundaries play a crucial role in initiating and sustaining the convective flow. In this regard, these materials behave very differently from ordinary fluids where buoyancy forces can drive convection. Convection appears to be directly linked to the interaction of the beads with the container wall, i.e. higher wall friction increases convection.

However, in both cases, the size ratio and the time dependence of the phenomenon is not quite understood. In general, the speed of the segregation is profoundly dependent upon the vibration frequency. For each system there is an optimum frequency for the segregation process, but this frequency varies from system to system. This is still an unsolved problem and the dependence of the segregation on the frequency is yet not fully understood (Vanel et al. 1997).

Segregation Driven by Horizontal Vibration. A relatively unexplored question is the behaviour of particles that are subject to horizontal shaking. During vibration in horizontal direction convective motion on the surface can be observed. This can give rise to a "reverse Muesli effect", where the large particles, initially, move towards the bottom of the particle bed (Liffman et al. 1997). In fact, horizontal shaking can produce two sets of convection cells in a box of granular material. The upper set of convection cells arises due to surface-wave effects, with material convecting up the side walls and down at the center of the heap. The lower convection cells move in opposite direction, i.e., down the side walls and up in the middle of the box. They arise due to the, gravity induced, downward movement of particles in the intermittent gaps that form between the heap and the side walls.

This convective behaviour give rise to a "reverse Muesli effect", where a large particle placed at the top-center of the pile will convect downwards until it reaches the upwelling of the lower convection rolls. For "small" heaps, the large particle will actually reach the bottom. In both cases, the surface-wave convection may cause a "large" particle to return to the top of the pile.

Stratification (Segregation by Pouring). Segregation also occurs when a mixture of particles is poured onto a horizontal plate. A spontaneous stratification appears, with alternating layers of small and large particles parallel to the surface of the sand pile. Whether this stratification appears only in two dimensions (Makse et al. 1997) or also in three dimension (Julien et al. 1997) still needs to be clarified. This mechanism could play a paramount role in the formation of thick stratified sediment deposits, which can be seen nowadays in sand stone rocks. Additionally, there is an overall tendency for the large and small particle to segregate spontaneously in different regions of the heap.

Hopper. In conical hoppers, most of the segregation occurs during the filling process through flow at the free surface when particles are poured into the hopper from the top. Small particles will accumulate in the central region whereas larger particles can mostly be found close to the side walls due to their higher mobility (Brown 1939, Standish 1985).

Two basic types of motion can be distinguished in hopper: (i) funnel–flow behaviour where the flow mostly occurs in a central region which might lead to the formation of rat holes (Johanson 1978) and (ii) mass-flow behaviour where flow occurs everywhere in the bin which can lead to *first-in-first-out* characteristics (Jenike 1964). The latter operational mode can also lead to a re-mixing of particles and thus reduce segregation.

Radial Segregation in a Horizontal Rotating Drum. Radial segregation in rotating drums takes place on very short time scales (Donald and Roseman 1962). Usually the drum is roughly about half-filled and rotated along the cylinder axis about the horizontal (see Fig.8). When the drum rotates, most particles can be viewed as being part of a solid block rotating upwards. On top of it, a fluidized layer is formed with downward flowing particles. For a small angular velocity, ω, there are distinct avalanches which are well separated. For increasing angular velocity, the avalanches follow each other more rapidly and finally exhibit a continuous downwards flow for a particular range of the angular velocity ω. For even higher speeds, the particles are centrifuged to the drum wall, but no segregation is observed in this regime. The kinematics of the segregation happens only in the shear flow along the surface, in that the small particles percolate between the larger ones in the flow and get trapped by the solid block before they can reach the cylinder wall. Due to the continuous solid block rotation, a core of small particles at the center of the drum below the surface flow is formed. This is commonly referred to as *radial segregation*. In two-dimensional systems, this was well observed experimentally (Clément et al. 1995, Cantelaube and Bideau 1995). Using discrete element simulation one can define an order parameter, q, that allows to quantify the amount of segregation. It can be used to compare the segregation speed and the final amount of segregation, q_∞, for different systems and materials directly (Dury and Ristow 1997). The width of the fluidized layer plays an important role and it is found that with increasing width, q_∞ will decrease since more and more portions of the already segregated regions will be destroyed again. The segregation speed also increases with the rotation speed of the drum which corresponds to an increase in width of the fluidized layer as well.

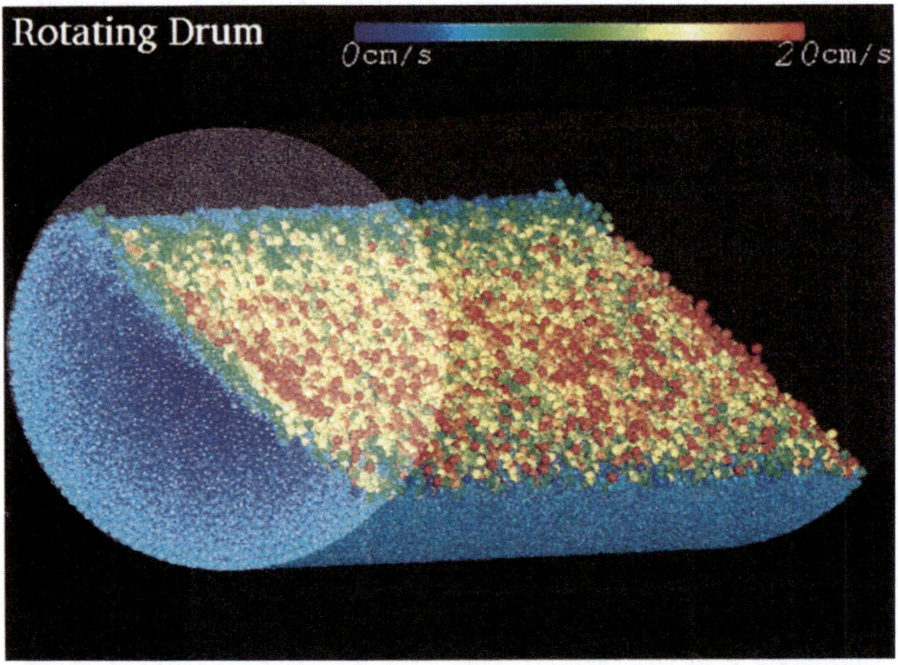

Fig. 8. Picture of the drum looking from the side. The particle velocities are color coded.

However in three spatial dimensions, radial segregation is more easily to achieve, since the voids between the particles are connected by a network and small particles can traverse more easily through it than the large particles, which will lead to a better segregation. Also in three dimensions, small particles colliding with larger ones can be deflected parallel to the direction of the rotational axis and therefore the velocity in direction of the downwards flow is reduced. Hence the particles have more time to segregate until they hit the wall (Dury and Ristow 1997). To verify this, an extended three-dimensional drum was simulated, containing a binary mixture of a total of more then 280 000 of small and large particles having a size ratio of 1.65:1, and a number ratio of 1:1, see Fig. 8. In Fig. 9, the segregation parameter is shown as a function of the drum rotations for two different widths of the fluidized layer. The upper curve corresponds to a narrow layer and shows a very fast segregation whereas the lower curve corresponds to a much wider layer and shows a significantly reduced segregation speed. A clear saturation is not yet visible but by inspecting a cross-section for the fast segregating system after two drum rotations, shown in Fig. 10, a nearly complete radial segregation of the smaller dark particles to the center is clearly visible. These preliminary results indicate that our hypothesis about the influence of the dimensionality on the segregation speed seems justified but for a more conclusive judgment, we have to rotate longer and wait for a saturation of the order parameter q.

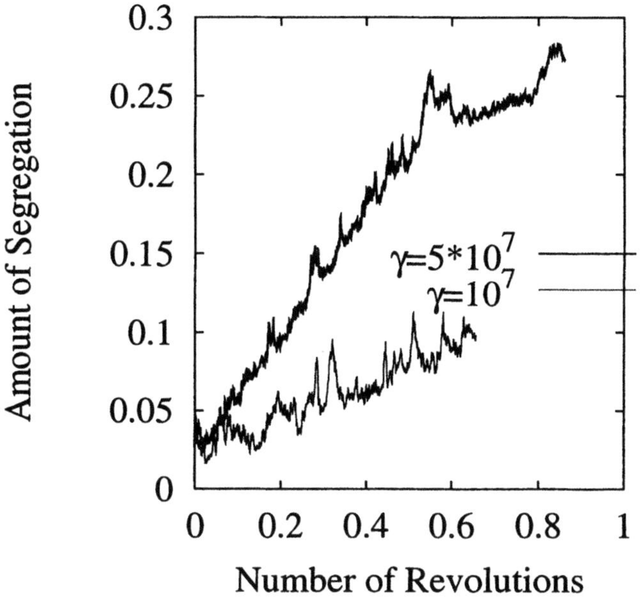

Fig. 9. Time evolution of the segregation order parameter for two different widths of the fluidized layer; narrow: upper curve, wide: lower curve.

Another point which has been observed is, that depending on the filling fraction there is also geometrical mixing (Metcalfe et al. 1995) which competes with the radial segregation. For a half-filled drum, mixing is minimal and segregation should therefore be maximal, whereas for a roughly quarter-filled drum, mixing is maximal and therefore segregation should be less pronounced than in the half-filled case. This interplay of mixing and segregation was recently studied numerically (Dury and Ristow 1998). It was found that the fastest and most complete segregation is achieved for a filling fraction of about 60% rather than the predicted 50%. This difference can be explained by the width of the fluidized layer since the numerical work was done in the continous flow regime.

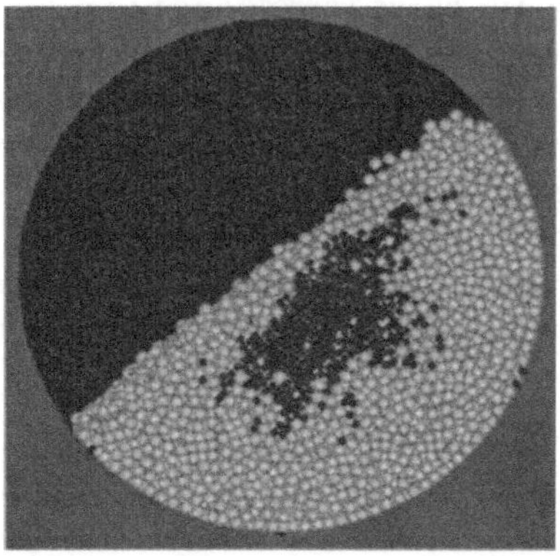

Fig. 10. Snapshot after two rotations for the faster segregating system leading to a concentration of small particles in the central region.

Axial Segregation in a Horizontal Rotating Drum. On the other hand, axial segregation happens on a much longer time scale than radial segregation. Axial bands are typically formed within minutes to hours. Also in contrast to the radial segregation, not all polydisperse systems show axial segregation; it is still an unsolved question whether a polydisperse mixture of particles will eventually segregate or not. The phenomenon of axial segregation has been long known (Donald and Roseman 1962) and an example is shown in Fig. 11. But the origin of the bands (Nakagawa 1994, Zik et al. 1994, Hill and Kakalios 1995) and whether they are stable or not (Choo et al. 1997) is still debated.

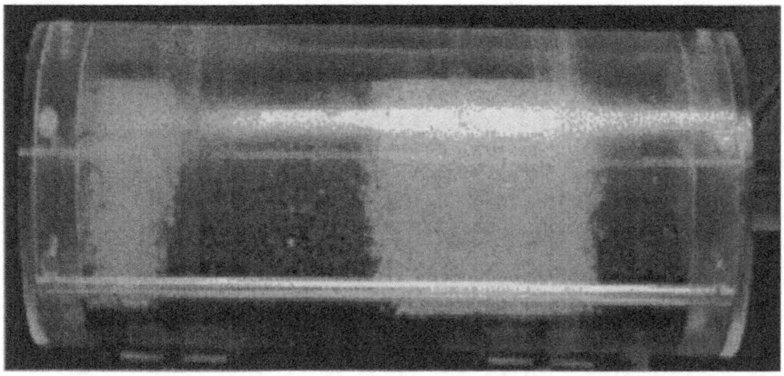

Fig. 11. Photo of a 50:50 volume mixture of poppy and mustard seeds after rotating for 1 hour at 30 rpm. A nearly perfect axial pattern is clearly visible.

One suggested mechanism builds on the observation that axial segregation occurs when the smaller particles have a higher angle of repose (the angle of repose is the angle between the horizontal and the surface of the granular mixture, Das Gupta et al. 1991). Due to local concentration fluctuations there will be regions with less small particles and therefore with a lower angle of repose. Now the larger particles from the neighboring sides with the higher angle of repose will go into this region and therefore enlarge the fluctuation. This systematic self-concentrating effect leads to zones with no large particles and zones with a very high percentage, eventually 100%, of large particles. Another possible mechanism is due to the percolation of small particles in the solid block. This might become the dominating mechanism when the radial segregation has set in. In that case, only few small particles will be present in the fluidized surface layer and the particle transport of the smaller species has to go through the radially segregated core. But more work needs to be done to fully understand the interplay of these mechanisms.

In order to investigate the band formation, we prepared different binary mixtures, 50:50 by volume, of various round particles, e.g. glass beads, pharmaceutical pills, mustard and poppy seeds. No band formation could be observed using the pharmaceutical pills and the most stable configuration was obtained by using small mustard seeds with a diameter of 1.7 mm and poppy seeds with a diameter of 1 mm. The drum was 26 cm long, 7 cm in diameter and half-filled with the initially well mixed particles. In Fig. 12, we show the concentration of poppy seeds along the rotation axis in weight percentage.

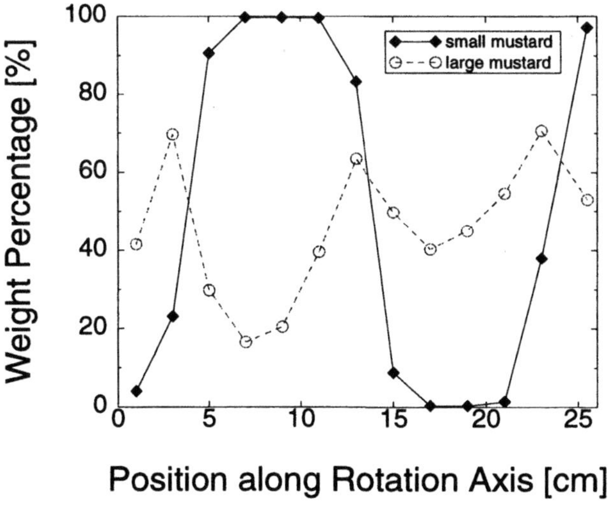

Fig. 12. Weight percentage of poppy seeds in a 50:50 mixture by volume with mustard seeds of different sizes. Axial bands are much more pronounced for small mustard seeds.

Two curves are shown: Filled diamonds correspond to the small mustard seeds and give an average value of 49.7%, whereas the open circles correspond to the of slightly heavier larger mustard seeds with a diameter of 2.5 mm and an corresponding average value of 45.7%. For this experiment the drum was rotated for one hour at 30 rpm and the mixture of small mustard seeds with poppy seeds shows a very sharp and pure stable 4 band pattern. The mixture of large mustard seeds with poppy seeds did not show very pure bands, i.e. the concentration is not close to 0 or 100%, the number of bands is between 5 and 6, and was not as stable as the previous combination of the mixture.

Ripple Formation Induced by Water Shear Flow

One of the most obvious examples of pattern–forming systems in nature are the ripples and dunes formed in sand, either by the wind, by steady (i. e. rivers) or oscillatory flow (i. e. surface waves). Apart from the basic instability triggering the ripple formation the dynamic behaviour such as the travel velocity of the established ripples is not understood.

In order to investigate this transition we have designed an annular shaped channel which is sketched in Fig. 13. The annular channel has an inner diameter of 292 mm and a channel width of 46 mm. The thickness of the inner and outer cylinder made of Perspex (Röhm) forming the annular channel is 4 mm. The channel is 60 mm high and is filled with glass beads of $150-250\mu$m diameter. The distance between the sand surface and the rotor disk, later referred to as h, is $18-21$ mm. The channel walls are made of Perspex and fixed on a plate of aluminium. This arrangement is placed in an aquarium filled with water up to 10 cm above the upper edge of the channel. The experiment were performed at room temperature. Above the channel there is a rotor dish of $D_r = 290$mm diameter made of Perspex which is connected to a motor and gear unit (Mattke Antriebstechnik). The rotor disk reaches 4 mm into the channel. The reducing gear (Stöber) with a reduction of 10.3 is used to have the motor run in the area of smallest wow and flutter. The frequency of rotation of the motor is set by a controlling unit which is driven via the serial interface by a PC (90 MHz Pentium). The variations of the frequency of rotation is less than 0.1% according to the manufacturer. A CCD–camera (Sony) fixed in the laboratory frame and controlled by a frame grabber card (Leutronvision) is observing a window of 100 mm in width and a height capable of covering the interface curve between the sand and the water flowing above it. With a constant sampling rate $f_s = \frac{1}{T}$, the boundary between sand and water is detected with the help of the gradient in contrast between the bright sand and the darker water. The parametrized boundary is stored for the purpose of further evaluation.

In Fig. 14 the dependence of the amplitude, the velocity and the total number of the ripples on the Reynolds number is given. In order to obtain these data all measurements were performed with a flat sand surface and zero rotation of the rotor disk as starting conditions. The rotor is speeded up

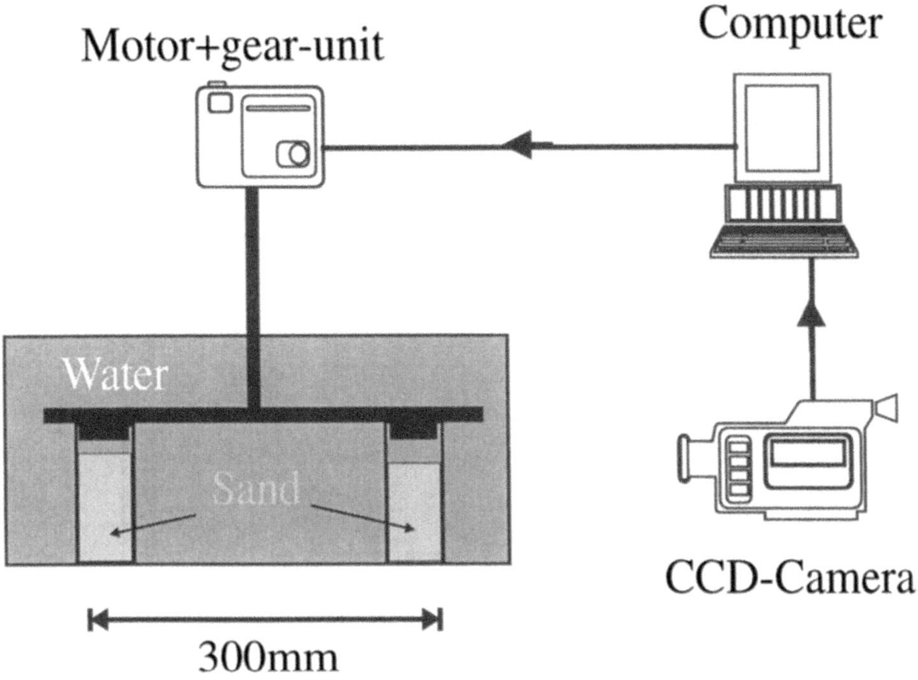

Fig. 13. Experimental setup for water shear flow.

within a few seconds to a certain frequency. This frequency corresponds to the Reynolds number on the abscissa of Fig. 14 via $Re = \frac{\rho u h}{\mu}$. Here, u denotes the maximal azimuthal velocity $u = \pi f D_r$ of the rotor disk of diameter D_r rotating at the frequency f and h is the distance between the flat sand surface and the rotor disk. The values for the density ρ and the viscosity μ of water are taken at room temperature. For each Reynolds number we wait for a time t_w of half a day before starting the data acquisition. The amplitude, which is plotted in the lower part, is determined by the difference between the maximal and the minimal amplitude of a ripple $(A_{max} - A_{min})/2$. For each Reynolds number this procedure is repeated at least 100 times. The error bars are given by the standard deviation $\sigma = \frac{1}{N} \sum_{i=1}^{N} (A_i - \bar{A})^2$. The solid line is a fit of a square root function (expected for a supercritical bifurcation) to the lower eight data points. The travel velocity of the ripples shown in the middle part of Fig. 14 was determined from the shift L of two consecutive boundary curves via a correlation function. As the sample time between two consecutive data sets is T, one can calculate the travel velocity $v = \frac{L}{T}$. Again the error bars give the standard deviation. In the upper part of Fig. 14 the

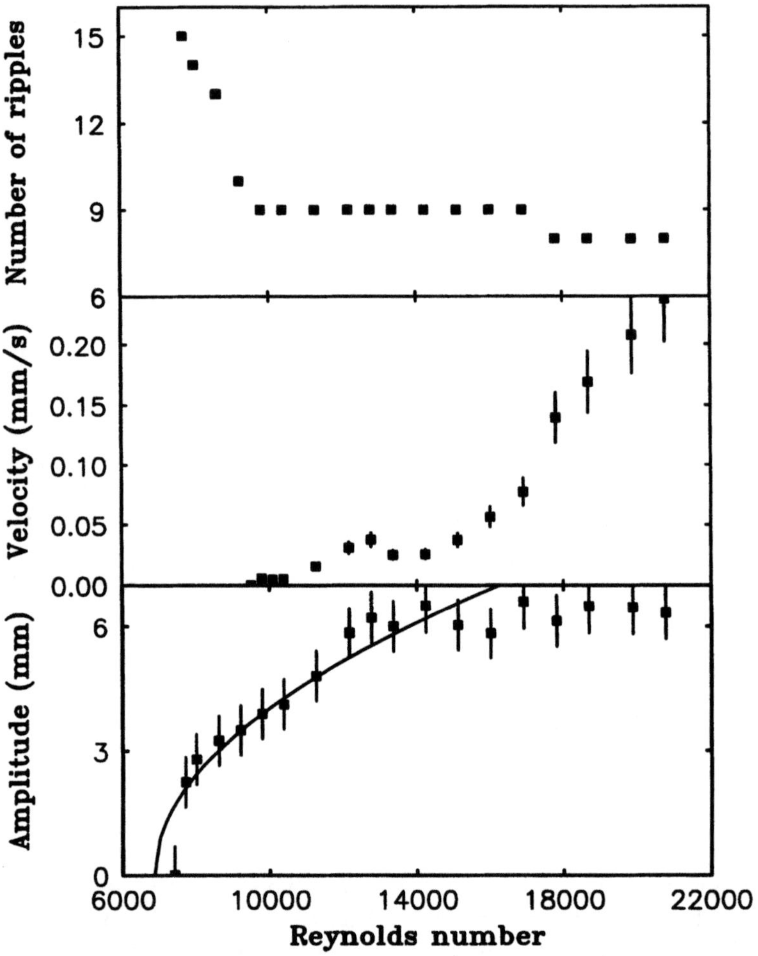

Fig. 14. The dependence of the amplitude (lower part), the velocity (in the middle) and the number of ripples (upper part) on the Reynolds number.

number of ripples along the circumference of the channel after running the experiment with a fixed frequency of rotation for time t_w is shown. This quantity was determined by visual inspection.

On the basis of Fig. 14 we interpret the transition from the initially flat to the rippled sand surface as a supercritical bifurcation taking place at a Reynolds number of 6860. We have not been able so far to measure the details of the transition because the time necessary to establish ripples all over the circumference increases rapidly as the critical Reynolds number is approached. In this regime no velocity data are available because the possibly very small movement is beyond our experimental resolution. A clearly measurable velocity sets in at a Reynols number of about 9500. This increase of the speed is accompanied by a dramatic change in the slope in the upper curve. We cannot offer an explanation for this observation. The second transition takes place at a Reynolds number of about 14000, where again a qualitative change in the velocity occurs and the amplitude saturates.

All three observations need to be investigated in more detail. Our current work focusses on the transition from a flat sand surface to a ripple pattern, however.

Pattern Formation under the Influence of Horizontal Vibration

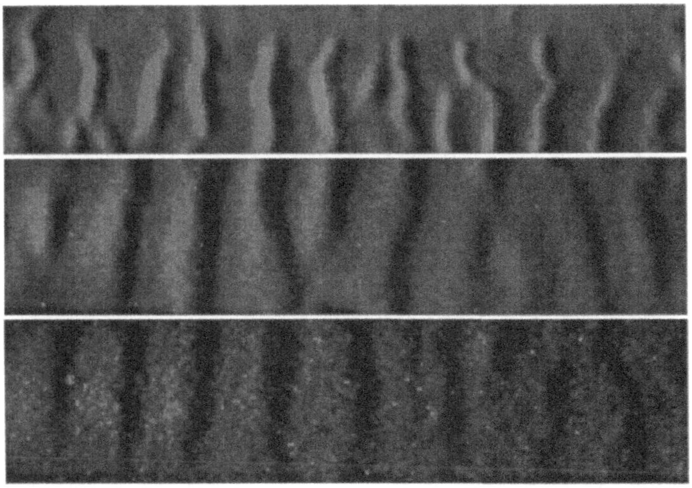

Fig. 15. Patterns generated during horizontal vibration. Dumping height between one and three layers of granulate, 25 cm of a channel with a total length of 35 cm are visible, driving amplitude: 2 cm, driving frequency: 4 Hz, material: glass spheres with different diameters: a) $150\mu m \leq D \leq 160\mu m$, b) $210\mu m \leq D \leq 250\mu m$, c) $430\mu m \leq D \leq 600\mu m$.

The behaviour of granular matter under the influence of periodic forces has so far mostly been studied on systems which are vibrating parallel to the gravitation field (Melo et al. 1994, Jaeger et al. 1994, Luding et al. 1994). Only a few works focus on horizontal driven systems. Evesque (1992) and

Rosenkranz and Pöschel (1997) reported about an experiment with horizontal vibrated granular matter and large dumping heights. Above a given threshold of vibrating amplitude they observed a small convexity of the free surface of the pile and a convection flow.

We are dealing with the behaviour of only a few layers of granulate after fluidization. For a pattern formation under horizontal vibration, which is stable in space and time, one can use vibrating plates or channels. In Fig. 15 some images of patterns are shown. There are many parameters which may influence the pattern, e.g. the properties of the granulate, the roughness of the container surface, the driving amplitude and frequency. The influence of some of these parameters on the pattern formation have been studied qualitatively. We find that already a monolayer or even less than a monolayer of granulate is sufficient for cluster formation. For a characterisation of the density of the granulate in such a system we use the "solid fraction" (McNamara and Young 1996). Because of the relatively small number of particles that exist in such a system it could be simulated numerically by means of molecular dynamics. A theoretical description with similar approaches as used by Sela et al. (1996), McNamara and Young (1996), Spahn et al. (1997), and Grossmann et al. (1997) seems to be possible.

For generating the vibration and capturing the pictures we used the setup described in section **3.1**. As container for the granulate we used boxes at 4 cm width and with a length between 35 and 47 cm in the direction of vibration. The container was made of aluminium and had a plastic foil with a roughness of 7 μm glued on its bottom. A CCD–camera observed the container from above, i.e. perpendicular to the direction of vibration. We used grazing incidence of parallel light along the direction of the oscillatory motion to visualise the modulation of the surface. When using less than one monolayer of granulate we illuminate with divergent light from a point source, which is perpendicular to the channel, to visualize the fluctuation of the grain density. For the determination of space–time–diagrams we measure the intensity along one or more lines of a frame. The measurement can be taken at any stage of the motion. For time–resolved measurements of cluster formation and cluster interaction a high–speed–camera (1000 frame/s, 5400 frames) was used. The material used in preliminary measurements was the same as in section 3.1. Later we used real granulate (grains of seed) and monodisperse steel spheres with polished surface.

Influence of the Channel Walls on the Patterns. The influence of the channel walls can be seen in Fig. 16. The wavelength seems to rise with the channel width. The channel with the largest width has a wavelength close to the one formed in a layer that has no contact with the boundaries at all. This channel width was used in the following experiments, i.e. the influence of the channel walls parallel to the direction of vibration has only little influence in our experiments.

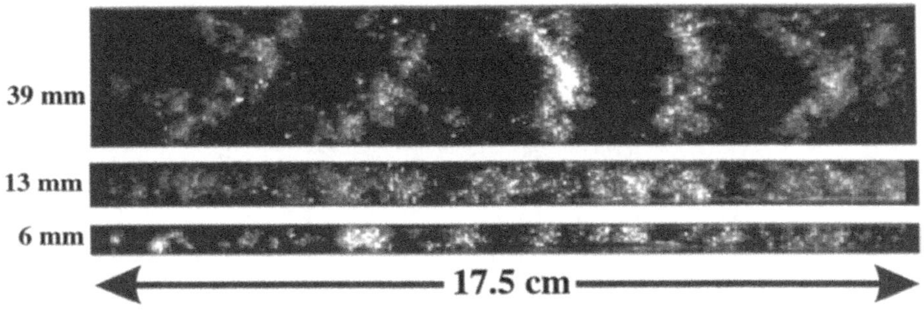

Fig. 16. Influence of the channel-width on the pattern. Driving amplitude: 1.8 cm, driving frequency: 3.3 Hz. Material: Monolayer of glass spheres, 0.6 mm diameter.

Influence of the Roughness of the Channel. The influence of the surface roughness was studied by preparing the bottom of the channels with glass spheres of different radii (Fig. 17). With rising roughness of the surface a pattern formation was prevented. If the ratio of the radius of the moving spheres to the radius of the ones glued on the bottom of the surface is around 10, pattern formation occurs. This indicates that the possibility for rolling or sliding motion of the spheres is a prerequisite for pattern formation. For great dumping heights only the upper layers of the granulate are fluidized. The lower layers act as a solid–like surface with the roughness of the granulate and thereby prevent a pattern formation. The influence of the roughness on the motion may be demonstrated by "cooling" a fluidized granulate by decreasing the driving frequency. Fig. 17 shows the patterns consisting of clusters of a real granulate (grains of seed) in three synchronously vibrated channels of different roughness after such a "cooling". A large roughness hinders the motion in the co–moving system, and decreases the distance of the clusters observed after the "cooling" process.

Influence of the Driving Amplitude. The influence of the driving amplitude on the pattern may be demonstrated by the "cooling" process and is shown in Fig. 18. The distance of the cluster after "solidification" increases with the driving amplitude, i.e. with the amplitude of the motion of grains in the co–moving system. The significant difference in the graphs of the seed grains and the glass spheres is not surprising because of the difference in all relevant properties of the two materials. Nevertheless, the behaviour of a real granulate is not totally different from the behaviour of the glass spheres in this experiment.

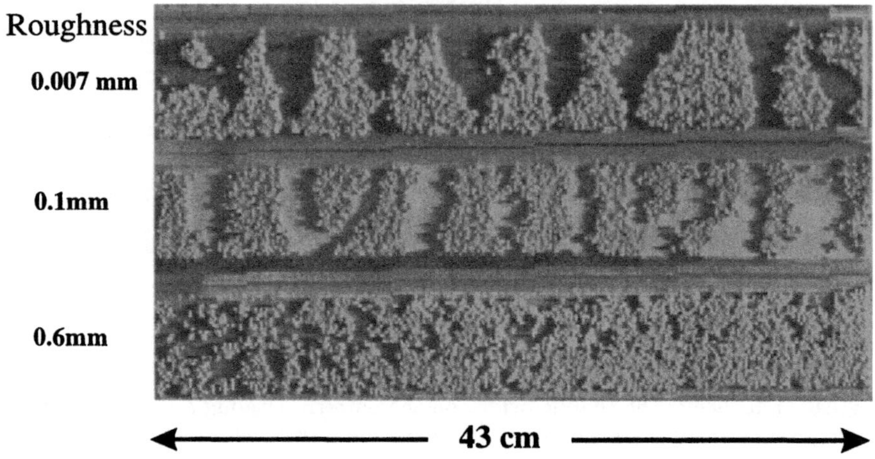

Fig. 17. Influence of the roughness of the channel on pattern. Material: Mustard seeds (1-2 mm diameter), driving amplitude: 2 cm, solid fraction: 0.44

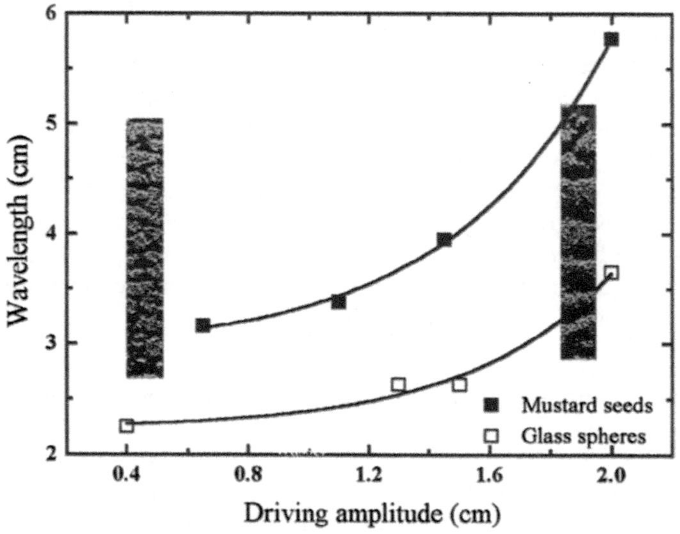

Fig. 18. Influence of the amplitude on clustering observed by "cooling". The images in the figure are taken from mustard seeds.

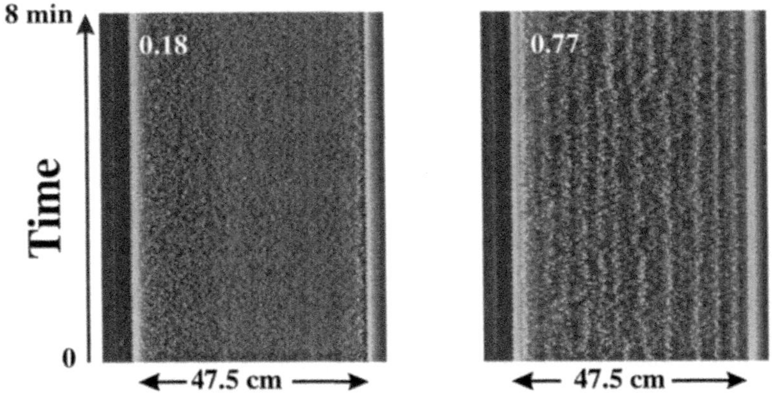

Fig. 19. Influence of the solid fraction on pattern formation. Material: Glass spheres of 1 mm diameter. Frequency: 4.5 Hz, amplitude: 1.2 cm.

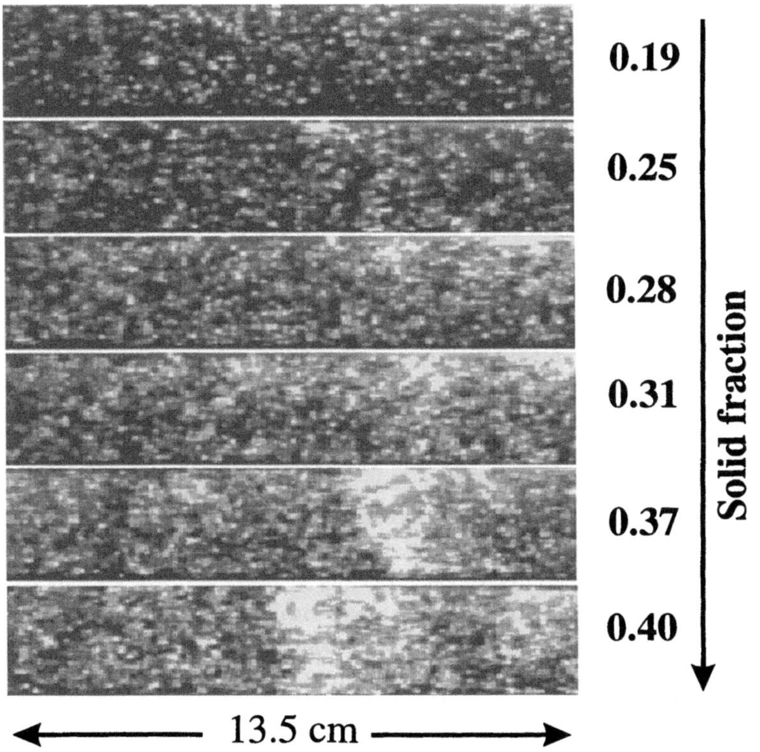

Fig. 20. High speed images below and above the transition point to clustering. Material: Glass spheres 1 mm diameter. Frequency: 3.5 Hz, amplitude: 1.2 cm.

Mechanism of Pattern Formation. In searching for a mechanism for the pattern formation we focus on a 2-dimensional system similar to those in Fig. 17 and Fig. 18. This means that we use a system with less than one monolayer. Depending on the solid fraction we have recorded space time diagrams from the intensity of one line from the images (Fig. 19). We used gracing incidence for the light. Above a critical value of the "solid fraction" we observe patterns, which are nearly stable in space and time.

By using an illumination similar to the one in Fig. 17 one can capture a representative number of single grains. We demonstrate, that the patterns are periodical changes in the granulate density (Fig. 20). By starting the vibration with grains distributed uniformly in the channel one can identify the clustering by comparing consecutive images. Pattern formation (clustering) starts above a critical value for the solid fraction. Near the critical value the interaction of clusters is observable. The exact determination of the transition point from a statistical independent motion to clustering will require the determination of the local position of the spheres by image processing.

Pattern Formation under the Influence of Circular Vibration

In this section we now extend the above–mentioned linear horizontal vibration, where stripe–like patterns were found, to an orbital vibration. It is the typical motion which can be found in the late stages of never–ending boring dinner conversations. Under such circumstances people often start swirling their glasses in which some table water, wine, beer or even cognac might be left. In most cases the performed motion ends up to be an almost perfect orbital vibration which is horizontally aligned by the table. For a physicist this situation yields the advantage that he or she can focus on phenomena that arise in this kind of driven system. In a cognac glass i. g. gravity driven finger patterns appear in the region where the evaporated alcohol again condenses. In the context of this presentation we study a system which is unlikely to be present after dinner: a granular material. For a thin layer of sand it is found that it behaves like a solid at low rotation frequencies f_d. As f_d exceeds a critical value the granular material fluidizes, which may lead to spiral patterns as shown in Fig. 21.

To uncover the underlying physics of this system we restrict our examinations to a monolayer of spheres and small numbers of particles N with a rather large diameter D. Here, we review measurements that were performed in two different configurations: a circular dish and an annular channel. A transition where the sense of rotation of the cluster changes due to the number of particles is found in both: For small N the particles follow the direction of the swirling motion (rotation) whereas the sense of rotation inverses (reptation) when N exceeds a critical number. The experimental results are compared with a two–dimensional molecular dynamics simulation in the case of the circular container, and with a one–dimensional model for the annular channel.

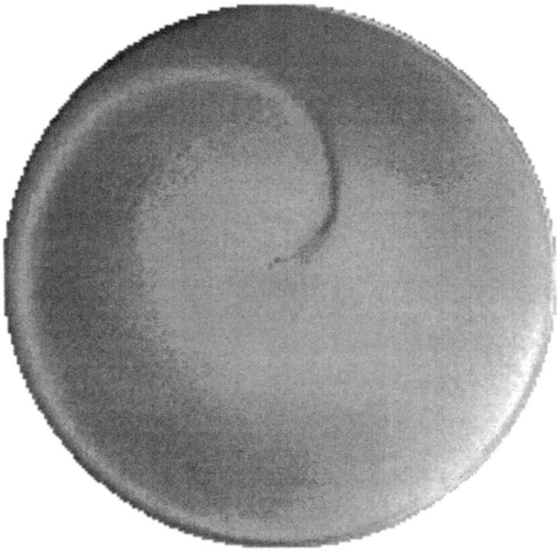

Fig. 21. Inverse image of a sand spiral obtained by swirling a thin layer of sand (size of circular container: 90 mm, frequency of driving: 3.63 Hz, amplitude of driving: 12.70 mm, diameter of sand beads: 50-150 μm, mass of sand: 13.2 g).

Experimental Setup. To investigate the behaviour of spheres under horizontal orbital vibration we used an adjustable reciprocating orbital shaker (Thermolyne AROS 160). In this apparatus every point (x, y) on the swirling table moves counterclockwise on a circle in the laboratory frame: $x = A_d \cos(2\pi f_d t)$ and $y = A_d \sin(2\pi f_d t)$, where A_d is the driving amplitude of the shaker, f_d the driving frequency and t the time. In the following experiments A_d is set to two different values: 9.53 mm and 12.70 mm. f_d is in the range from 0 to 4.5 Hz. This kind of orbital vibration yields no center of rotation and every point feels the same centrifugal acceleration. As can be seen in Fig. 22, the setup is fixed on a heavy marble table to avoid internal vibrations. As containers we used a Petri dish and an annular channel. The Petri dish has an inner radius of 45 mm. The annular channel is milled out of Perspex. The outer radius of the channel is 50 mm and the width 10.5 mm, respectively. A CCD–camera (Sony XC–77RR–CE) on top of the setup allows visualization of the particle dynamics in the co–moving frame. A ring–shaped lamp (Schott KL 1500 electronic) and an annular neon light (Philips TLE Cool White) illuminates the scene. For the investigations in the circular container we use ceramic spheres of diameter $d = 12$ mm. The experiments in the annular channel are performed with precision spheres of different materials: bronze (10.00 $\pm$ 0.11 mm), brass (10.00 $\pm$ 0.11 mm), Polyurethane

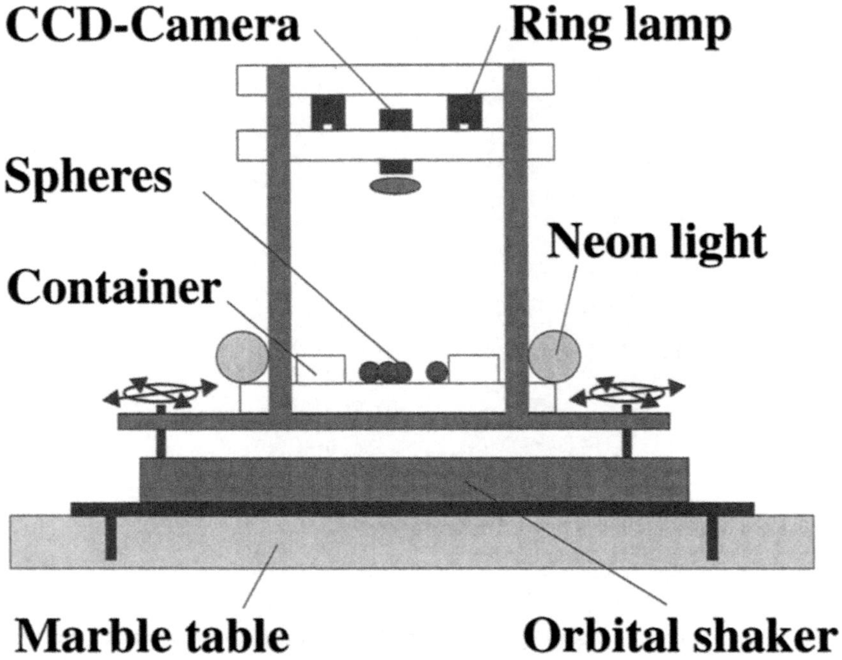

Fig. 22. Experimental setup for circular vibration.

(PUR, (9.85–10.10 mm)) and two types of Soda lime glass (glass): polished (10.00 ± 0.02 mm) and deadened (10.00 ± 0.02 mm).

Experimental Results for the Circular Dish. We now present results on the dynamics of the ceramic spheres in the circular dish. A_d was set to 9.53 mm and f_d to 2.5 Hz. Fig. 23(a) – Fig. 23(f) show the temporal evolution of 26 spheres in the co–moving frame. The momentary acceleration is indicated by the white arrow in each image. The arrow is obtained by drawing a line from a marker (small white cross) through the center of the Petri dish. The marker is fixed on a glass plate which is stationary in the laboratory frame and in between the dish and the camera. The position of the glass plate is adjusted in such a way that the marker circles around the center of the container when visualized in the co–moving frame. The snapshots were taken every 80 ms which corresponds to a time interval of $(5f_d)^{-1}$. Thus, Fig. 23(a) and Fig. 23(f) belong to the same phase of the swirling cycle. By tracing the position of a single sphere in the cluster one finds that the sphere moves in the direction of the orbital shaking.

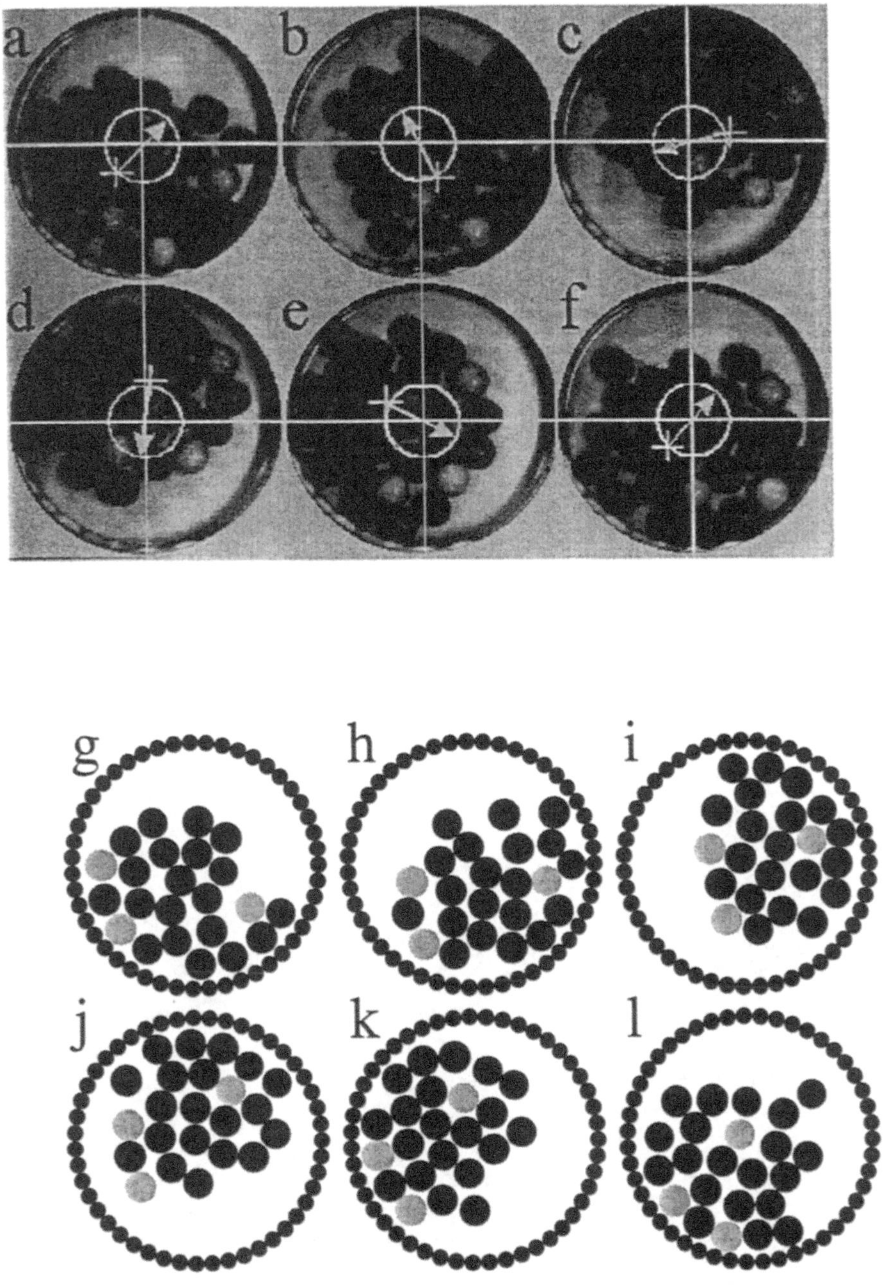

Fig. 23. Temporal behaviour of 26 spheres during one cycle of the orbital vibration (Scherer et al. 1996). The arrow indicates the direction of the momentary acceleration in the experimental snapshots ((a) – (f)). The lower six figures ((g) – (l)) correspond to the molecular dynamics simulation. Frequency of driving: 2.5 Hz, diameter of spheres: 12 mm.

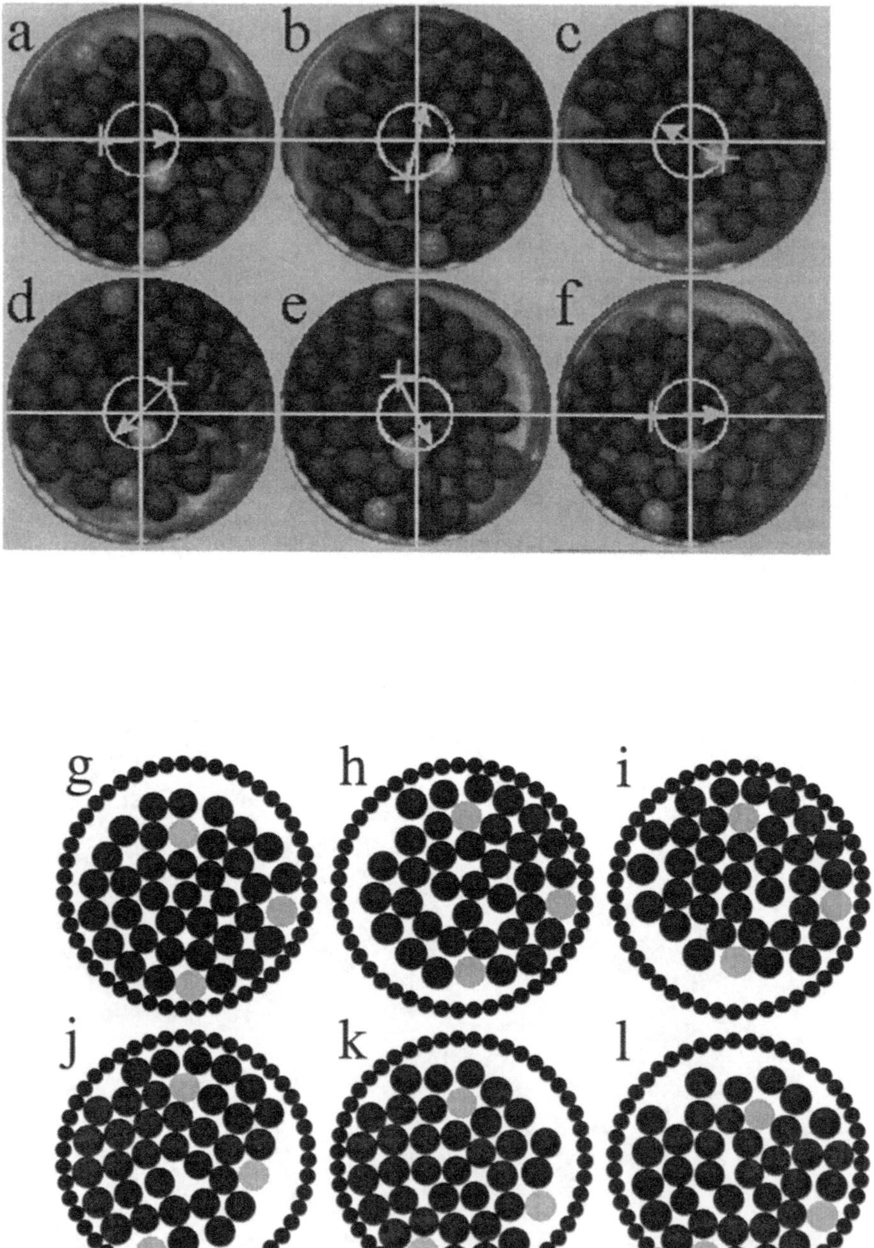

Fig. 24. Temporal behaviour of 37 spheres (Scherer et al. 1996). (a) – (f): experiment, (g) – (l): molecular dynamics simulation.

This behaviour is observed for all spheres and results in a global rotation of the cluster in the same sense as the external driving. This state we call rotation-mode. The experimental system has been compared to a two–dimensional molecular dynamics simulation. For details on the numerical procedure see the paper Scherer et al. (1996). Fig. 23(g)–Fig. 23(l) shows the same qualitative behaviour as the experiment.

For a higher number of spheres we observe a change in the dynamics of the cluster. Fig. 24(a)–Fig. 24(l) show the behaviour of 37 spheres for both the experiment and the numerical simulation. Here, the spheres move clockwise despite the counterclockwise motion of the shaking. We call this different state reptation-mode.

To quantify the dynamics of the cluster we measured the time needed by a single sphere close to the boundary of the Petri dish to circumnavigate the container. This gives a frequency of rotation f_s as a function of the number of particles N. The open circles in Fig. 25 represent these measurements. It can be seen that when N exceeds 24 f_s decreases. At a critical particle number of 32 we observe a change in the sense of rotation of the cluster. Increasing the number of particles gives a higher frequency of rotation whereas the motion again slows down at very high particle numbers. We compare this findings with measurements obtained by the molecular dynamics simulation. Here, all positions and velocities of the spheres were taken into account and the frequency of rotation of all particles with respect to the center of mass of the cluster was calculated. The measurements are indicated by solid squares in Fig. 25. For sufficiently small N it shows a good quantitative agreement with the experimental findings. For small particle numbers the experimental method for measuring the rotation frequency of the cluster breaks down because a single particle does not stay close to the boundary of the Petri dish. It has the tendency to migrate inwards.

Experimental Results for the Annular Channel. In a swirled annular channel it is found that there exists the same transition from rotation to reptation as seen in the circular dish (Scherer et al. 1997). Here we present measurements on the influence of the sphere material and the driving frequency for both modes, rotation and reptation. Figure 26 shows the response of the frequency of rotation on increasing the driving frequency for a particle number of 10. At this stage the one–dimensional array of particles is in the rotation mode. There is no significant influence of the material of the spheres: f_s increases linearly as f_d is set to higher values. In contrast, a cluster of 25 spheres shows a different behaviour (Fig. 27). The array is now in the reptation mode. While increasing the driving frequency the spheres revolve the channel with a higher frequency. This behaviour does not seem to depend on the material at low frequencies. But it breaks down when f_d exceeds a critical value of 2.5 Hz. At this stage deadened glass slows down while the frequency of rotation of PUR still increases.

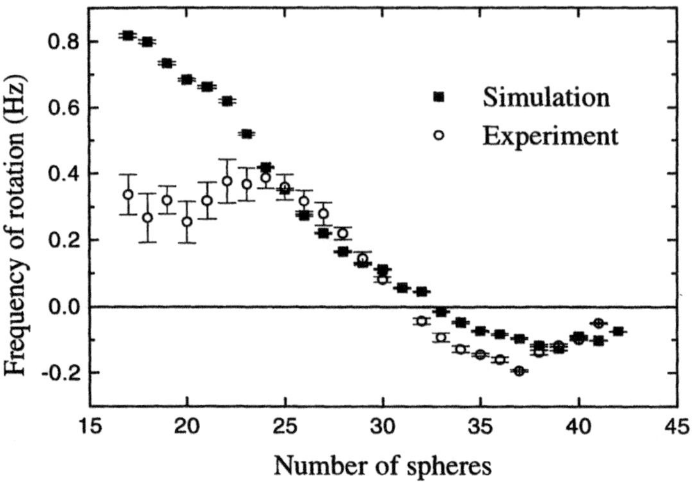

Fig. 25. Frequency of rotation of the cluster (Scherer et. al 1996). The solid squares show the results of the molecular dynamics simulation, while the open circles represent the experimental measurements. For a critical number of 32 spheres there is a reversal in the sense of rotation of the cluster.

To get a closer picture of the dynamics of the spheres we visualize the trace of the center of each sphere. Fig. 28 shows space–time evolutions for four different particle numbers. As particles bronze spheres were used. The amplitude of driving is set to 9.53 mm; the frequency of driving is 2.5 Hz. The horizontal axes of each image belongs to the space coordinate and the vertical axes represent time, respectively. Time evolves from bottom to top. The centers of mass of the spheres appear black. At low particle numbers ($N = 10$) we notice that there are almost no collisions. Each particle has a slow translatory motion which is superimposed by a fast back and forth motion of the sphere. Near the critical point where the transition from rotation to reptation ($N = 25$) occurs we observe strong fluctuations in the motion of the spheres. At higher solid fractions the spatial–temporal behavior becomes more ordered. Fig. 28 gives an example for $N = 24$. The highest possible number of particles in this experimental case is $N = 28$. Here, the frequency of rotation is very slow; for a limited part of the period of the external driving the particles even move backwards. The dynamics of a single sphere can be

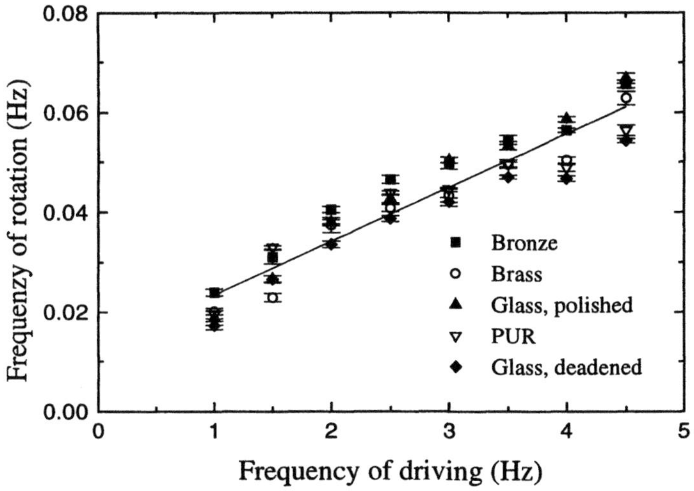

Fig. 26. Dependence of the frequency of rotation of the spheres on the driving frequency for different materials (Scherer et al. 1997). The number of spheres in the annular channel is 10. The line serves as a guide to the eye.

described by a one dimensional model,

$$\ddot{\phi} + \eta\dot{\phi} + \epsilon \sin(\phi - t) = 0, \tag{2}$$

where ϕ is the position of the sphere in the channel, η an idealized viscous friction, and ϵ the ratio between the driving amplitude and the distance from the center to the middle of the annular channel. The time t has been made dimensionless by multiplication with the angular frequency of the driving.

It can be shown (Scherer et al. 1997, Scherer 1998) that for $\eta/\epsilon < 1$, which is the regime of the experimental setup, there exists a solution where the sphere has a slow drift superimposed by a fast oscillatory part,

$$\phi(t) = \frac{\epsilon^2}{2(1+\eta^2)}t + \frac{\epsilon}{\sqrt{\eta^2+1}} \sin(\phi - t + \arctan\eta). \tag{3}$$

This is in good agreement with the experimental findings. Measuring the drift velocity of the slow translatory motion and the amplitude of the fast oscillation give two independent methods to determine the coefficient of viscous friction η. For bronze spheres, which are swirled with a driving amplitude of 9.53 mm and a driving frequency of 2.5 Hz (see Fig. 28), we have obtained $\eta = 0.7$ for both cases.

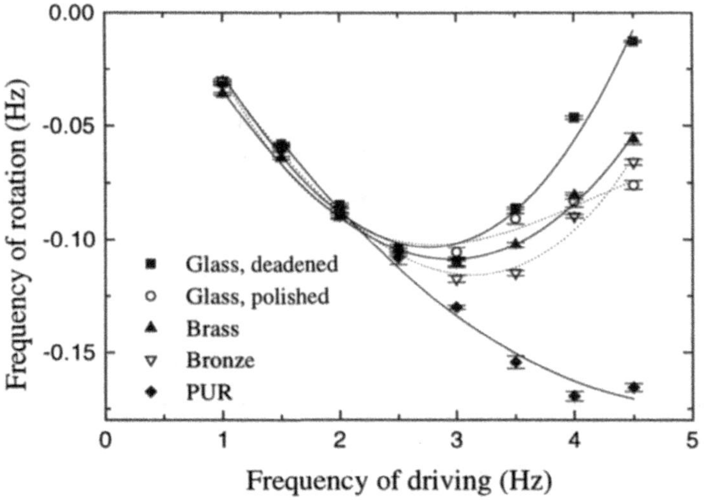

Fig. 27. Dependence of the frequency of rotation on the driving frequency for 25 spheres (Scherer et al. 1997). All curves have parabolic shape except for polished glass where a cubic fit is more reasonable.

We use Eq. (2) with $\epsilon = 0.21$ and $\eta = 0.7$ to simulate the behaviour of many particles[1] (Scherer et al. 1997). When there is a collision between two neighbouring particles 1 and 2, we calculate their new velocities v_1' and v_2' from

$$\begin{pmatrix} v_1' \\ v_2' \end{pmatrix} = \begin{pmatrix} 0 & \nu \\ \nu & 0 \end{pmatrix} \begin{pmatrix} v_1 \\ v_2 \end{pmatrix}, \tag{4}$$

where v_1 and v_2 are the velocities of both particles before the collision, and ν is the coefficient of restitution. Fig. 29 shows f_s normalized by f_d as a function of the number of spheres. The solid squares are the experimental values obtained for bronze spheres. The simulation starts with elastic collisions, which means $\nu = 1$ (solid circles). In disagreement with the experimental values, the frequency of rotation does not decrease when approaching the critical point where we observe experimentally a transition from rotation to reptation. Only for $N = 28$ we find a reversal in the sense of rotation. This behaviour changes dramatically when we set the coefficient of restitution to

[1] An interactive Java simulation of the numerical system can be found at URL http://comserv.urz.uni-magdeburg.de/~anp/safe/swirlsim/swirlsim.html. Various parameters like the coefficient of restitution or the number of spheres can be varied.

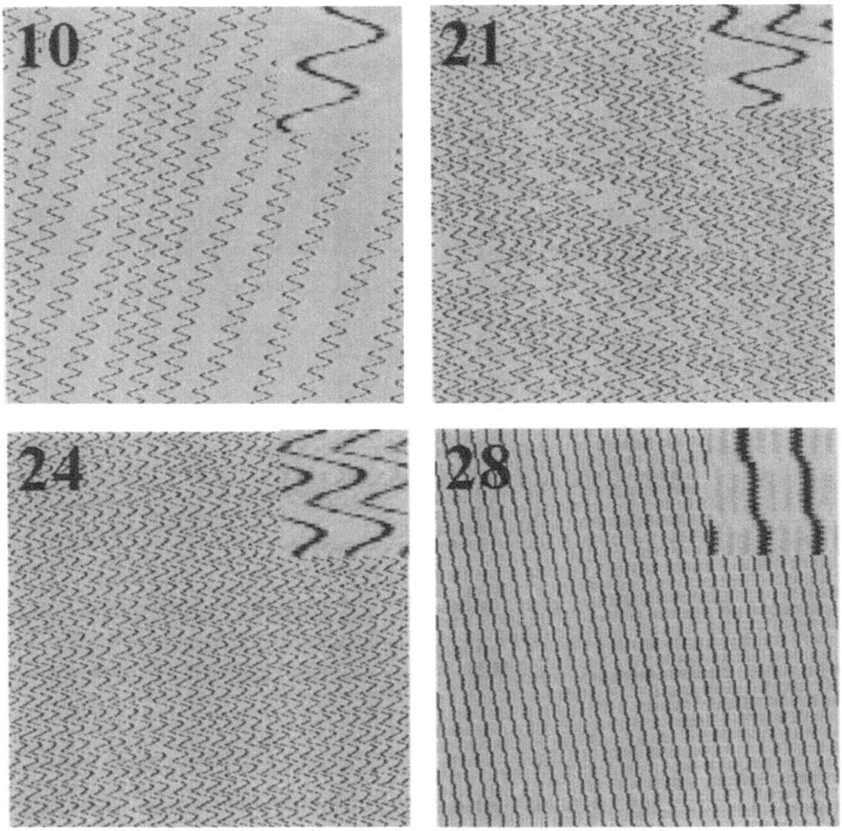

Fig. 28. Spatial–temporal evolution of the centers of the spheres (Scherer et al. 1997). The horizontal axes in each image is the space coordinate (0 to 2π) while the vertical axes shows the time (0 to 10.24 s). The number in the upper left corner denotes the number of spheres in the annular channel. In the upper right corner a magnified section is given. Material of the spheres: bronze, driving amplitude: 9.53 mm and driving frequency: 2.5 Hz.

0.99 (open triangles). Here, we already get the correct qualitative behaviour. Varying ν we find that there is a good agreement for 0.7 (open circles).

We have tested a collision model where energy dissipation is introduced by reducing the relative velocity of both particles and an additional conservation of momentum (Scherer et al. 1997). Even in the case of completely inelastic collisions we were not able to reproduce the experimental results. This is due to the fact that not only normal collisions are important for this process. The

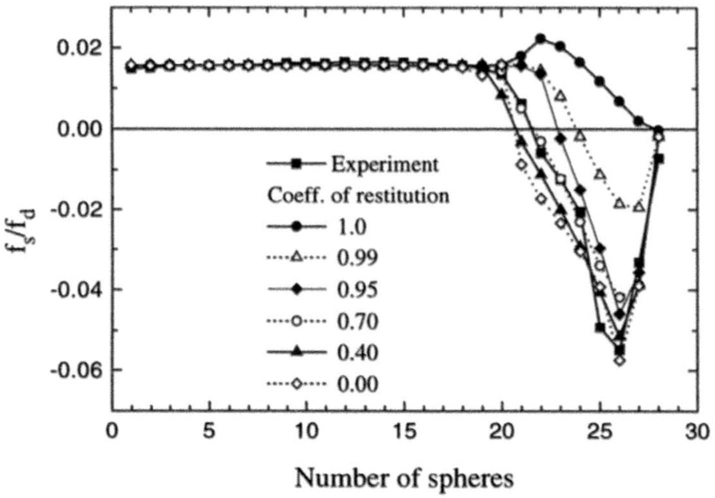

Fig. 29. Comparison of experimental results with numerical simulations of many particles in dependence of the coefficient of restitution (Scherer et al. 1997). The experimental conditions are the following: As sphere material we use bronze, $A_d = 9.53$ mm, and $f_d = 2.5$ Hz.

tangential shear at the contact points of the colliding spheres has to be taken into account as well. Thus, the dissipation of energy also depends on the spin of the particles and the collision process is governed by the absolute velocity of the spheres as modeled by Eq. (4).

In summary, we have presented measurements on the transition from rotation to reptation in granular material under horizontal orbital shaking. The experimental findings in the case of the circular dish are in good agreement with a two dimensional molecular dynamics simulation. This is of importance for processes in vibrational mills where the same effects occur (Yokoyama et al. 1996). To understand the underlying physics of this phenomena we have reduced our system to an annular channel. It was shown that material properties do not influence the dynamics in the rotation mode. In contrast we see a strong dependence on the sphere material in the reptation mode. This suggests that the granular dynamics is governed by binary collisions. We have confirmed this idea numerically. We first studied the dynamics of a single sphere. The results found by analysing a one–dimensional model were in good agreement with the experiment. A single sphere shows a slow translatory motion on which a fast back and forth motion of small amplitude is superimposed. We have used these results to simulate a many–particle situation. It was found that elastic collisions do not give a transition from rotation to reptation whereas inelastic collisions confirm the experimental values accurately. The degree of accuracy depends on the coefficient of restitution.

We plan to study pattern forming processes like the spiral seen in Fig. 21. The experimental results should give a hint on how to model the basic dynamics of granular materials.

Pattern Formation at the Interface of a Sedimenting Suspension

In many cases the air between granular particles can be neglected, and the granulate behaves in the same manner as it would under vacuum conditions. For reasons of convenience, most experiments are not performed in vacuum, however, and it is well known that the air might give rise to forces between the particles. These forces are much stronger if the air is replaced by water. If the resulting suspension is a dilute one, the characteristics of the granulate might become totally unimportant. The suspension behaves like a Newtonian fluid with a slightly increased viscosity, namely the viscosity coefficient increases by a fraction which equals 2.5 times the volume fraction Φ of the suspended spheres (Einstein 1906, Einstein 1911).

This is no longer true for more concentrated suspensions. Expanding the viscosity η in terms of Φ one finds the coefficient of Φ^2 to be dependent on the geometry of the flow (Batchelor and Green 1972, Batchelor 1977). This fact also explains why the result of viscosity measurements depend on the geometry of the apparatus (Pätzold 1980). If the volume fraction exceeds a certain threshold there occurs shear thickening i.e. the apparent viscosity increases with the shear rate. In this region, which was first described by Bagnold (1954), the suspension is clearly non–Newtonian. All these complications should be born in mind when interpreting the experimental results described below, namely the formation of patterns in a sedimenting suspension.

Measurements. The main problem for an experimental setup is how to prepare the initial situation with the denser material on top of the lighter one, and with a sufficiently flat interface between both phases. In our experiment we try to reach this situation by turning a closed Hele–Shaw–like cell around a horizontal axis (see Fig. 30).

The cell, a CCD–camera and a neon tube are fixed to a frame whose bottom bar is the rotation axis. This allows image analysis in the co–moving frame and ensures a homogeneous illumination at every stage of the pattern forming process. The length of the cell is 160 mm, the height 80 mm and the width 4 mm, respectively. The cell is filled with 2 g sand (which is a layer of about 2.5 mm height) and distilled water. As sand we used sieved spherical glass particles (Würth Ballotini MGL) with a diameter range of 71–80 μm. Its material density is given by 2.45 g/cm^3. The cell was rotated by

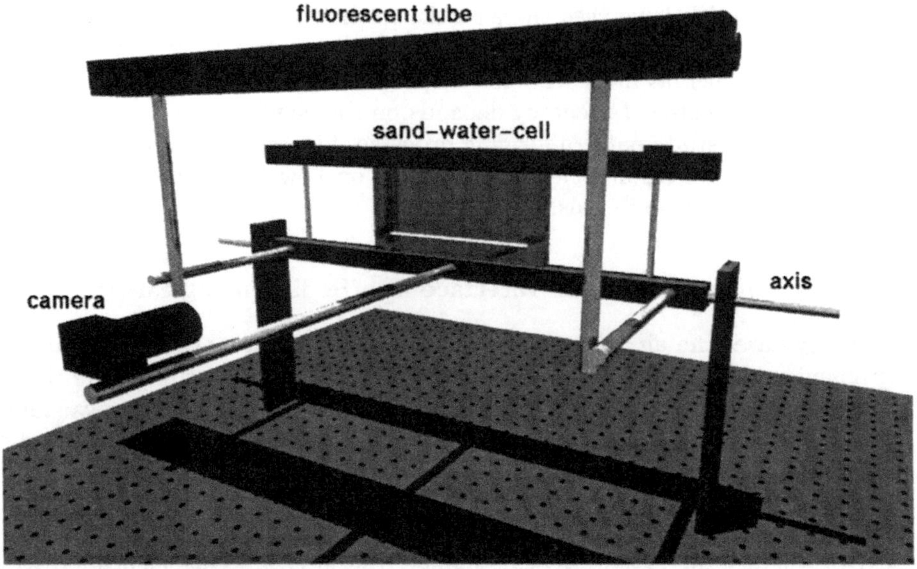

Fig. 30. Experimental setup for measuring sedimentation.

hand. To obtain reproducible results the vertical and horizontal acceleration is measured by acceleration sensors (ADXL05). When the cell passes a rotation angle of 170 degree, a light-gate triggers a number of snapshots. This moment defines the starting time for taking images every 20 ms for later analysis. The images have a dimension of 256 x 300 pixel. To achieve a reasonable resolution we focus only on a horizontal length of 61 mm at the middle of the cell. This gives a resolution of 4.9 pixel/mm. Fig. 31 shows five images of the sand-water cell sampled in time intervals of 60 ms. The first snapshot shows that the initial flat sand layer is modulated at small scales (Fig. 31(a)). These disturbances are enhanced and give rise to sand fingers as seen in Fig. 31(c). At later stages the fingers evolve to a mushroom-like pattern (Fig. 31(d) and Fig. 31(e)).

We have applied a threshold algorithm to obtain the water–sand interface. Starting at the bottom (water) and continuing to the top (sand) we look at every column of our digitized image to determine the point where the grey level exceeds a certain value. In this way we track down the interface of the pattern.

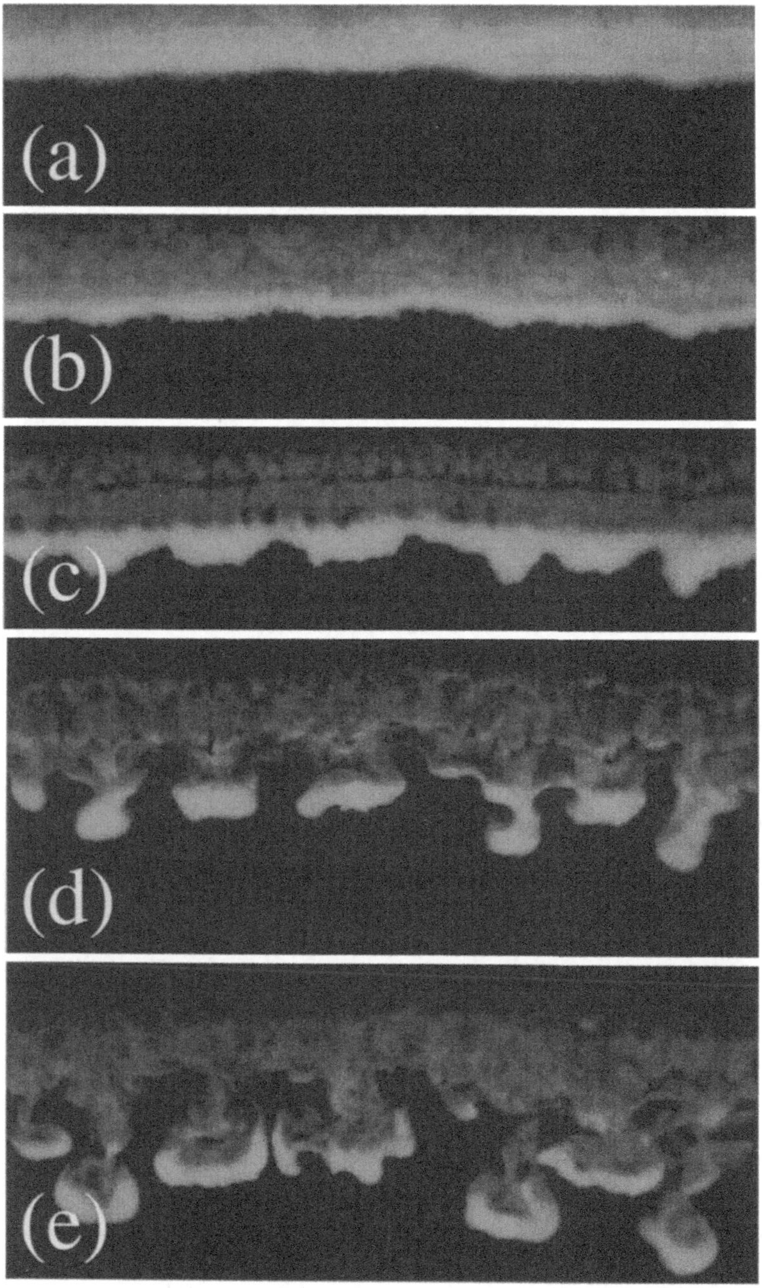

Fig. 31. Sand–water interfaces at a) 20 ms, b) 80 ms, c) 140 ms, d) 200 ms, and e) 260 ms after start of the experiment. The diameter range of the sand particles was 71-80 μm, and the total sand mass in the cell is 2 g. The frames show the middle part of the cell with a horizontal extension of 68 mm.

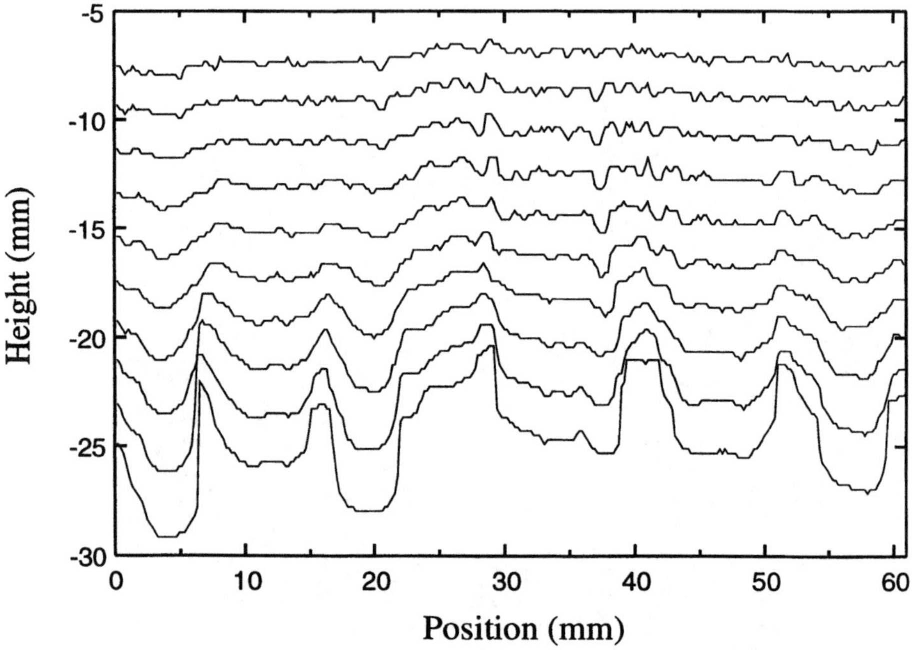

Fig. 32. Temporal evolution of the water–sand interfaces. The patterns are detected every 20 ms and plotted with a constant vertical offset of 1 mm. The experimental conditions are the same as in Fig. 31

Fig. 32 shows the temporal evolution of the front for the experiment in Fig. 31. Here the interfaces of all patterns are shown. While our detection method works well for the patterns during the first stages, namely small scale modulations and sand fingers, it gives erroneous results for the mushroom-like patterns. However, this is not crucial because those patterns are beyond the scope of the analysis that we want to perform here. A discrete Fourier transformation (DFT) gives the power spectrum of each interface. Fig. 33 shows the temporal evolution of the amplitude A of a typical mode. It is seen that A grows exponentially from the first image to $t = 200$ ms. By an exponential fit

$$A(k) = A_i(k)\exp(n(k)t) \tag{5}$$

we obtain the growth rate n for every wave number k in our spectra, where A_i is the initial amplitude.

In order to test the reliability of our experimental setup we have performed 100 independent runs. We only analysed image series for which the time for rotating the apparatus 180 degrees was smaller than 450 ms. These 42 measurements (average rotation time: 404 ms ± 31 ms) were analysed to obtain a mean growth rate for each wave number.

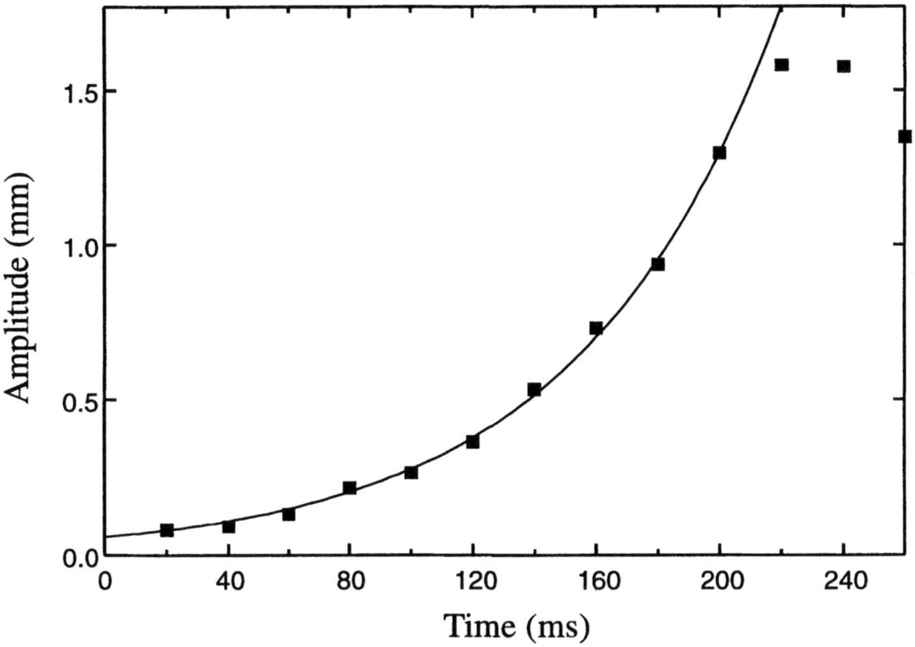

Fig. 33. Amplitude A obtained from a DFT–analysis for a typical wave number (here $k = 7\ \mathrm{cm}^{-1}$) in dependence on time t. The solid line is yielded by a fit according to Eq. (5). The data correspond to Fig. 32.

Results. The square dots in Fig. 34 show the measured dispersion relation. It is seen that the growth rates increase with increasing wave number until they saturate at larger k. In the case of large wave numbers we did not obtain exponential fits for all experimental runs. This is due to the fact that the amplitude is very small and that we approach the limit of the resolution of our image processing.

An obvious question concerning the underlying mechanism of the pattern formation in our system is: How do different material parameters affect the dispersion relation? We have carried out experiments with different values for the total mass and different size distributions. Using 2 g, 4 g, and 8 g of sand we observe a small shift of all growth rates towards larger values with increasing mass of sand. This effect is independent of the wave number. Using three different size distributions 56-63μm, 71-80μm and 90-100μm we found that the mean particle diameter does not have any significant influence on $n(k)$. As a common feature we find the same overall behaviour for all material sets: The growth rates increase for small k and reach a plateau for larger values of k.

In an attempt to understand the pattern formation of the sedimenting

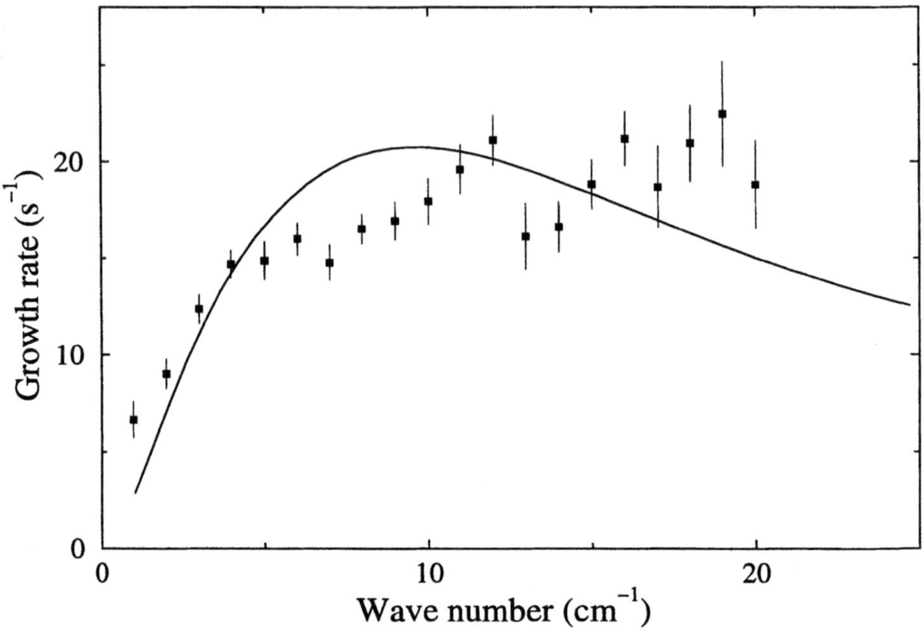

Fig. 34. Comparison of the experimental dispersion relation with theory. The square dots are the mean values of 42 experiments. The error bars give the standard deviation of the mean value. The solid line is yielded by a least square fit of the function obtained by a linear stability analysis.

suspension we considered a two–fluid model, where the suspension is seen as one fluid with a higher density than water, the second fluid. Within this model, our experiment is very reminescent of a classical hydrodynamic instability — the Rayleigh–Taylor instability. It occurs when a more dense fluid is stratified over a less dense one. The interface between the fluids is unstable, and the growth rate of a disturbance is a characteristic function of the wavelength and the fluid parameters. Following this idea, we have compared the experimentel dispersion relation with the results of a linear stability analysis of the two–fluid problem (Lange et al. 1998). The suspension is treated as a Newtonian fluid with a time dependent density, the initial density is set to be equal to that of the sedimented sand layer. The viscosity of the suspension is taken as a fit parameter. Under the assumption of vanishing surface tension a viscosity of about 100 times that of water is found. However, the theory is two–dimensional, while in the experiment due to the long rotation-time a three–dimensional flow ensues, which affects the appearing patterns. In addition, we feel that there is a strong need of more precise measurements of the volume fraction of the falling suspension in order to enhance the quality of the quantitative comparison.

To our knowledge there is only one other publication examining Rayleigh-Taylor instabilities with suspensions (Didwania and Homsy 1981). These authors were not able to measure any growth rates or dispersion relations. There are also two–dimensional numerical simulations of a sedimenting particle front (Manwart and Schwarzer 1998). As the growth law for the amplitudes of the power spectrum the authors find:

$$A(t) = at + \frac{a}{b}\exp(-bt) + c \tag{6}$$

This result is confirmed by an linear stability analysis where the density is taken to be constant in time. However, for our experimental data the exponential growth gives the better fit. It is obvious that additional work is needed in order to understand the nature of the pattern formation during sedimentation.

3.3 Summary and Outlook

We presented a number of experiments and numerical simulations suitable for quantitative comparison with theoretical descriptions of granular matter. Especially in the case where only a few particles were involved, we found a remarkable agreement with the numerical and analytical models. For most of the experiments presented here, however, a quantitative comparison has not yet been achieved. We expect to see a lot of progress within the next few years especially due to numerical methods. An overview on these methods is provided in the appendix.

3.4 Acknowledgement

CMD & GHR gratefully acknowledge the invitation of the "Particle Science and Technology Group" of M. Nakagara at Colorado School of mines to conduct the experiment on axial segregation, which was performed with the help of J. L. Moss. The Magdeburg Group would like to thank many members of the European TMR-network on "Patterns, Noise and Chaos" for stimulating discussions and enthusiastic support. This work was supported by the Deutsche Forschungsgemeinschaft through Re 588/11 and Ri 826/1 within the Schwerpunktprogramm on "Strukturbildung in dissipativen kontinuierlichen Systemen: Theorie und Experiment im quantitativen Vergleich".

Appendix: Numerical methods for simulating the dynamics of granular materials

Since there is no general theory for granular matter, ***numerical simulations*** are a useful tool for investigating idealized systems. Here they can provide a deeper insight into the systems. In numerical simulations we are looking for a passage from microscopic laws to macroscopic behaviour, i.e. the prediction of characteristic length and time scales. A big issue is the size of the investigated system. In order to model a realistic system at least 10^5 particles have to be used. Even though granular particles in vacuum only interact when they collide, since the particles are large enough not to feel the Van der Waals forces, it is a numerically very intensive job and can only be done with todays high performance computers. For the dynamics of granular media it is essential that the interaction is ***dissipative*** and ***nonlinear***. Various numerical techniques have been developed to address the difficulties in simulations of the dynamics of granular matter.

Monte Carlo Method

One widely used numerical technique to study particle motions is the so–called *Monte Carlo* method. The name stems from the extensive use of random numbers in the calculations which are used to find new particle positions which minimize the potential energy of the system. This method was introduced into the field of granular materials more than 10 years ago to study the size segregation of binary mixtures undergoing vertical vibrations (Rosato et al. 1986, Rosato et al. 1987). Hard spheres were used with a normal restitution coefficient of zero to avoid an increase in potential energy. It was successful in stressing the geometrical effects of the process. A scaling law, relating the segregation velocity to the amplitude of shaking and to the diameter ratio of the particles, was found as well (Devillard 1990). It also served in testing a segregation model in a ternary system (Rosato et al. 1991).

Cellular Automata

The use of ***cellular automaton*** models to study granular materials dates back to the introduction of the concept of ***self-organized criticality***, where sand pile avalanche statistics were used as one example (Bak et al. 1987). In these models space is discretized into cells which have the size of the particles and can either be occupied or empty. The particle dynamics are modeled by a set of particle collision rules which apply when certain conditions are fulfilled, i.e. the local surface angle (slope) exceeds a threshold value. Theses rules were later refined by deriving them from experiments to study the segregation process (Fitt and Wilmott 1992) during particle outflow from two-dimensional hoppers. It is also possible to choose collision rules which require the system to fulfill the Navier-Stokes equations, which is then called *lattice-gas* automaton (Flekkøy and Herrmann 1993). This seems to be a promising starting point to model granular materials when the rules are modified to include the important aspects of gravity, dissipation and static friction.

Method of steepest descent

This deterministic model, which is capable of dealing with a very large number of particles, was first described by Jullien et al. (1992). It is motivated by the method of ballistic deposition and should be understood as a *toy-model* to study the geometrical effects in granular materials. Particles follow the path of steepest descent and undergo a sequence of *rolling* and *falling* steps before they reach a local surface minimum. New particle contacts are treated as completely inelastic collisions. This model was used to model the segregation process in the growing heap of a pile with an impressive 10^7 and more particles (Meakin and Jullien 1992). The original model was extended by incorporating the exact form of the external excitation by Baumann et al. (1994) to study particle size segregation in rotating drums and was renamed *bottom-to-top-restructuring.*

Molecular Dynamics (Granular Dynamics or Discrete Element Method)

The aforementioned methods have several limitations: (i) No physical time scale enters the models since the collision time is assumed to be zero, therefore a direct connection of the update time and the physical time is missing which greatly reduces the scope of these approaches. (ii) These models are capable to study mostly geometrical effects. To overcome these difficulties and limitations, another method is most widely used to model the dynamics of granular materials called *discrete element method* (DEM) which is essentially a molecular dynamics method including the particle history (Cundall and Strack 1979).

Here, each particle i is approximated by a sphere with radius R_i, sketched in Fig. 35 and the wall is treated as a particle with infinite mass and radius. Only contact forces during collisions are considered and, depending on the material we want to simulate, the particle can or cannot rotate. For an example: in contrast to glass beads mustard seeds are slightly aspherical, so they tend not to rotate easily and the approximation of non-rotating particles is used. Whenever two particles i and j are closer than the sum of their radii particle j exerts a force on particle i and vice versa. The total contact forces can be split up into:

- Elastic, repulsive contact force in normal direction

$$f_{\text{el}}^{(ij)} = \tilde{Y}(m_{\text{eff}}, r_{\text{eff}}) \; ((R_i + R_j) - \mid \mathbf{r}_{ij} \mid)^{\alpha} \quad ,$$

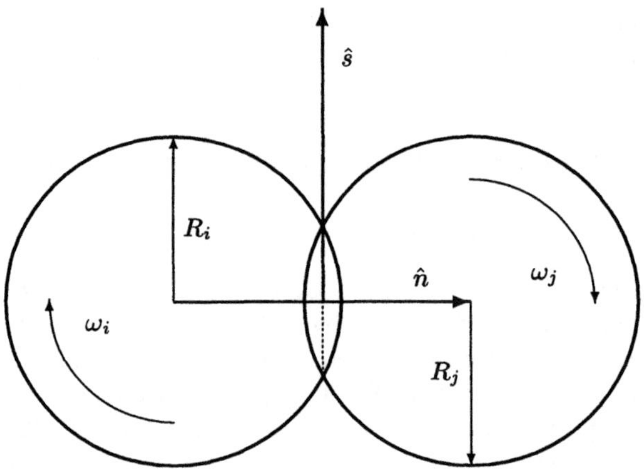

Fig. 35. Coordinate system during particle-particle collisions.

where $\tilde{Y}$ is the Young modulus of the investigated material and $|\mathbf{r}_{ij}|$ the distance between the centers of mass of particle i and particle j.
For $\alpha = 1$ we get a Hooke like force and for $\alpha = 1.5$ the Hertzian contact force.

– Dissipation in normal direction

$$f_{\text{diss}}^{(ij)} = -\tilde{\gamma}(m_{\text{eff}}, r_{\text{eff}})\ \mathbf{v}_{ij} \cdot \hat{n}\ ,$$

$\tilde{\gamma}(m_{\text{eff}}, r_{\text{eff}})$ is the dissipation coefficient and is determined by the coefficient of restitution ϵ_{res}. $\mathbf{v}_{ij} = \mathbf{v}_i + (\boldsymbol{\omega}_i \times \hat{n}) - (\mathbf{v}_j + (\boldsymbol{\omega}_j \times (-\hat{n})))$ denotes the relative velocity of the two colliding particles at the contact point, see Fig. 35. .

– Frictional forces in shear direction

1. Static friction (with static friction we get a finite angle of repose):

$$f_{\text{fric}}^{(ij)} = -k_s \int \mathbf{v}_{ij} \cdot \hat{s}(t) dt\ .$$

For the static shear force, we put upon contact a linear spring between the two particles which results in a restoring force, i.e. the friction is proportional to the displacement of the original contact points.

2. Dynamic friction:

$$f_{\text{fric}}^{(ij)} = -\gamma_s \mathbf{v}_{ij} \cdot \hat{s}(t)\ .$$

This is a viscous friction where the force is proportional to the relative velocity.

Both tangential forces are limited by the Coulomb criterion which states that the magnitude of the tangential force cannot exceed the magnitude of the normal force multiplied by the friction coefficient μ. Therefore, if the restoring force would be larger than Coulomb's frictional force we switch to the friction with Coulomb's Law and we get:

$$f_{\text{shear}}^{(ij)} = \text{sign}(f_{\text{fric}}^{(ij)}) \; \min(f_{\text{fric}}^{(ij)}, \mu \mid \mu f_{\text{el}}^{(ij)} + f_{\text{diss}}^{(ij)} \mid) \; .$$

The model parameters $(\tilde{Y}, \tilde{\gamma}, k_s(\gamma_s), \mu)$ have the following physical interpretation: $\tilde{Y}$ is related to the material stiffness, $\tilde{\gamma}$ to the energy loss during collisions (in experiments measured via the restitution coefficient) and the pair $k_s(\gamma_s)$ and μ controls the energy loss and static friction in the shear direction, e.g. via the surface roughness (Foerster et al. 1994). Commonly, μ is referred to as the Coulomb friction coefficient.

A detailed discussion of the different force laws is given by Schäfer et al. (1996), and a review of different applications using granular dynamics is given by Ristow (1994).

A technique very similar to the discrete element method are event driven simulations in which a variable time step is used which is well suited for dilute systems (Luding et al. 1994).

References

Allen M. P., Tildesley D. J. (1990): *Computer Simulation of Liquids* (Clarendon Press, Oxford 1990)

Bagnold R. A. (1954): Experiments on a Gravity-free Dispersion of Large Solid Spheres in a Newtonian Fluid Under Shear. Proc. R. Soc. **225**, 49–63

Bak P., Tang C., Wiesenfeld K. (1987): Self-organized criticality: an explanation of $1/f$ noise. Phys. Rev. Lett. **59**, 381–384

Batchelor G. K., Green J. T. (1972): Determination of the bulk stress in a suspension of spherical particles to order c^2. J. Fluid Mech. **56**, 401–427

Batchelor G. K. (1977): The effect of Brownian motion on the bulk stress in suspension of spherical particles. J. Fluid Mech. **83**, 97–117

Baumann G., Janosi, I., Wolf D. E. (1994): Particle trajectories and segregation in a two-dimensional rotating drum. Europhys. Lett. **27**, 203–208

Bizon C., Shattuck M. D., Swift J. B., McCormick W. D., Swinney H. L. (1998): Patterns in 3d vertically oscillated granular layers: simulation and experiment. Phys. Rev. Lett. **80**, 57–60

Bridgewater J. (1976): Fundamental powder mixing mechanisms. Powder Technol. **15**, 215–236

Brown R. L. (1939): The fundamental principles of segregation. J. Inst. Fuel. **13**, 15–19

Cantelaube F., Bideau D. (1995): Radial segregation in a 2d drum: an experimental analysis. Europhys. Lett. **30**, 133–138

Choo K., Molteno T. C. A., Morris S. W. (1997): Traveling granular segregation patterns in a long drum mixer. Phys. Rev. Lett. **79**, 2975–2978

Clément E., Rajchenbach J. (1991): Fluidization of a Bidimensional Powder. Europhys. Lett. **16**, 133–138

Clément E., Duran J., Rajchenbach J. (1992): Experimental study of heaping in a two–dimensional "sandpile". Phys. Rev. Lett. **69**, 1189–1192

Clément E., Rajchenbach J., Duran J. (1995): Mixing of a granular material in a bidimensional rotating drum. Europhys. Lett.**30**, 7–12

Cooke W., Warr S., Huntley J., Ball R. (1996): Particle size segregation in a two-dimensional bed undergoing vertical vibration. Phys. Rev. E **53**, 2812–2822

Cundall P., Strack O. D. L. (1979): A discrete numerical model for granular assemblies. Géotechnique **29**, 47–65

Das Gupta S., Khakar D. V., Bathia S. K. (1991): Axial segregation of particles in a horizontal rotating cylinder. Chem. Engineering Science **46**, 1513–1517

Devillard P. (1990): Scaling behaviour in size segregation ("Brazil Nuts"). J. Phys. France **51**, 369–373

Didwania A. K., Homsy G. M. (1981): Rayleigh-Taylor Instabilities in Fluidized Beds. Ind. Eng. Chem. Fundam. **20**, 318–323

Dippel S., Luding S. (1995): Simulation on size segregation: Geometrical effects in the absence of convection. J. Phys. I France **5**, 1527–1537

Donald M. B., Roseman B. (1962): Mixing and demixing of solid particles. Br. Chemical Engineering **7**, 749–753

Dury C. M., Ristow G. H. (1997): Radial segregation in a two-dimensional rotating drum. J. Phys. I France **7**, 737–745

Dury C. M., Ristow G. H. (1998): Competition of Mixing and Segregation in Rotating Cylinders. subm. to Phys. Fluids

Einstein A. (1906): Eine neue Bestimmung der Moleküldimensionen. Ann. Phys. **19**, 289–306

Einstein A. (1911): Berichtigung zu meiner Arbeit: Eine neue Bestimmung der Moleküldimensionen. Ann. Phys. **34**, 591–592

Esipov S. E., Pöschel T. (1997): The granular phase diagram. J. Stat. Phys. **86**, 1385–1395

Evesque P., Rajchenbach J. (1989): Instability in a sand heap. Phys. Rev. Lett. **62**, 44–46

Evesque P. (1992): Shaking dry powders and grains. Contemporary Physics **33**, 245–261

Fitt A. D., Wilmott P. (1992): Cellular-automaton model for segregation of a two-species granular flow. Phys. Rev. A **45**, 2383–2388

Flekkøy E. G., Herrmann H. J. (1993): Lattice Boltzmann models for complex fluids. Physica A **199**, 1–11

Foerster S. F., Louge M. Y., Chang H., Allia K. (1994): Measurements of the collision properties of small spheres. Phys. Fluids **6**, 1108–1115

Gallas J. A. C., Herrmann H. J., Sokołowski S. (1992): Molecular dynamics simulation of powder fluidization in two dimensions. Physica A **189**, 437–446

Grossmann E.L., Zhou T., Ben-Naim E. (1997): Towards granular hydrodynamics in two dimensions. Phys. Rev. E **55**, 4200–4206

Hill K. M., Kakalios J. (1995): Reversible axial segregation of rotating granular media. Phys. Rev. E **52**, 4393–4400

Ichiki K., Hayakawa H. (1995): Dynamical simulation of fluidized beds: Hydrodynamically interacting granular particles. Phys. Rev. E **52**, 658–670

Janssen H. A. (1895): Versuche über Getreidedruck in Silozellen. Z. Verein Deutsch. Ing. **93**, 1045–1049

Jaeger H. M., Knight J. B., Liu C. H., Nagel S. (1994): What is shaking in the sandbox? MRS Bulletin **19**, 25

Jaeger H. M., Nagel S. R., Behringer R. P. (1996): The Physics of Granular Materials, Physics Today **49**, 32

Jaeger H. M., Nagel S. R., Behringer R. P. (1996): Granular solids, liquids, and gases. Rev. Mod. Phys. **68**, 1259–1273

Jenike W. (1964): Why bins don't flow. Mech. Engineering **5**, 40–43

Johanson J. R. (1978): Particle segregation ... and what to do about it. Chem. Engineering **8**, 183–192

Julien P., Lan Y., Raslan Y. (1997): Experimental mechanics of sand stratification. In: Behringer R. P., Jenkins J. T. (eds.): *Powders & Grains 97* (Balkema, Rotterdam), 487–490

Jullien R., Meakin P., Pavlovitch A. (1992): Three-dimensional model for particle-size segregation by shaking. Phys. Rev. Lett. **69**, 640–643

Knight J., Jaeger H., Nagel S. (1993): Vibration-induced size seperation in granular media: The convection connection. Phys. Rev. Lett. **70**, 3728–3731

Lange A., Schröter M., Scherer M. A., Engel A., Rehberg I. (1998): Fingering Instability in a Water-Sand Mixture. submitted to Euro. Phys. J. B

Lee J. (1995): Scaling behaviour of granular particles in a vibrating box. Physica A **219**, 305–326

Liffman K., Metcalfe G., Cleary P. (1997): Convection due to horizontal shaking. In: Behringer R. P., Jenkins J. T. (eds.): *Powders & Grains 97* (Balkema, Rotterdam), 405–408

Luding S., Herrmann H. J., Blumen A. (1994): Simulation of twodimensional arrays of beds under external vibrations: Scaling behaviour. Phys. Rev. E **50**, 3100–3108

Luding S., Clément E., Blumen A., Rajchenbach J., Duran J. (1994): The onset of convection in molecular dynamics simulations of grains. Phys. Rev. E, **50**, R1762–R1765

Makse H. A., Cizeau P., Stanley H. E. (1997): Possible stratification mechanism in granular mixtures. Phys. Rev. Lett. **78**, 3298–3301

Manwart C., Schwarzer S. (1998): Response analysis for the early stage of sedimentation processes. Preprint.

Meakin P., Jullien R. (1992): Simple models for two and three dimensional particle size segregation. Physica A **180**, 1–18

Melo F., Umbanhowar P., Swinney H. L. (1994): Transition to parametric wave patterns in a vertically oscillated granular layer. Phys. Rev. Lett. **72**, 172–175

Melzer A., Homann A., Piel A. (1996): Experimental investigation of the melting transition of a plasma crystal. Phys. Rev. E **53**, 2757–2766

Metcalfe G., Shinbrot T., McCarthy J. J., Ottino J. M. (1995): Avalanche mixing of granular solids. Nature **374**, 39–41

Nakagawa M. (1994): Axial segregation of granular flows in a horizontal rotating cylinder. Chem. Engineering Science - Shorter Communications **49**, 2540–2544

McNamara S., Young W.R. (1996): Dynamics of a freely evolving, two-dimensional granular medium. Phys. Rev. E **53**, 5089–5100

Pätzold R. (1980): Die Abhängigkeit des Fließverhaltens konzentrierter Kugelsuspensionen von der Strömungsform: Ein Vergleich der Viskosität in Scher- und Dehnströmungen. Rheo. Acta **19**, 322–344

Ristow G. H. (1994): Granular dynamics: a review about recent molecular dynamics simulations of granular materials. In: Stauffer D., (ed.): *Annual Reviews of Computational Physics I* (World Scientific, Singapore), 275–308

Ristow G., Straßburger G., Rehberg I. (1997) Phase Diagram and Scaling of Granular Materials under Horizontal Vibrations. Phys. Rev. Lett. **79**, 833–836

Rosato A. D., Lan Y., Wang D. T. (1991): Vibratory particle size sorting in multicomponent systems. Powder Technol. **66**, 149–160

Rosato A. D., Prinz F., Strandburg K., Swendsen R. (1986): Monte carlo simulation of particulate matter segregation. Powder Technol. **49**, 59

Rosato A. D., Strandburg K. J., Prinz F., Swendsen R. H. (1987): Why the Brazil Nuts are on top: Size segregation of particulate matter by shaking. Phys. Rev. Lett. **58**, 1038–1040

Rosenkranz D.E., Pöschel T. (1997) Recurrent swelling of horizontally shaken granular material. Preprint cond-mat / 9712108.

Schäfer J., Dippel S., Wolf D. E. (1996): Force schemes in simulations of granular materials. J. Phys. I France **6**, 5–20

Scherer M. A., Buchholtz V., Pöschel T., Rehberg I. (1996): Swirling granular matter: from rotation to reptation. Phys. Rev. E **54**, R4560–R4563

Scherer M. A., Mahr T., Engel A., Rehberg I. (1997): Granular dynamics in a swirled annulus. subm. to Phys. Rev. E

Scherer, M. A. (1998): *Der Schwenkeffekt: Experimentelle Untersuchungen zur granularen Dynamik* (Dissertation, Universiät Magdeburg, Magdeburg)

Sela N., Goldhirsch I., Noskowicz S. H. (1996): Kinetik theoretical study of a simple sheared two-dimensional granular gas to Burnett order. Phys. Fluids **8**, 2337–2353

Spahn F., Schwarz U., Kurths J. (1997): Clustering of Granular Assemblies with Temperature Dependent Restitution under Keplerian Differential Rotation. Phys. Rev. Lett. **78**, 1596–1599

Standish N. (1985): Studies of size segregation in filling and emptying a hopper. Powder Technol. **45**, 43–56

Straßburger G., Betat A., Scherer M. A., Rehberg I. (1996): Pattern Formation by Horizontal Vibration of Granular Material. In: Wolf D. E., Schreckenberg M., Bachem A. (eds.): *Traffic and Granular Flow* (World Scientific), 329-333

Taguchi Y. H. (1993): $k^{-5/3}$ Power Spectrum in Powder — Turbulent flow in a Vibrated Bed: Numerical Results. Europhys. Lett. **24**, 203–209

Vanel L., Rosato A. D., Dave R. (1997): Rise–times regimes of a large sphere in vibrated bulk solids. Phys. Rev. Lett. **78**, 1255–1258

Warr S., Huntley J. M., Jacques G. T. H. (1995): Fluidization of a two–dimensional granular system: Experimental study and scaling behaviour. Phys. Rev. E **52**, 5583–5595

Williams J. C. (1976): The segregation of particulate materials. A review. Powder Technol. **15**, 245–251

Yokoyama T., Tamura K., Usui H., Jimbo G. (1996): Simulation of ball behavior in a vibration mill in relation with its grinding rate – effects of fractional ball filling and liquid viscosity. In: Forssberg E. (ed.): *Proceedings of the Eighth Symposium on Comminution, Stockholm, 1994* (Elsevier, Amsterdam) 413–425

Zik O., Levine D., Lipson S. G., Shtrikman S., Stavans J. (1994): Rotationally induced segregation of granular materials. Phys. Rev. Lett. **73**, 644–647

Spontaneous Generation of Magnetic Fields in the Laboratory

F.H. Busse[1] U. Müller[2], R. Stieglitz[2] and A. Tilgner[1]

[1] Institute of Physics, University of Bayreuth, D-95440 Bayreuth, Germany
[2] IATF, Forschungszentrum Karlsruhe, D-76021 Karlsruhe, Germany

Abstract. An introduction to the dynamo effect as a bifurcation phenomenon in moving, electrically conducting fluids is given and its importance as the origin of cosmic magnetic field is pointed out. The second part of this article is devoted to the planned experimental realization of the dynamo effect in the laboratory. Since liquid sodium must be pumped for this purpose through helical pipes at rates of about $100m^3/h$, the planned experiment has required several years of preparation. First results are expected after this article has gone to print.

1 Introduction

The transformation of mechanical into electromagnetic energy is a familiar process in daily life, the bicycle dynamo being a common example of this transformation. On a larger scale this process provides the basis for modern civilisation. The conversion of mechanical into electromagnetic energy is also the origin of the magnetism of celestial bodies and even of the galaxies. But there are some important differences. While the technical dynamo converts energy of simple rotational motion into electromagnetic energy through a complex distribution of electrical conductivity (the wiring of dynamos), the process of magnetic field generation in celestial bodies occurs in singly connected domains of rather uniform conductivity. This latter property has been used as an argument against the existence of homogeneous dynamos, and Cowling [?] has indeed been able to prove that an axisymmetric field can not be generated by motion in a fluid sphere of uniform electrical conductivity. Since the geomagnetic field is axisymmetric in first approximation Cowling's theorem had cast severe doubts on the feasibility of the dynamo mechanism as the origin of planetary and stellar magnetic fields. Only in 1958 proofs for working homogeneous dynamos have been obtained by Backus [?] and Herzenberg [?]. Sufficiently complex velocity fields are usually required to drive a homogeneous dynamo.

The homogeneous dynamo realized in nature and simple version of the technical dynamo as, for example, the disk dynamo shown in figure 1 have in common that they represent a bifurcation phenomenon. In contrast to other bifurcation problems considered in other chapters of this book, the solution bifurcating from the basic non-magnetic state is characterized by a new physical quantity, the magnetic field. In the following we consider this

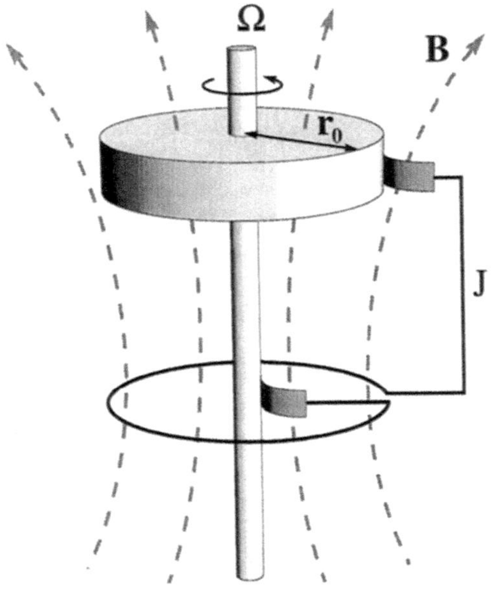

Fig. 1. Disk dynamo. A rotating metal disk is connected to an electric circuit through which the current may flow.

bifurcation problem first in the case of the disk dynamo and then in section 3 for a simple homogeneous dynamo driven by a spatially periodic velocity field. This latter problem provides the basis for the planned first laboratory realisation of a homogeneous dynamo to which the remaining sections of this chapter are devoted.

2 The Disk Dynamo

In the absence of a magnetic field the rotating metallic disk shown in figure 1 is governed by an equation of motion of the form

$$I\dot{\Omega} = -f\Omega + T \tag{1}$$

where I denotes the moment of inertia of the disk, Ω its angular velocity of rotation, and $f\Omega$ denotes the frictional torque assumed to be growing linearly with Ω. T is the prescribed applied torque. In the presence of magnetic field

$\boldsymbol{B}$ intersecting the disk an electromotive force between axis and rim of the disk will be generated leading to the potential

$$U = \int_0^{r_0} r\boldsymbol{B} \cdot \boldsymbol{\Omega} r dr \tag{2}$$

between rim and axis. This potential can be used to drive a current J through the electric circuit governed by the equation

$$L\dot{J} + RJ = U \tag{3}$$

where L and R are the inductivity and the Ohmic resistence of the circuit, respectively. When the magnetic field generated by the current in the circuit is identified with the magnetic field intersecting the disk, dynamo action results when the feedback is sufficiently strong. To study this possibility we note that U can be written in the form

$$U = M\Omega J/2\pi \tag{4}$$

where M is the mutual inductivity between circuit and disk and where now Ω is assumed to be a constant. After inserting expression (4) in (3) the differential equation (3) yields

$$J = J_0 \exp\left\{\left(\frac{\Omega M}{2\pi} - R\right) t/L\right\} \tag{5}$$

as solution indicating the condition

$$\frac{M\Omega}{R} > 2\pi \tag{6}$$

for dynamo action.

The exponentially growing magnetic field $\boldsymbol{B}$ corresponding to solution (5) becomes unphysical as soon as the Lorentz force $(\nabla \times \boldsymbol{B}) \times \boldsymbol{B})/\mu$ acting in the disk reaches a significant level. Here μ refers to the magnetic permeability of the disk. Accordingly relationship (1) must be replaced by an equation of the form

$$I\dot{\Omega} = -f\Omega + T - gJ^2 \tag{7}$$

where g is a positive constant. The dynamics of the system is now described by the combined equations (3), (4) and (7) and the steady state of the dynamo solution bifurcating from the non-magnetic state $\Omega = T/f$ is given by

$$J^2 = (T - f2\pi R/M)/g \tag{8}$$

provided the right hand side is positive.

3 A Homogeneous Dynamo

The disk dynamo in figure 1 is characterized by a simple form of motion, but a complex distribution of the electrical conductivity. In fact, if the winding sense of the circuit around the axis had been the other way around, the dynamo would not function since expression (4) would change sign and both terms in the argument of the exponential function (5) would be negative. It is thus not surprising that the existence of a dynamo in a singly connected domain of uniform conductivity had been doubtful for a long time. Here we wish to present briefly a simple spatially periodic dynamo which was first described by G.O. Roberts (1972).

We consider the velocity field of the form

$$\boldsymbol{v} = \nabla\psi \times \boldsymbol{k} + \boldsymbol{k}w \tag{9a}$$

with

$$\psi = A \sin \alpha x \sin \alpha y = w(x, y)A/C \tag{9b}$$

where $\boldsymbol{k}$ is the unit vector in the z-direction of the Cartesian system of coordinates. The equation of magnetic induction for the magnetic flux density $\boldsymbol{B}$ which follows from Maxwell's equation and Ohm's law in the magnetohydrodynamic approximation is given by

$$\frac{\partial}{\partial t}\boldsymbol{B} - \lambda\nabla^2\boldsymbol{B} = \nabla \times (\boldsymbol{v} \times \boldsymbol{B}) \tag{10}$$

where the magnetic diffusivity λ is defined by the inverse of the product of μ and the electrical conductivity σ. Since the velocity field (9a) does not depend on the time t nor on the z-coordinate the solution of equation (10) can be obtained in the form

$$\boldsymbol{B} = \boldsymbol{B}(x, y) \exp\{\mathrm{i}\gamma z + pt\} \tag{11}$$

where γ is the real wavenumber of the magnetic field in the z-direction. After separating $\boldsymbol{B}(x, y)$ into an average part $\bar{\boldsymbol{B}}$ and a fluctuating part $\check{\boldsymbol{B}}$,

$$\boldsymbol{B}(x, y) = \bar{\boldsymbol{B}} + \check{\boldsymbol{B}} \text{ with } \bar{\check{\boldsymbol{B}}} = 0 \tag{12}$$

where the average over x and y has been indicated by a bar we obtain from the average of equation (10) and by subtracting the latter from equation (10) the following two equations

$$(p + \gamma^2\lambda)\bar{\boldsymbol{B}} = \mathrm{i}\gamma\boldsymbol{k} \times \overline{(\boldsymbol{v} \times \check{\boldsymbol{B}})} \tag{13a}$$

$$(p - \lambda\nabla^2)\check{\boldsymbol{B}} = \bar{\boldsymbol{B}} \cdot \nabla\boldsymbol{v} - w\mathrm{i}\gamma\bar{\boldsymbol{B}} + \nabla \times (\boldsymbol{v} \times \check{\boldsymbol{B}}) - i\gamma\boldsymbol{k}\overline{(\boldsymbol{v} \times \check{\boldsymbol{B}})} \tag{13b}$$

We are interested in the limit

$$\gamma^2 \ll \alpha^2 \quad , \qquad p \approx \gamma^2 \tag{14a}$$

in which case the inequality

$$| \check{\boldsymbol{B}} | \ll | \bar{\boldsymbol{B}} | \tag{14b}$$

can be anticipated. Accordingly we neglect all but the first term on the right hand side of equation (13b) in first approximation and obtain

$$\check{\boldsymbol{B}} = \bar{\boldsymbol{B}} \cdot \nabla \boldsymbol{v} / 2\lambda\alpha^2 \tag{15}$$

After insertion of this result on the right hand side of equation (13a) we obtain

$$(p + \lambda\gamma^2)\bar{\boldsymbol{B}} = -i\gamma \frac{AC}{4\lambda} \boldsymbol{k} \times \bar{\boldsymbol{B}} \tag{16}$$

which is solved by

$$\bar{\boldsymbol{B}} = B_0(\boldsymbol{i} - \frac{\gamma AC}{| \gamma AC |} i\boldsymbol{j}), \tag{17a}$$

$$p = -\lambda\gamma^2 + \frac{| \gamma AC |}{4\lambda}, \tag{17b}$$

where $\boldsymbol{i}$ and $\boldsymbol{j}$ are the unit vectors in the x- and y-direction, respectively. Dynamo action thus occurs when the condition

$$R_A R_C > 4 \tag{18a}$$

is satisfied when the poloidal and the toroidal magnetic Reynolds numbers R_C and R_A are defined by

$$R_A \equiv | A | / \lambda \qquad R_C = | C | / \gamma\lambda \tag{18b}$$

Homogeneous dynamos typically depend on the interaction of two components of the velocity field and thus the dynamo condition depends on the product of two magnetic Reynolds numbers which have been defined for the "toroidal" component given by $\psi(x, y)$ with the length scale α^{-1} and for the poloidal component described by $w(x, y)$ with the magnetic length scale γ^{-1}. The latter length scale is restricted for all realistic dynamos. But as long as $| \gamma/\alpha | \ll 1$ the assumption (14b) corresponding to

$$R_A + \left| \frac{\gamma}{\alpha} \right| R_C \ll 1 \tag{19}$$

can be satisfied together with the dynamo condition (18a).

The structure of the mean magnetic field is shown in figure 2 together with the streamlines of the velocity field (9). The spiral staircase like structure of the mean magnetic field always exhibits the same screw sense as the velocity field. The dynamo action can be understood from the distortion of the magnetic field lines owing to the velocity field as indicated in figure 3. The field lines are distorted into a helix such that the upwards shifted part of the field lines from the plane $z = 0$ aquires a component in the direction of the mean field in the plane $z = \pi/2\gamma$. The distorted field line from the latter

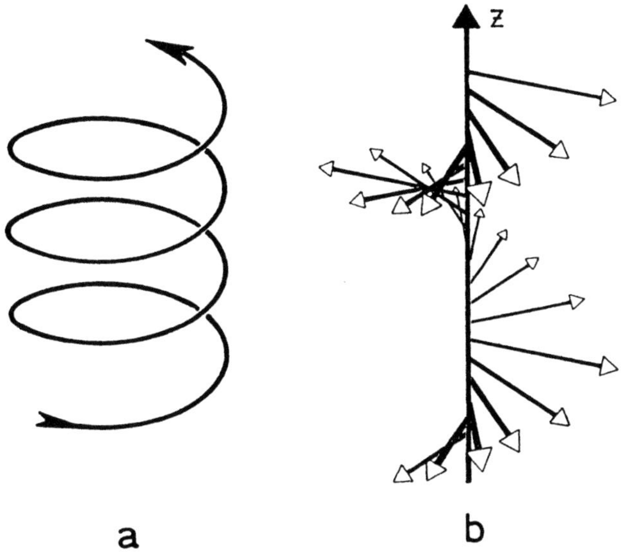

Fig. 2. Streamline of the velocity field for $AC > 0$ (a) and structure of the mean magnetic field (b).

level generates a component strengthening the field at the lower level $z = 0$. The distortions of the field lines thus exhibit the tendency to strengthen the field at the neighboring levels. To increase the magnetic flux density Ohmic diffusion is required, of course, which can not easily be modelled in the field line picture. But the latter provides a good understanding for the mutual reinforcement of the magnetic field lines at different levels. It is of interest to note that the screw sense of the distorted field lines is opposite to that of the velocity field.

Besides the particular solution (17a) discussed there are other types of magnetic fields possible as solutions of equ. (10), for example, those with a subharmonic structure. For details we refer to [5]. But the subharmonic solutions require higher values of the magnetic Reynolds numbers (18b) for dynamo action. The velocity field (9) is also the most effective one among the various periodic velocity fields studied in [4].

4 An experimental honogeneous dynamo

The flow described in the previous section underlies the design of an experiment which is currently under construction at the "Forschungszentrum Karlsruhe" [6]. The goal of the experiment is to experimentally demonstrate the dynamo effect with a conducting medium in a simply connected volume and to study the non-linear interactions between the generated magnetic field and the fluid.

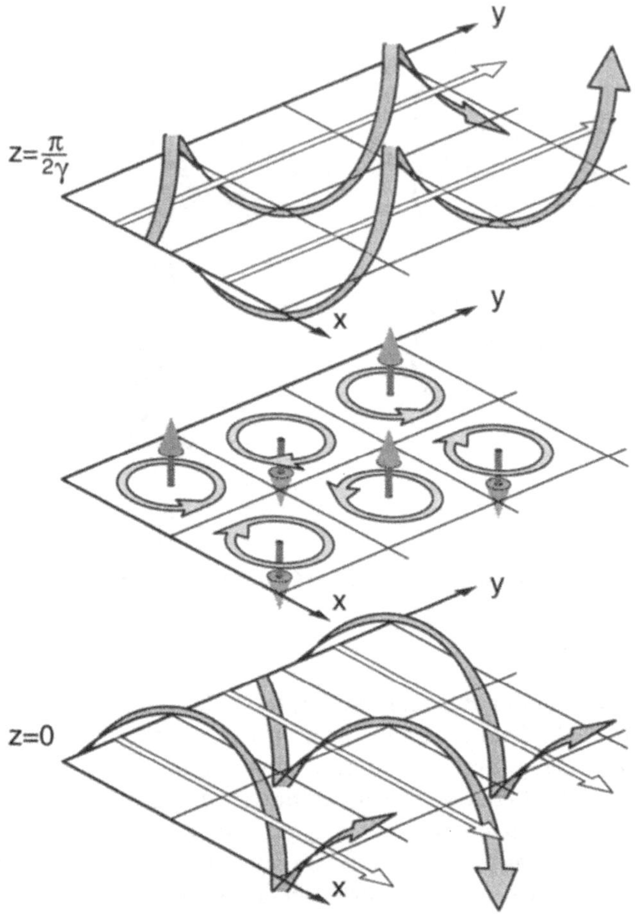

Fig. 3. Distortions of the mean magnetic field in the levels $z = \pi/2\gamma$ and $z = 0$ owing to the z-independent velocity field (9) indicated in the middle.

The first choice to be made in the design of an experimental dynamo concerns the fluid to be used. Liquid sodium appears to be the most favorable material available. It melts at 98° C, has a better conductivity than other liquid metals in the temperature range 100-150° C, and also has a lower density. An experiment with sodium can therefore be operated at modest temperatures, allows to reach higher magnetic Reynolds numbers than for example mercury or gallium and stores less kinetic energy. Furthermore, the kinematic viscosities of liquid sodium at around 120° C and water at room temperature are very similar. A sodium flow can thus be optimized through preliminary experiments with water in which the flow is conveniently visualized. Safety is a concern in operations with sodium. However, much experience has been gathered during the construction of sodium filled cooling circuits for fast breeder reactors so that both the technology and the know-how is available to handle sodium appropriately.

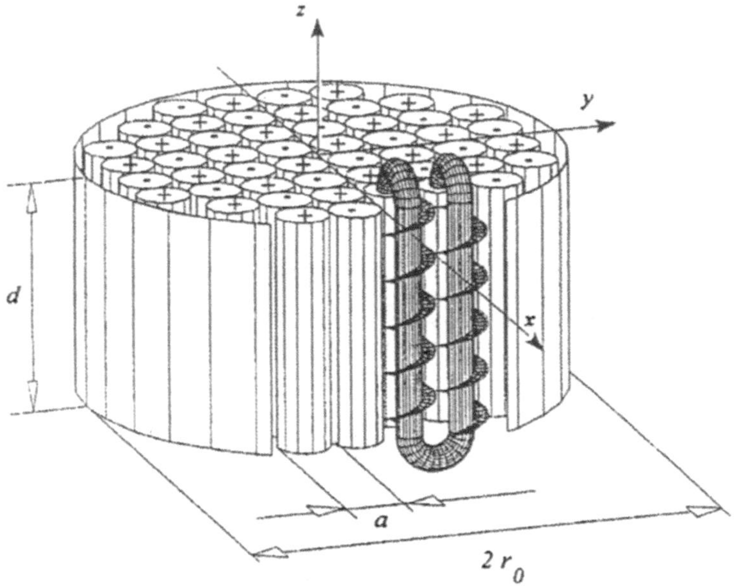

Fig. 4. Sketch of the dynamo experiment. The inner structure is shown for two cells in the front row. The signs indicate the direction of the sodium flow in individual cells.

The choice of the flow geometry is less straightforward than the choice of the fluid. The experimental setup that has been chosen following the proposal [7] is sketched in figure 4. The basic idea is to pump liquid sodium through a structure of pipes and helical blades which force the sodium to flow approximately as prescribed in equ. (9). 52 cells are contained in a cylindrical vessel of radius r_0 and height d. Each individual cell is made of two coaxial pipes. Sodium flows straight through the inner pipe whereas a helical sheet inserted in the space between the two pipes forces the sodium on a helical path there. The main features of the velocity field (9) are reproduced in the sense that a periodic helical flow is created with a larger pitch at the center of the cell than at the perimeter. In order to ensure electrical homogeneity of the experiment, the guiding structure is made of stainless steel which is a good conductor and resists the corrosive effect of sodium. The space outside the pipes but inside the cylindrical vessel is filled with stagnant sodium.

Fig. 5. Photograph of the dynamo cell during assembly.

Sodium needs to flow up and down in adjacent cells. The fluid is guided from one cell to a neighboring cell through U-bends connecting the inner pipes (two of which are shown in figure 4) whereas the helical parts are connected by a tube in the shape of an "S". Figure 5 shows a photograph of the experiment at an intermediate stage of the assembly. The S-shaped connectors can clearly be seen, whereas the U-bends are still missing. Further tubing between neighboring cells is included to allow convenient filling and complete emptying of the experiment.

In- and outlets leading sodium from the pumps to the dynamo cell and back are seen at the top of the picture. The entire sodium circuit is split into three separate loops, each equiped with an independent pump. One pump forces sodium through all 52 inner pipes. The helical parts are separated in two symmetric groups of 26 cells each located at $x < 0$ and $x > 0$ in figure 4. Both halves are fed by their own pump. The pressure losses are of course higher in the helical loops than in the central pipes. The separation into three circuits as described above distributes the pressure head roughly equally among the three pumps.

The dimensions in figure 4 are $r_0 = 0.94$m, $d = 0.9$m and $a = 0.21$m. These numbers are largely dictated by the characteristics of the available pumps, i.e. the pressure difference between the in- and outlet of a pump for a given volumetric flow. Other considerations concerning optimization emerged from numerical simulations are discussed in the next section and ref. [8].

Provisions are made for Hall probes which can be moved along the axis of the cylinder in order to measure the magnetic field inside the dynamo. A hole along the axis of the cylinder allows to wind meridional induction coils which run along the axis and close outside the cylinder. These coils together with an equatorial coil measure magnetic fields averaged over large cross sections of the flow. Other quantities of interest which will be recorded are the pressure applied by the pumps and the volumetric flows through the three circuits. A Helmholtz coil will also be placed next to the cylinder which allows to apply a nearly homogeneous magnetic field and to investigate the behavior of the flow before it actually starts to operate as a dynamo. Further details of the experimental setup are given in [9, 10].

5 Numerical simulations

Predictions of the critical magnetic Reynolds number in dependence of parameters like a, r_0 and d are necessary on which the design of the dynamo can be based. It is possible to extend the analysis of section 3, but one can also resort to direct numerical integration of equ. (10). A relatively high resolution is necessary to specify the velocity field $\boldsymbol{u}$. It is therefore advisable to use spectral methods which are known to yield the best accuracy per grid point. A major difficulty in simulating the experiment with any method is the cylindrical geometry which potentially leads to resolution problems at the corners. It has therefore been decided to simulate a flow in a cylinder which is enclosed in a sphere filled with stagnant sodium. The sphere itself is surrounded by electrical insulators.

The numerical method which produced the results presented below uses a decomposition of the magnetic field into poloidal and toroidal fields. The spatial dependence of each field is represented by spherical harmonics and Chebychev polynomials for the angular and radial variables, respectively. In order to find the fastest growing (or most slowly decaying) mode, equ. (10) has been integrated in time starting from initial seed fields which respect the symmetries of the problem, with some random noise added.

Features not included in the simulation are the pipes leading to the pumps and the anisotropy in electrical conductivity: because the stainless steel does not have exactly the same resistivity as the sodium, the average resistivity measured along the axis of the cylinder is less than in a perpendicular direction. On the other hand, the flow field can be specified at wish and does not need to be as smooth as the velocity field (9). In particular the flow at the ends of each cell from one cell to the next can be taken into account. The simulations have shown that the precise profile of the velocity field has a minor impact on the critical magnetic Reynolds number [11]. As suggested by section 3, only certain averages of the profile over individual cells are relevant.

Figure 6 gives an impression of the simulated magnetic field generated for a flow 1.6% above critical in a cylinder with $r_0 = d$ containing an 8×8 lattice of the velocity field (9) with $\frac{A\alpha}{C} = \frac{1}{\sqrt{2}}$. This comes close to the experimental geometry: figure 4 shows an 8×8 lattice of cells from which 3 cells are removed at each corner to make the structure fit into the cylinder.

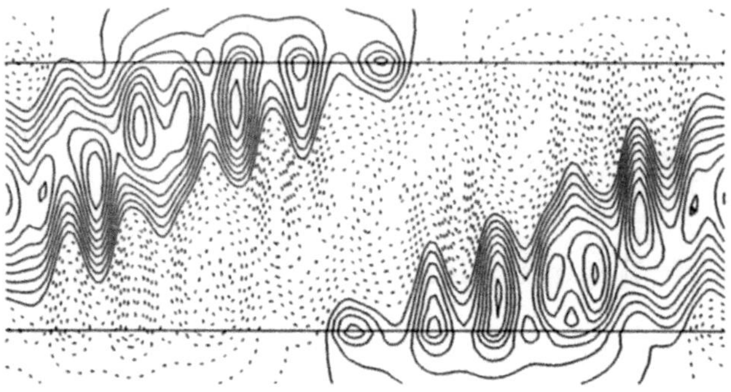

Fig. 6. Contour plot of B_ϱ on the cylindrical surface $\varrho = 0.79 r_0$ in cylindrical coordinates (ϱ, φ, z). φ increases from $-\pi$ to π in going from left to right ($\varphi = 0$ corresponds to the x-axis). The two horizontal lines locate the top and bottom lids of the simulated dynamo cell. Solid (dashed) contours indicate positive (negative) B_ϱ.

Figure 6 gives the component B_ϱ of the magnetic field in cylindrical coordinates (ϱ, φ, z) on a surface $\varrho = \text{const}$. One clearly recognizes the staircase structure predicted by (17a) with half a wavelength fitting in between the top and bottom boundaries. Fluctuations on the length scale of the cell size a are superposed on the average field. It now becomes interesting to compare the critical value of $R_A R_c$ given by (18a) with numerical simulations. Simulations yield critical values of $A\alpha$ and C. The definition of R_A and R_C is based on the length scale γ^{-1} which was taken as d/π when evaluating $R_A R_C$ for the numerical results. Figure 6 shows that this is a reasonable choice for γ. The numerical results for $R_A \cdot R_C$ are given in figure 7 for flow (9a) with $\frac{A\alpha}{C} = \frac{1}{\sqrt{2}}$ in a cylinder of aspect ratio 1 containing an $N \times N$ lattice. The approximations in section 3 are best realized for large N; where the separation of scales is most pronounced and the magnetic Reynolds number is smallest. A finite lattice invariably requires higher critical flow rates than predicted in (18a). Numerical simulations are thus preferred to make quantitative predictions. However, the main features of the solutions are already well captured by the simple model of section 3. Where comparisons appear to be possible, the results of the numerical computations also seem to agree reasonably well with computations [13] based on the mean field magnetohydrodynamic formulation for the α-effect [14].

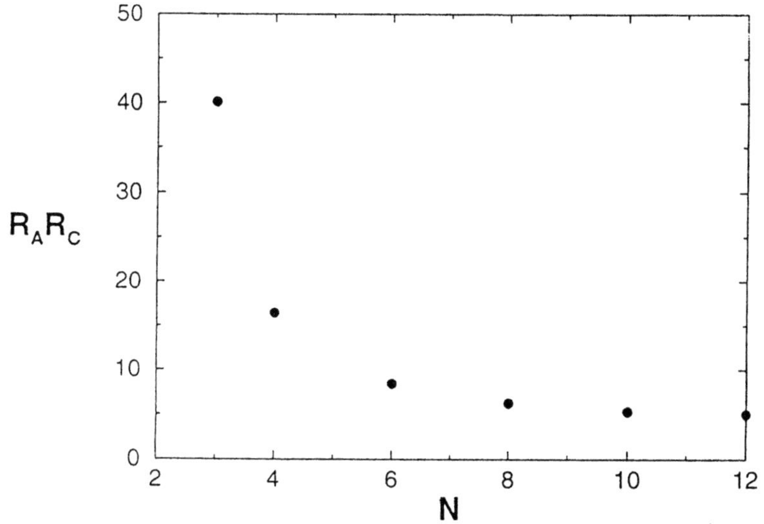

Fig. 7. Critical values of $R_A \cdot R_C$ obtained for lattices of size N (see text for remaining parameters).

6 Non-linear effects

Equation (10) allows magnetic fields that grow to infinity. In reality, the magnetic field saturates because it exerts a Lorentz force density $\frac{1}{\varrho\mu_0}(\nabla \times \boldsymbol{B}) \times \boldsymbol{B}$ on the fluid which needs to be overcome by the pumps in addition to the hydraulic friction force. When the Lorentz force contributes a significant fraction to the total pressure loss, oscillations may set in which transfer energy back and forth between the sodium flow and the magnetic field: a rising field slows down the flow, which decreases the field strength, which then allows the flow to accelerate again, etc. However, it is difficult to predict exactly how the pressure drop depends on $\boldsymbol{B}$ because the sodium flow in the pipes is highly turbulent. Turbulence has been irrelevant in previous sections because the value of λ in (10) far outweighs any "eddy diffusivity". Estimates of the latter [12] based on flow rates necessary for dynamo action yield only a 3 percent contribution to the magnetic diffusivity. The pressure losses on the other hand are entirely determined by the turbulence of the flow. A generated magnetic field may modify the small scale structures of the flow and change the relation between the pressure drop applied by the pumps and the volumetric flow rate. In a first approximation, one may assume that flow structures are unaffected by the magnetic field. In that case, the experiment should be able to generate fields of several hundred times the earth's magnetic field strength.

The most interesting non-linear aspect relates to transitions of the field between different spatial configurations. Such a behavior would be reminescent of the reversals of the earth's magnetic dipole field. It is quite possible that the experiment will show transitions owing to the degeneracy of the preferred mode. The field in figure 6 has a distinguished direction, for instance the line joining the maximum and minimum of B_ϱ in the equatorial plane. The flow pattern in figure 4 locks this direction either along the x- or the y-axis. Both axes are equivalent in theory. In reality, the degeneracy will be lifted by the in- and outflows and other minor asymmetries not included in the model. Since three independent pumps are available, it may be possible to fine tune their flow rates such that the two axes are just equally preferred and that the field switches between the two states because of non-linear interaction with the flow.

References

[1] Cowling, T.E., The magnetic field of sun spots, *Monthly Not. Roy. Astr. Sci.* **94**, 39-48, 1934

[2] Backus, G.E., A class of self sustaining dissipative spherical dynamos, *Ann. Phys.* **4**, 372-447, 1958

[3] Herzenberg, A., Geomagnetic dynamos, *Phil. Trans. Roy. Soc. London, Ser. A* **250**, 543-585, 1958

[4] Roberts, G.O., Dynamo action of fluid motions with two-dimensional periodicity, *Phil. Trans. Roy. Soc. London, Ser. A* **271**, 411-454, 1972

[5] Tilgner, A., and Busse, F.H., Subharmonic dynamo action of fluid motions with two-dimensional periodicity, *Proc. Roy. Soc. London* **A 448**, 237-244, 1995

[6] Busse, F.H., and Müller, U., The homogeneous dynamo: an analytical model and a planned experimental demonstration, pp. 149-163 in Proc. 2. Int. Conf. on Energy Transfer in MHd, Aussois, 1994

[7] Busse, F.H., Dynamo Theory of Planetary Magnetism and Laboratory Experiments, pp. 197-208 in "Evolution of Dynamical Structures in Complex Systems", R. Friedrich and A. Wunderlin, eds., Springer Proceedings in Physics, Vol. 6, 1992

[8] Tilgner, A., A kinematic dynamo with a small scale velocity field, *Phys. Rev. Lett. A* **226**, 75-79, 1997

[9] Busse, F.H., Müller, U., Stieglitz, R., and Tilgner, A., An Experimental Investigation of the Homogeneous Dynamo Effect, *Magnetohydrodynamics* **32**, 235-248, 1996

[10] Stieglitz, R,. and Müller, U., GEODYNAMO - Eine Versuchsanlage zum Nachweis des homogenen Dynamoeffektes, Bericht FZKA 5716, Forschungszentrum Karlsruhe, 1996

[11] Tilgner, A., Predictions on the behaviour of the Karlsruhe dynamo, *Acta Astron. et Geophys. Univ. Comenianae XIX* **51-62**, 1997

[12] Moffatt, H.K., Magnetic Field Generation in Electrically Conducting Fluids, Cambridge University Press, 1978

[13] Rädler, K.-H., Apstein, E., and Schüler, M., The α-effect in the Karlsruhe dynamo experiment, to appear in Proceedings of the PAMIR Conference, Assois, France, 1998

[14] Krause, F., and Rädler, K.-H., Mean-field Magnetohydrodynamics and Dynamo Theory, Oxford: Pergamon, 1980

From Single Cells to a Multicellular Organism: The Development of the Social Amoebae *Dictyostelium Discoideum*

B.N.Vasiev & C.J.Weijer

Department of Anatomy & Physiology, Wellcome Trust Building, University of Dundee, Dundee, DD1 5EH, UK

Abstract. One of the main goals of contemporary biology is to understand the development of multicellular organisms. This is a very complex process that involves cell division, cell death, cell differentiation and cell movement, which are all highly regulated in space and time. The co-ordination of these processes involves extensive cell-cell communication. Development of a vertebrate involves hundreds of cell types and thousands of signalling molecules, all interacting through positive and negative feedback pathways. Due to this complexity, it is very difficult to describe the development of these organisms in a meaningful physical sense. There are however much simpler organisms that show all of the essential characteristics of the development and are more accessible to such a description. *Dictyostelium discoideum* (Dd) is one such a system. Discoideum species are known as social amoebae, and their colonies are among the simplest and best studied objects of developmental biology. Under normal conditions *Dictyostelium* populations consist of groups of independent, single amoebae. However, under unfavourable conditions, when food becomes limiting, these amoebae become "social" and enter a complex developmental program. They collect into multicellular aggregates and differentiate into a number of cell types. These cell types organise themselves to form a fruiting body consisting of (dead) stalk cells and spores. The spores disperse and each spore can start a new colony under favourable conditions. Some of the features of Dd development are beginning to be understood both at the biological and formal level. Here we review *Dictyostelium* development and the progress made in the mathematical modelling of *Dictyostelium* morphogenesis.

1 Introduction

Most multicellular organisms develop from a single cell, the fertilised egg. This involves processes such as regulated cell division, cell death and cell differentiation in a multitude of cell types. These cell types may arise in situ at the right position at the right time of development or they may differentiate first as a precursor cell type and then move to their final destinations, as is the case in the formation of the nervous and immune system. These processes are highly complex and there is not yet any formalised understanding. For these reasons there has been a tendency to study the development of much simpler model organisms which still show some of the essential features of multicellular development. One prominent model organism is the social amoebae *Dictyostelium* discoideum. This is a very simple organism that on an evolutionary scale stands on the threshold between unicellular and multicellular life forms. Part of its life cycle it lives as single amoebae in the soil where it feeds on bacteria and multiplies by binary fission. However, upon starvation the cells aggregate to form a multicellular structure, the slug, consisting of up to 10^5 amoebae. The slug is motile and guided by photo and thermotaxis can move away to a place suitable to form a fruiting body. The fruiting body consists of a stalk, formed by dead vacuolated stalk cells, that supports a mass of spore cells. The spores can disperse and germinate, so that each gives rise to a single amoebae. The cells start to differentiate in stalk and spore cells during aggregation and a pre-pattern of prestalk and prespore cells is formed in the slug. The prestalk cells (which will later form the stalk) are localised in the tip of the slug while the prespore cells (transforming later to the spores) are localised in the back of the slug. This organism retains many of the commonly found elements of multicellular development such as differentiation of one cell type, the amoebae, into at least two cells types, the stalk cells and the spores. When the slug is cut into a prestalk and prespore piece both pieces will regulate and form perfectly normal proportioned fruiting bodies. This shows that the proportions of these cell types are strongly regulated by an elaborate feedback mechanism in which the cells signal each other all the time.

During the differentiation phase of the slime moulds, the cells do hardly divide (since there is no food). Therefore, morphogenesis results solely from the co-ordinated movement of individual cells. This co-ordination also requires extensive cell-cell signalling the mechanism of which will be reviewed extensively in this article.

In this article we will give an overview of the essential behaviour and mechanisms involved in the morphogenesis of this simple organism and describe the progress that has been made to understand the essential principles involved in a more theoretical formalism. We will finish by highlighting some unsolved problems, which may direct future research.

2 The life cycle

The life cycle of a developing *Dictyostelium* population is shown schematically in Fig.1. Normally slime moulds live as single amoebae in the soil. They feed on bacteria and divide by binary fission. The population multiplies and at a certain point in time will have depleted the food source. Starvation induces the activation of a developmental program in which the cells aggregate chemotactically to form a multicellular mass of cells containing $10^3 - 10^5$ cells. A few hours after the beginning of starvation cells begin to move synchronously towards the aggregation centre. This is the result of a chemotactic movement of the cells to signals emitted by the cells in the aggregation centre. The cells in the aggregation centre periodically emit a signal, 3'-5' cyclic-adenosinemonophosphate (cAMP), which is detected and relayed by surrounding cells. This leads to formation of propagating waves of cAMP, which instruct the cells to move towards the centre by chemotaxis. These waves can be seen as target or spiral dark-field waves (Figs.1,4). These dark-field waves are known to reflect changes in the cell shape: elongated moving cells form light areas, round-shaped non-moving cells – dark areas. Cells move during the rising phase of the cAMP waves towards the source of waves. Then the aggregating cells make contacts and form branching streams, in which the cells move towards the aggregation centre. Formation of these streams accelerates aggregation, since cells in streams move faster than individual crawling cells. All cells move towards the wave sources or aggregation centres. The presence of several competing centres divides the Dd population into domains (aggregation territories, Fig. 4), so that all cells in the same domain move to the same aggregation centre. The aggregation stage of Dd finishes when all cells collect in their respective centres and form multicellular aggregates (hemispherical mounds). In the mound, the cells start to differentiate. They transform into several prestalk types which will form different parts (the stalk, basal disk and upper and lower cup) of the fruiting body as well as into prespore cells which will continue to differentiate to form spores [1]. The cells differentiate in random positions in the late aggregate. The prestalk cells sort to the top of the mound to form the tip. The mound erects and extends up in the air to form the standing slug, which falls over and migrates away. The slug has a distinct polarity with a tip at the anterior end, which guides all its movement. The slug is photo and thermotactic, which allows it to move up towards the soil surface. There it transforms into a small fruiting body (up to 4mm high) consisting of a stalk supporting a spore mass. The spores disperse and under suitable conditions germinate to release amoebae and the whole cycle can start all over again.

There are several important problems concerning *Dictyostelium* development:

-How do single cells move chemotactically?

-Which signals control the aggregation process?

-Which signals control the differentiation of the cells?

-What controls cell sorting?

-How does the hemispherical mound form a cylindrical slug?

-How does the slug move?

-Which processes are responsible for the formation of the fruiting body?

All these problems are being addressed in experimental and theoretical studies and some of the mechanisms are beginning to be understood. The best understood part of the developmental cycle is the aggregation stage of development. The cells aggregate by chemotaxis to propagating waves of cAMP (Fig. 3). Cells in the aggregation centre start to produce the chemoattractant cAMP in a periodic fashion. They release the cAMP in the extracellular medium where it diffuses to neighbouring cells. These cells detect the signal

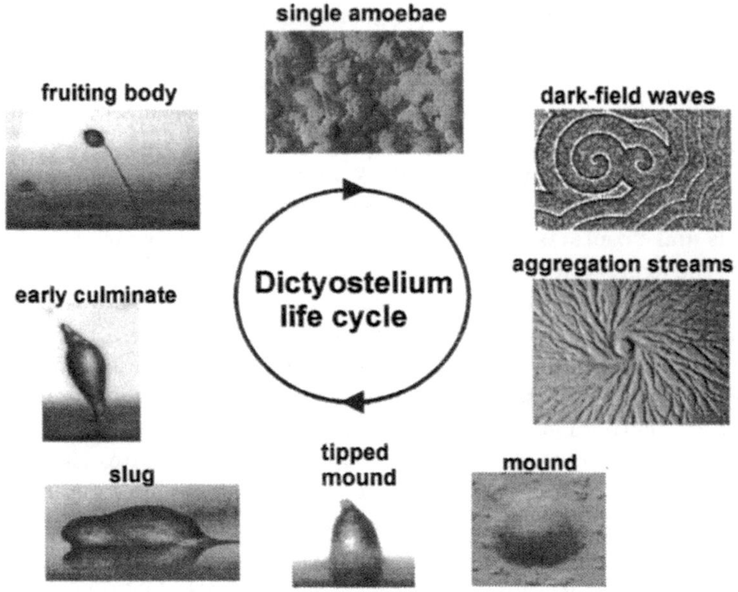

Fig. 1. Dictyostelium life cycle. Starving single amoebae aggregate towards propagating waves cyclic AMP. These waves co-ordinate chemotactic cell movement and can be visualised as darkfield waves. The cells form aggregation streams in which they move towards the aggregation centre to form multicellular aggregates (mounds). Each mound transforms into a migrating slug. The slug migrates to a place suitable to form a fruiting body. The fruiting body consists of a stalk supporting a spore mass. The spores disperse, germinate and each spore releases an amoeba, which can start a new colony.

by specific transmembrane cAMP receptors, which then in a relatively complex cascade of reactions, activate the enzyme adenylate cyclase, which starts to produce cAMP (Fig. 2). This cAMP is secreted to the outside where it can activate neighbouring cells again. After the cells have been stimulated, they become refractory for a while to further stimulation, which leads to a cessation of their relay response. Adaptation ensures the unidirectional wave propagation. Since the cells also secrete an enzyme cAMP phosphodiesterase, which degrades the cAMP, the levels of cAMP are going to fall after a while. This allows the cells to de-adapt to regain sensitivity to a new signal coming from the aggregation centre. This mechanism leads to the initiation of waves by the aggregation centre and their outward propagation (Fig. 3). The biochemical mechanisms underlying this signalling network are quite complex and involve many different proteins [2] (Fig. 2). The most important ones are depicted in the Figure 2. The cAMP signal is detected by transmembrane cAMP receptors, which upon binding of cAMP change their conformation and activate a hetero-trimeric G protein. This G protein undergoes a complex reaction cycle in which the GDP, which is normally bound to the Gα subunit, is exchanged for GTP. It dissociates splits in a $(\beta\gamma)$ complex and an activated (GTP bound) Gα2. The γ subunit has a fatty acid modification and stays bound to the membrane where the $(\beta\gamma)$ complex now binds another protein CRAC (cytosolic regulator of adenylate cyclase). This complex then stimulates adenylate cyclase leading to the synthesis of cAMP. It is not yet known how the cAMP is secreted but a major portion gets to the outside. Upon binding of cAMP, the receptor is phosphorylated by a receptor kinase on its cytoplasmic tail. The Gα2 subunit also gets phosphorylated. The consequence of phosphorylation is a decrease in the affinity for cAMP. Initially it was thought that the phosphorylation step was the adaptation step, however this does not seem to be so, since cells expressing cAMP receptors, in which the amino acids which are phosphorylated have been changed for ones that cannot be phosphorylated anymore, still adapt [3].

Local binding of cAMP also leads to the rapid local activation of the cell and the extension of pseudopodia in the direction of the cAMP signal [4-6]. This involves a complex biochemical machinery, which stimulates locally actin polymerisation and is catalysed by at least 20 different actin-binding proteins. The activity of these proteins is regulated in part by the cAMP signal associated with second messengers such as calcium, particular lipids and cGMP. It also involves the action of myosin motor molecules. It is not yet clear how this local activation process works. The local activation of the receptors has to be translated in a polarisation of the cell, i.e. in a formation of an extending front end and retracting tail end. This process seems to involve phosphorylation of myosin molecules. Myosin II molecules are preferentially found in the rear retracting end of the cell while several members of the myosin I class of motor protein are found in the extending part of the cell [7-9]. There have been few models for the local activation of the cytoskeleton during chemotaxis, although kinetic models for stimulated actin polymerisation in other chemotactic cells are beginning to appear [10-12].

Initially Dictyostelium cells are not excitable and then as a result of the starvation program the cells start to express essential components of the cAMP relay system such as cAMP receptors, G proteins and the enzyme that synthesises cAMP, adenylyl cyclase. The expression of these genes is under the control of cAMP pulses so that there is a feedback system between the cAMP signal and the components responsible for the cAMP oscillations [14]. This is a very interesting situation and obviously much more complex as that found in physical or chemical systems. This also implies that the system is never the same kinetically at any given moment of time, the consequences of which have not yet been fully investigated. Initially the cells are not excitable, they then become excitable and finally some, or all cells, become autonomous cAMP oscillators. The exact process of wave formation has not yet been studied in great detail experimentally. Since the waves control and direct cell movement,

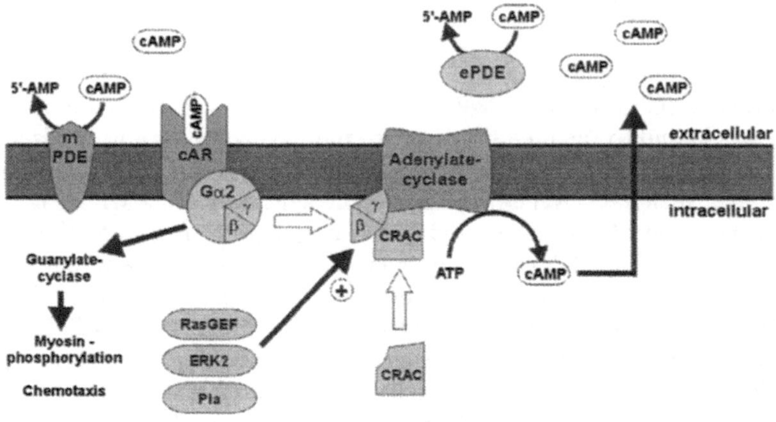

Fig. 2. Figure 2. Diagram of the cAMP signalling system of aggregation competent Dictyostelium cells. Indicated are the biochemical components known to play an essential role in the cAMP relay response. The cyclic AMP receptor (cAR), the G protein Ga2bg, the cytosolic regulator of adenylate cyclase (CRAC), the soluble secreted cAMP phosphodiesterase, the enzyme that degrades cAMP (ePDE), the membrane bound phosphodiesterase (mPDE) as well as the proteins that modulate the activation of cyclase activation (Ras GEF, the map kinase ERK2 and the gene product of the pianissimo gene (Pia) [2,13]. Some of these components may be involved in the adaption step.

the system is in a different state after every passage of a wave, a situation, which is also different from normal chemical and physical excitable systems.

The next two sections are devoted to the description of the biology and mathematics of the signalling and movement systems. They will lay the foundation for the detailed consideration of all stages of Dd development in later sections of the paper.

3 cAMP signalling system

We already know that aggregating cells communicate by means of travelling waves of cAMP. Such chemical waves are known to occur in oscillatory or excitable chemical systems (see, for example, [15]). At the very early stage of aggregation the waves are concentric and probably initiated by cells, which periodically emit cAMP pulses. We will call such cells "oscillatory" or cells with oscillatory signalling system. There is evidence that all cells are oscillatory [15]. However, frequency of oscillations is different from cell to cell and the concentric waves occurring during early aggregation are presumably initiated by cells having the highest frequencies. Another fact indicating the oscillatory behaviour of cells is that first a few waves in aggregation territories look very much like phase waves, which have irregular shapes and velocities (Fig. 4A,D). However, after short period of time the shapes and velocities of these waves stabilise, and phase waves transform into slower excitation waves (Fig. 4B,E). Such a transformation is known to happen in inhomogeneous oscillatory media [17,18]. There are many reasons why an aggregation territory can be inhomogeneous, i.e. cells have slightly different signalling systems or cell density varies over an aggregation territory. Spiral waves probably occur when concentric waves break up on these inhomogeneities, although it has been argued recently that small local fluctuation in the level of a phosphodiesterase inhibitor activity can also be responsible for the formation of spiral waves [19].

Over time a number of different mathematical models for the cAMP relay system have been proposed Some are just based on the general properties of chemical oscillators and some try to model the essentials of the biochemical mechanisms in more detail [20-23]. Detailed biochemical models tend to get quickly outdated as new experimental data accumulate. Let us consider in some detail the two most recent models for the cAMP oscillator. The first ("receptor box") model has been suggested by Martiel and Goldbeter [24,25]. It is based on the assumption that cAMP receptors detecting the level of extracellular cAMP can be in two states, active or inactive. Only active receptors after binding of cAMP molecules can activate adenylyl cyclase. Therefore, the transition between the states of receptors plays an essential role in the cells response. Although this assumption has been proven wrong by the most recent experiments (see above) the reduced Martiel-Goldbetter (MG) model remains the most popular among mathematicians modelling the cAMP signalling system. The Martiel-Goldbeter model describes the cAMP relay system of individual cells. It consists of three coupled (there is a two-variable version) non-linear equations which define the level of extracellular and intracellular cAMP, and the activation state of the cAMP receptors:

$$\frac{\partial c}{\partial t} = \left(\frac{k_t}{h}\right)\beta - k_e c; \tag{1}$$

$$\frac{\partial \beta}{\partial t} = q'\Phi(r, c) - k_t\beta - k_i\beta; \tag{2}$$

$$\frac{\partial r}{\partial t} = -f_1(c)r + f_2(c)(1-r); \tag{3}$$

where

$$\Phi(\rho,\gamma) = \frac{\lambda_1 + Y^2}{\lambda_2 + Y^2};\ Y = \frac{\rho\gamma}{1+\gamma};\ f_1(\gamma) = \frac{k_1 + k_2\gamma}{1+\gamma};\ f_2(\gamma) = \frac{k_{-1} + k_{-2}c\gamma}{1+c\gamma};$$

Equation (1) describes the change in the level of extra-cellular cAMP, c, over time. These changes are due to the secretion of intracellular cAMP, β: cAMP is transported over the cell membrane with the rate defined by k_t and is diluted in the extracellular medium by factor h representing ratio of the volume of extracellular solution to the total volume of cells. The second is cAMP hydrolysis by phosphodiesterase outside the cell, which is assumed to be proportional (k_e) to cAMP concentration. Equation (2) defines the changes in the level of intracellular cAMP, β as depending on three processes, which are taken into account by the three terms in the right hand side of (2). The first is the synthesis of intracellular cAMP by adenylate cyclase in response to a cAMP stimulus (it depends on the level of extracellular cAMP, c and the state of cAMP receptors, r). The second is a loss of intracellular cAMP due to secretion (same as first term in (1)). The third term denotes the intracellular hydrolysis of cAMP, which differs from second term in (1) by the rate of hydrolysis. Equation (3) reflects changes in the state of the receptors. It defines the change in the relative number of activated (r) and inactivated ($1 - r$) receptors resulting from their transition governed by the level of extracellular cAMP. To describe cAMP waves propagating through a population of cells, a diffusion term for extracellular cAMP has to be included in (1).

A second model has been introduced by Tang and Othmer [26]. This model postulates other pathways for excitation and adaptation of a cell's cAMP signalling system. It is based on the mammalian paradigm for membrane receptor mediated signalling pathways. In this model the activation and adaptation are mediated by stimulatory and inhibitory G proteins. In short, when the receptor binds cAMP it can activate the α subunit of a stimulatory G protein. This in turn activates adenylyl cyclase and causes the production of cAMP. Adaptation involves the slower receptor activation of an inhibitory G protein, which inhibits cyclase activation. Here, as well as in the case of the MG model, there are assumptions, which are not in agreement with newer biochemical data or not confirmed experimentally. For example, the α subunit of stimulatory G protein is not involved in the activation of adenylyl cyclase and there is only little evidence for the involvement of an inhibitory subunit. The final system of equations is more complex as the Martiel Goldbeter model and therefore this model is not used as extensively by other groups. Both models mentioned above are able to capture the essential elements of the cAMP oscillator and describe oscillations in cAMP level in the cell suspensions of isolated cells as well as cAMP wave propagation in a dispersed cell population [27,28]. However, there is a clear scope for improvement, i.e. modelling the detailed biochemical mechanism.

Both models basically describe excitable and/or oscillatory media and are qualitatively similar to the prototype FitzHugh-Nagumo (FHN) system, which is widely used for the study of general properties of excitable, oscillatory or trigger media [15]. Finally, since the MG and TO models are not completely satisfactory, using the FHN model for qualitative description of cell signalling system can also be considered as a good option [29,30]. The FHN model is described by the following equations:

$$\partial g/\partial t = D\Delta g - k_g(g-a)(g-1) - k_r r \tag{4}$$

$$\partial r/\partial t = (g - r)/\tau \quad (5)$$

Here g is assumed to define the level of extra-cellular cAMP, and r - the recovery variable responsible for the adaptation step. The last serves the same role as the proportion of active and inactive cAMP receptors (in MG model) or activated $G\alpha_i$ subunits of the inhibitory G- protein (in TO model).

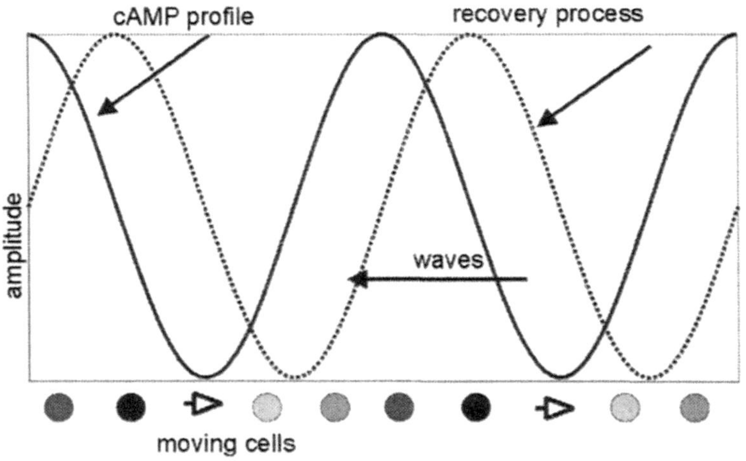

Fig. 3. Dictyostelium cells communicate over long distances by chemical (cAMP) waves. The wave propagation in a one-dimensional medium is schematically shown to occur from right to left, the cells (circles and arrows) move from left to right. A black spot is a resting cell and arrow indicates a fast moving cell. The x-axis represents distance, while y-axis corresponds to the level of cAMP (solid line) and adaptation (dotted line). According to experimental data [31] cAMP profiles should be very close to sinusoidal. At the early stage of aggregation the level of cAMP varies between 10^{-10} and 10^{-6} M. During development both these values presumably increase. This happens due to accumulation of cells in densely packed aggregates and is accompanied by switching of cells to lower affinity cAMP receptors [2]. Cells move chemotactically during the rising phase of the cAMP waves, in the direction opposite to that of cAMP wave propagation.

4 Chemotactic cell movement

cAMP waves not only propagate through the cell population but also co-ordinate their movement. cAMP orients the direction of otherwise randomly moving cells. There is strong evidence that cells detect the gradient of cAMP over their length [32,33,34]. There is also evidence that cells use the temporal derivative of cAMP and only move up the gradient as long as the cAMP level is rising [35,36]. This is a result of adaptation of the chemotactic response and inhibits the cells from turning around and chasing the waves outwards again once they have passed. There is a good experimental evidence for an adaptation step in the chemotactic response. It appears to be mediated by an adaptation of the cGMP signalling system that leads to chemotactic cell movement [2].

A number of mathematical models have been proposed to describe chemotactic cell movement. The oldest and best known one is the Keller-Segel model [37] describing a cell flux, J, as a function of cell density, ρ , and concentration of cAMP, c.

$$J = -D(c)\nabla\rho + \chi(c)\rho\nabla c \tag{6}$$

The first term on the right hand side describes random cell movement (the velocity can depend on the level of cAMP) and the second term describes the directed motion of the cells along the cAMP gradient.

There are also a number of models where chemotactic cell motion is described in an axiomatic way, as rules for motion of cells in a concentration field of cAMP. A few sets of such rules have been used successfully to model cell aggregation [23,38-40,41-43]. Most recently chemotactic cell motion in multicellular aggregates was successfully modelled by assuming that a group of cells can be considered as a fluid whose behaviour can be described by the Navier-Stokes equations, [44,30]. The equation for the velocity of cells then assumes the following form:

$$\rho[\partial \boldsymbol{V}/\partial t + (\boldsymbol{V}\nabla)\boldsymbol{V}] = \boldsymbol{F}_{ch} + +F_{fr} + \eta\Delta\boldsymbol{V} + \xi \mathrm{grad}\, div\boldsymbol{V} + F_{ad} - \mathrm{grad}p \tag{7}$$

where $\boldsymbol{V}$ is the velocity and r is a density of cells; p is a pressure in an aggregate developing due to differences in the velocities of cells, $\boldsymbol{F}_{ch}$ is a force exerted by the cells in response to the chemoattractant gradient; the second and third terms in the right hand side of (7) take into account the mechanical (viscous) cell-cell interactions. Once we know the cell flow velocities (equation 6 or 7) we can compute the evolution of cell density in an aggregation territory using the equation for the conservation of mass:

$$\partial\rho/\partial t = D_\rho \Delta\rho - div(\rho\boldsymbol{V}) \tag{8}$$

Here we assume that the coefficient for random cell movement is constant (Fig. 5,6).

All models developed so far for chemotaxis are phenomenological i.e. they describe the phenomenology of chemotactic motion rather than the mechanisms of chemotaxis. To obtain a more mechanistic model of chemotaxis it will be necessary to develop more detailed models for amoeboid cell motility first [45] and then incorporate these in the description of the behaviour of populations of cells. This is a clear goal for the near future.

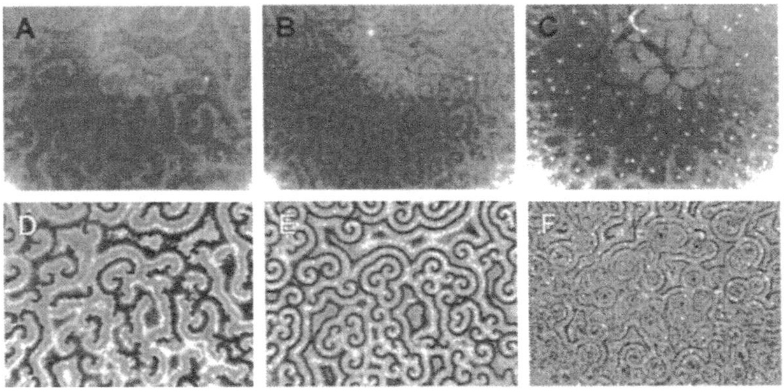

Fig. 4. Development of optical density waves during the aggregation of Dictyostelium cells strain DH1. A-C show original images as recorded under a macroscopic darkfield [46] at three successive times in development 4,5 and 6 hrs of development. A, D show the early darkfield waves, during the time that the phase waves are about to disappear. The wave fronts are still ragged. B,D show images when the excitation waves are fully developed, the wave fronts are smooth. C,F show images at a time where the cells start to organise themselves into streams and aggregation territories start to develop. The sections shown are 2.5 cm wide. The lower series of images are obtained after enhancement of the images shown in A-C by a rolling subtraction of three successive images taken 10 seconds apart.

5 Aggregation

Dictyostelium aggregation has been the subject of many modelling attempts. There are two main problems associated with the early aggregation phase of Dd development. The first concerns the mechanism of the formation, interaction, and geometry of the cAMP waves. The second is the mechanism responsible for the formation of aggregation streams. The modelling of the aggregation waves is mostly dealt with as purely being a consequence of the cell signalling system. The formation of streaming patterns is clearly more complicated and depends on properties of both, the signalling system and the chemotactic response of the cell.

One of the first models devoted to Dd aggregation was developed by Novak and Seelig [38]. It describes the cAMP signalling system and chemotaxis by simple rules and they were able to obtain simple aggregation of patterns of discrete cells in computer simulations. However, no clear aggregation streams were observed in their model. Streams were obtained a few years later in simulations by MacKay [23] using a more complex discrete model. These and other model simulations [47,48] showed that accumulation of cells in aggregation centres is easy to understand and to simulate, i.e. the cells simply collect at the location of waves source. However, the mechanisms of stream formation remained unclear for a much longer period. The first theoretical papers on the mechanisms responsible for aggregation and stream formation were devoted to the analytical estimation of the stability of equation (6), see for instance [37]. "Chemotactic instability" introduced in this ref. could be used to explain accumulation of cells in aggregation centres but not to explain the formation of streaming patterns. A large step forward was made by Nanjundiah [49] and later by Levine [50] who showed the existence of a "streaming instability" which results from the influence of cell density on the effectiveness of the cell signalling system. This has been shown in more detail in [29] where it was stated that streams form when the velocity of the cAMP waves increases with an increase in the local cell density. This statement contradicted experimental data that failed to show a dependence of the velocity of cAMP waves on cell density [51] as well as some theoretical considerations [52,53]. However, an increase in the velocity of cAMP waves with an increase in cell density agrees with numerical results obtained from a variety of models (convection-reaction-diffusion model [29] cellular-automata model [40] and hybrid model [39]) and newer experiments [39]. Stream formation can be explained qualitatively as following. Local accumulation of cells will, due to the dependence of the rate of cAMP accumulation on cell density, result in a local speeding up of the wave propagation at regions of high cell density. This local deformation of the wave front will lead to the attraction of even more cells to this region and finally to the formation of bifurcating aggregation streams. However, many researchers are still trying to formulate new explanations [54-56,57].

One area of research concerns itself with the formation of wave patterns in aggregation territories. There is little doubt that an aggregating population can be viewed as an inhomogeneous oscillatory medium where the sources of concentric waves are cells whose frequencies of oscillation are higher than frequencies of other cells (see, for example [18,58]). Describing population of aggregating Dictyostelium cells as an excitable system and the waves propagating through fields of cells as excitation waves also works well [27] and can easily describe the formation of spiral waves. While concentric waves of cAMP initiate at spots of local high frequency in aggregation territories, the mechanisms of spiral waves formation are less clear. In both [40] and [29] it was suggested that concentric waves can result from breakpoints due to inhomogeneities of the system (i.e. in areas of lower local density of cells) and spirals develop from these break points. In [59] a more complicated approach (positive feedback between the cAMP level and local excitability of the medium) to get formation of spirals in homogeneous medium was developed. The problem of the interaction between concentric and spiral waves has been studied in [60,61]. The main result of these papers can be formulated in general terms of excitable media. Spirals are dominant over concentric waves and remove the latter from the medium when the excitability of the medium is high. Under these conditions spiral will run at their maximal frequency in agreement with early experimental data [62-64]. A decrease in the excitability of the medium can alter the situation so that, concentric waves will become dominant and will replace spirals in the medium. Another interesting observation is that the spiral wavelength decreases continuously during development. This is the reason that initially during aggregation the spiral waves can organise aggregation territories of several centimetres in diameter and later in development also much smaller mounds, which are typically a few hundred microns in diameter. This change in spiral size has been partially attributed to the change in cell density during aggregation leading to an increase in oscillation frequency and a dispersion mediated decrease in wave propagation velocity [65,66], which has also been found in model calculations [30,39]. However in reality, this effect is quite large, the decrease in spiral wavelength has been estimated to be a factor of 20 [67] and dispersion is not enough to account for this. It is most likely caused by the feedback of the cAMP pulses on the increase of the expression of the components of the cAMP oscillator (cAMP receptors, adenylate cyclase etc [14]). One model has recently tried to incorporate this feedback [68] and found that it can control the dominance of established spiral centres.

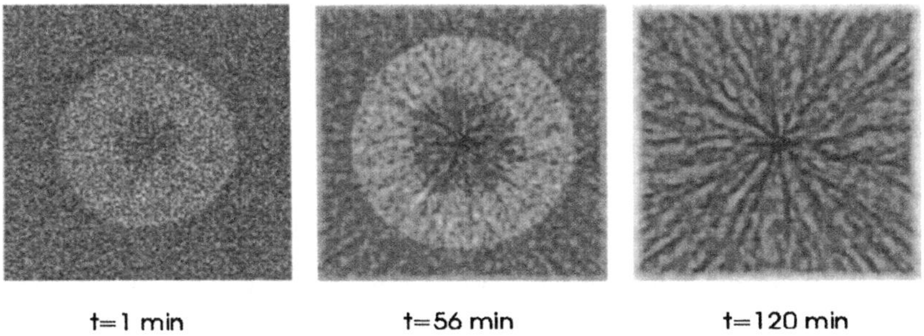

Fig. 5. Formation of streaming patterns in convection-reaction-diffusion system. Two- variable version of Martiel-Goldbeter model has been used two simulate cAMP waves and equation of mass conservation based on the Keller-Segel equation for chemotactic cell flows – to describe cell motion. Random numbers between 0 and 1 in each grid point gives the initial distribution of cell density. Concentric waves of cAMP were initiated every 6 minutes by stimulating the central area of medium. The cAMP wave (white) is superimposed on the pattern of amoebae density (various shades of grey).

6 Mound stage of development

When all the cells accumulate at the aggregation centre they form a mound and development enters a new phase. The most important problems concerning the mound are the following: How do the cells in the mound signal each other? How do the cells move? What is responsible for the formation and shape of the mound? What controls the differentiation of cells and how do the differentiated cells sort out? Finally, which mechanisms are responsible for the transformation of the mound into a slug?

There is good evidence, that the Dictyostelium mound is still organised by cAMP waves [69]. In the mound, clear optical density (OD) waves can be seen to propagate and these waves can be interfered with by periodic microinjection of cAMP pulses with a glass microelectrode. This suggests that the OD waves are carried by cAMP. Furthermore, mutants expressing cAMP receptors of different affinities show different wave patterns (Weijer et al, unpublished observations). The geometry of the waves is strain dependent and can be either concentric, single- or multi-armed spiral waves. Multi-armed spirals are often observed in mounds and the mechanisms for their formation and their significance are not yet clear [67,69]. These diverse geometry's of the signals lead to a variety of complex cell motion patterns. Since cell movement is always opposite to the direction of signal propagation [67,70] cell movement in mounds organised by concentric waves is directed towards the organising centre and slow. In the case of spiral waves, cell movement is rotational and fast.

While aggregation patterns were an object of intensive theoretical investigations for almost 30 years, formation of mound wasn't considered in any detail until recently. Last year however 4 publications dealing with the simulation of mound formation appeared [71,30,43,41]. All of them are very interesting and treat the problem from different points of view. The model by Savill and Hogeweg [71] is based on the assumption that the main reason for cell motility is the tendency to decrease a cell's free surface energy, which is defined by its adhesive properties to other cells and to the extra cellular matrix is strongly effected by the level of cAMP. Using this model authors were able to simulate all stages of Dd development from single cells to a crawling slug. According to this model, the main force responsible for mound formation is cell-cell adhesion. Another important outcome is that differential cell-cell adhesion results in cell sorting in the mound.

The model developed by Vasiev et al [30] was based on the assumption that the chemotactic movement of the cells can be viewed as a fluid flow and described by the Navier-Stokes equations. Using this hydrodynamic approach, it was possible to simulate the formation of streaming patterns and mound formation (Fig.6). It was found that hemispherical mounds form even in absence of cell-cell adhesion. The forces maintaining the mound are an inward directed force (accumulation by chemotaxis) and a pressure, which prevents too close compaction of the chemotactically moving cells. A further two papers [41,43] deal with hybrid models, in which a number of cells were modelled to move according to a number of rules (different rules in each model) in a continuum field of cAMP. The model by Bretschneider et al [43] reinforces the importance of pressure in the process of mound formation,

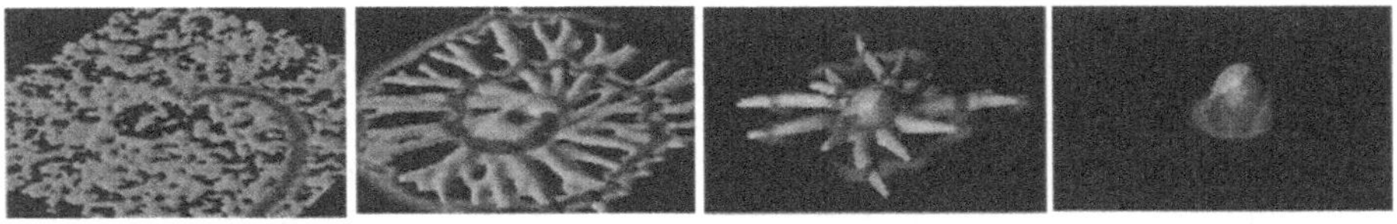

Fig. 6. Formation of the streaming patterns and the mound in hydrodynamic model [30]. Cell density is shown as a white iso-surface ($\rho = 0.5$) and the cAMP concentrations are mapped on this surface from low cAMP (white) to high cAMP (black). The initial density of cells was zero everywhere in 3d-space except for the bottom plane. A random number varying between 0 and 1 represented the cell density in each grid of this plane so that average density in this plane was equal to 0.5. In response to cAMP spiral wave cells move and form aggregation streams (t =40-200 min) and then mound (t =250 min) which represents a stable solution of the system.

while another [41] stresses the importance of adhesive forces in maintaining the mound. There is however an important advantage in the results obtained using the pressure based models. In these models the cells continue to move inside the mound (i.e. rotate along mound's vertical axis in response to spiral wave), while in both adhesion based models the cells stop to move after the formation of the mound. In reality, both adhesion and pressure affect the formation of the mound. A further development of the hybrid model by Bretschneider et al [72] incorporates both pressure and adhesion.

Probably the most important events during the mound stage of Dd development are cell differentiation and sorting [73]. In the mound cells differentiate in prestalk and prespore cells. The differentiation of cells happens at random positions, but then prestalk cells sort towards the top of the mound to form a tip [74]. The mound then contracts at the base while extending up in the air to form a standing slug.

The control of differentiation has been the object of several theoretical studies [75-78]. Since cell differentiation in a mound happens independently from their location, the theory of dissipative structures developed by Gierer and Meinhard, and applied for the process of cell differentiation in hydra [79,80] does not work here. Let us briefly introduce two of the most recent models for cell differentiation in mound as they look to be most promising. Schaap and co-authors [81] developed a model where detailed biological information about the interactions of the morphogens cAMP, ammonia, DIF (differentiation-inducing factor) on differentiated cells are taken into account to estimate the dynamics of the different cell types over time. Different cells in the population produce these factors and different cell types show different responsiveness to these factors (morphogens). By variation of model parameters, they were able to get experimentally observed ratio of fractions of differentiated cells with reasonable kinetics. A completely different model has been proposed by Mizuguchi and Sano [82]. The units (cells) are each considered to be described by a system of coupled activator-inhibitor (FitzHugh-Nagumo) equations. By introducing a global coupling between equations describing each unit, the authors where able to get a separation of all units into two groups representing the two different cell types. When the coupling is strong enough, there are two stationary solutions of the system describing the state of cells, and a portion of cells get one solution while others adopt the second. This description does however not correspond to any known biochemical mechanism.

Once the cell types have formed they have to sort to form the axial pattern of cell types found in the slug. Sorting of cells in mound or slug has also been the subject of extensive theoretical considerations [83-85]. The main problems to be solved are what mechanism is responsible for cell sorting and how is the final pattern stabilised? According to several theoretical investigations, the force that drives cell sorting is differential adhesion and the final pattern of cell sorting is defined by the relative adhesion forces between different cell types and the cells and the substratum [66,80]. In order to sort out, the different cell types should move differentially inside the mound. This differential motion can be caused by cell type specific differences in the motive force generated by prestalk and prespore cells, caused by differential chemotaxis (different cell types exert different forces due to cell type specific differences in their cytoskeleton) or by differential adhesion. We have shown that cell sorting can be driven only by differential chemotaxis [86]. According to these model calculations sorting can be achieved if the following conditions are satisfied:

1. A scroll wave of cAMP is rotates in the mound and the filament coincides with the vertical axis of a mound
2. Prestalk and prespore cells have different excitability: amount of cAMP in each pulse produced by prestalk cells is higher than by prespore cells.
3. Prestalk and prespore cells move chemotactically to cAMP signals as during aggregation. However, prestalk cells move faster in response to the cAMP waves than prespore cells.
4. Movement of cells can be treated as a fluid flow and described by the Navier-Stokes equations.

The model used in this investigation is an expansion of the hydrodynamic model used in [40]. Here we consider a mound as consisting of a mixture of two liquids which, correspond to prestalk and prespore cells. The motion of each liquid is defined by the momentum balance equation:

$$(\partial/\partial t+\boldsymbol{V}_i\cdot\nabla)(\alpha_i\boldsymbol{V}_i) = \boldsymbol{F}_i+\eta\nabla(\alpha_i\nabla\boldsymbol{V}_i)-\alpha_i\mathrm{grad}p+(-1)^i\psi\alpha_1\alpha_2(\boldsymbol{V}_1-\boldsymbol{V}_2) \tag{9}$$

where $i = 1$ or 2 is an index representing prestalk and prespore cells, α_i are the corresponding volume fractions, $\boldsymbol{V}_i$ the velocities and $\boldsymbol{F}_i$ the chemotactic forces. The second term on the right hand side defines the cell-cell interactions between the cells of the same type; the last term defines the interactions between the different cell types. Pressure is defined by the condition of total incompressibility of the mound. We neglect terms containing density derivatives and use the reduced version of equation (9):

$$\partial \boldsymbol{V}/\partial t + (\boldsymbol{V}\nabla)\boldsymbol{V} = \boldsymbol{F}_{ch} + \boldsymbol{F}_{fr} + \eta \Delta \boldsymbol{V} - \text{grad}p \tag{10}$$

We assumed that the force exerted by prestalk cells in response to the chemoattractant cAMP is stronger than that of prespore cells. The FitzHugh-Nagumo equations describe the cell signalling system, where we assume that the excitability of prestalk cells (k_g in (4)) is higher that for prespore cells, in agreement with previous experimental data.

Our computations show that starting from a random distribution of cell types in the mound one can obtain a spatially separated pattern of cell types as well as tip formation. In response to scroll waves rotating along the vertical axis of the hemispherical mound, the liquid begins to rotate in the opposite direction. The faster moving fluid (prestalk cells) accumulates in the centre and top of the mound (the pressure developing in the centre of the mound drives the prestalk cells to the top of the mound and gives rise to asymmetry along vertical axis). Since the faster fluid is more excitable, separation of the fluids leads to the mound becoming inhomogeneous with respect to its excitability so that the topside of the mound is more excitable than the bottom. This in turn results in a change of the cAMP wave shape i.e. the scroll becomes twisted and gets a component of velocity directed downwards to the substratum, along the vertical mound's axis. Therefore, the cells experience a chemotactic force with an upward component, which results in the further collection of faster moving cells on the top. Finally all the faster cells collect at the top of the mound and form a tip, very similar to what happens in real mounds.

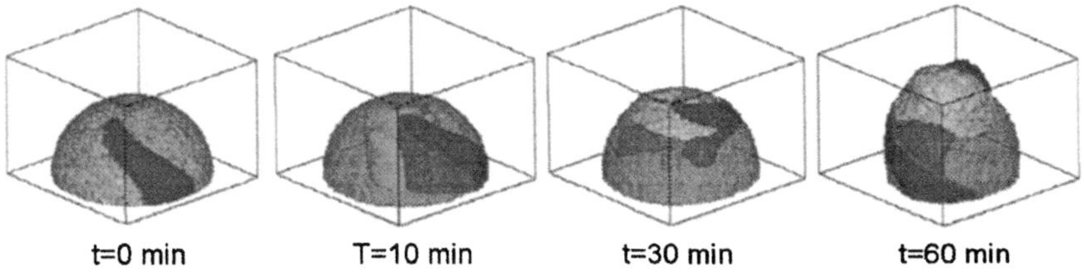

Fig. 7. Cell sorting in the hydrodynamic model. The mound consists of 20% of prestalk cells (light grey) and 80% prespore cells (grey). The cell types differ in their chemotactic response and excitability [86]. The prestalk cells move faster and are more excitable than prespore cells. Initially the mound is a hemisphere in which a cAMP scroll wave (dark grey) rotates clockwise, and both cell types are mixed randomly. Affected by the cAMP waves the cells move and sort, such that the prestalk cells collect at the top of the mound and form a tip.

We found that sorting of cells in the mound results in a transformation of the hemispherical mound into an elongated slug. According to our numerical simulations, the mechanisms responsible for this transformation are the following: 1. The chemotactic cell flow in a mound occurring in response to a scroll wave of cAMP always transforms the hemispherical aggregate into cylindrical one. 2. The geometry of this cylinder depends on the excitability of the cells: cylinders formed by cells that are more excitable are thinner. 3. Since the more excitable prestalk cells form the tip, it differs in geometry from the rest of the mound and has a smaller cross-section. 4. Since the twisted scroll originates in the tip and prestalk cells move to the top of the mound, the mound becomes elongated to form a standing slug (Fig.8).

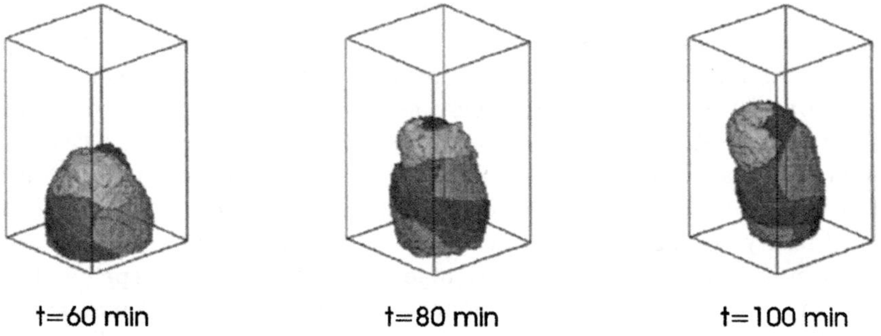

Fig. 8. Transformation of the mound into the standing slug in hydrodynamic model. The mound shown in Fig.7 has been placed in larger medium and allowed to evolve further.

7 Slug stage of development

The standing slug falls over and starts to crawl to form a migrating slug [87,88]. The anterior of the slug (about 20% of its volume) consists of prestalk cells while the rest of its volume consists of the prespore cells. A slime sheath consisting of 50% cellulose and 50% glycoproteins surrounds the slug, which gives the slug some mechanical stability.

The mechanism of slug movement is an interesting problem since it represents one of the simplest kinds of motion exhibited by any multicellular organism. According to experimental data, cells in the back of the slug move forward in the direction of slug migration in a periodic manner [89,90,91,92]. This indicates that they, most probably, move chemotactically in response to propagating waves of cAMP (or other chemoattractant). Tracking of cells shows that cells in the slug tip rotate around the slug's long axis. Assuming that cells move in response to propagating wave of chemoattractant one can conclude that this wave should have a shape of scroll at the top of the mound and be shaped as planar waves in the back of the slug (Fig.9). Computer simulations showed that the conversion of a scroll into a series of planar waves could occur when there is a substantial difference in excitability between the prestalk and prespore cell zones of the slug [93,94]. If this difference is not very large there is no transformation of scroll to planar waves, instead the scroll wave is only twisted, rotating around slug's axis. The waves propagate from the tip to the back. Cells in the back should exhibit a rotational forward motion in response to these waves. This is the case observed in another slime mould strain Dictyostelium mucoroides [88].

A further important point is the differentiation of the cells in the slug [1,95]. New cell types occur in the slug stage of Dd morphogenesis: pstA cells form in the anterior outer part of the prestalk zone; pstO cells at the boundary between prestalk and prespore cells; and pstB cells in the central core of the prestalk zone. The assumption about twisted scroll of cAMP rotating inside the slug helps in understanding this differentiation pattern. It is known that prespore genes need cAMP for their induction and stabilisation; expression of the prestalk specific ecmB gene by the pstB cells is inhibited by high concentrations of extra cellular cAMP while ecmA expression by pstA cells requires high concentrations of cAMP [96]. Computer simulations show, that the core of the scroll wave in the prestalk zone is a region of low average extracellular cAMP [94], exactly the condition which facilitates the expression of the stalk specific ecmB gene in the central core of the prestalk zone. Despite the complex mode of wave propagation it gives rise to a relatively simple spatial pattern of average cAMP, which can be read out by the cells in different positions in the slug to stabilise the differentiated state of the cells in the slug.

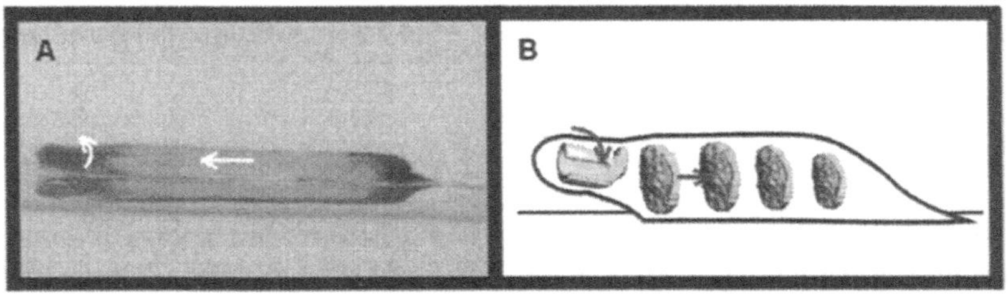

Fig. 9. Model for wave propagation in slugs. A: neutral red stained slug. Cells in the anterior tip exhibit rotational movement around slug's axis, while all other cells move progressively along this axis. B: model for waves in the slug.

Probably the most important property of the slug is its ability to crawl. There were a few attempts to model migration of the slug [44,72,97]. Slug migration must result from the co-ordinated motion of its constitutive cells and most models assume that this motion is chemotactic. Odell and Bonner [44] were the first to propose that the flow of tissue in a slug could be modelled as the flow of a viscous fluid under the influence of a chemoattractant. In addition, one main assumption was that all cells produced and destroyed a chemical, which modulated the chemotactic response of all cells, resulting in a gradient from the centre to the outside. This results in fountain like cell flow patterns along the long axis of the slug, which are not in agreement with experimental observations [91].

Fig. 10. Migrating slug in a model system [72]. Side view of a migrating slug at three successive points of time is shown. Slug is moving from the right to the left; its tip (formed by prestalk cells) is depicted in grey. The track of one (arbitrarily chosen) prestalk cell is shown. The spiral-shaped trajectory indicates that the cell is moving in rotational fashion. Slug migration is driven by cAMP waves of the shape shown in Fig.9B.

More recent attempts to simulate slug migration [72] are more successful, since they show reasonable patterns of cell movement inside the moving slug. However, all these models fail to give a satisfactory description of the mechanism of cell movement inside the slug, i.e. it is not clear how the cells gain traction. Obviously this traction comes from the substrate, and therefore might be proportional to the area of slug- substrate contact. There are experimental data that are in agreement with this notion [98]. However, a better understanding of the force transmission by cells during slug migration can only be achieved based on a good model for individual cell motion.

8 Culmination

The ability of the slug to crawl is very important. Under natural conditions, it migrates up from the leave litter in the soil to the surface before to enter a culmination phase when the slug transforms into a fruiting body. The cellular basis of culmination is not yet well investigated. During culmination the prestalk cells from the tip of the slug start to move downward through the middle of the slug in an inverse fountain like process [99,100]. As soon as the prestalk cells touch the substrate they undergo the final differentiation step into stalk cells. These cells are highly vacuolated cells that have a rigid cell wall and resemble plant cells in structure. During the culmination process cells move on top of the stalk where they also vacuolate. By doing so, the stalk extends from the top end since more and more cells crawl on top of it. The other cells move upward along the stalk as it extends. By this mechanism all prespore cells are lifted in the air. These cells perform a final differentiation step into spore cells. They are attached to the stalk by structures made from prestalk cells called the upper and lower cup. The spores can then disperse and wait for favourable conditions to germinate and release amoebae again so that the life cycle can start all over again. Since the cellular basis of the culmination process is not yet understood, i.e. there is especially a lack of understanding of the signals that control this process there have been no serious attempts to model this stage of development. This clearly has to be seen as a future goal but has to await further experimentation first.

9 Conclusions and outlook

Except for the culmination, all stages of Dictyostelium development have been intensively studied using mathematical methods and physical concepts. Since during all parts of the Dictyostelium life cycle the cells communicate by propagating chemical waves of cAMP, a population of these cells can be considered as forming a non-linear dynamical system, which can be described as an excitable medium. Due to the chemotactic movement of cells, the geometry of this excitable medium changes over time. It transforms from a 2-dimensional to a 3- dimensional medium. The change of shape of the different three-dimensional media is accompanied by complex transformations of the chemical waves in the system. In the course of the development phase waves transform into excitation waves, concentric waves give rise to spirals, single-armed spirals transform into multi-armed spirals and scrolls, which in turn twist and even give rise to planar waves. These waves co-ordinate cell movement and the combination of both processes leads to morphogenesis.

To conclude let us note here some of the main unsolved problems concerning Dictyostelium development: -We know that cells communicate by means of chemical waves of cAMP, which result from co-ordinated cell-cell signalling. The detailed biochemical network of this process is still not yet completely known. This clearly has to be the subject of further experimentation and modelling efforts. For instance, it is now well known that there are different cAMP receptors of different affinity that are expressed in different phases of the life cycle [2]. Their precise role is still the subject of both further experimental and theoretical investigations. –The detailed process of chemotaxis at the single cell level is still poorly understood. We know that Dictyostelium cells react chemotactically to cAMP, i.e. can move up a cAMP gradient. Many questions remain unanswered. How do the cells detect these gradients and translate them into cell movement? Where and how are the traction forces generated? How are contacts with the substrate made and released how are all these processes regulated and coordinated in a precise spatio-temporal manner? The physical basis of force generation by individual cells in multicellular tissues is very important but as yet also only poorly understood and investigated. This is clearly a very important area of investigation. –The interaction between the wave propagation system and the regulation of the gene expression of the components involved is also a very important area of investigation and has only been explored to a limited extent. This introduces further feedbacks into the system and can give rise to new types of behaviour. Furthermore, the fact that all cells are different and express different numbers of important macromolecules leads to a large heterogeneity, i.e. no two cells are alike. This again is very different from most chemical and physical excitable systems and will give rise to further heterogeneity and new solutions to the underlying dynamical systems. –Most models for cell sorting have dealt only with two cell types but there are likely to be more which all differ in the movement and signalling parameters. This has to be explored further as well. The cell type proportioning mechanism has to be integrated in the models for cell movement in mounds and slugs. – Finally the genome of Dictyostelium will soon be sequenced and it is to be expected that all of it approximately 10.000-15.000 genes will be known. The big challenge will be to understand this very complex highly integrated and compartmentalised dynamical system.

Acknowledgements

This work was supported by the DFG and the BBSRC.

References

[1] Williams, J., Current Opinion in Genetics & Development **5**(4), 426-431 (1995).
[2] Parent, C.A., and P.N. Devreotes, Ann. Rev. Biochem. **65**, 411-440 (1996).
[3] Kim, J.Y., et al., Journal of Biological Chemistry **272**(43), 27313-27318 (1997).
[4] Gerisch, G., et al., Curr. Biol. **5**, 1280-1285 (1995).
[5] Westphal, M., et al., Curr. Biol. **7**(3), 176-183 (1997).
[6] Noegel, A.A., et al., pp. 116-126 in The Cytoskeleton (45. Colloquium - Mosbach 1994), B.M. Jockusch, E. Madelkow, and K. Weber, eds., Springer-Verlag: Berlin (1995)
[7] Moores, S.L., J.H. Sabry, and J.A. Spudich, Proc. Nat. Acad. Sci. USA **93**, 443-446 (1996).
[8] Spudich, J.A., et al., CSH Symp. Quant. Biol. **60**, 783-791 (1995).
[9] Springer, M.L., B. Patterson, and J.A. Spudich, Development **120**, 2651-2660 (1994).
[10] Mogilner, A., and G. Oster, Biophysical Journal **71**(6), 3030-3045 (1996).
[11] Mogilner, A., and G. Oster, European Biophysics Journal With Biophysics Letters **25**, 46-53 (1996).

[12] Alt, W., Biofizika **41**(1), 169-171 (1996).
[13] Chen, M.Y., Y. Long, and P.N. Devreotes, Genes & Development **11**(23), 3218-3231 (1997).
[14] Gerisch, G., Annu. Rev. Biochem. **56**, 853-879 (1987).
[15] Holden, A.V., M.M., and H. Othmer, Ed. Nonlinear Wave Processes in Excitable Media, Plenum Press, New York (1991).
[16] Gerisch, G., et al., J. Exp. Biol. **81**, 33-47 (1979).
[17] Field, R.J., M.B., ed. Wiley, New York (1984).
[18] Kapral, K., K.S., ed. Kluwer: Dordrecht (1994).
[19] Palsson, E,. and E.C. Cox, Proc. Nat. Acad. Sci. USA **93**, 1151-1155 (1996).
[20] Cohen, M.S., J. Theor. Biol. **69**, 57-85 (1977).
[21] Cohen, M.S., J. Theor. Biol. **72**, 231-255 (1978).
[22] Barchilon, M., and L.A. Segel, J. Theor. Biol. **133**, 437-446 (1988).
[23] MacKay, S.A., J. Cell. Sci. **33**, 1-16 (1978).
[24] Martiel, J.-L., and A. Goldbeter, Biophys. J. **52**, 807-828 (1987).
[25] Goldbeter, A., "Biochemical oscillations and cellular rhythms. The molecular bases of periodic and chaotic behaviour", Cambridge University Press 605 (1996).
[26] Tang, Y.H., and H.G. Othmer, Math. Biosci. **120**, 25-76 (1994).
[27] Tyson, J., et al., Physica D **34**, 193-207 (1989).
[28] Tang, Y.H., and H.G. Othmer, Phil. Trans. Roy. Soc. London B **349**, 179-195 (1995).
[29] Vasiev, B.N., P. Hogeweg, and A.V. Panfilov, Phys. Rev. Lett. **73**, 3173-3176 (1994).
[30] Vasiev, B., F. Siegert, and C.J. Weijer, J. Theor. Biol. **184**(4), 441 (1997).
[31] Devreotes, P.N., M.J. Potel, and S.A. MacKay, Dev. Biol., **96**, 405-415 (1983).
[32] Mato, J.M., et al., Proc. Natl. Acad. Sci. USA, **72**, 4991-4993 (1975).
[33] Gerisch, G., Biol. Cellulaire **32**, 61-68 (1978).
[34] Caterina, M.J. and P.N. Devreotes, FASEB J. **5**, 3078-3085 (1991).
[35] Wessels, D., J. Murray, and D.R. Soll, Cell Motil. Cytoskel. **23**, 145-156 (1992).
[36] Wessels, D., et al., Molecular Biology of the Cell **7**(SS), 134 9-1349 (1996).
[37] Keller, E.F. and L.A. Segel, J. Theor. Biol. **26**, 399-415 (1970).
[38] Novak, B. and F.F. Seelig, J. Theor. Biol. **56**, 301-327 (1976).
[39] van Oss, C., et al., J. Theor. Biol. **181**(3), 203-213 (1996).
[40] Vasieva, O.O., et al., J. Theor. Biol. **171**, 361-368 (1994).
[41] Levine, H., L. Tsimring, and D. Kessler, Physica D **106**(3-4), 375-388 (1997).
[42] Levine, H., Physica A **249**(1-4), 53-63 (1998).
[43] Bretschneider, T., B. Vasiev, and C.J. Weijer, J. Theor. Biol. **189**(1), 41 (1997).
[44] Odell, G.M. and J.T. Bonner, Phil. Trans. R. Soc. Lond. B **312**, 487-525 (1986).
[45] Evans, E., Biophysical Journal **64**, 1306-1322 (1993).
[46] Siegert, F. and C. Weijer, J. Cell Sci. **93**, 325-335 (1989).
[47] Parnas, H. and L.A. Segel, J. Theor. Biol. **71**, 185-207 (1978).
[48] Agarwal, P., Computer Applications in the Biosciences **10**(6), 647-655 (1994).
[49] Nanjundiah, V., J. Theor. Biol. **42**, 63-105 (1973).
[50] Levine, H. and W. Reynolds, Phys. Rev. Lett. **66**, 2400-2403 (1991).
[51] Gross, J.D., M.J. Peacey, and D.J. Trevan, J. Cell Sci. **22**, 645-656 (1976).
[52] Cohen, M.H. and A. Robertson, J. Theor. Biol. **31**, 119-130 (1971).
[53] Hashimoto, Y., M.H. Cohen, and A. Robertson, J. Cell Sci. **19**, 215-229 (1975).
[54] Hofer, T., J.A. Sherratt, and P.K. Maini, Physica D **85**(3), 425-444 (1995).

[55] Hofer, T., J.A. Sherratt, and P.K. Maini, Proc. Roy. Soc. London B **259**(1356), 249-257 (1995).

[56] Hofer, T. and P.K. Maini, Phys. Rev. E **56**(2), 2074-2080 (1997).

[57] Dallon, J.C. and H.G. Othmer, Phil. Trans. Roy. Soc. London Bs **352**(1351), 391-417 (1997).

[58] Kuramoto, Y., "Chemical Oscillations, Waves and Patterns", Berlin: Springer (1984).

[59] Levine, H., et al., Proc. Nat. Acad. Sci. USA **93**(13), 6382-6386 (1996).

[60] Aranson, I., H. Levine, and L. Tsimring, Phys. Rev. Lett. **76**(7), 1170- 1173 (1996).

[61] Palsson, E., et al., Proc. Nat. Acad. Sci. USA **94**(25), 13719-13723 (1997).

[62] Durston, A.J., J. Theor. Biol. **42**, 483-504 (1973).

[63] Durston, A.J., Dev. Biol. **37**, 225-235 (1974).

[64] Durston, A.J., Dev. Biol. **38**, 308-319 (1974).

[65] Foerster, P., S. Muller, and B. Hess, Development **109**, 11-16 (1990).

[66] Siegert, F., C.J. Weijer, in Oscillations and Morphogenesis, L. Rensing, Editor., Marcel Dekker, Inc: New York, Basel, Hong Kong. 133-152 (1993).

[67] Rietdorf, J., F. Siegert, and C.J. Weijer, Developmental Biology **177**(2), 427-438 (1996).

[68] Levine, H., et al., Proc. Natl. Acad. Sci. USA **93**, 6382-6386 (1996).

[69] Siegert, F. and C.J. Weijer, Curr. Biol. **5**, 937-943 (1995).

[70] Alcantara, F. and M. Monk, J. Gen. Microbiol. **85**, 321-334 (1974).

[71] Savill, N.J. and P. Hogeweg, J. Theor. Biol. **184**(3), 229-235 (1997).

[72] Bretschneider, T., B. Vasiev, and C.J. Weijer, Proc. Roy. Soc. B, submitted.

[73] Takeuchi, I., in Cell-Cell Interactions in Early Development, J. Gerhart , Editor, Wiley-Liss: New York. 249-259 (1991).

[74] Early, A., T. Abe, and J. Williams, Cell **83**, 91-99 (1995).

[75] Bonner, J.T., Quart. Rev. Biol. **32**, 232-246 (1957).

[76] Cohen, M.H. and A. Robertson, in "Cell Differentiation", R. Harris, P. All in, and D. Viza, Editors, Munkgaard: Copenhagen, 35-45 (1972).

[77] Belintsev, B.N., I.D. Yudin, and M.V. Volkenstein, in "Dokl. Akad. Nauk SSSR", 470-474 (1985).

[78] Inouye, K., in "Devel. Growth Differ., 406 (1987).

[79] Gierer, A., Naturwissenschaften **68**(5), 245-251 (1981).

[80] Meinhardt, H., "Models of Biological Pattern Formation", London: Academic Press, 1-189 (1982).

[81] Schaap, P., Y.H. Tang, and H.G. Othmer, Differentiation **61**(2), 141-141 (1996).

[82] Mizuguchi, T., S.M., Phys. Rev. Lett. **75**(5), 966-969 (1995).

[83] Meinhardt, H., Differentiation **24**, 191-202 (1983).

[84] Sekimura, T., M. Mimura, and E. Teramoto, in "Devel. Growth Differ., 406 (1987).

[85] Umeda, T., Bull. Math. Biol. **55**, 451-464 (1993).

[86] Vasiev, B., and W., C.J., Biophysical J., submitted.

[87] Breen, E.J., and K.L. Williams, Dev. Genet. **9**, 539-548 (1988).

[88] Dormann, D., F. Siegert, and C.J. Weijer, Development **122**, 761-769 (1996).

[89] Durston, A.J., F. Vork, and C. Weinberger, in "Biophysical and Biochemical Information Transfer in Recognition", J.G. Vassileva-Popova and E.V. Jensen, Editors, Plenum: New York, 693-708 (1979).
[90] Durston, A.J., and F. Vork, J. Cell Sci. **36**, 261-279 (1979).
[91] Siegert, F., and C.J. Weijer, Proc. Nat. Acad. Sci. USA **89**, 6433-6437 (1992).
[92] Abe, T., et al., Cell **77**, 687-699 (1994).
[93] Steinbock, O., et al., Proc. Nat. Acad. Sci. USA **90**, 7332-7335 (1993).
[94] Bretschneider, T., F. Siegert, and C.J. Weijer, Proc. Nat. Sci. USA **92**, 4387-4391 (1995).
[95] Jermyn, K.A., K.T. Duffy, and J.G. Williams, Nature **340**, 144-146 (1989).
[96] Berks, M., and R.R. Kay, Development **110**, 977-984 (1990).
[97] Umeda, T.A., Bull. Math. Biol. **51**, 485-500 (1989).
[98] Inouye, K., Protoplasma **121**, 171-177 (1984).
[99] Jermyn, K.A., and J.G. Williams, Development **111**, 779-787 (1991).
[100] Sternfeld, J., W.R. Arch. Dev. Biol. **201**, 354-363 (1992).

Druck: Strauss Offsetdruck, Mörlenbach
Verarbeitung: Schäffer, Grünstadt

Lecture Notes in Physics

For information about Vols. 1–477
please contact your bookseller or Springer-Verlag

Vol. 478: G. Sierra, M. A. Martín-Delgado (Eds.), Strongly Correlated Magnetic and Superconducting Systems. Proceedings, 1996. VIII, 323 pages. 1997.

Vol. 479: H. Latal, W. Schweiger (Eds.), Perturbative and Nonperturbative Aspects of Quantum Field Theory. Proceedings, 1996. X, 430 pages. 1997.

Vol. 480: H. Flyvbjerg, J. Hertz, M. H. Jensen, O. G. Mouritsen, K. Sneppen (Eds.), Physics of Biological Systems. From Molecules to Species. X, 364 pages. 1997.

Vol. 481: F. Lenz, H. Grießhammer, D. Stoll (Eds.), Lectures on QCD. VII, 276 pages. 1997.

Vol. 482: X.-W. Pan, D. H. Feng, M. Vallières (Eds.), Contemporary Nuclear Shell Models. Proceedings, 1996. XII, 309 pages. 1997.

Vol. 483: G. Trottet (Ed.), Coronal Physics from Radio and Space Observations. Proceedings, 1996. XVII, 226 pages. 1997.

Vol. 484: L. Schimansky-Geier, T. Pöschel (Eds.), Stochastic Dynamics. XVIII, 386 pages. 1997.

Vol. 485: H. Friedrich, B. Eckhardt (Eds.), Classical, Semi-classical and Quantum Dynamics in Atoms. VIII, 341 pages. 1997.

Vol. 486: G. Chavent, P. C. Sabatier (Eds.), Inverse Problems of Wave Propagation and Diffraction. Proceedings, 1996. XV, 379 pages. 1997.

Vol. 487: E. Meyer-Hofmeister, H. Spruik (Eds.), Accretion Disks – New Aspects. Proceedings, 1996. XIII, 356 pages. 1997.

Vol. 488: B. Apagyi, G. Endrédi, P. Lévay (Eds.), Inverse and Algebraic Quantum Scattering Theory. Proceedings, 1996. XV, 385 pages. 1997.

Vol. 489: G. M. Simnett, C. E. Alissandrakis, L. Vlahos (Eds.), Solar and Heliospheric Plasma Physics. Proceedings, 1996. VIII, 278 pages. 1997.

Vol. 490: P. Kutler, J. Flores, J.-J. Chattot (Eds.), Fifteenth International Conference on Numerical Methods in Fluid Dynamics. Proceedings, 1996. XIV, 655 pages. 1997.

Vol. 491: O. Boratav, A. Eden, A. Erzan (Eds.), Turbulence Modeling and Vortex Dynamics. Proceedings, 1996. XII, 245 pages. 1997.

Vol. 492: M. Rubí, C. Pérez-Vicente (Eds.), Complex Behaviour of Glassy Systems. Proceedings, 1996. IX, 467 pages. 1997.

Vol. 493: P. L. Garrido, J. Marro (Eds.), Fourth Granada Lectures in Computational Physics. XIV, 316 pages. 1997.

Vol. 494: J. W. Clark, M. L. Ristig (Eds.), Theory of Spin Lattices and Lattice Gauge Models. Proceedings, 1996. XI, 194 pages. 1997.

Vol. 495: Y. Kosmann-Schwarzbach, B. Grammaticos, K.M. Tamizhmani (Eds.), Integrability of Nonlinear Systems. VII, 380 pages. 1997.

Vol. 496: F. Lenz, H. Grießhammer, D. Stoll (Eds.), Lectures on QCD. VII, 483 pages. 1997.

Vol. 497: J. P. Greve, R. Blomme, H. Hensberge (Eds.), Stellar Atmospheres: Theory and Observations. VIII, 352 pages. 1997

Vol. 498: Z. Horváth, L. Palla (Eds.), Conformal Field Theories and Integrable Models. Proceedings, 1996. X, 251 pages. 1997.

Vol. 499: K. Jungmann, J. Kowalski, I. Reinhard, F. Träger (Eds.), Atomic Physics Methods in Modern Research, IX, 448 pages. 1997.

Vol. 500: D. Joubert (Ed.), Density Functionals: Theory and Applications, XVI, 194 pages. 1998.

Vol. 501: J. Kertész, I. Kondor (Eds.), Advances in Computer Simulation. VIII, 166 pages. 1998.

Vol. 502: H. Aratyn, T. D. Imbo, W.-Y. Keung, U. Sukhatme (Eds.), Supersymmetry and Integrable Models. Proceedings, 1997. XI, 379 pages. 1998.

Vol. 503: J. Parisi, S. C. Müller, W. Zimmermann (Eds.), A Perspective Look at Nonlinear Media. From Physics to Biology and Social Sciences. VIII, 372 pages. 1998.

Vol. 504: A. Bohm, H.-D. Doebner, P. Kielanowski (Eds.), Irreversibility and Causality. Semigroups and Rigged Hilbert Spaces. XIX, 385 pages. 1998.

Vol. 505: D. Benest, C. Froeschlé (Eds.), Impacts on Earth. XVII, 223 pages. 1998.

Vol. 506: D. Breitschwerdt, M. J. Freyberg, J. Trümper (Eds.), The Local Bubble and Beyond. Proceedings, 1997. XXVIII, 603 pages. 1998.

Vol. 507: J. C. Vial, K. Bocchialini, P. Boumier (Eds.), Space Solar Physics. Proceedings, 1997. XIII, 296 pages. 1998.

Vol. 508: H. Meyer-Ortmanns, A. Klümper (Eds.), Field Theoretical Tools for Polymer and Particle Physics. XVI, 258 pages. 1998.

Vol. 509: J. Wess, V. P. Akulov (Eds.), Supersymmetry and Quantum Field Theory. Proceedings, 1997. XV, 405 pages. 1998.

Vol. 510: J. Navarro, A. Polls (Eds.), Microscopic Quantum Many-Body Theories and Their Applications. Proceedings, 1997. XIII, 379 pages. 1998.

Vol. 511: S. Benkadda, G. M. Zaslavsky (Eds.), Chaos, Kinetics and Nonlinear Dynamics in Fluids and Plasmas. Proceedings, 1997. VIII, 438 pages. 1998.

Vol. 512: H. Gausterer, C. Lang (Eds.), Computing Particle Properties. Proceedings, 1997. VII, 335 pages. 1998.

Vol. 513: A. Bernstein, D. Drechsel, T. Walcher (Eds.), Chiral Dynamics: Theory and Experiment. Proceedings, 1997. IX, 394 pages. 1998.

Vol. 514: F. W. Hehl, C. Kiefer, R. Metzler, Black Holes: Theory and Observation. Proceedings, 1997. XV, 519 pages. 1998.

Vol. 515: C.-H. Bruneau (Ed.), Sixteenth International Conference on Numerical Methods in Fluid Dynamics. Proceedings. XV, 568 pages. 1998.

Monographs

For information about Vols. 1–10 please contact your bookseller or Springer-Verlag

GPSR Compliance

The European Union's (EU) General Product Safety Regulation (GPSR) is a set of rules that requires consumer products to be safe and our obligations to ensure this.

If you have any concerns about our products, you can contact us on ProductSafety@springernature.com

In case Publisher is established outside the EU, the EU authorized representative is:

Springer Nature Customer Service Center GmbH
Europaplatz 3
69115 Heidelberg, Germany

Batch number: 10704025

Printed by Printforce, the Netherlands